The Plant Family Fabaceae

Mirza Hasanuzzaman · Susana Araújo ·
Sarvajeet Singh Gill
Editors

The Plant Family Fabaceae

Biology and Physiological Responses to Environmental Stresses

Springer

Editors
Mirza Hasanuzzaman
Department of Agronomy
Faculty of Agriculture
Sher-e-Bangla Agricultural University
Dhaka, Bangladesh

Susana Araújo
Instituto de Tecnologia Química e Biológica
António Xavier
Universidade Nova de Lisboa
Oeiras, Portugal

Sarvajeet Singh Gill
Centre for Biotechnology
Maharshi Dayanand University
Rohtak, Haryana, India

ISBN 978-981-15-4751-5 ISBN 978-981-15-4752-2 (eBook)
https://doi.org/10.1007/978-981-15-4752-2

This Springer imprint is published by the registered company Springer Nature Singapore Pte Ltd.
The registered company address is: 152 Beach Road, #21-01/04 Gateway East, Singapore 189721, Singapore

Preface

The Food and Agriculture Organization of United Nations projects that global population will increase 10.8 billion by 2080, representing an increase by 47% in the future.[1] In such context, worldwide agricultural and food production systems are being challenged to provide enough food to meet the growing demands. Nowadays, agriculture has also to address the sustainable use of existing natural resources while tackling the challenges associated with the impact of climate change.

The cultivation of *Fabaceae* plants, commonly known as legumes, emerges as one of the relevant approaches to tackle these challenges. This plant family with 800 genera and 20,000 species,[2] being the second most economically important family of plants after *Poaceae*, encloses many benefits for agricultural and food sustainability. Due to their ability to fix atmospheric nitrogen by establishing symbiotic associations with nitrogen-fixing microorganisms, the cultivation of legumes is a sustainable option to reduce the use of fertilizers and production costs, while contributing to improve overall soil conditions. Importantly, several grain legumes are important sources of vegetable protein for humans or important role as forage for animals. Consequently, a growing body of research has been also devoted to study aspects associated with their nutritional quality and health-promoting effects.

Similarly, to other crops, legume physiology and yield are severely affected by non-optimal environmental conditions. Abiotic stresses such as drought, salinity, extreme temperatures, nutrient deficiencies, or toxicities have been reported to cause crucial losses on legume growth and productivity. Although relevant fundamental knowledge underlying the adaptative responses of legumes to abiotic constraints and their genetic basis have been elucidated, more research needs to be done to translate these findings into improved elite lines that can contribute to achieve food security. Recent advances and developments in molecular, biotechnological, and breeding tools have contributed to ease and wider this mission. Still,

[1]FAO (2018).

[2]Lewis et al. (2005).

the ongoing goal is to develop legumes not able to cope with environmental stresses, but still with considerable yield and quality.

The Plant Family Fabaceae—Biology and Physiological Responses to Environmental Stresses combines a group of 20 chapters written by worldwide researchers to provide novel information, regarding the major physiological, metabolic, cellular, and molecular processes, as well as the genetic basis and diversity, associated with abiotic stress responses. This book includes both several chapters addressing general and unique aspects and questions of legume Biology and worldwide impact, and a considerable number of chapters devoted to the effects of environmental stresses have on legume responses. A special focus is provided on running crop breeding and state-of-the-art biotechnological approaches to breed abiotic stress resistance traits into modern crop varieties, highlighting their achievements and still open challenges.

We would like to give special thanks to the authors for their outstanding and timely work in producing such fine chapters. We are highly thankful to Dr. Mei Hann Lee (Senior Editor, Life Science), Springer, Japan, for her prompt responses during the acquisition. We are also thankful to Arulmurugan Venkatasalam, Project Coordinator of this book, and all other editorial staffs for their precious help in formatting and incorporating editorial changes in the manuscripts. Special thanks to Dr. Md. Mahabub Alam, Department of Agronomy, Sher-e-Bangla Agricultural University, Bangladesh, for his generous help in formatting the manuscripts. We believe that this book is useful for undergraduate and graduate students, teachers, and researchers, particularly from the fields of the plant science, botany or agronomy, environmental science, biotechnology, and food science.

Dhaka, Bangladesh — Mirza Hasanuzzaman
Oeiras, Portugal — Susana Araújo
Rohtak, India — Sarvajeet Singh Gill

References

FAO (2018) The future of food and agriculture—alternative pathways to 2050. Rome, p 224

Lewis G, Schrire B, Mackinder B, Lock M (2005) Legumes of the world. Royal Botanic Gardens, Kew, UK

Contents

Editors and Contributors

About the Editors

Dr. Mirza Hasanuzzaman is Professor of Agronomy at Sher-e-Bangla Agricultural University, Dhaka, Bangladesh. He received his Ph.D. on 'Plant Stress Physiology and Antioxidant Metabolism' from the United Graduate School of Agricultural Sciences, Ehime University, Japan, with Japanese Government (MEXT) Scholarship. Later, he completed his postdoctoral research in Center of Molecular Biosciences (COMB), University of the Ryukyus, Okinawa, Japan, with 'Japan Society for the Promotion of Science (JSPS)' postdoctoral fellowship. Subsequently, he joined as Adjunct Senior Researcher at the University of Tasmania with Australian Government's Endeavour Research Fellowship. He has been devoting himself in research in the field of crop science, especially focused on Environmental Stress Physiology since 2004. He published over 100 articles in peer-reviewed journals and books. He has edited 12 books and written 35 chapters on important aspects of plant physiology, plant stress responses, and environmental problems in relation to plant species. These books were published by the internationally renowned publishers. He is a research supervisor of undergraduate and graduate students and supervised 20 MS students so far. He is editor and reviewer of more than 50 peer-reviewed international journals and recipient of 'Publons Global Peer Review Awards 2017, 2018, and 2019.' He is active member of about 40 professional societies and acting as Publication

Secretary of Bangladesh Society of Agronomy. He has been honored by different authorities due to his outstanding performance in different fields like research and education. He received The World Academy of Sciences (TWAS) Young Scientist Award 2014. He attended and presented 25 papers and posters in national and international conferences in different countries (USA, UK, Germany, Australia, Japan, Austria, Sweden, Russia, etc.).

Dr. Susana Araújo graduated in Applied Plant Biology (2000) at the Faculdade de Ciências da Universidade de Lisboa, Portugal. In 2007, she obtained her Ph.D. in Biology at the Instituto de Tecnologia Química e Biológica António Xavier da Universidade Nova de Lisboa (ITQB NOVA), Portugal. During her Ph.D. thesis, she developed *Medicago truncatula* lines expressing a stress-related gene and studied their response to water deficit. After a postdoctoral fellowship at ITQB NOVA, she moved to the Tropical Research Institute in Lisbon (2009–2014) to continue her research on legume adaptation to water deficit. In 2015, she moved to the University of Pavia (UNIPV, Italy) as a senior researcher. Under the scope of the PRIMTECH project, she studied the molecular mechanisms behind seed germination and priming. This subject, among others, was further studied after returning to ITQB NOVA as postdoctoral researcher (2016–2018). Presently, she is a researcher at the Plant Cell Biotechnology Laboratory of ITQB NOVA, being an integrated member of the research unit "GREEN-IT—Bioresources for Sustainability." One of her goals is to develop legumes able to cope and yield within the current climate change scenario. Her main research aims to uncover molecular and physiological mechanisms by which legume plants grow and respond to abiotic stresses. Recently, her research focused on seed biology, with running research on seed development and germination. She has been delivering invited lectures and classes in these topics, while supervising several Ph.D., M.Sc., and B.Sc students. She co-authored 42 manuscripts in peer-reviewed international journals, 11 chapters, edited one book on seed technology, and has contributed with several communications in international meetings. She is a member of the International

Legume Society (ILS) and Portuguese Society of Plant Physiology. Since 2015, she is part of the Editorial Board of Frontiers in Plant Science, as Associate Editor of the Plant Breeding section, while being referee for several international scientific journals.

Dr. Sarvajeet Singh Gill has completed his Ph.D. at the age of 28 from AMU, Aligarh, and postdoctoral studies from International Centre for Genetic Engineering and Biotechnology (ICGEB). He is working as Assistant Professor of agriculture biotechnology at Centre for Biotechnology, Maharshi Dayanand University. He in collaboration with Dr. Narendra Tuteja (postdoctoral guide) conferred the novel function of DNA helicase (PDH45) in stress tolerance. He has published >100 papers in the journals of international repute, edited >28 books, and has been serving as an editorial board member in the journals of international repute. He has been conferred with 2017 Research Excellence and Citation Award from Clarivate Analytics (Web of Science).

Contributors

Gulraiz Ahmad State Key Laboratory of Grassland Agro-ecosystem, College of Pastoral Agriculture Science and Technology, Lanzhou University, Lanzhou, Gansu, China

Muhammad Ahmad Department of Agronomy, University of Agriculture, Faisalabad, Pakistan

Faical Akaichi Land Economy, Environment and Society, Scotland's Rural College (SRUC), Edinburgh, UK

Afsana Hoque Akhi Molecular Breeding Lab, Plant Breeding Division, Bangladesh Agricultural Research Institute, Gazipur, Bangladesh

Sadia Sabrina Alam Molecular Breeding Lab, Plant Breeding Division, Bangladesh Agricultural Research Institute, Gazipur, Bangladesh

Subham Anurag Department of Environmental Sciences, Central University of Jharkhand, Brambe, Ranchi, Jharkhand, India

Susana Araújo Instituto de Tecnologia Química e Biológica António Xavier, Universidade Nova de Lisboa (ITQB NOVA), Oeiras, Portugal

Bálint Balázs Environmental Social Science Research Group (ESSRG), Budapest, Hungary

Alma Balestrazzi Department of Biology and Biotechnology, "Lazzaro Spallanzani"-University of Pavia, Pavia, Italy

Saon Banerjee AICRP on Agrometeorology, Bidhan Chandra Krishi Viswavidyalaya, Kalyani, West Bengal, India

Christine Oré Barrios Land Economy, Environment and Society, Scotland's Rural College (SRUC), Edinburgh, UK

Rajan Bhatt Regional Research Station, Kapurthala, India;
Punjab Agricultural University, Ludhiana, Punjab, India

Tasnim Farha Bhuiyan Department of Agricultural Botany, Sher-e-Bangla Agricultural University, Dhaka, Bangladesh

M. H. M. Borhannuddin Bhuyan Citrus Research Station, Bangladesh Agricultural Research Institute, Jaintapur, Sylhet, Bangladesh

Tiziana Centofanti Environmental Social Science Research Group (ESSRG), Budapest, Hungary

Caspar C. C. Chater Departamento de Biología Molecular de Plantas, Instituto de Biotecnología, Universidad Nacional Autónoma de México, Cuernavaca, Mexico; Department of Molecular Biology and Biotechnology, University of Sheffield, Sheffield, UK

Sandeep Chatterjee Department of Environmental Sciences, Central University of Jharkhand, Brambe, Ranchi, Jharkhand, India

Arzu Çığ Department of Horticulture, Siirt University, Siirt, Turkey

Fatih Çığ Department of Field Crops, Faculty of Agriculture, Siirt University, Siirt, Turkey

Alejandra A. Covarrubias Departamento de Biología Molecular de Plantas, Instituto de Biotecnología, Universidad Nacional Autónoma de México, Cuernavaca, Mexico

Shainandni Dogra Punjab Biotechnology Incubator, Mohali, India

Rafael D. C. Duarte CBQF—Centro de Biotecnologia E Química Fina—Laboratório Associado, Universidade Católica Portuguesa, Escola Superior de Biotecnologia, Porto, Portugal

Anamika Dubey Metagenomics and Secretomics Research Lab., Department of Botany, Dr. Harisingh Gour University (A Central University), Sagar, MP, India

Ana Sofia Duque Instituto de Tecnologia Química e Biológica António Xavier (ITQB NOVA), Green-it Unit, Oeiras, Portugal

Ayman EL Sabagh Department of Agronomy, Faculty of Agriculture, Kafrelsheikh University, Kafr El-Shaikh, Egypt;
Department of Field Crops, Faculty of Agriculture, Siirt University, Siirt, Turkey

Murat Erman Department of Field Crops, Faculty of Agriculture, Siirt University, Siirt, Turkey

Shah Fahad Department of Agriculture, University of Swabi, Khyber Pakhtunkhwa, Pakistan

Muhammad Farooq Department of Crop Sciences, College of Agricultural and Marine Sciences, Sultan Qaboos University, Muscat, Oman;
Department of Agronomy, University of Agriculture, Faisalabad, Pakistan;
The UWA Institute of Agriculture and School of Agriculture, & Environment, the University of Western Australia, Perth, WA, Australia

Maham Fatima Department of Environmental Sciences, COMSATS University Islamabad, Vehari Campus, Pakistan

Helena Ferreira Universidade Católica Portuguesa, Escola Superior de Biotecnologia, CBQF - Centro de Biotecnologia e Química Fina—Laboratório Associado, Porto, Portugal

Dibakar Ghosh ICAR-Directorate of Weed Research, Jabalpur, Madhya Pradesh, India

Ana Gomes Universidade Católica Portuguesa, Escola Superior de Biotecnologia, CBQF - Centro de Biotecnologia e Química Fina—Laboratório Associado, Porto, Portugal

Gurinderjit Singh Goraya Himachal Pradesh Forest Department, Former PCCF & HoFF, Himachal Pradesh, Talland, Shimla, HP, India

Sabine Gruber University of Hohenheim, Institute of Crop Production (340a), Stuttgart, Germany

Michael A. Grusak USDA-ARS Edward T. Schafer Agricultural Research Center, Fargo, ND, USA

Carla Gualtieri Department of Biology and Biotechnology, "Lazzaro Spallanzani"-University of Pavia, Pavia, Italy

Muhammad Bilal Hafeez College of Agronomy, Northwest A&F University, Yangling, China

Abdul Hannan Seed Technology Division, Bangladesh Agricultural Research Institute, Rajshahi, Bangladesh

Mirza Hasanuzzaman Department of Agronomy, Faculty of Agriculture, Sher-e-Bangla Agricultural University, Dhaka, Bangladesh

Akbar Hossain Bangladesh Wheat and Maize Research Institute (BWMRI), Dinajpur, Bangladesh

Pete Iannetta Ecological Sciences, James Hutton Institute, Dundee, Scotland, UK

Muhammad Ikram Department of Agronomy, Faculty of Agricultural Sciences and Technology, Bahauddin Zakariya University, Multan, Pakistan

Shahid Iqbal Department of Agronomy, Muhammad Nawaz Shareef University of Agriculture, Multan, Pakistan

M Sohidul Islam Department of Agronomy, Hajee Mohammad Danesh Science and Technology University, Dinajpur, Bangladesh

Tofazzal Islam Institute of Biotechnology and Genetic Engineering (IBGE), Bangabandhu Sheikh Mujibur Rahman Agricultural University, Gazipur, Bangladesh

Euan K. James Ecological Sciences, James Hutton Institute, Dundee, Scotland, UK

Vaneet Jishtu Himalayan Forest Research Institute, Shimla, HP, India

Anestis Karkanis Department of Agriculture Crop Production and Rural Environment, University of Thessaly, Volos, Greece

Alison Karley Ecological Sciences, James Hutton Institute, Dundee, Scotland, UK

Mohammad Latif Khan Metagenomics and Secretomics Research Lab., Department of Botany, Dr. Harisingh Gour University (A Central University), Sagar, MP, India

Ashwani Kumar Metagenomics and Secretomics Research Lab., Department of Botany, Dr. Harisingh Gour University (A Central University), Sagar, MP, India

Susana T. Leitão Instituto de Tecnologia Química e Biológica António Xavier, Universidade Nova de Lisboa (ITQB NOVA), Oeiras, Portugal

Anca Macovei Department of Biology and Biotechnology, "Lazzaro Spallanzani"-University of Pavia, Pavia, Italy

Jubayer Al Mahmud Department of Agroforestry and Environmental Science, Faculty of Agriculture, Sher-e-Bangla Agricultural University, Dhaka, Bangladesh

Marta Maluk Ecological Sciences, James Hutton Institute, Dundee, Scotland, UK

Muhammed Khairujjaman Mazumder Plant Stress and Metabolomics Laboratory, CIL, Assam University, Silchar, Assam, India

Ram Swaroop Meena Department of Agronomy, Institute of Agricultural Sciences, BHU, Varanasi, Uttar Pradesh, India

Md. Motiar Rohman Molecular Breeding Lab, Plant Breeding Division, Bangladesh Agricultural Research Institute, Gazipur, Bangladesh

Debojyoti Moulick Plant Stress and Metabolomics Laboratory, CIL, Assam University, Silchar, Assam, India

Muhammad Mubeen Department of Environmental Sciences, COMSATS University Islamabad, Vehari Campus, Pakistan

Akash Nag Department of Environmental Sciences, Central University of Jharkhand, Brambe, Ranchi, Jharkhand, India

Kamrun Nahar Department of Agricultural Botany, Sher-e-Bangla Agricultural University, Dhaka, Bangladesh

Poornima K. Narayana Indian Institute of Pulses Research, Kanpur, India;
Department of Plant and Soil Sciences and Gund Institute for the Environment, University of Vermont, Burlington, VT, USA

Wajid Nasim Department of Agronomy, University College of Agriculture and Environmental Sciences, The Islamia University of Bahawalpur (IUB), Bahawalpur, Pakistan

Georgia Ntatsi Laboratory of Vegetable Production Athens, Department of Crop Science, Agricultural University of Athens, Athens, Greece

Andrea Pagano Department of Biology and Biotechnology, "Lazzaro Spallanzani"-University of Pavia, Pavia, Italy

Shyam Narain Pandey Department of Botany, University of Lucknow, Lucknow, India

Kishore C. S. Panigrahi School of Biological Sciences, National Institute of Science Education and Research, Jatni, Bhubaneswar, Odisha, India

Khursheda Parvin Laboratory of Plant Stress Responses, Department of Applied Biological Sciences, Faculty of Agriculture, Kagawa University, Kita-Gun, Kagawa, Japan;
Department of Horticulture, Faculty of Agriculture, Sher-e-Bangla Agricultural University, Dhaka, Bangladesh

Elisabete Pinto Universidade Católica Portuguesa, Escola Superior de Biotecnologia, CBQF - Centro de Biotecnologia e Química Fina—Laboratório Associado, Porto, Portugal

Jorge A. Pinto Paiva Instituto de Biologia Experimental e Tecnológica (IBET), Oeiras, Portugal;
Instituto de Tecnologia Química e Biológica António Xavier (ITQB NOVA), Green-it Unit, Oeiras, Portugal;
Institute of Plant Genetic Polish Academy of Science, Poznań, Poland

Jose A. Polania Departamento de Biología Molecular de Plantas, Instituto de Biotecnología, Universidad Nacional Autónoma de México, Cuernavaca, Mexico

Richa Rai Department of Botany, Rameshwar College, Babasaheb Bhimrao Ambedkar Bihar University, Muzaffarpur, Bihar, India

Idupulapati M. Rao International Center for Tropical Agriculture (CIAT), Cali, Colombia;
Agricultural Research Service, Department of Agriculture, Plant Polymer Research Unit, National Center for Agricultural Utilization Research, Peoria, IL, USA

Disna Ratnasekera Department of Agricultural Biology, Faculty of Agriculture, University of Ruhuna, Matara, Sri Lanka

Ali Raza Oil Crops Research Institute, Chinese Academy of Agricultural Sciences (CAAS), Wuhan, China

Md. Rezwan Molla Plant Genetic Resources Centre, Bangladesh Agricultural Research Institute, Gazipur, Bangladesh

Diego Rubiales Institute for Sustainable Agriculture, CSIC, Córdoba, Spain

Bedabrata Saha School of Biological Sciences, National Institute of Science Education and Research, Jatni, Bhubaneswar, Odisha, India

Purabi Saikia Department of Environmental Sciences, Central University of Jharkhand, Brambe, Ranchi, Jharkhand, India

Suman Samanta AICRP on Agrometeorology, Bidhan Chandra Krishi Viswavidyalaya, Kalyani, West Bengal, India

Carla S. Santos CBQF—Centro de Biotecnologia e Química Fina—Laboratório Associado, Universidade Católica Portuguesa, Escola Superior de Biotecnologia, Porto, Portugal

Subhas Chandra Santra Department of Environmental Science, University of Kalyani, Kalyani, West Bengal, India

Dimitrios Savvas Laboratory of Vegetable Production Athens, Department of Crop Science, Agricultural University of Athens, Athens, Greece

Kanval Shaukat Department of Botany, University of Balochistan, Quetta, Pakistan

Shailesh Shrestha Land Economy, Environment and Society, Scotland's Rural College (SRUC), Edinburgh, UK

Luiza Toma Land Economy, Environment and Society, Scotland's Rural College (SRUC), Edinburgh, UK

Ehsan Valiollahi Instituto de Biologia Experimental e Tecnológica (IBET), Oeiras, Portugal;
Instituto de Tecnologia Química e Biológica António Xavier (ITQB NOVA), Green-it Unit, Oeiras, Portugal

Albert Vandenberg Department of Plant Sciences, University of Saskatchewan, Saskatoon, Canada

Marta W. Vasconcelos CBQF—Centro de Biotecnologia e Química Fina—Laboratório Associado, Universidade Católica Portuguesa, Escola Superior de Biotecnologia, Porto, Portugal

Maria Carlota Vaz Patto Instituto de Tecnologia Química e Biológica António Xavier, Universidade Nova de Lisboa (ITQB NOVA), Oeiras, Portugal

Eric J. B. von Wettberg Department of Plant and Soil Sciences and Gund Institute for the Environment, University of Vermont, Burlington, VT, USA;
Mathematical Biology Laboratory, Peter the Great St. Petersburg Polytechnic University, St. Petersburg, Russia

Allah Wasaya College of Agriculture, BZU, Layyah, Pakistan

Michael Williams Department of Botany, School of Natural Sciences, Trinity College Dublin, Dublin, Ireland

Tauqeer Ahmad Yasir College of Agriculture, BZU, Layyah, Pakistan

Noreen Zahra Department of Botany, University of Agriculture, Faisalabad, Pakistan

General Aspects

The Biology of Legumes and Their Agronomic, Economic, and Social Impact

Marta W. Vasconcelos, Michael A. Grusak, Elisabete Pinto, Ana Gomes, Helena Ferreira, Bálint Balázs, Tiziana Centofanti, Georgia Ntatsi, Dimitrios Savvas, Anestis Karkanis, Michael Williams, Albert Vandenberg, Luiza Toma, Shailesh Shrestha, Faical Akaichi, Christine Oré Barrios, Sabine Gruber, Euan K. James, Marta Maluk, Alison Karley, and Pete Iannetta

Abstract Intensive agriculture and meat-based westernized diets have brought a heavy environmental burden to the planet. Legumes, or pulses, are members of the large Fabaceae (*Leguminosae*) family, which comprise about 5% of all plant species. They are ancient crops whose popularity both for farmers and consumers has gone

---

M. W. Vasconcelos (✉) · E. Pinto · A. Gomes · H. Ferreira
Universidade Católica Portuguesa, Escola Superior de Biotecnologia, CBQF - Centro de Biotecnologia e Química Fina—Laboratório Associado, Rua Diogo Botelho 1327, 4169-005 Porto, Portugal
e-mail: mvasconcelos@porto.ucp.pt

M. A. Grusak
USDA-ARS Edward T. Schafer Agricultural Research Center, 1616 Albrecht Blvd. North, Fargo, ND 58102-2765, USA

B. Balázs · T. Centofanti
Environmental Social Science Research Group (ESSRG), Budapest, Hungary

G. Ntatsi · D. Savvas
Laboratory of Vegetable Production Athens, Department of Crop Science, Agricultural University of Athens, Athens, Greece

A. Karkanis
Department of Agriculture Crop Production and Rural Environment, University of Thessaly, Volos, Greece

M. Williams
Department of Botany, School of Natural Sciences, Trinity College Dublin, Dublin, Ireland

A. Vandenberg
Department of Plant Sciences, University of Saskatchewan, Saskatoon, Canada

L. Toma · S. Shrestha · F. Akaichi · C. O. Barrios
Land Economy, Environment and Society, Scotland's Rural College (SRUC), West Mains Road, Edinburgh EH9 3FH, UK

S. Gruber
University of Hohenheim, Institute of Crop Production (340a), 70599 Stuttgart, Germany

E. K. James · M. Maluk · A. Karley · P. Iannetta
Ecological Sciences, James Hutton Institute, Dundee, Scotland, UK

M. Hasanuzzaman et al. (eds.), *The Plant Family Fabaceae*,
https://doi.org/10.1007/978-981-15-4752-2_1

through several stages of acceptance, and in recent years, legumes have regained their luster. This is due to a global understanding that: (1) farming systems need to promote biodiversity, (2) biological nitrogen fixation is an important tool to reduce the application of external chemical inputs, namely in the form of nitrogen fertilizers, and that (3) plant-based foods have fewer adverse environmental effects per unit weight, per serving, per unit of energy, or per protein weight than do animal source foods, across various environmental indicators. Legumes play a key role in answering these three global challenges and are pivotal actors in the diversification and sustainable intensification of agriculture, particularly in light of new and urgent challenges such as climate change. In this chapter, we showcase the importance of legumes as contemporary agents of change, whose impacts start in the field, but then branch out into competitive global economies, modernized societies, and ultimately, improved food security and human health.

Keywords Biodiversity · Biological nitrogen fixation · Nutrition and health · Pulses · Sustainability

1 Introduction

The word legume comes from the Latin word *legumen* which can be translated to "seeds harvested in pods." In many parts of the world, such as in Canada, Bangladesh, or India, the world pulse is used when referring to legume grains, especially those with a low content in fat. Legumes or pulses have accompanied farmers since the Neolithic revolution, the very onset of farming practices of mankind. Pea (*Pisum sativum*), lentil (*Lens culinaris*), chickpea (*Cicer arietinum*), and bitter vetch (*Vicia ervilia*) belong to the "Big Eight," that package of "founder crops" which have been domesticated in the Fertile Crescent during the 10th and 9th millennia BCE (Asouti and Fuller 2013). Legumes were domesticated alongside grasses as early as 10,000 years ago (Hancock 2012). Among the earliest legume crops were chickpea, garden pea, and lentil (Sprent 2009; Hancock 2012; Smýkal et al. 2015). The domestication of other important legumes followed later on in different regions of the world, for example, soybean in east Asia (Sedivy et al. 2017), Azuki bean (*Vigna angularis*) in west Asia (Lee 2012), or common bean (*Phaseolus vulgaris*) in Mesoamerica (Lopez et al. 2013). The cultivation of soybean [*Glycine max* (L.) Merrill] dates from China in around 2500 BCE, being now spread throughout the world mostly due to its elevated protein content of the seeds that can reach almost 40%. Despite its particular worldwide importance, soybean is heavily reliant on inoculation to bring it into profitable use in non-native countries like Brazil (Alves et al. 2003).

It is thought that the introduction of legumes into cropping systems in Europe (before the tenth century) enabled an improvement in soil quality and provided nourishment to populations, relieving famine and improving overall population growth.

More than 820 million people have insufficient food and many more consume low-quality diets that cause micronutrient deficiencies. This has contributed to a substantial rise in the incidence of diet-related noncommunicable diseases. All legumes offer a high level of protein in above- and belowground biomass, particularly in grains, in comparison to other crops such as cereals. They are self-supporters of nitrogen fertilization through atmospheric nitrogen fixation in root nodules in symbiosis with soil bacteria from the families *Rhizobium*, *Bradyrhizobium,* and others. The genetic regulation of these processes has been intensively investigated and various forward- and reverse-genetic approaches have identified nearly 200 genes required for symbiotic nitrogen fixation in legumes (Roy et al. 2019).

In times with low availability of meat, pulses—legumes with predominant grain usage—provided a valuable source of proteins for the human diet. The biblical tale of Esau who sold his birthright to his younger brother Jacob for the price of a lentil stew illustrates the estimation of the pulses in early societies. In the past, meat was often unavailable to common people, i.e., for the majority of ancient societies. Pulses, therefore, were a sufficient alternative to meat for a healthy and whole food diet. The traditional Milpa cropping system, a combination of maize, beans, and squash, is a good example for the integration of legumes in sustainable cropping systems and in the whole food human diet (Altieri et al. 2011). It integrates physiological and morphological benefits of crops, including pulses, at the field, and offers a balanced food composition for human consumption with beans as the main provider of protein. The Milpa system originated from Mesoamerica and has spread to many tropical and subtropical regions across the world because of its benefits. Meanwhile, it can be considered a model for innovative cropping systems today and in the future.

In the middle of the twentieth century, pulses disappeared more and more from the menu in the industrial countries and as well from cropping systems at the same time. Pulses were considered to be an old-fashioned food, with nonnutritive compounds such as lectins, alkaloids, saponins, or phytates (Muzquiz et al. 2012), with lengthy time-consuming preparation methods and some causing intestinal irritations. Finally, meat was available to all social classes. There are additionally some agricultural challenges of legume growing: they have lower yields and lower economic value in comparison to cereals. For example, a farmer in temperate Europe (France, Germany, Poland) can achieve a yield of 4.8–7.6 t ha^{-1} winter wheat and only 2.7–3.5 t ha^{-1} dry pea (FAO Stat 2019).

Recently, there seems to be a return to the value of pulses. Concerns about ecological impacts of meat production, ethical concerns in terms of animal welfare, and considerations for human health (Chai et al. 2019; Hagmann et al. 2019) have promoted an interest in a more sustainable plant-based food production, with legumes as a substantial contributor. In times of public discussions about the loss of biodiversity, sustainable agriculture, and climate change, a renaissance of legumes in agricultural systems seems a reasonable and promising way to design the future of our planet.

2 The Biology of Legumes

2.1 Taxonomy and Morphology

The Earth currently has almost 400,000 species of plants. About 5% of plant species are members of the large plant family Fabaceae (*Leguminosae*) which produce their protein-rich seeds within simple dehiscent dry fruits botanically known as legumes (commonly known as pods). The Fabaceae family includes 770 genera and nearly 20,000 worldwide distributed species (LPWG 2017). The Fabaceae evolved to have root systems that enable symbiotic relationships with various species of soil bacteria that are capable of fixing atmospheric nitrogen, thereby providing a basic biological source of nitrogenous compounds such as proteins and their biochemical derivatives. Legume species are very diverse and are adapted to almost all terrestrial ecosystems in the form of trees, shrubs, vines, and annual herbs. Legume flowers characteristically have five petals that have evolved to a wide range of characteristic sizes, shapes, and colors. Legume species can be self-pollinating, cross-pollinating or both. The traditional classification of *Fabaceae* into the three subfamilies, *Caesalpinioideae*, *Mimosoideae*, and *Papilionoideae*, has been revised by The Legume Phylogeny Working Group (LPWG 2017) and Sprent et al. (2017). A new subfamily classification presented by LPWG (2017) divides the Leguminosae into six subfamilies: *Detarioideae* (84 genera; ca. 760 species; Pantropical), *Cercidoideae* (12 genera; ca. 335 species; Pantropical, *Cercis* warm temperate), *Duparquetioideae* (1 genus; 1 species; West and West-central Africa), *Dialioideae* (17 genera; ca. 85 species; Pantropical), *Caesalpinioideae* (148 genera; ca. 4400 species; Pantropical, some temperate), and *Papilionoideae* (503 genera; ca 14,000 species; cosmopolitan). The previous subfamily *Mimosoideae* has been incorporated into the Caesalpinioideae as the mimosoid clade. Species from *the Detarioideae, Cercidoideae, Duparquetioideae,* and *Dialioideae* are all non-nodulators. Nodulation has been confirmed in only eight genera in the *Caesalpinioideae* sensu stricto subfamily. Most, but not all mimosoids and papilionoids can nodulate (Sprent et al. 2017).

The *Caesalpinoideae* subfamily is highly variable, mostly trees and shrubs with zygomorphic asymmetrical flowers. The mimosoid clade are adapted to tropical and subtropical climates and exist mostly in the form of trees and shrubs. Their flowers are symmetric with valvate petals and have large numbers of prominent stamens. The *Papilionoideae* is the largest, most widely adapted and diverse legume subfamily. Their floral morphology (standard, wings, and keel petals) is demonstrated by that of the widely known species (bean, pea, and soybean) that have edible pods and seeds used in food systems as vegetables and dry seeds. The members of this ecologically diverse group include trees, shrubs, and herbs.

2.2 Nodulation

Legumes form symbiotic relationships with nitrogen-fixing bacteria (rhizobia), most of which belong to the genera *Rhizobium*, *Bradyrhizobium*, *Mesorhizobium*, *Ensifer* (Sinorhizobium), and *Azorhizobium* in the Alphaproteobacteria (Denison and Okano 2003; Tampakaki et al. 2017a, b; Ferguson et al. 2019) and in the genera *Paraburkholderia*, *Cupriavidus,* and *Trinickia* in the Betaproteobacteria (Gyaneshwar et al. 2011; Estrada-de los Santos et al. 2018). The infection of roots by rhizobia results in the formation on roots (and occasionally stems) of unique organs called nodules (Ferguson et al. 2013) in which the biological nitrogen fixation process takes place (Ferguson et al. 2019). In this process, the bacterial enzyme nitrogenase catalyzes the reduction of atmospheric N_2 to ammonia (Howard and Rees 1996), which is a plant-available N form.

The nodulation process in many, but not all, legumes follows root infection by efficient compatible rhizobia strains. The root infection causes the curling of the root hairs that entrap the rhizobia and then, after the formation of infection threads through these structures, the bacteria enter the root cells (Peleg-Grossman et al. 2007; Fournier et al. 2015). According to Oldroyd and Downie (2008), the induction of cortical cell divisions is necessary for the nodule's morphogenesis. The bacteria within the nodule cells are a differentiated symbiotic form of rhizobia called bacteroids. Each bacteroid is surrounded by the symbiosome (or peribacteroid) membrane (Denison and Okano 2003; Peleg-Grossman et al. 2007). In the initiation of the rhizobia-legume symbiosis, several compounds (e.g., Nod factors and flavonoids) are implicated. Nod factors are lipochitooligosaccharides secreted by rhizobia that are involved in the initiation of cell divisions in the cortex, which leads to root hair curling and the formation of infection threads (Ibáñez and Fabra 2011; Murray 2011). Flavonoids produced by legume roots activate NodD proteins and consequently the expression of the Nod genes that are implicated in the synthesis of Nod factors (del Cerro et al. 2017).

Taken together, this chemical cross-talk between the rhizobia and the host legume allows the latter to impose a degree of stringency on which bacteria can enter and form a symbiotic nodule, but as it is based on nod genes rather than *nif* genes, it cannot guarantee that the symbiosis will be effective, and hence compatible, but "cheating" rhizobia are considered to be a significant problem for nodulated legumes (Sprent et al. 2017).

Oxygen plays a significant role in nitrogen fixation because an adequate supply of oxygen in the nodules is needed by bacteroids for respiration (Denison and Okano 2003). See review by Minchin et al. (2008). It is also important to point out that the nitrogen fixation process is characterized by high energy (ATP) demands (Rutten and Poole 2019), because the reduction of 1 molecule of N_2 to ammonia utilizes at least 16 molecules of ATP (Maier 2004). These energy requirements are covered by the respiration of bacteroids (Miller et al. 1988), but sufficient transport of carbohydrates to the roots is needed to maintain sufficiently high respiration rates. Nevertheless, excessive oxygen concentrations can inactivate the nitrogenase (Denison and Okano 2003), and thus the protein leghaemoglobin (Lb) is also an important component of the nodules, as it acts as an oxygen carrier that facilitates a controlled flux of oxygen to the nitrogen-fixing bacteroids (Denison and Okano 2003). Furthermore, Lb protects nitrogenase from being inactivated by free oxygen, while the Lb-bound oxygen is accessible to bacteroids (Abdelmajid et al. 2008; Rutten and Poole 2019). The internal red-pink color of the nodules is due to the presence of leghaemoglobin (Rejili et al. 2012). Abdelmajid et al. (2008) linked higher nitrogen fixation capacity with a higher accumulation of leghaemoglobin in the nodules.

3 Agronomic Impact

3.1 Nitrogen Supply via Biological Nitrogen Fixation (BNF)

Over the past decades, the excessive application of inorganic nitrogen fertilizers has resulted in groundwater contamination with nitrates (Lv et al. 2019). Groundwater pollution via leaching of these pollutants (NO_3^-) is one of the most serious environmental problems and is positively related to high nitrogen fertilization rates (Vinod et al. 2015; Zheng et al. 2019). Thus, to reduce the groundwater pollution with nitrates, it is important to reduce the excess application of inorganic fertilizers in agricultural fields and/or to apply organic nitrogen sources such as compost or manure. The use of legumes as green manures or the inclusion of legumes in crop rotation systems is alternative to inorganic nitrogen fertilizers that can contribute to higher crop yields and improved soil quality (Castro et al. 2017; Ntatsi et al. 2018).

Biological N fixation by legumes (e.g., faba bean, lentil, pea, chickpea, alfalfa, red clover etc.) ranges from 21 to 389 kg ha^{-1} (Table 1) (Cazzato et al. 2012; Nimmo et al. 2013; Büchi et al. 2015; Hossain et al. 2016; Snapp et al. 2017; Akter et al. 2018; da Silva Júnior et al. 2018; Dhamala et al. 2018; Ntatsi et al. 2018; Pampana et al. 2018; Ntatsi et al. 2019).

The N_2-fixation capacity of legumes (e.g., the proportion of N derived from the atmosphere [%Ndfa] and biomass productivity) is mainly dependent on plant species, genotypes, symbiotic bacteria (e.g., *Rhizobium* spp.) strains, and environmental conditions (Büchi et al. 2015; Hossain et al. 2016; Akter et al. 2018; Ntatsi et al. 2018; Benjelloun et al. 2019; Ntatsi et al. 2019).

Despite the fact that legumes contribute to nitrogen enrichment of soil BNF, it is worth mentioning that their over-frequent use of these plant species can also lead to nitrate leaching (De Notaris et al. 2018; Hansen et al. 2019). Thus, it is important to optimize the use of legumes (e.g., appropriate crop rotation sequences, mixtures of legumes, and nonlegumes) in order to reduce the risk of nitrate leaching (Hansen et al. 2019; Rakotovololona et al. 2019).

Table 1 Biological nitrogen fixation (BNF) capacity (kg ha^{-1}) of commonly cultivated legumes

Common name	Scientific name	BNF (kg ha^{-1})	Cultivation area	References
Faba bean	*Vicia faba* L.	118.6–311	Greece, Italy	Ntatsi et al. (2018), Pampana et al. (2018)
Pea	*Pisum sativum* L.	36.6–125.3	Canada, Greece	Hossain et al. (2016), Ntatsi et al. (2019)
Common vetch	*Vicia sativa* L.	107–131	Switzerland	Büchi et al. (2015)
Grass pea	*Lathyrus sativus* L.	101–149	Switzerland	Büchi et al. (2015)
White lupin	*Lupinus albus* L.	53.1–64.1	Italy	Cazzato et al. (2012)
Chickpea	*Cicer arietinum* L.	21.0–103.6	Canada	Hossain et al. (2016)
Lentil	*Lens culinaris* Med.	23.0–86.8	Switzerland, Canada	Büchi et al. (2015), Hossain et al. (2016)
Common bean	*Phaseolus vulgaris* L.	16.3–71.9	Canada	Akter et al. (2018)
Cowpea	*Vigna unguiculata* (L.) Walp.	36–75	Brazil	da Silva Júnior et al. (2018)
Soybean	*Glycine max* (L.) Merr.	90–95	USA	Snapp et al. (2017)
Alfalfa	*Medicago sativa* L.	103–209	Canada, China	Nimmo et al. (2013)
Egyptian clover	*Trifolium alexandrinum* L.	35–59	Switzerland	Büchi et al. (2015)
Red clover	*Trifolium pretense* L.	35.4–389	Denmark, USA	Snapp et al. (2017), Dhamala et al. (2018)

3.2 Pre-crop Benefits Through a Combination of Residual Nitrogen and Break-Crop Effects

Legume cropping, including rotation, intercropping, green manure, and legume-enriched pastures, shows significant advantages over nonlegume systems in terms of fertilizer use and hence emissions of the greenhouse gases CO_2 and N_2O (Jensen and Hauggaard-Nielsen 2003). Grain and forage legumes, by virtue of their symbiosis with N_2-fixing bacteria, can reduce the need for N fertilizer application. If legume cropping becomes more widely adopted, this could reduce the demand for manufactured fertilizer (Jensen et al. 2012). In terms of soil N inputs from BNF, an approximate value of 9 kg N mineralized per ton of stubble may be possible for grain legume crops, with higher transfer values being recorded for forage legume systems—15 to 20 kg N per tonne (Peoples et al. 2004, 2009, 2017). Typical rates of BNF for grain and forage legumes are between 100 and 200 kg shoot N ha^{-1} per year or growing season (Peoples et al. 2019).

Reduced fertilizer usage associated with legume cropping is only suitable after the successful establishment of the root-nodule symbiosis, adequate levels of BNF, and appropriate crop management practices to maintain N_2 fixation. This may involve inoculation of plants with appropriate strains of rhizobia to improve nitrogen fixation, carrying a cost in terms of energy and GHG emissions, and careful monitoring of soil N. Liming of soils is important too in maintaining N_2 fixation, N_2 fixation having the potential to acidify unbuffered soils and hence inhibit nitrogenase activity. This acidifying activity has the potential also to mineralize inorganic phosphate and reduce the requirement for P fertilizer addition (Williams et al. 2017).

Skowronska and Filipek (2014), in their review of life cycle analysis studies on fertilizer manufacture, provide illustrative data on the extent of GHG savings possible through reduced fertilizer production. Depending on the type of N fertilizer, the combined GHG cost of production, packaging, and delivery ranges from 1.9 to 6.3 kg CO_2e (carbon dioxide equivalent) kg^{-1} fertilizer The GHG cost for P fertilizer is considerably less, 0.6–1.66 kg CO_2e kg^{-1} fertilizer, with manufacture of calcium carbonate for soil amendment accounting for 0.15 kg CO_2e kg^{-1} (Skowronska and Filipek (2014).

Calculation of the reduction in field GHG emissions possible with legume cropping is problematic given the wide variance in data available due to differing crops, soils, climate, management, and most significantly the type of measurement and the time course of measurements employed. Using values averaged across 67–71 site-years of data, Peoples et al. (2019) report an overall reduction in N_2O emissions for legume crops compared with N fertilized crops and pastures of approximately 59%, assuming N_2O emissions of 0.47 t CO_{2e} ha^{-1} for legume crops and 1.16 t CO_{2e} ha^{-1} for N fertilized crops and pastures.

3.3 *Increased Crop Diversification and Biodiversity*

Modern intensive agricultural systems are relatively simplified, focussing on a small number of crop species, often in monocultures, and reliant on mineral fertilizers and chemical crop protection to maximize their productivity. Heterogeneous crop systems, however, can show improved production efficiency, yield stability, and resilience to environmental stresses. Legume crops have great potential for optimizing these benefits, whether by increasing the diversity of crops within the crop rotation sequence or as components of crop species mixtures. The positive contribution of legumes to diversification arises directly from legume-specific traits and indirectly from their reduced reliance on agronomic inputs. This is underpinned primarily by the ability of legumes to fix atmospheric nitrogen into nitrogen-rich organic compounds—as well as their capacity to capitalize on generating a symbiotic- or facilitative-microbiome in the rhizosphere (Chen et al. 2019).

Nonlegume crops show up to 30% greater biomass production in a legume-supported rotation: the benefit of BNF is estimated to be maximal when grain and forage legume crops are present in half of the years of the crop sequence (Iannetta et al. 2016). Despite this, grain legume production in Europe is falling (Magrini et al. 2016) and is characterized by only a few legume crop species, which has constrained progress in legume crop improvement. There are ample opportunities to diversify the range of legume crops and take advantage of adaptive traits in of orphan legume species to improve, for example, their resilience to biotic and abiotic stresses (Cullis and Kunert 2017). By enhancing soil nutrient supply and function, legumes can promote nutrient acquisition by nonleguminous plants growing in a species mixture. Legume crops exhibit multiple traits that complement or facilitate the growth of nonlegume species, leading to more efficient use of resources. This allows greater productivity and profitability per unit land area to be achieved with intercropping compared with monocultures (Martin-Guay et al. 2018). Floral resource provision in legume-supported mixtures, along with increased canopy and root system heterogeneity, and reduced reliance on agronomic inputs, can promote the abundance and activity of beneficial organisms, which facilitate ecosystem services such as pollination, nutrient cycling, and suppression of pests, diseases, and weeds (Everwand et al. 2017).

4 Economic Impact

Understanding the economics of legume systems requires an analysis of the factors influencing the equilibrium between supply (farmers) and demand (consumers) and the interlinkages along the supply chain, while assessing the impact of any shocks to the system on other aspects such as trade and environment. As is the case of any other agricultural industry, but even more so due to their benefits to the public good (European Parliament 2013), changes to the equilibrium between the supply and demand of legumes translates into wider long-term effects, and as such, an analysis of legume production (e.g., assessment of farm profitability) is incomplete and potentially incorrect if not coordinated with an analysis of demand, and of the corresponding ripples on trade, environment, and health.

The economics of legumes in the European Union (EU) shows a production trend closely correlated to the different types of subsidies and payments linked to the Common Agricultural Policy (CAP) reforms, and the global market prices for fertilisers (Fig. 1). Linked to similar factors affecting livestock production, feed demand mirrors production trends for dry pulses (only starting in the 1980s in the case of beans). Imports mirror the demand for processing in the case of soybeans, and mirror the production trends for peas and, respectively, lentils, lupins and other pulses, while showing an opposite trend to the domestic production of beans (following the rise in the 1980s in its use for feed and food).

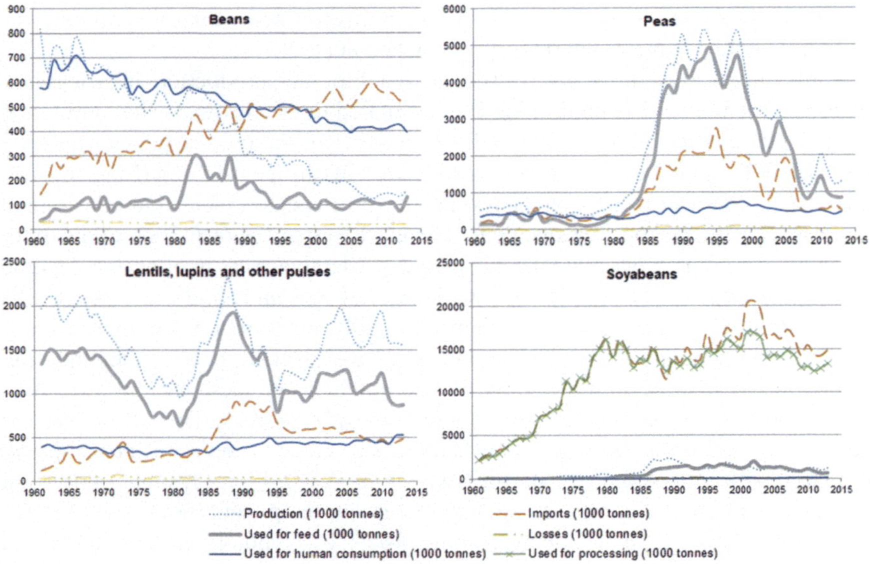

Fig. 1 Legumes production and imports, uses and losses in the European Union. *Source* Own creation based on FAOSTAT data (extracted November 2019)

The cultivation of dry pulses (i.e., grain legumes except soybeans) in EU countries is significantly more frequent in: regions with higher receipt of voluntary coupled CAP support to protein crops; regions with higher shares of organic farming; regions with a more important role of legume consumption in regional diets; regions with relatively deep soils and; regions displaying lower competition for land use with sunflower. Livestock density and share of irrigable agricultural areas are significantly negatively correlated with the share of dry pulses. Up to a certain temperature sum maximum, also higher temperature sums seem to be beneficial for the cultivation of dry pulses. In contrast to dry pulses, regional soybean shares in the arable area are positively correlated with a region's distance to the next main port and with the share of irrigable agricultural area. Agglomeration and spillover effects may matter (i.e., farms located in the neighborhood of dry legume producers are also more likely to commence cultivation of dry legumes; regions with a high share of dry pulses tend to be close to each other), as in the case of dry pulses where a significant spatial lag coefficient was found. Such effects, however, are likely to be effective on a spatial scale smaller than the regional level. Potentially significant causal factors, which have not been tested due to poor data availability, that may be positively linked to legume production include proximity to processing facilities and trading companies and access to extension services and regional networks and training programs (Oré Barrios et al. 2020). Other factors well acknowledged in the literature (European Parliament 2013) with positive causal effects on the cultivation of legumes are market factors, i.e., producer prices for outputs (pulses) and inputs (nitrogen fertilisers).

The economic circumstances of farms cultivating legumes are linked to some of the factors mentioned above, e.g., larger organic farms show higher profitability. There are also indirect economic benefits of legume cultivation such as the lower cost of agricultural inputs (nitrogen and tillage cost saving) and yield effects on other crops (e.g., cereals included in the rotation). While profits and the economic sustainability of the farm are necessary, they may not always be sufficient and farmers' decision-making may be influenced by noneconomic factors, such as their perceptions of how what they create affects other issues beyond the farm gate, such as the environment and human health.

Similarly, consumers' choices may be influenced by environmental and health concerns as opposed to purely economic reasons and assessing the weight of the different attributes of choice would help predict sustainable changes in shopping habits and subsequent consumption patterns. A study on the own-price elasticities of legumes shows that consumer's demand can only change significantly if factors other than price are considered, such as provision of targeted campaigns and better communication of legumes' health and environmental benefits, better availability of healthy convenience foods, access to information on cooking, and easy recipes (Akaichi 2019).

As represented in Fig. 1, while the consumption trend for beans has been in a stable decrease, the consumption of peas, lentils, lupins, and other dry pulses shows a gentle but steady increase, likely correlated to a slow change in consumer diets reflecting a healthier pattern.

While slow, changes in consumption patterns to include more legumes are apparent and need to be translated into production patterns. The current EU demand for legumes is met partly by domestic production, partly by imports (Fig. 1) and equilibrium analysis is necessary to assess the sustainability of the whole sector when faced with shocks such as price fluctuations in the context of higher dependency on imports, or changes in environmental policies leading to stronger incentives to EU producers and thus a larger share of the demand being met by local production.

5 Social Impact

5.1 Nutrition and Food Security

Food insecurity is a reality for millions of people and households, especially in poor and developing countries (FAO 2019). Recent data reveals that over 2 billion people around the world do not have regular access to safe, nutritious and sufficient food, including eight percent of the population in Northern America and Europe (FAO 2019). Indeed, more than 820 million people in the world were still hungry in 2018. Such living conditions increase the risk of malnutrition and ultimately impair the health of populations (FAO 2019). It is recognized that lack of protein-energy intake, as well as micronutrient deficiencies, are major undernutrition triggers, both frequently associated with more severe food insecurity states (FAO 2019; Webb et al. 2018). Grain legumes could be part of the solution for these problems; however, over time they have been significantly depreciated within human diets, and legume crops are yet greatly under-cultivated (Foyer et al. 2016). In fact, legumes are relatively invisible actors of our food system.

A 2015 Joint Research Center policy foresight assessing the role of EU policies for global food security called for a "Common Food Systems Policy," but failed to even mention protein crops, pulses, or legumes (Maggio et al. 2015). Whereas unique agro-ecological benefits of legumes are gaining more recognition and are slowly being acknowledged in food policy debates, their impact is still at the small-scale of home-grown legume production and consumption. Nevertheless, a recent public food procurement mandate in Portugal (Graça et al. 2018) successfully increased home-grown legume consumption. On the large scale, only the cultivation of non-GMO soybean for feed has increased in Italy, Germany, France, and Poland due to multiple support policies. In other geographical frameworks, such as Canada, cost-benefit analyses have revealed the positive impact of legume consumption (100 g cooked legumes/day by 50% of the population) in combination with a low glycaemic index or high fiber diet on healthcare costs. In addition, human productivity costs (reductions of roughly $370 million Canadian dollars), particularly related with cardiovascular disease and type 2 diabetes (Abdullah et al. 2017), have driven the 2019 Canada's Food Guide to emphasize plant-based protein foods within the "protein foods" group. Beans and lentils have been placed at the top of the list, before nuts and other seeds and animal protein products (meat, poultry, fish, eggs, and dairy foods) (https://food-guide.canada.ca/en/).

Historically, food and agricultural policy often lagged behind nutrition science, though more recently this position has started to change. Legumes may already be leading a green food revolution (Tenkouano 2011), because they have been identified as critical to provide nutrients and balanced diets and provide nutritional security with minimal use of resources, as well as to facilitate social-eating when cultivated in small areas within backyards or home-, school- and community-gardens (Keatinge et al. 2012). Their high protein content (17–30%) (Boye et al. 2010) associated with relevant nutritional richness (Marinangeli et al. 2017), turns grain legumes into better affordable nutritive options, comparatively to more expensive animal-based protein food sources, such as meat or dairy products, which may be less achievable among food insecure contexts. Legumes are also important food sources of slowly digested complex carbohydrates (~50–65%) (Havemeier et al. 2017) and fiber (~30/100 g, with low glycaemic index; dry weight) (Tosh and Yada 2010), as well as minerals (Grela and Samoli 2017) like magnesium, iron, potassium, phosphorus or zinc and several complex B vitamins (Mudryj et al. 2014), namely B1, B6, and B9. On the other hand, they possess low energy density in terms of fat (1.3 kcal/g cooked), providing mostly mono- and polyunsaturated fats (Grela and Samoli 2017).

Grain legumes are also relevant dietary sources of health-protective bioactive compounds (Singh et al. 2016). Last but not least, grain legumes hold versatile technological and cooking properties providing excellent opportunities among food industry to be used in the production of several convenience value-added food products, like flours, snacks, infant, or sports foods (Asif et al. 2013). The possibility to store grain legumes for long periods of time without altering their nutritional value results in one of their best features helping minimize food waste and therefore food insecurity. Although diet is the most apparent link by which agro-food systems affect our health, the role of legumes to provide solutions to the double burden of inadequate dietary intake (undernutrition) and excess food intake (overnutrition) in an unequal world is not widely considered or understood. Still, there seems to be a consensus on a sustainable and nutrient-rich plant-based diet (ovo-lacto-vegetarian and pescetarian) that may provide optimal synergy between nutrition health and environmental sustainability (Springmann et al. 2016; van Dooren et al. 2017). Indeed, markets for plant protein and fiber-based diets are rapidly growing (Logatcheva and van Galen 2015). However, without appropriate and careful reframing, public health improvement and environmental sustainability arguments in favor of a diet-change will not be enough to engage stakeholders and beneficiaries (c.f. de Boer and Aiking 2017) and to achieve a paradigm shift (Mason and Lang 2017).

5.2 Mitigating Effect on Climate Change

The food habits of the 7.7 billion people who inhabit our planet have been threatening all life domains, with extremely worrying expression at the climate change level (Macdiarmid and Whybrow 2019). Animal-based foods are a significant component of food production worldwide and meat or meat products are major dietary protein sources, especially across more westernized countries (Willett et al. 2019). Nevertheless, livestock production accounts for considerably high amounts of total greenhouse gas (GHG) emissions and other pollutants, together with increasing demands for scarce water resources and the promotion of soil erosion (Godfray et al. 2018). If such production and consumer patterns persist, it is expected that by 2050, there will be a 50–80% increase in GHG emissions and a ~13–66% expansion in land used for crop production (Clark et al. 2019), both associated with increased threats to biodiversity (Tilman et al. 2017).

Globally, this will also translate into ~15% more water use, as well as ~50% and ~100% more nitrogen and phosphorous fertilizer use, respectively (Clark et al. 2019). According to the literature, the production of plant-based foods has the lowest environmental impact, achieving for example 25–150 times less GHG emissions than ruminants produce for meat production (Clark et al. 2019). As such, a dietary shift toward more plant-based diets, providing more eco-friendly protein food sources, is being suggested (Willett et al. 2019). Grain legumes have caught the public's attention over the past few years, being considered as nutritious animal food alternatives and highlighted for their key role within sustainable food production systems (Calles et al. 2019b). Among several important features, the atmospheric nitrogen fixation capacity of legumes reduces the need for chemical fertilizer use during crop cultivation, helping reduce GHG production, like carbon dioxide (CO_2) and nitrous oxide

(N_2O). In fact, it appears that by substituting meat with grain legumes could lead to a reduction of up to 74% in GHG emissions, enabling the achievement of the 2020 target for the US (Harwatt et al. 2017).

Grain legume crops are also able to release high-quality organic matter into the soil and facilitate nutrient circulation in the soil, as well as promote water retention minimizing fossil energy inputs in the agricultural food production chain (Stagnari et al. 2017). Indeed, the water footprint per gram of protein for grain legumes seems 1.5 times smaller than for milk, eggs and chicken meat (Mekonnen and Hoekstra 2012). When beef production is considered, such difference becomes six times less, again in favor of grain legumes (Mekonnen and Hoekstra 2012). It is possible then that the increase in production and consumption of grain legumes could be a cornerstone to ensure food and nutritional security, in light of ongoing global climate change (Willett et al. 2019).

Trying to empower this message, the United Nations declared 2016 as the International Year of Pulses (Calles et al. 2019a). In the same year, the slogan for the celebration of the World Food Day was "Climate is changing. Food and agriculture must too" (FAO 2016). Since then, grain legumes have been in the spotlight across worldwide climate change mitigation strategies (Willett et al. 2019). Likewise, a recent World Bank report states that fruits, vegetables, and legume-based products should be supported at the expense of cereals, palm oil, and sugar, while the subsidies and price support mechanisms for unhealthy ingredients should be abandoned. However, to find pathways to more sustainable agro-food systems and innovative policy solutions, greater civil society engagement and more effective public-sector research and education efforts are required (Abarca-Gómez et al. 2017).

5.3 Cultural Valorisation

Traditional food products are naturally linked with local resources and cultural heritage of involved territories. Indeed, gastronomy and several cultural practices related to food represent a distinctive element between different populations. All over the world, it is possible to find a high number of traditional dishes containing grain legumes prepared and cooked in different ways and with unique organoleptic properties, combinations that people easily associate with comfort, societal wellness, and festive food (Polak et al. 2015). In this regard, FAO's Information Network on Post-harvest Operations maintains an updated database of more than 850 recipes from more than 50 countries, where not only the traditional recipes are preserved for posterity but also less common ingredients are presented and promoted (FAO 2015).

Cooking with pulses offers several advantages that go from their affordability for family budgets (food security asset as previously mentioned), their long shelf-life where nutritional value is preserved throughout, to their important organoleptic versatility where savory yet subtle tastes enable their harmonious inclusion in a wide range of cuisines and flavor profiles. Kidney beans, black beans, pigeon peas, chickpeas, and lentils are featured regularly in legume recipes supporting their popularity for cuisine management and making these grain legumes a staple of many diets. Such high versatility enables their inclusion in all types of servings, including entrees, soup bases, side dishes, salads, stews, and desserts (Figueira et al. 2019).

Across the globe, consumption of traditional legume foods is popular in Brazil (black beans with rice) (FAO 2015), in India (dhal, pappadums) (Appel 2005; Misra 2011), in Mexico (refried kidney beans or chili) (FAO 2015) in Middle Eastern countries (hummus, falafel, nakee, bajelah, fasolia) (Kamboj and Nanda 2018; Alalwan et al. 2017), and in Mediterranean countries (navy bean soup, bean stew "feijoada," fave bianche) (Lăcătușu et al. 2019; Renna et al. 2015). In addition, legume flours are traditionally incorporated into many different foods either as batter for vegetables (onions, leeks, aubergine) or as ingredients for savory snacks, that when combined with fried whole grain legumes (Alalwan et al. 2017), may be consumed as a mid-morning or a mid-afternoon meal.

New foods using grain legumes are emerging on the market and may be an interesting alternative to increase their consumption; nevertheless, efforts need to continue to be made in order to promote traditional dishes and associated nutritional value. A balance between old and new is undoubtedly the most favorable means to preserve the heritage and health of a population.

5.4 *Increasing the Social Acceptance of Legumes*

More and more, populations should be aware of the impact of their choices, namely dietary choices, on the planet's sustainability. The grain legumes should be considered as good options even though these foods continue to be unpopular, especially in developed countries (Perignon et al. 2017). Some misconceptions related to their consumption and subsequent gastrointestinal problems or presumed impact on weight gain, the time needed for their soaking and cooking, and the perceived lack of appealing preparations certainly contribute to their scarce utilization. The promotion of legumes should start among children for several reasons: this food group is unpopular among children and, consequently, their consumption is scarce at this age; dietary changes are easily implemented among children, since their food habits are not so fixed and evidence suggests that early food patterns tend to be maintained throughout life. On the other hand, the issues related to a "healthy planet" are motivating targets for children (Sadegholvad et al. 2017; Smith et al. 2016).

Taste modulation starts very early in the lifecycle and grain legumes could be gradually introduced as early as the eighths or ninth month of life (Fewtrell et al. 2017). However, since the insertion of the child within the family diet is crucial, it is important that these foods are offered regularly and the parents also eat them. Their inclusion in school meals is also fundamental, inclusively as a partial substitute of animal protein. Schools are the ideal place to widely spread concepts regarding nutritional and ecological benefits of legumes, having as major targets the children and their relatives (Smith et al. 2016). Issues related with the practicability of grain legume cooking could be easily overlapped with canned alternatives or even with pre-preparation and freezing in portions. Innovative foods containing grain legumes and ready-to-eat options are also a good solution.

6 Conclusions

Our global society faces several challenges that are negatively impacting the health of our planet, our people, and our agricultural economies. Increases in atmospheric CO_2 and other greenhouse gases are altering our climate in ways that are straining our food systems. Our growing human population is worsening our ability to meet food security and nutrition needs. And various production and processing costs are limiting the economic potential of different players in our food value chain. While legume crops cannot solve all of these issues, we have attempted to show how an expanded utilization of legumes in our food production systems could supply a number of benefits to mitigate these problems.

Our future challenge, then, is to scale up the use of legumes in a way that achieves the most social, economic, and environmental benefits. This will require a change in society's acceptance of legumes and a willingness to divert some cereal-based or meat-based production systems to legume systems and primarily to grain legumes. Fortunately, the benefits gained from legumes are highly integrated and complementary, providing clear value to all members of the legume supply chain (producers, food industry, and consumers). This should facilitate increased acceptance, but a concerted effort among agronomists, nutritionists, environmentalists, and others will be needed to educate and promote the societal benefits of legumes. A reshaping of public and policy maker's opinions will be critical to ensure that legumes are elevated to a new level in our global food network.

Acknowledgements This research is supported by the EU Horizon 2020 Research and Innovation Program project "TRansition paths to sUstainable legume based systems in Europel" (TRUE, www.true-project.eu) funded under Grant Agreement number 727973. MV, CS, EV, EP, HG, AP, AD are also supported by the Foundation for Science and Technology (FCT) under project UID/Multi/50016/2013. MAG is supported by the US Department of Agriculture through project 3060-21650-001-00D. The contents of this publication do not necessarily reflect the views or policies of the US Department of Agriculture, nor does mention of trade names, commercial products, or organizations imply endorsement by the US Government. Scotland's Rural College (SRUC) is supported by the Scottish Government's Rural and Environmental Science and Analytical Services (RESAS), a Division of the Scottish Government. The James Hutton Institute (PI, MM, AK) is supported by the Scottish Government's Rural and Environmental Science and Analytical Services (RESAS), a Division of the Scottish Government.

References

Abarca-Gómez L, Abdeen ZA, Hamid ZA et al (2017) Worldwide trends in body-mass index, underweight, overweight, and obesity from 1975 to 2016: a pooled analysis of 2416 population-based measurement studies in 128·9 million children, adolescents, and adults. Lancet 390(10113):2627–2642

Abdelmajid K, Karim BH, Chedly A (2008) Symbiotic response of common bean (*Phaseolus vulgaris* L.) to iron deficiency. Acta Physiol Plantarum 30:27–34

Abdullah M, Marinangeli CPF, Jones PJH, Carlberg JG (2017) canadian potential healthcare and societal cost savings from consumption of pulses: a cost-of-illness analysis. Nutrients 9:793

Akaichi F (2019) British consumers' demand for legumes. In: Presentation at the 'Transition paths to sustainable legume-based systems in Europe' TRUE annual meeting, Porto, July 2019

Akter Z, Pageni BB, Lupwayi NZ, Balasubramanian PM (2018) Biological nitrogen fixation by irrigated dry bean (*Phaseolus vulgaris* L.) genotypes. Can J Plant Sci 98(5):1159–1167

Alalwan TA, Mandeel QA, Al-Sarhani L (2017) Traditional plant-based foods and beverages in Bahrain. J Ethnic Foods 4:274–283

Altieri MA, Funes-Monzote MR, Peterson P (2011) Agroecologically efficient agricultural systems for smallholder farmers: contributions to food sovereignty. Agron Sustain Dev 32(1):1–13

Alves BRJ, Boddey RM, Urquiaga S (2003) The success of BNF in soybean in Brazil. Plant Soil 252:1–9

Appel LJ (2005) Effects of protein, monounsaturated fat, and carbohydrate intake on blood pressure and serum lipids: results of the omniheart randomized trial. JAMA J Am Med Assoc 294:2455–2464

Asif M, Rooney LW, Ali R, Riaz MN (2013) Application and opportunities of pulses in food system: a review. Crit Rev Food Sci Nutr 53(11):1168–1179

Asouti E, Fuller DQ (2013) A contextual approach to the emergence of agriculture in southwest Asia: reconstructing early Neolithic plant-food production. Curr Anthropol 54(3):299–345

Benjelloun I, Thami Alami I, Douira A, Udupa SM (2019) Phenotypic and genotypic diversity among symbiotic and non-symbiotic bacteria present in chickpea nodules in Morocco. Front Microbiol 10:1885

Boye J, Zare F, Pletch A (2010) Pulse proteins: processing, characterization, functional properties and applications in food and feed. Food Res Int 43(2):414–431. https://doi.org/10.1016/j.foodres.2009.09.003

Büchi L, Gebhard CA, Liebisch F, Sinaj S, Ramseier H, Charles R (2015) Accumulation of biologically fixed nitrogen by legumes cultivated as cover crops in Switzerland. Plant Soil 393:163–175

Calles T, Del Castello R, Baratelli M, Xipsiti M, Navarro DK (2019a) The international year of pulses: final report Rome. Available from http://www.wipo.int/amc/en/mediation/rules

Calles T, Xipsiti M, del Castello R (2019b). Legacy of the international year of pulses. Environ Earth Sci 78:124. https://link.springer.com/article/10.1007/s12665-019-8106-6

Castro RE, Sierra A, Mojica JE, Carulla JE, Lascano CE (2017) Effect of species and management of legumes used as green manures in the quality and yield of a forage crop used in livestock systems in the dry tropics. Archivos de Zootecnia 66:99–106

Cazzato E, Laudadio V, Stellacci AM, Ceci E, Tufarelli V (2012) Influence of sulphur application on protein quality, fatty acid composition and nitrogen fixation of white lupin (*Lupinus albus* L.). Eur Food Res Technol 235:963–969

Chai BC, van der Voort JR, Grofelnik K, Eliasdottir HG, Klöss I, Perez-Cueto FJA (2019) Which diet has the least environmental impact on our planet? A systematic review of vegan, vegetarian and omnivorous diets. Sustainability 11(15):4110. https://doi.org/10.3390/su11154110

Chen J, Shen W, Xu H, Li Y, Luo T (2019) The composition of nitrogen-fixing microorganisms correlates with soil nitrogen content during reforestation: a comparison between legume and non-legume plantations frontiers in microbiology, vol10, p 508. https://doi.org/10.3389/fmicb.2019.00508

Clark M, Science NR, Paul S, States U (2019) Changing dietary patterns as drivers of changing environmental impacts. In: Encyclopedia of food security and sustainability, vol 1. Elsevier, pp 172–177. http://dx.doi.org/10.1016/B978-0-08-100596-5.21992-6
Cullis C, Kunert KJ (2017) Unlocking the potential of orphan legumes. J Exp Bot 68:1895–1903
da Silva Júnior EB, Favero VO, Xavier GR, Boddey RM, Zilli JE (2018) Rhizobium inoculation of cowpea in Brazilian cerrado increases yields and nitrogen fixation. Agron J 110:722–727
de Boer J, Aiking H (2017) Pursuing a low meat diet to improve both health and sustainability: how can we use the frames that shape our meals? Ecol Econ 142:238–248
De Notaris C, Rasmussen J, Sørensen P, Olesen JE (2018) Nitrogen leaching: a crop rotation perspective on the effect of N surplus, field management and use of catch crops. Agric Ecosyst Environ 255:1–11
del Cerro P, Pérez-Montaño F, Gil-Serrano A, López-Baena FJ, Megías M, Hungria M, Ollero FJ (2017) The Rhizobium tropici CIAT 899 NodD2 protein regulates the production of Nod factors under salt stress in a flavonoid-independent manner. Sci Rep 7:46712. https://doi.org/10.1038/srep46712
Denison RF, Okano Y (2003) Leghaemoglobin oxygenation gradients in alfalfa and yellow sweetclover nodules. J Exp Bot 54:1085–1091
Dhamala NR, Rasmussen J, Carlsson G, Søegaard K, Eriksen J (2018) Effects of including forbs on N2-fixation and N yield in red clover-ryegrass mixtures. Plant Soil 424:525–537
Estrada-de los Santos P, Palmer M, Chávez-Ramírez B, Beukes C, Steenkamp ET, Briscoe L, Khan N, Maluk M, Lafos M, Humm E, Arrabit M, Crook M, Gross E, Simon MF, Dos Reis Junior FB, Whitman WB, Shapiro N, Poole PS, Hirsch AM, Venter SN, James EK (2018) Whole Genome Analyses Suggests that Burkholderia sensu lato Contains Two Additional Novel Genera (Mycetohabitans gen. nov., and Trinickia gen. nov.): Implications for the Evolution of Diazotrophy and Nodulation in the Burkholderiaceae. Genes 9:389
European Parliament's Committee on Agriculture and Rural Development (2013) Environmental role of protein crops in the new common agriculture policy. In: Brussels: European Parliament. Retrieved from http://www.legumefutures.de/images/The_environmental_role_of_protein_crops_in_the_new_Common_Agricultural_Policy.pdf
Everwand G, Cass S, Dauber J, Williams M, Stout J (2017) Legume Crops and Biodiversity. In: Murphy-Bokern D, Stoddard FL, Watson CA (eds) Legume-supported cropping systems. CABI International, UK, pp 55–69
FAO, IFAD, UNICEF, WFP and WHO (2019) The state of food security and nutrition in the world 2019. Safeguarding against economic slowdowns and downturns. Rome, FAO. Licence: CC BY-NC-SA 3.0 IGO
Ferguson B, Lin M-H, Gresshoff PM (2013) Regulation of legume nodulation by acidic growth conditions. Plant Signal Behav 8(3):e23426. https://doi.org/10.4161/psb.23426
Ferguson B, Mens C, Hastwell A, Zhang M, Su H, Jones C, Chu X, Gresshoff P (2019) Legume nodulation: the host controls the party. Plant Cell Environ 42:41–51
Fewtrell M, Bronsky J, Campoy C et al (2017) Complementary feeding: a position paper by the European Society for Paediatric Gastroenterology, Hepatology, and Nutrition (ESPGHAN) committee on nutrition. J Pediatr Gastroenterol Nutr 64(1):119–132
Figueira N, Curtain F, Beck E, Grafenauer S (2019) Consumer understanding and culinary use of legumes in Australia. Nut 11(7):1575
Food and Agriculture Organization (2015) INPHO cookbook http://www.fao.org/in-action/inpho/resources/cookbook/en
Food and Agriculture Organization (2016) The climate is changing. Food and agriculture must too. Available from https://login.libproxy.rpi.edu/login?qurl=https%3a%2f%2fsearch.proquest.com %2fdocview%2f1868155818%3faccountid%3d28525%250A, http://sfx-serv.lib.rpi.edu/locater ?url_ver=Z39.88-2004%26rft_val_fmt=info:ofi/fmt:kev:mtx:journal%26genre=unknown%26 sid=ProQ:ProQ%3Aabiglobal%26atitle=The+climate+
Food and Agriculture Organization Stat (2019) Food and agricultural organization of the United Nations, data. http://www.fao.org/faostat/en/#data/QC. Last access 14 Nov 2019

Fournier J, Teillet A, Chabaud M, Ivanov S, Genre A, Limpens E, Carvalho-Niebel FD, Barker D (2015) Remodeling of the infection chamber before infection thread formation reveals a two-step mechanism for rhizobial entry into the host legume root hair. Plant Physiol 167:1233–1242

Foyer CH, Cooper JW, Fodor N, Colmer TD, Considine MJ, Lam H-M et al (2016) Neglecting legumes has compromised human health and sustainable food production. Nat Plants 2(8):1–10. https://doi.org/10.1038/nplants.2016.112

Graça P, Gregório MJ, de Sousa SM, Brás S, Penedo T, Carvalho T et al (2018) A new interministerial strategy for the promotion of healthy eating in Portugal: implementation and initial results. Health Res Policy Syst 16(1):102

Grela ER, Samoli W (2017) Content of minerals and fatty acids and their correlation with phytochemical compounds and antioxidant activity of leguminous seeds. Biol Trace Elem Res 180:338–348

Gyaneshwar P, Hirsch AM, Moulin L, Chen WM, Elliot GN, Bontemps C, Estrada de Los Santos P, Gross E, dos Reis Junior FB, Sprent JI, JPW Y, James EK (2011) legume-nodulating beta-proteobacteria:diversity, host-range and future prospects. Mol Plant Microb Interact 24:1276–1288

Hagmann D, Siegrist M, Hartmann C (2019) Meat avoidance: motives, alternative proteins and diet qualityin a sample of Swiss consumers. Public Health Nutr 22(13):2448–2459

Hancock JF (2012) Plant evolution and the origin of crop species. CABI. James Hancock, Michigan State University, USA, p 256

Hansen S, Frøseth RB, Stenberg M, Stalenga J, Olesen J, Krauss M, Radzikowski P, Doltra J, Nadeem S, Torp T, Pappa V, Watson CA (2019) Reviews and syntheses: review of causes and sources of N_2O emissions and NO_3 leaching from organic arable crop rotations. Biogeosciences 16:2795–2819

Harwatt H, Sabaté J, Eshel G, Soret S, Ripple W (2017) Substituting beans for beef as a contribution toward US climate change targets. Clim Change 143(1–2):261–270

Havemeier S, Erickson J, Slavin J (2017) Dietary guidance for pulses: the challenge and opportunity to be part of both the vegetable and protein food groups. Ann N Y Acad Sci 1392(1):58–66

Hossain Z, Wang X, Hamel C, Morrison MJ, Gan Y (2016) Biological nitrogen fixation by pulse crops on semiarid Canadian prairies. Can J Plant Sci 97:119–131

Howard JB, Rees DC (1996) Structural basis of biological nitrogen fixation. Chem Rev 96:2965–2982

Iannetta PPM, Young M, Bachinger J, Bergkvist G, Doltra J, Lopez-Bellido RJ, Monti M, Pappa VA, Reckling M, Topp CFE, Walker RL, Rees RM, Watson CA, James EK, Squire GR, Begg GS (2016) A comparative nitrogen balance and productivity analysis of legume and non-legume supported cropping systems: the potential role of biological nitrogen fixation. Frontiers Plant Sci 7:1700. https://doi.org/10.3389/fpls.2016.01700

Ibáñez F, Fabra A (2011) Rhizobial Nod factors are required for cortical cell division in the nodule morphogenetic programme of the Aeschynomeneae legume Arachis. Plant Biol 13:794–800

Jensen ES, Hauggaard-Nielsen H (2003) How can increased use of biological N2 fixation in agriculture benefit the environment? Plant Soil 252:177–186

Jensen ES, Peoples MB, Boddey RM, Gresshoff PM, Hauggaard-Nielsen H, Alves BJR, Morrison MJ (2012) Legumes for mitigation of climate change and the provision of feedstock for biofuels and biorefineries. A Rev Agric Sust Dev 32:329–364. https://doi.org/10.1007/s13593-011-0056-7

Kamboj R, Nanda V (2018) Proximate composition, nutritional profile and health benefits of legumes—a review. Legum Res Int J 41:325–332

Keatinge JD, Chadha ML, Hughes JDA, Easdown WJ, Holmer RJ, Tenkouano A, Yang RY, Mavlyanova R, Neave S, Afari-Sefa V, Luther G (2012) Vegetable gardens and their impact on the attainment of the millennium development goals. Biol Agric Hort 28(2):71–85

Lăcătușu CM, Grigorescu ED, Floria M, Onofriescu A, Mihai BM (2019) The mediterranean diet: from an environment-driven food culture to an emerging medical prescription. Int J Env Res Pub Health 16(6):942

Lee GA (2012) Archeological perspectives on the origins of azuki (*Vigna angularis*). Holocene 23(3):453–559

Logatcheva K, van Galen MA (2015) Primary food processing; cornerstone of plant-based food production and the bio-economy in Europe. Wageningen, LEI Wageningen UR (University and Research centre), LEI Report 2015-121. 42 pp

Lopez VH, Vargas-Vasquez MLP, Muruaga JS, Mayek-Pérez YN (2013) Origin, domestication and diversification of common beans: advances and perspectives. Revista fitotecnia mexicana publicada por la Sociedad Mexicana de Fitogenética 36(2):95–104

Godfray HCJ, Aveyard P, Garnett T, Hall JW, Key TJ, Lorimer J, et al. (2018) Meat consumption, health, and the environment. Science 361(6399): eaam5324. http://dx.doi.org/10.1126/science.aam5324

LPWG—The Legume Phylogeny Taxonomy Group (2017) A new subfamily classification of the Leguminosae based on a taxonomically comprehensive phylogeny. the legume phylogeny working group. Taxon 66:44–77

Lv H, Lin S, Wang Y, Lian X, Zhao Y, Li Y, Du J, Wang Z, Wang J, Butterbach-Bahl K (2019) Drip fertigation significantly reduces nitrogen leaching in solar greenhouse vegetable production system. Environ Poll 245:694–701

Macdiarmid JI, Whybrow S (2019) Nutrition from a climate change perspective. Proc Nutr Soc 78(3):380–387. https://doi.org/10.1017/S0029665118002896

Maggio A, Criekinge TV, Malingreau JP (2015) Global food security 2030: assessing trends with a view to guiding future EU policies EUR 27252 EN—Joint Research Centre—Foresight and behavioural insights unit, 39 pp. http://dx.doi.org/10.2788/5992

Magrini M-B, Anton M, Cholez C, Corre-Hellou G, Duc G, Jeuffroy M-H, Meynard J-M, Pelzer E, Voisin A-S, Walrand S (2016) Why are grain-legumes rarely present in cropping systems despite their environmental and nutritional benefits? Analyzing lock-in in the French agrifood system. Environ Econ 126:152–162. https://doi.org/10.1016/j.ecolecon.2016.03.024

Maier RJ (2004) Chapter 5: Nitrogen fixation and respiration: two processes linked by the energetic demands of nitrogenase. In: Zannoni D (eds) Respiration in Archaea and Bacteria. Advances in photosynthesis and respiration, vol 16. Springer, Dordrecht

Marinangeli CPF, Curran J, Barr SI, Slavin J, Puri S, Swaminathan S et al (2017) Enhancing nutrition with pulses: defining a recommended serving size for adults. Nut Rev 75(12):990–1006

Martin-Guay M-C, Paquette A, Dupras J, Rivest D (2018) The new green revolution: sustainable intensification of agriculture by intercropping. Sci Tot Env 615:767–772

Mason P, Lang T (2017) Sustainable diets. Routledge, London. https://doi.org/10.4324/9781315802930

Mekonnen MM, Hoekstra AY (2012) A global assessment of the water footprint of farm animal products. Ecosyst 15:401–415

Miller RW, McRae DG, Al-Jobore A, Berndt WB (1988) Respiration supported nitrogenase activity of isolated Rhizobium meliloti bacteroids. J Cell Biochem 38:35–49

Minchin FR, James EK, Becana M (2008) oxygen diffusion, production of reactive oxygen and nitrogen species, and Antioxidants in Legume Nodules M.J. In: Dilworth et al (eds), Nitrogen-fixing Leguminous Symbioses. Springer, pp 321–362

Misra R (2011) Indian foods: AAPI's guide to nutrition, health and diabetes, 2nd edn. Allied Publishers Private Limited, USA. ISBN 978-81-8424-687-2

Mudryj AN, Yu N, Aukema HM (2014) Nutritional and health benefits of pulses. Appl Physiol Nutr Metab 39(11): 1197–204. http://www.nrcresearchpress.com/doi/abs/10.1139/apnm-2013-0557

Murray JD (2011) Invasion by invitation: rhizobial infection in legumes. Mol Plant-Mic Int 24:631–639

Muzquiz M, Varela A, Burbano C, Cuadrado C, Guillamón E, Pedrosa MM (2012) Bioactive compounds in legumes: pronutritiveand antinutritive actions. Implications for nutrition and health. Phytochem Rev 11:227–244

Nimmo J, Lynch DH, Owen J (2013) Quantification of nitrogen inputs from biological nitrogen fixation to whole farm nitrogen budgets of two dairy farms in Atlantic Canada. Nut Cyc Agroec 96:93–105

Ntatsi G, Karkanis A, Yfantopoulos D, Olle M, Travlos I, Thanopoulos R, Bilalis D, Bebeli P, Savvas D (2018) Impact of variety and farming practices on growth, yield, weed flora and symbiotic nitrogen fixation in faba bean cultivated for fresh seed production. Acta Agric Scand Sect B Plant Soil Sci 38:619–630

Ntatsi G, Karkanis A, Yfantopoulos D, Pappa V, Konosonoka IH, Travlos I, Bilalis D, Bebeli P, Savvas D (2019) Evaluation of the field performance, nitrogen fixation efficiency and competitive ability of pea landraces grown under organic and conventional farming systems. Arch Agron Soil Sci 65:249–307

Oldroyd GED, Downie JA (2008) Coordinating nodule morphogenesis with rhizobial infection in legumes. Ann Rev Plant Biol 59:519–546

Oré Barrios C, Mäurer E, Lippert C (2020) Factors determining the spatial distribution of grain legume cultivation in the EU. In: Schriften der Gesellschaft für Wirtschafts- und Sozialwissenschaften des Landbaues e.V., vol 55 (in press)

Pampana S, Masoni A, Mariotti M, Ercoli L, Arduini I (2018) Nitrogen fixation of grain legumes differs in response to nitrogen fertilization. Exp Ag 54:66–82

Peleg-Grossman S, Volpin H, Levine A (2007) Root hair curling and Rhizobium infection in Medicago truncatula are mediated by phosphatidylinositide-regulated endocytosis and reactive oxygen species. J Exp Bot 58:1637–1649

Peoples MB, Angus JF, Swan AD, Dear BS, Hauggard-Nielsen H, Jensen ES, Ryan MH, Virgona JM (2004) Nitrogen dynamics in legume-based pasture systems. In: Mosier AR, Syers K, Freney JR (eds) Agriculture and the nitrogen cycle. Island Press, Washington DC, pp 245–260

Peoples MB, Brockwell J, Herridge DF, Rochester IJ, Alves BJR, Urquiaga S, Boddey RM, Dakora FD, Bhattarai S, Maskey SL, Sampet C, Rerkasem B, Khans DF, Hauggaard-Nielsen H, Jensen ES (2009) The contributions of nitrogen-fixing crop legumes to the productivity of agricultural systems. Symbiosis 48:1–17. https://doi.org/10.1007/BF03179980

Peoples MB, Swan AD, Goward L, Kirkegaard JA, Hunt JR, Li GD, Schwenke GD, Herridge DF, Moodie M, Wilhelm N, Potter T, Denton MD, Browne C, Phillips LA, Khan DF (2017) Soil mineral nitrogen benefits derived from legumes and the comparisons of the apparent recovery of legume or fertiliser nitrogen by wheat. Soil Res 55:600–615

Peoples MB, Hauggaard-Nielsen H, Huguenin-Elie O, Jensen ES, Justes E, Williams M (2019) The contributions of legumes to reducing the environmental risk of agricultural production. In: Lemaire G, Carvalho PCDF, Kronberg S, Recous S (eds) Agroecosystem diversity: reconciling contemporary agriculture and environmental quality. Elsevier, Academic, Cambridge, pp 123–143. ISBN 978-01-2811-050-8

Perignon M, Vieux F, Soler LG, Masset G, Darmon N (2017) Improving diet sustainability through evolution of food choices: review of epidemiological studies on the environmental impact of diets. Nut Rev 75(1):2–17

Polak R, Phillips EM, Campbell A (2015) Legumes: health benefits and culinary approaches to increase intake. Clin Diab 33(4):198–205

Rakotovololona L, Beaudoin N, Ronceux A, Venet E, Mary B (2019) Driving factors of nitrate leaching in arable organic cropping systems in Northern France. Agric Ecosys Environ 272:38–51

Rejili M, Mahdhi M, Fterich A, Dhaoui S, Guefrachi I, Abdeddayem R, Mars M (2012) Symbiotic nitrogen fixation of wild legumes in Tunisia: soil fertility dynamics, field nodulation and nodules effectiveness. Agric Ecosys Environ 157:60–69

Renna M, Rinaldi VA, Gonnella M (2015) The mediterranean diet between traditional foods and human health: the culinary example of Puglia (Southern Italy). Int J Gastr Food Sci 2:63–71

Roy S, Liu W, Nandety RS, Crook AD, Mysore KS, Pislariu CI, Frugoli JA, Dickstein R, Udvardi MK (2019) Celebrating 20 years of genetic discoveries in legume nodulation and symbiotic nitrogen fixation. Plant Cell https://doi.org/10.1105/tpc.19.00279. (Epub ahead of print)

Rutten PJ, Poole PS (2019) Oxygen regulatory mechanisms of nitrogen fixation in rhizobia. Adv Micro Phys 75:325–389

Sadegholvad S, Yeatman H, Parrish AM, Worsley A (2017) What should be taught in secondary schools' nutrition and food systems education ?Views from prominent food-related professionals in Australia. Nutrients 9(11):1207

Sedivy J, Wu F, Hanzawa Y (2017) Soybean domestication: the origin, genetic architecture and molecularbases. New Phytol 214:539–553

Singh B, Pal J, Khetan S, Singh N (2016) Bioactive constituents in pulses and their health benefits. J Food Sci Technol 4(4):858–870

Skowronska M, Filipek T (2014) Life cycle assessment of fertilizers: a review. Int Agrophys 28:101–110

Smith D, Riddle L, Kerr S, Atterberry K, Lanigan J, Miles C (2016) Barriers and opportunities to serving pulses in school meals in Washington schools. J Child Nutr Manag 40(1):8

Smýkal P, Coyne CJ, Ambrose MJ, Maxted N, Schaefer H, Blair MW, Berger J, Greene SL, Nelson MN, Besharat N, Vymyslický T, Toker C, Saxena RK, Roorkiwal M, Pandey MK, Hu J, Li YH, Wang LX, Guo Y, Qiu LJ, Redden RJ, Varshney RK (2015) Legume crops phylogeny and genetic diversity for science and breeding. Crit Rev Plant Sci 34:43–104. https://doi.org/10.1080/07352689.2014.897904

Snapp S, Wilke B, Gentry LE, Zoellner D (2017) Compost legacy down-regulates biological nitrogen fixation in a long-term field experiment. Agron J 109(6):2662–2669

Sprent JI (2009) Global distribution of legumes. In: Sprent JI (ed) Legume nodulation: a global perspective. Wiley, NJ, USA, pp 35–50. ISBN 978-1-405-18175-4

Sprent JI, Ardley J, James EK (2017) Biogeography of nodulated legumes and their nitrogen-fixing symbionts. New Phytol 215:40–56. https://doi.org/10.1111/nph.14474

Springmann M, Godfray HCJ, Rayner M, Scarborough P (2016) Analysis and valuation of the health and climate change co-benefits of dietary change. Proc Nat Acad Sci 113(15):4146–4151

Stagnari F, Maggio A, Galieni A, Pisante M (2017) Multiple benefits of legumes for agriculture sustainability: an overview. Chem Biol Techn Agr 4:2. https://doi.org/10.1186/s40538-016-0085-1

Tampakaki AP, Fotiadis CT, Ntatsi G, Savvas D (2017a) A novel symbiovar (aegeanense) of the genus Ensifer nodulates Vigna unguiculata. J Sci Food Agr 97:4314–4325

Tampakaki AP, Fotiadis CT, Ntatsi G, Savvas D (2017b) Phylogenetic multilocus sequence analysis of indigenous slow-growing rhizobia nodulating cowpea (*Vigna unguiculata* L.) in Greece. Syst App Mic 40:179–189

Tenkouano A (2011) The nutritional and economic potential of vegetables. State of the World, pp 27–38

Tilman D, Clark M, Williams DR, Kimmel K, Polasky S, Packer C (2017) Future threats to biodiversity and pathways to their prevention. Nature 546(7656):73–81

Tosh SM, Yada S (2010) Dietary fibres in pulse seeds and fractions: characterization, functional attributes, and applications. Food Res Int 43(2):450–460. https://doi.org/10.1016/j.foodres.2009.09.005

van Dooren C, Douma A, Aiking H, Vellinga P (2017) Proposing a novel index reflecting both climate impact and nutritional impact of food products. Ecol Econ 131:389–398

Vinod PN, Chandramouli PN, Kochc Manfred (2015) Estimation of nitrate leaching in groundwater in an agriculturally used area in the State Karnataka, India, using existing model and GIS. Aquat Procedia 4:1047–1053

Webb P, Stordalen GA, Singh S, Wijesinha-Bettoni R, Shetty P, Lartey A (2018) Hunger and malnutrition in the 21st century. BMJ 361:1–5

Willett W, Rockström J, Loken B, Springmann M, Lang T, Vermeulen S et al (2019) Food in the Anthropocene: the EAT–*Lancet* Commission on healthy diets from sustainable food systems. Lancet 393:447–492

Williams M, Roth B, Pappa V, Rees R (2017) Nitrogen and phosphorous losses from legume based agriculture. In: Murphy-Bokern D, Stoddard F, Watson C (eds) Legumes in cropping systems. CABI, pp 37–54. ISBN: 978-17-8639-178-0

Health Canada Canada's Dietary Guidelines. Available online https://food-guide.canada.ca/en/

Zheng J, Qu Y, Kilasara MM, Mmari WN, Funakawa S (2019) Nitrate leaching from the critical root zone of maize in two tropical highlands of Tanzania: effects of fertilizer-nitrogen rate and straw incorporation. Soil Tillage Res 194:104295. https://doi.org/10.1016/j.still.2019.104295

Tropical Legumes: Status, Distribution, Biology and Importance

Purabi Saikia, Akash Nag, Subham Anurag, Sandeep Chatterjee, and Mohammad Latif Khan

Abstract Legumes have unusual flower structure, podded fruit, ability of 88% species to form nodules with *Rhizobia* to fix atmospheric nitrogen, protein-rich physiology, secondary metabolism, etc. Ca. 19,400 species under 730 genera of legumes include important grains, pastures, timbers, energy crops, and agroforestry species. Grain and forage legumes are grown on 12–15% of the Earth's arable surface, and grain legumes alone contribute 33% of the total dietary protein nitrogen (N) needs of human. Biodegradable plastics, oils, gums, dyes, and inks are some of the important industrial legumes-based produces. Besides, galactomannan gums and isoflavones are other important legume products, used as a thickener in sizing textiles, paper and pill formulation, as well as in folk medicines to reduce the risks of cancer and serum cholesterol. Agriculturally important legumes can fix Ca. 40–60 million MT of N_2 year^{-1}, whereas 3–5 million MT of N_2 used to be fixed in natural ecosystems. It requires adventurous breeding programs to develop disease resistance varieties, enhanced nitrogen fixation, and tolerance to soil constraints as well as better use of marker-assisted technologies to enhance the yields of legume in changing environmental conditions to fulfil the food security of the growing populations.

Keywords Diversity · Uses · Biological nitrogen fixation · Agroforestry crops

1 Introduction

Legumes (members of Fabaceae) are the second largest plant group after grasses (members of Poaceae) in terms of their importance to humans (NAS 1979). It is the third largest angiosperm family with about 19,400 species under 730 genera (Lewis

P. Saikia (✉) · A. Nag · S. Anurag · S. Chatterjee
Department of Environmental Sciences, Central University of Jharkhand, Brambe, Ranchi, Jharkhand 835205, India
e-mail: purabi.saikia83@gmail.com

M. L. Khan
Department of Botany, Dr. Harisingh Gour Vishwavidyalaya (A Central University), Sagar, Madhya Pradesh 470003, India

M. Hasanuzzaman et al. (eds.), *The Plant Family Fabaceae*,
https://doi.org/10.1007/978-981-15-4752-2_2

et al. 2005a, b) includes important grain, timber, bioenergy crops, pasture, and agroforestry species. It has some specialized features including unusual flower structure, podded fruit, ability of 88% of the species to form nodules with nitrogen-fixing bacteria, collectively called *Rhizobia* to fix atmospheric N_2 by a symbiotic association, protein-rich physiology, secondary metabolism, etc. (de Faria et al. 1989). Legumes are cosmopolitan in distribution and are found nearly everywhere in the world (NAS 1979) with the highest diversity in tropical and subtropical regions. Fabaceae showed high diversity in all three main tropical vegetation types including tropical rain forests, dry forests, and woody savannas (Sarkinen et al. 2011). It occurs in a great variety of vegetation formations from tropics to arctic zones, seashore to alpine habitats, and in rain forests, mangroves, peat swamp forests, seasonal forests, savannas, and deserts (Prado 2000; Lewis et al. 2005a, b). Succulent (a semi-arid, seasonally dry tropical forest, and bushland biome), grass (grass-rich woodland and savanna biome), rainforest (a tropical wet forest biome), and temperate (in both the Northern and Southern Hemispheres) are the four major biomes of Fabaceae been grouped by Schrire et al. (2009). Many tropical species of legumes play a significant role in the human life as it forms the most vital and valued food sources due to their protein-rich physiology. It has considerable importance in agriculture due to its ability to occupy different habitats and with diverse life forms (Yahara et al. 2013). Legumes are extensively grown in dry or semi-arid regions of the world, usually under rain fed rather than irrigated agriculture. Under optimal environmental condition, most of the nitrogen demand of soil can be supplied by biological nitrogen fixation leads to increased nitrogen resources in soil (Unkovich and Pate 2000). The production of a high nodule number is relevant for the production of high grain yield and protein. Genotype of the crop, *Rhizobium* strains, and the environment can affect the efficiency of the symbiosis and consequently the crop yield (Ayisi et al. 2000). Fabaceae contains many unique chemicals, especially in the seeds (Bisby 1994). Products (primary or secondary metabolites) of certain plants of Fabaceae are reported for their cytotoxicity against human cancer cells (Roberts et al. 2001). The American Nutraceutical Associations said that several nutraceuticals found in legumes including vitamin B6, calcium, magnesium, sodium, zinc, copper, and manganese (Roberts et al. 2001). Fabaceae includes many invasive species, posing serious economic threats to the native biodiversity (Bradshaw et al. 2008), and they are key invaders in several continents (Archer 1994; Lewis et al. 2009) and oceanic islands (Caetano et al. 2012) having strong impacts on savanna and grassland ecosystems (Scholes and Archer 1997; Chaneton et al. 2004). Thus, there is a need of detail study of legume diversity and uses to understand the present status, distribution, biology and importance of legumes to widen the genetic base and to support legume breeding programs to meet the demand of the increasing human populations in the changing climatic conditions.

2 Legume Biology and Diversity

Legumes (Fabaceae) are fast growing, multipurpose plants with great economic and ecological significance. It is a monophyletic family with three sub-families, Caesalpinioideae, Mimosoideae, and Papilionoideae divided on the basis of floral characteristics. Caesalpinioideae is the smallest sub-family (Mabberley 1997; Lewis et al. 2005a, b) comprised of Ca. 2250 species followed by Mimosoideae with Ca. 3270 species, and Papilionoideae is the largest legume sub-family with Ca. 13,800 species (Table 1). Caesalpinioideae and Mimosoideae have tropical distribution, whereas Papilionoideae is cosmopolitan. Caesalpinioideae is relatively rich in Africa, and Ca. 40% of the species are found in the two contrasting biomes, grass and rainforest, whereas Mimosoideae in the New World and in the grass biome and Papilionoideae in the Asia–Pacific region including Australia and in the temperate biome (Schrire et al. 2005). Caesalpinioideae typically bears large, showy flowers, and Mimosoideae has small flowers aggregated into heads or spikes, and papilionoideae bears characteristic "pea" flowers with vexillary aestivation (Lewis et al. 2005a, b). Fabaceae harbours extremely diversified life forms, including annual plants to herbaceous perennials,

Table 1 Species diversity of legumes as reported by various authors

Fabaceae	Almost 18,000–19,000 species belonging to 670–750 genera	Polhill et al. (1981)
	Ca. 18,000 species classified into ≈650 genera	Polhill and Raven (1981)
	Ca. 19,400 species of Ca. 730 genera	Lewis et al. (2005a, b)
	Ca. 19,500 species and Ca. 751 genera	Lewis et al. (2005a, b)
Papilionoideae	About 14,000 species of 476 genera	Lewis et al. (2005a, b)
	Approximately 12,000 species and 440 genera	Polhill and Raven (1981)
	Ca. 13,800 species of 478 genera	Lewis et al. (2005a, b)
	Ca. 14,000 species	Ahmed and Hasan (2014)
	Ca. 12,000 species	Ibeawuchi (2007)
Mimosoideae	Ca. 3000 species with 77 genera	Lewis et al. (2005a, b)
	Ca. 3270 species	Lewis et al. (2005a, b)
	Ca. 2900 species	Ahmed and Hasan (2014)
	Ca. 2800 species	Ibeawuchi (2007)
Caesalpinioideae	Approximately 3000 species and 162 genera	Lewis et al. (2005a, b)
	Ca. 2250 species of 171 genera	Lewis et al. (2005a, b)
	Ca. 2800 species	Ahmed and Hasan (2014), Ibeawuchi (2007)

woody shrubs, canopy trees, vines, and aquatic plants (Lewis et al. 2005a, b) which are basically deciduous in nature. Legumes bear typically alternate compound leaves composed of three to many leaflets arranged along a central stalk. In some herbaceous, climbing species of legumes (*Pisum sativum* L.), some of the leaflets are modified into spirally winding, clinging organs known as tendrils. Fabaceae is highly diversified in functional traits of leaves, stems, flowers, fruits, and seeds (Kattge et al. 2011). Flowers of Fabaceae are generally animal pollinated and thus sensitive to pollinator loss (Proctor et al. 1996).

3 Cultivated Legumes

Legumes were cultivated before 500 BC as confirmed by archaeological evidence, and perhaps, soybean is the oldest cultivated legume cultivated first time in China dated back to 3000 BC (Onwueme and Sinha 1991). Legumes cultivation can improve soil physical condition by producing copious amount of polysaccharides by *Rhizobia* besides fixed nitrogen (N) (Ojeniyi 2002). Tree legumes may provide N-rich mulch for cropping systems, living fences, shade trees for plantation crops and living trellises for climbing crops with fruit and vegetables for human consumption (Gutteridge and Shelton 1993). Effective managing of N_2 in the environment is one of the driving forces behind agricultural sustainability (Graham and Vance 2000). Most of the rural Indian populations depend on crop and animal mixed farming system to fulfil their livelihood as well as sustenance needs. Tree legumes are cultivated in home gardens in humid tropics (Saikia et al. 2012), and they are lopped periodically to provide mulch for vegetables and fruit trees. *Moringa oleifera* is one of the dominant tree legumes cultivated in entire India for its edible beans used a highly protein rich, fibrous vegetables. *Sesbania grandiflora* is another edible tree legumes commonly cultivated in Asia as its flowers and leaves are used for human consumption (Osman and Anderson 1990). Tree fallow and alley cropping is growing day by day due to food security and increasing soil degradation in Africa and Asia (Franzluebbers et al. 1998; Sanchez 2002). The Sloping Agricultural Land Technology (SALT) is using tree legumes significantly to reduce soil erosion and restoring degraded hilly lands with profitable farming system (Tacio et al. 1987).

4 Uses of Tropical Legumes

Legumes are known for its multipurpose utilities with great economic, ecological and social significance. It includes many useful plants such as crops, vegetables, timber, ornamentals, and medicinal plants (Gepts et al. 2005; Saslis-Lagoudakis et al. 2011). Legumes can be milled into flour, used to make bread, doughnuts, tortillas, chips, spreads, and extruded snacks or used in liquid form to produce milks, yogurt, and infant formula (Garcia et al. 1998) in addition to traditional food, fuelwood, timber, forage uses, soil conservation and fertility improvement (Table 2).

Table 2 List of some common legumes of various uses along with its area of origin and IUCN status as per the IUCN red list of threatened species

Use category	Species	Native	Exotic	IUCN red list category	Area of origin
Agroforestry crops	*Inga oerstediana* Benth.		+	LC	Mexico
	Sesbania punicea (Cav.) Benth.		+	LC	Brazil
	Sesbania bispinosa (Jacq.) W. Wight	+		LC	Indian subcontinent
	Leucaena leucocephala (Lam.) de Wit		+	CD	Mexico
	Cassia siamea (Lam.) H. S. Irwin and Barneby		+	LC	Sri Lanka
	Mimosa scabrella Benth.		+		Brazil
	Sesbania grandiflora (L.) Pers.	+			Indian subcontinent
Bioenergy	*Crotalaria juncea* L.	+			Indian subcontinent
	Vigna unguiculata (L.) Walp.		+		Africa
	Lablab purpureus (L.) Sweet		+		Africa
	Macrotyloma uniflorum Verdc.	+		LC	Indian subcontinent
	Phaseolus lunatus L.		+		Central America
	Pongamia pinnata (L.) Pierre	+		LC	Indian subcontinent

(continued)

Table 2 (continued)

Use category	Species	Native	Exotic	IUCN red list category	Area of origin
Medicinal/industrial	*Cassia occidentalis* (L.) Rose	+			Pantropical
	Desmodium triquetrum (L.) DC.	+			Indian subcontinent
	Pterocarpus marsupium Roxb.	+		NT	Indian subcontinent
	Astragalus glycyphyllos L.		+	LC	Europe
	Pongamia pinnata (L.) Pierre	+		LC	Indian subcontinent
	Dalbergia sissoo DC.	+			Indian subcontinent
	Saraca indica auct. non L.	+			Indian subcontinent
	Glycyrrhiza glabra L.		+	LC	France
Grains and forage	*Desmodium triquetrum* (L.) DC.	+			Indian subcontinent
	Lablab purpureus (L.)Sweet		+		Africa
	Vigna unguiculata (L.) Walp.		+		Africa
	Sesbania grandiflora (L.) Pers.	+			Indian subcontinent
	Cassia siamea (Lam.) H. S. Irwin and Barneby	+		LC	Pantropical
	Cassia occidentalis (L.) Rose	+			Pantropical

(continued)

Table 2 (continued)

Use category	Species	Native	Exotic	IUCN red list category	Area of origin
	Leucaena leucocephala C. E. Hughes		+		Mexico
	Glycine max (L.) Merr.		+		Eastern Asia, Australia
	Arachis hypogaea L.		+		South America
	Phaseolus vulgaris L.		+		Mexico
	Pisum sativum L.		+	LC	Netherlands
	Cicer arietinum L.		+		Iran, Iraq
	Vicia faba L.		+		Afghanistan, Iran
	Cajanus cajanifolius (Haines) Maesen	+		NT	Indian subcontinent
Ecosystem services	*Anthyllis vulneraria* Cullen and Pinto da Silva		+	DD	Europe, North Africa
	Coronilla emerus L.		+		Europe
	Lotus delortii F. W. Schultz		+		France

LC least concern, *NT* near threatened, *DD* data deficient
Sources The IUCN Red List of Threatened Species, Available at https://www.iucnredlist.org

5 Grain and Forage Legumes

Grain and forage legumes are grown on 12–15% of the Earth's arable surface (Ca. 180 million ha) contributing to 27% of the world's primary crop production. Grain legumes alone contributed 33% of the dietary protein nitrogen (N) needs of humans (Vance et al. 2000). The protein content of legumes depends on cultivar (legume variety), harvest date, and local environmental conditions (Berkenkamp and Meeres 1987). Grain legumes are rich source of dietary protein (Duranti and Gius 1997) especially for the Worlds' vegetarian populations. In the nineties, Onwueme and Sinha (1991) reported that *Glycine max* (L.) Merr. (Soybean) is the most important edible legume produced throughout the world as a source of high quality protein followed by *Arachis hypogaea* L. (Groundnut), *Lens esculenta* (Lentils). Nevertheless, this trend has changed recently, with Phaseolus vulgaris ranking on the third place (Duc et al. 2015). Soybean and peanut are the sources of more than 35% of the world's processed vegetable oil as well as rich sources of dietary protein for the chicken and pork industries (Graham and Vance 2003). In terms of percentage of legume protein N, bean (*Phaseolus vulgaris*), pea (*P. sativum*), chickpea (*Cicer arietinum*), broad bean (*Vicia faba*), pigeon pea (*Cajanus cajan*), cowpea (*Vigna unguiculata*), and lentils (*L. esculenta*) constitute the primary dietary legumes (NAS 1994).

Almost 80% of the global arable land dedicated to production of animal feed (FAO 2017). Tropical forage legumes have considerable potential to contribute towards environment-friendly agricultural productivity and sustainable livestock production because of their interaction with plants, soil, animals, and the atmosphere through nitrogen cycle. Forage legumes are required to maintain animal health (Wattiaux and Howard 2001) and productivity as well as milk and meat productivity. Meat and dairy production are almost dependent upon quality forage legumes and grasses as both are considered as primary food sources of livestock. Forage legumes perform an important role in energy and protein supply for livestock (Eskandari et al. 2009) and therefore considered as the foundation of dairy and meat production (Russelle 2001). Intercropping of cereals and legumes in forage production has been recognized as an important strategy for the improvement of protein content as well as productivity of fodders (Herbert et al. 1984). Animals need nutrient-rich fodders to meet their nutrient requirement to express their full genetic production potential (Singh et al. 2018). India is one of the leading milk producing countries in the world, but the productivity of milk per animal is very low. The major reason for the low animal as well as milk productivity is the deficiency of quality fodders. Addition of faba bean can improve the quality of wheat forage as it contains on average 2.1 times greater protein content that of wheat (Berkenkamp and Meeres 1987).

6 Leguminous Bioenergy Crops

Suitable bioenergy crop should have high biomass and energy yields, reduced agricultural inputs in terms of labour and fertilizers, and low-contaminant compositions (McKendry 2002; Cantrell et al. 2009). Legumes are considered as an important bioenergy crop due to their high biomass yield and no N-fertilizer requirements and grown in entire Southeast region. Tree legumes including *Leucaena leucocephala*, *Gliricidia sepium*, and *Acacia* spp. are used as fuelwood besides its timber and other uses and can be important energy sources for households in developing countries like India (Ryan 1994). The legume tree *Pongamia pinnata* can be considered as future energy crops, particularly in relation to their impact on nitrogen inputs and the net energy balance for biofuel production. It is gaining acceptance as a biofuel feedstock as it produces profuse quantities of seeds that are rich in oil (about 35–43%) composed predominantly of fatty acids (oleic, palmitic, stearic, and linoleic) (Scott et al. 2008). *Pongamia* oil contains about 50–55% oleic acid (C18:1) desirable for biodiesel production and about 7–10% palmitic (C16:0) and stearic acid (C18:0). Soybeans are regarded as feedstock in biorefinery for biofuel and/or biomaterials production (Brehmer et al. 2008). Several US cities and states now require that fleet vehicles be powered in part by biodiesel fuel from soybean. Alfalfa stems produce a lower output of bio-oil with higher energy content but slightly higher amounts of biochar than soybean (Boateng et al. 2008).

7 Industrial and Medicinal Legumes

Many legumes are used for nutraceutical and pharmaceutical purposes, whereas others contain medically important phytochemicals (primary or secondary metabolites) with potential therapeutic effects. Phytochemicals from legumes proved as useful for their anti-oxidant, anticancer, antiviral, anti-microbial, anti-diabetic, anti-allergenic, anti-inflammatory, anti-fungal activities (Tyler et al. 1981). Primary metabolites from legumes are used mainly as industrial raw materials, foods, or food additives and include products such as vegetable oils, fatty acids (for making soaps and detergents), and carbohydrates such as sucrose, starch, pectin, and cellulose (Morris 2003). Biodegradable thermoplastics produced from polyhydroxybutyrate (Paetau et al. 1994), oils, gums, dyes, and inks (Morris 1997) are some of the important industrial produces prepared from legumes. Besides, galactomannan gums and isoflavones derived from *Cyamopsis* spp. and *Sesbania* spp. are used as a thickener in sizing textiles and paper also used in pill formulation (Graham and Vance 2003). Many legumes have been used in folk medicine (Duke 1992) and isoflavones derived from soybeans, and other legumes have the potential to reduce the risks of cancer and to lower serum cholesterol (Kennedy 1995). *Mucuna pruriens* is one of the popular herbaceous legumes of India used in more than 200 indigenous drug formulations (Oudhia 2001). Higher intake of legume may be relevant for the prevention of the occurrence of several diseases due to the hypocholesterolaemic and anticarcinogenic activity of health-promoting compounds found in legume seeds (Vaz Patto et al. 2015). Legumes can modify cardiovascular risk factor including, hypertension, dyslipidemia, oxidative stress, obesity, and others (Morris 2003). Alfalfa sprouts (*Medicago sativa*) and soybeans contain phytoestrogens helps in reducing menopause symptoms and contributes to bone health in women due to the presence of isofavonoids (Morris 2003).

8 Legumes in Soil Health and Ecosystem Services

Nitrogen is the primary nutrient limiting plant productivity in most of the natural ecosystems (Seastedt and Knapp 1993; Vitousek et al. 1997). Legumes play an important role in colonizing fire-prone disturbed ecosystems through their symbiotic nitrogen fixation abilities (Arianoutsou and Thanos 1996), but the rate of N_2 fixation in such environments is often low. Legumes relied more on symbiotically fixed N_2 and result in higher N_2 availability for non-legumes (Temperton et al. 2007). Maximum N_2-fixation rates also depend on soil phosphorus availability (Kennedy and Cocking 1997), which in turn may limit ecosystem productivity and N_2 accumulation (Vitousek et al. 2001). Nitrogen fixers and fixation rates may vary within and across tropical forests as it composed of a variety of habitats (Harms et al. 2001). Symbiotic nitrogen fixers are most abundant in arid biomes with alkaline soils and higher temperatures (Steidinger et al. 2019). Legume abundance influences the rate of carbon and nitrogen accumulation in ecosystems (Knops et al. 2002). Atmospheric CO_2 enrichment and N_2 fixation are major ecological concern and are likely to have opposing effects in natural ecosystems. Many Fabaceae trees grow slowly and produce very heavy wood (e.g. *Dalbergia* L. f.) with slow decomposition rate and contribute to carbon storage in ecosystems (Weedon et al. 2009). N_2-fixing trees can sequester 0.11–0.07 $kgm^{-2}yr^{-1}$ of soil organic carbon (Resh et al. 2002). The ability of legumes to sequester carbon has been considered as a means to reduce atmospheric CO_2 levels. The presence of legumes often has a positive effect on ecosystem nitrogen pools which can significantly increase above-ground biomass (Spehn et al. 2002). Legumes are able of fix atmospheric nitrogen in the soil due to their symbiotic associations with bacteria viz. *Rhizobium* (NAS 1979; Ojeniyi 2002). Agriculturally important legumes can fix Ca. 40–60 million MT of N_2 $year^{-1}$, whereas 3–5 million MT of N_2 fixed in natural ecosystems (Smil 1999). Farming practices have strong influence on N_2 fixation, and it helps in make use of more economically viable and environmentally prudent N_2 fixation to enhance agricultural productivity and environmental sustainability (Peoples et al. 1995; Vance 2001). Vigorous tree legumes have the potential to accumulate 100–600 $kgNha^{-1}year^{-1}$ (NAS 1979). Trees legumes can also be used as a tool against desertification due to their colonizing potential in environmental stress conditions (Kaul 1970; Goor and Barney 1976).

9 Constrains of Legumes Production

Productivity of legumes has been declining for the last few decades despite a large growing area under legumes cultivation. Soil quality degradation, acidity, salinity, landslides in hills as well as draught in arid regions may limit the legume productivity. Drought problems for legumes productivity are likely to worsen with the expansion of water-stressed areas of the world from Ca. 28–30 countries to 50 countries by 2030 (Postel 2000). Diseases and pests are also major constraints to legume production, especially in tropical and subtropical regions of world due to higher rainfall and temperature. Important pathogens in common legumes include several viruses, root rots causing fungi, anthracnose, angular leaf spot, bean rust, white mould and web blight, *Ascochyta* blight and *Fusarium* wilt, and common bacterial blight and halo blight (Coyne et al. 2003; Miklas et al. 2003). Molecular marker-assisted research has enhanced the progress in disease resistance breeding in beans (Kelly et al. 2003).

Integrated approach including certified seed programs, fallow periods to reduce vector populations, ploughing to bury infected plant tissue, biological control of root disease as well as disease/pest resistance breeding, chemical application, and resistance breeding and cultivation of draught tolerant legumes with reduced leaf size and thick cuticle (Beaver et al. 2003; Coyne et al. 2003) are needed to enhance legumes productivity. Management of soil acidity is of critical importance to agriculture especially for temperate and tropical regions. It depends on acid-tolerant legume cultivars and *Rhizobia* as well as soil liming only to a pH at which Al and Mn are no longer toxic (Howieson et al. 2000).

10 Future Research Prospects

- Production of biotic or abiotic stresses tolerant cultivars with ability of adapting to changing climate is needed to meet the increasing demands for food, forage, timber, etc., through genetic and genomics-assisted breeding and biotechnological approaches.
- Adoption of legume-based pasture systems and increased knowledge of grazing management principles are necessary to enhance the livestock production as well fulfil the sustenance needs of livestock producers.
- There is a crucial need to produce drought and salinity tolerance legumes for arid and semi-arid agroecosystems and pests and disease-resistant varieties for tropics and subtropics.

11 Conclusion

Legumes play a critical role in natural ecosystems, agriculture, agroforestry, and industry as well as low nitrogen environments due to their ability to fix atmospheric nitrogen through symbiosis with *Rhizobia*. It requires adventurous breeding programs to develop varieties of disease resistance, enhanced nitrogen fixation, and tolerance to soil constraints as well as better use of marker-assisted technologies for the changing environmental conditions to enhance the yields of legume to fulfil the food as well as livelihood security of the growing populations. It also requires *Rhizobial* inoculation in the cultivated legumes for the small farmers which is considered as a low-cost technology. Tropical forage legumes have potential to contribute significantly to environment-friendly agricultural land use and sustainable livestock production. Landrace legumes serve as feed and food for animals and human, respectively, by producing quality starch and protein to sustain life. Efforts should be given towards discovery of more landraces from the wild to improve and integrate them into farming systems for sustained food, fodder, fibre, and medicine production across the globe. Farmers in tropical and subtropical regions must be educated to understand the importance of landrace legumes in agricultural productivity.

References

Ahmed S, Hasan MM (2014) Legumes: an overview. J Pharm Pharm Sci 2(1):34–38

Archer S (1994) Tree-grass dynamics in *Prosopis*–thorn shrub savanna parkland: reconstructing the past and predicting the future. Ecoscience 2:83–99

Arianoutsou M, Thanos CA (1996) Legumes in the fire-prone mediterranean regions: an example from Greece. Int J Wildland Fire 6:77–82

Ayisi KK, Nkgapele RJ, Dakora FD (2000) Nodule formation and function in six varieties of cowpea (*Vigna unguiculata* L. Walp.) grown in a nitrogen-rich field soil in South Africa. Symbiosis 28:17–31

Beaver JS, Rosas JC, Myers J, Acosta J, Kelly JD, Nchimbi-Msolla S, Misangu R, Bokosi J, Temple S, Arnaud-Santana E (2003) Contributions of the bean/cowpea CRSP program to cultivar and germplasm development in common bean. Field Crops Res 82(2–3):87–102

Berkenkamp B, Meeres J (1987) Mixtures of annual crops for forage in central Alberta. Can J Plant Sci 67:175–193

Bisby F (1994) Phytochemical dictionary of the Leguminosae. Chapman and Hall/CRC, London

Boateng AA, Mullen CA, Goldberg NM, Hicks KB, Jung HG, Lamb JF (2008) Production of bio-oil from alfalfa stems by fluidized-bed fast pyrolysis. Ind Eng Chem Res 47:4115–4122

Bradshaw CJA, Giam X, Tan HTW, Brook BW, Sodhi NS (2008) Threat or invasive status in legumes is related to opposite extremes of the same ecological and life-history attributes. J Ecol 96:869–883

Brehmer B, Struik PC, Sanders J (2008) Using an energetic and exergetic life cycle analysis to assess the best application of legumes within a bio based economy. Biomass Bioenerg 32:1175–1186

Caetano S, Currat M, Pennington RT, Prado DE, Excoffier L, Naciri Y (2012) Recent colonization of the Galapagos by the tree *Geoffroea spinosa* Jacq. (Leguminosae). Molec Ecol 21:2743–2760

Cantrell KB, Stone KC, Hunt PG, Ro KS, Vanotti MB, Burns JC (2009) Bioenergy from coastal Bermuda grass receiving subsurface drip irrigation with advance-treated swine wastewater. Bioresour Techno l 100:3285

Chaneton EJ, Mazia CN, Machera M, Uchitel A, Ghersa CM (2004) Establishment of honey locust (*Gleditsia triacanthos*) in burned Pampean grasslands. Weed Technol 18:1325–1329

Coyne DP, Steadman JR, Godoy-Lutz G, Gilbertson R, Arnaud-Santana EA, Beaver JS, Myers JR (2003) Contributions of the bean/cowpea CRSP to the management of bean diseases. Field Crops Res 82:155–168

de Faria SM, Lewis GP, Sprent JI, Sutherland JM (1989) Occurrence of nodulation in the leguminosae. New Phytol 111:607–619

Duc G, Agrama H, Bao S, Berger J, Bourion V, De Ron AM, Gowda CLL, Mikic A, Millot D, Singh KB, Tullu A, Vandenberg A, Vaz Patto MC, Warkentin TD, Zong X (2015) Breeding annual grain legumes for sustainable agriculture: new methods to approach complex traits and target new cultivar ideotypes. Crit Rev Plant Sci 34(1–3):381–411

Duke JA (1992) Handbook of legumes of economic importance. Plenum Press, New York

Duranti M, Gius C (1997) Legume seeds: protein content and nutritional value. Field Crops Res 53:31–45

Eskandari H, Ghanhari A, Javanmard A (2009) Intercropping of cereals and legumes for forage production. Not Sci Biol 1(1):07–13

FAO (2017) Food and agriculture organization of the United Nations. Animal Production, Rome

Franzluebbers K, Hossner LR, Juo ASR (1998) Integrated nutrient management for sustained crop production in sub-saharan agriculture: a review. Trop Soils Texas AM Univ Technol Bull: 98–103

Garcia MC, Marina ML, Laborda FT, Orre M (1998) Chemical characterization of commercial soybean products. Food Chem 62:325–331

Gepts P, Beavis WD, Brummer EC, Shoemaker RC, Stalker HT, Weeden NF, Young ND (2005) Legumes as a model plant family: genomics for food and feed report of the cross-legume advances through genomics conference. Plant Physiol 137:1228–1235

Goor AY, Barney CW (1976) Forest tree planting in arid zones, 2nd edn. Ronald Press, New York

Graham PH, Vance CP (2000) Nitrogen fixation in perspective: an overview of research and extension needs. Field Crops Res 65:93–106

Graham PH, Vance CP (2003) Legumes: importance and constraints to greater use. Plant Physiol 131:872–877

Gutteridge RC, Shelton HM (1993) The scope and potential of tree legumes in agroforestry. Agrofor Syst 23:177–194

Harms KE, Condit R, Hubbell SP, Foster RB (2001) Habitat associations of trees and shrubs in a 50-ha neotropical plot. J Ecol 89:947–959

Herbert SJ, Putnam DH, Creighton JF (1984) Forage yields intercropped corn and soybean in various planting patterns. Agron J 76:507–510

Howieson JG, O'Hara GW, Carr SJ (2000) Changing roles for legumes in mediterranean agriculture: developments from an Australian perspective. Field Crops Res 65:107–122

Ibeawuchi II (2007) Landrace legumes: synopsis of the culture, importance, potentials and roles in agricultural production systems. J Biol Sci 7:464–474

Kattge J, Díaz S, Lavorel S, Prentice IC, Leadley P, Boenisch G, Garnier E, Westoby M, Reich PB, Wright IJ et al (2011) TRY-a global database of plant traits. Global Change Biol 17:2905–2935

Kaul RN (1970) Afforestation in arid zones. Dr. W. Junk, The Hague, Netherlands

Kelly JD, Gepts P, Miklas PN, Coyne DP (2003) Tagging and mapping of genes and QTL and molecular marker-assisted selection for traits of economic importance in bean and cowpea. Field Crops Res 82:135–154

Kennedy AR (1995) The evidence for soybean products as cancer preventative agents. J Nutr 125:733–743

Kennedy IR, Cocking EC (eds) (1997) Biological nitrogen fixation: the global challenge and future needs. In: rockefeller foundation bellagio conference proceedings. SUN Fix Press, University of Sydney, Sydney, pp 83

Knops J, Bradley KL, Wedin DA (2002) Mechanisms of plant species impacts on ecosystem nitrogen cycling. Ecol Lett 5:454–466

Lewis GF, Schrire BD, Mackinder BA, Lock JM (eds) (2005a) Legumes of the world. Richmond, Kew, Royal Botanic Gardens

Lewis G, Schrire B, Mackinder B, Lock M (2005b) Legumes of the world. Royal Botanic Gardens, Kew, p 577

Lewis JP, Noetinger S, Prado DE, Barberis IM (2009) Woody vegetation structure and composition of the last relicts of Espinal vegetation in subtropical Argentina. Biodivers Conserv 18:3615–3628

Mabberley DJ (1997) The plant-book: a portable dictionary of the vascular plants. Cambridge University Press, Cambridge, p 857

McKendry P (2002) Energy production from biomass (part 1): overview of biomass. Bioresour Technol 83(1):37–46

Miklas PN, Coyne DP, Grafton KF, Mutlu N, Reiser J, Lindgren D, Singh SP (2003) A major QTL for common bacterial blight resistance derives from the common bean great northern landrace cultivar Montana No.5. Euphytica1 31: 137–146

Morris JB (1997) Special purpose legume genetic resources conserved for agricultural, industrial and pharmaceutical use. Econ Bot 51:251–263

Morris JB (2003) Bio-functional legumes with nutraceutical, pharmaceutical, and industrial uses. Econ Bot 57(2):254–261

NAS (1994) Biological nitrogen fixation. National Academy Press, Washington, DC

NAS (1979) Tropical legumes resources for the future. National Academy Press, Washington, DC

Ojeniyi SO (2002) Soil management, natural resources and environment. Adeniran Press, Ibadan, p 30
Onwueme IC, Sinha TD (1991) Field crop production in the tropical Africa CTA. Ede Netherlands, pp 1–319
Osman AM, Anderson DMW (1990) Food and other uses. In: Evans DO, Macklin B (eds) Perennial *Sesbania* production and use. Nitrogen Fixing Tree Association, Hawaii, pp 26–28
Oudhia P (2001) Records of *Aphis craccivora* Koch (Hemipteraaphididea) on medicinal crop *Mucuna pruriens* L. Chhattisgarh (India). Insect Environ 7:24–36
Paetau I, Chen CZ, Jane JL (1994) Biodegradable plastic made from soybean products: 1. Effect of preparation and processing on mechanical properties and water absorption. Ind Eng Chem Res 33:1821–1827
Peoples MB, Herridge DF, Ladha JK (1995) Biological nitrogen fixation: an efficient source of nitrogen for sustainable agricultural production. Plant Soil 174:3–28
Polhill RM, Raven PH (eds) (1981) Advances in legume systematics. Part I, Royal Botanic Gardens, Kew, UK
Polhill RM, Raven PH, Stirton CH (1981) Evolution and systematics of the Leguminosae. In: Polhill RM, Raven PH (eds) Advances in Legume Systematics Part 1. Royal Botanic Gardens, Kew, UK, pp 1–26
Postel SL (2000) Entering an era of water scarcity. Ecol Appl 10:941–948
Prado DE (2000) Seasonally dry forests of tropical South America: from forgotten ecosystems to a new phytogeographic unit. Edinburgh J Bot 57:437–461
Proctor M, Yeo P, Lack A (1996) The natural history of pollination. Harper Collins, London
Resh SC, Binkley D, Parrotta JA (2002) Greater soil carbon sequestration under nitrogen-fixing trees compared with *Eucalyptus* species. Ecosystems 5:217–231
Roberts AJ, O'Brien ME, Subak G, Sharpe MS (2001) Nutraceuticals. The complete encyclopedia of supplements, herbs, vitamins, and healing foods. Penguin Putnam, New York
Russelle M (2001) Alfalfa. Am Sci 89:252–259
Ryan PA (1994) The use of tree legumes for fuelwood production. In: Gutteridge RC, Shelton HM (eds), Forage tree legumes in tropical agriculture. Chapter 5.1 CABI, Wallingford
Saikia P, Choudhary BI, Khan ML (2012) Floristic composition and plant utilization pattern in homegardens of Upper Assam. India. Tropical Ecology 53(1):105–118
Sanchez PA (2002) Soil fertility and hunger in Africa. Science 295:2019–2020
Sarkinen T, Iganci JR, Linares-Palomino R, Simon MF, Prado D (2011) Forgotten forests-issues and prospects in biome mapping using seasonally dry tropical forests as a case study. BMC Ecol 11:27
Saslis-Lagoudakis CH, Klitgaard BB, Forest F, Francis L, Savolainen V, Williamson EM, Hawkins JA (2011) The use of phylogeny to interpret cross-cultural patterns in plant use and guide medicinal plant discovery: An example from *Pterocarpus* (Leguminosae). PLoS ONE 6:22275
Scholes RJ, Archer S (1997) Tree-grass interactions in savannas. Annual Rev Ecol Evol Syst 28:517–544
Schrire BD, Lavin M, Barker NP, Forest F (2009) Phylogeny of the tribe Indigofereae (Leguminosae-Papilionoideae): Geographically structured more in succulent-rich and temperate settings than in grass-rich environments. Amer J Bot 96:816–852
Schrire BD, Lewis GP, Lavin M (2005) Biogeography of the Leguminosae. In: Lewis G, Schrire BD, Mackinder B, Lock M (eds) Legumes of the world, Richmond, UK: Royal Botanic Gardens, Kew, pp 21–54
Scott PT, Pregelj L, Chen N, Hadler JS, Djordjevic MA, Gresshoff PM (2008) *Pongamia pinnata*: an untapped resource for the biofuels industry of the future. Bioenergy Res 1:2–11
Seastedt TR, Knapp AK (1993) Consequences of non equilibrium resource availability across multiple time scales: the transient maxima hypothesis. Am Nat 141:621–633
Singh T, Ramakrishnan S, Mahanta SK, Tyagi VC, Roy AK (2018) Tropical forage legumes in india: status and scope for sustaining livestock production. Chapter 8, Forage groups, IntechOpen, UK, pp 123–143

Smil V (1999) Nitrogen in crop production. Global Biogeochem Cycles 13:647–662
Spehn EM, Scherer-Lorenzen M, Schmid B, Hector A, Caldeira MC, Dimitrakopoulos PG, Finn JA, Jumpponen A, O' Donnovan G, Pereira JS, Schulze ED, Troumbis AY, Korner C (2002) The role of legumes as a component of biodiversity in a cross-European study of grassland biomass nitrogen. Oikos 98:205–218
Steidinger BS, Crowther TW, Liang J, et al (2019) Climatic control of decomposition drive the global biogeography of forest tree symbioses. Nature 569:404–408
Tacio HD, Watson HR, Laquihon WA (1987) Nitrogen fixing trees as multipurpose species for soil conservation. In: Withington D, Mac Dicken KG, Sastry CB, Adams NR (eds) Multipurpose tree species for small farm use. Winrock International, USA, IDRC, Ottawa, pp 114–121
Temperton VM, Mwangi PN, Bernhard M, Scherer-Lorenzen Buchmann SN (2007) Positive interactions between nitrogen-fixing legumes and four different neighbouring species in a biodiversity experiment. Oecologia 151:190–205
Tyler RT, Youngs CG, Sosulski FW (1981) Air classification of legumes: separation efficiency, yield, and composition of the starch and protein fractions. Cereal Chem 58:144–147
Unkovich MJ, Pate JS (2000) An appraisal of recent field measurements of symbiotic N_2 fixation by annual legumes. Field Crops Res 65:211–222
Vance CP (2001) Symbiotic nitrogen fixation and phosphorus acquisition: plant nutrition in a world of declining renewable resources. Plant Physiol 127:390–397
Vance CP, Graham PH, Allan DL (2000) Biological nitrogen fixation: phosphorus Ba critical future need? In: Pederosa FO, Hungria M, Yates MG, Newton WE (eds) Nitrogen fixation from molecules to crop productivity. Kluwer Academic Publishers, Dordrecht, The Netherlands, pp 509–518
Vaz Patto MC, Amarowicz R, Aryee ANA, Boye JI, Chung H-J, Martín-Cabrejas MA, Domoney C (2015) Achievements and challenges in improving the nutritional quality of food legumes. Crit Rev Plant Sci 34(1–3):105–143
Vitousek PM, Aber JD, Howarth RW, Likens GE, Matson PA, Schindler DW, Schlesinger WH, Tilman D (1997) Human alteration of the global nitrogen cycle: sources and consequences. Ecol Appl 7:737–750
Vitousek PM, Harrington RA, Fownes JH (2001) Production and resource use efficiencies in N- and P-limited tropical forests: a comparison of responses to long-term fertilization. Ecosystems 4:646–657
Wattiaux MA, Howard TM (2001) Technical dairy guide: nutrition and feeding. University of Wisconsin
Weedon JT, Cornwell WK, Cornelissen JHC, Zanne AE, Wirth C, Coomes DA (2009) Global meta-analysis of wood decomposition rates: a role for trait variation among tree species? Ecol Lett 12:45–56
Yahara T, Javadi F, Onoda Y, de Queiroz LP, Faith DP, Prado DE, Akasaka M, Kadoya T, Ishihama F, Davies S, Slik JWF, Yi T, Ma K, Bin C, Darnaedi D, Pennington RT, Tuda M, Shimada M, Ito M, Egan AN, Buerki S, Raes N, Kajita T, Vatanparast M, Mimura M, Tachida H, Iwasa Y, Smith GF, Victor JE, Nkonki T (2013) Global legume diversity assessment: concepts, key indicators, and strategies. Taxon 62:249–266

Nitrogen Fixation of Legumes: Biology and Physiology

Ali Raza, Noreen Zahra, Muhammad Bilal Hafeez, Muhammad Ahmad, Shahid Iqbal, Kanval Shaukat, and Gulraiz Ahmad

Abstract Legumes (Leguminosae or Fabaceae) are considered the second most cultivated crop, covering 14% of the total cultivated land worldwide. Legumes considered as the major source of important oils, micronutrients, vegetable proteins, fiber, and minerals for both livestock and human consumption. Besides, they provide nitrogen (N) to agroecosystem through their exclusive capability to fix atmospheric N in symbiotic relationship with soil rhizobia. The symbiotic association among legumes and rhizobial bacterium occurs in the root nodules of legume where they access to atmospheric N. Hence, it is of great benefit for us to understand the mechanism of symbiotic nitrogen fixation (SNF) and their effects on the neighboring environment. The excessive application of chemical fertilizers has resulted in an alarming disturbance of the N cycle, e.g., buildup of nitrates in soil and water and atmosphere contamination with nitrogen oxides. In order to maintain a sustainable agricultural system, substitutes for chemical fertilizers should be sought on an emergency basis. Biological nitrogen fixation (BNF), known as a microbiological process, is used by

A. Raza (✉)
Oil Crops Research Institute, Chinese Academy of Agricultural Sciences (CAAS), Wuhan 430062, China
e-mail: alirazamughal143@gmail.com

N. Zahra
Department of Botany, University of Agriculture, Faisalabad 38040, Pakistan

M. B. Hafeez
College of Agronomy, Northwest A&F University, Yangling 712100, China

M. Ahmad
Department of Agronomy, University of Agriculture, Faisalabad 38040, Pakistan

S. Iqbal
Department of Agronomy, Muhammad Nawaz Shareef University of Agriculture, Multan 60000, Pakistan

K. Shaukat
Department of Botany, University of Balochistan, Quetta, Pakistan

G. Ahmad
State Key Laboratory of Grassland Agro-ecosystem, College of Pastoral Agriculture Science and Technology, Lanzhou University, Lanzhou 730020, Gansu, China

M. Hasanuzzaman et al. (eds.), *The Plant Family Fabaceae*,
https://doi.org/10.1007/978-981-15-4752-2_3

legumes that transform atmospheric N into a plant in its usable form, which can be this alternative. Mineral N shortage is a recurring character of arid and semi-arid soils. Consequently, it is understood that BNF is a sustainable and environmentally friendly substitution to chemical fertilizers. There exists convincing proof that some non-leguminous crops, in some cases, may benefit from association with diazotrophs. Considerably, a natural association between plants mainly gains the potential benefit from N fixation and bacteria, which is seldom manageable as, part of agricultural practices. Especially in drylands, these associations are hard to maintain and are very unreliable. Nonetheless, through recent advancements in induction of nodular structures on roots of several cereal crops, i.e., wheat and rice, the possibility of dependable symbiosis with free-living diazotrophs, e.g., azospirilla and rhizobia may ultimately be achieved. Nevertheless, in this chapter, we have described the biology and physiology of N fixation by legumes.

Keywords Bacteria · Host-plant · Legumes · Nodules · Nitrogen fixation · *Rhizobia* · Symbiotic relationship

1 Introduction and Evolution

Nitrogen (N) comprises about 78% of the earth's atmosphere. It occurs as a dimolecule (N_2), and unfortunately, it is an unavailable form to the plants, animals, and microorganisms. All living organisms use N in the form of ammonia to manufacture various macromolecules vital to life, like amino acids, proteins nucleic acids, etc. (Mariotti et al. 2008; Lu and Tian 2017; Adams et al. 2018a). N is the limiting reactant for an agricultural point of view. To compensate for its deficiency, farmers use an immoderate quantity of fertilizers, which not only increases environmental pollution but also increased the input costs (Adams et al. 2018a; Salim and Raza 2020). Very few microorganisms called diazotrophs can shatter the bonds among two N atoms and make them in an available form that can be utilized by the living organisms. This conversion reaction is highly exothermic and occurs only in anoxic (absence of oxygen) conditions as the enzyme responsible (nitrogenase) is reactive to oxygen and is deactivated by oxygen. Ammonia production is potentially toxic and is immediately converted to amides and amino acids to form proteins, nucleic acids (Bittsánszky et al. 2015).

BNF is the process of atmospheric N_2 fixation by converting it into readily available form through the symbiotic association (Yang et al. 2017). Only the N-fixating bacteria, i.e., rhizobia, α-proteobacteria (Sørensen and Sessitsch 2007), mediate BNF. These bacteria can make N available for plants when they live in closed associations. Several SNF bacteria exist, having a great impact on agricultural and natural ecosystems (Fig. 1). However, legumes are considered as the major contributors for usable N inputs for making a symbiotic relationship with rhizobia (Valentine et al. 2018).

Legumes (members of family Leguminosae or Fabaceae) are the second major group of agriculturally important crops around the globe and covers about 14% of the

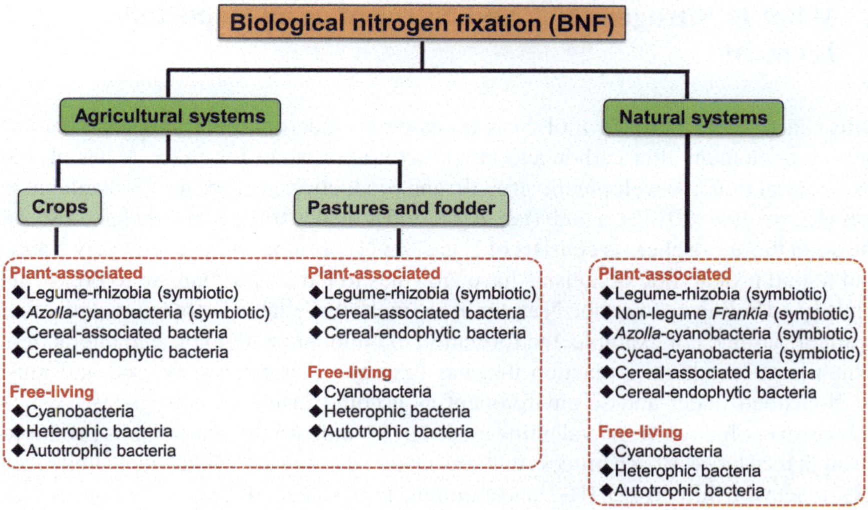

Fig. 1 Nitrogen-fixing organisms exist in both agricultural and natural systems associated with the nitrogen cycle

total cultivation land. The family includes plants that are used for food, fodder, medicinal, oil, and ornamental purposes; besides these, they have unique ability to fix the N and providing it into an available form for agroecosystem. Some important species include *Phaseolus vulgaris*, *Vicia faba*, *Pisum sativum*, *Arachis hypogea*, *Glycine max*, etc. (Afkhami et al. 2018; Valentine et al. 2018). The symbiotic association between legumes and the rhizobial bacteria occurs in the root nodules (Karmakar et al. 2015). Moreover, the symbiotic association and N fixation by legumes are highly sensitive to various environmental and edaphic factors. Therefore, exploring and understanding the N fixation by legumes responds to the ambient environmental conditions that are important for the preservation of the ecosystem.

Current studies prove that legumes emerged about sixty million years ago (Lavin et al. 2005). Interestingly, root nodule symbiosis is absent in most of the species of each legume subfamily (Griesmann et al. 2018). Soltis et al. (1995) hypothesized that nodulation was evolved independently within Rosid clade. Specific genetic characters led to the gain and loss of function, which is still undiscovered (Doyle 2016). Legumes have three subfamilies, and among these families, nodulation is very common in Papilionoideae, and the least common in Mimosoideae and rare in Caesalpinioideae, and their order is consistent, from which their lineage is evolved (Allen and Allen 1981).

2 What Is Nitrogen (N)? And Nitrogen as an Important Element

Nitrogen (N≡N; 225 kcal mol^{-1}) is an essential macronutrient and second most important element after carbon and input factor after water for crops. N has played an essential role in development, growth, and production in crops as it is available in soil (Krapp et al. 2014; Lu and Tian 2017; Vidal et al. 2014). Nevertheless, 78% of the air in the atmosphere is consists of N gas. As N_2 presents, here is relatively stable, and not all prokaryotic organisms have enzymes to have a mechanism to break that triple bond holing up together. N atoms are to make it available for reaction with other available atoms. For example, the formation of ammonia with hydrogen interaction, which is an exothermic reaction that has (energy), if a catalyst is used and must be controlled under anoxic environment as it could inactivate nitrogenase, which is an irreversible reaction (Valentine et al. 2018). Industrially ammonia is produced through the Haber–Bosch process that was invented in since 1904 in which dinitrogen (N_2) react with dihydrogen (H_2), and ammonia is produced; this process requires high temperature and pressure with efficient catalysts (Amar et al. 2011; Nielsen 1981; Wang et al. 2018).

The total dry weight of the plant is comprised of almost 1–5% of N, which is an important constituent in enzymes, amino acids, chlorophyll, co-enzymes, nucleic acids, proteins, phytochrome, and some secondary metabolites. That is why plant growth is determined by the amount of N available in the roots. The N_2 present in atmospheric is only absorbed by plants having bacteria for N fixation in symbiotic relation (Hawkesford et al. 2012; O'Brien et al. 2016). N regeneration occurs in two forms, first inorganic: oxides (NO_2 and NO_3) and ammonia, and second organic urea, nucleic acids, amino acids, enzymes, and proteins (Aluwihare and Meador 2008). N most preferred form that is absorbed by the plant is nitrate and ammonia also goes for the agricultural and cereal crops with exception to paddy, which has the most preference for ammonia only (Gioseffi et al. 2012; Paungfoo-Lonhienne et al. 2012).

The overall consumption by crops of applied N fertilizer is almost 40–50% (Adams et al. 2018b; Sylvester-Bradley and Kindred 2009). The significant efforts made for the enhancement of crop efficiency by N appliance has not improved further even after half a century has passed (Adams et al. 2018b; Cassman et al. 2002). The excessive utilization of N has caused many problems regarding the environment (Adams et al. 2018a). N fertilizers are not the only reason for increased eutrophication in aquatic ecosystem but also are the leading cause of ozone layer depletion in the stratosphere and global acidification (Gruber and Galloway 2008). The adverse environmental effects it causes and a significant cost for agriculture have made it important to understand the plant sensing and up taking a mechanism for N. To reveal the mechanisms and all steps in the process could be the early steps to sustain a strategy to improve the N utilization by crops for better and efficient agriculture (Adams et al. 2018a; Raza et al. 2019).

N is not the only required essential element, but it is taken from soil by the plants. The high yielding agriculture crops not only take up N from soil but also are dependent on the fertilizers, but a plant absorbs not all the applied fertilizers as a large proportion leaches down in the soil. This N that is leached into to environment could cause problems for example air and water pollution (Adams et al. 2018b; Hirel et al. 2011).

3 Nitrogen Fixation and Types of Nitrogen Fixation

The ideal growth of a plant heavily depends upon the adequate supply of fixed N. The most common growth limiting factors are fixed N and water. Even though N is present of about 80% of the atmospheric gases but unfortunately, plants cannot uptake it directly that they needed for appropriate development and growth. The unavailability of N is because it is present in an inert form (Dalton and Kramer 2007) and cannot be used by the number of organisms, so it needs to be fixed before assimilation. Ammonium (NH_4^+) and nitrate (NO_3) are the common fixed forms of N. This conversion needs an enormous amount of energy to split the triple bond of N atoms which encompass gaseous nitrogen (N_2), is known as N fixation.

Biogeochemical are cycles, which involved in N fixation. It includes the transformation of atmospheric N into ammonia (De Bruijn 2015). Hermann Hellriegel (1831–1895), a renowned German agricultural chemist, revealed that leguminous plants took atmospheric N and replenished the soil with the ammonium through N fixation. There are certain other components responsible for the conversion of NH_4^+ (ammonium) again into nitrogen gas (N_2) and then again goes to the atmosphere through nitrification (NH_4^+ into NO_3^-) and denitrification is a conversion of nitrate into N_2 gas and gaseous nitrogen oxides. Naturally, there are different ways to fix N like, one way is by lightning (atmospheric fixation), second way is the industrial fixation is known as the Haber–Bosch process in which atmospheric N is fixed into ammonia by utilizing the huge amount of energy and the third one is BNF. These processes are involved in fixing almost 380 Tg N y^{-1} (teragrams of N per year) totally (Galloway et al. 2008).

In the agriculture system, the N is supplied in available form, which is produced through the Haber–Bosch process, which transforms atmospheric N into ammonia by utilization of huge amount of energy either sometimes, through the mining of mineral deposits (such as rocks comprising of potassium or sodium nitrate). Chemical N fixation accounts for the 85 million metric tons of N y^{-1}. This fixed ammonia can be applied in the anhydrous form to the soil. The environmental and energetic cost of this process is very prodigious whether these fertilizers produced through a chemical reaction or mining process. So, there is another process in which the environmental cost and energy demands are not prodigious; that process is known as BNF (De Bruijn 2015; Herridge et al. 2008). BNF is an essential life-supporting phenomenon that accounts for the most fixed N supplier to endure life. The N fixation in BNF is carried out by the nitrogenase enzyme complex ("diazotrophs") that uses energy in the form of ATP or electrons (reductants) supplied by the respirational process. The resulting poisonous N is then readily converted into amides or amino acids by diazotrophic bacteria to produce proteins. When they die then this N is released into the atmosphere and mineralized and uptaken by some other plants. Since it is difficult to quantify the amount of N fixed biologically, approximations propose that possibly 150 (million metric tons) of N is being fixed in all over the world yearly, or chemical industry which is fixed it about twice the amount (Herridge et al. 2008; Peoples et al. 1995).

There are several organisms that govern this process, including (Eubacteria, Archaea, and actinomycetes). The point to be noted is that only prokaryotes can fix N, and there is no evidence of eukaryotes to fix N yet, and there are several claims that many bacterial genera can fix N (Sprent and Sprent 1990; Young 1992).

3.1 Atmospheric Nitrogen Fixation (ANF)

In the atmosphere, the N is fixed via abiotic processes like lightning or volcanic activity that releases energy. That energy is used to drive several reactions used to form N compounds like NO_2 for atmospheric fixation (Noxon 1976). The oxidized produced by these processes occurs in the atmosphere, dissolved in rainwater, and went back to the soil in NH_3 form (Noxon 1976). About 12% of the total N fixed came through atmospheric fixation annually (Galloway et al. 2004; Mori et al. 1998). The contribution of ANF is almost 5 (Tg N y^{-1}) into global N cycles (Galloway et al. 2008). NO dissolve in the rain to form nitrates hence come to earth (acid rains) as HNO_3. Below equations showing the synthesis if NH_4^+ and NO_3^-.

(i) $\mathbf{NH_4^+}$

$$H_2O\ \underline{\text{Lightening}}\ H^+ + OH^-$$

$$N_2 + H^+ \rightarrow NH_3/NH_4$$

(ii) $\mathbf{NO_3^-}$

$$N + O_2\ \underline{\text{Lightening}}\ NO_3$$

3.2 Biological Nitrogen Fixation (BNF)

N is an essential element in the lifecycle of several living organisms after oxygen, carbon, and hydrogen because it is the part of biochemical compounds like nucleotide amino acids, proteins, nucleic acids, and phosphates and BNF provides an enroot to N for plants (Vance 2002). According to Galloway et al. (2004), the involvement of BNF in naturally terrestrial ecosystems has been assessed at 107 Tg N y^{-1} and

195 Tg N y^{-1}. According to Cleveland et al. (1999), whereas input of BNF to agricultural system varies from 40 Tg N y^{-1} (Galloway et al. 2008) to 50–70 Tg N y^{-1} (Herridge et al. 2008). The addition of organic N in food is very important for the working of photoautotrophic organisms to fix N, such as terrestrial plants, cyanobacteria, and algae. Therefore, this process is ignited by primary producers who take N as nitrate from the environment, reduce it into ammonia, and then produce amino acids by assimilation of ammonia into organic acids. So, N has different forms to pass through biosphere as NH_3, NH_4 the most reduced forms to NO_3 to most oxidized forms via a biochemical cycle also involved living organisms as described in Fig. 5.

3.2.1 Nitrogen-Fixing Organisms

Despite that huge percentage of N in the environment, metabolically its availability is deficient because of the presence of N (triple covalent bond) that makes it an inert material. The ability of N fixation is present in a few members of prokaryotes, containing members of two kingdoms, bacteria, and archaea. Amid bacteria, the ability to fix N_2 has extensive taxonomic dissemination and includes proteobacteria, clostridia, chlorobi, cyanobacteria, and actinobacteria. Proteobacteria includes alpha-proteobacteria (*Methylobacterium*, *Ochrobactrum*, *Devosia*, *Phyllobacterium*) and beta-proteobacteria (*Ralstonia*, *Burkholderia*, *Cupriavidus*) (Sprent 2009; Willems 2006). Some of them need plant associations, and some are free-living bacteria that can speed up the reduction of N into ammonia (NH_3) with the help of enzymes. In all ecosystems, those diazotrophic or N-fixing microorganisms are present abundantly as symbiotic or symbiotic organisms, which are associated with several different plant species as shown in Table 1; and the biological organisms found in agricultural and natural systems are illustrated in Fig. 1.

3.3 Symbiotic Nitrogen Fixation (SNF)

The life-governing phenomenon of SNF has got attention due to its increasing contribution to agriculture productivity. N-fixing bacteria present in numerous phyla (Boyd and Peters 2013), and many of the individuals from that phyla are recognized to involve in fixation of N through symbiosis along with plants (Hardoim et al. 2015). Mutually, plants have established several ways to cope with and support diazotrophs, so they obtain atmospheric N (Doty et al. 2009). The key element between a plant host and bacterial symbiont is a nutrient exchange and falls into three broad groups, established on the grade of sensitivity and microbe's interdependency and plants: unfastened relationship along with free-living organisms which fix N, endosymbiosis intercellular and endophytic relationships.

The life-governing phenomenon of SNF has got attention due to its increasing contribution to agriculture productivity. The easiest way to fix N through symbiosis is the interactions with associative N-fixing bacteria and plants that are believed as a

Table 1 Classification of nitrogen-fixing microorganisms

Nitrogen-fixing organisms classification								
Asymbiotic (fix N without the help of a host)			Symbiotic (fix N with the help of a host)					
			Symbiotic–nodulating				Symbiotic non-nodulating	
Bacteria		Blue-green algae	Bacteria		BGA	Actinomycetes	Bacteria	Blue-green algae
Aerobic	Anaerobic	BGA	Root nodulating	Leaf nodulating	Associated with gymnosperms Nodulation on surface Roots-light required	Associated with woody angiosperms, e.g., *Alnus* (Alder)		
Azotobacter, *Azospirillum*, *Beijerinckia*	*Clostridium*, *Rhodospirillum rubrum*, *Chromatium*	*Anabaena*, *Nostoc*	*Rhizobium*	*Burkholderia species*			*Azotobacter paspali*, *Azospirillum brasilense*, *Spirillum lipoferum* Found in pastures of tropical and semi tropical C_4 grasses	Blue-green algae Associated with fern *Azolla* (rice) and fungi (Lichens)

part of PGPR (plant growth-promoting rhizobacteria), including: (1) a relationship in both plant hosts and diazotrophs. So diazotrophs have been separated into two distinct categories: PGPR (plant growth-promoting rhizobacteria) and root bacteria present in a nodule of the root. *Frankia* and rhizobia are grouped in root nodule bacteria. Rhizobia α- and β-proteobacteria has an association with leguminous plants, which is symbiotic, while *Frankia* has an actinorhizal relationship with plants. *Parasponia* also makes a relationship with alpha-proteobacteria and form nodulation. *Nostoc* (N-fixing cyanobacteria) also forms an endosymbiotic association with some plants. PGPRs contain actinobacteria, proteobacteria α-, β-, and γ-proteobacteria bacilli, and cyanobacteria. Cereals also form both associative and endophytic relationship along with numerous PGPRs. Some cyanobacteria are present inside tissues of plant are termed as endophytes and (2) relationship in both a nodulating plant and diazotrophic bacterial cell for the period of N fixation through symbiosis so rhizobia encourage the nodule development in legumes using both nod factor-independent and nod factor-dependent processes (Ahemad and Kibret 2014; Bakker et al. 2007).

The most famous symbiosis in agriculture is between leguminous plants and rhizobia. This process of symbiosis maximizes the fixed N_2 from the nodules on plant roots (where rhizobia are present) to the plant. Give in return, plant serves them as a home by providing protection and supplying them energy-rich food (Ahemad and Kibret 2014). The activity of nitrogenase is severely affected by the presence of oxygen because it might denature the enzyme, but nodules give them an oxygen-free environment for the proper activity. While complete anoxia is not suitable but there is a need for oxygen for the respiration of microbes and formation of energy to derive N fixation, and that is possible due to the presence of leghemoglobin; a protein that supplies the optimum amount of oxygen to run the process. Association is classically symbiotic and a vastly particular mechanism both fine plant hosts and bacteria (Dénarié et al. 1992). The infection process starts inside the legume–rhizobia association after the adherence of bacteria and becomes the reason for root tip hair curling (Dazzo and Gardiol 1984), and becomes infection threads. These infection threads become the way to penetrate bacteria inside the root cortex, which leads bacteria to the cortical cells.

Additional N-fixing, a symbiotic relationship comprises liverworts, bryophytes, mosses, lichens, cycads, (ferns) other non-leguminous angiosperms, and pteridophytes. Non-leguminous root nodules (called actinorhizae) are found on several spp. Such as alder and *Myrica*, and also present on huge trees like Casuarina (Sprent and Sprent 1990), dispersed all over both climates of temperate and tropical bacteria present within nodules has been classified under genetic name *Frankia* and in individuals of the bacterial family, Streptomyces. These associations between *Frankia* are called and non-legumes actinorhizal associations. The formation and working for the effectiveness of symbiosis is reliant on genetic determinants in plants and bacteria. These steps of penetration, recognition, stimulation of host cell division, and distinction operate the entire harmonious endosymbiosis.

3.4 Free-Living Nitrogen Fixation

Although the emphasis of agriculture is largely on SNF, there are several microbes, which are associated neither with plants nor with animals as they live freely and fix N by their own and are present within soils but they are free from the direct effect of roots of plant (Rousk et al. 2016). The first free-living was *Clostridium pasteurianum*, an anaerobic microorganism found in 1901. After that, an aerobic free-living *Azotobacter chroococcum* and many others were discovered. During the enzymatic level, the free-living and SNF processes are almost the same. However, freely living, N-fixing microorganisms are extensively dispersed within soils, but they seldom reach to the amount of N fixed by the symbiotic associations (Ashby 1907; Elmerich 2007; Franche et al. 2009; Lema et al. 2012).

While one anticipates that a highly effective symbiotic association can fix 100–200 kg ha^{-1} of N annually and only 1–5 kg ha^{-1} per year of N fixed is being fixed by the free-living microorganisms (Reed et al. 2011; Zackrisson et al. 2004), these free-living microbes are very diverse physiologically and are ubiquitous in aquatic and terrestrial ecosystems (Reed et al. 2011). As there is a deficiency of C and N in most soils and the N fixation is limited due to less contact with energy sources, which is the substrates to produce ATP and for the proper functioning and synthesis of nitrogenase, there must be a proper supply of micronutrients (Reed et al. 2011).

The drastic oxygen sensitivity of nitrogenase leads to limit the BNF by the free-living microbes, so this problem has been at least partially reduced due to various diazotrophs contributing to the fixation of N by symbioses (Olivares et al. 2013). When an organism competes for nutrient, then antagonistic interaction like parasitism and competition decreased the N_2 quantity that they fix (Bashan and Holguin 1997; Bashan et al. 2004). Though over-all conviction is that freely living diazotrophs does not add vast amounts of N, which is fixed to most terrestrial ecosystems, possibly 3–5 kg per hectare per year (Newton 2007), within some temperate and tropical forest ecosystems, their cumulative N involvement is considered to be significant (Gehring et al. 2005; Reed et al. 2008).

3.5 Associative Nitrogen Fixation (ANF)

Unfortunately, the most important crops like wheat, corn, rice do not have a nodule based symbiotic relationship, and they do not have any partner to facilitate N fixation. Therefore, in such crops, N fixation is also possible normally by associative microbes through associative symbiosis. In ANF, the association between plants and diazotrophs is very casual, and this association lacks lasting interdependence (Jones et al. 2003). ANF and freely living fixation are tough to separate from each other within rhizosphere (Reed et al. 2011). At this time, we assign to any fixation of N within rhizosphere of a plant-like ANF but accept that some fixation of N might happen due to diazotrophs which are free living. ANF has been long-reported within sugarcane (Döbereiner et al. 1972); and many other tropical kinds of grass (Reis et al. 2001; Thilakarathna et al. 2016) whereas microbes, which fix N, are nearly related with surfaces of root (Van Dommelen and Vanderleyden 2007) as well as endophytic (Rothballer et al. 2007).

Though ANF happens in a variety of temperate and tropical spp., vast ambiguity left near its impact on N budgets in ecosystem and plant, rates of seasonal changes, and its reaction toward adding N externally, particularly within temperate regions (Bottomley and Myrold 2007; Herridge et al. 2008). Switchgrass, a C_4 grass was predominating in the area of North historically. The process of conversion of N_2 into NH_3 by a loose microbial-plant association is known as ANF. The association among grasses of tropics like *Digitaria* and *Paspalum* and bacteria such as *Azospirillum brasilense* and *Azotobacter paspali* is an example of associative symbiosis (Döbereiner et al. 1972). *Paspalum*, on external surface, the N_2-fixing bacteria are present in a mucilaginous sheath within the root. *Digitaria–Azospirillum* relationships include roots assault but no development of nodule. The amount of advantage that plants can get from such association is yet to be discovered. *Azospirillum lipoferum* is a temperate "cousin" of *A. brasilense* and is termed as partners with some sorghum and corn cultivars, but its outcome might be negligible (Boddey and Dobereiner 1988, 1995).

Further official relationships of two *Acetobacter*, *Herbaspirillum*, and *diazotrophicus* spp. along with sugarcane (Boddey and Dobereiner 1988) gratify the fixation of N and required for the plant growth. They become established in sugarcane stem, but they do not form any structure as the nodulation of root of leguminous plants does. Moreover, the Acetobacter adores the lavishness of sugar to promote growth vigorously and fixed the N. A portion of fixed N compounds produced by *Acetobacter* is provided into and spread within the sugarcane stem. In crops, fixation of as much as one hundred and fifty-kilogram N per hectare has been documented. Likewise, *Azoarcus* spp. associated with Kallar grass and perhaps along with rice (James et al. 2000b) also specifies a substantial agronomic and economic potential (Chalk 2016; James et al. 2000b).

4 Key Factors Regulating the Nitrogen Fixation Activity

4.1 Energy and Fixation of Nitrogen

N fixation is an energy-dependent procedure in the presence of free-living microbes, which might get energy from the chemical source (non-photosynthetic), while photosynthetic utilizes light energy. Higher energy is required for the disassociation of triple bonds present between two atoms of dinitrogen and converted into ammonia (Figg et al. 2012; Qiu et al. 2018). As compared to associative diazotrophs free-living contributes less fixed N for agriculture sector, but they contribute a healthy amount of N to the arid land (Sorochkina et al. 2018) and the whole ecosystem (Unkovich and Baldock 2008). In this relationship of providing energy and getting fixed N in return plants give nutrition and habitat to the diazotrophs and in return they get fixed N from this association and this fixed N (ammonia) is transferred to plants to fulfill its nutritional requirements by synthesizing different compounds like enzymes, amino acids, nucleic acid, chlorophyll, and proteins. Schubert (1982) has stated that the energy requirement estimation is heavily dependent on analytical data and results of experiment. However, energy demand to fix the triple bonded dinitrogen molecules is nearly same as needed for the assimilation of nitrate (Kennedy and Tchan 1992) that is another main mean of N supply for a number of field crops, except rice. The determination of the energy required for N fixation is a complicated process because it is very nearly linked with the growth of bacteria that lives freely.

Approximations of ATP/N, molar ratios attained through equating molar growth yields of cultures which is supplied by NH^+ and N which are the N sources that ranges from four to five for *A. chroococcum* (Dalton and Postgate 1969), 29 for *Klebsiella pneumoniae* (Hill 1976), to 20 for *Clostridium pasteurianum* (Daesch and Mortenson 1968). In vitro studies, mostly along nitrogenase from numerous bacteria specify the least requirement of two electrons transferred by using four to five molecules of ATP (ATP/2e value). These values depend upon the temperature, pH, and the fraction of the constituent proteins of nitrogenase (Orme-Johnson and Davis 1977; Postgate 2012). In the case of energy requirements, the N_2 fixation should not exist an evolutionary obstacle to plants. Eight ATP molecules are obligatory to fix every 1/2 molecule of dinitrogen having a triple bond between them reduced to NH_3 by the bacterial nitrogenase. Further, six ATP molecules are needed as a reductant, consequence in elaborating that 14 ATP molecules are required per NH_3 produced. The assimilation of N is done via reduction of nitrate in the plants that need 12 ATP molecules to give one molecule of NH_3 (Postgate 1992; Cocking 2009). Therefore, reduction in the form of energy usage, N fixation is slightly more demanding as compared to nitrate (Postgate 1974). When a comparison is made between the growing proportion of different legumes on nitrate and N, the energetics requirements are being estimated and that there is no such difference for energy requirements in both nitrate and N applied in crops (Kennedy and Tchan 1992).

The efflux of the CO_2 approach was used to estimate the energy requirement in nitrate and N applied field and found that significant efflux was being observed in the dinitrogen applied crops that means they have more energy demand as compared to other and looking for more carbohydrates (Schubert 1982; Silsbury 1977). No one focused on a fact that in the nitrate applied experiment plants will produce bicarbonate instead of C2 as in N-applied crop. Therefore, if we take into account this metabolic difference of sources of N that will disappear the difference between energy demands, nevertheless, in case of direct access engineered diazotrophs to the abundant energy source, then the proposal to engineer the non-legumes based on the energy required to fix N will be queried. Root infection substantially in the cortex through organisms such as *Azospirillum* would be illustrated by an example. For the effectiveness of N_2 fixation, though not only penetration will work but also further colonization is needed diazotrophs and some symbiosis sufficiently (Kennedy and Tchan 1992).

4.2 Oxygen and Nitrogen Fixation

A possibly harmful element in outspreading the array of NFS (nitrogen-fixing systems) is the familiar as this enzyme is susceptible to O_2, which causes a great instability for the N-fixing enzyme complex (nitrogenase enzyme complex) (Pauly et al. 2006). Nitrogenase is highly sensitive to oxygen (Seefeldt et al. 2018). That is why this enzyme cannot be utilized by all genera of aerobic organisms. As it is a complex metallo-enzyme that has a major share of Fe protein that works as ATP dependent reductase for the Mo–Fe complex that serves to give a catalytic place for the reduction of dinitrogen molecule. Fe protein provides an electron to the Mo–Fe complex as it was reduced by the ferredoxin. Dinitrogen will accept these electrons that lead to the synthesize of ammonia at catalytic site of MoFe protein. Fe protein is known for having a [4Fe–4S] cluster that bridges the NifH promoters. This [4Fe–4S] cluster makes it highly sensitive toward the oxygen (Nomata et al. 2006). In diazotrophs, this sensitivity toward oxygen can be regulated through several mechanisms (Yamamoto et al. 2009). To minimize this sensitivity, they enhance the consumption of oxygen through increasing respiration; the cellular location of nitrogenase is also crucial in regards, some oxygen protecting proteins presence, buffering of oxygen as by leghemoglobin (Clara and Knowles 1984). At the same time, oxygen is obligatory for the mechanism of N fixation as it is necessary for the formation of ATP through oxidative phosphorylation. Oxygen also has a role in the development of nodules. The high O_2 ingestion of nodules, essential for provision of ATP energy, might produce reactive oxygen species (ROS). Rhizobia has a number of defense mechanisms that can minimize the toxic impact of these ROS, comprising ROS scavengers and reductants, which seems vital for regular nodule evolution (Tavares et al. 2007).

4.3 Leghemoglobin and Nitrogen Fixation

Nodules have to play a vital role to ensure all physiological prerequisites for N fixation in bacteroids; (1) the N fixation as it needs oxygen for the proper functioning of nitrogenase by proving the energy generated from the oxidative phosphorylation for N fixation on the same time at more or less on the same location; (2) it has to protect the nitrogenase enzyme from the oxygen. So, the low concentration of oxygen in the nodules can be achieved by: (i) in the inner nodule cortex there must be an O_2 diffusion hurdle in thickly packed cells, and (ii) enhance the rate of respiration in the bacteroids. The mechanisms behind the activity of well-packed diffusion barrier are yet to be discovered, but it can rapidly adjust the oxygen concentration outside and inside the nodule cortex (Vessey et al. 2005). This barrier consists of five-layer thickness and the spaces between the cells might fill with water or air. Meanwhile, the diffusion coefficient in air for oxygen is 10^4 times greater as compared to water, so the water barrier is more desirable mean of lowering oxygen concentration in the interior of nodules (Blevins 1989). Oxygen dissemination rates can more be prejudiced by the volume of intercellular H_2O (Denison 1992); and intercellular glycoprotein (James et al. 2000a). Nevertheless, in the case of insufficient supply of oxygen, leghemoglobin takes part to make possible the sufficient supply of oxygen to the bacteriod. It is an important constituent of the nodule proteins as it contributes about 5% of total nodule proteins, being encoded by the four lb genes. Its porphyrin ring is identical to cytochrome having Fe as a central atom and it does best to bind oxygen from the intercellular area and transporting it along the concentration gradient of oxyleghemoglobin to a great affinity cbb3-type cytochrome oxidase within bacteroids (Denison and Okano 2003; Preisig et al. 1996). In cytoplasm of infected plant cells, leghemoglobin is present in milli-molar amount, and it plays its role to buffer the oxygen in nano-molar concentration that will maintain high oxygen flux for the proper phosphorylation and avoids the nitrogenase inactivation (Franssen et al. 1995).

4.4 Carbohydrates and Nitrogen Fixation

The N fixation in the heterotrophs heavily depends upon the availability of carbohydrates that serves as an energy source. As these heterotrophs are in efficient using these carbohydrates, they ultimately need large quantities to fix a considerable amount of N that exhibits massive competition among non-N and N-fixing forms (Stewart 1969). The efficiency of some organisms to fix mg of N by per g consumption of carbohydrates is reported as for *Azotobacter* 10–20 mg of N_2 g^{-1} of carbohydrates, *Klebsiella* 5 mg of N_2 g^{-1} of carbohydrates, *Bacillus* 12 mg of N_2 g^{-1} of carbohydrates and *Clostridium* 2–27 mg of N_2 g^{-1} of carbohydrates (Stewart 1969). The availability of carbohydrates in the soil to fix N is very unequal, so sometimes, several energy sources like organic acids or alcohols might be used (Mus et al. 2016). Aerobic microbes present in large numbers in low oxygen, and here, they use carbohydrates more efficiently. The highest N_2 fixing rate is present in upper layers of the soil even in the absence of photosynthetic N_2 fixing microbes, (Jaiyebo and Moore 1963). Ruinen (1965) have described that *Beijerinckia* and *Azotobacter*, which lives on leaves which provide carbohydrates via exudations in tropical trees, along with rhizosphere and soil. Gibson (1966) has discovered that in the course of active fixation of N, the need of carbohydrate to fix N_2 and nodule maintenance varies from 3.2 mg carbohydrate per 1 mg of fixed N that is C.R. = 3.2:1 down to −3.0:1, for that standard being comparative to the carbohydrate prerequisite to assimilate the combined N through plants which were in control.

4.5 Nodulation

Nevertheless, dalbergoid legumes are monophyletic clades, contain aeschynomenoid nodules (Lavin et al. 2001), that are distinguished by the lack of transcellular tissues, and uninfected cells in infected areas (Doyle and Luckow 2003). Indeterminate nodules form in tribe Fabaceae and Trifolieae, which are characterized by the presence of peripheral vasculature ontology (Magne et al. 2018) and central and apical meristem zone (Franssen et al. 2015). Moreover, they also possess root apical meristem (RAM) such as the meristematic zone (Franssen et al. 2015). Determinate nodule formation occurs in tribes Loteae and Phaseoleae (Guinel 2009). Figure 2 shows the currently recognized variety of legume nodules morphology. For more detail about morphology of legume nodules, read the article published by Schwartz et al. (2013), Sprent et al. (1989, 2013), WEBB and Sheehy (1991). Figure 3 indicates the phylogenetic relationship among the broad diversity of nodulating and non-nodulating Fabids as well as non-Fabids (Gauthier-Coles et al. 2018). Nevertheless, below we have discussed the advancement in root, stem, and para-nodules.

4.5.1 Root Nodules

Several higher plants made close associations with N-fixing microbes. Therefore, this symbiotic association made plants independent of soil N use. Among the gymnosperm cycads, coralloid roots harbor symbiotic cyanobacteria (Gehringer et al. 2010). The N-fixing capability among the flowering plants evolved in several times in one clade, the Rosid clade (Soltis et al. 1995). For N fixation, Fabales make associations with single-celled gram-negative bacteria called *Rhizobium* species, and others like Fagales, Cucurbitales, and Rosales-made associations along gram-positive filamentous actinobacteria belonging from genus Frankia (Vessey et al. 2005). Genes in bacteria known for their host-specific infection are known as nod (nodulation) genes, which produces nod-factor (Gibson et al. 2008). A distinctive group of LysM plant receptor kinases documents nod factors (Lohmann et al. 2010). The entry of rhizobial cells is called as a crack-entry (Sprent and James 2007). The infection in legumes proceeds via infection threads, which are transcellular and intercellular channels, and at the same time in roots, cortex cells are encouraged to split/divide and develop the nodule tissues. The physiology and anatomy of root nodules together replicate an abundance of structural and metabolic interaction among plant and microbial symbionts (Lodwig et al. 2003). The specialized bacteroids cells fix N that is wrapped in cytoplasmic compartments. The physiology and anatomy of root nodules together replicate a high degree of structural and metabolic interaction among plant and microbial symbionts (Lodwig et al. 2003). The physiology of root nodules is integrated with different growth phases of the plant (Okamoto et al. 2008). There is a huge diversity of shape of root nodules among different symbiosis (Sprent and James 2007). In legume nodules, the infected tissue is central, and the vasculature is peripheral, whereas, in Frankia-induced non-legume nodules, the infected tissue is always located outside the root endodermis.

The symbiotic association between legumes and the rhizobial bacteria occurs in the root nodules. Invasion of bacteroids initiated with signaling exchange between the saprophytes free-living bacteria and legumes in the rhizosphere. Legumes root

Fig. 2 Morphology of legume nodules. **a** Determinate, desmodioid. **b** Determinate, aeschynomenoid. **c** Indigoferoid. **d** Lupinoid. **e** Indeterminate unbranched nodules. **f** Indeterminate having one or more than one branches. **g** Indeterminate with many branches. **h** Woody. Note that many nodules of types (**f**) and (**g**) are unbranched when young. Adapted from Sprent et al. (2013) with permission from the publishers and the authors of the paper

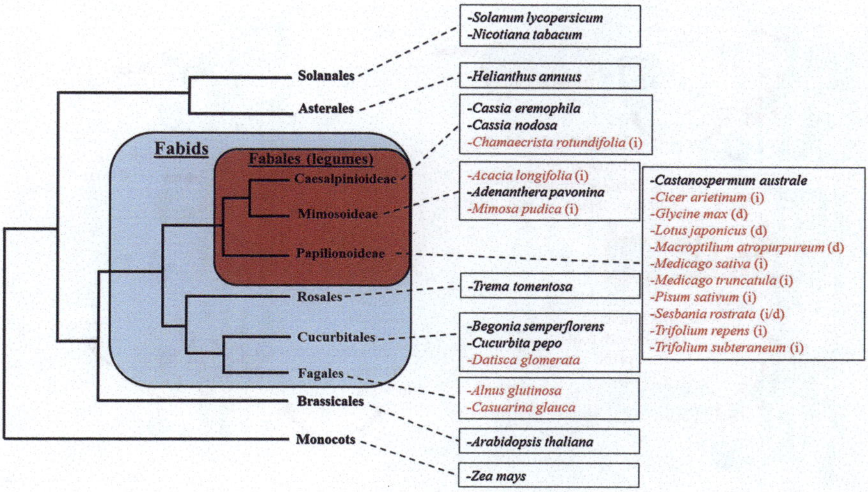

Fig. 3 Relative phylogenetic relationship among subject species. Species in red showing nodulating plant species. (i) Indicating the legumes making indeterminate nodules; (d) indicating the legumes making determinate nodules. Adapted from Gauthier-Coles et al. (2018)

exudates such as betains and flavonoids, which work as signaling molecules (Karmakar et al. 2015). The type of signaling molecules is specific to different rhizobia and among legume species (Downie 1994). Representation of nodule formation in legumes by Rhizobia is illustrated in Fig. 4. These flavonoids are detected by NodD transcriptional protein present in rhizobia (Peters et al. 1986). Complex hormones signaling are involved in nodule organogenesis, which includes suppression of auxin transport toward the cortex and activation of cytokinin (Oldroyd et al. 2011). Infection is started with cell wall breakup at the connection site of plant and bacterial infection foci (Muñoz et al. 1998). Rhizobia multiply at the tip of IT and move toward the cortex by the process of endocytosis, thereby resulted in the acquisition of symbiosome membrane (SM), which enclosed the rhizobial cells (Limpens et al. 2005).

4.5.2 Stem Nodules

These are the nodules present on the stems in submerged plants. A rare phenomenon known as nodulation in the stem has been found in some genera of hydrophytic like *Sesbania* and *Aeschynomene*. Alazard (1985) and Loureiro et al. (1994) revealed that there are three species that are reported to have stem nodules. These nodules can be formed ion the aerial parts of the plants, including stem and adventitious roots as on Neptunia. As stem nodule, derive carbon from the photosynthetic process, and in return, bacteroid fixes N. As stem nodulation is different from root, nodulation, their

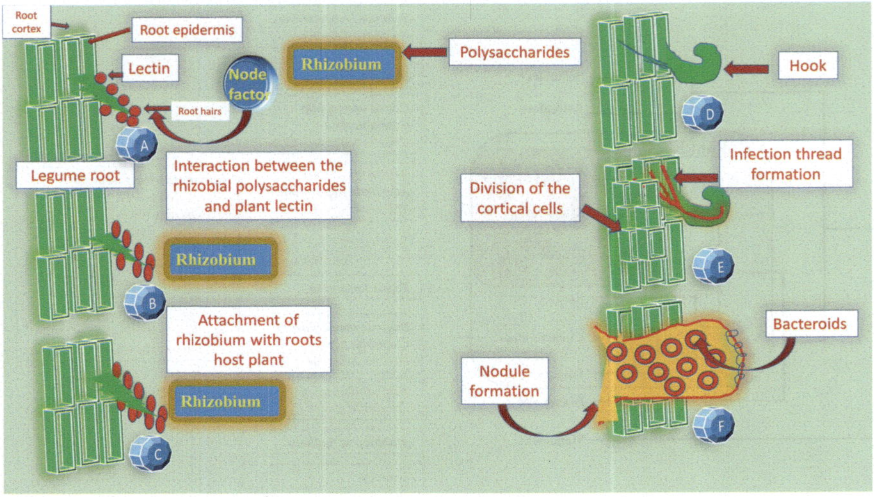

Fig. 4 Representation of nodule formation in legumes by Rhizobia. (A) Plant secretes Lectin to attract the *Rhizobium*, (B) interaction between rhizobia and legume roots, (C) attachment of *Rhizobium* with plant roots, (D) plant root curling to trap the bacteria, (E) formation of infection thread and division of cortical cells, and (F) nodule formation for nitrogen fixation

N-fixing bacteria are also different. Unlike root nodules, an *Azorhizobium* microbe fixes N in stem nodules.

4.5.3 Para-Nodules

Nie et al. (1992) stated that structures like nodules produced through 2,4-D might be colonized due to diazotrophs. The name *para-nodule* (p-nodule) was given by Tchan and Kennedy for defining the chemically persuaded nodule; subsequently, it varies when compared with the nodules of the legume, which occurs naturally (Kennedy et al. 1990). By using 2,4-D, the development of para-nodules can be induced, but the activity of nitrogenase was not noticed (Bender et al. 1990). Similarly, Al-Mallah et al. (1989) and Cocking et al. (1990) were able to induce the development of nodule on non-legumes by treating roots with a cell wall degrading enzyme mixture followed by inoculation with rhizobia in the presence of polyethylene glycol. Moreover, Jing et al. (1990) used a mutant of R. *Sesbania* for the induction of nodulation on the surface of rice roots. Nevertheless, in all these approaches, only a minimum activity of nitrogenase was attained within these systems.

5 Legumes and Nitrogen Fixation

The phenomenon of BNF has got attention due to its increasing contribution to agriculture productivity. Figure 5 depicts some advantages of the SNF made legumes and illustrates the N-fixing process integrated with rhizobia in legumes. In this section, we discussed the particular aspects of N fixation by some of the most important legumes (Table 2).

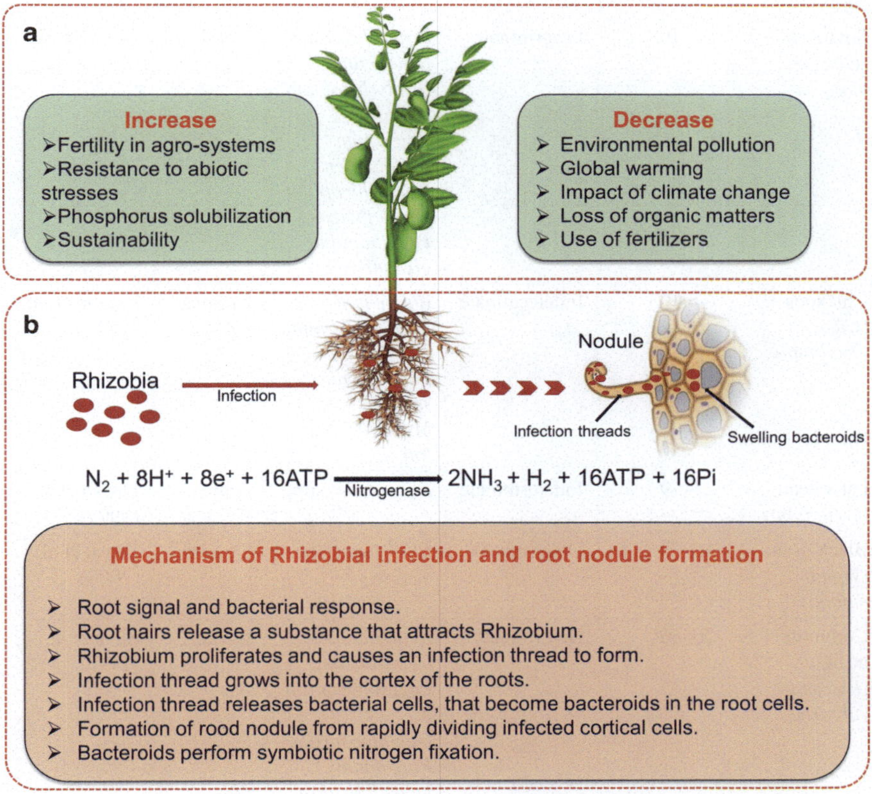

Fig. 5 (A) Some advantages of nitrogen fixation by legumes. (B) Schematic scheme illustrating the nitrogen-fixing process integrated with rhizobia in legumes. Rhizobia impose the roots of legumes and create nodules. Throughout the process of biological nitrogen fixation (BNF) by nodules, N_2 is brought down to two NH_3 molecules via rhizobial nitrogenase. However, H_2 is the byproduct of the symbiotic nitrogen fixating process

Table 3.2 Summary of nitrogen fixation by grain and forage legumes

Grain legume	N-fixing ability (kg ha^{-1})	Type of nodules	Strains of symbionts	Export product	References
Groundnut (*Arachis hypogaea*)	150–200	Determinate	*Rhizobium*, *Bradyrhizobium*	Ureides	Toomsan et al. (1995)
Pigeon pea (*Cajanus cajan*)	120–170	Indeterminate	*Bradyrhizobium*	Ureides	Adu-Gyamfi et al. (1997)
Soybean (*Glycine max*)	71–108	Determinate	*Bradyrhizobium japonicum*, *Bradyrhizobium elkanii*, *Rhizobium radiobacter*, *Bradyrhizobium diazoefficiens*, *Paenibacillus glycanilyticus*	Ureides	Gogoi et al. (2018)
Chickpea (*Cicer arietinum*)	64–103	Indeterminate	*Rhizobium* spp., *Mesorhizobium ciceri*, *Paenibacillus* sp., *Mesorhizobium loti*	Ureides Amides	Fatima et al. (2008)
Faba bean (*Vicia faba*)	23–79	Indeterminate	*Rhizobium* spp.	Amides	Danso et al. (1987)
Black gram (*Vigna mungo*)	16–79	Determinate	*Rhizobium* spp.	Ureides	Hayat et al. (2008)
Common bean (*Phaseolus vulgaris*)	20–60	Determinate	*R. lusitanum*, *R. mesosinicum*, Leguminosarum, *R. leguminosarum*, *Rhizobium leguminosarum* bv. *phaseoli*	Ureides	Silva et al. (1993)
Mung bean (*Vigna radiata*)	19–54	Determinate	*Rhizobium* spp.	Ureides	Hayat et al. (2008)
Pea (*Pisum sativum*)	46	Indeterminate	*Rhizobium leguminosarum*	Amides	Bhatia et al., (2001)

(continued)

Table 3.2 (continued)

Grain legume	N-fixing ability (kg ha^{-1})	Type of nodules	Strains of symbionts	Export product	References
Cowpea (*Vigna unguiculata*)	14–35	Determinate	*Azotobacter*, *Bacillus*, *Enterobacter*, *Pseudomonas*, *Serratia*, and *Azospirillum*	Ureides	Okereke and Ayama (1992)
Lentil (*Lens culinaris*)	8–14	Indeterminate	*Rhizobium etli*, *Rhizobium leguminosarum*	Amide	Cowell et al. (1989)
Forage legumes					
Hyacinth bean (*Lablab purpureus*)	214.6	Determinate	*Bradyrhizobium* spp.	Ureides	Haque and Lupwayi (2000)
Hairy vetch (*Vicia dasycarpa*)	163.0	Indeterminate	*Rhizobium leguminosarum* bv. *viciae*	Amides	Haque and Lupwayi (2000)
Medicago (*Medicago scutellata*)	140.1	Indeterminate	*Ensifer meliloti*	Amide	Haque and Lupwayi (2000)
Tufted vetch (*Vicia atropurpurea*)	103.6	Indeterminate	*Rhizobium leguminosarum* bv. *viciae*	Amide	Haque and Lupwayi (2000)
Clover (*Trifolium steudneri*)	44.7	Indeterminate	*Rhizobium leguminosarum* bv. *trifolii*	Amide	Haque and Lupwayi (2000)

5.1 Peanut

Arachis hypogaea L. (peanut) fixes the N across the seasons of the year (Devi et al. 2010). Peanut fixes 100–190 kg N ha^{-1}, which is quite comparable to pea, chickpea, and soybean (Pattee and Stalker 1995). N fixation in peanut contains a unique infection mechanism through which bacteria enter in crack entry/intercellular spreading (Boogerd and van Rossum 1997). In this infection type, rhizobia invasion occurs through epidermal cells of lateral emerging root, and then bacteria proliferate into the cortical cells (Nievas et al. 2012). Peanut nodule is spherical and has no permanent meristematic tissues, but central tissues have consisted of tightly packed bacteroids that take part in N fixation (Dupont et al. 2012).

5.2 Chickpea

A gram-negative and N-fixing bacterium known as *Mesorhizobium ciceri* forms nodules in chickpea roots. *M. ciceri*, under symbiotic association with chickpeas can fix around 120 kg N ha^{-1} (Herridge et al. 1995). However, nodulated chickpeas can fix less, with values around 103 kg N ha^{-1} according to Kurdali (1996). Nevertheless, Turpin et al. (2002) documented that chickpea can fix about 80–135 kg N ha^{-1}. In this species, the major constraint in efficient N fixation is the lower availability of inorganic phosphate (Pi) (Esfahani et al. 2016).

5.3 Soybean

Soybean is enriched with a higher level of protein content that can fulfill a significant part of human and animal nutritional diet. To make an excessive amount of proteins, soybean plants required a lot of N content that was fixed with the help of bacteroids (Bergersen 1969). Globally, it is described that *Bradyrhizobium japonicum* fixes 450 kg N ha^{-1} within symbiosis with soybean, representing 90% of the N requirement. Soybean is economically suitable for low N-containing areas (Zablotowicz and Reddy 2004). This species forms determinate nodules (Ferguson et al. 2011), whose life is a few weeks. After that, it becomes senescence and new nodule forms on the direction of the new root system (Gresshoff and Rolfe 1978). Espinosa-Victoria et al. (2000) observed senescence in a nine-week older nodule in soybean. In soybean, ureides act as primary N transport compounds (Tegeder 2014).

5.4 Pea

P. sativum forms a symbiotic relationship with *Rhizobium leguminosarum* bacteria. Pea forms indeterminate nodules, which have cylindrical shapes (Dupont et al. 2012). According to Bhatia et al. (2001), pea can fix 46 N kg ha^{-1}, and according to Adu-Gyamfi et al. (2007), pigeon pea can fix 168–280 kg N ha^{-1}. Asparagine functions as a dominant N transport compound in pea (Tegeder 2014), and much lesser extent, other amino acids; exported via phloem and xylem to other plant parts. According to Borisov et al. (2003), pea plants go through different nodule organogenesis.

5.5 Alfalfa

Alfalfa (*Medicago sativa*) forms a symbiotic association with *Ensifer* (*Sinorhizobium*) (a genus of N-fixing bacteria (rhizobia)) and can fix 200–400 N kg h^{-1} y^{-1}

(Mouradi et al. 2018). Its nodules are mostly spherical and consisted of markedly definite functional and developmental zones (Vasse et al. 1990). Asparagine utilized as dominant primary N transport compound in alfalfa (Tegeder 2014). Alfalfa as temperate legumes makes indeterminate nodules (Van Rhijn and Vanderleyden 1995). Usually, its nodules are pink in color due to leghemoglobin (Barsch et al. 2006).

5.6 Clover

Trifolium (Clover) is usually inoculated with *Rhizobium leguminosarum* bv. *trifolii* (Yates et al. 2011). Clover forms indeterminate nodules, which are well organized and cylindrical (Dupont et al. 2012). Clover can fix 44.7 kg N ha^{-1} (Haque and Lupwayi 2000). Amides are the main transporting compound in clover species, according to Gordon et al. (1995).

5.7 Common Bean

Common bean (*P. vulgaris* L.) has a low capacity to fix N than other legumes (Olivera et al. 2004; Isoi and Yoshida 1991). *Rhizobium leguminosarum* bv. *phaseoli* forms nodules in common bean (Mhamdi et al. 2002). According to Gualtieri and Bisseling (2000), *P. vulgaris* nodules matured within few weeks and appeared spherical due to non-persistent meristematic characteristics. Fernández-Luqueño et al. (2008) reported that during nodule senescence some morphology changes are taking place in two studied cultivars of *P. vulgaris* (Bayomex and Cacahuate 72) inoculated with *Rhizobium etli* strain CE-3. These include a change in protein and starch granules and their distribution, disintegration of central tissues, breakdown of a cell wall, and cellular collapse of infected zone. Ureides utilized as dominant primary N transport compound in alfalfa (Tegeder 2014).

5.8 Lupin

Only a few studies are based on N fixation of white lupin as compared to other legumes (Unkovich et al. 2010). Andreeva (1998) reported that the number of dead nodules increased at seed filling and the pod developmental stage of lupin. According to Sulas et al. (2016), lupins can fix 160–300 kg N ha^{-1}. Lupin nodules are documented as a subtype of intermediate nodules known as lupinoid nodules; as these nodules grow, form broadening connection with roots, causing both nodules and connections to surround root (Corby et al. 1983). Moreover, the other distinguished feather of lupinoid nodules is that their initiation takes place in outer cortex, and the basal laterally located meristem has bacteroids colonies, and the infected central tissues do not show any uninfected interstitial cells and typical zonation (Golinowski et al. 1987). However, in white lupin James et al. (1997) observed infection thread-like structure. Lupin nodule-forming bacteria are classified in genus *Bradyrhizobium* (Stępkowski et al. 2005).

6 Conclusion

There is a reliable prediction that the world population will increase more rapidly in the upcoming years. This prediction must be utilized in a parallel way to boost agricultural production to firm enough food for the increasing population. The excessive application of N fertilizers causes unequaled disturbance in the N cycle, documented by the immoderate amassing of various nitrates in the water and soil, and NO in the environment. Legumes are considered the second major crop and covering 14% of the total cultivated land worldwide. Legumes contribute a great many functions and ecosystem services toward the sustainability of different agricultural systems in different parts of the globe. Usually, they provide N to agroecosystem through their exclusive capability to fix atmospheric N in a symbiotic relationship with soil rhizobia. The ideal growth of a plant heavily depends upon the adequate supply of fixed N. The most common growth limiting factors are fixed N and water. Legumes contain the significant potential to be used in different agricultural practices, e.g., soil stabilization and reforestation. The symbiotic association among legumes and rhizobial bacterium occurs in the root nodules of legume where they access to atmospheric N. Hence, it is of great benefit for us to understand the mechanism of SNF and their resulting effects on the neighboring environment. However, SNF heavily depends on the selection of rhizobia. Nevertheless, the natural association between plants mainly gains the potential benefit from N fixation and bacteria, which is seldom manageable as part of agricultural practices. Especially in drylands, these associations are hard to maintain and are very unreliable. BNF by legumes is considered as an imperishable and ecofriendly choice to chemical fertilizers.

Several factors play a vital role in the regulation of N fixation by legumes, which include energy, oxygen, carbohydrates, leghemoglobin, and nodulation. On the other hand, water cycling among the plant shoots and nodules plays a vital part to comfort the high N fixation. Therefore, the selection of host-plant genotype is recommended as a critical step to improve the efficiency of N fixation by grain and forage legumes. In short, N fixation legumes play a crucial role in sustainable agriculture, and mainly, their utilization in the recovery of negligible areas, notable has a very fruitful future.

Acknowledgements The authors are grateful to all the researchers whose contributions have been cited in this chapter, which have helped us to prepare this chapter. Further, we are thankful to publishers and authors to grant us permission to use illustration from published papers.

References

Adams MA, Buchmann N, Sprent J, Buckley TN, Turnbull TL (2018a) Crops, nitrogen, water: are legumes friend, foe, or misunderstood ally? Trends Plant Sci 23:539–550
Adams MA, Buckley TN, Salter WT, Buchmann N, Blessing CH, Turnbull TL (2018b) Contrasting responses of crop legumes and cereals to nitrogen availability. New Phytol 217:1475–1483
Adu-Gyamfi JJ, Ito O, Yoneyama T, Katayama K (1997) Nitrogen management and biological nitrogen fixation in sorghum/pigeonpea intercropping on Alfisols of the semi-arid tropics. Soil Sci Plant Nutr 43:1061–1066
Adu-Gyamfi JJ, Myaka FA, Sakala WD, Odgaard R, Vesterager JM, Høgh-Jensen H (2007) Biological nitrogen fixation and nitrogen and phosphorus budgets in farmer-managed intercrops of maize–pigeonpea in semi-arid southern and eastern Africa. Plant Soil 295:127–136
Afkhami ME, Luke Mahler D, Burns JH, Weber MG, Wojciechowski MF, Sprent J, Strauss SY (2018) Symbioses with nitrogen-fixing bacteria: nodulation and phylogenetic data across legume genera. Ecology 99:502
Ahemad M, Kibret M (2014) Mechanisms and applications of plant growth promoting rhizobacteria: current perspective. J King Saud Univ Sci 26:1–20
Alazard D (1985) Stem and root nodulation in Aeschynomene spp. Appl Environ Microbiol 50:732–734
Allen ON, Allen EK (1981) The Leguminosae, a source book of characteristics, uses, and nodulation. University of Wisconsin Press, Madison, WI, USA
Al-Mallah MK, Davey MR, Cocking EC (1989) Formation of nodular structures on rice seedlings by rhizobia. J Exp Bot 40:473–478
Aluwihare LI, Meador T (2008) Chemical composition of marine dissolved organic nitrogen. In: Nitrogen in the marine environment. Academic Press, Cambridge, pp 95–140
Amar IA, Lan R, Petit CT, Tao S (2011) Solid-state electrochemical synthesis of ammonia: a review. J Solid State Electrochem 15:1845
Andreeva I (1998) Senescence of legume nodules. Russ J Plant Physiol 45:101–112
Ashby S (1907) Some observations on the assimilation of atmospheric nitrogen by a free living soil organism—*Azotobacter chroococcum* of Beijerinck. J Agric Sci 2:35–51
Bakker PA, Raaijmakers JM, Bloemberg GV, Höfte M, Lemanceau P, Cooke M (2007) New perspectives and approaches in 'plant growth-promoting rhizobacteria' research—foreword. Eur J Plant Pathol 119:241–242
Barsch A, Tellström V, Patschkowski T, Küster H, Niehaus K (2006) Metabolite profiles of nodulated alfalfa plants indicate that distinct stages of nodule organogenesis are accompanied by global physiological adaptations. Mol Plant-Microbe Interact 19:998–1013
Bashan Y, Holguin G (1997) Azospirillum–plant relationships: environmental and physiological advances (1990–1996). Can J Microbiol 43:103–121
Bashan Y, Holguin G, De-Bashan LE (2004) Azospirillum-plant relationships: physiological, molecular, agricultural, and environmental advances (1997–2003). Can J Microbiol 50:521–577
Bender G, Preston L, Barnard D, Rolfe B (1990) Formation of nodule-like structures on the roots of the non-legumes rice and wheat. In: Nitrogen fixation: achievements and objectives, p 825
Bergersen FJ (1969) Nitrogen fixation in legume root nodules: biochemical studies with soybean. Proc R Soc Lond Ser B Biol Sci 172:401–416
Bhatia CR, Nichterlein K, Maluszynski M (2001) Mutations affecting nodulation in grain legumes and their potential in sustainable cropping systems. Euphytica 120:415–432
Bittsánszky A, Pilinszky K, Gyulai G, Komives T (2015) Overcoming ammonium toxicity. Plant Sci 231:184–190
Blevins DG (1989) An overview of nitrogen metabolism in higher plants. In: Plant nitrogen metabolism. Springer, Boston, MA, pp 1–41
Boddey RM, Dobereiner J (1988) Nitrogen fixation associated with grasses and cereals: recent results and perspectives for future research. Plant Soil 108:53–65

Boddey RM, Dobereiner J (1995) Nitrogen fixation associated with grasses and cereals: recent progress and perspectives for the future. In: Nitrogen economy in tropical soils. Springer, Dordrecht, pp 241–250

Boogerd FC, van Rossum D (1997) Nodulation of groundnut by Bradyrhizobium: a simple infection process by crack entry. FEMS Microbiol Rev 21:5–27

Bottomley PJ, Myrold DD (2007) Biological N inputs. In: Soil microbiology, ecology and biochemistry. Elsevier, Academic Press, Amsterdam, pp 365–387

Borisov AY, Madsen LH, Tsyganov VE, Umehara Y, Voroshilova VA, Batagov AO, Sandal N, Mortensen A, Schauser L, Ellis N, Tikhonovich IA (2003) The Sym35 gene required for root nodule development in pea is an ortholog of Nin from *Lotus japonicus*. Plant Physiol 131:1009–1017

Boyd E, Peters JW (2013) New insights into the evolutionary history of biological nitrogen fixation. Front Microbiol 4:201

Cassman KG, Dobermann A, Walters DT (2002) Agroecosystems, nitrogen-use efficiency, and nitrogen management. AMBIO J Hum Environ 31:132–141

Chalk PM (2016) The strategic role of 15N in quantifying the contribution of endophytic N_2 fixation to the N nutrition of non-legumes. Symbiosis 69:63–80

Clara RW, Knowles R (1984) Superoxide dismutase, catalase, and peroxidase in ammonium-grown and nitrogen-fixing *Azospirillum brasilense*. Can J Microbiol 30:1222–1228

Cleveland CC et al (1999) Global patterns of terrestrial biological nitrogen (N_2) fixation in natural ecosystems. Global Biogeochem Cycles 13:623–645

Cocking EC, Al-Mallah MK, Benson E, Davey MR (1990) Nodulation of non-legumes by rhizobia. In: Nitrogen fixation. Springer, Boston, MA, pp 813–823

Cocking EC (2009) The challenge of establishing symbiotic nitrogen fixation in cereals. Nitrogen Fixation Crop Prod 52:35–63

Corby HDL, Polhill RM, Sprent JI (1983) Taxonomy. In: Broughton WJ (ed) Nitrogen fixation. Legumes, vol 3. Oxford University Press, Oxford, UK, pp 1–35

Cowell LE, Bremer E, Van Kessel C (1989) Yield and N, fixation of pea and lentil as affected by intercropping and N application. Can J Soil Sci 69:243–251

Daesch G, Mortenson LE (1968) Sucrose catabolism in *Clostridium pasteurianum* and its relation to N_2 fixation. J Bacteriol 96:346–351

Dalton DA, Kramer S (2007) Nitrogen-fixing bacteria in non-legumes. In: Plant-associated bacteria. Springer, Dordrecht, pp 105–130

Dalton H, Postgate J (1969) Growth and physiology of *Azotobacter chroococcum* in continuous culture. Microbiology 56:307–319

Danso SKA, Zapata F, Hardarson G (1987) Nitrogen fixation in fababeans as affected by plant population density in sole or intercropped systems with barley. Soil Biol Biochem 19:411–415

Dazzo F, Gardiol A (1984) Host specificity in Rhizobium-legume interactions. In: Genes involved in microbe-plant interactions. Springer, Vienna, pp 3–31

De Bruijn FJ (2015) Biological nitrogen fixation. In: Principles of plant-microbe interactions. Springer, Cham, pp 215–224

Dénarié J, Debellé F, Rosenberg C (1992) Signaling and host range variation in nodulation. Ann Rev Microbiol 46:497–531

Denison RF (1992) Mathematical modeling of oxygen diffusion and respiration in legume root nodules. Plant Physiol 98:901–907

Denison RF, Okano Y (2003) Leghaemoglobin oxygenation gradients in alfalfa and yellow sweetclover nodules. J Exp Bot 54:1085–1091

Devi MJ, Sinclair TR, Vadez V (2010) Genotypic variability among peanut (*Arachis hypogea* L.) in sensitivity of nitrogen fixation to soil drying. Plant Soil 330:139–148

Döbereiner J, Day J, Dart P (1972) Nitrogenase activity in the rhizosphere of sugar cane and some other tropical grasses. Plant Soil 37:191–196

Doty SL et al (2009) Diazotrophic endophytes of native black cottonwood and willow. Symbiosis 47:23–33

Downie JA (1994) Signalling strategies for nodulation of legumes by rhizobia. Trend Microbiol 2:318–324

Doyle JJ (2016) Chasing unicorns: nodulation origins and the paradox of novelty. Am J Bot 103:1865–1868

Doyle JJ, Luckow MA (2003) The rest of the iceberg. Legume diversity and evolution in a phylogenetic context. Plant Physiol 131:900–910

Dupont L, Alloing G, Pierre O, El Msehli S, Hopkins J, Hérouart D, Frendo P (2012) The legume root nodule: from symbiotic nitrogen fixation to senescence. In: Senescence. IntechOpen, pp 137–168

Elmerich C (2007) Historical perspective: from bacterization to endophytes. In: Associative and endophytic nitrogen-fixing bacteria and cyanobacterial associations. Springer, Dordrecht, pp 1–20

Esfahani MN et al (2016) Adaptation of the symbiotic Mesorhizobium–chickpea relationship to phosphate deficiency relies on reprogramming of whole-plant metabolism. Proc Nat Acad Sci USA 113:E4610–E4619

Espinosa-Victoria D, Vance CP, Graham PH (2000) Host variation in traits associated with crown nodule senescence in soybean. Crop Sci 40:103–109

Fatima Z, Bano A, Sial R, Aslam M (2008) Response of chickpea to plant growth regulators on nitrogen fixation and yield. Pak J Bot 40:2005–2013

Ferguson BJ, Foo E, Ross JJ, Reid JB (2011) Relationship between gibberellin, ethylene and nodulation in *Pisum sativum*. New Phytol 189:829–842

Fernández-Luqueño F, Dendooven L, Munive A, Corlay-Chee L, Serrano-Covarrubias LM, Espinosa-Victoria D (2008) Micro-morphology of common bean (*Phaseolus vulgaris* L.) nodules undergoing senescence. Acta Physiol Plant 30:545–552

Figg TM, Holland PL, Cundari TR (2012) Cooperativity between low-valent iron and potassium promoters in dinitrogen fixation. Inorg Chem 51:7546–7550

Franche C, Lindström K, Elmerich C (2009) Nitrogen-fixing bacteria associated with leguminous and non-leguminous plants. Plant Soil 321:35–59

Franssen H et al (1995) Plant genes involved in root-nodule development on legumes. Philos Trans R Soc Lond B Biol Sci 350:101–107

Franssen HJ, Xiao TT, Kulikova O, Wan X, Bisseling T, Scheres B, Heidstra R (2015) Root developmental programs shape the *Medicago truncatula* nodule meristem. Development 142:2941–2950

Galloway JN et al (2004) Nitrogen cycles: past, present, and future. Biogeochemistry 70:153–226

Galloway JN et al (2008) Transformation of the nitrogen cycle: recent trends, questions, and potential solutions. Science 320:889–892

Gauthier-Coles C, White R, Mathesius U (2018) Nodulating legumes are distinguished by a sensitivity to cytokinin in the root cortex leading to pseudonodule development. Front Plant Sci 9:1901

Gehring C, Vlek PL, de Souza LA, Denich M (2005) Biological nitrogen fixation in secondary regrowth and mature rainforest of central Amazonia. Agric Ecosyst Environ 111:237–252

Gehringer MM, Pengelly JJ, Cuddy WS, Fieker C, Forster PI, Neilan BA (2010) Host selection of symbiotic cyanobacteria in 31 species of the Australian cycad genus: Macrozamia (Zamiaceae). Mol Plant Microbe Interact 23:811–822

Gibson A (1966) The carbohydrate requirements for symbiotic nitrogen fixation: a "whole-plant" growth analysis approach. Aust J Biol Sci 19:499–516

Gibson KE, Kobayashi H, Walker GC (2008) Molecular determinants of a symbiotic chronic infection. Ann Rev Genet 42:413–441

Gioseffi E, Neergaard Ad, Schjørring JK (2012) Interactions between uptake of amino acids and inorganic nitrogen in wheat plants. Biogeosciences 9:1509–1518

Gogoi N, Baruah KK, Meena RS (2018) Grain legumes: impact on soil health and agroecosystem. In: Legumes for soil health and sustainable management. Springer, Singapore, pp 511–539

Golinowski W, Kopcinska J, Borucki W (1987) The morphogenesis of lupine root nodules during infection by *Rhizobium lupini*. Acta Soc Bot Pol 56:687–703

Gordon AJ, Thomas BJ, James CL (1995) The location of sucrose synthase in root nodules of white clover. New Phytol 130:523–530
Gresshoff PM, Rolfe BG (1978) Viability of Rhizobium bacteroids isolated from soybean nodule protoplasts. Planta 142:329–333
Griesmann M et al (2018) Phylogenomics reveals multiple losses of nitrogen-fixing root nodule symbiosis. Science 361:eaat1743
Gruber N, Galloway JN (2008) An Earth-system perspective of the global nitrogen cycle. Nature 451:293
Gualtieri G, Bisseling T (2000) The evolution of nodulation. In: Plant molecular evolution. Springer, Dordrecht, pp 181–194
Guinel FC (2009) Getting around the legume nodule: I. The structure of the peripheral zone in four nodule types. Botany 87:1117–1138
Haque I, Lupwayi NZ (2000) Nitrogen fixation by annual forage legumes and its contribution to succeeding wheat in the Ethiopian Highlands. J Plant Nutr 23:963–977
Hardoim PR et al (2015) The hidden world within plants: ecological and evolutionary considerations for defining functioning of microbial endophytes. Microbiol Mol Biol Rev 79:293–320
Hawkesford M, Horst W, Kichey T, Lambers H, Schjoerring J, Møller IS, White P (2012) Functions of macronutrients. In: Marschner's mineral nutrition of higher plants. Elsevier, Academic Press, Amsterdam, pp 135–189
Hayat R, Ali S, Ijaz SS, Chatha TH, Siddique MT (2008) Estimation of N_2-fixation of mung bean and mash bean through xylem ureide technique under rainfed conditions. Pak J Bot 40:723–734
Herridge DF, Marcellos H, Felton WL, Turner GL, Peoples MB (1995) Chickpea increases soil-N fertility in cereal systems through nitrate sparing and N_2 fixation. Soil Biol Biochem 27:545–551
Herridge DF, Peoples MB, Boddey RM (2008) Global inputs of biological nitrogen fixation in agricultural systems. Plant Soil 311:1–18
Hill S (1976) The apparent ATP requirement for nitrogen fixation in growing *Klebsiella pneumoniae*. Microbiology 95:297–312
Hirel B, Tétu T, Lea PJ, Dubois F (2011) Improving nitrogen use efficiency in crops for sustainable agriculture. Sustainability 3:1452–1485
Isoi T, Yoshida S (1991) Low nitrogen fixation of common bean (*Phaseolus vulgaris* L.). Soil Sci Plant Nutr 37:559–563
Jaiyebo E, Moore A (1963) Soil nitrogen accretion under different covers in a tropical rain-forest environment. Nature 197:317–318
James EK, Minchin FR, Iannetta PPM, Sprent JI (1997) Temporal relationships between nitrogenase and intercellular glycoprotein in developing white lupin nodules. Ann Bot 69:493–503
James EK, Iannetta PPM, Deeks L, Sprent JI, Minchin FR (2000a) Detopping causes production of intercellular space occlusions in both the cortex and infected region of soybean nodules. Plant Cell Environ 23:377–386
James EK, Gyaneshwar P, Barraquio WL, Mathan N, Ladha JK (2000b) Endophytic diazotrophs associated with rice. In: The quest for nitrogen fixation in rice. International Rice Research Institute, Philippines, pp 119–140
Jing Y, Li G, Jin G, Shan X, Zhang B, Guan C, Li J (1990) Rice root nodules with acetylene reduction activity. In: Nitrogen fixation achievements and objectives. Springer, Berlin, p 829
Jones DL, Farrar J, Giller KE (2003) Associative nitrogen fixation and root exudation—What is theoretically possible in the rhizosphere? Symbiosis 35:19–38
Karmakar K, Rana A, Rajwar A, Sahgal M, Johri BN (2015) Legume-rhizobia symbiosis under stress. In: Plant microbes symbiosis: applied facets. Springer, Berlin, pp 241–258
Kennedy IR, Tchan Y-T (1992) Biological nitrogen fixation in non-leguminous field crops: recent advances. Plant Soil 141:93–118
Kennedy I, Zeman A, Tchan Y, New P, Sriskandarajah S, Nie Y (1990) Biological nitrogen fixation and prospects for yield increases in wheat. In: Transactions 14th international congress of soil science, Kyoto, Japan, August 1990, vol III, pp 146–151
Krapp A et al (2014) Nitrate transport and signalling in Arabidopsis. J Exp Bot 65:789–798

Kurdali F (1996) Nitrogen and phosphorus assimilation, mobilization and partitioning in rainfed chickpea (*Cicer arietinum* L.). Field Crop Res 47:81–92

Lavin M, Pennington RT, Klitgaard BB, Sprent JI, de Lima HC, Gasson PE (2001) The dalbergioid legumes (Fabaceae): delimitation of a pantropical monophyletic clade. Am J Bot 88:503–533

Lavin M, Herendeen PS, Wojciechowski MF (2005) Evolutionary rates analysis of Leguminosae implicates a rapid diversification of lineages during the tertiary. Syst Biol 54:575–594

Lema KA, Willis BL, Bourne DG (2012) Corals form characteristic associations with symbiotic nitrogen-fixing bacteria. Appl Environ Microbiol 78:3136–3144

Limpens E, Mirabella R, Fedorova E, Franken C, Franssen H, Bisseling T, Geurts R (2005) Formation of organelle-like N_2-fixing symbiosomes in legume root nodules is controlled by DMI2. Proc Nat Acad Sci 102:10375–10380

Lodwig EM et al (2003) Amino-acid cycling drives nitrogen fixation in the legume–Rhizobium symbiosis. Nature 422:722

Lohmann GV et al (2010) Evolution and regulation of the *Lotus japonicus* LysM receptor gene family. Mol Plant Microbe Interact 23:510–521

Loureiro MF, De Faria S, James E, Pott A, Franco A (1994) Nitrogen-fixing stem nodules of the legume, *Discolobium pulchellum* Benth. New Phytol 128:283–295

Lu CC, Tian H (2017) Global nitrogen and phosphorus fertilizer use for agriculture production in the past half century: shifted hot spots and nutrient imbalance. Earth Syst Sci Data 9:181

Magne K, George J, Berbel Tornero A, Broquet B, Madueño F, Andersen SU, Ratet P (2018) *Lotus japonicus* NOOT-BOP-COCH-LIKE 1 is essential for nodule, nectary, leaf and flower development. Plant J 94:880–894

Mariotti F, Tomé D, Mirand PP (2008) Converting nitrogen into protein—beyond 6.25 and Jones' factors. Crit Rev Food Sci Nutr 48:177–184

Mhamdi R, Laguerre G, Aouani ME, Mars M, Amarger N (2002) Different species and symbiotic genotypes of field rhizobia can nodulate *Phaseolus vulgaris* in Tunisian soils. FEMS Microbiol Ecol 41:77–84

Mori M, Hori K, Akashi M, Hori M, Sato Y, Nishida M (1998) Fixation of atmospheric nitrogen: synthesis of heterocycles with atmospheric nitrogen as the nitrogen source. Angew Chem Int 37:636–637

Mouradi M, Farissi M, Bouizgaren A, Qaddoury A, Ghoulam C (2018) *Medicago sativa*-rhizobia symbiosis under water deficit: physiological, antioxidant and nutritional responses in nodules and leaves. J Plant Nutr 41:384–395

Muñoz JA, Coronado C, Pérez-Hormaeche J, Kondorosi A, Ratet P, Palomares AJ (1998) MsPG3, a *Medicago sativa* polygalacturonase gene expressed during the alfalfa–*Rhizobium meliloti* interaction. Proc Nat Acad Sci 95:9687–9692

Mus F et al (2016) Symbiotic nitrogen fixation and the challenges to its extension to nonlegumes. Appl Environ Microbiol 82:3698–3710

Newton WE (2007) Physiology, biochemistry, and molecular biology of nitrogen fixation. In: Biology of the nitrogen cycle. Elsevier, Amsterdam, pp 109–129

Nie Y, Vesk M, Kennedy I, Sriskandarajah S, Lane E, Tchan Y (1992) Structure of 2, 4-D induced paranodules with Rhizobium on wheat. Phytochemistry 11:67–73

Nielsen A (1981) Ammonia synthesis: exploratory and applied research. Catal Rev 23:17–51

Nievas F, Bogino P, Nocelli N, Giordano W (2012) Genotypic analysis of isolated peanut-nodulating rhizobial strains reveals differences among populations obtained from soils with different cropping histories. Appl Soil Ecol 53:74–82

Nomata J, Mizoguchi T, Tamiaki H, Fujita Y (2006) A second nitrogenase-like enzyme for bacteriochlorophyll biosynthesis reconstitution of chlorophyllide a reductase with purified x-protein (bchX) and yz-protein (bchY-bchZ) from *Rhodobacter capsulatus*. J Biol Chem 281:15021–15028

Noxon J (1976) Atmospheric nitrogen fixation by lightning. Geophys Res Lett 3:463–465

O'Brien JA, Vega A, Bouguyon E, Krouk G, Gojon A, Coruzzi G, Gutiérrez RA (2016) Nitrate transport, sensing, and responses in plants. Mol Plant 9:837–856

Okamoto S, Ohnishi E, Sato S, Takahashi H, Nakazono M, Tabata S, Kawaguchi M (2008) Nod factor/nitrate-induced CLE genes that drive HAR1-mediated systemic regulation of nodulation. Plant Cell Physiol 50:67–77

Okereke GU, Ayama N (1992) Sources of nitrogen and yield advantages for monocropping and mixed cropping with cowpeas (*Vigna unguiculata* L.) and upland rice (*Oryza sativa* L.). Biol Fertil Soils 13:225–228

Oldroyd GE, Murray JD, Poole PS, Downie JA (2011) The rules of engagement in the legume-rhizobial symbiosis. Ann Rev Genet 45:119–144

Olivares J, Bedmar EJ, Sanjuán J (2013) Biological nitrogen fixation in the context of global change. Mol Plant Microbe Interact 26:486–494

Olivera M, Tejera N, Iribarne C, Ocana A, Lluch C (2004) Growth, nitrogen fixation and ammonium assimilation in common bean (*Phaseolus vulgaris*): effect of phosphorus. Physiol Plant 121:498–505

Orme-Johnson W, Davis L (1977) Current topics and problems in the enzymology of nitrogenase. Ref 12:15

Pattee HE, Stalker HT (1995) Advances in peanut science. American Peanut Research and Education Society, Stillwater, OK

Pauly N et al (2006) Reactive oxygen and nitrogen species and glutathione: key players in the legume–Rhizobium symbiosis. J Exp Bot 57:1769–1776

Paungfoo-Lonhienne C, Visser J, Lonhienne TG, Schmidt S (2012) Past, present and future of organic nutrients. Plant Soil 359:1–18

Peoples M, Herridge D, Ladha J (1995) Biological nitrogen fixation: an efficient source of nitrogen for sustainable agricultural production? In: Management of biological nitrogen fixation for the development of more productive and sustainable agricultural systems. Springer, Berlin, pp 3–28

Peters NK, Frost JW, Long SR (1986) A plant flavone, luteolin, induces expression of *Rhizobium meliloti* nodulation genes. Science 233:977–980

Postgate J (1974) Prerequisites for biological nitrogen fixation in free-living heterotrophic bacteria. In: Biology of nitrogen fixation. North Holland, Amsterdam, pp 663–686

Postgate J (1992) The Leeuwenhoek lecture, 1992. Bacterial evolution and the nitrogenfi xing plant. Philos Trans R Soc Lond B Biol Sci 338:409–416

Postgate J (2012) The chemistry and biochemistry of nitrogen fixation. Springer Science and Business Media, Berlin

Preisig O, Zufferey R, Thöny-Meyer L, Appleby CA, Hennecke H (1996) A high-affinity cbb3-type cytochrome oxidase terminates the symbiosis-specific respiratory chain of *Bradyrhizobium japonicum*. J Bacteriol 178:1532–1538

Qiu W et al (2018) High-performance artificial nitrogen fixation at ambient conditions using a metal-free electrocatalyst. Nat Commun 9:3485

Raza A, Razzaq A, Mehmood SS, Zou X, Zhang X, Lv Y, Xu J (2019) Impact of climate change on crops adaptation and strategies to tackle its outcome: A review. Plants 8:34

Reed SC, Cleveland CC, Townsend AR (2008) Tree species control rates of free-living nitrogen fixation in a tropical rain forest. Ecology 89:2924–2934

Reed SC, Cleveland CC, Townsend AR (2011) Functional ecology of free-living nitrogen fixation: a contemporary perspective. Ann Rev Ecol Evol Syst 42:489–512

Reis VM, dos Reis FB Jr, Quesada DM, de Oliveira OC, Alves BJ, Urquiaga S, Boddey RM (2001) Biological nitrogen fixation associated with tropical pasture grasses. Funct Plant Biol 28:837–844

Rothballer M, Schmid M, Hartmann A (2007) Diazotrophic bacterial endophytes in Gramineae and other plants. In: Prokaryotic symbionts in plants. Springer, Berlin, pp 273–302

Rousk K, Sorensen PL, Michelsen A (2016) Nitrogen transfer from four nitrogen-fixer associations to plants and soils. Ecosystems 19:1491–1504

Ruinen J (1965) The phyllosphere: III. Nitrogen fixation in the phyllosphere. Plant Soil 22:375–394

Salim N, Raza A (2020) Nutrient use efficiency (NUE) for sustainable wheat production: a review. J Plant Nutr 43:297–315

Schubert KR (1982) The energetics of biological nitrogen fixation. American Society of Plant Physiologists, Maryland, pp 1–30

Schwartz A et al (2013) *Bacillus simplex*—a little known PGPB with anti-fungal activity—alters pea legume root architecture and nodule morphology when coinoculated with *Rhizobium leguminosarum* bv. *viciae*. Agronomy 3:595–620

Seefeldt LC, Hoffman BM, Peters JW, Raugei S, Beratan DN, Antony E, Dean DR (2018) Energy transduction in nitrogenase. Acc Chem Res 51:2179–2186

Silsbury J (1977) Energy requirement for symbiotic nitrogen fixation. Nature 267:149

Silva PMD, Tsai SM, Bonetti R (1993) Response to inoculation and N fertilization for increased yield and biological nitrogen fixation of common bean (*Phaseolus vulgaris* L.). Plant Soil 152:123–130

Soltis DE, Soltis PS, Morgan DR, Swensen SM, Mullin BC, Dowd JM, Martin PG (1995) Chloroplast gene sequence data suggest a single origin of the predisposition for symbiotic nitrogen fixation in angiosperms. Proc Nat Acad Sci 92:2647–2651

Sørensen J, Sessitsch A (2007) Plant-associated bacteria-lifestyle and molecular interactions. In: Modern soil microbiology. CRC Press, Boca Raton, pp 211–236

Sorochkina K, Ayuso SV, Garcia-Pichel F (2018) Establishing rates of lateral expansion of cyanobacterial biological soil crusts for optimal restoration. Plant Soil 429:199–211

Sprent JI (2009) Legume nodulation: a global perspective. Wiley, Hoboken

Sprent JI, James EK (2007) Legume evolution: where do nodules and mycorrhizas fit in? Plant Physiol 144:575–581

Sprent JI, Sprent P (1990) Nitrogen fixing organisms: pure and applied aspects. Chapman and Hall Ltd., UK

Sprent J, Sutherland M, De Faria S (1989) Structure and function of nodules from woody legumes. AGRIS

Sprent J, Ardley J, James E (2013) From North to South: a latitudinal look at legume nodulation processes. South Afr J Bot 89:31–41

Stępkowski T, Moulin L, Krzyżańska A, McInnes A, Law IJ, Howieson J (2005) European origin of Bradyrhizobium populations infecting lupins and serradella in soils of Western Australia and South Africa. Appl Environ Microbiol 71:7041–7052

Stewart WDP (1969) Biological and ecological aspects of nitrogen fixation by free-living microorganisms. Proc R Soc Lond B Biol Sci 172:367–388

Sulas L, Canu S, Ledda L, Carroni AM, Salis M (2016) Yield and nitrogen fixation potential from white lupine grown in rainfed Mediterranean environments. Sci Agric 73:338–346

Sylvester-Bradley R, Kindred DR (2009) Analysing nitrogen responses of cereals to prioritize routes to the improvement of nitrogen use efficiency. J Exp Bot 60:1939–1951

Tavares F, Santos CL, Sellstedt A (2007) Reactive oxygen species in legume and actinorhizal nitrogen-fixing symbioses: the microsymbiont's responses to an unfriendly reception. Physiol Plant 130:344–356

Tegeder M (2014) Transporters involved in source to sink partitioning of amino acids and ureides: opportunities for crop improvement. J Exp Bot 65:1865–1878

Thilakarathna M, Papadopoulos Y, Rodd A, Grimmett M, Fillmore S, Crouse M, Prithiviraj B (2016) Nitrogen fixation and transfer of red clover genotypes under legume–grass forage based production systems. Nutr Cycl Agroecosyst 106:233–247

Toomsan B, McDonagh JE, Limpinuntana V, Giller KE (1995) Nitrogen fixation by groundnut and soyabean and residual nitrogen benefits to rice in farmers' fields in Northeast Thailand. Plant Soil 175:45–56

Turpin JE, Herridge DF, Robertson MJ (2002) Nitrogen fixation and soil nitrate interactions in field-grown chickpea (*Cicer arietinum*) and fababean (*Vicia faba*). Aust J Agric Res 53:599–608

Unkovich M, Baldock J (2008) Measurement of asymbiotic N_2 fixation in Australian agriculture. Soil Biol Biochem 40:2915–2921

Unkovich M, Baldock J, Peoples M (2010) Prospects and problems of simple linear models for estimating symbiotic N_2 fixation by crop and pasture legumes. Plant Soil 329:75–89

Valentine AJ, Benedito VA, Kang Y (2018) Legume nitrogen fixation and soil abiotic stress: from physiology to genomics and beyond. Ann Plant Rev Online 42:207–248
Vance CP (2002) Root-bacteria interactions: symbiotic N_2 fixation. In: Plant roots. CRC Press, Boca Raton, pp 1298–1343
Van Dommelen A, Vanderleyden J (2007) Associative nitrogen fixation. In: Biology of the nitrogen cycle. Elsevier, Amsterdam, pp 179–192
van Rhijn P, Vanderleyden J (1995) The Rhizobium-plant symbiosis. Microbiol Mol Biol Rev 59:124–142
Vasse J, De Billy F, Camut S, Truchet G (1990) Correlation between ultrastructural differentiation of bacteroids and nitrogen fixation in alfalfa nodules. J Bacteriol 172:4295–4306
Vessey JK, Pawlowski K, Bergman B (2005) Root-based N_2-fixing symbioses: legumes, actinorhizal plants, Parasponia sp. and cycads. In: Root physiology: from gene to function. Springer, Berlin, pp 51–78
Vidal EA, Moyano TC, Canales J, Gutiérrez RA (2014) Nitrogen control of developmental phase transitions in *Arabidopsis thaliana*. J Exp Bot 65:5611–5618
Wang L, Xia M, Wang H, Huang K, Qian C, Maravelias CT, Ozin GA (2018) Greening ammonia toward the solar ammonia refinery. Joule 2:1055–1074
Webb J, Sheehy J (1991) Legume nodule morphology with regard to oxygen diffusion and nitrogen fixation. Ann Bot 67:77–83
Willems A (2006) The taxonomy of rhizobia: an overview. Plant Soil 287:3–14
Yamamoto H, Kurumiya S, Ohashi R, Fujita Y (2009) Oxygen sensitivity of a nitrogenase-like protochlorophyllide reductase from the cyanobacterium *Leptolyngbya boryana*. Plant Cell Physiol 50:1663–1673
Yang Y et al (2017) Characterization of genetic basis on synergistic interactions between root architecture and biological nitrogen fixation in soybean. Front Plant Sci 8:1466
Yates RJ, Howieson JG, Reeve WG, O'Hara GW (2011) A re-appraisal of the biology and terminology describing rhizobial strain success in nodule occupancy of legumes in agriculture. Plant Soil 348:255
Young J (1992) Phylogenetic classification of nitrogen-fixing organisms. BNF 1544:43–86
Zablotowicz RM, Reddy KN (2004) Impact of glyphosate on the *Bradyrhizobium japonicum* symbiosis with glyphosate-resistant transgenic soybean. J Environ 33:825–831
Zackrisson O, DeLuca TH, Nilsson M-C, Sellstedt A, Berglund L (2004) Nitrogen fixation increases with successional age in boreal forests. Ecology 85:3327–3334

Nitrogen Fixation of Legumes Under the Family Fabaceae: Adverse Effect of Abiotic Stresses and Mitigation Strategies

Ayman EL Sabagh, Akbar Hossain, M Sohidul Islam, Shah Fahad, Disna Ratnasekera, Ram Swaroop Meena, Allah Wasaya, Tauqeer Ahmad Yasir, Muhammad Ikram, Muhammad Mubeen, Maham Fatima, Wajid Nasim, Arzu Çığ, Fatih Çığ, Murat Erman, and Mirza Hasanuzzaman

Abstract Nitrogen (N) is an essential element of the building blocks of almost all plant structures and a vital component of chlorophyll, enzymes, and proteins. It is used in relatively large amounts than other plant nutrients. Therefore, N has been

---

A. EL Sabagh (✉)
Department of Agronomy, Faculty of Agriculture, Kafrelsheikh University, Kafrelsheikh, Egypt
e-mail: ayman.elsabagh@agr.kfs.edu.eg

A. Hossain (✉)
Bangladesh Wheat and Maize Research Institute (BWMRI), Dinajpur 5200, Bangladesh
e-mail: akbarhossainwrc@gmail.com; tanjimar2003@yahoo.com

M. S. Islam
Department of Agronomy, Hajee Mohammad Danesh Science and Technology University, Dinajpur 5200, Bangladesh

S. Fahad
Department of Agriculture, University of Swabi, Khyber Pakhtunkhwa, Pakistan

D. Ratnasekera
Department of Agricultural Biology, Faculty of Agriculture, University of Ruhuna, Matara, Sri Lanka

R. S. Meena
Department of Agronomy, Institute of Agricultural Sciences, BHU, Varanasi, Uttar Pradesh 221005, India

A. Wasaya · T. A. Yasir
College of Agriculture, BZU, Bahadur Sub-campus, Layyah, Pakistan

M. Ikram
Department of Agronomy, Faculty of Agricultural Sciences and Technology, Bahauddin Zakariya University, Multan, Pakistan

M. Mubeen · M. Fatima
Department of Environmental Sciences, COMSATS University Islamabad, Vehari Campus, Pakistan

M. Hasanuzzaman et al. (eds.), *The Plant Family Fabaceae*,
https://doi.org/10.1007/978-981-15-4752-2_4

recognized as the most limiting nutrient in crop production systems. Several transformation processes are involved in the nitrogen cycle. Among them, biological nitrogen fixation is an environmentally friendly natural resource for sustainable agricultural systems. Recently, the reports directed to a decrease in agricultural dependence on symbiotic nitrogen fixation due to abiotic stresses. Therefore, abiotic stresses are a topic that increasingly occupies the attention of the world is still a matter of debate. Although physiological mechanisms are affected in more intense abiotic stresses, most research efforts have focused on the study of these processes. In legume plants being grown under symbiotic conditions, one of the primary effects of abiotic stresses is a decline in the rates of symbiotic nitrogen fixation (SNF). In the present chapter, we summarize our current understanding of the factors that are affected by SNF in legumes. Finally, an overview of the available resources and applications of the physiological system for understanding the complex responses of legumes to abiotic stresses is provided. The overall conclusion was that all physiological mechanisms are important in understanding the regulation of N_2 fixation and its response to abiotic stresses.

Keywords Abiotic stresses · Fabaceae family · Legumes · Symbiotic nitrogen fixation

Abbreviations

ARA	Acetylene reduction activity
ATP	Adenosine triphosphate
BNF	Biological nitrogen fixation
DON	Dissolved organic nitrogen
DST	Decreased salt tolerance
EPS	Exopolysaccharides
EON	Extractable organic nitrogen
HSP	Heat shock proteins
Lb	Leghemoglobin

W. Nasim
Department of Agronomy, University College of Agriculture and Environmental Sciences, The Islamia University of Bahawalpur (IUB), Bahawalpur, Pakistan

A. Çığ
Department of Horticulture, Siirt University, Siirt, Turkey

A. EL Sabagh · F. Çığ · M. Erman
Department of Field Crops, Faculty of Agriculture, Siirt University, Siirt, Turkey

M. Hasanuzzaman
Department of Agronomy, Faculty of Agriculture, Sher-e-Bangla Agricultural University, Sher-e-Bangla Nagar, Dhaka 1207, Bangladesh

N	Nitrogen
NH_3	Ammonia
NH_4^+	Ammonium
NO_3	Nitrate
N_2O	Nitrous oxide
NUE	Nitrogen use efficiency
$NADP^+$/NADPH	Nicotinamide adenine dinucleotide phosphate
PMN	Potentially mineralizable N
ROS	Reactive oxygen species
SNF	Symbiotic nitrogen fixation
TSN	Total soluble nitrogen

1 Introduction

Nitrogen (N) has been recognized as the most limiting nutrient in crop production. It is a crucial nutrient for plants as it is the major component of the chlorophyll molecule and an essential element of the building blocks of almost all plant structures and a vital component of enzymes, and proteins (Hofman and Cleemput 2004). For example, N is directly linked with the optimization of crop quality through enriching protein which plays a significant role in many biochemical activities in plants by acting as a structural unit as well as enzymes (Grant and Flaten 1998). Among others, N is the most imperative plant nutrient for crop production and being used in relatively large amounts by all living things (Campbell et al. 1986; Hodges 2010). It occupies a unique position among essential nutrients because agricultural crops require a more enormous amount of N to achieve maximum yield through stimulating root growth and crop development as well as uptake of the other nutrients (Hofman and Cleemput 2004; St. Luce et al. 2011). However, N concentration in soil and plant is continuously declining due to lack of consideration associated with intensive cropping (Ghosh et al. 2003). Hence, scientists in the world are more concentrated on research upon N fixation and plant nutrition. Also, the role of human activities in the global nitrogen cycle has become a topic of increasing research attention (Galloway et al. 1995).

In most ecosystems, nitrogen is the element that regulates net plant primary production (Lambers et al. 1998). However, nitrogen is also a big issue for environmental sustainability because leaching of nitrate (NO_3) can reduce groundwater quality, and nitrous oxide (N_2O) emission—a significant contributor to greenhouse gas emission and global warming (Campbell et al. 1995). Consequently, N management plays a vital role in improving crop yield and quality, environmental safety, and ultimately the economics of crop production (Campbell et al. 1995). But its content is declined which owing to continuous crop cultivation without giving due attention to nitrogen requirement (Ghosh et al. 2003). Mechanistic knowledge of the soil N cycle is critical in understanding the behavior of ecosystems and their responses to natural and anthropogenic mediated change (Jones et al. 2004).

Imbalanced uses of chemical fertilizers have led to creating the problem in the environment (Sun et al. 2017). For example, greenhouse gas such as nitrous oxide (N_2O) is produced by microbial action on inorganic fertilizers in the soil which causes a reduction of the ozone layer, as the ozone layer serves as a safeguard against the harmful effect of UV-rays. Although the application of nitrogenous fertilizers increases the crop yield (carbohydrate), the excessive use of these fertilizers causes the negative effects on the environment as well as decreases the potassium (K) content in food grains. K is an important element for human health which controls the rise in blood pressure and also avoids the risks of heart attack (American Heart Association 2016). Furthermore, the excessive use of chemical fertilizers is the foundation of the ammonia (NH_3) emissions (Sheppard et al. 2010), and are currently responsible for 75% of global NH_3 emissions (Hamaoui-Laguel et al. 2014). Therefore, it is an urgent need to lower the adverse environmental impacts of agricultural chemical fertilization. Different initiatives are in progress, aiming to improve N nutrition and N use efficiency (NUE) in plants, such as the manipulation of plant N metabolism. The biological N fixation (BNF) is a promising alternative to improve N nutrition, as the use of inoculants of diazotrophic bacteria in agriculture has been proven to enhance N availability and uptake, to promote plant growth, to increase biomass, and to keep the plants healthy (Adesemoye and Kloepper 2009). The main benefit of using legumes is to ensure that the BNF which, in turn, decreases the amount of nitrogen (N) fertilizer needed for the succeeding crop (Kumar et al. 2018).

In legume plants grown under symbiotic conditions, one of the primary effects of abiotic stresses is the decline in the rates of symbiotic N fixation (SNF). In the present chapter, we summarized the current understanding of the factors that are affected by SNF in legumes. Finally, an overview of the available resources and applications of the physiological system for understanding the complex responses of legumes to abiotic stresses stress is provided.

2 Transformation Processes that Are Involved in the N Cycle

2.1 *N Transformation*

Many transformation processes of N are engaged in the soil. These transformation processes are called the N cycle. The knowledge of these processes is important for the management of N and fertilizer. In agricultural and natural ecosystems, N is very dynamic and subjected to several processes. There are a few different processes involved in N cycling (Fig. 4.1), which are described below:

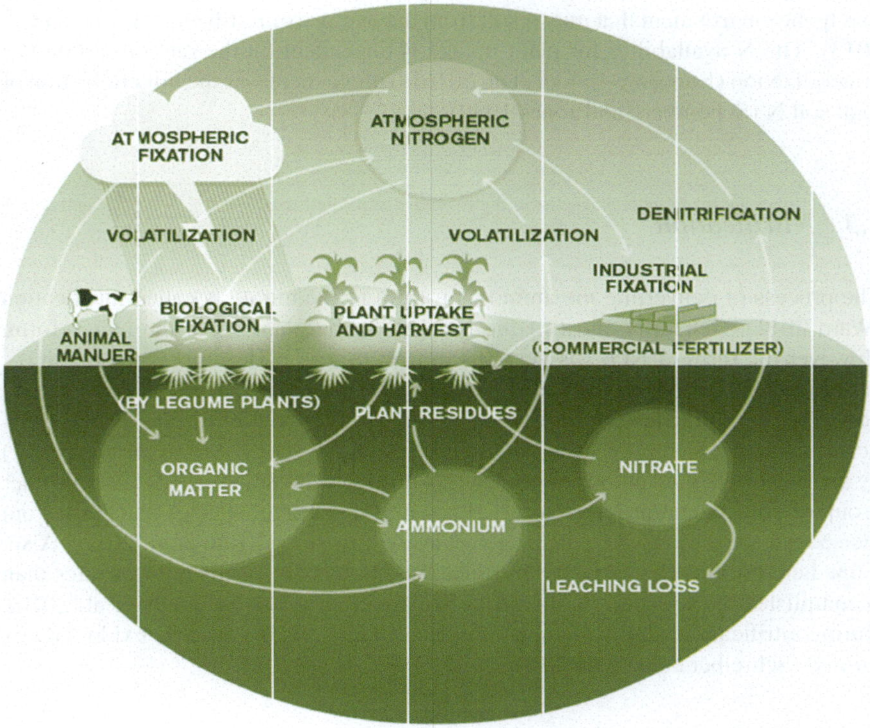

Fig. 4.1 Transformation processes that are involved in the N cycle

2.2 *Mineralization*

This a process by which N from organic matter in the soil is converted to ammonium (NH_4^+). The three major forms of N involved in biological N are proteins (amino acids), plant cell wall components (amino sugars, such as cellulose and chitin), and nucleic acids. Mineralization of organic matter is the process of biological degradation of proteins, amino sugars, and nucleic acids to the ammonium form. Plants can uptake either NH_4^+ or NO_3^-, depending on the species and environment. When plants uptake N as NH_4^+ form, it is quickly transformed into NO_3^-) or NH_3 (Boudsocq et al. 2012).

Nitrogen (N) released by mineralization is often a vital source of N for plant growth and strongly depends on soil factors (D'Haene et al. 2008). Generally, N in crop residues is in the organic form which is not directly available for plant growth, and this organically bound N is only available for crops or microbial growth through N mineralization during the decomposition of crop residues (Lupwayi et al. 2006; Van Den Bossche et al. 2009). Therefore, an accurate estimate of potentially available N in soil is essential to ensure optimum crop yield and quality and to minimize nutrient loss to the environment that may result from the overuse of fertilizers (Bordoloi et al. 2013). The N availability for plant uptake is dependent on the rate of carbon (C) mineralization (Kaneez-e-Batool et al. 2016). Tillage increases the mineralization of total soil N (Schomberg and Jones 1999).

2.3 *Nitrification*

The process of converting the ammonium form to the nitrate is called nitrification (Ward 2011). Generally, most of the N taken up by plants is in the NO_3^- form. However the form of NO_3^- is very susceptible to loss by leaching because NO_3^- is negatively charged, and soil also has a negative charge, NO_3^- is not chemically bonded to the soil tended to each other (Azam and Farooq 2003; Ward 2011). Nitrification is a two-step process, in the first step, NH_4^+ is converted to nitrite (NO_2^-) by a group of obligate autotrophic bacteria known as nitrosomonas species. Another group of obligate autotrophic bacteria known as nitrobacter species that carries out the second step, where NO_2 is further converted to NO_3^- (Rao et al. 2017). Also, a few heterotrophs can carry out nitrification, but usually at much lower rates than accomplished by the autotrophic bacteria (Sun et al. 2006; Schläppy et al. 2010). During nitrification, slight amounts of nitrous oxide (N_2O) and nitric oxide (NO) is formed (Schreiber et al. 2009, 2012).

2.4 *Denitrification*

Denitrification is an anaerobic biological process that is the opposite of the nitrification process (Munch and Velthof 2007; Bouwman et al. 2013). There are two types of denitrification: (i) biological denitrification—biochemical reduction of NO_3^- N to gaseous compounds. During biological denitrification, NO_3^- and NO_2^- are reduced to N oxides (NO, N_2O) and molecular N (N_2) by microorganisms and these gaseous products are not available for plant uptake, and (ii) chemo denitrification—refers to the same reduction pattern and end products, but it is not carried out by microorganisms (Spanning et al. 2005; Munch and Velthof 2007; Bouwman et al. 2013). This denitrification mainly occurs in the subsoil and acid conditions. This type of denitrification is more crucial to NO_3^- pollution in deep groundwater (Shivran et al. 2006).

2.5 *Immobilization*

The process by which the available ammonium form is consumed by microorganisms, while becomes unavailable to the plant is called N immobilization. This process is the opposite of mineralization and mainly occurs when soil microorganisms absorb inorganic N and make it immobilize for plant uptake. The C:N ratio of soil organic materials vastly influences the immobilization process. For example, the C:N ratio > 25:1 is good for immobilization while the C:N ratio below 20:1 could be optimum for mineralization (Tully and Ryals 2017).

2.6 *Nitrogen Fractions*

2.6.1 Total Soil Nitrogen

The total soil N content is generally mirrors of total soil organic C as the N cycle is inextricably linked to the C cycle (Bradford and Peterson 2000).

2.6.2 Total Soluble Nitrogen (TSN)

The term TSN can be used as soluble organic N (SON) or extractable organic N (EON) or dissolved organic N (DON). Recently, Xiang et al. (2008) proposed that it would be the more appropriate term to say EON, which is obtained by extraction rather than SON. It is defined as organic N present in dissolved form in soil solution (Murphy et al. 2000). Dissolved organic N is defined as the fraction of soil organic N, which is collected in situ using a lysimeter, rhizon or suction cup, among other devices, and where no extractant is used (Ros et al. 2009). Dissolved organic N originates from many compounds which enter the soil from a range of sources including dry and wet deposition, litter-fall, root, and microbial exudation, turnover of roots and organisms, urine and feces, and organic fertilizer additions to soil (Haynes 2005; Kalbitz et al. 2000). Among these sources, root exudates and turnover are likely to be the most important in many agricultural systems mainly where inorganic fertilizers are added (Christou et al. 2006). In most ecosystems, N plays a vital role in mineralization, leaching and plant uptake (Jones et al. 2004). Altering land use induces a quantitative change of soil water-soluble organic matter, but knowledge of such change is still limited (Zhao et al. 2012). Water-soluble organic matter contains labile organic carbon and N, which is very sensitive to soil management (Zhang et al. 2011).

2.6.3 Potentially Mineralizable Nitrogen (PMN)

The active fraction of soil organic N is considered as a potentially mineralizable N (PMN) which is mainly responsible for the release of mineral N through microbial action, and it is composed of a heterogeneous array of organic substrates including microbial biomass, residues of recent crops, and humus (Curtin and Campbell 2007). Theoretically, PMN is the amount of N that will mineralize in infinite time at optimum temperature and moisture (Curtin and Campbell 2007). Franzluebbers et al. (1995) reported that PNM at 0–5 cm depth was higher in NT (no-tillage) soils than in CT (conventional tillage) and with rotated than continuous sorghum, and increased with increasing cropping intensity after nine years in south-central Texas.

2.6.4 $KMnO_4$ Extractable-N

$KMnO_4$ extractable-N as a predictor of N availability in soils. Alkaline $KMnO_4$ is a quick and reliable indicator of indigenous soil N supply in soils under a rice-wheat cropping system (Nayyar et al. 2006).

2.6.5 Amine Nitrogen

Amino sugar N has been considered as potentially labile N pools in the soil (Khan et al. 2001; Mulvaney et al. 2001). Amino sugar N appears to be the more sensitive indicator of N availability. Amelung et al. (1999) sampled soils along a climosequence from Saskatchewan-Canada to Texas-USA and found that amino sugar N concentration varied predictably with mean annual temperature and precipitation. Zhang et al. (1999) noted that cultivation reduces 50% amino sugar N concentration in the soil, like many other fractions. Mulvaney et al. (2001) have concept amino sugar N in the soil would serve as an identifiable and easily mineralizable N pool.

3 Prospect of Legume-Rhizobium Symbiosis in Agriculture

The biological N fixation (BNF) is a promising alternative to improve N nutrition, as the use of inoculants of diazotrophic bacteria in agriculture has been proven to enhance N availability and uptake, to promote plant growth, to increase biomass, and to keep the plants healthy (Adesemoye and Kloepper 2009). The associative and endophytic diazotrophic bacteria naturally colonize and contribute with fixed N to several economically important plant species, comprising a natural system to be explored. However, the mechanisms regulating this plant–bacteria association are still not clear; thus, a better understanding of the mechanisms is necessary to allow improvement and manipulation of this association, and possibly an extension of it to non-natural hosts. Quantitative analyses of BNF and plant growth promotion demonstrated that plant and bacterial genotypes are important factors in controlling the efficiency of the association (Carvalho et al. 2011).

In this context, one challenge in this area is the determination of the best combination of diazotrophic bacteria and plant varieties to obtain the maximum benefit from this association in agriculture. A considerable effort should be made to understand the molecular and genetic factors controlling all steps of the association: recognition, colonization, N fixation, and plant growth promotion. Several advances came from genomic approaches, and integrative gene expression maps are being generated for some plant species colonized with associative and endophytic diazotrophic bacteria. Possible regulatory mechanisms involved were identified, and functional analyses are now necessary. Also, it is important to determine common regulatory pathways governing a fruitful association with diazotrophic bacteria, as well as those specific to plant–bacteria genotypes.

N status might act as a critical signal-regulating and integrating various metabolic processes that occur during association with diazotrophic bacteria. Besides directly providing ammonium to plants, the associative and endophytic diazotrophic bacteria enhance N uptake of inoculated plants, an effect that could be important for improving N Use Efficiency (NUE). Nevertheless, high N levels inside plants seem to signal feedback control, negatively regulating BNF, and bacterial colonization (Carvalho et al. 2014; Sapountzis 2016). High N in soil could contribute to an increase in N levels in plants, activating this negative feedback control. Therefore, a clear understanding of the mechanisms in N regulation during plant interaction with associative and endophytic diazotrophic bacteria could provide tools to maximize the benefits for crop production.

4 BNF Inputs to Agricultural Systems

While the Haber-Bosch N fertilizer production system is considered to have ensured crop production to fed millions of people throughout the world, it is not without the following major concerns: the several-fold increase in reactive N cycling throughout the ecosphere; the relative lack of efficiency of reactive N under agricultural use, which may lead to major ecological issues of water contamination and eutrophication; and the demand for fossil fuel, which is generally demanded as natural gas (Rockström et al. 2009; Good and Beatty 2011; Kim et al. 2011; Sutton et al. 2011). In contrast, even with all these caveats, the growing population and the desired increase in income will only demand higher levels of food production, particularly of meat and dairy products (Godfray et al. 2010). Meeting these demands will be impossible without a reliable N source. While chemically fixed fertilizer will necessarily be a part of the various options deployed by agricultural and soil scientists throughout the world, and increased reliance on BNF is one of the significant and sustainable alternatives for both maintaining and/or increasing agricultural yield and reducing both environmental and economic concerns linked to N fertilizer use (Doane et al. 2009). One point to keep in mind is that these alternatives are not either/or solution sets and should not be thought of as such. Both other options are simply tools to increase agricultural yield to allow human resources needs to be fulfilled in such a way that future generations will have access to at least the same pool of natural resources as previous ones did. Unfortunately, even though obtaining global N fertilizer estimates is relatively easy, getting estimates for biologically fixed N is not easy (Herridge et al. 2008; Peoples et al. 2009). One of the first reasons for this difficulty is the sheer number of possible biological systems, which all have different BNF capabilities, natural ranges, cultivated ranges, areas, and potential yields. Burris (1981) has been quoted by Herridge et al. (2008) as having said that "potential authors could use a variety of methods to fill in the values in the N cycle, from gazing at crystal balls, consulting sages to cranking out computer-generated random numbers." However, the most common method is a literature review, and choosing the numbers thought to be a more logical approach (Figueiredo et al. 2013).

The difficulty in estimating global BNF can be roughly divided into the following several different reasons:

1. Methodological problems in field-scale BNF estimation
2. Highly variable BNF rates, which are strongly affected by environmental and agricultural concerns
3. Difficulty in estimating individual cropping systems, worldwide distribution, and cultivated areas.

The first and second reasons intermingle with the high variation in BNF rates and lead to highly variable estimates, for example, for soybeans (*Glycine max*), which range from 0 to 450 kg shoot N ha^{-1} according to different sources cited by Peoples et al. (2009). Besides, another major problem is that the root system is routinely not included in the BNF estimates, which may lead to soybean going from a net exporter (Di Ciocco et al. 2011) to a net fixer of soil N (Singh et al. 2004). When considering the importance and number of agricultural systems in which soybean participates, this change may have significant effects on the overall N balance. Even with all these caveats, it is still highly important to achieve a total estimate, which has been well executed in several reports on a global or national scale (Herridge et al. 2008; Peoples et al. 2009). One common approach has been to obtain an estimate for the average BNF rate per hectare and then multiply by another estimate of the total cultivated area of the specific system. This approach may also be performed with some subdivision as exemplified by the recognition of different BNF rates for Argentinean-, Brazilian-, Chinese-, and North American-grown soybeans (Okogun et al. 2005; Hungria et al. 2006; Oberson et al. 2007; Schipanski et al. 2010; Di Ciocco et al. 2011). As indicated by the previous examples, a further point to consider is that BNF estimation is much more common for the legume-rhizobia symbiosis than for other systems, such as grass-endophyte associations or Azolla-cyanobacteria symbiosis, which results from a better knowledge of the system, a much more precise estimate of occurrence, a vastly higher number of punctual experiments from which to derive raw data, and the sheer importance of the legume-rhizobia symbiosis in world agriculture. Most estimates for global BNF are 120 Tg of N $year^{-1}$ (Herridge et al. 2008) and legume-rhizobia symbiosis accounting for approximately a sixth of that estimation (Peoples et al. (2009). These values are less than those estimated for the N fertilizer industry of ca. 140 Tg of N $year^{-1}$ (Canfield et al. 2010). In contrast, while most estimates indicate that approximately 1–2% of global energy consumption is directly linked to N fertilizer production. There should also be a lower NOx emission derived from BNF, and thus lower glasshouse effects, because all the N is in an organic form by definition and, therefore, should not be available for denitrification most of the time (Jensen et al. 2012). This consideration is also important when considering that there is an international tendency to demand more sustainable agriculture with less resource consumption for a given yield level, which could be maximized through further use of BNF as a rule in agricultural systems (Wilkins 2008), and if non-BNF advantages of legume inclusion are considered, such as the reduction in disease incidence or nutrient mining of deeper layers allowed by their higher root system (Köpke and Nemecek 2010). All these advantages require a greater need for management knowledge, which is one of the primary reasons for the greater reliance on N fertilizer as an alternative. The greater need for management knowledge is due to the highly localized effects of environment and cropping systems, which demand a high level of experience of the farmer to maximize their efficiency as exemplified by the variable effects of species and cultivars of legumes seeded with barley, the effect of a fertility gradient on BNF from legume-grass mixtures (Schipanski and Drinkwater 2012), or the impact of cutting management and Desmodium species in a mixture with corn (Kifuko-Koech et al. 2012).

5 Nitrogen Fixation in Legumes as Influenced by Abiotic Stresses

The rate of N fixation was affected by many different physiological and environmental factors in soil, such as temperature, water holding capacity, water stress, salinity, N level, pH, and other nutrients. Many of these factors, including temperature, affect many aspects of N fixation and assimilation, as well as factors such as respiratory activity, gaseous diffusion, and solubility of dissolved gases, which ultimately affect plant growth (Coskan and Dogan 2011).

5.1 Soil Water Status and Nitrogen Fixation

The development of soil microbes is affected by ground moisture via absorption mechanisms, water flow, and accumulation of nutrients. Soil moisture affects the microbial population in the soil, including the rhizobium population, decreased osmotic pressure under crucial resistance levels and partially affecting photosynthesis, root structure, and exudations (Manzoni et al. 2012; Van Horn et al. 2014). Poor legume nodule formation is probably due to a decline in landmass rhizobium levels during and after the dry spell in the tropical climate. It is not appropriate to disregard the effect of land-moisture efficiency on photosynthesis and vitality, and therefore also impacting in nodule formation. Rhizobium has established a range of pathways to respond to moisture stress, mainly through the biosynthetic aggregation of inorganic and organic dissolved salts. For example, *R. meliloti* eliminates osmoregulation stress-induced development suppression by producing suitable dissolved salts such as potassium ions, dopamine, arginine, nicotinamide methionine, betaine proline, trehalose, and dipeptide, *N*-acetyl glutamine amide (Boscari et al. 2002). Many integrated solutions would be utilized as development resources of either N or C, indicating that its catabolism may be controlled during aqueous pressure to avoid deterioration.

Water stress is one of the main factors influencing the mutualistic relationship of N-fixing rhizobium and legume hosts. It was previously reported that some mechanisms are disturbed in the physiochemical reactions of mutualistic N-fixing to stress caused by moisture deficiency, namely C shortage and C nodulation metabolites, oxygen (O_2) constraint, and responded directive through the accumulation of N fixation products. During drought, compared with drought-tolerant genotypes, the amount of 14C released for nodules is about four times that of drought-resistant fixed N_2 genotypes as reported by Serraj (2003). Initially, plants produced a higher concentration of organic compounds that help plants to survive against drought stress (Ulrich et al. 2019). But later on, field evidence suggested that organic compounds produced by plants in response to drought stress inhibit the enzymatic activity which leads to a reduced rate of the biological N fixation process (Serraj 2003; Streeter 2003; Ulrich et al. 2019). The conclusion is that every one of three physiological mechanisms are vital for understanding the regulation of N_2 fixation and its response to soil drying. Water deprivation is a status that causes dehydration in plants (Kozlowski and Pallardy 1997). The principal mechanism through that plants develops when in drought is to manage water loss. Water loss from plants is modulated by morphological modification or physiological adaptation. Morphological adaptation of plants includes leaf rolling, stomatal response, while physiological adaptation includes metabolism at low leaf wetness, enzymatic and non-enzymatic antioxidants in plant cells and secretion of stress response plant hormones (Taiz and Zeiger 2002). The area of arid or semi-arid land is growing worldwide, as soils grow to be salinized thanks to intensive irrigation and inflated temperatures, thus limiting crop yields.

5.2 *Salinity: Physiological Impacts on Legume Nitrogen Fixation*

Abiotic stresses threats agriculture and leading to the deterioration of the environment, which resulted in crop loss global (EL Sabagh et al. 2015, 2019). Salt stress impairs plant metabolism and structure. Salinity has devastating impacts on legumes by affecting germination and vegetative and reproductive growth, and to biological N fixation capacity (Van Hoorn et al. 2001). In this scenario, proteins play a crucial role in salt tolerance, acclimatization, and cellular adjustment. In legumes, salt stress has significantly limited productivity as an adverse effect on the growth of the host plant, root nodule bacteria and their symbiotic development and N fixation capacity (Zahran 1999). Legume plants are classified into sensitive or moderately sensitive according to their degree of sensitivity level. Legumes such as *Medicago truncatula*,

Medicago sativa, or *Pisum sativum* very sensitive to high salinity while *Vicia faba*, *G. max*, *Phaseolus vulgaris*, or *A. hypogea* are more salt-tolerant (Swaraj and Bishnoi 1999).

In rhizobium-legume symbiosis, the process of nitrogen fixation is highly related to the physiological status of the host plant (Zahran 1999). As said previously, symbiotic N fixation is dependent on the host plant genotype, *Rhizobium* strain, and the interaction of these symbionts with pedo-climatic factors and environmental conditions. There are two phases of salinity: the early phase (osmotic phase), the response of plants under saline conditions due to outside the root zone, whereas late phase (ionic phase), where the toxicity of the accumulation of salt inside the plant (Chen et al. 2014). Salinity imposes three primary stresses on plants. A combination of these three factors adversely affects the physiological, biochemical, and molecular changes in the plant process (Edelstein et al. 2011).

Soil salinity causes impacts on rhizobial biology, as well as legumes physiology. Most of the rhizobia are sensitive to the harmful effects of high osmolarity (Talibart et al. 1997). The sensitivity of different rhizobia species and strain to salt stress varies considerably, and some of these bacterial strains can persist and survive in saline soils (Van Hoorn et al. 2001). Morphological alterations of bacteria under salt stress include the expansion of cell size, distortion of the cell envelope, and change of rhizobial cell structure (Vanderlinde et al. 2010). The same author also stated that besides the surface components, several symbiotic bacteria might suffer additional changes in the saline environment: alterations such as a decrease in the synthesis of exopolysaccharides. Further, there are some reports on changes in the length of the side chain of lipopolysaccharide (Soussi et al. 2001), and deletion of periplasmic oligosaccharide involved in bacterial osmotic adaptation (Ghittoni and Bueno 1995).

In legumes, higher sensitivity to salinity is observed during seedling and developmental stages than the germination stage (Bayuelo-Jiménez et al. 2002). Also, salinity affects nodulation and the final yield. Early studies uncovered that salinity interferes with N uptake and biological N fixation, which limits N supply in legumes (Essa 2002). Nodules are vital for N fixation in legumes but nodulation is susceptible to salinity. Salinity affects the infection process by inhibiting root hair growth and by decreasing the number of nodules per plant and the amount of N fixed per unit weight of nodules (Zahran 1999; Bruning and Rozema 2013). Legumes are grown in saline environments often exhibit reduced number and weight of root nodules (Swaraj and Bishnoi 1999). A significant reduction in N fixation and total N content in snap bean plants grown under saline conditions has been recorded (Campbell et al. 1986). The negative effect of salinity on rhizobium–legume symbiosis is reflected in the inhibition of specific nitrogenase activity. Thus, in saline soils, the yield of leguminous crops is decreased due to the lack of successful symbiosis mainly because of drop-in bacteroid respiration and in the leghemoglobin (Lb) content (Delgado et al. 1994). For instance, previous studies on salinity and legumes revealed that salinity substantially affects the formation of the nodules in pigeon pea (Manchanda and Garg 2008) and faba bean (Rabie and Almadini 2005) leading to senescence of premature nodules in soybean (Sarath et al. 1986). The treatment of *Cajanus cajan* (L.) Millsp. plants with NaCl resulted in early greening of the nodules, a decline in nodule Lb, decreased acetylene reduction activity (ARA) and reduced N content of leaf and stem (Swaraj et al. 2000; Manchanda and Garg 2008). Salt tolerance was linked with the higher stability of O_2 nodule conductance of the tolerant rhizobia when increasing soil salinity (L'taief et al. 2007). Bolaños et al. (2006) stated that nodule number and weight diminished in salt-stressed nodules of pea plants, and most of them appeared pale in contrast with the control pink nodules. These results reveal that the effectiveness of symbiotic interaction under salinity depended on the symbiotic efficacy of the isolates under standard conditions and no correlation with the source of root nodule bacteria (soil or nodule) or their salt tolerance.

A multifaceted sequence of events coordinated by the host and bacterial signaling underlie and lead to the development of symbiotic interaction. Rhizobia respond to flavonoids exuded by the leguminous roots by secreting lipo-chito oligosaccharides nodulation factors (NOD), the crucial molecular signal to trigger the nodulation program in a compatible host (Bruning and Rozema 2013). Salinity can negatively affect this signal exchange (Miransari and Smith 2009). According to Estévez et al. (2009), salinity stress also increases the biosynthesis of nod factors (NFs) and changes their structure with a high number of different new biologically active NFs being generated. Primary NF-triggered responses causing root hair deformation involved several changes as intra- and extra-cellular alkalinization, phosphatidic acid, and diacylglycerol formation, and accumulation of reactive O_2 species (Cooper 2007). It was observed that, in the absence of flavonoid inducers, a high concentration of NaCl induced nodulation genes (nod genes) and the production of NFs (Guasch-Vidal et al. 2013). Pre-incubation of *Bradyrhizobium japonicum* with the molecular signal genistein (NF inducers) was described as a method to alleviate the stressful effects of high salinity on soybean and *B. japonicum* symbiosis (Miransari and Smith 2009).

Accumulation of low molecular compounds as a result of salt in the external root environment was found, but they are not interfering with normal biochemical reactions. Disaccharides, sugar alcohols, amino acids, and sulfonium compounds are some examples of the low molecular compounds accumulated under salt stress (Iqbal et al. 2015; Slama et al. 2015). The accumulation compatible osmolytes help to raise the osmotic pressure to counteract the high concentration of inorganic salts in the vacuole and the root medium and act as an energy sink, such as C and N_2 source. Additionally, these compounds also protect some structures from denaturation (Iqbal et al. 2015; Slama et al. 2015). Sucrose metabolism decreases under salt stress via lowering the supply of dicarboxylic acids by bacteroids. It is mainly associated with a lower content of malate in the nodule cytosol (Liu et al. 2018). Trehalose (a non-reducing sugar), is abundant in root nodules (López et al. 2008). During nodule senescence, 84% of sucrose content reduced while trehalose is remaining constant. Meanwhile, trehalose acts as an osmoprotectant (Müller et al. 2001; Vogel et al. 2001). Khan et al. (2017) reported that salt stress declined photosynthesis, decreased the sugar contents by 22–47%, total sugar in stems, leaves, and developing pods was increased by increasing sucrose infusion of salt-treated chickpea plants. In salt-stressed chickpea plants, sucrose infusion increased dry mass (2.6 fold), pod numbers (3.8 fold), seed number (6.5 fold), and seed yield (10.4 fold).

Amino acids, such as alanine, arginine, glycine, serine leucine, and valine, together with the amino acid, proline, and the non-protein amino acids, citrulline and ornithine have been accumulated plant under salt stress (Annunziata et al. 2017, 2019). Under salt stress, legume plants accumulate high levels of proline to adjust osmotic potential. High levels of proline in nodular tissues were found in *M. truncatula* (Chun et al. 2018), and *L. japonicus* plants. *M. sativa* accumulates a large amount of proline betaine, and proline as osmoprotectants. Proline–betaine within nodules interfere with N fixation (Trinchant et al. 2004). In soybean nodules, proline level increases resulting in high ratios of $NADP^+$/NADPH (nicotinamide adenine dinucleotide phosphate) and activation of pentose phosphate pathway and final production of purine. Derivate of proline act as the transporters of fixed N_2 (Miransari et al. 2013).

Nodulation is highly sensitive to salt stress, climatic conditions, soil properties, and growth stage of the plant. Previous studies uncovered pigeon pea and faba bean substantially effect by the salt stress and led to senescence of nodules (Cordovilla et al. 1995). The curly root hair formation causes inhibition of bacteria colonization, eventually reducing the infection rate, weight, and the number of nodule formation. Chickpea cultivar inoculated with *Rhizobium* strain and irrigated with saline water significantly decreased total nodule number per plant and average nodule weight also reported by Flowers et al. (2010) and Egamberdieva et al. (2014). Microscopic observations showed loss of turgor of the peripheral cells, loss of rigidity of the cell wall, inhibition of nitrogenase activity because of the reduction of the protein leghemoglobin by oxidative stress and by the carbohydrate content of both cytosol and the bacteroids (López-Gómez et al. 2016). *P. vulgaris* nodules inoculated with salt-tolerant rhizobium tropically wild-type strains showed a reduction of the nitrogenase activity rather than its decreased salt-tolerance mutant derivatives (Manchanda and Garg 2008; Latef and Ahmad 2015).

Early studies revealed that salt stress interferes with nitrogen uptake, which limits N supply in legumes. For instance, the reduction of NO_3 supply from the soil causes a reduced protein content of legumes including mung bean, chickpea and faba bean (Ghassemi-Golezani et al. 2010; Stagnari et al. 2017). Further, salt stress decreased the amino acid, carbohydrates, polysaccharides, and protein contents in legumes such as mung bean which interfere with photosynthesis, nutrient uptake and nutritional imbalance, and final yield (Farooq et al. 2017). The reduction of nodule-soluble protein may result from the decrease in protein synthesis and increase of protease activity in the cytosol. The *P. sativum* L. cv. Lincoln plants subjected to 50 mM sodium chloride stress for three weeks, showed a significant reduction in acetylene reduction activity (ARA) and leghemoglobin (Lb) content in isolated nodules. NaCl reduced ARA activity by 77% and Lb by 50% in pea nodules treated with 50 mM NaCl (Delgado et al. 1994). In chickpea plants inoculated with *Mesorhizobium ciceri* strain Ch-191 and challenged with salt stress during the vegetative stage, it was reported that growth was inhibited by the highest NaCl, whereas inhibition of nodulation and N fixation even at the lowest NaCl concentration (Soussi et al. 1998). Salt-tolerant or salt-sensitive *Rhizobium leguminosarum* strains were inoculated to legumes plants under salt stress and reported that higher levels of nodule mass and N fixation decline in response to salinity (Cordovilla et al. 1999; Dong et al. 2017). According to Sadiki and Rabih (2001), 200 Moroccan accessions of chickpea were evaluated for salt stress for dry weight, nodule mass, nitrogenase activity, and total N under NaCl treatments and revealed MCA 103, MCA 131, and MCA 250 were selected as salt SNF tolerant lines. The efficiency of symbiotic interaction under salinity condition depends on the symbiotic efficiency of the isolates, and it did not correlate with the source of root nodule bacteria or their salt tolerance (Roumiantseva et al. 2006).

A said previously, rhizobial strains are also sensitive to salinity, which severely affect the symbiotic N fixation and thus limit the legume productivity. Salinity limits the growth of rhizobia, which reduces rhizobial cell number, restrict root infection, inhibit nodule development, and impairing nodule activity (Alexandre and Oliveira 2013). Different genotypes show variation in salt tolerance. *Rhizobium japonicum* strain USDA191 is more tolerant to 0.4 M NaCl treatment than is the strain USDA110 (Yelton et al. 1983) whereas *Mesorhizobium* strains are more sensitive to salt stress than are strains of *Rhizobium* and *Sinorhizobium* but are more tolerant than strains of *Bradyrhizobium* (Alexandre and Oliveira 2013). Therefore isolation and evaluating salt-tolerant rhizobia are key aspects for alleviating the adverse effect of salinity, particularly in arid and semi-arid regions.

Rhizobium has a different internal porousness, measured by its ability to grow at totally different concentrations of binary compounds (two-element salt). *B. japonicum*, *Rhizobium etli* and *R.* are salt-sensitive, and their growth is totally reserved at a hundred-millimeter binary compound (Boncompagni et al. 1999); *Arthrobacter*, *Rhizobium tropici* IIB, and *S. fredii* are moderately sensitive, their growth at two hundred millimeters binary compound is reserved, while *Aiolopus strepens* and *Agrobacterium tumefactiens* are strongly salt resistant and grow at three hundred millimeters binary compound (Bernard et al. 1986). *Rhizobium* spp. isolated from nodules of Hedysarum, Leucaena, and Prosopis plants will tolerate up to five hundred millimeter binary compounds (Zhang et al. 1991). Rhizobia survive in stress due to the expression of NaCl-responsive loci leading to the accumulation of osmoregulators, enhanced production of exopolysaccharides, ROS-scavenging enzymes, and heat shock proteins and chaperones (Vriezen et al. 2007).

5.3 N Fixation in Legumes as Influenced by High-Temperature Stress and Soil Temperature

In general, plants experience the situation of heat stress when they expose to high-temperature events. The impact of heat stress (herein referring to the one due to high temperature) depends on the intensity, exposure duration of the stress and the growth stage of the crop. Heat stress that often co-occurs with drought is reported to affect various crops leading to a severe reduction in yield potential (Bita and Gerats 2013). Indeed, the temperature is one of the major factors affecting seed yield and quality in legumes (Ruelland and Zachowski 2010; Christophe et al. 2011). Numerous reports highlight the effects of high-temperature stress in the reproductive phase of legumes such as chickpea (Kaushal et al. 2013; Kumar et al. 2013), pea (Guilioni et al. 1997), common bean (Gross and Kigel 1994; Vara Prasad et al. 2002), mungbean (Tzudir et al. 2014; Bindumadhava et al. 2016) and cowpea (Ahmed et al. 1992).

Further, the high-temperature stress was found to be one of the foremost factors affecting the symbiotic N fixation of legume crops. Keerio and Wilson (1998) and Keerio (2001) reported that N fixation in legumes was severely inhibited at high temperatures. These reports highlighted that when the temperature exceeds 35 °C in the root zone, this could affect N fixation while N fixation severely inhibited at 45 °C in soybean cultivars (Keerio 2001). High soil temperature is one of the severe factors, which can restrain the development of N-fixing association in arid and semi-arid regions. The persistence of rhizobia in the soil is profoundly affected by high temperatures because it causes detrimental effects (Niste et al. 2013). The most favorable temperature range for optimum growth of rhizobia is 28–31 °C, and many of rhizobia are unable to grow at 38 °C (Graham 1992). High soil temperature could adversely affect both free and symbiotic rhizobia (Zahran 1999). However, certain rhizobial strains isolated from Acacia are reported to survive at high temperatures reached up to 44 °C (Zahran et al. 1994). High temperature can influence the exchange of molecular signals between rhizobia and the symbiotic partners (Sadowsky 2005). Moreover, it is also stated that high temperatures can inhibit bacterial adherence to root hairs of legumes, bacteroid differentiation, nodule structure, and function of legume root nodules (Zahran 1999; Alexandre and Oliveira 2013). Sudden temperature fluctuations are known to stimulate the synthesis of heat shock proteins (HSP), which can play a defensive role by tolerating heat effects without alteration of the inner cell temperature (Xu et al. 2011). Though most bacteria have only a less number of HSP, rhizobia appear having exclusion of HSP (Alexandre and Oliveira 2013).

High soil temperatures in tropical and subtropical regions are a significant problem for the biological N fixation of legumes (Michiels et al. 1994). High root temperatures strongly affect microbial inoculation and N_2 fixation in several legume species, including soybean (Munevar and Wollum 1982), guar [*Cyamopsis tetregonoloba* (L.)] (Arayankoon et al. 1990), peanut (Kishinevsky et al. 1992), cowpea (Rainbird et al. 1983) and beans (Piha and Munnus 1987; Hungria and Franco 1993).

However, critical temperatures for BNF appeared to be species dependent. For example, N_2 fixation is 30 °C for clover and pea and range between 35 and 40 °C for soybean, guar, peanut, and cowpea (Michiels et al. 1994), for common beans (*Phaseolus* spp.) 25–30 °C being hindered by root temperatures between 30 and 33 °C (Piha and Munnus 1987). Many early reports showed that high soil temperature inhibits legume biological N fixation by affecting on nodulation, nodule establishment, and nitrogenase activity (Roughley and Dart 1970; Whitehead 1995). Soil temperature in the root zone is one of the controlling factors for nodulation and nodule establishment; thus, attention is necessary on the minimum and maximum soil temperatures, and the optimum range of temperatures, which are favorable for N fixation. Responses of N_2 fixation in legume species to soil temperature were extensively studied and are summarized in Table 4.1.

It was further shown that higher root temperatures affect disease, N fixation, and leguminous crop development (Kasper et al. 2019) and even have a significant impact

Table 4.1 Summary of soil optimum temperature range and minimum and maximum temperatures for N_2 fixation in different legume species

Plant species	Minimum (°C)	Optimum (°C)	Maximum (°C)	References
Alfalfa (*Medicago sativa* L.)	2	2–25		Waughman (1977)
	2	35	40	Dart and Day (1971)
Guar (*Cyamopsis tetragonoloba* Guar)			37–40	Pate (1961)
Barrel medic (*Medicago truncatula* Gaertn.)	2	20	35	Dart and Day (1971)
Narrowleaf lupin (*Lupinus angustifolius* L.)	10	20–30		Halliday (1975)
Big-leaved lupin (*Lupinus polyphyllus* Lindl.)	1.5	25		Waughman (1977)
Birdsfoot trefoil (*Lotus corniculatus* L.)	10	25–27	35	Rao (1977)
			40	Michiels et al. (1994)
Common vetch (*Vicia sativa* L.)	2	20	40	Dart and Day (1971)
Purple vetch (*Vicia atropurpurea* Desf.)		24		Pate (1961)
Cowpea (*Vigna unguiculata* (L.) Walp)	5	40		Dart and Day (1971)
	2	20–30	40	Dart and Day (1971)
Faba bean (*Vicia faba* L.)	5	20		Waughman (1977)
Common bean (*Phaseolus vulgaris* L.)			35	Piha and Munnus (1987)
	5	15–25		Halliday (1975)
Field/garden pea (*Pisum sativum* L.)	0.5	25		Waughman (1977)
		20–26		Lie (1971)
Peanut (*Arachis hypogaea* L.)			40	Kishinevsky et al. (1992)

(continued)

Table 4.1 (continued)

Plant species	Minimum (°C)	Optimum (°C)	Maximum (°C)	References
	2	25	35	Dart and Day (1971)
Soybean (*Glycine max* L.)		27		Kuo and Boersma (1971)
		20–25		Lindemann and Ham (1979)
	5	20–35	40	Dart and Day (1971)
	2	5–30		Dart and Day (1971)
Red clover (*Trifolium pratense* L.)		12–26		Small and Joffe (1968)
Subterranean clover (*Trifolium subterraneum* L.)	5	12–32		Gibson (1971)
		15		Roughley and Dart (1970)
	5	13–26		Halliday and Pate (1976)
		26		Small and Joffe (1968)
Arrowleaf clover (*Trifolium vesiculosum*)		25		Schomberg and Weaver (1992)
White clover (*Trifolium repens* L.)		21		Masterson and Murphy (1976)
	9			Frame and Newbould (1986)

on species and varietal interactions. The optimal temperatures interaction between legume and the rhizobial combination is about 30 °C for clover and pea, between 35 and 40 °C for soybean, peanut, and cowpea, and between 25 and 30 °C for popular legumes (Long 2001; Reynolds-Henne et al. 2010). LaFavre and Eaglesham (1986) reported that nodulation of soybean was strikingly inhibited at 42 and 45 °C during 12-h and 9-h days, respectively. Moreover, the same study indicated that no correlation was found between the ability of plant strains to grow at high temperatures under in vitro culture and its ability to nodulate at high temperatures (ElSheikh and Wood 1989). Piha and Munnus (1987) observed that in response to high temperature, bean nodules formed were small with low specific nitrogenase activity at 35 °C temperature.

The acetylene reduction assay has been extensively used in research to measure N fixation (Witty and Minchin 1988; Vessey 1994). Minchin et al. (1983) reported that the nodulated roots of some legumes were decreased the rate of ethylene production by 40–60 when exposed to acetylene. By contrary, Mederski and Streeter (1977) found a 60% decline in nitrogenase activity in soybean, which occurred at the end of 6 days' continuous exposure to an acetylene-air mixture. Further, acetylene reduction by N-fixing organisms occurred over a wide range of temperatures with maximum activity between 24 and 33 °C but rapidly declining at higher temperatures (Dart and Day 1971). Sinclair and Weisz (1985) reported that the rate of acetylene reduction in soybean increased with soil temperatures up to 30 °C and declined slightly up to 34 °C and reduced drastically above this temperature. In addition, some reports suggested that the activity of leghaemoglobin in transporting oxygen to the respiratory sites without inhibiting the oxygen-sensitive nitrogenase enzyme system could be also affected by high temperatures (Brun 1978). Heat waves could regulate the diffusion of oxygen into the nodules, and high temperatures could affect the role of leghaemoglobin in delivering oxygen to the respiratory sites without inhibiting the oxygen-sensitive nitrogenase enzyme system (Brun 1978). Hartwig et al. (1987); Vessey et al. (1988) reported that nitrogenase is limited by oxygen supply and thereby ATP (Adenosine triphosphate) availability rather than by reductant reduced ferredoxin availability in clover and soybean (Table 4.2).

The chaperones are characterized as folding modulators, in configuring and stabilizing a wide range of polypeptides presented in the wrong conformational structure (Alexandre and Oliveira 2013). The chaperone genes dnaKJ and groESL in thermo-tolerant and thermosensitive isolate of rhizobia were studied to understand the molecular basis of high-temperature stress tolerance by comparing the expression levels (Alexandre and Oliveira 2013). Nandal et al. (2005) stated that the mutants of *Rhizobium* sp. strain, tolerant to high temperature, exhibited a different protein profile compared to the wild-type at high temperature and showed over-expressed proteins and new proteins as well. Moreover, this study also showed that exopolysaccharides (EPSs) and lipopolysaccharides (LPSs) and protein synthesis in the mutant strains were found less thermosensitive than in the parent strain and the role of Hsps synthesized by the heat-resistant mutants at 43 °C can relevant for the thermostolerance found in mutants. However, this overproduction of protein was confirmed by other studies in mutant strains as DnaK (Alexandre and Oliveira 2013; Abd-Alla et al. 2014a, b), in chickpea rhizobia as GroEL (Rodrigues et al. 2006) and also in *Mesorhizobium* strains (Laranjo and Oliveira 2011). All these studies suggested the possible molecular mechanisms behind the high-temperature effects on BNF associated with legumes.

5.4 N Fixation in Legumes as Influenced by Soil Acidity and Alkalinity

Soil acidity has long been known to decrease symbiotic N fixation in legumes under acidic conditions. The excess of H^+ alters cell-membranes permeability while

Table 4.2 Genetic engineering of rhizobia to obtain abiotic stress tolerance and enhanced N fixation

Host Rhizobium	Enhanced trait or effect	Proposed mechanism	References
S. meliloti	Nodulation, Nase activity, plant N content and growth in alfalfa	Increase in Nod factor production and the regulation of N fixation	Castillo et al. (1999)
	Delayed nodule senescence in alfalfa nodulated plants N fixation with enhanced cadmium tolerance in alfalfa nodulated plants.	Enhanced ROS Detoxification Changes in antioxidant metabolism	Redondo et al. (2009) Shvaleva et al. (2010)
	N fixation in nodulated alfalfa subjected to salt stress	Osmolytes accumulation proline betaine and glycine betaine in bacteroids under osmotic stress	Boscari et al. (2006)
	The advantage over natural rhizobial populations to infect and nodulate alfalfa under drought stress	Oxidation of proline from alfalfa root exudates to glutamate as the energy source for rhizobia	Van Dillewijn et al. (2001, 2002)
R. tropici	Symbiotic performance and dry weight of Phaseolus vulgaris nodulated plants	Block of glycogen synthesis and putative pleiotropic effects	Marroquí et al. (2001)
R. etli	Nitrogenase activity and total N content in nodulated bean plants	Stimulated respiratory efficiency in free-living rhizobia and bacteroids	Ramírez et al. (1999)
	Enhanced nodule occupancy in host *Phaseolus vulgaris*	Antibiotic activity	Robleto et al. (1997, 1998)
R. etli ntr C mutant	N fixation in nodulated Phaseolus vulgaris plants	Enhanced ATP supply to nitrogenase	Soberón et al. (1999)

(continued)

Table 4.2 (continued)

Host Rhizobium	Enhanced trait or effect	Proposed mechanism	References
R. leguminosarum *S. meliloti*	N fixation in *Vicia hirsuta* nodulated plants Tolerance to heat, cold, salt and UV-irradiation stresses in free-living rhizobia N-fixing activity, stem dry weight, and salt tolerance in *M. truncatula* nodulated plants	Enhanced IAA synthesis General hormone re-modulation in the plant	Camerini et al. (2008), Imperlini et al. (2009)
R. leguminosarum	Dry seed weight yield in nodulated Phaseolus vulgaris plants	Antibiotic activity and recycling and oxidation of hydrogen derived from nitrogenase activity	Iniguez et al. (2004)

inducing cation efflux, impairing plant nutritional status and growth (Bambara and Ndakidemi 2010; Abd-Alla et al. 2014a, b). Liming is often required to neutralize the undesired effect of high H^+ in the soil, and therefore, to solve the negative influence of acidic soil on plant growth (Ferreira et al. 2016).

6 Strategies to Enhance Legumes Nitrogen Fixation

6.1 Using the Conservation Agricultural Practices for Improving Rhizobium-Legume Symbiosis

The amount of nitrogen that is fixed by legumes crop is reduced by environmental factors like rainfall and temperature (Streeter 2003). Although the Legume family consists of about 18,000 species, only small numbers of legume species are used as crops and pastures, being critical food and feed crops worldwide (Herridge et al. 2008). Higher amounts and percentages of the proportion of nitrogen derived from the atmosphere (% Ndfa) have been observed in the grain of pea under a no-tillage system compared with conventional tillage field within a 4-year rotation experiment (Matus et al. 1997). Indeed, the water content in the root zone during nodule formulation control nitrogen fixation, gas exchange, and nodule activity (Weisz and Sinclair 1988; Liu et al. 2012). Reduced crop growth and development under water tress may also secondarily affect %Ndfa by lowering plant demand for N (Chalk et al. 2010). The percent N derived from the atmosphere by legumes was a higher increase in the landscape with low inorganic soil N (Sullivan et al. 2007; Singh et al. 2018). Doran (1980) observed that soils under zero-tillage management had lower soil nitrate compared with soils under conventional tillage. Indeed, the reduced levels of soil nitrate in soils under zero-tillage management were associated with increased N fixation by soybean (Hughes and Herridge 1989; Wheatley et al. 1995).

The CT practices typically leave less than 15% residue on the surface, while conservation tillage practices often leave more than 30% of residue as a soil cover at the time of planting of the next crop (Sullivan et al. 2007; Singh et al. 2018). Zero tillage often results in increased soil water content (Lafond et al. 1992), and this reduced water stress may increase N fixation in grain legumes (Bennet and Albretch 1984). N fixation by lentil grown by using zero tillage management was 10% higher than when cultivated using conventional tillage, whereas N fixation by pea was 31% higher using zero tillage. Pea fixed more N than lentil under zero-tillage management, while no differences in N fixation by pea and lentil were observed under conventional tillage as reported by Herridge et al. (2008). Zero tillage retains crop residues on the soil surface, and this often results in lower available soil N content in the rooting zone as compared to conventional tillage (Doran 1980).

The minimum/zero-tillage practices and maintaining the crop residues on soil surface play an important role to sustain soil fertility, improving water use efficiency, physical conditions of soils and enhance crop productivity (Dalal and Chan 2000).

Overall, the no-tillage enhanced the N_2 fixation as compared to tillage treatment (Mohammad et al. 2010), ultimately increased the growth and yield of growing crops. It is due to a significant proportion of N (58.20–76.51%) was derived by mungbean through symbiotic atmospheric N fixation plus crop residues retained under the no-tillage treatment (Mohammad et al. 2010). It has been demonstrated the merits and demerits of both conventional tillage and no-tillage impact on the amount of N fixed, growth, and yield of a crop. It has also been reported that the effect of tillage methods depends on crop species, climate, site, and time of tillage (Martinez et al. 2008).

Roots near the surface derive a higher proportion of their N from the symbiosis with rhizobia (Lory et al. 1992). While deeper nodules account for relatively low levels of N_2 fixation (McDermott and Graham 1989; Tajima et al. 2007). Tillage practices may also interact with the method of inoculant application, which itself may impact nodulation patterns, nitrogen fixation, as well as legume growth and yield (Denton et al. 2017). Conservation tillage systems that provide limited soil mixing may have lower nutrient levels at depth, which may increase the N_2 fixing activity of nodules within these regions because the N would rarely reach levels that inhibit N_2 fixation (Virginia et al. 1986).

Antagonism between nitrate concentration in the soil solution and the N_2 fixation process in nodules is the primary constraint the crop faces in terms of increasing N uptake (Streeter and Wong 1988). High levels of soil mineral N as NO_3 form can reduce the amount of N_2 fixed by legumes (Høgh-Jensen and Schjoerring 1994; Thilakarathna et al. 2016) as it costs less energy for legumes to take up N from the soil than fix N biologically from the atmosphere (Cannell and Thornley 2000). The number and dry weight of nodules decrease with increasing mineral N contents (Koutroubas et al. 1998), and the percentage of N derived from atmosphere decreases, with a concomitant increase in the percentage of N derived from fertilizer (George and Singleton 1992).

7 Conclusion

Earlier findings in the chapter revealed that N is essential for plants, belonging to almost all structures of plant and a vital constituent of chlorophyll, enzymes, and proteins. As a result, N is needed relatively large amounts than other nutrients. Numerous transformation processes that are involved in the N cycle contributes to the uptake N by plants. Among them, symbiosis biological N fixation is an environmentally friendly natural resource of N for sustainable agricultural systems. Recent research findings found a decreasing trend of symbiotic N fixation in agricultural systems as a result of adverse abiotic stresses. Therefore, new and growing research efforts focusing on the effects of abiotic stresses on the symbiotic N fixation are needed to cope with their adverse effects on the physiological mechanisms of plants. In legume plants grown under symbiotic conditions, one of the primary effects of abiotic stresses is a decline in the rates of symbiotic biological N fixation. In the present chapter, we summarized the current understanding of the abiotic stresses which are responsible for symbiotic biological N fixation in legumes.

Conflict of interest The authors declared no conflict of interest.

References

Abd-Alla MH, El-Enany AW, Nafady NA, Khalaf DM, Morsy FM (2014a) Synergistic interaction of *Rhizobium leguminosarum* bv. *viciae* and arbuscular mycorrhizal fungi as a plant growth promoting biofertilizers for faba bean (*Vicia faba* L.) in alkaline soil. Microbiol Res 169(1):49–58

Abd-Alla MH, Issa AA, Ohyama T (2014b) Impact of harsh environmental conditions on nodule formation and dinitrogen fixation of legumes. In: Advances in biology and ecology of nitrogen fixation, Takuji Ohyama. IntechOpen https://doi.org/10.5772/56997

Adesemoye AO, Kloepper JW (2009) Plant–microbes interactions in enhanced fertilizer-use efficiency. Appl Microbiol Biotechnol 85:1–12

Ahmed FE, Hall AE, DeMason DA (1992) Heat injury during floral development in cowpea (*Vigna unguiculata*). Am J Bot 79:784–791

Alexandre A, Oliveira S (2013) Response to temperature stress in rhizobia. Crit Rev Microbiol 39(3):219–228

Amelung W, Zhang X, Flach KW, Zech W (1999) Amino sugars in native grassland soils along a climosequence in North America. Soil Sci Soc Am J 63(1):86–92

American Heart Association (2016) How potassium can help control high blood pressure. National Center, 7272 Greenville Ave., Dallas, TX 75231. Available link: https://www.heart.org/en/health-topics/high-blood-pressure/changes-you-can-make-to-manage-high-blood-pressure/how-potassium-can-help-control-high-blood-pressure

Annunziata MG, Ciarmiello LF, Woodrow P, Maximova E, Fuggi A, Carillo P (2017) Durum wheat roots adapt to salinity remodeling the cellular content of nitrogen metabolites and sucrose. Front Plant Sci 7:2035. https://doi.org/10.3389/fpls.2016.02035

Annunziata MG, Ciarmiello LF, Woodrow P, Dell'Aversana E, Carillo P (2019) Spatial and temporal profile of glycine betaine accumulation in plants under abiotic stresses. Front Plant Sci 10:230. https://doi.org/10.3389/fpls.2019.00230

Arayankoon T, Schomberg HH, Weaver RW (1990) Nodulation and N_2 fixation of guar at high root temperature. Plant Soil 126:209–213

Azam F, Farooq S (2003) Nitrification inhibition in soil and ecosystem functioning—an overview. Pak J Biol Sci 6(6):528–535

Bambara S, Ndakidemi PA (2010) The potential roles of lime and molybdenum on the growth, nitrogen fixation and assimilation of metabolites in nodulated legume: a special reference to *Phaseolus vulgaris* L. Afr J Biotechnol 9(17):2482–2489

Bayuelo-Jiménez JS, Craig R, Lynch JP (2002) Salinity tolerance of *Phaseolus* species during germination and early seedling growth. Crop Sci 42:1584–1594. https://doi.org/10.2135/cropsci2002.1584

Bennet JM, Albretch SL (1984) Drought and flooding effects on N_2 fixation, water relations and diffusive resistance of soybean. Agron J 76:735–740

Bernard T, Pocard JA, Perround B, Le Rudulier D (1986) Variations in the response of salt-stressed Rhizobium strains to betaines. Archiv Microbiol 143(4):359–364

Bindumadhava H, Nair RM, Nayyar H (2016) Salinity and high temperature tolerance in mungbean [*Vigna radiata* (L.) Wilczek] from a physiological perspective. Front Plant Sci 7:957. https://doi.org/10.3389/fpls.2016.00957

Bita CE, Gerats T (2013) Plant tolerance to high temperature in a changing environment: scientific fundamentals and production of heat tolerance crops. Front Plant Sci 4:273. https://doi.org/10.3389/fpls.2013.00273

Bolaños L, Martín M, El-Hamdaoui A, Rivilla R, Bonilla I (2006) Nitrogenase inhibition in nodules from pea plants grown under salt stress occurs at the physiological level and can be alleviated by B and Ca. Plant Soil 280:135–142. https://doi.org/10.1007/s11104-005-2853-8

Boncompagni E, Østerås M, Poggi MC, le Rudulier D (1999) Occurrence of choline and glycine betaine uptake and metabolism in the family Rhizobiaceae and their roles in osmoprotection. Appl Environ Microbiol 65(5):2072-2077

Bordoloi LJ, Singh AK, Kumar M, Patiram Hazarika S (2013) Microbial biomass nitrogen as an index of N availability in acidic soils of North East India. Indian J Hill Farm 26(1):22–28

Boscari A, Mandon K, Dupont L, Poggi MC, Le Rudulier D (2002) BetS is a major glycine betaine/proline betaine transporter required for early osmotic adjustment in *Sinorhizobium meliloti*. J Bacteriol 184:2654–2663

Boscari A, Van de Sype G, Le Rudulier D, Mandon K (2006) Overexpression of BetS, a Sinorhizobium meliloti high-affinity betaine transporter, in bacteroids from *Medicago sativa* nodules sustains nitrogen fixation during early salt stress adaptation. Mol Plant-Microbe Interact 19:896–903. https://doi.org/10.1094/MPMI-19-0896

Boudsocq S, Niboyet A, Lata JC, Raynaud X, Loeuille N, Mathieu J, Blouin M, Abbadie L, Barot S (2012) Plant preference for ammonium versus nitrate: a neglected determinant of ecosystem functioning? Am Nat 180(1):60–69. https://doi.org/10.1086/665997

Bouwman AF, Beusen AH, Griffioen J, Van Groenigen JW, Hefting MM, Oenema O, Van Puijenbroek PJ, Seitzinger S, Slomp CP, Stehfest E (2013) Global trends and uncertainties in terrestrial denitrification and N_2O emissions. Philos Trans R Soc B: Biol Sci 368(1621):20130112

Bradford JM, Peterson GA (2000) Conservation tillage. In: Sumner ME (ed) Handbook of soil science. CRC Press, Boca Raton, pp 247–269

Brun WA (1978) Assimilation. In: Soybean physiology, agronomy, and utilization. Academic, New York, pp 45–76

Bruning B, Rozema J (2013) Symbiotic nitrogen fixation in legumes: perspectives for saline agriculture. Environ Exp Bot 92:134–143. https://doi.org/10.1016/j.envexpbot.2012.09.001

Burris RH (1981) The nitrogen cycle the nitrogen cycle: a royal society discussion held on 17 and 18 June 1981 WDP Stewart T. Rosswall

Camerini S, Senatore B, Lonardo E, Imperlini E, Bianco C, Moschetti G, Rotino GL, Campion B, Defez R (2008) Introduction of a novel pathway for IAA biosynthesis to rhizobia alters vetch root nodule development. Archiv Microbiol 190:67–77. https://doi.org/10.1007/s00203-008-0365-7

Campbell CA, Zentner RP, Dormaar JF, Voroney RP (1986) Land quality, trends and wheat production in Western Canada. p. 318–353. Wheat production in Canada: a review. In: Proceedings of Canadian Wheat Production Symposium, Div. of Ext. and Community Relation, Saskatoon. University of Saskatchewan, Saskatoon, SK. Canada, 3–5 Mar 1986

Campbell CA, Myers RJK, Curtin D (1995) Managing nitrogen for sustainable crop production. Fertilizer Res 42(1–3):277–296

Canfield D, Glazer AN, Falkowski PD (2010) The evolution and future of earth's nitrogen cycle. Science 330:192–196

Cannell MGR, Thornley JHM (2000) Modelling the components of plant respiration: some guiding principles. Ann Bot 85:45–54

Carvalho TLG, Ferreira PCG, Hemerly AS (2011) Sugarcane genetic controls involved in the association with beneficial endophytic nitrogen fixing bacteria. Trop Plant Biol 4:31–41

Carvalho TL, Balsemão-Pires E, Saraiva RM, Ferreira PC, Hemerly AS (2014) Nitrogen signalling in plant interactions with associative and endophytic diazotrophic bacteria. J Exp Bot 65(19):5631–5642

Castillo M, Flores M, Mavingui P, Martínez-Romero E, Palacios R, Hernández G (1999) Increase in the alfalfa nodulation, nitrogen fixation and plant growth by specific DNA amplification in Sinorhizobium meliloti. Appl Environ Microbiol 65:2716–2722

Chalk PM, Alves BJR, Boddey RM, Urquiaga S (2010) Integrated effects of abiotic stresses on inoculant performance, legume growth and symbiotic dependence estimated by 15 N dilution. Plant Soil 328:1–16

Chen L, Luo S, Li X, Wan Y, Chen J, Liu C (2014) Interaction of Cd-hyperaccumulator Solanum nigrum L. and functional endophyte *Pseudomonas* sp. Lk9 on soil heavy metals uptake. Soil Biol Biochem 68:300–308

Christophe S, Jean-Christophe A, Annabelle L, Alain O, Marion P, AnneSophie V (2011) Plant N fluxes and modulation by nitrogen, heat and water stresses: a review based on comparison of legumes and non-legume plants. In: Shanker A, Venkateswarlu B (eds) Abiotic stress in plants–mechanisms and adaptations. Intech Open Access Publisher, Rijeka, pp 79–118

Christou M, Avramides EJ, Jones DL (2006) Dissolved organic nitrogen dynamics in a Mediterranean vineyard soil. Soil Biol Biochem 38(8):2265–2277

Chun J, Oren A, Ventosa A, Christensen H, Arahal DR, da Costa MS, Rooney AP, Yi H, Xu XW, De Meyer S, Trujillo ME (2018) Proposed minimal standards for the use of genome data for the taxonomy of prokaryotes. Int J System Evol Microbiol 68(1):461–466

Cooper JE (2007) Early interactions between legumes and rhizobia: Disclosing complexity in a molecular dialogue. J Appl Microbiol 103:1355–1365. https://doi.org/10.1111/j.1365-2672.2007.03366.x

Cordovilla M, Ligero F, Lluch C (1995) Influence of host genotypes on growth, symbiotic performance and nitrogen assimilation in faba bean (*Vicia faba* L.) under salt stress. Plant Soil 172(2):289–297

Cordovilla MD, Berrido SI, Ligero F, Lluch C (1999) Rhizobium strain effects on the growth and nitrogen assimilation in *Pisum sativum* and *Vicia faba* plant growth under salt stress. J Plant Physiol 154(1):127–131

Coskan A, Dogan K (2011) Symbiotic nitrogen fixation in soybean. Soybean Physiol Biochem 307:167–182

Curtin D, Campbell CA (2007) Mineralizable nitrogen. In: MCarter MR, Gregorich EG (eds) Soil sampling and methods of analysis, 2nd edn. CRC Press, Taylor & Francis Group, Boca Raton, pp 33487–2742

Dart P, Day J (1971) Effects of incubation temperature and oxygen tension on nitrogenase activity of legume root nodules. Plant Soil 35:167–184

Delgado MJ, Ligero F, Lluch C (1994) Effects of salt stress on growth and nitrogen fixation by pea, faba-bean, common bean and soybean plants. Soil Biol Biochem 26:371–376. https://doi.org/10.1016/0038-0717(94)90286-0

Denton M, Phillips L, Peoples M, Pearce D, Swan A, Mele P, Brockwell J (2017) Legume inoculant application methods: effects on nodulation patterns, nitrogen fixation, crop growth and yield in narrow-leaf lupin and faba bean. Plant Soil 419:25–39

D'Haene K, Vandenbruwane J, De Neve S, Gabriels D, Salomez J, Hofman G (2008) The effect of reduced tillage on nitrogen dynamics in silt loam soils. Eur J Agron 28(3):449–460

Di Ciocco C, Penón E, Coviella C, López S, Díaz-Zorita M, Momo F, Álvarez R (2011) Nitrogen fixation by soybean in the Pampas: relationship between yield and soil nitrogen balance. Agrochimica 55:305–313

Doane TA, Horwath WR, Mitchell JP, Jackson J, Miyao G, Brittan K (2009) Nitrogen supply from fertilizer and legume cover crop in the transition to no-tillage for irrigated row crops. Nutr Cycl Agroecosyst 85:253–262

Dong R, Zhang J, Huan H, Bai C, Chen Z, Liu G (2017) High salt tolerance of a Bradyrhizobium strain and its promotion of the growth of *Stylosanthes guianensis*. Int J Mol Sci 18(8):1625. https://doi.org/10.3390/ijms18081625

Doran JW (1980) Soil microbial and biochemical changes associated with reduced tillage. Soil Sci Soc Am J 44:765–771

Edelstein M, Plaut Z, Ben-Hur M (2011) Sodium and chloride exclusion and retention by non-grafted and grafted melon and Cucurbita plants. J Exp Bot 62(1):177–184. https://doi.org/10.1093/jxb/erq255

Egamberdieva D, Shurigin V, Gopalakrishnan S, Sharma R (2014) Growth and symbiotic performance of chickpea (*Cicer arietinum*) cultivars under saline soil conditions. J Biol Chem Res 1–10

EL Sabagh A, Omar AE, Saneoka H, Barutçular C (2015) Comparative physiological study of soybean (*Glycine max* L.) cultivars under salt stress. YYU J Agric Sci 25:269–284

EL Sabagh A, Hossain A, Islam MS, Barutçular C, Ratnasekera D, Kumar N, Teixeira da Silva JA (2019) Sustainable soybean production and abiotic stress management in saline environments: a critical review. Aust J Crop Sci 13(2):228–236

ElSheikh EA, Wood M (1989) Response of chickpea and soybean rhizobia to salt: influence of carbon source, temperature and pH. Soil Biol Biochem 21(7):883–887

Essa TA (2002) Effect of salinity stress on growth and nutrient composition of three soybean. J Agron Crop Sci 93:86–93

Estévez J, Soria-Díaz ME, De Córdoba FF, Morón B, Manyani H, Gil A, Thomas-Oates J, Van Brussel AAN, Dardanelli MS, Sousa C, Megías M (2009) Different and new Nod factors produced by *Rhizobium tropici* CIAT899 following Na^+ stress. FEMS Microbiol Lett 293:220–231. https://doi.org/10.1111/j.1574-6968.2009.01540.x

Farooq M, Gogoi N, Hussain M, Barthakur S, Paul S, Bharadwaj N, Migdadi HM, Alghamdi SS, Siddique KH (2017) Effects, tolerance mechanisms and management of salt stress in grain legumes. Plant Physiol Biochem 118:199–217

Ferreira TC, Aguilar JV, Souza LA, Justino GC, Aguiar LF, Camargos LS (2016) pH effects on nodulation and biological nitrogen fixation in *Calopogonium mucunoides*. Brazilian J Bot 39(4):1015–1020

Figueiredo MVB, Mergulhão ACES, Sobral JK, Lira Jr MA, de Araújo ASF (2013) Biological nitrogen fixation: importance, associated diversity, and estimates. In: Arora NK (ed) Plant microbe symbiosis: fundamentals and advances, pp 267–289. https://doi.org/10.1007/978-81-322-1287-4_10

Flowers TJ, Gaur PM, Gowda CL, Krishnamurthy L, Samineni S, Siddique KH, Turner NC, Vadez V, Varshney RK, Colmer TD (2010) Salt sensitivity in chickpea. Plant, Cell Environ 33(4):490–509

Frame J, Newbould P (1986) Agronomy of white clover. In: advances in agronomy, vol 40, Academic Press, pp 1–88

Franzluebbers AJ, Hons FM, Zuberer DA (1995) Soil organic carbon, microbial biomass, and mineralizable carbon and nitrogen in sorghum. Soil Sci Soc Am J 59(2):460–466

Galloway JN, Schlesinger WH, Levy II A, Michaels A, Schnoor JL (1995) Nitrogen fixation: Anthropogenic enhancement-environmental response. Global Biogeochem Cycles 9:235–252

George T, Singleton PW (1992) Nitrogen assimilation traits and dinitrogen fixation in soybean and common bean. Agron J 84:1020–1028

Ghassemi-Golezani K, Taifeh-Noori M, Oustan S, Moghaddam M, Seyyed-Rahmani S (2010) Oil and protein accumulation in soybean grains under salinity stress. Not Sci Biol 2(2):64–67

Ghittoni NE, Bueno MA (1995) Peanut rhizobia under salt stress: Role of trehalose accumulation in strain ATCC 51466. Can J Microbiol 41:1021–1030. https://doi.org/10.1139/m95-141

Ghosh PK, Dayal D, Mandal KG, Wanjari RH, Hati KM (2003) Optimization of fertilizer schedules in fallow and groundnut-based cropping systems and an assessment of system sustainability. Field Crops Res 80(2):83–98

Gibson AH (1971) Factors in the physical and biological environment affecting nodulation and nitrogen fixation by legumes. Plant Soil 35(1):139–152

Godfray HCJ, Beddington JR, Crute IR, Haddad L, Lawrence D, Muir JF, Pretty J, Robinson S, Thomas SM, Toulmin C (2010) Food security: the challenge of feeding billion people. Science 327:812–818

Good AG, Beatty PH (2011) Fertilizing nature: a tragedy of excess in the commons. PLoS Biol 9:100–124

Graham PH (1992) Stress tolerance in Rhizobium and Bradyrhizobium, and nodulation under adverse soil conditions. Can J Microbiol 38:475–484

Grant CA, Flaten DN (1998) Fertilizing for protein content of wheat. In: Proceedings of Canadian Wheat Protein Symposium, Saskatoon. University of Saskatchewan, Saskatoon, 9–10 Mar 1998

Gross Y, Kigel J (1994) Differential sensitivity to high temperature of stages in the reproductive development of common bean (*Phaseolus vulgaris* L.). Dierential sensitivity to high temperature of stages in the reproductive development of common bean (*Phaseolus vulgaris* L.). Field Crops Res 36:201–212 https://doi.org/10.1016/0378-4290(94)90112-0

Guasch-Vidal B, Estévez J, Dardanelli MS, Soria-Díaz ME, Fernández De Córdoba F, Balog CIA, Manyani H, Gil-Serrano A, Thomas-Oates J, Hensbergen PJ, Deelder AM, Megías M, Brussel AAN (2013) High NaCl concentrations induce the nod genes of Rhizobium tropici CIAT899 in the absence of flavonoid inducers. Mol Plant-Microbe Interact 26:451–460. https://doi.org/10.1094/MPMI-09-12-0213-R

Guilioni L, Wery J, Tardieu F (1997) Heat stress-induced abortion of buds and flowers in pea: is sensitivity linked to organ age or to relations between reproductive organs? Ann Bot 80:159–168. https://doi.org/10.1006/anbo.1997.0425

Halliday J (1975) An interpretation of seasonal and short term fluctuations in nitrogen fixation, PhD thesis, University of Western Australia, Perth

Halliday J, Pate JS (1976) The acetylene reduction assay as a means of studying nitrogen fixation in white clover under sward and laboratory conditions. Grass Forage Sci 31(1):29–35

Hamaoui-Laguel L, Meleux F, Beekmann M, Bessagnet B, Génermont S, Cellier P, Létinois L. (2014) Improving ammonia emissions in air quality modelling for France. Atmosph Environ 92:584–595

Hartwig U, Boller B, Nösberger J (1987) Oxygen supply limits nitrogenase activity of clover nodules after defoliation. Ann Bot 59(3):285–291

Haynes RJ (2005) Labile organic matter fractions as central components of the quality of agricultural soils: an overview. Adv Agron 85:221–268

Herridge DF, Peoples MB, Boddey RM (2008) Global inputs of biological nitrogen fixation in agricultural systems. Plant Soil 311(1–2):1–8

Hodges SC (2010) Soil fertility basics, NC Certified Crop Advisor Training, Soil Science Extension, North Carolina State University

Hofman G, Cleemput OV (2004) Soil and plant nitrogen. International Fertilizer Industry Association (IFA), Paris, France, pp 1–48

Høgh-Jensen H, Schjoerring JK (1994) Measurement of biological dinitrogen fixation in grassland: comparison of the enriched 15 N dilution and the natural 15 N abundance methods at different nitrogen application rates and defoliation frequencies. Plant Soil 166(2):153–163

Hughes RM, Herridge DF (1989) Effect of tillage on yield, nodulation and nitrogen fixation of soybean in far north-coastal New South Wales. Aust J Exp Agric 29(5):671–677

Hungria M, Franco AA (1993) Effects of high temperature on nodulation and nitrogen fixation by *Phaseolus vulgaris* L. Plant Soil 149:95–102

Hungria M, Franchini JC, Campo RJ, Crispino CC, Moraes JZ, Sibaldelli RNR, Mendes IC, Arihara J (2006) Nitrogen nutrition of soybean in Brazil: contributions of biological N_2 fixation and N fertilizer to grain yield. Can J Plant Sci 86:927–939

Imperlini E, Bianco C, Lonardo E, Camerini S, Cermola M, Moschetti G, Defez R (2009) Effects of indole-3-acetic acid on Sinorhizobium meliloti survival and on symbiotic nitrogen fixation and stem dry weight production. Appl Microbiol Biotechnol 83:727–738. https://doi.org/10.1007/s00253-009-1974-z

Iniguez AL, Robleto EA, Kent AD, Triplett EW (2004) Significant yield increase in *Phaseolus vulgaris* obtained by inoculation with a trifolitoxin-producing, Hup + strain of *Rhizobium leguminosarum* bv. *Phaseoli*, Crop Management. https://doi.org/10.1094/CM-2004-0301-07-RV. Available at http://www.Plantmanagementnetwork.org/sub/cm/review/2004/yield/

Iqbal N, Nazar R, Khan NA (eds) (2015) Osmolytes and plants acclimation to changing environment: emerging omics technologies. Springer, Berlin. https://doi.org/10.1007/978-81-322-2616-1

Jensen ES, Peoples MB, Boddey RM, Gresshoff PM, Henrik HN, Alves BJR, Morrison MJ (2012) Legumes for mitigation of climate change and the provision of feedstock for biofuels and biorefineries. A review. Agron Sustain Dev 32:329–364

Jones DL, Shannon D, Murphy DV, Farrar J (2004) Role of dissolved organic nitrogen (DON) in soil N cycling in grassland soils. Soil Biol Biochem 36(5):749–756

Kalbitz K, Solinger S, Park JH, Michalzik B, Matzner E (2000) Controls on the dynamics dissolved organic matter in soils: a review. Soil Sci 165(4):277–304

Kaneez-e-Batool N, Shahzad T, Mahmood F, Hussain S, Riaz M, Maqbool Z, Anwar F, Rehman K, Rashid MI (2016) Carbon mineralization in response to nitrogen and litter addition in surface and subsoils in an agroecosystem. Archiv Agron Soil Sci 62(9):1285–1292. https://doi.org/10.1080/03650340.2016.1145792

Kasper S, Christoffersen B, Soti P, Racelis A (2019) Abiotic and biotic limitations to nodulation by leguminous cover crops in South Texas. Agriculture 9(10):209. https://doi.org/10.3390/agriculture9100209

Kaushal N, Awasthi R, Gupta K, Gaur P, Siddique KH, Nayyar H (2013) Heat-stress-induced reproductive failures in chickpea (*Cicer arietinum*) are associated with impaired sucrose metabolism in leaves and anthers. Funct Plant Biol 40:1334–1349. https://doi.org/10.1071/FP13082

Keerio IM (2001) Nitrogenase activity of soybean root nodules inhibited after heat stress. J Appl Sci 1(3). https://doi.org/10.3923/jas.2001.297.300

Keerio MI, Wilson JM (1998) Effects of high temperatures on nitrogen fixation in soybean cultivars. In: Proceedings of 7th international symposium on nitrogen fixation with non-legumes, NIBGE Faisalabad, Pakistan, 16–21 Oct 1996. Developments in plant and soil science, vol 79. Kluwer Academic Publishers, London

Khan SA, Mulvaney RL, Hoeft RG (2001) A simple soil test for detecting sites that are nonresponsive to Nitrogen fertilization. Soil Sci Soc Am J 65(6):1751–1760

Khan HA, Siddique KH, Colmer TD (2017) Vegetative and reproductive growth of salt-stressed chickpea are carbon-limited: sucrose infusion at the reproductive stage improves salt tolerance. J Exp Bot 68(8):2001–2011

Kifuko-Koech M, Pypers P, Okalebo JR, Othieno CO, Khan ZR, Pickett JA, Kipkoech AK, Vanlauwe B (2012) The impact of *Desmodium* spp. and cutting regimes on the agronomic and economic performance of Desmodium—maize intercropping system in western Kenya. Field Crop Res 137:97–107

Kim TW, Lee K, Najjar RG, Jeong H-D, Jeong HJ (2011) Increasing N abundance in the Northwestern Pacific Ocean due to atmospheric nitrogen deposition. Science 334:505–509

Kishinevsky BD, Sen D,Weaver RW (1992) Effect of high root temperature on Bradyrhizobium-peanut symbiosis. Plant Soil 143:275–282

Köpke U, Nemecek T (2010) Ecological services of faba bean. Field Crop Res 115:217–233

Koutroubas SD, Papakosta DK, Gagianas AA (1998) The importance of early dry matter and nitrogen accumulation in soybean yield. Eur J Agron 9:1–10

Kumar S, Thakur P, Kaushal N, Malik JA, Gaur P, Nayyar H (2013) Effect of varying high temperatures during reproductive growth on reproductive function, oxidative stress and seed yield in chickpea genotypes during heat sensitivity. Arch Agron Soil Sci 59:823–843. https://doi.org/10.1080/03650340.2012.683424

Kumar S, Meena RS, Lal R, Yadav GS, Mitran T, Meena BL, Dotaniya ML, EL-Sabagh A (2018) Role of legumes in soil carbon sequestration. In: Meena RS et al (eds) Legumes for soil health and sustainable management. Springer, Singapore, pp 109–138

Kuo T, Boersma L (1971) Soil water suction and root temperature effects on nitrogen fixation in soybeans. Agron J 63(6):901–904

Kozlowski TT, Pallardy SG (1997) Growth control in woody plants. Elsevier Science, Ltd., The Boulevard, Langford Lane, Kidlington OX5 1GB UK. p 641

L'taief B, Sifi B, Zaman-Allah M, Drevon JJ, Lachaâl M (2007) Effect of salinity on root-nodule conductance to the oxygen diffusion in the *Cicer arietinum–Mesorhizobium ciceri* symbiosis. J Plant Physiol 164:1028–1036. https://doi.org/10.1016/j.jplph.2006.05.016

LaFavre AK, Eaglesham ARJ (1986) The effects of high temperature on soybean nodulation and growth with different strains of bradyrhizobia. Can J Microbiol 32:22–27

Lafond GP, Loeppky H, Derksen DA (1992) The effects of tillage systems and crop rotations on soil water conservation, seedling establishment and crop yield. Can J Plant Sci 72:103–125

Lambers H, Chapin SF, Pons T (1998) Plant physiological ecology. Springer, New York

Laranjo M, Oliveira S (2011) Tolerance of Meso-Rhizobium type strains to different environmental stresses. Antonie Van Leeuwenhoek 99:651–662

Latef AA, Ahmad P (2015) Legumes and breeding under abiotic stress: an overview. In: Ahmad P (ed) Legumes under environmental stress: yield, improvement and adaptations. Wiley, New York, p 315

Lie TA (1971) Symbiotic nitrogen fixation under stress conditions. Plant Soil 35(1):117–127

Lindemann WC, Ham GE (1979) Soybean plant growth, nodulation, and nitrogen fixation as affected by root temperature. Soil Sci Soc Am J 43(6):1134–1137

Liu L, Zhang Q, Hu L, Tang J, Xu L, Yang X, Yong JW, Chen X (2012) Legumes can increase cadmium contamination in neighboring crops. PLoS ONE 7(8):e42944

Liu A, Contador CA, Fan K, Lam HM (2018) Interaction and regulation of carbon, nitrogen, and phosphorus metabolisms in root nodules of legumes. Front Plant Sci 9:1860. https://doi.org/10.3389/fpls.2018.01860

Long SR (2001) Genes and signals in the Rizobium-legume symbiosis. Plant Physiol 125:69–72

López M, Tejera NA, Iribarne C, Lluch C, Herrera-Cervera JA (2008) Trehalose and trehalase in root nodules of *Medicago truncatula* and *Phaseolus vulgaris* in response to salt stress. Physiol Plant 134(4):575–582

López-Gómez M, Cobos-Porras L, Prell J, Lluch C (2016) Homospermidine synthase contributes to salt tolerance in free-living *Rhizobium tropici* and in symbiosis with *Phaseolus vulgaris*. Plant Soil 404(1–2):413–425

Lory JA, Russell MP, Heichel GH (1992) Quantification of symbiotically fixed nitrogen in soil surrounding alfalfa roots and nodules. Agron J 84:1033–1040

St. Luce M, Whalen JK, Ziadi N, Zebarth BJ (2011) Nitrogen dynamics and indices to predict soil nitrogen supply in humid temperate soils. Adv Agron 55–102

Lupwayi NZ, Clayton GW, O'Donovan JT, Harker KN, Turkington TK, Soon YK (2006) Nitrogen release during decomposition of crop residues under conventional and zero tillage. Can J Soil Sci 86(1):11–19

Manchanda G, Garg N (2008) Salinity and its effects on the functional biology of legumes. Acta Physiol Plant 30(5):595–618. https://doi.org/10.1007/s11738-008-0173-3

Manzoni S, Schimel JP, Porporato A (2012) Responses of soil microbial communities to water stress: results from a meta-analysis. Ecology 93(4):930–938

Marroquí S, Zorreguieta A, Santamaría C, Temprano F, Soberón M, Megías M, Downie JA (2001) Enhanced symbiotic performance by *Rhizobium tropici* glycogen synthase mutants. J Bacteriol 183:854–864. https://doi.org/10.1128/jb.183.3.854-864.2001

Martinez E, Fuentes JP, Silva P, Valle S, Acevedo E (2008) Soil physical properties and wheat root growth as affected by no till and conventional tillage systems in a Mediterranean environment of Chile. Soil Tillage Res 99:232–244

Masterson CL, Murphy PM (1976) Application of the acetylene reduction technique to the study of nitrogen fixation by white clover in the field. Cambridge University Press, Cambridge, UK p 584

Matus A, Derksen DA, Walley FL, Loeppky HA, van Kessel C (1997) The influence of tillage and crop rotation on nitrogen fixation in lentil and pea. Can J Plant Sci 77:197–200

McDermott TR, Graham PH (1989) *Bradyrhizobium japonicum* inoculant mobility, nodule occupancy, and acetylene reduction in soybean root system. Appl Environ Microbiol 55:2493–2498

Mederski HJ, Streeter JG (1977) Continuous, automated acetylene reduction assays using intact plants. Plant Physiol 59(6):1076–1081

Michiels J, Verreth C, Vanderleyden J (1994) Effects of temperature stress on bean-nodulating Rhizobium strains. Appl Environ Microbiol 60(4):1206–1212

Minchin FR, Witty JF, Sheehy JE, Müller M (1983) A major error in the acetylene reduction assay: decreases in nodular nitrogenase activity under assay conditions. J Exp Bot 34(5):641–649

Miransari M, Smith DL (2009) Alleviating salt stress on soybean (*Glycine max* (L.) Merr.)—*Bradyrhizobium japonicum* symbiosis, using signal molecule genistein. Eur J Soil Biol 45:146–152. https://doi.org/10.1016/j.ejsobi.2008.11.002

Miransari M, Riahi H, Eftekhar F, Minaie A, Smith DL (2013) Improving soybean (*Glycine max* L.) N_2 fixation under Stress. J Plant Growth Regulat 32(4):909–921

Mohammad W, Shehzadi S, Shah SM, Shah Z (2010) Effect of tillage and crop residues management on mungbean (*Vigna radiata* (L.) Wilczek) crop yield, nitrogen fixation and water use efficiency in rainfed areas. Pak J Bot 42(3):1781–1789

Müller J, Aeschbacher RA, Wingler A, Boller T, Wiemken A (2001) Trehalose and trehalase in Arabidopsis. Plant Physiol 125(2):1086–1093

Mulvaney RL, Khan SA, Hoeft RG, Brown HM (2001) A soil organic nitrogen fraction that reduces the need for nitrogen fertilization. Soil Sci Soc Am J 65(4):1164–1172

Munch JC, Velthof GL (2007) Denitrification and agriculture. In: Bothe H, Ferguson SJ, Newton WE (eds) Biology of the nitrogen cycle. Elsevier, Amsterdam, pp 331–341. https://doi.org/10.1016/B978-044452857-5.50022-9

Munevar F, Wollum AG (1982) Response of soybean plants to high root temperature as affected by plant cultivar and Rhizobium strain. Agron J 74:138–142

Murphy DV, Macdonald AJ, Stockdale EA, Goulding KWT, Fortune S, Gaunt JL, Poulton PR, Wakefield JA, Webster CP, Wilmer WS (2000) Soluble organic nitrogen in agricultural soils. Biol Fertil Soils 30:374–387

Nandal K, Sehrawat AR, Yadav AS, Vashishat RK, Boora KS (2005) High temperature-induced changes in exopolysaccharides, lipopolysaccharides and protein profile of heat-resistant mutants of *Rhizobium* sp. (Cajanus). Microbiol Res 160:367–373

Nayyar A, Bijay S, Yadvinder S (2006) Nitrogen-supplying capacity of soils for rice and wheat and nitrogen availability indices in soils of northwest India. Commun Soil Sci Plant Anal 37(7–8):961–976

Niste M, Vidican R, Pop R, Rotar I (2013) Stress factors affecting symbiosis activity and nitrogen fixation by Rhizobium cultured in vitro. ProEnvironment/ProMediu 6(13):42–45

Oberson A, Nanzer S, Bosshard C, Dubois D, Mäder P, FrossarD E (2007) Symbiotic N_2 fixation by soybean in organic and conventional cropping systems estimated by 15 N dilution and 15 N natural abundance. Plant Soil 290:69–83

Okogun JA, Sanginga N, Abaidoo R, Dashiell KE, Diels J (2005) On-farm evaluation of biological nitrogen fixation potential and grain yield of Lablab and two soybean varieties in the northern Guinea savanna of Nigeria. Nutr Cycl Agroecosyst 73:267–275

Pate JS (1961) Temperature characteristics of bacterial variation in legume symbiosis. Nature 192(4803):637–639

Peoples MB, Brockwell J, Herridge DF, Rochester IJ, Alves BJR, Urquiaga S, Boddey RM, Dakora FD, Bhattarai S, Maskey SL, Sampet C, Rerkasem B, Khan DF, Hauggaard-Nielsen H, Jensen ES (2009) The contributions of nitrogen-fixing crop legumes to the productivity of agricultural systems. Symbiosis 48:1–17

Piha MI, Munnus DN (1987) Sensitivity of the common bean (*Phaseolus vulgaris* L.) symbiosis to high soil temperature. Plant Soil 98:183–194

Rabie GH, Almadini AM (2005) Role of bioinoculants in development of salt-tolerance of *Vicia faba* plants under salinity stress. Afr J Biotechnol 4:210–222. https://doi.org/10.5897/AJB2005.000-3041

Rainbird RM, Atkins CA, Pate JS (1983) Effect of temperature on nitrogenase functioning in cowpeanodules. Plant Physiol 73(2):392–394

Ramírez M, Valderrama B, Arredondo-Peter R, Soberón M, Mora J, Hernández G (1999) Rhizobium etli genetically engineered for the heterologous expresionexpression of *Vitreoscilla* sp. hemoglobin: effects on free-living and symbiosis. Mol Plant-Microbe Interact 12:1008–1015. https://doi.org/10.1094/MPMI.1999.12.11.1008

Rao VR (1977) Effect of root temperature on the infection processes and nodulation in Lotus and Stylosanthes. J Exp Bot 28(2):241–259

Rao AS, Jha P, Meena BP, Biswas AK, Lakaria BL, Patra AK (2017) Nitrogen processes in agroecosystems of India. In: The Indian Nitrogen Assessment. Elsevier, Amsterdam, pp 59–76. https://doi.org/10.1016/B978-0-12-811836-8.00005-7

Redondo FJ, Coba de la Peña T, Morcillo CN, Lucas MM, Pueyo JJ (2009) Overexpression of flavodoxin induces changes in antioxidant metabolism leading to delayed senescence and starch accumulation in alfalfa root nodules. Plant Physiol 149:1166–1178. https://doi.org/10.1104/pp.108.129601

Reynolds-Henne CE, Langenegger A, Mani J, Schenk N, Zumsteg A, Feller U (2010) Interactions between temperature, drought and stomatal opening in legumes. Environ Exp Bot 68(1):37–43

Robleto EA, Scupham AJ, Triplett EW (1997) Trifolitoxin production in *Rhizobium etli* strain CE3 increases competitiveness for rhizosphere growth and root nodulation of *Phaseolus vulgaris* in soil. Mol Plant-Microbe Interact 10:228–233. https://doi.org/10.1094/MPMI.1997.10.2.228

Robleto EA, Kmiecik K, Oplinger ES, Nienhuis J, Triplett EW (1998) Trifolitoxin production increases nodulation competitiveness of Rhizobium etli CE3 under agricultural conditions. Appl Environ Microbiol 64:2630–2633

Rockström J, Steffen W, Noone K, Persson A, Chapin FS III, Lambin EF, Lenton TM, Scheffer M, Folke C, Schellnhuber HJ, Nykvist B, de Wit CA, Hughes T, Van der Leeuw S, Rodhe H, Sörlin S, Snyder PK, Constanza R, Svedin U, Falkenmark M, Karlberg L, Corell RW, Fabry VJ, Hansen J, Walker B, Liverman D, Richardson K, Crutzen P, Foley JA (2009) A safe operating space for humanity. Nature 461:472–475

Rodrigues C, Laranjo M, Oliveira S (2006) Effect of heat and pH stress in the growth of chickpea mesorhizobia. Curr Microbiol 53:1–7

Ros GH, Hoffland E, Kessel VC, Temminghoff EJM (2009) Extractable and dissolved soil organic nitrogen—a quantitative assessment. Soil Biol Biochem 41(6):1029–1039

Roughley RJ, Dart PJ (1970) Growth of *Trifolium subterraneum* L. selected for sparse and abundant nodulation as affected by root temperature and Rhizobium strain. J Exp Bot 21(3):776–786

Roumiantseva ML, Andronov EE, Kurchak ON (2006) Diversity of soil microbial community towards alfalfa root nodule bacteria. In: Proceedings of the 7th European nitrogen fixation conference, p 107

Ruelland E, Zachowski A (2010) How plants sense temperature. Environ Exp Bot 69:225–232. https://doi.org/10.1016/j.envexpbot.2010.05.011

Sadiki M, Rabih K (2001) Selection of chickpea (*Cicer arietinum* L.) for yield and symbiotic nitrogen fixation ability under salt stress. Agron EDP Sci 21(6–7):659–666 https://doi.org/10.1051/agro:2001158

Sadowsky MJ (2005) Soil stress factors influencing symbiotic nitrogen fixation. In: Werner D, Newton WE (eds) Nitrogen fixation research in agriculture, forestry. Springer, Dordrecht, pp 89–102

Sapountzis, P (2016) Potential for nitrogen fixation in the fungus-growing termite symbiosis. Front Microbiol 7:1993. https://doi.org/10.3389/fmicb.2016.01993

Sarath G, Pfeiffer NE, Sodhi CS, Wagner FW (1986) Bacteroids are stable during dark-induced senescence of soybean root nodules. Plant Physiol 82:346–350. https://doi.org/10.1104/pp.82.2.346

Schipanski ME, Drinkwater LE (2012) Nitrogen fixation in annual and perennial legume-grass mixtures across a fertility gradient. Plant Soil 357:147–159

Schipanski ME, Drinkwater LE, Russelle MP (2010) Understanding the variability in soybean nitrogen fixation across agroecosystems. Plant Soil 329:379–397

Schläppy ML, Schöttner SI, Lavik G, Kuypers MM, de Beer D, Hoffmann F (2010) Evidence of nitrification and denitrification in high and low microbial abundance sponges. Mar Biol 157(3):593–602

Schomberg HH, Weaver RW (1992) Nodulation, nitrogen fixation, and early growth of arrowleaf clover in response to root temperature and starter nitrogen. Agron J 84(6):1046–1050

Schomberg HH, Jones OR (1999) Carbon and nitrogen conservation in dryland tillage and cropping systems. Soil Sci Soc Am J 63(5):1359–1366

Schreiber F, Loeffler B, Polerecky L, Kuypers MM, De Beer D (2009) Mechanisms of transient nitric oxide and nitrous oxide production in a complex biofilm. The ISME J 3(11):1301–1313

Schreiber F, Wunderlin P, Udert KM, Wells GF (2012) Nitric oxide and nitrous oxide turnover in natural and engineered microbial communities: biological pathways, chemical reactions, and novel technologies. Front Microbiol 3:372. https://doi.org/10.3389/fmicb.2012.00372

Seneviratne M, Gunaratne S, Bandara T, Weerasundara L, Rajakaruna N, Seneviratne G, Vithanage M (2016) Plant growth promotion by *Bradyrhizobium japonicum* under heavy metal stress. South Afr J Bot 105:19–24

Serraj R (2003) Effects of drought stress on legume symbiotic nitrogen fixation: physiological mechanisms. Ind J Exp Biol 42(10):1136–1141

Sheppard SC, Bittman S, Bruulsema TW (2010) Monthly ammonia emissions from fertilizers in 12 Canadian Ecoregions. Can J Soil Sci 90(1):113–127

Shivran HS, Kumar D, Singh RV (2006) Improvement of water quality through biological denitrification. J Environ Sci Eng 48(1):57–60

Shvaleva A, Coba de la Peña T, Rincón A, Morcillo CN, García de la Torre VS, Lucas MM, Pueyo JJ (2010) Flavodoxin overexpression reduces cadmium-induced damage in alfalfa root nodules. Plant Soil 326:109–121. https://doi.org/10.1007/s11104-009-9985-1

Sinclair TR, Weisz PR (1985) Response to soil temperature of dinitrogen fixation (acetylene reduction) rates by field-grown soybeans 1. Agron J 77(5):685–688

Singh M, Kundu S, Biswas AK, Saha JK, Tripathi AK, Acharya CL (2004) Quantification of N_2 fixation and annual N benefit from N_2 fixation in soybean accrued to the soil under soybean-wheat continuous rotation. J Plant Nutr Soil Sci-Z Pflanzenernahr Bodenkd 167:577–583

Singh RP, Gupta SC, Yadav AS (2008) Effect of levels and sources of phosphorus and PSB on growth and yield of blackgram (*Vigna mungo* L. Hepper). Legume Res 31:139–141

Singh BP, Setia R, Wiesmeier M, Kunhikrishnan A (2018) Agricultural management practices and soil organic carbon storage. In: Soil carbon storage, 1 Jan 2018. Academic, New York, pp 207–244

Slama I, Abdelly C, Bouchereau A, Flowers T, Savouré A (2015) Diversity, distribution and roles of osmoprotective compounds accumulated in halophytes under abiotic stress. Ann Bot 115(3):433–447

Small IGC, Joffe A (1968) Physiological studies on the genus Trifolium with special reference to the South African species. II. Influence of root temperature on growth, nodulation and symbiotic nitrogen fixation. South Afr J Agric Sci 11:41–56

Soberón M, López O, Morera C, de Lourdes Girard M, Tabche ML, Miranda J (1999) Enhanced nitrogen fixation in a Rhizobium etli ntrC mutant that overproduces the *Bradyrhizobium japonicum* symbiotic terminal oxidase cbb3. Appl Environ Microbiol 65:2015–2019

Soussi M, Ocana A, Lluch C (1998) Effects of salt stress on growth, photosynthesis and nitrogen fixation in chick-pea (*Cicer arietinum* L.). J Exp Bot 49(325):1329–1337

Soussi M, Santamaría M, Ocaña A, Lluch C (2001) Effects of salinity on protein and lipopolysaccharide pattern in a salt-tolerant strain of *Mesorhizobium ciceri*. J Appl Microbiol 90:476–481. https://doi.org/10.1046/j.1365-2672.2001.01269.x

Spanning R, Delgado M, Richardson D (2005) The nitrogen cycle: denitrification and its relationship to N_2 fixation. In: Nitrogen fixation: origins, applications, and research progress, pp pp 277–342. https://doi.org/10.1007/1-4020-3544-6_13

Stagnari F, Maggio A, Galieni A et al (2017) Multiple benefits of legumes for agriculture sustainability: an overview. Chem Biol Technol Agric 4:2. https://doi.org/10.1186/s40538-016-0085-1

Streeter JG (2003) Effects of drought on nitrogen fixation in soybean root nodules. Plant Cell Environ 26(8):1199–1204

Streeter J, Wong PP (1988) Inhibition of legume nodule formation and N_2 fixation by nitrate. Crit Rev Plant Sci 7:1–23

Sullivan DG, Truman CC, Schomberg HH, Endale DM, Franklin DH (2007) Potential impact of conservation tillage on conserving water resources in Georgia. J Soil Water Conserv 62(3):145–152

Sun YJ, Zuo JE, Yang Y, Lu YQ, Xing W, Bu DH (2006) Community structure of nitrification bacteria in aerobic short-cut nitrification granule. Huan jing ke xue = Huanjing kexue 27(9):1858–1861

Sun WL, Zhao YG, Yang M (2017) Microbial fertilizer improving the soil nutrients and growth of reed in degraded wetland. In: IOP conference series: earth and environmental science, vol 69, 3rd international conference on advances in energy, environment and chemical engineering 26–28 May 2017, Chengdu, China, 69(1):012062. IOP Publishing. https://doi.org/10.1088/1755-1315/69/1/012062

Sutton MA, Oenema O, Erisman JW, Leip A, Van Grinsven H, Winiwarter W (2011) Too much of a good thing. Nature 472:159–161

Swaraj K, Bishnoi NR (1999) Effect of salt stress on nodulation and nitrogen fixation in legumes. Ind J Exp Biol 37(09):843–848

Swaraj K, Dhandi S, Sheokhand S, Sheoran IS (2000) Nodule functioning and recovery under salt stress in Pigeonpea (*Cajanus cajan* L. Millsp). J Plant Biol 27:19–26

Tajima R, Lee ON, Abe J, Lux A, Morita S (2007) Nitrogen-fixing activity of root nodules in relation to their size in peanut (*Arachis hypogaea* L.). Plant Prod Sci 10(4):423–429

Taiz L, Zeiger E (2002) Photosynthesis: physiological and ecological considerations. Plant Physiol 9:172–174

Talibart R, Jebbar M, Gouffi K, Pichereau V, Gouesbet G, Blanco C, Bernard T, Pocard J (1997) Transient accumulation of glycine betaine and dynamics of endogenous osmolytes in salt-stressed cultures of sinorhizobium meliloti. Appl Environ Microbiol 63(12):4657–4663

Thilakarathna MS, McElroy MS, Chapagain T, Papadopoulos YA, Raizada MN (2016) Below-ground nitrogen transfer from legumes to non-legumes under managed herbaceous cropping systems. A review. Agron Sustain Develop 36(4):58. https://doi.org/10.1007/s13593-016-0396-4

Trinchant JC, Boscari A, Spennato G, Van de Sype G, Le Rudulier D (2004) Proline betaine accumulation and metabolism in alfalfa plants under sodium chloride stress. Exploring its compartmentalization in nodules. Plant Physiol 135(3):1583–1594

Tully K, Ryals R (2017) Nutrient cycling in agroecosystems: balancing food and environmental objectives. Agroecol Sustain Food Syst 41(7):761–798. https://doi.org/10.1080/21683565.2017.1336149

Tzudir L, Bera PS, Chakraborty PK (2014) Impact of temperature on the reproductive development in mungbean (*Vigna radiata*) varieties under different dates of sowing. Int J Bioresour Stress Manage 5:194–199. https://doi.org/10.5958/0976-4038.2014.00555.7

Ulrich DEM, Sevanto S, Ryan M, Albright MBN, Johansen RB, Dunbar JM (2019) Plant–microbe interactions before drought influence plant physiological responses to subsequent severe drought. Sci Rep 9:249. https://doi.org/10.1038/s41598-018-36971-3

Van Den Bossche A, De Bolle S, De Neve S, Hofman G (2009) Effect of tillage intensity on N mineralization of different crop residues in a temperate climate. Soil Tillage Res 103(2):316–324

Van Dillewijn P, Soto MJ, Villadas PJ, Toro N (2001) Construction and environmental release of a *Sinorhizobium meliloti* strain genetically modified to be more competitive for alfalfa nodulation. Appl Environ Microbiol 67:3860–3865. https://doi.org/10.1128/AEM.67.9.3860-3865.2001

Van Dillewijn P, Soto MJ, Villadas PJ, Toro N (2002) Effect of a *Sinorhizobium meliloti* strain with a modified putA gene on the rhizosphere microbial community of alfalfa. Appl Environ Microbiol 67:4201–4208. https://doi.org/10.1128/aem.68.9.4201-4208.2002

Van Hoorn JW, Katerji N, Hamdy A, Mastrorilli M (2001) Effect of salinity on yield and nitrogen uptake of four grain legumes and on biological nitrogen contribution from the soil. Agric Water Manag 51:87–98. https://doi.org/10.1016/S0378-3774(01)00114-7

Van Horn DJ, Okie JG, Buelow HN, Gooseff MN, Barrett JE, Takacs-Vesbach CD (2014) Soil microbial responses to increased moisture and organic resources along a salinity gradient in a polar desert. Appl Environ Microbiol 80(10):3034–3043

Vanderlinde EM, Harrison JJ, Muszyński A, Carlson RW, Turner RJ, Yost CK (2010) Identification of a novel ABC transporter required for desiccation tolerance, and biofilm formation in *Rhizobium leguminosarum* bv. *viciae* 3841. FEMS Microbiol Ecol 71:327–340. https://doi.org/10.1111/j.1574-6941.2009.00824.x

Vara Prasad PV, Boote KJ, Allen LH, Thomas JM (2002) Effects of elevated temperature and carbon dioxide on seed-set and yield of kidney bean (*Phaseolus vulgaris* L.). Glob Change Biol 8:710–721. https://doi.org/10.1046/j.1365-2486.2002.00508.x

Vessey JK (1994) Measurement of nitrogenase activity in legume root nodules: in defense of the acetylene reduction assay. Plant Soil 158(2):151–162

Vessey JK, Walsh KB, Layzell DB (1988) Can a limitation in pholem supply to nodules account for the inhibitory effect of nitrate on nitrogenase activity in soybean? Physiol Plant 74(1):137–146

Virginia RA, Jenkins MB, Jarrell WM (1986) Depth of root symbiont occurrence in soil. Biol Fert Soil 2:127–130

Vogel G, Fiehn O, Jean-Richard-dit-Bressel L, Boller T, Wiemken A, Aeschbacher RA, Wingler A (2001) Trehalose metabolism in Arabidopsis: occurrence of trehalose and molecular cloning and characterization of trehalose-6-phosphate synthase homologues. J Exp Bot 52(362):1817–1826

Vriezen JAC, de Bruijn FJ, Nüsslein KR (2007) Responses of 1040 rhizobia to desiccation in relation to osmotic stress, oxygen, 1041 and temperature. Appl Environ Microbiol 73:3451–3459

Ward BB (2011) Nitrification: an introduction and overview of the state of the field. In: Nitrification. American Society of Microbiology, pp 3–8. https://www.asmscience.org/content/book/10.1128/9781555817145.ch01?crawler=true#

Waughman GJ (1977) The Effect of Temperature on Nitrogenase Activity. J Exp Bot 28(4):949–960

Weisz PR, Sinclair TR (1988) Soybean nodule gas permeability, nitrogen fixation and diurnal cycles in soil temperature. Plant Soil 109:227–234

Wheatley DM, Macleod DA, Jessop RS (1995) Influence of tillage treatments on N_2 fixation of soybean. Soil Biol Biochem 27(4–5):571–574

Whitehead DC (1995) Legumes: biological nitrogen fixation and interaction with grasses. In: Whitehead DC (ed) Grassland nitrogen, CAB international, Wallingford, UK, pp 36–57

Wilkins RJ (2008) Eco-efficient approaches to land management: a case for increased integration of crop and animal production systems. Philos Trans R Soc B Biol Sci 363:517–525

Witty JF, Minchin FR (1988) Measurement of nitrogen fixation by the acetylene reduction assay; myths and mysteries. In: Nitrogen fixation by legumes in Mediterranean agriculture. Springer, Dordrecht, pp 331–344

Xiang SR, Doyle A, Holden PA, Schimel JP (2008) Drying and rewetting effects on C and N mineralization and microbial activity in surface and subsurface California grassland soils. Soil Biol Biochem 40(9):2281–2289

Xu Y, Zhan C, Huang B (2011) Heat shock proteins in association with heat tolerance in grasses. Int J Proteom, 11, Article ID 529648. https://doi.org/10.1155/2011/529648

Yelton MM, Yang SS, Edie SA, Lim ST (1983) Characterization of an effective salt-tolerant, fast-growing strain of *Rhizobium japonicum*. Microbiol 129(5):1537–1547

Zahran HH (1999) Rhizobium-legume symbiosis and nitrogen fixation under severe conditions and in an arid climate. Microbiol Mol Biol Rev 63(4):968–989

Zahran HH, Räsänen LA, Karsisto M, Lindström K (1994) Alteration of lipopolysaccharide and protein profiles in SDS-PAGE of rhizobia by osmotic and heat stress. World J Microbiol Biotechnol 10(1):100–105

Zhang XP, Karsisto M, Harper R, Lindstrom K (1991) Diversity of Rhizobium bacteria isolated from the root nodules of leguminous trees. Int J Syst Bacteriol 41:104–113

Zhang X, Amelung W, Yuan Y, Samson-Liebig S, Brown L, Zech W (1999) Land-use effects on amino sugers in particle size fractions of an Arguidoll. Appl Soil Ecol 11:271–275

Zhang M, He Z, Zhao A, Zhang H, Endale DM, Schomberg HH (2011) Water-extractable soil organic carbon and nitrogen affected by tillage and manure application. Soil Sci 176(6):307–312

Zhao A, Zhang M, He Z (2012) Characteristics of soil water-soluble organic C and N under different land uses in Alaska. Soil Sci 177(12):683–694

Genetic Engineering and Genome Editing for the Improvement of Fabaceae for Abiotic Stress Tolerance

Ehsan Valiollahi, Jorge A. Pinto Paiva, and Ana Sofia Duque

Abstract According to the new WMO Global Climate Report (released on March 2019), the years 2015–2018 were the four warmest years on record. Moreover, the alternation of heat waves and drought with the occurrence of extreme events become a challenge for agriculture in forthcoming years. Plants from Fabaceae family are the third largest family of flowering plants and second in economic importance, being cultivated for food and animal feed. They also provide the benefits of nitrogen fixation due to Legume-Rhizobium symbiotic relationship. However, their productivity is limited by different biotic and abiotic constraints, and the improvement of Fabaceae for stress tolerance is an important requirement regarding the future climate scenario. In this chapte we will focus on works related to the genetic engineering and genome editing for the improvement of Fabaceae plants regarding their tolerance to abiotic stresses. As candidates for genetic engineering purposes we can find numerous abiotic stress-related genes (e.g., *LEAs*, *ZIPs*, *HSPs*) and transcriptional factors (e.g., *MYBs*, *ABFs*, *HSFs*); that have been used with some degree of success either in model or other important grain and forage legumes. Recently, considerable research has revealed the active role of micro-RNAs (miRNAs) in many aspects of plant development and in the response to various environmental stress conditions. Some examples of their usage for engineering for abiotic stress tolerance in legumes and their possible future exploitation will be addressed. Finally, we will provide a resume on genome editing techniques related to stress tolerance and discussed forthcoming applications in Fabaceae improvement.

Ehsan Valiollahi and Jorge A. Pinto Paiva are equally contributing authors.

E. Valiollahi · J. A. Pinto Paiva
Instituto de Biologia Experimental e Tecnológica (IBET), Apartado 12, 2781-901 Oeiras, Portugal

E. Valiollahi · J. A. Pinto Paiva · A. S. Duque (✉)
Instituto de Tecnologia Química e Biológica António Xavier (ITQB NOVA), Green-it Unit, Apt. 127, 2781-901 Oeiras, Portugal
e-mail: sduque@itqb.unl.pt

J. A. Pinto Paiva
Institute of Plant Genetic Polish Academy of Science, ul. Strzeszyńska 34, 60-479 Poznań, Poland

M. Hasanuzzaman et al. (eds.), *The Plant Family Fabaceae*,
https://doi.org/10.1007/978-981-15-4752-2_5

Keywords Fabaceae · Abiotic stress · Tolerance · Transgenic plants · miRNAs · Genome editing

1 Introduction

Pulses and forage legumes (Leguminosae or Fabaceae) well known for their nutritional and health benefits may play important roles on mitigation of climate change impact and on the contribution to the sustainability of Agriculture (Araújo et al. 2015). Main advantages of legumes in terms of soil conservation and sustainability are related to their ability to fix atmospheric nitrogen due to Legume-Rhizobium symbiotic relationship. They are also responsible for providing the main source of vegetable protein for human diet and are essential for livestock feed (as forage production). In developing countries, legumes could play a major role in the eradication of protein–energy malnutrition (Maphosa and Jideani 2017). Nevertheless, legume productivity might be limited by different biotic and abiotic constraints. According to the 25th World Meteorological Organization (WMO) "Statement on the State of the Global Climate in 2018" report (WMO 2019) the exposure of the agricultural sector to climate extremes would threaten to reverse the gains made in ending hunger and malnutrition. Indeed, the alternation of heat waves and drought with the occurrence of extreme events, like flooding, extreme rainfall, and extratropical storms, constitute a challenge for agriculture in the forthcoming years. Thus, the improvement of Fabaceae for stress tolerance is an important requirement regarding the current and future climate scenario. In this context, it is crucial to provide deep insights into the molecular responses to stress, including the function of stress-responsive genes and their effects on phenotypic changes, allowing the development of biotechnological tools to accelerate breeding of legumes crops for stress resistance (Kudapa et al. 2013; Duque et al. 2013; Deshmukh et al. 2014; Nadeem et al. 2019).

Currently, abiotic stress-responsive genes can be grouped in genes involved in protecting cells from stress by producing important metabolic proteins, and in genes involved in regulatory pathways and signal transduction during the perception and response to stress (Duque et al. 2013). The first group of stress-responsive genes include those with a putative function in stress tolerance by coding for important proteins such as chaperones, membrane proteins (e.g., water channel proteins and potassium transporters), detoxification enzymes, or enzymes involved in the synthesis of osmoprotectants or compatible solutes. A second group of stress-responsive genes comprises those coding for protein factors involved in the further regulation of signal transduction or gene expression such as transcription factors (Agarwal et al. 2018).

Recently, plant non-coding RNA (long-ncRNAs and small-ncRNA) have also been considered major regulators of stress response gene expression mechanisms in both biotic (e.g., virus, bacteria, and fungus pathogens) and abiotic stresses (e.g., drought, heat, salinity, cold, and nutrient starvation). Among small non-coding RNA, micro-RNAs (miRNAs) represent a class of 20- to 24-nucleotides small non-coding molecules which regulate gene expression at the posttranscriptional level, by binding to the coding sequence (CDS) or untranslated region (UTR) of specific mRNAs, targeting them for cleavage or by directing translational repression at the mRNA level (Zhang et al. 2006; Ding et al. 2012; Alptekin et al. 2017; Wang et al. 2018b, 2019). MiRNAs biogenesis initiates from the transcription of miRNA-coding genes by the RNA polymerase II activity in the nucleus into long primary miRNAs (pri-miRNAs), which are then processed by an enzyme DICER-LIKE1 (DCL1) into a precursor miRNA (pre-miRNA) of 80–500 nt (Liu et al. 2018). Pre-miRNA is unstable in the nucleus and is processed into guide (miRNA) and passenger (miRNA*) strands duplex, by DCL1, and transported outside the nucleus by the protein HASTY (HST; a homolog of Exportin 5) (Park et al. 2005; Pareek et al. 2015). Exported miRNA-miRNA* duplexes are processed by helicases to produce mature miRNAs, which are incorporated into the RNA-induced silencing complex (RISC) to repress the expression of the target genes by perfect or near perfect pairing of miRNAs and their target sites (Bartel 2004; Sanei and Chen 2015) (see Fig. 1). The discovery and understanding of this natural conserved defense mechanism against double-stranded RNA (dsRNA) has transformed the aforementioned into a powerful tool for genetic engineering and functional genomics studies. Several conserved miRNAs are important modulators of the expression of various abiotic stress-related genes which are significantly altered (upregulated or downregulated) in stress conditions (Shriram et al. 2016; Chauhan et al. 2017).

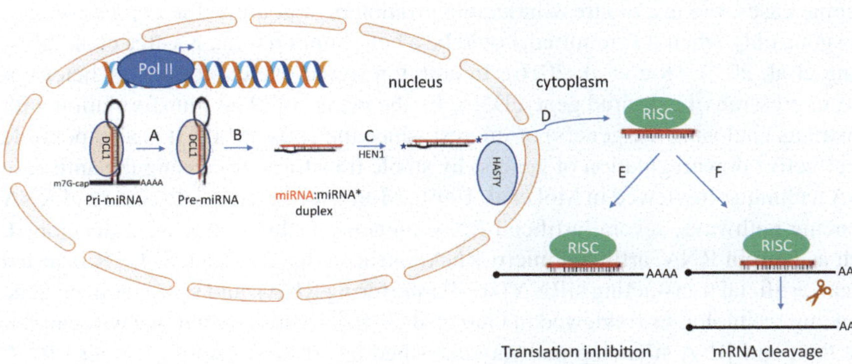

Fig. 1 Schematic representation of plant miRNA biogenesis and miRNA roles in the RNAi mechanisms. MiRNA genes begin to be transcribed by RNA polymerase II as long primary transcripts (pri-miRNA) that are stabilized by the addition of a 5′ 7-methylguanosine cap and a 3′ polyadenylated tail (Rogers and Chen 2013). These pri-miRNAs fold into stem-loop structure that can be recognized and processed into a precursor miRNA (pre-miRNA) of about 80–500 nucleotides in length, by Dicer-like 1 (DCL1) enzyme with RNase III endonucleases activity (A). Then, DCL1 generates miRNA-miRNA* duplex (B) and the 3′terminal nucleotide of each strand needs to be methylated by the Hua Enhancer1 (HEN1) RNA methyltransferase (Park et al. 2005; Yu et al. 2005) (C). Methylated miRNA-miRNA* duplex is exported to the cytoplasm with the assistance of HASTY (D). After the removal of miRNA*, mature miRNA is incorporated into the RISC (a complex ribonucleoprotein containing the Argonaute (AGO) protein, responsible for mRNA cleavage), and act by inhibiting the translation (E) or by modulating the cleavage of target mRNA (F)

Plant genetic engineering has been claimed as the fastest agricultural technique for crop improvement (Zhou and Luo 2013). Potent biotechnological advancements especially in post-genomic era are developing infrastructures for sustainable agricultural crop development to bypass the environmental hurdles challenges (Xu et al. 2019).

Two main transformation strategies have been employed for genetic engineering using expression constructs with the gene/cDNA of interest to be integrated into the plant genome. The most used method for foreign gene transfer into plant cells and for the subsequent regeneration of transgenic plants is based on the natural *Agrobacterium tumefaciens* system; but particle bombardment has also been extensively exploited for plant transformation especially in recalcitrant species or when the aim is the DNA integration into the plastid genome (plastid transformation) (review in Duque et al. 2013). *Agrobacterium*-mediated transformation approach is been mainly used in dicotyledonous species (dicots), and particle bombardment of DNA coated gold or tungsten beads, preferably in monocotyledonous species (monocots) (Gomes et al. 2019). Additionally, by using *Agrobacterium rhizogenes* (revised as *Rhizobium rhizogenes*; Young et al. 2001), it is also possible to induce the production of transformed hairy roots and obtain composite plants, rather than transgenic whole plants, as further described in this chapter in An et al. (2017) work in soybean. Moreover, one key factor to take into consideration in engineering of abiotic stress-tolerant plants is the choice of the promoters. The cauliflower mosaic virus (CaMV) 35S promoter has been the most commonly used for dicots crops. However, in some cases, the use of stress-inducible promoters, that allow the expression of a transgene only when it is required, could be a better option (e.g., Kasuga et al. 2004; Wang et al. 2013; Hou et al. 2016). In addition to the overexpression or heterologous expression of a desired gene/cDNA, by the means of plant transformation with constructs enclosing the gene(s) of interest; since the early 1990s, it is also possible the effective downregulation of gene(s) by stable transformation using the antisense RNA technique (reviewed in Mol et al. 1990). Moreover, since the discovery of RNA silencing pathways, several artificial RNA silencing technologies were developed, such as hairpin RNA, artificial micro-RNA, intrinsic direct repeat, 3′ UTR inverted repeat, artificial trans-acting siRNA (small interfering RNA), and virus-induced gene silencing technologies (reviewed in Guo et al. 2016). In this chapter, we will emphasize the plant RNA silencing pathway classified as posttranscriptional micro-RNA pathway, whose bases are shown in Fig. 1. Since plant miRNAs are assigned to modulate stress responses and adaptation processes, the miRNA-based biotechnology has huge potential for improvement of crop tolerance to abiotic stresses (Budak and Zhang 2017).

Recently, the advent of the genome editing (GE) technology opened new possibilities of engineering any genomic sequence in a crop plant, by efficient targeted precise mutagenesis (including gene knockouts, knockdowns, modifications, and the repression and activation of target genes) (Khandagale and Nadaf 2016). While there is still a dynamic debate between stakeholders, including regulators, scientists, industry, and non-governmental organizations, about considering GE technology as non-GM (non-genetically modified) (e.g., Eckerstorfer et al. 2019); it is undoubtedly that GE has a vital role for future utilization in molecular breeding to enhance abiotic stress tolerance. Even though only few studies have adopted GE for editing for stress tolerance in Fabaceae, a brief introduction to GE (especially CRISPR/Cas9 technology) and some recent examples will be provided in Sect. 4 of this chapter.

To conclude, in this chapter, we will focus on works related to genetic engineering for the improvement of Fabaceae for abiotic stresses tolerance. Some examples of stress-related genes and stress-related miRNAs, their usage, and possible future exploitation for genetic engineering will be addressed. Finally, we will provide a resume on genome editing techniques and discussed forthcoming applications in Fabaceae abiotic stress improvement.

2 Stress-Responsive Candidates for Genetic Engineering Purposes

The development of new high-throughput methodologies with reduced costs to access the transcriptome and proteome, and the release of plant genome sequences for several species contributed to the fast increase knowledge on candidate responsive genes to a variety of abiotic stresses, such as heavy metals, flood, osmotic and ionic stress, drought, heat or cold stresses, that can be further used for genetic engineering purposes [see the recent reviews by Ramalingam et al. (2015), Jha et al. (2019), Nadeem et al. (2019)]. Ramalingam et al. (2015) discussed the use of proteomics and metabolomics approaches to understand the stress response mechanisms in Fabaceae, as well as cellular and developmental processes, including symbiosis. These authors presented an overview of important proteins that are differentially expressed in response to various stresses in some model and crop legumes (see tables in Ramalingam et al. 2015, and references therein). Jha et al. (2019) review the current progress in "omics" advances to understanding the genetic makeup of salinity tolerance and for improving salt tolerance of grain legumes. More recently, Nadeem et al. (2019) reviewed the progresses on the research in drought stress in legumes, focusing also on the "omics" approaches and providing a resume of genes that could be putatively explored for improving drought resistance in legumes.

Bashir et al. (2019) identified a list of genes considered potential candidates for the development of drought and cadmium-tolerant soybean (*Glycine max* L. (Merrill) cultivars). The genes involved in abscisic acid (ABA) degradation, gibberellin and salicylic acid biosynthesis, hydrogen peroxide (H_2O_2) production and Cd transport were upregulated, while those involved in zeatin riboside (ZR), indole 3-acetic acid (IAA), and methyl jasmonate (MeJA) biosynthesis were downregulated, under Cd and drought stress. Wang et al. (2018a) used RNA-seq to determine the transcriptional responses of soybean to heat, drought, and combined stress and identified potential new targets for enhancing soybean stress tolerance. In wild soybean (*Glycine soja* Siebold & Zucc.), major intrinsic proteins like Aquaporins (AQPs), responsive to salt and drought stresses, were identified through genome-wide screening, and their mode of action elucidated using yeast two-hybrid assay and bimolecular fluorescence complementation analysis (BiFC) (Zhang et al. 2017).

An integrated approach of proteomics and computational genetic modification effectiveness analysis (a mathematical model constructed for metabolic simulation) was applied to explore flood-tolerant genes in soybean (Wang et al. 2018c). Protein levels of glyceraldehyde-3-phosphate dehydrogenase (GAPDH), aconitase 1, and 2-oxoglutarate dehydrogenase (2-OGDH) were higher in flood-tolerant soybean plants compared to wild-type plants, under flood conditions. These three proteins were considered good candidates to promote flood tolerance and the respective set of genes potential candidates for genetically modified soybean (Wang et al. 2018c).

A comparative transcriptomic analysis based on full-length transcript sequencing was performed to evaluate the contribution of osmotic and ionic stress toward the salinity tolerance in roots of *Medicago sativa* L. (Luo et al. 2019). Within the alfalfa transcriptome, these authors found 188 TFs in 32 TF families present as differentially expressed genes (DEGs) under the two evaluated stresses; among those were important TFs families such as the GRAS, MYB, AP2-EREBP, NAC, WRKY, and bHLH. These key osmotic and ionic stress-related genes were considered good candidates for future studies as potential targets to improve resistance to salinity stress via genetic engineering (Luo et al. 2019). In common bean, *Phaseolus vulgaris* L., 27 Teosinte-branched1/Cycloidea/Pcf (TCP) proteins were characterized by in silico analysis of transcriptome, and their involvement in the response to salinity confirmed by comparing salt-tolerant and sensitive genotypes under salt stress (İlhan et al. 2018). Interestingly, these authors also found that 20 TCP genes were predicted to be targeted by miRNAs (among those miRNAs, the miR319 was the highly represented regulatory miRNA to target TCP transcripts). Jha et al. (2019) explored germplasm resources to identify salinity stress-responsive candidate genes using transcriptome analysis. They presented a list of salinity tolerance genes and their putative function for salinity tolerance in soybean, a list of QTLs/genes conferring salinity tolerance in various grain legumes, and also a list of differentially expressed genes related to salinity stress tolerance.

In the model legumes, *Medicago truncatula* and *Lotus japonicus*, transcriptomic analysis allowed the identification of a high number of differentially expressed genes during the early response of nodules to water deprivation, highlight genes related to hormone metabolism and signal transduction, and genes encoding key enzymes in ABA biosynthesis (Sańko-Sawczenko et al. 2019). Genes encoding late embryogenesis abundant (LEA) glycoproteins and serine-rich proteins; and a variety of transcription factors (TFs) encoding genes, belonging to myeloblastosis oncogene (MYB) and to the ethylene response factor (ERF) subfamily of AP2 (AP2/ERF), basic Helix-Loop-Helix, Constans-like, basic region/leucine zipper motif (bZIP), WRKY, C2H2-type zinc finger and NAC families, were also found (Sańko-Sawczenko et al. 2019).

Recently, NGS-based (next-generation sequencing-based) small RNA sequencing (sRNA-seq) revealed the importance of these approaches for understanding the importance of small RNAs (sRNAs), and more specifically of miRNAs, on cellular responses in legumes when exposed to abiotic stress conditions (Abdelrahman et al. 2018). Using RNA-seq, Xu et al. (2016) identified 51 chilling-responsive miRNAs in *Glycine max*. The targets of these cold stress-responsive miRNAs were enriched in oxidation–reduction, signal transduction, and metabolic process functional categories. Furthermore, 23 drought-responsive miRNA families were also reported in soybean. Among those, 11 families were upregulated, while the other 12 were downregulated under stress conditions (Zheng et al. 2016). Additionally, a genome-wide study of the miRNA expression dynamics in the cell division zone of chickpea (*Cicer arietinum*) root apex under water-deficit and salinity stresses revealed possible miRNA-mediated modulation of root development and root hair formation under these conditions (Khandal et al. 2017).

Deep sequencing (also referred as high-throughput sequencing) of *Vicia faba* sRNA transcriptome led to the identification of 527 and 693 salt stress-responsive miRNAs, in salt-sensitive (Hassawi-3) and salt-tolerant (ILB4347) genotypes, respectively. Functional annotation of stress-responsive miRNAs target genes demonstrated their role in regulation of salt stress-related biological pathways; including the ABC transporter pathway, MAPK signaling pathway, plant hormone signal transduction, and the phosphatidylinositol signaling system (Alzahrani et al. 2019). High-throughput sequencing of *M. truncatula* seedlings showed that 81 and 129 miRNAs were stress-responsive under salt and alkali treatment, respectively. Among them, 101 miRNAs were significantly upregulated in at least one of the treatment groups and 35 miRNAs (including mtr-miR156 and mtr-miR171 families, and mtr-miR159a) were overexpressed in both salt and alkali treatments. In contrast, 67 miRNAs were downregulated in one of the treatments. Pathway analysis revealed that salt/alkali stress-related miRNAs participate in regulation of various biosynthesis and metabolic pathways (Cao et al. 2018). In addition, the soybean miRNOme (defined as the full spectrum of miRNAs in a genome) profiling in aluminum toxicity conditions revealed that 32 miRNAs were differentially expressed during aluminum stress. Overexpressed miRNAs in tolerant (BX10) and susceptible (BD2) genotypes regulate the expression of genes related to amino acid metabolism and secondary metabolism, respectively, and indicated miRNA-mediated root elongation in tolerant genotype (Huang et al. 2018).

3 Genetic Engineering for Abiotic Stress Tolerance in Fabaceae

Genetic engineering strategies are being used to characterize the function of several candidate genes involved in abiotic stress response. In this section, we will focus on the most recent studies based on the genetic engineering for abiotic stress tolerance in Fabaceae, emphasizing different examples, both in terms of transformed species and of the function of the chosen genes.

3.1 Improving Tolerance to Abiotic Stress in Grain and Forage Legumes

Protective Proteins and Chaperones

Here, we present some examples of proteins with recognized protective function (e.g., LEA and ZFNs), under both normal and extreme conditions; including those considered as plant molecular chaperones. Being molecular chaperones defined as a family of unrelated classes of proteins that mediate the correct assembly of other polypeptides (Ellis 1991), as is the case of some heat shock proteins (HSPs) (Kim et al. 2013; Park and Seo 2015; Trivedi et al. 2016).

A late embryogenesis abundant (LEA) protein from the desert grass *Cleistogenes songorica* was inserted into alfalfa (*Medicago sativa* L.) via agrobacterium-mediated transformation (Zhang et al. 2016). The expression of the *CsLEA* gene in alfalfa enhanced drought and salt tolerance by improving growth and relative water content (RWC), protection of the membrane (evaluated by the malondialdehyde (MDA) content; a parameter related to the degree of cell membrane damage), and increase in compatible solutes (namely the osmoprotectant proline). Transgenic alfalfa overexpressing Cs*LEA* had decreased Na^+ and increased K^+ levels, resulting in a higher K^+/Na^+ ratio compared to non-transgenic plants, suggesting that was able to regulate K^+/Na^+ homeostasis to reduce the ion toxicity (Zhang et al. 2016). Moreover, transgenic alfalfa plants presented higher chlorophyll fluorescence (increased Fv/Fm; a parameter related to the photochemical efficiency of the PS II), chlorophyll content, and net photosynthetic rate, compared to the control. Plants were tested with a period of 15d of drought (by water withdrawn) and 10d of salt stress, followed by re-watering. The transgenic plants were able to return to a normal state, though the WT plants could not be recovered.

Also in alfalfa, the overexpression of a *Glycine soja* zinc finger protein1 (GsZFP1), under the control of the cauliflower mosaic virus (CaMV) 35S promoter, resulted in enhanced drought and salt tolerance (Tang et al. 2013). After a drought treatment of 20d, three transgenic lines performed better than the WT plants; and when were re-watered the survival rates, fresh weights, root lengths, and shoot heights of the transgenic lines were greater than those of the WT plants. Under high salinity stress (plants irrigated with 250 mM of NaCl solutions for 17d) the relative membrane permeability and MDA of WT plants were greater than those of the lines overexpressing GsZFP1. Conversely, more free proline and protective soluble sugars accumulated in the transgenic alfalfa than in the WT plants (Tang et al. 2013). An example using a HSP molecular chaperone will be provided in the Genome editing for stress tolerance 4.2 subsection.

Membrane Proteins

$Na^+(K^+)/H^+$ antiporters (NHXs) are ubiquitous membrane proteins that play important roles in maintaining the cellular pH and $Na^+(K^+)$ homeostasis (Wu et al. 2016). In plants, the Na^+/H^+ antiporter in the plasma and vacuolar membrane pumps excessive Na^+ from the cytosol or compartmentalizes it into tonoplasts to maintain higher

K^+/Na^+ ratios in the cytoplasm, protecting cells from sodium toxicity (Almeida et al. 2017). In order to obtain a salt-tolerant alfalfa plants, Zhang et al. (2014a, b) transferred the halophyte *Salicornia europaea* L. Na^+/H^+ antiporter gene, *SeNHX1*, using the Agrobacterium-mediated transformation method. Upon a salt stress treatment (0.6% NaCl) during 21 days, the authors observed that the amount of chlorophyll decrease and MDA contents in transgenic overexpressing *SeNHX1* plants were lower; while proline levels and also the activities of superoxide dismutase (SOD), peroxidase (POD), and catalase (CAT) were higher, compared to wild-type plants. In this way, the *SeNHX1* transgenic alfalfa plants showed enhanced tolerance to salt through improved photosynthesis and membrane protection, and also osmotic adjustment and scavenging of free radicals by the increase in proline content, and SOD, POD, and CAT activities (Zhang et al. 2014a, b). Another antiporter gene, the *AtNHX5* gene from the model plant Arabidopsis, was used to transform soybean, and the resulting transgenic plants were also tolerant to the presence of 300 mM NaCl, when compared to the non-transgenic plants (Wu et al. 2016). In a more recent work, the *AtNHX1* Na^+/H^+ antiporter overexpression, together with the inoculation with the stress-resistant nitrogen-fixing strain *Sinorhizobium meliloti* B401, resulted in higher forage production and increased nutrient quality in alfalfa under salt stress (Stritzler et al. 2018).

Aquaporins (AQP) are membrane proteins that belong to the major intrinsic protein (MIP) superfamily and play an important role in plant water relations (Kapilan et al. 2018). AQPs facilitate water movement by regulating root hydraulic conductivity under diverse stress conditions such as salt and water stresses (Shekoofa and Sinclair 2018). Plant AQPs are classified into five main subfamilies including plasma membrane intrinsic proteins (PIPs), tonoplast intrinsic proteins (TIPs), nodulin 26 like intrinsic proteins (NIPs), small basic intrinsic proteins (SIPs), and X intrinsic proteins (XIPs) (Kapilan et al. 2018). Soybean composite plants (obtained by *Agrobacterium rhizogenes* transformation) and transgenic whole plants, both transformed with the *Panax ginseng* tonoplast intrinsic protein1 (*PgTIP1*) gene, had enhanced salt and drought tolerance by maintaining homeostasis of water, salt ions, and reactive oxygen species (ROS) (An et al. 2017, 2018).

Osmoprotectants

Transgenic strategies can also be used for the overexpression of enzymes/proteins involved in the biosynthesis of osmoprotectants or compatible solutes, such as sugars, polyols, amino acids (e.g., proline), polyamines, glycine betaine, and other related compounds (Singh et al. 2015).

As an example, glycine betaine (GB) accumulates in many plants under stress conditions, is considered non-toxic at high concentrations, and is responsible for raising the cytoplasm osmotic pressure. GB is synthetized from two different substrates, choline and glycine. The pathway of GB biosynthesis from choline involves one or two enzymes, depending on the biological system. In higher plants, animals and *E. coli* is a two-step reaction catalyzed by the choline mono-oxygenase (CMO) or choline dehydrogenase (CDH), and the NAD^+-dependent betaine aldehyde dehydrogenase (BADH) (Fariduddin et al. 2013). In contrast, in the soil bacterium *Arthrobacter* sp., a single-step reaction catalyzed by choline oxidase (COD) is responsible for the production of GB (Fariduddin et al. 2013). The choline dehydrogenase (CDH) and choline oxidase (COD) overexpression, in tobacco and Arabidopsis, are some examples of the transgenic approaches used for the accumulation of GB for the purpose of obtaining salt and freezing tolerant plants (Lilius et al. 1996; Hayashi et al. 1997; Sakamoto et al. 2000; He et al. 2001). In *M. sativa*, Li et al. (2014) generated transgenic plants expressing the *cod*A cDNA derived from *Arthrobacter globiformis* under the control of the oxidative stress-inducible sweet potato peroxidase (SWPA2) promoter (Kim et al. 2011). A targeting signal to the chloroplast was included by the presence of an N-terminal transit peptide (TP) in the used construct. Salt and drought stress analysis was performed in whole 4-week-old plants, by irrigation treatment with 250 mM NaCl solution or by withholding the water supply, respectively (Li et al. 2014). The tolerance of transgenic plants to salt stress was estimated based on the contents of GB, chlorophyll, and MDA in leaves, after the stress treatment. For the drought stress free proline content and relative water content (RWC) were accessed in the leaves after 5 days of water withholding. As expected, the GB content in transgenic plants significantly increased under both stresses due to the induction of *cod*A transcriptional expression under the control of the stress-inducible SWPA2 promoter. All transgenic plants showed enhanced tolerance to drought stress by maintaining high RWC and increased levels of GB and proline compared to non-transgenic (NT) plants (Li et al. 2014). Moreover, two transgenic lines maintained higher chlorophyll contents and lower MDA levels, compared to NT plants. More recently, in mung bean (*Vigna radiata* L. Wilczek), a grain legume cultivated in equatorial and semi-tropical regions of the Indian subcontinent and in Southeast Asian countries, Baloda et al. (2017) induced salt stress tolerance by introducing the *cod*A bacterial gene using the plasmid pGAH/*codA* (Hayashi et al. 1997). A simple disk leaf assay was used on primary transformants to evaluate salt stress tolerance, and authors reported that leaf disks from non-transgenic plants showed complete senescence while those from transgenic plants remained green after a 4d treatment with 50–200 mM NaCl (Baloda et al. 2017).

Other interesting biomolecules used for plant engineering toward abiotic stress tolerance are polyamines (PAs). PAs are low-molecular-weight, positively charged, aliphatic amines found widespread in living organisms. They are synthesized from the decarboxylation of the amino acids arginine, ornithine, methionine, and lysine. Putrescine (Put), spermidine (Spd), and spermine (Spm) are the plant most abundant PAs, and their levels increase under several environmental disturbing conditions, including drought, high salinity, and exposure to extreme temperatures (heating or freezing) (Araújo et al. 2019). Regarding the modulation of the PAs biosynthetic pathways, we have several works describing an increased abiotic stress tolerance in different plant species by the means of regulating the arginine decarboxylase (ADC) gene expression; a key enzyme responsible for Put synthesis (e.g., Alcázar et al. 2010; Alet et al. 2011; Espasandin et al. 2014; Duque et al. 2016; Espasandin et al. 2018; Hidalgo-Castellanos et al. 2019). Put is then converted into Spd by spermidine synthase (SPDS), and Spd is further converted into Spm or thermospermine (tSpm), by Spm synthase (SPMS) and tSpm synthase (ACL5), respectively (reviewed in Araújo et al. 2019). Plants from the model legume *Medicago truncatula* cv. Jemalong constitutively expressing the *Adc* gene from *Avena sativa* showed higher leaf accumulation of Put, Spd and norspermidine (NorSpd), when compared to non-transformed plants (Duque et al. 2016). When subjected to severe water deficit (SWD), by water withholding, the transgenic plants showed increased photosynthetic parameters, specifically, leaf internal CO_2 concentration (*Ci*), net CO_2 assimilation rate (*A*), transpiration (*E*), and stomatal conductance (*gs*). Besides, transgenic plants that recovered from SWD showed higher seed yield compared to non-transformed plants, suggesting a benefit of increased PA levels in preserving legume grain yield when exposed to drought stress (Duque et al. 2016). Recently, using those transgenic *M. truncatula* ADC lines, Hidalgo-Castellanos et al. (2019) found that the expression of the *Adc* gene increased the nodule biomass by favoring *Sinorhizobium meliloti* root colonization and/or nodule organogenesis, which led to an increment in the nitrogen fixation rate (NFR) per plant. Regarding the nodule metabolism, under salt stress, ADC transgenic plants accumulated higher amounts of proline and Spm in nodules compared to non-transformed plants (Hidalgo-Castellanos et al. 2019).

Detoxification

In this subsection, we will focus on some specific examples concerning the heterologous expression of enzymes considered important as detoxifying agents. As an example, we have ascorbate peroxidases, which play important protective roles in higher plants preventing stress toxic effects by scavenging reactive oxygen species (ROS) (Singh et al. 2014b).

In Peanut (*Arachis hypogaea* L.), an economically and agriculturally important crop, the yield and quality are negatively affected by drought and salinity (Sun et al. 2013). The peroxisomal ascorbate peroxidase gene (*SbpAPX*), from the extreme halophyte *Salicornia brachiate*, under the control of the CaMV35S promoter (Singh et al. 2014b) was used for peanut Agrobacterium-mediated transformation (Singh et al. 2014a). For salt stress treatments 4-week-old transgenic peanut plants were watered with 0–150 mM NaCl for 2 weeks, and the phenotypical differences were recorded. Transgenic peanut plants expressing *SbpAPX* performed better than WT plants, since the total chlorophyll content, shoot length, fresh weight, and relative water content increased concomitantly with salt stress. Furthermore, transgenic plants continued normal growth and completed life cycle and presented lower electrolyte leakage (EL) during the course of the experiment (Singh et al. 2014b).

Another interesting approach is plant engineering toward the overproduction of reactive carbonyl species (RCS)-detoxifying enzymes; with the example of *Vigna mungo* (blackgram) transformation with the Aldose Reductase *ALDRXV4* cDNA from *Xerophyta viscosa* (Singh et al. 2016). RCS species are α-ß unsaturated aldehydes and ketones produced from lipid peroxides, with action in environmental stress (Mano 2012). The enzyme ALD (belonging to the aldo-keto reductase superfamily) catalyzes the reduction of carbonyl metabolites and play an important role in the osmoprotection and detoxification of RCS in plant cells (Mano 2012; Singh et al.

2016). In this experiment, transgenic ALDRXV4 plants showed improved tolerance to various environmental stresses [including drought, salt, methyl viologen (MV) and H_2O_2 induced oxidative stress] (Singh et al. 2016). Transgenic plants presented increased aldose reductase activity, higher sorbitol content, and less accumulation of the toxic methylglyoxal, either under stress and non-stress conditions (drought and salinity). Additionally, accumulation of ROS remarkably decreased in the transgenic lines, compared with the non-transformed plants (Singh et al. 2016)

Transcription Factors

Transcription Factors (TFs) interact with *cis*-elements in the promoter regions of abiotic stress-related genes and upregulate the expression of different responsive genes resulting in abiotic stress tolerance. In this way, TF-based genetic engineering for stress tolerance has enormous potential and transcription factors (TFs) like Apetala2/ethylene-responsive factor (AP2/ERF), and homologous of myelocytomatosis (MYC) and myeloblastosis (MYB) oncogenes, NAC, WRKY, basic leucine zipper (bZIP), protein kinases (MAPK, CDPK), and protein phosphatases, among others, are important candidates for this approach (reviewed in Agarwal et al. 2018). Numerous stress-responsive TFs have been exploited to genetic engineering for abiotic stress tolerance in Fabaceae, among those we can highlight common TFs, such as MYB, AP2, BZIP, GRAS, and Zinc finger (see Table 1). In this chapter, we present few selected examples of application of TF legume engineered for increase abiotic stress response.

AP2/ERFs, especially those in DREB (dehydration-responsive element-binding) and ERF (ethylene-responsive factors) subfamilies, are ideal candidates for crop improvement because their overexpression enhances tolerance to multiple abiotic stress tolerance such as salt, drought, heat, and cold, as well as resistances several diseases in the transgenic plants (Xu et al. 2011; Deokar et al. 2015; Debbarma et al. 2019).

In early studies, genetic engineering using DREB transcription factors was accomplished in the attempt to improve salt tolerance in the forage legumes *M. sativa* and *M. truncatula* (Jin et al. 2010; Li et al. 2011). In the first study (Jin et al. 2010), a soybean DREB orthologue (GmDREB1) was used to transform alfalfa plants under the control of the Arabidopsis Rd29A promoter. Plants were exposed to salt stress by watering with concentrations up to 400 mM NaCl for 60 days. Transgenic plants showed significantly higher tolerance when compared to non-transformed plants, considering the chlorophyll fluorescence values, electrolytic leakage (EL) rates, and contents of proline and total soluble sugars. Additionally, the transcript level of the proline biosynthetic enzyme Δ1-pyrroline-5-carboxylate synthase (P5CS), important in determining proline levels in plants, was upregulated by the stress-inducible expression of *GmDREB1* (Jin et al. 2010). In the second case (Li et al. 2011), the *MtCBF4*, belonging to the AP2-EREBP transcription factor family, was overexpressed in *M. truncatula* and enhanced salt tolerance was achieved; associated with induced expression level of corresponding downstream genes containing DRE elements, specifically *MtCAS15* and *MtCAS31* (Li et al. 2011). Lately, in peanut, transgenic lines expressing *AtDREB1A* transcription factor showed improved tolerance to both salt and drought stresses (Sarkar et al. 2014). The experiments were conducted with 18-day-old plants, and drought and salt stress were imposed by polyethylene glycol (PEG) treatment (0, 10, 15, 20% PEG) or NaCl-induced salinity stress (0, 100, 150, and 200 mM NaCl), respectively. At 20% PEG, three transgenic lines exhibited significantly improved growth parameters, including pod weight, shoot weight, total biomass, and kernel weight, root and shoot length, and an improved root-shoot ratio, compared to non-transformed ones. Similarly, higher root weight, root-shoot ratio, and harvest-index (Donald 1962) were observed for transgenic plants across the various concentrations of salinity with the increased exposure time (Sarkar et al. 2014). It is well documented that root growth is usually less affected by water deficit than shoot growth, making the root system an essential part of drought avoidance mechanism (Purushothaman et al. 2017); and consequently, an increased root-shoot ratio is commonly observed under stress situations. Other important growth parameters such as proline content, total chlorophyll content, osmotic potential, EL, and RWC were also increased in those transgenic peanut lines (Sarkar et al. 2014).

While AP2/ERF transcription factors are ABA independent TFs, others, for example, basic leucine zipper (bZIP) TFs are ABA dependent, meaning that those TFs are responsible for regulating many ABA-responsive genes by interaction with specific ABA-responsive elements (ABREs). ABREs are cis-acting elements containing a (C/T)ACGTGGC consensus sequence that is present in the promoter region of ABA-regulated genes that are induced by abiotic stresses and also mediate defense responses (e.g., Uno et al. 2000; Agarwal and Jha 2010; Hoang et al. 2017). Homodomain-leucine zipper (HD-Zip) genes noticeable increase after ABA exposure (Chew et al. 2013). *M. sativa* plant overexpressing the Arabidopsis enhanced drought Tolerance1 (*AtEDT1*) cDNA, that encodes a protein of the HD-ZIP TF family, were subjected to drought in both laboratory and field trials (Zheng et al. 2017). These researches observed that compared with wild-type, transgenic plants had higher survival rates and biomass, but reduced water loss under drought stress. Moreover, *EDT1* transgenic alfalfa plants had increased stomatal size (but reduced stomatal density) larger root systems and reduced membrane permeability and MDA content. However, chlorophyll, soluble sugar, and proline content were higher. As a consequence of this improvement transgenic alfalfa plants grew better in 2-y field trials and showed enhanced growth performance (Zheng et al. 2017).

Finally, we present an example regarding the overexpression of a MYB-type transcription factor; a class of TFs also reported to be engaged in ABA-mediated gene expression (Fujita et al. 2011). *GmMYB84*-overexpressing soybean plants exhibited enhanced drought resistance with a higher survival rate, longer primary roots, greater proline and ROS contents, higher antioxidant enzyme activities (POD, CAT, and SOD), lower dehydration rate, and reduced MDA. Wang et al. (2017) also described that ROS levels were related to the modulation of primary root elongation and demonstrated that GmMYB84 were able to bind to the promoter of *GmRBOHB-1* and *GmRBOHB-2* genes (Respiratory burst oxidase homolog protein Bs- proteins involved in ROS production). Table 1 summarizes several studies based on the genetic engineering for the improvement of Fabaceae for abiotic stress tolerance.

Table 1 Examples of genetic engineering of Fabaceae plants toward tolerance to environmental stresses

Transformed plant	Stress related genes	Promoter/methodology	Abiotic Stress	Effect on phenotype and responses to environmental stresses	References
Medicago sativa (Alfalfa)	*GsZFP1* (Cys2/His2-type zinc finger protein gene from *Glycine soja*)	CaMV 35S (cauliflower mosaic virus 35S constitutive promoter); *Agrobacterium tumefaciens*	Drought and salt stress	After recovering from water-deficit stress alfalfa transgenic lines presented higher survival rates, fresh weights, root lengths, and shoot height. Under high salinity stress, transgenic alfalfa showed greater relative membrane permeability and malondialdehyde (MDA) content. Higher free proline and soluble sugars accumulation was also observed	Tang et al. (2013)

(continued)

Table 1 (continued)

Transformed plant	Stress related genes	Promoter/methodology	Abiotic Stress	Effect on phenotype and responses to environmental stresses	References
Medicago sativa (Alfalfa)	*CodA* (choline oxidase cDNA from *Arthrobacter globiformis*)	Expression of *codA* in chloroplasts; SWPA2 (sweet potato oxidative stress-inducible promoter); *A. tumefaciens* (EHA105 strain)	Drought and salt stress	Transgenic plants showed enhanced tolerance to NaCl and drought stress at the whole-plant level. Under salt stress plants maintained higher chlorophyll contents and lower MDA levels. Under drought stress conditions, plants maintained a high relative water contents (RWC) and increased levels of glycinebetaine and proline, compared to control plants	Li et al. (2014b)

(continued)

Table 1 (continued)

Transformed plant	Stress related genes	Promoter/methodology	Abiotic Stress	Effect on phenotype and responses to environmental stresses	References
Medicago sativa (Alfalfa)	*AtNDPK2* (Arabidopsis nucleoside diphosphate kinase 2 gene)	SWPA2; *A. tumefaciens* (EHA105)	Multiple abiotic stresses	Transgenic plants showed enhanced tolerance to high temperature, NaCl, and drought stress at the whole-plant level. In addition, exhibited better growth through increased expression of several auxin-related indole acetic acid (*IAA*) genes	Wang et al. (2014)
Medicago sativa (Alfalfa)	*IbOr* (Sweet potato Orange gene)	SWPA2; *A. tumefaciens* (EHA105)	Multiple abiotic stresses	Transgenic plants exhibited less damage in response to methyl viologen (MV)-mediated oxidative stress and salt stress; and also exhibited enhanced tolerance to drought stress, along with higher total carotenoid levels (*IbOr* previously reported to increase carotenoid accumulation)	Wang et al. (2015)

(continued)

Table 1 (continued)

Transformed plant	Stress related genes	Promoter/methodology	Abiotic Stress	Effect on phenotype and responses to environmental stresses	References
Medicago sativa (Alfalfa)	*rstB* (salt tolerance B gene from *Sinorhizobium fredii*)	CaMV 35S; *A. tumefaciens* (LBA4404)	Salt stress	Enhanced resistance to salt-shock treatment was observed in *rstB* transgenic T0 plants. Transgenic second-generation (T1) seeds, showed improved germination rates and better growth under salt stress conditions. Hindered Na^+, but enhanced $Ca2^+$ accumulation was also observed in salt stressed plants	Zhang and Wang (2015)

(continued)

Table 1 (continued)

Transformed plant	Stress related genes	Promoter/methodology	Abiotic Stress	Effect on phenotype and responses to environmental stresses	References
Medicago sativa (Alfalfa)	*AtABF3* (Abcisic-acid responsive element-binding factor 3 (bZIP TF) from Arabidopsis)	SWPA2; *A. tumefaciens* (GV3101)	Multiple abiotic stresses	Under normal conditions, transgenic plants showed smaller leaf size. However, under drought they exhibited better growth, reduced transpiration rate (*E*) and lower reactive oxygen species (ROS). Under salt stress transgenic plants performed better, showing higher chlorophyll levels and lower MDA. Transgenic alfalfa also showed enhanced resistance to methyl viologen (MV)-mediated oxidative stress	Wang et al. (2016)

(continued)

Table 1 (continued)

Transformed plant	Stress related genes	Promoter/methodology	Abiotic Stress	Effect on phenotype and responses to environmental stresses	References
Medicago sativa (Alfalfa)	*CsLEA* (*C. songorica* late embryogenesis abundant glycoprotein)	CAM 35S; *A. tumefaciens* (GV3101)	Drought and salt stress	Transgenic plants grew better than control plants after imposed drought and salt stresses. Those plants were able to maintain a higher RWC, higher shoot biomass, fewer photosystem changes, decreased membrane injury, and lower osmotic stress injury	Zhang et al. (2016)

(continued)

Table 1 (continued)

Transformed plant	Stress related genes	Promoter/methodology	Abiotic Stress	Effect on phenotype and responses to environmental stresses	References
Medicago sativa (Alfalfa)	*ZxNHX* and *ZxVP1-1* (tonoplast NXH and H+ -PPase genes from xerophyte *Z. xanthoxylum*)	CAM 35S; *A. tumefaciens* (GV3101)	Drought and salt stress	Transgenic alfalfa plants, co-expressing ZxNHX and ZxVP1-1, grew better, with greater plant height and dry mass, under both normal or stress conditions (NaCl or water-deficit). They exhibited enhanced photosynthesis capacity by maintaining higher net photosynthetic rate (*A*), stomatal conductance (*gs*), and water use efficiency (WUE), compared to control	Bao et al. (2016)

(continued)

Table 1 (continued)

Transformed plant	Stress related genes	Promoter/methodology	Abiotic Stress	Effect on phenotype and responses to environmental stresses	References
Vigna mungo (Blackgram)	*ALDRXV4* (aldose reductase *RXV4* gene form *Xerophyta viscosa*)	CaMV 35S; *A. tumefaciens* (GV3101)	Multiple abiotic stresses	Transgenic plants showed improved tolerance to various environmental stresses; including drought, salt, MV and H_2O_2 induced oxidative stress. Those plants presented increased aldose reductase activity, higher sorbitol content and less accumulation of the toxic methylglyoxal, under stress and non-stress conditions (drought and salinity). The accumulation of ROS was also remarkably decreased in transgenic lines	Singh et al. (2016)

(continued)

Table 1 (continued)

Transformed plant	Stress related genes	Promoter/methodology	Abiotic Stress	Effect on phenotype and responses to environmental stresses	References
Vigna radiata (Mung bean)	*CodA* (choline oxidase cDNA from *A. globiformis*)	CaMV 35S; *A. tumefaciens* (EHA101 strain)	Salt stress	Leaf disks from non-transgenic plants showed complete senescence, while those from transgenic plants remained green after a 4d salt stress treatment	Baloda et al. (2017)
Arachis hypogaea (Peanut)	*AtDREB1A* (Dehydration-Responsive Element-Binding 1A gene from Arabidopsis)	RD29A (*stress-inducible* Arabidopsis *rd29A promoter)*; *A. tumefaciens*	Drought and salt stress	Transgenic peanut lines showed improved tolerance to both stresses, by delayed and less severe wilting of the leaves, and by improved growth parameters, such as: proline content, total chlorophyll content, osmotic potential, electrolytic leakage and RWC	Sarkar et al. (2014)

(continued)

Table 1 (continued)

Transformed plant	Stress related genes	Promoter/methodology	Abiotic Stress	Effect on phenotype and responses to environmental stresses	References
Arachis hypogaea (Peanut)	*SbpAPX* (peroxisomal ascorbate peroxidase gene from *Salicornia brachiata*)	CaMV35S; *A. tumefaciens* (LBA4404)	Salt stress	Higher chlorophyll content, shoot lengths, fresh weights (shoot and root) and RWC was observed in transgenic lines under salt stress compared to control plants	(Singh et al. 2014b)

(continued)

Table 1 (continued)

Transformed plant	Stress related genes	Promoter/methodology	Abiotic Stress	Effect on phenotype and responses to environmental stresses	References
Arachis hypogaea (Peanut)	*AtHDG11* (Arabidopsis homeodomain-leucine zipper TF)	RD29A; *A. tumefaciens* (EHA105)	Drought and salt stress	Transgenic plants with improved water use efficiency traits, such as: longer root system, reduced stomatal density, higher chlorophyll content, increased specific leaf area, improved photosynthetic rates, and increased intrinsic instantaneous WUE, under both drought and salt stress conditions. Additional upregulation of known stress-responsive genes, antioxidative enzymes and free proline, in transgenic plants	Banavath et al. (2018)

(continued)

Table 1 (continued)

Transformed plant	Stress related genes	Promoter/methodology	Abiotic Stress	Effect on phenotype and responses to environmental stresses	References
Glycine max (Soybean)	*GmACP1* (Acid phosphatase gene from soybean)	CaMV 35S; *Agrobacterium rhizogenes* (K599)	Low-P stress	Overexpression of GmACP1 in soybean hairy roots increased Phosphorus (P) efficiency by 11–20% relative to the control, and contributed to soybean tolerance to low-P stress.	Zhang et al. (2014a, b)
Glycine max (Soybean)	*AtNHX5* (Na + (K +)/H + antiporter gene from Arabidopsis)	*2x* CaMV 35S; *A. tumefaciens*	Salt stress	After salt treatment the transgenic plants had a higher content of free proline and lower content of MDA, compared to the control, and were more tolerant to the NaCl stress. AtNHX5 possibly working by efficiently transporting Na^+ and K^+ ions from the roots to the leaves	Wu et al. (2016)

(continued)

Table 1 (continued)

Transformed plant	Stress related genes	Promoter/methodology	Abiotic Stress	Effect on phenotype and responses to environmental stresses	References
Glycine max (Soybean)	*Ncl* (salt tolerance gene from soybean)	CaMV 35S; *A. tumefaciens*	Salt stress	Overexpression of *Ncl* in transgenic soybean lines enhanced salt tolerance by improving leaf chlorophyll (SPAD measurement) and shoot dry weight. Higher expression of the *Ncl* gene in the root resulted in lower accumulations of Na^+, K^+, and Cl^- in the shoot under salt stress	Do et al. (2016)

(continued)

Table 1 (continued)

Transformed plant	Stress related genes	Promoter/methodology	Abiotic Stress	Effect on phenotype and responses to environmental stresses	References
Glycine max (Soybean)	*GmMYB84* (R2R3-MYB TF from soybean)	CaMV35S; *A. tumefaciens*	Drought stress	Transgenic overexpressing soybean lines exhibited enhanced drought resistance with a higher survival rate, longer primary root length, and greater proline and ROS contents. Additionally, higher antioxidant enzyme activities [peroxidase (POD), catalase (CAT) and superoxide dismutase (SOD)], lower dehydration rate and reduced MDA content, were also observed	Wang et al. (2017)

(continued)

Table 1 (continued)

Transformed plant	Stress related genes	Promoter/methodology	Abiotic Stress	Effect on phenotype and responses to environmental stresses	References
Glycine max (Soybean)	*GmFDL19* (bZIP transcription factor from soybean)	CaMV35S; *A. tumefaciens*	Drought and salt stress	The relative plant height (RPH) and relative shoot dry weight (RSDW) of transgenic plants were significantly higher than those of the control after PEG and salt treatments. Germination rate and plant height of transgenic soybean were also significantly higher after various salt treatments	Li et al. (2017)
Glycine max (Soybean)	*PgTIP* (*Panax ginseng* tonoplast intrinsic protein1)	CaMV35S; *Agrobacterium rhizogenes* and *A. tumefaciens*	Drought and salt stress	Soybean composite plants (*Agrobacterium rhizogenes* transformation) and transgenic whole plants (*A. tumefaciens* transformation), both transformed with the (*PgTIP1*) gene, had enhanced salt and drought tolerance by maintaining homeostasis of water, salt ions and ROS species	An et al. (2017), (2018)

(continued)

Table 1 (continued)

Transformed plant	Stress related genes	Promoter/methodology	Abiotic Stress	Effect on phenotype and responses to environmental stresses	References
Glycine max (Soybean)	*AtABF3* (ABRE-binding factor (ABF) from Arabidopsis)	CaMV35S; *A. tumefaciens* (EHA105)	Drought and salt stress	Some transgenic lines showed lower ion leakage and higher chlorophyll contents. Water loss by transpiration and higher stomata closure was also observed in transgenic plants. Moreover, pods of non-transgenic plants showed severe withering and most of the pods failed to set normal seeds	Kim et al. (2018)

(continued)

Table 1 (continued)

Transformed plant	Stress related genes	Promoter/methodology	Abiotic Stress	Effect on phenotype and responses to environmental stresses	References
Cicer arietinum (Chickpea)	CaHDZ12 (HD-Zip (I) gene from chickpea)	CaMV35S; Constructs in antisense orientation [CaHDZ12(-)]; *A. tumefaciens* (LBA4404)	Drought and salt stress	Silencing of CaHDZ12 in chickpea resulted in increased sensitivity to salt and drought stresses. CaHDZ12-silenced plants displayed severe symptoms of wilting and damage, and produced higher H_2O_2 and O_2^-, than the control plants upon stress treatments. The antioxidant enzymes (CAT, APX and SOD) were also significantly reduced	Sen et al. (2017)
Cicer arietinum (Chickpea)	*AtDREB1A* (Dehydration-Responsive Element-Binding 1A gene from Arabidopsis)	RD29A; *A. tumefaciens*	Drought stress	Under drought stress transgenic plants showed better root and shoot partitioning, and presented higher transpiration efficiency (measured by vapor pressure deficit-VPD), compared to control	Anbazhagan et al. (2014)

(continued)

Table 1 (continued)

Transformed plant	Stress related genes	Promoter/methodology	Abiotic Stress	Effect on phenotype and responses to environmental stresses	References
Lotus tenuis (Narrowleaf trefoil)	*OatADC (Arginine Decarboxylase gene from Avena sativa)*	*pRD29A; A. tumefaciens*	Drought stress	Transgenic plants showed increased putrescine (Put) levels, improved cellular hydration and increased root growth. Put also controls the level of ABA by modulating ABA synthesis at the transcriptional leve	Espasandin et al. (2014)
Lotus tenuis (Narrowleaf trefoil)	*OatADC (Arginine Decarboxylase gene from A. sativa)*	*pRD29A; A. tumefaciens*	Salt stress	Improved tolerance to salt stress in transgenic lines by reduction in shoot biomass and increase in root growth in response to stress (improved root-shoot ratio), and increased osmotic adjustment via proline production	Espasandin et al. (2018)

(continued)

Table 1 (continued)

Transformed plant	Stress related genes	Promoter/methodology	Abiotic Stress	Effect on phenotype and responses to environmental stresses	References
Medicago truncatula (barrel medic)	*OatADC* *(Arginine Decarboxylase gene from A. sativa)*	*2x* CaMV 35S; *A. tumefaciens* (EHA105)	Drought stress	Transgenic plants with elevated Put and spermidine (Spd) levels. Under water-deficit stress transgenic lines presented higher leaf relative water content (RWC) and increased photosynthetic parameters, namely, higher leaf *Ci* (internal CO_2 concentration), *A* and *gs*. higher seed yield upon stress	Duque et al. (2016)

(continued)

Table 1 (continued)

Transformed plant	Stress related genes	Promoter/methodology	Abiotic Stress	Effect on phenotype and responses to environmental stresses	References
Medicago truncatula (barrel medic)	*MtWRKY76* (WRKY TF from *M. truncatula*)	*2x* CaMV 35S; *A. tumefaciens*	Drought and salt stress	Transgenic *MtWRKY76* plants presented higher survival rates and lower relative electrolyte leakage upon water withdrawn. Higher levels of several abiotic stress-inducible genes were also observed in transgenic lines under drought stress. Under salt treatment transgenic lines have higher root and plant dry weights, compared to control	Liu et al. (2016)
Medicago truncatula (barrel medic)	*OatADC (Arginine Decarboxylase gene from A. sativa)*	*2x* CaMV 35S; *A. tumefaciens* (EHA105)	Salt stress	Overexpression of the *Adc* gene increased the nodule biomass by favoring *Sinorhizobium meliloti* root colonization and supporting an increment in the Nitrogen Fixation Rate (NFR). Under salt stress, ADC transgenic plants accumulated higher amounts of proline and Spm in nodules, compared to control plants	Hidalgo-castellanos et al. (2019)

3.2 MiRNA-Based Biotechnology for Abiotic Stress Tolerance

In this subsection, we present few examples of functional characterization of miRNAs involved in abiotic stress response in Legumes. For genetic engineering purposes, the RNAi (RNA interference) mechanism is the usual choice (a biological process involved in sequence-specific gene regulation driven by the introduction of dsRNA that degrade mRNA molecules and decrease their activity by preventing translation, via posttranscriptional gene silencing (see Fig. 1); or that can inhibit gene expression via the pre-transcriptional silencing mechanism, through DNA methylation at genomic positions complementary to siRNA or miRNA molecules (Hirai and Kodama 2008). Plant transformation, using the RNAi mediated gene silencing technique, comprises the identification of the target gene(s), followed by the generation of RNAi constructs cassettes, plant transformation, and finally screening and evaluation of the obtained traits (Kamthan et al. 2015). An example of an RNAi cassette includes inverted repeat sequences against the target gene and a spacer fragment; the resultant transcript forms a hairpin-like structure that is processed into siRNAs that further operate within the RNAi pathway. Briefly, the RNAi construct produces long dsRNAs that, depending on the construct, can be processed by two DCL enzymes, resulting either in the endonucleolytic cleavage of the homologous mRNA (DCL4/AGO1 processing) or the methylation of cytosine in the complementary genomic DNA sequence (DCL3/AGO4 processing) (Hirai and Kodama 2008). The RNAi technology is considered precise, efficient, stable, and even better than the antisense technology (Saurabh et al. 2014; Guo et al. 2016). Nowadays, the vectors used for RNAi constructs are based on gateway vectors like pHELLSGATE, pSTARGATE, and pWATERGATE (Saurabh et al. 2014). In many RNAi constructs constitutive promoters like CaMV 35S and plant ubiquitin promoter (Ubi) are, among others, efficient in order to drive foreign gene expression in plant cells.

Arshad et al. (2017) showed a significant improvement in drought tolerance in alfalfa by overexpressing miR156, which regulates major transcription factors belonging to the SQUAMOSA PROMOTER BINDING PROTEIN-LIKE (SPL) family of stress-related genes. Three *SPL* target genes (*SPL6*, *SPL12*, and *SPL13*) were previously found to be downregulated by miR156 in alfalfa (Aung et al. 2015). In Arshad et al. (2017) work, a *SPL13*-RNAi construct was made using the pHELLSGATE12 vector and the gateway system (Invitrogen). Transgenic plants were obtained using the *Agrobacterium tumefaciens* strain GV3101. Resulting miR156 overexpression alfalfa genotypes (miR156OE) exhibited reduced water loss and enhanced root growth, and also maintained higher stomatal conductance compared to the wild type under drought stress.

In the same way, drought stress-induced tolerance was observed in chickpea transgenic lines overexpressing miR408 through the regulation of copper accumulation (Hajyzadeh et al. 2015). Under drought conditions (17d treatment of water deficiency), the overexpression of miR408 resulted in the repression of the plantacyanin transcript (a miR408 target, coding for a blue copper protein) and led to the induction of DREB and six others drought-responsive genes in the transgenic plants, when compared to control (Hajyzadeh et al. 2015).

In another work, by transforming soybean with *Agrobacterium rhizogenes*, the miR172c was found to be a positive regulator of root plastic development and salt tolerance (Sahito et al. 2017). The root system of soybean plants overexpressing miR172c demonstrated enhanced salt tolerance when compared to the control plants. Overexpression of miR172a in soybean plants improved salt tolerance through cleaving AP2/EREBP-type transcription factor gene (SSAC1). This gene is inhibitor of thiamine biosynthesis gene (THI1), which encodes a positive regulator of salt tolerance (Pan et al. 2016). In contrast, a functional analysis of miR399a (a salt-responsive miRNA in the root meristem) in *A. rhizogenes* transformed soybean plants revealed that overexpressing miR399a roots were more sensitive to salt stress, when compared to control roots transformed with the empty vector (Sun et al. 2016).

A resume table with examples of Fabaceae transformation using miRNA technology, and consequent phenotypic alterations observed under stress conditions, is presented below (Table 2).

Table 2 Examples of Fabaceae transformation using miRNA technology and functional characterization of phenotypic alterations through abiotic stress

Transformed plant	miRNA	Promoter/transformation technique	miRNA target	Effect on abiotic stresses	References
Medicago sativa (alfalfa)	miR156	CaMV35S/*A. tumefaciens* (LBA4404)	SPLs (Squamosa promoter binding protein-like)	Adaptive responses leading to drought tolerance	Aung et al. (2015), Arshad et al. (2017)
Cicer arietinum (chickpea)	miR408	CaMV35S/*A. tumefaciens* (LBA4404)	Plantacyanin (blue copper family protein)	Induction of drought tolerance	Hajyzadeh et al. (2015)
Glycine max (soybean)	miR172c	CaMV35S/*A. rhizogenes* (K599)	NNC1 (Nodule Number Control 1)	Increased root plasticity and salt tolerance	Sahito et al. (2017)
Glycine max~ (soybean)	miR399a	CaMV35S/*A. rhizogenes* (K599)	Multicopper oxidases (*Glyma08G359400;* Growth-regulating factor (*Glyma03G021900*)	Increased sensitivity to salt stress	Sun et al. (2016)
Glycine max (soybean)	miR172a	CaMV35S/*A. rhizogenes* (K599)	SSAC1 (AP2/EREBP-type transcription factor gene)	Enhanced tolerance to salt stress	Pan et al. (2016)

4 Genome Editing for Abiotic Stress Tolerance in Fabaceae

4.1 A Brief Introduction to Genome Editing Technology

The foundation of genome editing technique relies on the efficiency of targeting genome alterations mainly based on the induction of DNA double-stranded breaks (DSBs) or adjacent single-stranded breaks (SSBs). Since all organisms have to deal with DNA-damaging factors, understanding the different repair pathways is a prerequisite to control and enhance targeted gene modifications (Schmidt et al. 2019). The DNA repair by endonucleases can occur by non-homologous end joining (NHEJ) repair, or homology-dependent repair (HDR) (Puchta 2005). In plants, the DSB reparation is typically performed by the way of NHEJ, a very efficient repair mechanism but also susceptible to frequent mutation errors due to nucleotide insertions and deletions (indels) (Puchta 2005). HDR is considered the dominant mechanism but requires the initial formation of single-stranded DNA (ssDNA) for pairing with homologous DNA template sequences, thus resulting in a low efficiency process (Decottignies 2013). Simplified models of NHEJ and HDR methods for DSB repair are schematically represented in Fig. 2.

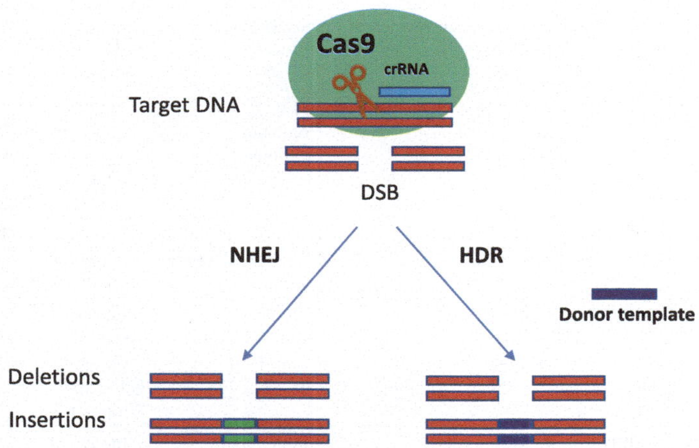

Fig. 2 Resumed schemes of CRISPR/Cas9 mediated DNA double-stranded breaks (DSBs) and repair mechanisms. The CRISPR-associated enzyme Cas9 achieves site-specific genomic engineering by introducing a double-stranded break (DSB) at the chromosomal site specified by the guide RNA (gRNA) (an upstream region called protospacer adjacent motif (PAN) is involved in the recognition of the target sequence by Cas9). The sequences of the spacer are further used as templates to produce short CRISPR RNAs (crRNAs) complementary to the target DNA region. Cleavage of both strands of DNA target is then accomplished. The DNA repair by endonucleases can occur by non-homologous end joining (NHEJ), or by homology-dependent repair (HDR) (Adapted from Tang et al. 2019)

Combining the knowledge about DNA repair and the specific induction of DSBs using programmable site-specific nucleases provided a powerful tool for successful genome engineering. Synthetic site-specific nucleases are able to induce one or more DSBs at defined loci, predominantly repaired via NHEJ (Schmidt et al. 2019). Examples of artificial site-specific nucleases included Zink finger nucleases (ZFNs), transcription activator-like effector nucleases (TALENs), and clustered regularly interspaced short palindromic repeats (CRISPR) associated with a nuclease protein (CRISP/CAS) (reviewed in Gaj et al. 2013; Wright et al. 2014). Since 2013, the CRISPR/Cas system was considered a breakthrough and has been used for gene editing (adding, interrupting or altering the sequence of specific genes) and gene regulation in various species. The CRISPR-Cas system relies on two main components: a guide RNA (gRNA) and CRISPR-associated (Cas) nuclease. By providing the cell with Cas9 protein and the appropriate guide RNA, the organism's genome can be cut at any desired location and this technology also enables simultaneous editing of several sites (Cong et al. 2013). The gRNA includes a sequence complementary that recognizes the target DNA region, the CRISPR-RNA (crRNA) containing 17–20-nucleotides (nt), and a transactivating crRNA (tracrRNA) that function as a binding support between the crRNA and Cas9 enzyme. Recognition of a target sequence by Cas9 also involves the identification of an upstream protospacer adjacent motif (PAM) (Cromwell et al. 2018). Although there are various types of CRISPR/Cas systems, the most commonly used system for genome editing is the Cas9 nuclease from the type II CRISPR/Cas9 system of *Streptococcus pyogenes*, which requires an NGG (N, any nucleotide; G, guanine) PAM sequence for DNA targeting (Liu et al. 2017). A simplified model of the CRISPR/Cas9 mechanism is represented on Fig. 2.

4.2 Genome Editing for Stress Tolerance: Some Examples

The use of genome editing, and more specifically the use of the CRISPR/Cas9 efficient methodology for abiotic stress tolerance in legumes, is still scarce. Similarly, for other crops only few studies can be found in the literature concerning the use of genome editing for abiotic stress tolerance (reviewed in Debbarma et al. 2019). Successful reports on legume genome editing using the CRISPR/Cas9 system were accomplished in soybean (Cai et al. 2015, 2018; Bao et al. 2019). As an example, the knockout of soybean gene *GmFT2a* (*FT-Flower Locus T*) associated with flowering time, using the CRISPR/Cas9 methodology, resulted in *GmFT2a* mutants showing delayed flowering (Cai et al. 2018). Likewise, in *Glicine max* and *Medicago truncatula* the CRISPR/Cas9 and TALENs systems were also used to generate heritable mutations for genes encoding proteins of the machinery involved in small RNA processing (Curtin et al. 2018). This study validated the use of these two different genome engineering methodologies to generate a mutant collection with utility for future studies on small RNA processing in crop legume.

Regarding stress tolerance studies, recently, the roles of GmHsp90A2 (Heat shock protein 90s) in soybean were investigated by the use of stable transgenic lines overexpressing *GmHsp90A2* and mutant lines generated by the CRISPR/Cas9-mediated genome editing system (Huang et al. 2019). The Hsp90s is a conserved and abundant molecular chaperone active in the protective stress response; and in a previous work the GmHsp90A2 overexpression in the model *Arabidopsis thaliana* resulted in increased tolerance to heat stress (Xu et al. 2013). In Huang et al. (2019) work the CRISPR/Cas9 knockout of the endogenous soybean *GmHsp90A2* gene was confirmed up to the T1 generation. In overexpressing GmHsp90A2 soybean plants they found less heat stress-mediated damage (decreases chlorophyll loss and less MDA accumulation) and maintenance of plant growth, while in CRISPR/Cas9 knockout plants there found opposite effects, an accelerated chlorophyll loss and lipid peroxidation.

Even though only few scientific reports on the use of genome editing for targeted and precise editing for abiotic stress tolerance in Fabaceae could be found, the potential of these methodologies is enormous for future utilization in legume molecular breeding.

5 Conclusions

Fabaceae plants are responsible for providing the main source of vegetable proteins for human diet and are essential for livestock feed; however, legume productivity is compromised by different abiotic constraints. Grain and forage legumes are very sensitive to abiotic stresses, especially during the reproductive period (during flowering and pod filling), and the establishment and functionality of Legume-Rhizobium symbiosis is also critically affected. Besides, under abiotic stress situations, legumes become more vulnerable to plant pathogens (such as fungi, bacteria, viruses, nematodes, insects, and mites). As a consequence, there is a reduction in plant yield, decreased productivity, and also the decline of soil nitrogen fixation.

Implementation of molecular tools in breeding programs will facilitate the development of tolerant varieties to sustain food security under abiotic stress conditions. Consequently, research to understand the complex molecular mechanisms of plant responses to abiotic stress is essential to develop stress-tolerant crops using genetic engineering or genome editing technologies. The identification and further study of stress-responsive candidates, based on high-throughput omics methodologies, will not only contribute to increase the knowledge on stress response mechanisms, but will also allow the identification of future candidates for genetic engineering or genome editing purposes.

In this chapter, we presented an overview of the recent applications of genetic engineering and genome editing for the improvement of legumes toward abiotic stress tolerance. We highlighted several examples of the use of stress-related genes, including transcriptional factors, for genetic engineering of model and important grain and forage legumes. We have also focused on the application of miRNAs posttranscriptional gene regulation mechanism for crop improvement. Indeed, miRNA-based biotechnology could be an innovative and efficient approach for Fabaceae genetic engineering. Although only a small numbers of studies have adopted genome editing technology for improving abiotic stress in legumes, there are an enormous potential in these new molecular breeding approaches (specially the CRISPR/Cas9 tool) for an efficient targeted precise mutagenesis, and for the development of legume cultivars with increased tolerance to abiotic stresses.

Acknowledgements Financial support from FCT (Fundação para a Ciência e Tecnologia, Lisbon, Portugal) is acknowledged through the research unit "GREEN-it: Bioresources for Sustainability" (UID/Multi/04551/2019), and through ASD Ph.D. holders DL57 research contracts. We also acknowledge the Program for Scientific Cooperation FCT/Poland 2017/2018. EV thanks the post-doc fellowship in the frame of FCT funded project PTDC/AGR-FOR/0931/2014.

References

Abdelrahman M, Jogaiah S, Burritt DJ, Tran L-SP (2018) Legume genetic resources and transcriptome dynamics under abiotic stress conditions. Plant, Cell Environ 41:1972–1983. https://doi.org/10.1111/pce.13123

Agarwal PK, Jha B (2010) Transcription factors in plants and ABA dependent and independent abiotic stress signalling. Biol Plant 54:201–212. https://doi.org/10.1007/s10535-010-0038-7

Agarwal P, Pradeep K, Agarwal PK, Gohil D (2018) Transcription factor-based genetic engineering for salinity tolerance in crops. In: Kumar V, Wani SH, Suprasanna P, Tran L-SP (eds) Salitity responses and tolerance in plants: targeting sensory, transport and signaling mechanisms. Springer International Publishing AG, Cham

Alcázar R, Planas J, Saxena T, Zarza X, Bortolotti C, Cuevas J, Bitrián M, Tiburcio AF, Altabella T (2010) Putrescine accumulation confers drought tolerance in transgenic Arabidopsis plants overexpressing the homologous Arginine decarboxylase 2 gene. Plant Physiol Biochem 48:547–552. https://doi.org/10.1016/j.plaphy.2010.02.002

Alet AI, Sanchez DH, Cuevas JC, del Valle S, Altabella T, Tiburcio AF, Marco F, Ferrando A, Espasandín FD, González ME, Carrasco P, Ruiz OA (2011) Putrescine accumulation in *Arabidopsis thaliana* transgenic lines enhances tolerance to dehydration and freezing stress. Plant Signal Behav 6:278–286. https://doi.org/10.4161/psb.6.2.14702

Alzahrani SM, Alaraidh IA, Khan MA, Migdadi HM, Alghamdi SS, Alsahli AA (2019) Identification and characterization of salt-responsive MicroRNAs in *Vicia faba* by high-throughput sequencing. Genes 10:303. https://doi.org/10.3390/genes10040303

Alptekin B, Akpinar B, Budak H (2017) A comprehensive prescription for plant miRNA identification. Front Plant Sci 7:2058. https://doi.org/10.3389/fpls.2016.02058

Almeida DM, Margarida Oliveira M, Saibo NJM (2017) Regulation of Na+ and K+ homeostasis in plants: towards improved salt stress tolerance in crop plants. Genet Mol Biol 40:326–345. https://doi.org/10.1590/1678-4685-gmb-2016-0106

An J, Hu Z, Che B, Chen H, Yu B, Cai W (2017) Heterologous expression of *Panax ginseng* PgTIP1 confers enhanced salt tolerance of soybean cotyledon hairy roots, composite, and whole plants. Front Plant Sci 8:1232. https://doi.org/10.3389/fpls.2017.01232

An J, Cheng C, Hu Z, Chen H, Cai W, Yu B (2018) The *Panax ginseng* PgTIP1 gene confers enhanced salt and drought tolerance to transgenic soybean plants by maintaining homeostasis of water, salt ions and ROS. Environ Exp Bot 155:45–55. https://doi.org/10.1016/j.envexpbot.2018.06.025

Anbazhagan K, Bhatnagar-Mathur P, Vadez V, Dumbala SR, Kishor PBK, Sharma KK (2014) DREB1A overexpression in transgenic chickpea alters key traits influencing plant water budget across water regimes. Plant Cell Rep 34:199–210. https://doi.org/10.1007/s00299-014-1699-z

Araújo SS, Beebe S, Crespi M, Delbreil B, González EM, Gruber V, Lejeune-Henaut I, Link W, Monteros MJ, Prats E, Rao I, Vadez V, Vaz Patto MC (2015) Abiotic stress responses in legumes: strategies used to cope with environmental challenges. Crit Rev Plant Sci 34(1–3):237–280

Araújo SS, Wendt dos Santos AL, Duque AS (2019) Engineering polyamines metabolic pathways for abiotic stress tolerance in plants, In: Hossain MA, Kumar V, Burritt D, Fujita M, Mäkelä PSA (eds) Osmoprotectant-mediated abiotic stress tolerance in plants. Springer International Publishing Cham. https://doi.org/10.1007/978-3-030-27423-8_14

Arshad M, Feyissa BA, Amyot L, Aung B, Hannoufa A (2017) MicroRNA156 improves drought stress tolerance in alfalfa (*Medicago sativa*) by silencing SPL13. Plant Sci 258:122–136. https://doi.org/10.1016/j.plantsci.2017.01.018

Aung B, Gruber MY, Amyot L, Omari K, Bertrand A, Hannoufa A (2015) MicroRNA156 as a promising tool for alfalfa improvement. Plant Biotechnol J 13:779–790. https://doi.org/10.1111/pbi.12308

Baloda A, Madanpotra S, Aiwal PK (2017) Transformation of mungbean plants for salt and drought tolerance by introducing a gene for an osmoprotectant glycine betaine. J Plant Stress Physiol 3:5. https://doi.org/10.19071/jpsp.2017.v3.3148

Banavath JN, Chakradhar T, Pandit V, Konduru S, Guduru KK, Akila CS, Podha S, Puli COR (2018) Stress inducible overexpression of AtHDG11 leads to improved drought and salt stress tolerance in peanut (*Arachis hypogaea* L.). Front Chem 6:34. https://doi.org/10.3389/fchem.2018.00034

Bao A, Chen H, Chen L, Chen S, Hao Q, Guo W, Qiu D, Shan Z, Yang Z, Yuan S, Zhang C, Zhang X, Liu B, Kong F, Li X, Zhou X, Tran LSP, Cao D (2019) CRISPR/Cas9-mediated targeted mutagenesis of GmSPL9 genes alters plant architecture in soybean. BMC Plant Biol 19:1–12. https://doi.org/10.1186/s12870-019-1746-6

Bao AK, Du BQ, Touil L, Kang P, Wang QL, Wang SM (2016) Co-expression of tonoplast Cation/H+ antiporter and H+ -pyrophosphatase from xerophyte *Zygophyllum xanthoxylum* improves alfalfa plant growth under salinity, drought and field conditions. Plant Biotechnol J 14:964–975. https://doi.org/10.1111/pbi.12451

Bartel DP (2004) MicroRNAs: genomics, biogenesis, mechanism, and function. Cell 116:281–297. https://doi.org/10.1016/s0092-8674(04)00045-5

Bashir W, Anwar S, Zhao Q, Hussain I, Xie F (2019) Interactive effect of drought and cadmium stress on soybean root morphology and gene expression. Ecotoxicol Environ Saf 175:90–101. https://doi.org/10.1016/j.ecoenv.2019.03.042

Budak H, Zhang B (2017) MicroRNAs in model and complex organisms. Funct Integr Genomics 17:121–124. https://doi.org/10.1007/s10142-017-0544-1

Cai Y, Chen L, Liu X, Guo C, Sun S, Wu C, Jiang B, Han T, Hou W (2018) CRISPR/Cas9-mediated targeted mutagenesis of GmFT2a delays flowering time in soya bean. Plant Biotechnol J 16:176–185. https://doi.org/10.1111/pbi.12758

Cai Y, Chen L, Liu X, Sun S, Wu C, Jiang B, Han T, Hou W (2015) CRISPR/Cas9-mediated genome editing in soybean hairy roots. PLoS ONE 10:1–13. https://doi.org/10.1371/journal.pone.0136064

Cao C, Long R, Zhang T, Kang J, Wang Z, Wang P, Sun H, Yu J, Yang Q (2018) Genome-wide identification of microRNAs in response to salt/alkali stress in *Medicago truncatula* through high-throughput sequencing. Int J Mol Sci 19:4076. https://doi.org/10.3390/ijms19124076

Chauhan S, Yogindran S, Rajam MV (2017) Role of miRNAs in biotic stress reactions in plants. Ind J Plant Physiol 22:514–529. https://doi.org/10.1007/s40502-017-0347-3

Chew W, Hrmova M, Lopato S (2013) Role of homeodomain leucine zipper (HD-Zip) iv transcription factors in plant development and plant protection from deleterious environmental factors. Int J Mol Sci 14:8122–8147. https://doi.org/10.3390/ijms14048122

Cong L, Ran FA, Cox D, Lin S, Barretto R, Habib N, Hsu PD, Wu X, Jiang W, Marraffini LA, Zhang F (2013) Multiplex genome engineering using CRISPR/Cas systems. Sci 339 (6121):819–823. https://doi.org/10.1126/science.1231143

Cromwell CR, Sung K, Park J, Krysler AR, Jovel J, Kim SK, Hubbard BP (2018) Incorporation of bridged nucleic acids into CRISPR RNAs improves Cas9 endonuclease specificity. Nat Commun 9:1–11. https://doi.org/10.1038/s41467-018-03927-0

Curtin SJ, Xiong Y, Michno JM, Campbell BW, Stec AO, Čermák T, Starker C, Voytas DF, Eamens AL, Stupar RM (2018) CRISPR/Cas9 and TALENs generate heritable mutations for genes involved in small RNA processing of *Glycine max and Medicago truncatula*. Plant Biotechnol J 16:1125–1137. https://doi.org/10.1111/pbi.12857

Debbarma J, Sarki YN, Saikia B, Boruah HPD, Singha DL, Chikkaputtaiah C (2019) Ethylene response factor (ERF) family proteins in abiotic stresses and CRISPR–Cas9 genome editing of ERFs for multiple abiotic stress tolerance in crop plants: a review. Mol Biotechnol 61:153–172. https://doi.org/10.1007/s12033-018-0144-x

Decottignies A (2013) Alternative end-joining mechanisms: a historical perspective. Front Genet 4:1–7. https://doi.org/10.3389/fgene.2013.00048

Deokar AA, Kondawar V, Kohli D, Aslam M, Jain PK, Karuppayil SM, Varshney RK, Srinivasan R (2015) The CarERF genes in chickpea (Cicer arietinum L.) and the identification of CarERF116 as abiotic stress responsive transcription factor. Funct Integr Genomics 15(1):27–46. https://doi.org/10.1007/s10142-014-0399-7

Deshmukh R, Sonah H, Patil G, Chen W, Prince S, Mutava R, Vuong T, Valliyodan B, Nguyen HT (2014) Integrating omic approaches for abiotic stress tolerance in soybean. Front Plant Sci 5:244. https://doi.org/10.3389/fpls.2014.00244

Ding J, Zhou S, Guan J (2012) Finding MicroRNA targets in plants: current status and perspectives. Genomics Proteomics Bioinform 10:264–275. https://doi.org/10.1016/j.gpb.2012.09.003

Do TD, Chen H, Hien VTT, Hamwieh A, Yamada T, Sato T, Yan Y, Cong H, Shono M, Suenaga K, Xu D (2016) Ncl synchronously regulates Na+, K+, and Cl- in soybean and greatly increases the grain yield in saline field conditions. Sci Rep 6:19147. https://doi.org/10.1038/srep19147

Donald CM (1962) In search of yield. J Aust Inst Agric Sci 28:171–178

Duque AS, de Almeida AM, da Silva AB, da Silva JM, Farinha AP, Santos D, Fevereiro P, Araújo SS (2013) Abiotic stress responses in plants : unraveling the complexity of genes and networks to survive. In: Vahdati K, Leslie C (eds) Abiotic stress—plant responses and applications in agriculture as (InTech). Rijeka, pp 49–101. https://doi.org/10.5772/52779

Duque AS, López-Gómez M, Kráčmarová J, Gomes CN, Araújo SS, Lluch C, Fevereiro P (2016) Genetic engineering of polyamine metabolism changes *Medicago truncatula* responses to water deficit. Plant Cell, Tissue Organ Cult 127:681–690. https://doi.org/10.1007/s11240-016-1107-1

Eckerstorfer MF, Engelhard M, Heissenberger A, Simon S, Teichmann H (2019) Plants developed by new genetic modification techniques-comparison of existing regulatory frameworks in the EU and Non-EU countries. Front Bioeng Biotechnol 7. https://doi.org/10.3389/fbioe.2019.00026

Ellis RJ (1991) Molecular chaperones interesting. Annu Rev Biochem 60:321–347

Espasandin FD, Calzadilla PI, Maiale SJ, Ruiz OA, Sansberro PA (2018) Overexpression of the arginine decarboxylase gene improves tolerance to salt stress in lotus tenuis plants. J Plant Growth Regul 37:156–165. https://doi.org/10.1007/s00344-017-9713-7

Espasandin FD, Maiale SJ, Calzadilla P, Ruiz OA, Sansberro PA (2014) Transcriptional regulation of 9-cis-epoxycarotenoid dioxygenase (NCED) gene by putrescine accumulation positively modulates ABA synthesis and drought tolerance in *Lotus tenuis* plants. Plant Physiol Biochem 76:29–35. https://doi.org/10.1016/j.plaphy.2013.12.018

Fariduddin Q, Varshney P, Yusuf M, Ali A, Ahmad A (2013) Dissecting the role of glycine betaine in plants under abiotic stress. Plant Stress 7:8–18

Fujita Y, Fujita M, Shinozaki K, Yamaguchi-Shinozaki K (2011) ABA-mediated transcriptional regulation in response to osmotic stress in plants. J Plant Res 124:509–525. https://doi.org/10.1007/s10265-011-0412-3

Gaj T, Gersbach CA, Barbas CF (2013) ZFN, TALEN, and CRISPR/Cas-based methods for genome engineering. Trends Biotechnol 31:397–405. https://doi.org/10.1016/j.tibtech.2013.04.004

Gomes C, Oliveira F, Vieira SI, Duque AS (2019) Prospects for the production of recombinant therapeutic proteins and peptides in plants: Special focus on angiotensin I-converting enzyme. In: Jamal F (ed) Genetic engineering—an insight into the strategies and applications (InTech). https://doi.org/10.5772/intecopen.84419

Guo Q, Liu Q, Smith NA, Liang G, Wang MB (2016) RNA silencing in plants: mechanisms, technologies and applications in horticultural crops. Curr Genomics 17(6):476–489. https://doi.org/10.2174/1389202917666160520103

Hajyzadeh M, Turktas M, Khawar KM, Unver T (2015) miR408 overexpression causes increased drought tolerance in chickpea. Gene 555:186–193. https://doi.org/10.1016/j.gene.2014.11.002

Hayashi H, Alla Mustardy L, Deshnium P, Ida M, Murata N (1997) Transformation of *Arabidopsis thaliana* with the codA gene for choline oxidase; accumulation of glycinebetaine and enhanced tolerance to salt and cold stress. Plant J 12:133–142. https://doi.org/10.1046/j.1365-313x.1997.12010133.x

He P-M, Zhang D-B, Liang W-Q, Yao Q-H, Zhang R-X (2001) Expression of choline oxidase gene (codA) enhances salt tolerance of the tobacco. Sheng Wu Hua Xue Yu Sheng Wu Wu Li Xue Bao (Shanghai) 33:519–524

Hidalgo-Castellanos J, Duque AS, Burgueño A, Herrera-Cervera JA, Fevereiro P, López-Gómez M (2019) Overexpression of the arginine decarboxylase gene promotes the symbiotic interaction *Medicago truncatula-Sinorhizobium meliloti* and induces the accumulation of proline and spermine in nodules under salt stress conditions. J Plant Physicol 241:153034. https://doi.org/10.1016/j.jplph.2019.153034

Hirai S, Kodama H (2008) RNAi vectors for manipulation of gene expression in higher plants. Open Plant Sci J 2:21–30. https://doi.org/10.2174/1874294700802010021

Hoang XLT, Nhi DNH, Thu NBA, Thao NP, Tran L-SP (2017) Transcription factors and their roles in signal transduction in plants under abiotic stresses. Curr Genomics 18:483–497. https://doi.org/10.2174/1389202918666170227150057

Hou J, Jiang P, Qi S, Zhang K, He Q, Xu C, Ding Z, Zhang K, Li K (2016) Isolation and functional validation of salinity and osmotic stress inducible promoter from the maize type-II H+ -pyrophosphatase gene by deletion analysis in transgenic tobacco plants. PLoS ONE 11:1–23. https://doi.org/10.1371/journal.pone.0154041

Huang SC, Lu GH, Tang CY, Ji YJ, Tan GS, Hu DQ, Cheng J, Wang GH, Qi JL, Yang YH (2018) Identification and comparative analysis of aluminum-induced microRNAs conferring plant tolerance to aluminum stress in soybean. Biol Plant 62:97–108. https://doi.org/10.1007/s10535-017-0752-5

Huang Y, Xuan H, Yang C, Guo N, Wang H, Zhao J, Xing H (2019) GmHsp90A2 is involved in soybean heat stress as a positive regulator. Plant Sci 285:26–33. https://doi.org/10.1016/j.plantsci.2019.04.016

İlhan E, Büyük İ, İnal B (2018) Transcriptome —scale characterization of salt responsive bean TCP transcription factors. Gene 642:64–73. https://doi.org/10.1016/j.gene.2017.11.021

Jha UC, Bohra A, Jha R, Parida SK (2019) Salinity stress response and 'omics' approaches for improving salinity stress tolerance in major grain legumes. Plant Cell Rep 38:255–277. https://doi.org/10.1007/s00299-019-02374-5

Jin T, Chang Q, Li W, Yin D, Li Z, Wang D, Liu B, Liu L (2010) Stress-inducible expression of GmDREB1 conferred salt tolerance in transgenic alfalfa. Plant Cell, Tissue Organ Cult 100:219–227. https://doi.org/10.1007/s11240-009-9628-5

Kamthan A, Chaudhuri A, Kamthan M, Datta A (2015) Small RNAs in plants: recent development and application for crop improvement. Front Plant Sci 6:1–17. https://doi.org/10.3389/fpls.2015.00208

Khandagale K, Nadaf AB (2016) Genome editing for targeted improvement of plants. Plant Biotechnol Rep 10:327–343

Khandal H, Parween S, Roy R, Meena MK, Chattopadhyay D (2017) MicroRNA profiling provides insights into post-transcriptional regulation of gene expression in chickpea root apex under salinity and water deficiency. Sci Rep. https://doi.org/10.1038/s41598-017-04906-z

Kapilan R, Vaziri M, Zwiazek JJ (2018) Regulation of aquaporins in plants under stress. Biol Res 51:1–11. https://doi.org/10.1186/s40659-018-0152-0

Kasuga M, Miura S, Shinozaki K, Yamaguchi-Shinozaki K (2004) A combination of the Arabidopsis DREB1A gene and stress-inducible rd29A promoter improved drought- and low-temperature stress tolerance in tobacco by gene transfer. Plant Cell Physiol 45:346–350. https://doi.org/10.1093/pcp/pch037

Kim HJ, Cho HS, Pak JH, Kwon T, Lee J-H, Kim D-H, Lee DH, Kim C-G, Chung Y-S (2018) Confirmation of drought tolerance of ectopically expressed AtABF3 gene in soybean. Mol Cells 41:413. https://doi.org/10.14348/molcells.2018.2254

Kim YE, Hipp MS, Bracher A, Hayer-Hartl M, Ulrich Hartl F (2013) Molecular chaperone functions in protein folding and proteostasis> Annu Rev Biochem. https://doi.org/doi:10.1146/annurev-biochem-060208-092442

Kim YH, Kim MD, Choi YI, Park SC, Yun DJ, Noh EW, Lee HS, Kwak SS (2011) Transgenic poplar expressing Arabidopsis NDPK2 enhances growth as well as oxidative stress tolerance. Plant Biotechnol J 9:334–347. https://doi.org/10.1111/j.1467-7652.2010.00551.x

Kudapa H, Ramalingam A, Nayakoti S, Chen X, Zhuang W-J, Liang X, Kahl G, Edwards D, Varshney RK (2013) Functional genomics to study stress responses in crop legumes: progress and prospects. Funct Plant Biol 40:1221. https://doi.org/10.1071/fp13191

Li D, Zhang Y, Hu X, Shen X, Ma L, Su Z, Wang T, Dong J (2011) Transcriptional profiling of *Medicago truncatula* under salt stress identified a novel CBF transcription factor MtCBF4 that plays an important role in abiotic stress responses. BMC Plant Biol 11. https://doi.org/10.1186/1471-2229-11-109

Li H, Wang Z, Ke Q, Ji CY, Jeong JC, Lee HS, Lim YP, Xu B, Deng XP, Kwak SS (2014) Over-expression of codA gene confers enhanced tolerance to abiotic stresses in alfalfa. Plant Physiol Biochem. https://doi.org/10.1016/j.plaphy.2014.10.010

Li Y, Chen Q, Nan H, Li X, Lu S, Zhao X, Liu B, Guo C, Kong F, Cao D (2017) Overexpression of GmFDL19 enhances tolerance to drought and salt stresses in soybean. PLoS ONE 12:e0179554. https://doi.org/10.1371/journal.pone.0179554

Lilius G, Holmberg N, Bülow L (1996) Enhanced NaCl stress tolerance in transgenic tobacco expressing bacterial choline dehydrogenase. Nat Biotechnol 14:177–180. https://doi.org/10.1038/nbt0296-177

Liu L, Zhang Z, Dong J, Wang T (2016) Overexpression of MtWRKY76 increases both salt and drought tolerance in *Medicago truncatula*. Environ Exp Bot 123:50–58. https://doi.org/10.1016/j.envexpbot.2015.10.007

Liu X, Xie C, Si H, Yang J (2017) CRISPR/Cas9-mediated genome editing in plants. Methods 121–122:94–102. https://doi.org/10.1016/j.ymeth.2017.03.009

Liu H, Yu H, Tang G, Huang T (2018) Small but powerful: function of microRNAs in plant development. Plant Cell Rep 37:515–528. https://doi.org/10.1007/s00299-017-2246-5

Luo D, Zhou Q, Wu Y, Chai X, Liu W, Wang Y, Yang Q, Wang Z, Liu Z (2019) Full-length transcript sequencing and comparative transcriptomic analysis to evaluate the contribution of osmotic and ionic stress components towards salinity tolerance in the roots of cultivated alfalfa (*Medicago sativa* L.). BMC Plant Biol 19:32. https://doi.org/10.1186/s12870-019-1630-4

Mano J (2012) Reactive carbonyl species: their production from lipid peroxides, action in environmental stress, and the detoxification mechanism. Plant Physiol Biochem 59:90–97. https://doi.org/10.1016/j.plaphy.2012.03.010

Maphosa Y and Jideani VA (2017) The role of legumes in human nutrition. In Hueda MC (ed) Nutrition in functional food- improve health through adequate food (InTech). Rijeka, pp 103–121. http://dx.doi.org/10.5772/intechopen.69127

Mol JNM, van der Krol AR, van Tunen AJ, van Blokland R, de Lange P, Stuitje AR (1990) Regulation of plant gene expression by antisense RNA. FEBS Lett 268:427–430. https://doi.org/10.1016/0014-5793(90)81298-3

Nadeem M, Li J, Yahya M, Sher A, Ma C, Wang X, Qiu L, Nadeem M, Li J, Yahya M, Sher A, Ma C, Wang X, Qiu L (2019) Research progress and perspective on drought stress in legumes: a review. Int J Mol Sci 20:2541. https://doi.org/10.3390/ijms20102541

Pan JW, Tao JJ, Cheng TC, Bian XH, Wei W, Zhang WK, Ma B, Chen SY, Zhang JS (2016) Soybean miR172a improves salt tolerance and can function as a long-distance signal. Mol Plant 9:1337–1340. https://doi.org/10.1016/j.molp.2016.05.010

Pareek M, Yogindran S, Mukherjee SK, Rajam MV (2015) Plant MicroRNAs: biogenesis, functions, and applications. In: Bahadur B et al (eds) Plant biology and biotechnology. Plant genomics and biotechnology. Springer India, New Delhi, pp 639–661

Park CJ, Seo YS (2015) Heat shock proteins: A review of the molecular chaperones for plant immunity. Plant Pathol J 31:323–333. https://doi.org/10.5423/ppj.rw.08.2015.0150

Park MY, Wu G, Gonzalez-Sulser A, Vaucheret H, Poethig RS (2005) Nuclear processing and export of microRNAs in arabidopsis. Proc Natl Acad Sci 102:3691–3696. https://doi.org/10.1073/pnas.0405570102

Puchta H (2005) The repair of double-strand breaks in plants: mechanisms and consequences for genome evolution. J Exp Bot 56:1–14. https://doi.org/10.1093/jxb/eri025

Purushothaman R, Krishnamurthy L, Upadhyaya HD, Vadez V, Varshney RK (2017) Genotypic variation in soil water use and root distribution and their implications for drought tolerance in chickpea. Funct Plant Biol 44:235–252. https://doi.org/10.1071/fp16154

Ramalingam A, Kudapa H, Pazhamala LT, Weckwerth W, Varshney RK (2015) Proteomics and metabolomics: two emerging areas for legume improvement. Front Plant Sci 6:1–21. https://doi.org/10.3389/fpls.2015.01116

Rogers K, Chen X (2013) Biogenesis, turnover, and mode of action of plant MicroRNAs. Plant Cell 25:2383–2399. https://doi.org/10.1105/tpc.113.113159

Sahito ZA, Wang L, Sun Z, Yan Q, Zhang X, Jiang Q, Ullah I, Tong Y, Li X (2017) The miR172c-NNC1 module modulates root plastic development in response to salt in soybean. BMC Plant Biol. https://doi.org/10.1186/s12870-017-1161-9

Sakamoto A, Valverde R, Alia Chen TH, Murata N (2000) Transformation of arabidopsis with the codA gene for choline oxidase enhances freezing tolerance of plants. Plant J 22:449–453. https://doi.org/10.1046/j.1365-313x.2000.00749.x

Sanei M, Chen X (2015) Mechanisms of microRNA turnover. Curr Opin Plant Biol 27:199–206. https://doi.org/10.1016/j.pbi.2015.07.008

Sańko-Sawczenko I, Łotocka B, Mielecki J, Rekosz-Burlaga H, Czarnocka W (2019) Transcriptomic changes in *Medicago truncatula* and *Lotus japonicus* root nodules during drought stress. Int J Mol Sci 20:1204. https://doi.org/10.3390/ijms20051204

Sarkar T, Thankappan R, Kumar A, Mishra GP, Dobaria JR (2014) Heterologous expression of the AtDREB1A gene in transgenic peanut-conferred tolerance to drought and salinity stresses. PLoS ONE 9:e110507. https://doi.org/10.1371/journal.pone.0110507

Saurabh S, Vidyarthi AS, Prasad D (2014) RNA interference: concept to reality in crop improvement. Planta 239:543–564. https://doi.org/10.1007/s00425-013-2019-5

Schmidt C, Pacher M, Puchta H (2019) DNA break repair in plants and its application for genome engineering. In: Kumar S, Barone P, Smith M (eds) Transgenic plants. Methods in molecular biology, vol 1864. Humana Press, New York, NY

Sen S, Chakraborty J, Ghosh P, Basu D, Das S (2017) Chickpea WRKY70 regulates the expression of a homeodomain-leucine zipper (HD-Zip) I transcription factor CaHDZ12, which confers abiotic stress tolerance in transgenic tobacco and chickpea. Plant Cell Physiol 58:1934–1952. https://doi.org/10.1093/pcp/pcx126

Shekoofa A, Sinclair T (2018) Aquaporin activity to improve crop drought tolerance. Cells 7:123. https://doi.org/10.3390/cells7090123

Shriram V, Kumar V, Devarumath RM, Khare TS, Wani SH (2016) MicroRNAs as potential targets for abiotic stress tolerance in plants. Front Plant Sci. https://doi.org/10.3389/fpls.2016.00817

Singh M, Kumar J, Singh S, Singh VP, Prasad SM (2015) Roles of osmoprotectants in improving salinity and drought tolerance in plants: a review. Rev Environ Sci Biotechnol 14:407–426. https://doi.org/10.1007/s11157-015-9372-8

Singh N, Mishra A, Jha B (2014a) Ectopic over-expression of peroxisomal ascorbate peroxidase (SbpAPX) gene confers salt stress tolerance in transgenic peanut *(Arachis hypogaea)*. Gene 547:119–125. https://doi.org/10.1016/j.gene.2014.06.037

Singh N, Mishra A, Jha B (2014b) Over-expression of the peroxisomal ascorbate peroxidase (SbpAPX) gene cloned from halophyte *Salicornia brachiata* confers salt and drought stress tolerance in transgenic tobacco. Mar Biotechnol 16:321–332. https://doi.org/10.1007/s10126-013-9548-6

Singh P, Kumar D, Sarin NB (2016) Multiple abiotic stress tolerance in *Vigna mungo* is altered by overexpression of ALDRXV4 gene via reactive carbonyl detoxification. Plant Mol Biol 91:257–273. https://doi.org/10.1007/s11103-016-0464-9

Statement on the State of the Global Climate in 2018 (2019) World Meteorological Organization WMO-No. 1233, ISBN 978-92-63-11233-0

Stritzler M, Elba P, Berini C, Gomez C, Ayub N, Soto G (2018) High-quality forage production under salinity by using a salt-tolerant AtNXH1-expressing transgenic alfalfa combined with a natural stress-resistant nitrogen-fixing bacterium. J Biotechnol 276–277:42–45. https://doi.org/10.1016/j.jbiotec.2018.04.013

Sun L, Hu R, Shen G, Zhang H (2013) Genetic engineering peanut for higher drought- and salt-tolerance. Food Nutr Sci 04:1–7. https://doi.org/10.4236/fns.2013.46a001

Sun Z, Wang Y, Mou F, Tian Y, Chen L, Zhang S, Jiang Q, Li X (2016) Genome-wide small RNA analysis of soybean reveals auxin-responsive microRNAs that are differentially expressed in response to salt stress in root apex. Front Plant Sci. https://doi.org/10.3389/fpls.2015.012

Tang L, Cai H, Ji W, Luo X, Wang Z, Wu J, Wang X, Cui L, Wang Y, Zhu Y, Bai X (2013) Overexpression of GsZFP1 enhances salt and drought tolerance in transgenic alfalfa (*Medicago sativa* L.). Plant Physiol Biochem 71:22–30. https://doi.org/10.1016/j.plaphy.2013.06.024

Tang X-D, Gao F, Liu M-J, Fan Q-L, Chen D-K, Ma W-T (2019) Methods for enhancing clustered regularly interspaced short palindromic repeats/cas9-mediated homology-directed repair efficiency. Frontiers Genetics 10:551. https://doi.org/10.3389/fgene.2019.00551

Trivedi DK, Huda KMK, Gill SS, Tuteja N (2016) Molecular chaperone: structure, function, and role in plant abiotic stress tolerance. Abiotic Stress Response Plants 135–154. https://doi.org/10.1002/9783527694570.ch7

Uno Y, Furihata T, Abe H, Yoshida R, Shinozaki K, Yamaguchi-Shinozaki K (2000) Arabidopsis basic leucine zipper transcription factors involved in an abscisic acid-dependent signal transduction pathway under drought and high-salinity conditions. Proc Natl Acad Sci U S A 97:11632–11637. https://doi.org/10.1073/pnas.190309197

Wang JY, Lai LD, Tang SM, Li QL (2013) Constitutive and salt-inducible expression of SlBADH gene in transgenic tomato (*Solanum lycopersicum* L. cv. Micro-Tom) enhances salt tolerance. Biochem Biophys Res Commun 432:262–267. https://doi.org/10.1016/j.bbrc.2013.02.001

Wang A, Hu J, Gao C, Chen G, Wang B, Lin C, Song L, Ding Y, Zhou G (2019) Genome-wide analysis of long non-coding RNAs unveils the regulatory roles in the heat tolerance of Chinese cabbage (*Brassica rapa* ssp. chinensis). Sci Rep. https://doi.org/10.1038/s41598-019-41428-2

Wang L, Liu L, Ma Y, Li S, Dong S, Zu W (2018a) Transcriptome profilling analysis characterized the gene expression patterns responded to combined drought and heat stresses in soybean. Comput Biol Chem 77:413–429. https://doi.org/10.1016/j.compbiolchem.2018.09.012

Wang X, Sakata K, Komatsu S (2018b) An integrated approach of proteomics and computational genetic modification effectiveness analysis to uncover the mechanisms of flood tolerance in soybeans. Int J Mol Sci 19. https://doi.org/10.3390/ijms19051301

Wang W, Liu D, Zhang X, Chen D, Cheng Y, Shen F (2018c) Plant MicroRNAs in cross-kingdom regulation of gene expression. Int J Mol Sci 19:2007. https://doi.org/10.3390/ijms19072007

Wang N, Zhang W, Qin M, Li S, Qiao M, Liu Z, Xiang F (2017) Drought tolerance conferred in soybean (*Glycine max*. L) by GmMYB84, a novel R2R3-MYB transcription factor. Plant Cell Physiol 58:1764–1776. https://doi.org/10.1093/pcp/pcx111

Wang Z, Ke Q, Kim MD, Kim SH, Ji CY, Jeong JC, Lee HS, Park WS, Ahn MJ, Li H, Xu B, Deng X, Lee SH, Lim YP, Kwak SS (2015) Transgenic alfalfa plants expressing the sweetpotato orange gene exhibit enhanced abiotic stress tolerance. PLoS One 10. https://doi.org/10.1371/journal.pone.0126050

Wang Z, Li H, Ke Q, Jeong JC, Lee HS, Xu B, Deng XP, Lim YP, Kwak SS (2014) Transgenic alfalfa plants expressing AtNDPK2 exhibit increased growth and tolerance to abiotic stresses. Plant Physiol Biochem. https://doi.org/10.1016/j.plaphy.2014.08.025

Wang Z, Su G, Li M, Ke Q, Kim SY, Li H, Huang J, Xu B, Deng XP, Kwak SS (2016) Overexpressing Arabidopsis ABF3 increases tolerance to multiple abiotic stresses and reduces leaf size in alfalfa. Plant Physiol Biochem. https://doi.org/10.1016/j.plaphy.2016.09.020

Wright DA, Li T, Yang B, Spalding MH (2014) TALEN-mediated genome editing: prospects and perspectives. Biochem J 462:15–24. https://doi.org/10.1042/bj20140295

Wu XX, Li J, Wu XD, Liu Q, Wang ZK, Liu SS, Li SN, Ma YL, Sun J, Zhao L, Li HY, Li DM, Li WB, Su AY (2016) Ectopic expression of Arabidopsis thaliana Na+(K +)/H+ antiporter gene, AtNHX5, enhances soybean salt tolerance. Genet Mol Res 15. https://doi.org/10.4238/gmr.15027483

Xu J, Hou Q-M, Khare T, Verma SK, Kumar V (2019) Exploring miRNAs for developing climate-resilient crops: a perspective review. Sci Total Environ 653:91–104. https://doi.org/10.1016/j.scitotenv.2018.10.340

Xu J, Xue C, Xue D, Zhao J, Gai J, Guo N, Xing H (2013) Overexpression of GmHsp90s, a heat shock protein 90 (Hsp90) gene family cloning from soybean, decrease damage of abiotic stresses in *Arabidopsis thaliana*. PLoS ONE 8:1–14. https://doi.org/10.1371/journal.pone.0069810

Xu S, Liu N, Mao W, Hu Q, Wang G, Gong Y (2016) Identification of chilling-responsive microRNAs and their targets in vegetable soybean (*Glycine max* L.). Sci Rep. https://doi.org/10.1038/srep26619

Xu Z-S, Chen M, Li L-C, Ma Y-Z (2011) Functions and application of the AP2/ERF transcription factor family in crop improvement. J Integr Plant Biol 53:570–585. https://doi.org/10.1111/j.1744-7909.2011.01062.x

Young JM, Kuykendall LD, Martínez-Romero E, Kerr A, Sawada H (2001) A revision of Rhizobium Frank 1889, with an emended description of the genus, and the inclusion of all species of Agrobacterium conn 1942 and Allorhizobium undicola de Lajudie et al. 1998 as new combinations: R*hizobium radiobacter, R. rhizogenes, R. rubi*. Int J Syst Evol Microbiol 51:89–103. https://doi.org/10.1099/00207713-51-1-89

Yu B, Yang ZY, Li JJ, Minakhina S, Yang MC, Padgett RW, Steward R, Chen XM (2005) Methylation as a crucial step in plant microRNA biogenesis. Science 307:932–935. https://doi.org/10.1126/science.1107130

Zhang B, Pan X, Cobb G, Anderson T (2006) Plant microRNA: a small regulatory molecule with big impact. Dev Biol 289:3–16. https://doi.org/10.1016/j.ydbio.2005.10.036

Zhang D, Song H, Cheng H, Hao D, Wang H, Kan G, Jin H, Yu D (2014a) The acid phosphatase-encoding gene GmACP1 contributes to soybean tolerance to low-phosphorus stress. PLoS Genet 10:e1004061. https://doi.org/10.1371/journal.pgen.1004061

Zhang L, Niu Y, Huridu H, Hao J, Qi Z, Hasi A (2014b) Salicornia europaea L. Na+/H+ antiporter gene improves salt tolerance in transgenic alfalfa (*Medicago sativa* L.) The SeNHX1 gene improves salt tolerance in transgenic alfalfa. Genet Mol Res 13:5350–5360. https://doi.org/10.4238/2014.July.24.14

Zhang DY, Kumar M, Xu L, Wan Q, Huang YH, Xu ZL, He XL, Ma JB, Pandey GK, Shao HB (2017) Genome-wide identification of Major Intrinsic Proteins in *Glycine soja* and characterization of GmTIP2;1 function under salt and water stress. Sci Rep 7. https://doi.org/10.1038/s41598-017-04253-z

Zhang WJ, Wang T (2015) Enhanced salt tolerance of alfalfa (*Medicago sativa*) by rstB gene transformation. Plant Sci 234:110–118. https://doi.org/10.1016/j.plantsci.2014.11.016
Zhang J, Duan Z, Zhang D, Zhang J, Di H, Wu F, Wang Y (2016) Co-transforming bar and CsLEA enhanced tolerance to drought and salt stress in transgenic alfalfa (*Medicago sativa* L.). Biochem Biophys Res Commun 472:75–82. https://doi.org/10.1016/j.bbrc.2016.02.067
Zheng G, Fan C, Di S, Wang X, Xiang C, Pang Y (2017) Over-expression of arabidopsis EDT1 gene confers drought tolerance in Alfalfa (*Medicago sativa* L.). Front Plant Sci 8:2125. https://doi.org/10.3389/fpls.2017.02125
Zheng Y, Hivrale V, Zhang X, Valliyodan B, Lelandais-Brière C, Farmer AD, May GD, Crespi M, Nguyen HT, Sunkar R (2016) Small RNA profiles in soybean primary root tips under water deficit. BMC Syst Biol. https://doi.org/10.1186/s12918-016-0374-0
Zhou M, Luo H (2013) MicroRNA-mediated gene regulation: potential applications for plant genetic engineering. Plant Mol Biol 83:59–75. https://doi.org/10.1007/s11103-013-0089-1

GWAS and Genomic Approaches in Legumes, an Expanding Toolkit for Examining Responses to Abiotic Stresses

Poornima K. Narayana and Eric J. B. von Wettberg

Abstract The early twenty-first century has seen a rapid development of new high-throughput DNA sequencing technologies and related "-omics" approaches that are rapidly revolutionizing many areas of biology. These developments have made it possible to perform previously impossible studies of the genetic basis of abiotic stress tolerance in a range of organisms. Here, we explore some of these developments with regards to previously understudied or "orphan" legume crops, many of which have a long history of providing food and nutritional security to people living in semiarid regions with limited soil fertility.

Keywords Genomics · High-throughput sequencing · Orphan crops · Grain legumes

1 Introduction

Across the biological sciences, the advent of genomics and related technologies has had immense impacts on the nature of questions asked by researchers. Here, we examine the contribution of some of these new "-omics" technologies, which when applied to DNA are sometimes variously called "next-generation sequencing" (NGS),"massively parallel sequencing," "third-generation sequencing" and "high-throughput genotyping" to our understanding of abiotic stress tolerance in legumes (e.g., Schadt et al. 2010; Moorthie et al. 2011; Reuter et al. 2015; Bleidorn 2016). We give particular attention to emerging work on crop legumes and emphasize work that is increasingly feasible in previously understudied minor legumes of the developing

P. K. Narayana
Indian Institute of Pulses Research, Kanpur, India

P. K. Narayana · E. J. B. von Wettberg
Department of Plant and Soil Sciences and Gund Institute for the Environment, University of Vermont, Burlington, VT, USA

E. J. B. von Wettberg (✉)
Mathematical Biology Laboratory, Peter the Great St. Petersburg Polytechnic University, 29 Polytechnicheskaya, 195251 St. Petersburg, Russia
e-mail: Eric.Bishop-von-Wettberg@uvm.edu

M. Hasanuzzaman et al. (eds.), *The Plant Family Fabaceae*,
https://doi.org/10.1007/978-981-15-4752-2_6

world, sometimes known as "orphan crops" (e.g., Naylor et al. 2004; Langridge and Fleury 2011; Varshney et al. 2012a, b). This revolution has already been widely reviewed (e.g., above, as well as Bauchet et al. 2019, for recent reviews specific to legumes), so we focus on shared approaches that we view as most useful to stress tolerance across legumes.

Part of our rationale for focusing on cultivated legumes is their importance to global food and nutritional security. Traditional crop breeding based on phenotypic selection can be enhanced by adopting "omics" approaches. Since legumes have long been considered the "slow runners" of the Green Revolution (Borlaug 1975), there is a perceived need to improve legume yields in challenging agroecological settings. These "-omics" approaches have led to the discovery of genes and pathways controlling specific traits and enhance selection strategies, which we will selectively review. The rapid advances in "omics" technologies provide an opportunity to generate new datasets for crop species, including those previously neglected, and improve them in the face of a rapidly changing climate and shifting global market trends (Gupta et al. 2014; Singh et al. 2015). In addition, the high-throughput genotyping technologies that are at the heart of the genomics "revolution" enable the screening of large germplasm collections to identify novel alleles from diverse sources, thus offering a major expansion in the variation available for breeding (e.g., Langridge and Fleury 2011).

In the era of advanced high-throughput technologies, we have the capacity to integrate genomics (DNA sequence variation), transcriptomics (RNA expression), proteomics (protein expression), interactomics (interactions of proteins and other molecular structures), metabolomics (cellular metabolites), lipidomics (lipids) and microbiomics (presence and abundance of associated microbes). This integration of immense datasets into a meaningful outcome of the cellular mechanisms is referred to as the multi-omics approach (e.g., Beale et al. 2016; Yugi et al. 2016). As "single-omics" approaches can be limited to correlational inference and may generally end up identifying the consequential changes rather than the causative ones, integrating the single levels to generate a global view will be increasingly important.

The pace of development of "-omics" tools and resources has been expedited by remarkable technological advances in next-generation sequencing platforms and high-throughput genotyping assays. Increasingly, the deployment of these approaches is constrained by quality of associated phenotypic data. In crop breeding, such as for abiotic stress-tolerant legumes, this is often germplasm such as landraces and wild relatives from genebanks. The extent of genetic variation present in the germplasm largely attributes to its extent of success of breeding efforts and genomics efforts to find genes underlying traits. Remarkable changes have also been and are being witnessed in downstream deployments of the genetic markers/QTLs in crop improvement programs as a result (e.g., Kumar et al. 2011). The aspects of high throughput and automation by identification of SNPs throughout the genome sequence have played important roles. Due to recently introduced sequencing-cum-genotyping methods like restriction site associated-DNA(RAD) sequencing (Davey et al. 2011, 2013), genotyping by sequencing (GBS, Elshire et al. 2011), diversity array sequencing (DArT-Seq, Grzebelus 2015) and whole-genome resequencing

(WGRS, Turner et al. 2010), a major shift has been revealed in the methods used for discovery and mapping of DNA markers.

Notably, thousands of DNA markers could be discovered and mapped in a short and increasingly cost-effective one-step process using the above referenced NGS-based methods. Consequently, we can construct high and ultrahigh density recombination maps not only for the major crop species with reference genome sequence but also for the crops where no reference genome is yet available (Poland et al. 2012). Recently available semi-/fully automated HTP genotyping systems have allowed accurate and rapid scoring of several hundreds to thousands of genetic markers. These include large-scale SNP genotyping systems like Illumina GoldenGate (GG)/Infinium (e.g., Kassa et al. 2012) and moderate-scale assays such as MassARRAY and iPLEX, (e.g., Jurinke et al. 2002; Saxena et al. 2014), TaqmanSNPlex (e.g., Tobler et al. 2005), VeraCode (Lin et al. 2009) and KASPar assays (Kumar et al. 2012). Furthermore, due to recently introduced sequencing-cum-genotyping methods like restriction site associated-DNA (RAD) sequencing, genotyping by sequencing (GBS) and WGRS, a major shift has been revealed in the methods used for discovery and mapping of DNA markers.

2 Association Genetics

In the previous decades, many efforts to find cultivated legume accessions with high abiotic stress tolerance utilized biparental linkage mapping, or QTL mapping, as a means to find genes for stress tolerance (e.g., Tanksley and McCouch 1997; Elnaccash and Tonsor 2010). Borrowing from human medicinal genomics, these approaches have been increasingly complemented by genome-wide association mapping (e.g., Atwell et al. 2010; Varshney et al. 2017a, b). Both of these approaches can be extremely powerful, but do have limitations that can be exacerbated by the nature of the germplasm available to researchers (e.g., Vilhjálmsson and Nordborg 2013). In conjunction with the technological advancements, the concept of biparental linkage mapping is also changing to multiparent-based mapping like multiparent advanced generation intercrosses (MAGIC, Cavanagh et al. 2008) and nested association mapping (NAM, Yu et al. 2008) to enable reaping maximum benefits from the recently available HTP genotyping/sequencing and phenotyping platforms. The highly saturated recombination maps, thus developed for these populations, can reveal the important genomic regions underlying economically important traits. Aside from traditional QTL mapping, these complex mapping resources create new possibilities for applying genome-wide association studies (GWAS) and more importantly, joint linkage–LD analysis for a much comprehensible genetic investigation of complex traits (Cavanagh et al. 2008). They have particular power for harnessing the diversity of crop wild relatives (CWR) and diverse landrace germplasm in breeding programs (e.g., Tanksley and McCouch 1997; Warschefsky et al. 2014).

3 Transcriptional, Proteomic, Metabolomic and Ionomic Profiling

A number of technologies to assess gene expression have arisen over the past two decades. Transcriptional profiling with microarrays was the most popular means of characterizing gene expression before the advent of RNA sequencing (e.g., Benedito et al. 2008). Serial analysis of gene expression (SAGE) and massively parallel signature sequencing (MPSS) are older but established alternatives to microarrays as they do not depend on prior probe selection (e.g., Matsumura et al. 1999; Jain 2011). For example, large-scale transcriptome analysis in chickpea (*Cicer arietinum* L.) provides insights into the chickpea transcriptome and differential responses to drought and also enabled development of gene-based markers to accelerate breeding (Hiremath et al. 2011). Differential protein abundance has traditionally been analyzed using 2D-PAGE with 2D difference gel electrophoresis (DIGE) (Issaq and Veenstra 2008). Non-gel-based proteomics approaches such as multidimensional protein identification technology (MudPIT, Washburn et al. 2001), iTRAQ (Ross et al. 2004; Aggarwal et al. 2006) or metabolic labeling and MALDI-TOF (e.g., Wittmann and Heinzle 2001) also followed. The first shotgun proteomics study in wheat gave some important insights into the protein response to drought, also identifying largest number of wheat proteins in a single proteomics experiment (Ford et al. 2011). Metabolite profiling in plants includes mainly NMR, gas chromatography (GC)-MS and liquid chromatography (LC)-MS (Schauer and Fernie 2006). Metabolite composition analysis in tomato inbred lines indicated their association with whole-plant yield associated QTL (Schauer et al. 2006).

The ionome of a plant is the summation of many biological processes (Salt et al. 2008). A high-throughput ionomics platform offers a viable system for probing the multiple physiological and biochemical activities that affect the ionome, in tens of thousands of individuals (e.g., Baxter et al. 2008). Ionomics, in combination with other phenotyping platforms such as transcript profiling, proteomics and metabolomics, offers the potential to close the growing gap between our knowledge of genotype and the phenotypes it controls (Salt et al. 2008). These new high-throughput technologies have ultimately enabled increase in efficiency of selection and accelerate the rates of genetic gain in breeding populations.

Phenomics, a large-scale phenotyping, is a natural complement to genome sequencing, transcriptomics, etc., that has the potential to lead to rapid advances in biology (Finkel 2009; Furbank and Tester 2011). Plant phenomics approaches provide the quantitative phenotyping needed to elucidate the genetic bases for agricultural traits and to screen germplasm for genetic variation in form, function and performance. Often using sensitive cameras and scales, these approaches can measure whole-plant growth trajectories in different environmental conditions. These approaches vary from those that use custom platforms such as the Lemnatec system (Lemnatech, Aachen, Germany), to approaches that are more modular and take advantage of tools such as drones (e.g., Ahmed et al. 2019). This area is currently evolving as rapidly as any other "-omic," with rapid shifts in both technology and perhaps more importantly in tools to handle and effectively analyze these complex datasets (e.g., Gehan et al. 2017; Araus et al. 2018). One custom-built system, a combined lysimeter and scanner at the International Crop Research Institute for the Semi-Arid Tropics, is an example of an intermediate system specially built to examine precise phenotypic differences in response to water limitation (Vadez et al. 2017).

4 Genome Size

The *Leguminosae* involves legume family which is the second only to the grass family in economic and nutritional value (Graham and Vance 2003; Smýkal et al. 2015). About 17,000 species of legumes are found in nature, and many of these are grain, pasture and agroforestry species (Vance 2001). Currently, either partial or complete sequences exist for at least 99 higher plants (69 angiosperms), with numbers growing at an increasing rate. Sixteen legume species (https://legumeinfo.org/genomes) have been sequenced and data stored in the LIS system, although many more are subject to completed or ongoing genome sequencing (Bauchet et al. 2019). This is the largest cluster of genomic information for legumes, although many of these species have their own sites or can be found at Phytozome, and much of the information on LIS is contributed from other we list these in Table 1.

5 Genome-Wide and Omics Approaches in Grain Legumes: Examples for Uncovering the Basis of Plant Architecture, Morphological Traits, Stress Tolerance and Disease Resistance

The plant genome sequencing started in the late twenty-first century, and the first plant genome that was sequenced was *Arabidopsis thaliana* in October 2001 by the Arabidopsis genome initiative (2000), with a tiny size of 125 Mbp. The first crop legume to be sequenced was *Glycine max* (Schmutz et al. 2010), with a size of 1115 Mbp. With the improvement in recent technologies, even very large genome of crop plants that are polyploidy has been successful. For example, in early 2019, the 4 GB *Pisum* genome was finally published (Kreplak et al. 2019).

5.1 Soybeans

The relatively large genome sizes of many legumes make for a range of complications. For example, the paleopolyploid soybean has about 46,430 protein-coding

Table 1 Genomic information on selected crop legume and crop wild relatives

Legume	Genome (1N)	Chromosome No.	No. of genes	References	Link out from LIS to Genome project home
Arachis hypogaea	2.7	$2n = 4x = 40$	67,124	Bertioli et al. (2019)	https://peanutbase.org
Arachis duranensis	1.25	$2n = 2x = 20$	36,734	Bertioli et al. (2016)	https://www.peanutbase.org
Arachis ipaensis	1.56	$2n = 2x = 20$	41,840	Bertioli et al. (2016)	https://www.peanutbase.org
Cajanus cajan	8.33	$2n = 2x = 22$	48,680	Varshney et al. (2012a, b)	https://legumeinfo.org
Cicer arietinum (CDC Frontier, kabuli-type chickpea	7.38	$2n = 2x = 16$	28,269	Varshney et al. (2013)	https://legumeinfo.org
Cicer arietinum (ICC 4958, desi type chickpea)	5.2	$2n = 2x = 16$	27,571	Jain et al. (2013)	http://www.ncbi.nlm.nih.gov/bioproject?term=PRJNA78951
Glycine max (soybean)	11.03	$2n = 4x = 40$	46,430	Schmutz et al. (2010)	https://legumeinfo.org
Glycine latifolia	9039	$2n = 40$	54,475	Liu et al. (2018)	https://soybase.org
Lens culinaris (lentil)	41.16	$2n = 2x = 14$		Not yet published as of late 2019	https://knowpulse.usask.ca
Lotus japonicus	4.66	$2n = 2x = 16$	22,650	Sato et al. (2008)	http://www.plantgdb.org
Medicago truncatula	4.66	$2n = 2x = 16$	50,894	Young et al. (2011)	https://phytozome.jgi.doe.gov
Phaseolus vulgaris	5.88	$2n = 2x = 22$	27,197	Schmutz et al. (2014)	https://phytozome.jgi.doe.gov
Trifolium pratense	6.37	$2n = 2x = 14$	40,868	De Vega et al. (2015)	https://www.ncbi.nlm.nih.gov/assembly/GCA_900079335.1

(continued)

Table 1 (continued)

Legume	Genome (1N)	Chromosome No.	No. of genes	References	Link out from LIS to Genome project home
Vigna angularis	5.38	$2n = 2x = 22$	26,857 to 34,183	Kang et al. (2015), Yang et al. (2015)	http://plantgenomics.snu.ac.kr
Vigna radiata	5.79	$2n = 2x = 22$	22,427	Kang et al. (2014)	http://plantgenomics.snu.ac.kr
Vigna unguiculata	6.20	$2n = 2x = 22$	29,773	Lonardi et al. (2019)	http://harvest.ucr.edu

genes, 70% more than Arabidopsis and similar to the poplar genome (another polyploidy genome). The soybean genome has segments that are rich in repeats and low recombination heterochromatic regions surrounding the centromeres. The exon–intron structure of genes among soybean, poplar and grapevine are conserved with a high degree as in the case of position and phase conservation found broadly across angiosperms (Roy and Penny 2007). Each round of polyploidy results in loss of genes, and this rate of loss can be determined by analyzing the genome repeated duplications (Schmutz et al. 2010).

Complementary "-omics" approaches can help clarify the complexity of large genomes. For example, as a component of transcriptomics, RNA sequencing has been done for salt tolerance in soybean by Zeng et al. (2018). Yet another study has shown differential gene expression among soybean downy mildew resistant and susceptible plants using RNA-seq data (Dong et al. 2018). Also, a study of the cyst nematode infection in soybean and its transcriptomics and metabolomics revealed the co-regulation of gene expression and metabolites in soybean plants (Kang et al. 2018). Gupta et al. (2018) have used an integrated approach of proteomics and metabolomics to investigate their individual and combined (ABA + ET) signaling in soybean leaves.

5.2 *Pigeonpeas*

The draft genome of pigeon pea (*Cajanus cajan*) sequenced by Varshney et al. (2012a, b) with a genome size of 833 Mbp. Genome analysis predicted 48,680 genes. About 266 *Phaseoleae*-restricted ORFans have been identified out of the whole chunk of predicted protein-coding genes. 97 ORFs had significant similarity to ORFs from soybean leaving the 169 putative pigeon pea-specific ORFan genes which were found to contain few introns, short length and unusual GC content. A total of 23,410 SSR primer pairs were designed from 29,467 sequences that could be converted into genetic markers. Kudapa et al. (2012) have done the transcriptomic assembly for pigeon pea by analyzing 128.9 million short Illumina GA IIx single end reads, 2.19 million single end FLX/454 reads and 18,353 Sanger expressed sequence tags from more than 16 genotypes. About 6284 intron spanning markers (ISR) have been designed using the transcript reads. In yet another study, RNA-seq data was generated from embryo sac from the day of anthesis (0 DAA), seed and pod wall formation (5, 10, 20 and 30 DAA) of pigeon pea variety "Asha" (Pazhamala et al. 2017). Liu et al. (2010) have determined and quantified active phenolic compounds in pigeon pea leaves and its medicinal product using liquid chromatography-tandem mass spectrometry. Krishnan et al. (2017) have performed a high-resolution two-dimensional (2-D) electrophoresis followed by MALDI-TOF-TOFMS/MS analysis of these protein fractions which resulted in the identification of 373 pigeon pea seed proteins. In consensus with the reported high degree of synteny between pigeon pea and soybean genomes, there exists a large homology between pigeon pea and soybean seed proteins. A large amount of stress-related proteins may be due to adaptation to drought-prone environments that were found in this study. An in-depth validation of these genes would result in identification of drought resistance/tolerance genes that can be used for introgression into the cultivated lines either via molecular breeding approaches or molecular cloning of such genes.

5.3 Chickpeas

The next important pulse crop that was sequenced by two groups is chickpea (*Cicer arietinum*) in the year 2013. Varshney et al. (2013) sequenced the kabuli chickpea and Jain et al. (2013) sequenced the desi chickpea. About 738 Mb of whole-genome shotgun sequence was reported in kabuli chickpea CDC frontier which is resistant to several diseases and insects. It contains an estimated 28,269 genes, and approximately, half (49.41%) of the genome is composed of transposable elements and unclassified repeats which is comparable to other sequenced legumes. In comparison with the kabuli type, the desi chickpea reference assembly had 520 Mb with 27,571 genes and 210 Mb as repeat elements. About 274 million RNA-seq reads were identified in several tissue-specific and stress responsive genes. Nucleotide diversity analysis shows assessment of narrow genetic base within chickpea cultivars (Roorkiwal et al. 2014). RNA-seq was done for roots and shoots under desiccation, salinity and cold stresses, and about 11,640 transcripts showed response to at least one of the stress conditions (Garg et al. 2016). The highest number of genes were differentially expressed in response to salinity (5321), followed by cold (4145) and desiccation (4078) stresses. Pathway analysis in chickpea showed a total of 121 pathways representing 640 enzymes involved in biosynthesis, metabolism, signaling and cell differentiation which were found to be activated. RNA-seq of leaf tissues from two contrasting chickpea genotypes at leaf apical meristem stage showed 1562 genes that were differentially expressed in drought-tolerant genotype (Badhan et al. 2018). An integrative network analysis of wilt transcriptome in chickpea using cDNA microarray temporal datasets consisting of 1749 unigenes has shown that there are genotype-dependent regulatory hubs in immunity and susceptibility phenotypes (Ashraf et al. 2009). To enhance the crop productivity using molecular breeding approaches, the next-generation sequencing was used to determine the sequence of most gene transcripts and to identify drought-responsive genes and gene-based molecular markers (Hiremath et al. 2011). Varshney et al. (2013) generated a set of chickpea ESTs that serves as a high-quality resource of transcripts for gene discovery and development of functional markers associated with abiotic stresses like drought and salinity. Wild chickpea, *Cicer reticulatum* (PI489777), the progenitor of cultivated chickpea, was analyzed for its transcriptome using GS-FLX454 technology. This study generated 37,265 transcripts in total with an average length of 946 bp. A total of 4072 SSRs could be identified in these transcripts of which only 561 SSRs were polymorphic (Jhanwar et al. 2012). Molina et al. (2008) have used deep SuperSAGE to obtain the salt-responsive transcriptome of chickpea roots and nodules. The high carbohydrate and protein content of chickpea has made it very important to study transcriptome of its seed developmental stages. Deep sequencing of transcriptomes from four developmental stages of seeds gave about 1.3 million reads assembled into 51,099 unigenes by merging the de novo and reference assemblies (Pradhan et al. 2014). Identification of potential fungal pathogenicity factors has been done using RNA-seq and massive analysis of cDNA ends to produce comprehensive expression profiles of *Ascochyta* rabiei [teleomorph Didymella rabiei (Kovachevski) v. Arx] genes (Fondevilla et al. 2015). Proteomic analysis was done to understand the diversity and complexity of membrane proteins in chickpea. A proteome reference map has been developed using 2D-gel electrophoresis. MALDI-TOF/TOF and LC-ESI-MS/MS have led to the identification of 91 proteins involved in a variety of cellular functions (Jaiswal et al. 2012). Fusarium wilt caused by *Fusarium oxysporum* f. sp. *ciceri* in chickpea is another devastating disease exacerbated by warm temperatures, and thus, its metabolic profiling at several time points of resistance and susceptible cultivars has been done. Researchers have taken up multivariate analysis of the data (OPLS–DA) which revealed discriminating metabolites in chickpea root tissue after *foc* inoculation such as flavonoids and isoflavonoids along with their malonyl conjugates (Kumar et al. 2015).

5.4 Groundnuts/Peanuts

Sequencing combined with precise phenotyping of segregating populations and germplasm collections is expected to measure the accurate genetic diversity present in the germplasm as well as to identify the gene/nucleotide associated with the trait(s) (e.g., Warschefsky et al. 2014). For example, a genome-wide association study of peanuts identified 48 significant indels associated with seed oil content across five environments, while transcriptome data of 49 peanut cultivars was analyzed to identify 5458 differentially expressed genes for oil synthesis process (Wang et al. 2018).

PeanutDB is an integrated bioinformatics Web portal for *Arachis hypogaea* (Duan et al. 2012). Peanuts have a total of 5712 of 8505 proteins identified through isobaric tags for relative and absolute quantitation (iTRAQ) which is used for determining the protein profiles of lipid dynamics during peanut seed development and post-germination (Wang et al. 2016). Soil salinity has been a limiting factor for growth of peanuts, and thus, an insight into its plant physiology in response to salt stress and alleviation has been studied. On analyzing the results by GC/MS and RNA-seq methods, 92 metabolites among 391 varied in response to salt, and 42 metabolites responded to especially to recovery (Cui et al. 2018). Drought has been another challenge for groundnuts (e.g., Puppala et al., 2017). Breeding efforts have been difficult to develop peanuts of multiple resistance, high seed quality and yield due to narrow genetic base. In contrast, peanut wild relatives are considered to be a major source of resistance alleles, and genomics and transcriptomic studies of these wild genotypes can be used for introgression into cultivated lines for their improvement (Guimarães et al. 2017).

5.5 Field Peas

Although perhaps the original model plant, as Mendel's study system, peas have lagged behind other legumes in the genomic era due to their relatively large genome size. However, the recent publication of the *Pisum* genome (Kreplak et al. 2019) as well as a growing range of tools such as a single seed descent population for GWAS analysis (Holdsworth et al. 2017) is helping to catch peas up to other legumes in terms of genomic resources. Peas have a fascinating domestication history, with two independent domestications in the Fertile Crescent and Ethiopia (Trněný et al. 2018), and wild relatives with tolerance to a range of abiotic stresses. For example, comparison of wild and cultivated peas has shown that seed dormancy was lost during domestication but also varies in wild populations in geographic patterns suggesting a strong relationship with climatic patterns (Hradilová et al. 2019). Other studies provide new results for a better understanding of plant architecture genetic determinism and genetic interdependency of root disease resistance and root system architecture inheritance in peas, *Pisum sativum (*Desgroux et al. 2018). Pea lines with good levels of resistance to *Aphanomyces euteiches* and a large root system (larger number of roots and longer roots) were identified and could be useful for breeders to improve resistance to *A. euteiches* in pea varieties. A SNP marker, detected for both improved resistance to *A. euteiches* and high projected root area will be relevant for use in the marker-assisted selection of resistant varieties (Desgroux et al. 2018).

5.6 *Phaseolus Species*

The common bean (*Phaseolus vulgaris*) is one of the most important staples in many areas of the world, has a fascinating double domestication in Mesoamerica and the Andes and has had a complete genome sequence for several years (Schmutz et al. 2014). A large number of experiments have been designed on phenotypic and genetic characterization of unexplored bean germplasm to unlock the breeding potential of this crop and to develop new varieties able to adapt to changing climatic conditions. An attempt is made to study the genetic control of days to flowering using a whole-genome association approach on a panel of 192 highly homozygous common bean genotypes purposely developed from landraces using single seed descent (Raggi et al. 2019). Within this work, a dataset of about 50 k single-nucleotide polymorphisms (SNPs) was obtained by RAD sequencing producing strict quality control genotyping data of the above genotypes. A subsequent genome-wide association study revealed significant and meaningful associations between days to flowering and several SNP markers; seven genes are proposed as the best candidates to explain the detected associations (Raggi et al. 2019). More information on other research conducted in *P. vulgaris*, namely in drought responses, could be found in the Chap. 3 of this book.

5.7 *Lentils*

Lentil (*Lens culinaris*Medik.) is the world's fifth most important pulse crop. It is self-pollinated diploid ($2n = 14$) and has a relatively large genome size of approximately 4 GB. The next-generation sequencing (NGS) technology has been used to genotype the ICARDA lentil reference set and a larger Canadian diversity panel using the two-enzyme (PstI, MspI) genotyping-by-sequencing (GBS) method described by Poland and colleagues (Wong et al. 2015). This method has been shown to be capable of discovering and genotyping thousands of markers across the lentil genome (Wong et al. 2015). To focus on coding regions, a more accurate exome capture array has been developed (Ogutcen et al. 2018). The published genome is expected in 2020, although a current draft genome is already available (Table 1).

5.8 *Vigna Species*

Mung bean, adzuki bean and cowpea, all in the genus *Vigna,* all have completed draft genome sequences (Kang et al. 2014; Kang et al. 2015; Muñoz-Amatriaín et al. 2017). Mung bean [*Vigna radiata* (L.) R. Wilczek var. *radiata*] is an important grain legume globally, providing a high-quality plant protein source largely produced and consumed in South and East Asia. One of the recent studies to characterize a mung bean diversity panel conducted a pilot genome-wide association study of seed coat color with 466 cultivated accessions (Noble et al. 2018). In addition, this study genotyped 16 wild accessions for comparison, and in total, over 22,000 polymorphic genome-wide SNPs were identified and used to analyze the genetic diversity, population structure, linkage disequilibrium (LD) of mung bean. Polymorphism was lower in the cultivated accessions in comparison with the wild accessions, with average polymorphism information content values 0.174, versus 0.305 in wild mung bean. LD decayed in ~100 kb in cultivated lines, a distance higher than the linkage decay of ~60 kb estimated in wild mung bean. Four distinct subgroups were identified within the cultivated lines, which broadly corresponded to geographic origin and seed characteristics. In a pilot genome-wide association mapping study of seed coat color, five genomic regions associated were identified, two of which were close to seed coat color genes in other species. This mung bean diversity panel constitutes a valuable resource for genetic dissection of important agronomical traits to accelerate mung bean breeding (Noble et al. 2018). Two more recent papers have identified loci controlling flowering time, using the same DArT-seq markers (Sokolva et al., in press; Koslov et al., in press) and a set of accessions from the Vavilov seed bank in Russia. This collection, which includes pre-Green Revolution accessions from a range of drought-prone and cold-season environments, could be an excellent source of stress tolerance for mung bean breeding.

A novel genotyping assay for over 50,000 SNPs was recently to delineate genomic regions governing pod length in cowpea, *Vigna unguiculata* (Xu et al. 2017), a crop known for its capacity to tolerate drought-prone environments (Carvalho et al. 2017). Major, minor and epistatic QTLs were identified through QTL mapping. Seventy-two SNPs associated with pod length were detected by genome-wide association studies (GWAS). Over 50,000 SNPs were employed to delineate genomic regions governing pod length. Population stratification analysis revealed subdivision among a cowpea germplasm collection consisting of 299 accessions, which is consistent with pod length groups (Xu et al. 2017). A more recent study has used the same array to make a range of domestication-related traits (Lo et al. 2018). It is well equipped to facilitate future molecular breeding for drought tolerance and other abiotic stresses (Chamarthi et al. 2019).

Similar approaches are likely to emerge soon from research groups working with other *Vigna*, such as adzuki bean, which has a small genome (Kang et al. 2015; Yang et al. 2015). Adzuki beans are known to have feral forms (Yang et al. 2015), which may have evolved repeatedly and adapted to a range of different abiotic stresses. However, as a group of largely orphan legume crops, this has trailed work in cowpeas, the most widely used and study *Vigna* species. Furthermore, other *Vigna* crops, such as Bambara groundnut and Zombi peas, have fascinating traits such as geocarpic pods (fruit buried belowground) or tubers and have tractable genomes sizes approximately similar to cowpea, mung bean and adzukia bean. Other cultivated *Vigna* crops, such as moth bean and *V. aconitifolius*, are remarkably drought tolerant. With multiple completed genomes and a growing set of tools, the capacity to uncover the wide range of abiotic stress tolerances in this group is quite high.

5.9 *Faba Bean- the Ultimate Large Genome Challenge*

With no known compatible wild relative and an enormous 12 GB genome, Faba bean (*Vicia faba*) is perhaps the most challenges cultivated legume genome to sequence and assemble. Frost has been one of the important stresses in fava bean production in Europe. A recent attempt was made to identify and validate QTL associated with frost tolerance and use in marker-assisted selection (Sallam et al. 2016). QTL mapping and GWAS identified new putative QTL associated with promising frost tolerance and related traits. A set of 54 SNP markers common in both genetic backgrounds showed a high genetic diversity with polymorphic information content (PIC) ranging from 0.31 to 0.37 and gene diversity ranging from 0.39 to 0.50. This indicates that these markers may be polymorphic for many faba bean populations (Sallam et al. 2016). Cheap sequencing technologies have stimulated the production of deep transcriptome coverage from several tissue types and numerous distinct cultivars. This has permitted the reconstruction of the faba bean meta-transcriptome and has fueled development of extensive sets of simple sequence repeat and single-nucleotide polymorphism (SNP) markers (O'Sullivan and Angra 2016).

6 Concluding Remarks

The family of high-throughput approaches we have described here in the context of minor legume crops is part of a trend across biology where it is increasingly possible to do genomics research in non-model taxa in which relatively little work has been done previously. We have highlighted selected aspects of this work in a range of grain legumes, historically mostly orphan crops that were the "slow runners" of the Green Revolution (Borlaug 1975). As these technologies continue to grow in power and decline in cost, we expect to see more work focused on legume taxa with particular abiotic stresses, such as drought, low pH or salinity stress, which can be harnessed in agricultural programs. In the face of rapid climate change and anthropogenic land use shifts, coupled with widespread agricultural degradation of arable lands globally, such research is badly needed.

However, work done in two model legumes, *Medicago truncatula* and *Lotus japonicus*, remains important for driving forward our understanding across all legumes. Much of the foundational information needed for modern "-omics" approaches rests on a deep basis of knowledge built in model plants, such as the model legumes and Arabidopsis. Continued work in these groups has much to give to those looking to understand the basis of stress tolerance in other legumes (e.g., De Bruijn 2019).

Finally, tools are essential to the utilization of "-omics" approaches, particularly by researchers in parts of the global south who most immediately face the challenges of improving legume stress tolerance for food security. We have not focused on bioinformatic approaches here, as they were recently discussed by Bauchet et al.

(2019). However, our emphasis does not diminish the need for user-friendly tools to allow all researchers to benefit from the power of the next-generation sequencing and related "-omics" approaches.

Acknowledgements The authors thank Susana Araújo for the opportunity to contribute this chapter. Research on the genomics of stress tolerance done by von Wettberg and collaborators is supported by Russian Scientific Fund Project No. 18-46-08001 on the basis of a unique scientific installation «Collection of plant genetic resources VIR», by a cooperative agreement from the United States Agency for International Development under the Feed the Future Program AID-OAA-A-14-00008 to D. R. Cook and Co-PI E.J.B.v.W, by a grant from the US National Science Foundation Plant Genome Program under Award IOS-1339346 to D. R. Cook, and E. J. B. v. W; US NIFA grant # 2018-67013-27619 R. V. Penmetsa and EBvW, EJBvW is further supported by the USDA Hatch program through the Vermont State Agricultural Experimental Station. Funds from the Russian Scientific Project focused on environmental factors impacting the domestication of species from subtropical areas such as cultivated *Vigna* species.

References

Aggarwal K, Choe LH, Lee KH (2006) Shotgun proteomics using the iTRAQ isobaric tags. Briefings Funct Genomics 5(2):112–120

Ahmed I, Eramian M, Ovsyannikov I, van der Kamp W, Nielsen K, Duddu HS, Rumali A, Shirtliffe S, Bett K (2019) Automatic detection and segmentation of lentil crop breeding plots from multi-spectral images captured by UAV-mounted camera. In: 2019 IEEE winter conference on applications of computer vision (WACV). IEEE, New York, pp 1673–1681

Araus JL, Kefauver SC, Zaman-Allah M, Olsen MS, Cairns JE (2018) Translating high-throughput phenotyping into genetic gain. Trends Plant Sci 23(5):451–466

Atwell S, Huang YS, Vilhjálmsson BJ, Willems G, Horton M, Li Y, Meng D, Platt A, Tarone AM, Hu TT, Jiang R et al (2010) Genome-wide association study of 107 phenotypes in Arabidopsis thaliana inbred lines. Nature 465(7298):627

Arabidopsis Genome Initiative (2000) Analysis of the genome sequence of the flowering plant Arabidopsis thaliana. Nature 408(6814):796

Ashraf N, Ghai D, Barman P, Basu S, Gangisetty N, Mandal MK, Chakraborty N, Datta A, Chakraborty S (2009) Comparative analyses of genotype dependent expressed sequence tags and stress-responsive transcriptome of chickpea wilt illustrate predicted and unexpected genes and novel regulators of plant immunity. BMC Genom 10(1):415

Badhan S, Kole P, Ball A, Mantri N (2018) RNA sequencing of leaf tissues from two contrasting chickpea genotypes reveals mechanisms for drought tolerance. Plant Physiol Biochem 129:295–304

Bauchet G, Bett K, Cameron C, Campbell JD, Cannon EKS, Cannon SB, Carlson JW, Chan A, Cleary A, Close TJ, Cook DR, Cooksey AM, Coyne C, Dash S, Dickstein R, Farmer AD, Fernáandez-Baca D, Hokin S, Jones ES, Kang Y, Monteros MJ, Muñoz-Amatriain M, Mysore K, Pislariu CI, Richards C, Shi A, Town C, Udvardi M, von Wettberg EJ, Young ND, Zhao P (2019) The future of legume genetic data resources: challenges, opportunities, and priorities (In press, Legume Science)

Baxter IR, Vitek O, Lahner B, Muthukumar B, Borghi M, Morrissey J, Guerinot ML, Salt DE (2008) The leaf ionome as a multivariable system to detect a plant's physiological status. Proc Natl Acad Sci 105(33):12081–12086

Beale DJ, Karpe AV, Ahmed W (2016) Beyond metabolomics: a review of multi-omics-based approaches. In: Microbial metabolomics. Springer, Cham, pp 289–312

Benedito VA, Torres-Jerez I, Murray JD, Andriankaja A, Allen S, Kakar K, Wandrey M, Verdier J, Zuber H, Ott T, Moreau S (2008) A gene expression atlas of the model legume Medicagotruncatula. Plant J 55(3):504–513

Bertioli DJ, Cannon SB, Froenicke L, Huang G, Farmer AD, Cannon EK, Liu X, Gao D, Clevenger J, Dash S, Ren L (2016) The genome sequences of Arachis duranensis and Arachis ipaensis, the diploid ancestors of cultivated peanut. Nat Genet 48(4):438

Bertioli DJ, Jenkins J, Clevenger J, Dudchenko O, Gao D, Seijo G, Leal-Bertioli SCM, Ren L, Farmer AD, Pandey MK, Samoluk SS, Abernathy B, Agarwal G, Ballén-Taborda C, Cameron C, Campbell J, Chavarro C, Chitikineni A, Chu Y, Dash S, El Baidouri M, Guo B, Huang W, Kim KD, Korani W, Lanciano S, Lui CG, Mirouze M, Moretzsohn MC, Pham M, Shin JH, Shirasawa K, Sinharoy S, Sreedasyam A, Weeks NT, Zhang X, Zheng Z, Sun Z, Froenicke L, Aiden EL, Michelmore R, Varshney RK, Holbrook CC, Cannon EKS, Scheffler BE, Grimwood J, Ozias-Akins P, Cannon SB, Jackson SA, Schmutz J (2019). The genome sequence of segmental allotetraploid peanut Arachis hypogaea. Nat Genet 51(5):877–884. https://doi.org/10.1038/s41588-019-0405-z

Bleidorn C (2016) Third generation sequencing: technology and its potential impact on evolutionary biodiversity research. Syst Biodivers 14(1):1–8

Borlaug NE (1975) Building a protein revolution on grain legumes. Nutritional Improvement of Food Legumes by Breeding (No. B0035-R. CIMMYT)

Carvalho M, Lino-Neto T, Rosa E, Carnide V (2017) Cowpea: a legume crop for a challenging environment. J Sci Food Agric 97(13):4273–4284

Cavanagh C, Morell M, Mackay I, Powell W (2008) From mutations to MAGIC: resources for gene discovery, validation and delivery in crop plants. Curr Opin Plant Biol 11(2):215–221

Chamarthi SK, Belko N, Togola A, Fatokun CA, Boukar O (2019) Genomics-assisted breeding for drought tolerance in cowpea. In: Genomics assisted breeding of crops for abiotic stress tolerance, vol II. Springer, Cham, pp 187–209

Cui F, Sui N, Duan G, Liu Y, Han Y, Liu S, Wan S, Li G (2018) Identification of metabolites and transcripts involved in salt stress and recovery in peanut. Front Plant Sci 9:217

Davey JW, Hohenlohe PA, Etter PD, Boone JQ, Catchen JM, Blaxter ML (2011) Genome-wide genetic marker discovery and genotyping using next-generation sequencing. Nat Rev Genet 12(7):499

Davey JW, Cezard T, Fuentes-Utrilla P, Eland C, Gharbi K, Blaxter ML (2013) Special features of RAD Sequencing data: implications for genotyping. Mol Ecol 22(11):3151–3164

De Bruijn F (2019) The model legume *Medicago truncatula*, vols 1 & 2. Wiley-Blackwell, NJ, 1264 p

Desgroux A, Baudais VN, Aubert V, Le Roy G, de Larambergue H, Miteul H et al (2018) Comparative genome-wide-association mapping identifies common loci controlling root system architecture and resistance to *Aphanomyceseuteiches* in Pea. Front Plant Sci. 8: 2195. https://doi.org/10.3389/fpls.2017.02195

De Vega JJ, Ayling S, Hegarty M, Kudrna D, Goicoechea JL, Ergon Å, Rognli OA, Jones C, Swain M, Geurts R, Lang C (2015) Red clover (Trifoliumpratense L.) draft genome provides a platform for trait improvement. Sci Rep 5:17394

Dong H, Shi S, Zhang C, Zhu S, Li M, Tan J, Yu Y, Lin L, Jia S, Wang X, Wu Y (2018) Transcriptomic analysis of genes in soybean in response to *Peronospora manshurica* infection. BMC Genom 19(1):366

Duan X, Schmidt E, Li P, Lenox D, Liu L, Shu C, Zhang J, Liang C (2012) PeanutDB: an integrated bioinformatics web portal for Arachis hypogaea transcriptomics. BMC Plant Biol 12(1):94

Elnaccash TW, Tonsor SJ (2010) Something old and something new: wedding recombinant inbred lines with traditional line cross analysis increases power to describe gene interactions. PLoS ONE 5(4):e10200

Elshire RJ, Glaubitz JC, Sun Q, Poland JA, Kawamoto K, Buckler ES, Mitchell SE (2011) A robust, simple genotyping-by-sequencing (GBS) approach for high diversity species. PLoS ONE 6(5):e19379

Finkel E (2009) With 'phenomics,' plant scientists hope to shift breeding into overdrive. Sci, 325(5939):380–381

Fondevilla S, Krezdorn N, Rotter B, Kahl G, Winter P (2015) In planta identification of putative pathogenicity factors from the chickpea pathogen *Ascochyta rabiei* by de novo transcriptome sequencing using RNA-Seq and massive analysis of cDNA ends. Frontiers in microbiology 6:1329

Ford KL, Cassin A, Bacic AF (2011) Quantitative proteomic analysis of wheat cultivars with differing drought stress tolerance. Front Plant Sci 2:44

Furbank RT, Tester M (2011) Phenomics–technologies to relieve the phenotyping bottleneck. Trends Plant Sci 16(12):635–644

Garg R, Shankar R, Thakkar B, Kudapa H, Krishnamurthy L, Mantri N, Varshney RK, Bhatia S, Jain M (2016) Transcriptome analyses reveal genotype-and developmental stage-specific molecular responses to drought and salinity stresses in chickpea. Sci Rep 6:19228

Graham PH, Vance CP (2003) Legumes: importance and constraints to greater use. Plant Physiol 131(3):872–877

Gehan MA, Fahlgren N, Abbasi A, Berry JC, Callen ST, Chavez L, Doust AN, Feldman MJ, Gilbert KB, Hodge JG, Hoyer JS (2017) PlantCV v2: image analysis software for high-throughput plant phenotyping. PeerJ 5:e4088

Grzebelus D (2015) Diversity arrays technology (DArT) markers for genetic diversity. In: Genetic diversity and erosion in plants. Springer, Cham, pp 295–309

Guimarães PM, Brasileiro ACM, Mehta A, Araujo ACG (2017) Functional genomics in peanut wild relatives. In: The peanut genome. Springer, Cham, pp 149–164

Gupta DS, Gupta S, Nadarajan N (2014) Legumes in the omic era. Springer, Berlin

Gupta R, Min CW, Kramer K, Agrawal GK, Rakwal R, Park KH, Wang Y, Finkemeier I, Kim ST (2018) A multi-omics analysis of glycine max leaves reveals alteration in flavonoid and isoflavonoid metabolism upon ethylene and abscisic acid treatment. Proteomics 18(7):1700366

Hiremath PJ et al (2011) Large-scale transcriptome analysis in chickpea (*Cicer arietinum L.*), an orphan legume crop of the semi-arid tropics of Asia and Africa. Plant Biotechnol J 1:10

Holdsworth WL, Gazave E, Cheng P, Myers JR, Gore MA, Coyne CJ, McGee RJ, Mazourek M (2017) A community resource for exploring and utilizing genetic diversity in the USDA pea single plant plus collection. Horticulture Res 4:17017

Hradilová I, Duchoslav M, Brus J, Pechanec V, Hýbl M, Kopecký P, Smržová L, Štefelová N, Vaclávek T, Bariotakis M, Machalová J (2019) Variation in wild pea (*Pisum sativum* subsp. elatius) seed dormancy and its relationship to the environment and seed coat traits. Peer J 7:e6263

Issaq HJ, Veenstra TD (2008) Two-dimensional polyacrylamide gel electrophoresis (2D-PAGE): advances and perspectives. Biotechniques 44(5):697–700

Jain M (2011) Next-generation sequencing technologies for gene expression profiling in plants. Briefings Funct Genomics 11(1):63–70

Jain M, Misra G, Patel RK, Priya P, Jhanwar S, Khan AW, Shah N, Singh VK, Garg R, Jeena G, Yadav M (2013) A draft genome sequence of the pulse crop chickpea (*Cicer arietinum* L.). Plant J 74(5):715–729

Jaiswal DK, Ray D, Subba P, Mishra P, Gayali S, Datta A et al (2012) Proteomic analysis reveals the diversity and complexity of membrane proteins in chickpea (*Cicer arietinum* L.). Proteome Sci 10:59. http://doi.org/10.1186/1477–5956-10-59

Jhanwar S, Priya P, Garg R, Parida SK, Tyagi AK, Jain M (2012) Transcriptome sequencing of wild chickpea as a rich resource for marker development. Plant Biotechnol J 10(6):690–702

Jurinke C, van den Boom D, Cantor CR, Köster H (2002) The use of MassARRAY technology for high throughput genotyping. In: Chip technology. Springer, Berlin, pp 57–74

Kang YJ, Kim SK, Kim MY, Lestari P, Kim KH, Ha BK, Jun TH, Hwang WJ, Lee T, Lee J, Shim S (2014) Genome sequence of mungbean and insights into evolution within *Vigna* species. Nat Commun 5:5443

Kang YJ, Satyawan D, Shim S, Lee T, Lee J, Hwang WJ, Kim SK, Lestari P, Laosatit K, Kim KH, Ha TJ (2015) Draft genome sequence of adzuki bean, *Vigna angularis*. Sci Rep 5:8069

Kang W, Zhu X, Wang Y, Chen L, Duan Y (2018) Transcriptomic and metabolomic analyses reveal that bacteria promote plant defense during infection of soybean cyst nematode in soybean. BMC Plant Biol 18(1):86

Kassa MT, Penmetsa RV, Carrasquilla-Garcia N, Sarma BK, Datta S, Upadhyaya HD, Varshney RK, von Wettberg EJ, Cook DR (2012) Genetic patterns of domestication in pigeonpea (*Cajanuscajan* (L.) Millsp.) and wild Cajanus relatives. PloS one 7(6):e39563

Kreplak J, Madoui MA, Cápal P, Novak P, Labadie K, Aubert G, Bayer PE, Gali KK, Syme RA, Main D, Klein A (2019) A reference genome for pea provides insight into legume genome evolution. Nat Genet 51(9):1411–1422

Krishnan HB, Natarajan SS, Oehrle NW, Garrett WM, Darwish O (2017) Proteomic analysis of pigeonpea (Cajanuscajan) seeds reveals the accumulation of numerous stress-related proteins. J Agric Food Chem 65(23):4572–4581

Kudapa H, Bharti AK, Cannon SB, Farmer AD, Mulaosmanovic B, Kramer R, Bohra A, Weeks NT, Crow JA, Tuteja R, Shah T (2012) A comprehensive transcriptome assembly of pigeonpea (*Cajanuscajan* L.) using Sanger and second-generation sequencing platforms. Mol Plant 5(5):1020–1028

Kumar Y, Dholakia BB, Panigrahi P, Kadoo NY, Giri AP, Gupta VS (2015) Metabolic profiling of chickpea-Fusarium interaction identifies differential modulation of disease resistance pathways. Phytochemistry 116:120–129

Kumar J, Choudhary AK, Solanki RK, Pratap A (2011) Towards marker-assisted selection in pulses: a review. Plant Breeding 130(3):297–313

Kumar S, Banks TW, Cloutier S (2012) SNP discovery through next-generation sequencing and its applications. Int J Plant Genomics

Langridge P, Fleury D (2011) Making the most of 'omics' for crop breeding. Trends Biotechnol 29(1):33–40

Lin CH, Yeakley JM, McDaniel TK, Shen R (2009) Medium-to high-throughput SNP genotyping using VeraCode microbeads. In: DNA and RNA profiling in human blood. Humana Press, pp 129–142

Liu W, Kong Y, Zu Y, Fu Y, Luo M, Zhang L, Li J (2010) Determination and quantification of active phenolic compounds in pigeon pea leaves and its medicinal product using liquid chromatography–tandem mass spectrometry. J Chromatogr A 1217(28):4723–4731

Liu Q, Chang S, Hartman GL, Domier LL (2018) Assembly and annotation of a draft genome sequence for Glycine latifolia, a perennial wild relative of soybean. Plant J 95(1):71–85

Lo S, Muñoz-Amatriaín M, Boukar O, Herniter I, Cisse N, Guo YN, Roberts PA, Xu S, Fatokun C, Close TJ (2018) Identification of QTL controlling domestication-related traits in cowpea (*Vignaunguiculata* L. Walp). Sci Rep 8(1):6261

Lonardi S, Muñoz-Amatriaín M, Liang Q, Shu S, Wanamaker SI, Lo S, Tanskanen J, Schulman AH, Zhu T, Luo MC, Alhakami H (2019) The genome of cowpea (*Vignaunguiculata* [L.] Walp.). Plant J 98(5):767–782

Matsumura H, Nirasawa S, Terauchi R (1999) Transcript profiling in rice (*Oryza sativa* L.) seedlings using serial analysis of gene expression (SAGE). Plant J 20(6):719–726

Molina C, Rotter B, Horres R, Udupa SM, Besser B, Bellarmino L, Baum M, Matsumura H, Terauchi R, Kahl G, Winter P (2008) SuperSAGE: the drought stress-responsive transcriptome of chickpea roots. BMC Genom 9(1):553

Moorthie S, Mattocks CJ, Wright CF (2011) Review of massively parallel DNA sequencing technologies. The HUGO journal 5(1–4):1–12

Muñoz-Amatriaín M, Mirebrahim H, Xu P, Wanamaker SI, Luo M, Alhakami H, Alpert M, Atokple I, Batieno BJ, Boukar O, Bozdag S (2017) Genome resources for climate-resilient cowpea, an essential crop for food security. Plant J 89(5):1042–1054

Naylor RL, Falcon WP, Goodman RM, Jahn MM, Sengooba T, Tefera H, Nelson RJ (2004) Biotechnology in the developing world: a case for increased investments in orphan crops. Food Policy 29(1):15–44

Noble TJ, Tao Y, Mace ES, Williams B, Jordan DR, Douglas CA, Mundree SG (2018) Characterization of linkage disequilibrium and population structure in a mungbean diversity panel. Front Plant Sci 8:2102

Ogutcen E, Ramsay L, von Wettberg EB, Bett KE (2018) Capturing variation in Lens (Fabaceae): development and utility of an exome capture array for lentil. Appl Plant Sci 6(7):e01165

O'Sullivan DM, Angra D (2016) Advances in faba bean genetics and genomics. Front Genetics 7:150

Pazhamala LT, Purohit S, Saxena RK, Garg V, Krishnamurthy L, Verdier J, Varshney RK (2017) Gene expression atlas of pigeonpea and its application to gain insights into genes associated with pollen fertility implicated in seed formation. J Exp Bot 68(8):2037–2054

Poland JA, Brown PJ, Sorrells ME, Jannink JL (2012) Development of high-density genetic maps for barley and wheat using a novel two-enzyme genotyping-by-sequencing approach. PLoS ONE 7(2):e32253

Pradhan S, Bandhiwal N, Shah N, Kant C, Gaur R, Bhatia S (2014) Global transcriptome analysis of developing chickpea (*Cicer arietinum* L.) seeds. Front Plant Sci 5:698

Puppala N, Xavier A, Garg V, Nayak SN, Pandey MK, Varshney RK, Layland N, Payton P, Holbrook C (2017) Next generation sequencing approaches for understanding genetic mechanism of drought tolerance in Valencia Peanut

Raggi L, Caproni L, Carboni A, Negri V (2019) Genome-wide association study reveals candidate genes for flowering time variation in common bean (*Phaseolus vulgaris* L.). Front Plant Sci 10:962

Reuter JA, Spacek DV, Snyder MP (2015) High-throughput sequencing technologies. Mol Cell 58(4):586–597

Roorkiwal M, Von Wettberg EJ, Upadhyaya HD, Warschefsky E, Rathore A, Varshney RK (2014) Exploring germplasm diversity to understand the domestication process in Cicer spp. using SNP and DArT markers. PLoS One 9(7):e102016

Ross PL, Huang YN, Marchese JN, Williamson B, Parker K, Hattan S, Khainovski N, Pillai S, Dey S, Daniels S, Purkayastha S (2004) Multiplexed protein quantitation in Saccharomyces cerevisiae using amine-reactive isobaric tagging reagents. Mol Cell Proteomics 3(12):1154–1169

Roy SW, Penny D (2007) Patterns of intron loss and gain in plants: intron loss dominated evolution and genome-wide comparison of *O. sativa* and *A. thaliana*. Mol Biol Evol 24:171–181

Sallam A, Arbaoui M, El-Esawi M, Abshire N, Martsch R (2016) Identification and verification of QTL associated with frost tolerance using linkage mapping and GWAS in winter faba bean. Front Plant Sci 7:1098

Salt DE, Baxter I, Lahner B (2008) Ionomics and the study of the plant ionome. Annu Rev Plant Biol 59:709–733

Sato S, Nakamura Y, Kaneko T, Asamizu E, Kato T, Nakao M, Sasamoto S, Watanabe A, Ono A, Kawashima K, Fujishiro T (2008) Genome structure of the legume, Lotus japonicus. DNA Res 15(4):227–239

Schadt EE, Turner S, Kasarskis A (2010) A window into third-generation sequencing. Human Mol Genetics 19(R2):R227–R240

Saxena MS, Bajaj D, Kujur A, Das S, Badoni S, Kumar V, Singh M, Bansal KC, Tyagi AK, Parida SK (2014) Natural allelic diversity, genetic structure and linkage disequilibrium pattern in wild chickpea. PLoS ONE 9(9):e107484

Schauer N, Fernie AR (2006) Plant metabolomics: towards biological function and mechanism. Trends Plant Sci 11(10):508–516

Schauer N, Semel Y, Roessner U, Gur A, Balbo I, Carrari F, Pleban T, Perez-Melis A, Bruedigam C, Kopka J, Willmitzer L (2006) Comprehensive metabolic profiling and phenotyping of interspecific introgression lines for tomato improvement. Nat Biotechnol 24(4):447

Schmutz J, Cannon SB, Schlueter J, Ma J, Mitros T, Nelson W, Hyten DL, Song Q, Thelen JJ, Cheng J, Xu D (2010) Genome sequence of the palaeopolyploid soybean. Nature 463(7278):178

Schmutz J, McClean PE, Mamidi S, Wu GA, Cannon SB, Grimwood J, Jenkins J, Shu S, Song Q, Chavarro C, Torres-Torres M (2014) A reference genome for common bean and genome-wide analysis of dual domestications. Nat Genet 46(7):707

Singh B, Bohra A, Mishra S, Joshi R, Pandey S (2015) Embracing new-generation 'omics' tools to improve drought tolerance in cereal and food-legume crops. Biologiaplantarum 59(3):413–428

Smýkal P, Coyne CJ, Ambrose MJ, Maxted N, Schaefer H, Blair MW, Berger J, Greene SL, Nelson MN, Besharat N, Vymyslický T (2015) Legume crops phylogeny and genetic diversity for science and breeding. Crit Rev Plant Sci 34(1–3):43–104

Tanksley SD, McCouch SR (1997) Seed banks and molecular maps: unlocking genetic potential from the wild. Science 277(5329):1063–1066

Tobler AR, Short S, Andersen MR, Paner TM, Briggs JC, Lambert SM, Wu PP, Wang Y, Spoonde AY, Koehler RT, Peyret N (2005) The SNPlex genotyping system: a flexible and scalable platform for SNP genotyping. J Biomolecular Tech: JBT 16(4):398

Trněný O, Brus J, Hradilová I, Rathore A, Das R, Kopecký P, Coyne C, Reeves P, Richards C, Smýkal P (2018) Molecular evidence for two domestication events in the pea crop. Genes 9(11):535

Turner TL, Bourne EC, Von Wettberg EJ, Hu TT, Nuzhdin SV (2010) Population resequencing reveals local adaptation of *Arabidopsis lyrata* to serpentine soils. Nat Genet 42(3):260

Vadez V, Kholova J, Srikanth M, Rekha B, Tharanya M, Sivasakthi K, Alimagham M, Karthika G, Keerthi C (2017) LeasyScan: 3D scanning of crop canopy plus seamless monitoring of water use to harness the genetics of key traits for drought adaptation

Vance CP (2001) Symbiotic nitrogen fixation and phosphorus acquisition. Plant nutrition in a world of declining renewable resources. Plant Physiol 127(2):390–397

Varshney RK, Ribaut JM, Buckler ES, Tuberosa R, Rafalski JA, Langridge P (2012a) Can genomics boost productivity of orphan crops? Nat Biotechnol 30(12):1172

Varshney RK, Chen W, Li Y, Bharti AK, Saxena RK, Schlueter JA, Donoghue MT, Azam S, Fan G, Whaley AM, Farmer AD (2012b) Draft genome sequence of pigeonpea (Cajanuscajan), an orphan legume crop of resource-poor farmers. Nat Biotechnol 30(1):83

Varshney RK, Song C, Saxena RK, Azam S, Yu S, Sharpe AG, Cannon S, Baek J, Rosen BD, Tar'an B, Millan T (2013) Draft genome sequence of chickpea (*Cicer arietinum*) provides a resource for trait improvement. Nat Biotechnol 31(3):240

Varshney RK, Pandey MK, Puppala N (2017) Future prospects for peanut improvement. In: The peanut genome. Springer, Cham, pp 165–169

Varshney RK, Saxena RK, Upadhyaya HD, Khan AW, Yu Y, Kim C, Rathore A, Kim D, Kim J, An S, Kumar V, von Wettberg EJB et al (2017b) Whole-genome resequencing of 292 pigeonpea accessions identifies genomic regions associated with domestication and agronomic traits. Nat Genet 49(7):1082

Vilhjálmsson BJ, Nordborg M (2013) The nature of confounding in genome-wide association studies. Nat Rev Genet 14(1):1–2

Wang Y, Ma X, Zhang X, He X, Li H, Cui D, Yin D (2016) ITRAQ-based proteomic analysis of the metabolic mechanisms behind lipid accumulation and degradation during peanut seed development and postgermination. J Proteome Res 15(12):4277–4289

Wang X, Xu P, Yin L, Ren Y, Li S, Shi Y, Alcock TD, Xiong Q, Qian W, Chi X, Pandey MK (2018) Genomic and transcriptomic analysis identified gene clusters and candidate genes for oil content in peanut (*Arachis hypogaea* L.). Plant Mol Biol Rep, pp 1–12

Warschefsky E, Penmetsa RV, Cook DR, von Wettberg EJ (2014) Back to the wilds: tapping evolutionary adaptations for resilient crops through systematic hybridization with crop wild relatives. Am J Bot 101(10):1791–1800

Washburn MP, Wolters D, Yates JR III (2001) Large-scale analysis of the yeast proteome by multidimensional protein identification technology. Nat Biotechnol 19(3):242

Wittmann C, Heinzle E (2001) Application of MALDI-TOF MS to lysine-producing *Corynebacterium glutamicum*: a novel approach for metabolic flux analysis. Eur J Biochem 268(8):2441–2455

Wong MM, Gujaria-Verma N, Ramsay L, Yuan HY, Caron C, Diapari M, Vandenberg A, Bett KE (2015) Classification and characterization of species within the genus Lens using genotyping-by-sequencing (GBS). PLoS ONE 10(3):e0122025

Xu P, Wu X, Muñoz-Amatriaín M, Wang B, Wu X, Hu Y, Huynh BL, Close TJ, Roberts PA, Zhou W, Lu Z, Li G (2017) Genomic regions, cellular components and gene regulatory basis underlying pod length variations in cowpea (*V. unguiculata L. Walp*). Plant Biotechnol J 15(5):547–557. https://doi.org/10.1111/pbi.12639. Epub 2016 Oct 17. PMID: 27658053; PMCID: PMC5399003

Yang K, Tian Z, Chen C, Luo L, Zhao B, Wang Z, Yu L, Li Y, Sun Y, Li W, Chen Y (2015) Genome sequencing of adzuki bean (*Vigna angularis*) provides insight into high starch and low fat accumulation and domestication. Proc Natl Acad Sci 112(43):13213–13218

Young ND, Debellé F, Oldroyd GE, Geurts R, Cannon SB, Udvardi MK, Benedito VA, Mayer KF, Gouzy J, Schoof H, Van de Peer Y (2011) The Medicago genome provides insight into the evolution of rhizobial symbioses. Nature 480(7378):520

Yu J, Holland JB, McMullen MD, Buckler ES (2008) Genetic design and statistical power of nested association mapping in maize. Genetics 178(1):539–551

Yugi K, Kubota H, Hatano A, Kuroda S (2016) Trans-omics: how to reconstruct biochemical networks across multiple 'omic' layers. Trends Biotechnol 34(4):276–290

Zeng H, Zhang X, Zhang X, Pi E, Xiao L, Zhu Y (2018) Early transcriptomic response to phosphate deprivation in soybean leaves as revealed by RNA-sequencing. Int J Mol Sci 19(7):2145

Use of Osmolytes for Improving Abiotic Stress Tolerance in Fabaceae Plants

Md. Motiar Rohman, Md. Rezwan Molla, Afsana Hoque Akhi, Sadia Sabrina Alam, Abdul Hannan, and Mirza Hasanuzzaman

Abstract Fabaceae family is considered to be the second largest family of dicotyledonous plants with great economic importance which are terrestrial and widely distributed in temperature and tropical countries. Major cultivated economic plants in Fabaceae family include food, forage, and fodder crops, and often face different abiotic stresses like drought, salinity, extreme temperature, unpredictable flooding, heavy metals, radiation, unfavorable soil condition, malnutrition, etc. Among them, drought and salinity are most detrimental and cause crop losses about 50% of the total crop worldwide. Further, forthcoming climatic changes have made the agriculture production system more vulnerable to feed the fast-growing population. Higher production of reactive oxygen species (ROS) and methylglyoxal (MG) break down the cellular metabolic system and cause structural and functional losses of cell organelles like nucleic acid, proteins, lipids, pigments, etc. In recent, exogenous protectants such as osmolytes, plant hormones, antioxidants, signaling molecules, polyamines, and trace elements have drawn great attention to be effective to improve abiotic stress tolerance though osmotic adjustment, stabilization of proteins and membranes, induction of antioxidation, and gene expression. In this chapter, we accumulate the recent information on role of exogenous osmolytes in mitigating the adverse effect of abiotic stress in Fabaceae crops under important abiotic stresses which will be useful

Md. Motiar Rohman (✉) · A. H. Akhi · S. S. Alam
Molecular Breeding Lab, Plant Breeding Division, Bangladesh Agricultural Research Institute, Gazipur-1701, Bangladesh
e-mail: motiar_1@yahoo.com

Md. Rezwan Molla
Plant Genetic Resources Centre, Bangladesh Agricultural Research Institute, Gazipur-1701, Bangladesh

A. Hannan
Seed Technology Division, Bangladesh Agricultural Research Institute, Gazipur-1701, Rajshahi, Bangladesh

M. Hasanuzzaman
Department of Agronomy, Faculty of Agricultural University, Sher-e-Bangla Agricultural University, Sher-e-Bangla Nagar, Dhaka 1207, Bangladesh
e-mail: mhzsauag@yahoo.com

M. Hasanuzzaman et al. (eds.), *The Plant Family Fabaceae*,
https://doi.org/10.1007/978-981-15-4752-2_7

for their further development of economic uses in farmers' field to meet the future challenges in agriculture by abiotic stresses.

Keywords Abiotic stress · Salinity · Drought · Proline · Glycinebetaine · Trehalose · Osmotic stress

Abbreviations

$O_2^{\cdot-}$	Superoxide radical
$^{\cdot}OH$	Hydroxyl radical
@	At the rate of
μM	Micromole
$^1O_2$	Singlet oxygen
ABA	Abscisic acid
APX	Ascorbate peroxidase
ASA	Ascorbic acid
BADH	Betaine aldehyde dehydrogenase
BRRI	Bangladesh Rice Research Institute
CAT	Catalase
Cd	Cadmium
Chl	Chlorophyll
CKs	Cytokinins
CMO	Choline monooxygenase
CO_2	Carbon dioxide
COX	Choline oxidase
Cu	Copper
DHAR	Dehydro ascorbate reductase
$dS\ m^{-1}$	Deci Siemens per meter
DW	Dry weight
EBL	24-epibrassinolide
Fd	Ferredoxin
FW	Fresh weight
GB	Glycinebetaine
Glu	Glutamate
Gly-I	Glyoxalase-I
Gly-II	Glyoxalase-II
GPX	Glutathione peroxidase
GR	Glutathione reductase
gs	Stomatal conductance
GSA	Glutamic semialdehyde
GST	Glutathione *s*-transferase
H_2O_2	Hydrogen peroxide
JA	Jasmonic acid

K	Potassium
KIN	Kinetin
LAI	Leaf area index
LP	*Lolium perenne* leaf extract
LRWC	Leaf relative water content
MDA	Malondialdehyde
MDHAR	Monodehydro ascorbate reductase
MG	Methylglyoxal
mM	Millimolar
MPa	Megapascal
MSI	Membrane stability index
N	Nitrogen
NO	Nitric oxide
Orn	Ornithine
P	Phosphorus
P5CS	Pyrroline-5-carboxylate synthetase
PN	Net photosynthetic rate
POD	Peroxidase
Pro	Proline
Put	Putrescine
ROS	Reactive oxygen species
RWC	Relative water content
SA	Salicylic acid
SOD	Superoxide dismutase
Spd	Spermidine
Spm	Spermine
T6P	Trehalose-6-phosphate
TPP	Trehalose phosphate phosphatase
TPS	Phosphate synthase
TSS	Total soluble sugars
UV	Ultraviolet ray
WUE	Water use efficiency

1 Introduction

Stress is the overpowering pressure that affects the normal functions of individual life or the conditions in which plants are prevented from fully expressing their genetic potential for growth, development, and reproduction. In agriculture, stress is defined as a phenomenon that limits crop productivity or destroys biomass. Agriculture production and productivity are vulnerable to abiotic stresses. These stresses emerge due to drought, temperature extremes (heat, cold chilling/frost), radiation (UV, ionizing radiation), floods, edaphic factors which include chemical (nutrient deficiencies, excess of soluble salts, salinity, alkalinity, low pH/acid sulfate conditions, high pH and anion retention, calcareous or gypseous conditions, low redox, chemical contaminants—geogenic and xenobiotic), physical (high susceptibility to erosion, steep slopes, shallow soils, surface crusting and sealing, low water-holding capacity, impeded drainage, low structural stability, root-restricting layer, high swell/shrink potential), and biological (low or high organic contents) components where drought and salt stress are predominant many parts of the world. These stresses are the major challenges for production of crops, livestock, fisheries, and other commodities. Only 9% of the world's agricultural area is conducive for crop production, while 91% is under stresses which widely occur in combinations (Minhas et al. 2017).

Drought is considered as the major abiotic stress in worldwide. Drought stress limits crop productivity more severely than any other abiotic stress. Drought stress is supposed to cause global crop production losses of up to 30% by 2025, compared to present yield (according to "Water Initiative" report of "The World Economic Forum 2009" at Davos; Zhang 2011).

More than 45 million hectares of irrigated land have been occupied by salt which is more than 20% of total arable land. Every year, more than 1.5 million hectare area goes under current fallow due to salinity problem (Pitman and Läuchli 2002; Munns and Tester 2008), and increasing salinity problem is expected to result in up to 50% loss of cultivable lands by the middle of the twenty-first century (Mahajan and Tuteja 2005; Hasanuzzaman et al. 2013a, b).

Drought, salinity, and other abiotic stresses cause serious crop losses which are near about 50% of the total crop of the world (Acquaah 2007). Oxidative stress occurs within the plant cell during moisture-deficit condition due to reduction in stomatal conductivity, which restricts CO_2 influx into the leaves, inhibits RuBisCo activity, and disrupts energy balance and distribution during photosynthesis (Demirevska et al. 2010; Rapacz et al. 2010). As a consequence, these leading to the formation of reactive oxygen species (ROS) such as singlet oxygen ($^1O_2$), superoxide radical ($O_2^{\cdot-}$), hydrogen peroxide (H_2O_2), and hydroxyl radical ($^{\cdot}OH$) by enhanced leakage of electrons to molecular oxygen (Yadav 2010; Hasanuzzaman et al. 2014a, b, c).

Various subcellular organelles such as chloroplast, mitochondrion, and peroxisome are the common sites of ROS production. Normal concentration of ROS is necessary for inter- and intracellular signaling but overproduction of ROS can damage various cellular process such as lipid peroxidation, protein degradation, inactivation of enzymes, damage of nucleic acids, disruption in normal cell metabolism, and damage of cell membrane which leads to cell death (Ozkur et al. 2009). In addition, methylglyoxal (MG) is a potential cytotoxic compound produced under abiotic stress, which can react with and modify other molecules including DNA and proteins (Yadav et al. 2005a, b). Therefore, both ROS and MG must be kept in threshold level.

Plant itself tried to prevent oxidative injury by antioxidant defense system. ROS scavenging enzymes include superoxide dismutase (SOD), peroxidase (POD), catalase (CAT), glutathione peroxidase (GPX), glutathione *S*-transferase (GST), ascorbate peroxidase (APX), monodehydro ascorbate reductase (MDHAR), dehydro ascorbate reductase (DHAR), and glutathione reductase (GR) (Noctor et al. 2002). SOD deploys the primary protection to convert $O_2^{\cdot -}$ to H_2O_2, whereas POD, CAT, GPX, and APX decompose H_2O_2 to water. In ascorbate–glutathione cycle, MDHAR and DHAR play essential role in defense system against ROS through maintaining ascorbic acid (ASA), while GR maintains the reduced status of glutathione (GSH). Plant GSTs are known to function in herbicide detoxification, hormone homeostasis, vacuolar sequestration of anthocyanin, tyrosine metabolism, hydroxyperoxide detoxification, regulation of apoptosis, and in plant response to biotic and abiotic stress (Dixon et al. 2010). On the other hand, cytotoxic MG is detoxified via glyoxalase system which consists of two enzyme such as glyoxalase-I (Gly-I) and glyoxalase-II (Gly-II) which use glutathione (GSH) as a cofactor (Yadav et al. 2005a). Enhanced antioxidant and glyoxalase systems improve drought tolerance in many crop plants (Hasanuzzaman and Fujita 2011; Alam et al. 2013).

In recent, exogenous protectants such as osmolytes have been found to be effective in abiotic stress tolerant in plant. Exogenous use of the protectants plays a vital role in osmotic adjustment in the cytoplasm and can also stabilize proteins and membranes and increased antioxidant enzyme activities including detoxifying ROS and MG; therefore, osmoprotectants can play an important role in the adaptation of cells to various adverse environmental conditions (Tiwari et al. 2010; Alam et al. 2013; Cevic and Unyayar 2015; Rohman et al. 2015).

Fabaceae family is known as Papilionaceae, commonly known as Pea family. It is considered to be the second largest family of dicotyledonous plants. The plants of the family are terrestrial and widely distributed in temperature and tropical countries of the world. The plants are of great economic importance. The followings are the major cultivated economic plants in Fabaceae family:

Lentil (*Lens culinaris*), mung bean (*Vigna radiata* L.), common bean (*Phaseolus vulgaris* L.), faba bean (*Vicia faba*), green bean (*Phaseolus vulgaris*), soybean (*Glycine max*), pea (*Pisum sativum*), cowpea (*Vigna unguiculata*), pigeon pea (*Cajanus cajan*), grass pea (*Lathyrus sativus* L.) chickpea (*Cicer arietinum*), black gram (*Vigna mungo*), fenugreek (*Trigonella foenum-graecum*), lima bean (*Phaseolus lunatus*), and peanut/ground nut (*Arachis hypogaea*).

Alfalfa (*Medicago sativa*), sainfoin (*Onobrychis viciaefolia*), hyacinth bean (*Lablab purpureus*), sunn hemp (*Crotalaria juncea*), and white clover (*Trifolium repens* L.).

The economically important plants of Fabaceae family often face different abiotic stress conditions. Nowadays, different osmoprotectants are used to overcome the abiotic stresses for successful crop production. This chapter is our little effort to gather and present information highlighting the vital roles of osmoprotectants against devastating effects of abiotic stress in plants. Finally, we focus the use of exogenous osmolytes for improving abiotic stress tolerances in Fabaceae plants.

2 Osmolytes or Osmoprotectants

Osmolytes or osmoprotectants are small, electrically neutral nontoxic molecules at molar concentrations and highly soluble organic compounds that efficiently maintain osmotic balance and stabilize proteins and membranes under salt, drought, or other stress conditions (Yancey 1994). Most common osmoprotectants are:

Osmolytes: Proline, glycine betaine, and trehalose.
Plant hormones: Abscisic acid, indoleacetic acid, jasmonic acid, and salicylic acid.
Antioxidants: Ascorbic acid or ascorbate, glutathione, and tocopherol.
Signaling molecules: Nitric oxide and hydrogen peroxide.
Polyamines: Spermidine, spermine, and putrescine, and
Trace elements: Selenium and silicon.

3 Role of Osmolytes and Osmoprotectants in Abiotic Stress Tolerance in Different Crops Including the Important Crops Under Fabaceae

3.1 Osmolytes

Under abiotic stress condition, osmoprotectants serve to raise cellular osmotic pressure and also protect the cell constituents (Yancey 1994). The mechanisms of osmolytes are biological membrane protection, detoxification of toxic compounds such as ROS and MG, alleviation of ionic toxicity, protection of photosynthetic and mitochondrial structure, and metabolism. Moreover, the signaling role of osmolytes also designated as a vital stress-protective mechanism. The role of osmolytes for using abiotic stress tolerance in different crops including the important crops under Fabaceae (Pea family) is reviewed below.

Proline (Pro), an essential amino acid, is known to participate in the biosynthesis of primary metabolism during growth and development (Hare et al. 1999; Funck et al. 2012). Proline also accumulates in response to the imposition of a wide range of stress responses in plants. It has been well documented as an osmotic regulator helping in reduction of osmotic damage (Slama et al. 2008; Reddy et al. 2015). It is further hypothesized that accumulation of proline in leaves could possibly play a protection role aside from osmoregulation during water stress (Jungklang et al. 2015). Dien et al. (2019) suggested that osmotic adjustment is a biochemical mechanism that helps plants to acclimate to abiotic stress. One mechanism for osmotic adjustment is the accumulation of compatible solutes, such as the amino acid Pro. Stressful environment results in an overproduction of Pro in plants which is highly associated with stress tolerance development imparting cellular turgor or osmotic balance, stabilizing membranes, controlling reactive oxygen species (ROS) within tolerable range, thus preventing plant cell from oxidative damage. Proline is synthesized in plant via two pathways, viz. glutamate (Glu) and ornithine (Orn) pathways. However, glutamate (Glu) pathways take part mostly in Pro synthesis during osmotic stress (Lv et al. 2011; Witt et al. 2012). Glutamate is first converted into an intermediate glutamic semialdehyde (GSA) by the action of Δ1-pyrroline-5-carboxylate synthetase (P5CS) which is spontaneously cyclized into pyrroline-5-carboxylate (P5C). This intermediate P5C is finally converted into Pro which is catalyzed by Δ1-pyrroline-5-carboxylate reductase (P5CR). The potentiality of Pro synthesis is largely dependent on the activity and expression of the catalyzing enzymes.

Proline is a compatible osmolyte or osmoprotectant against salt stress (Matysik et al. 2002). Proline can induce the expression of salt stress-responsive proteins involved in plant adaptation to salt stress (Khedr et al. 2003). According to Hoque et al. (2008), salt stress increased protein carbonylation and contents of GSH and oxidized glutathione (GSSG) and enhanced the activities of GST and Gly-II enzymes in *Nicotiana tabacum* L. cv. BY-2. Proline showed important role for stabilizing the functional units of electron transport and stabilizing the membranes, proteins, and enzymes such as RuBisCo (Hamilton and Heckathorn 2001). It protects the photosynthetic apparatus by functioning as an oxygen radical scavenger (Heuer 2003). Exogenous Pro and glycine betaine (GB) application alleviated salt damages by improving antioxidant and glyoxalase system and improving physiological adaptation (Hasanuzzaman et al. 2014a). Another study reported by Sabagh (2015a) observed that exogenous osmoprotectants combined with compost will effectively solve seasonal salinity stress problem and are a good strategy to increase salinity resistance of soybean in the drylands.

The sugars are major organic compounds present in the plant world. These are basically products of photosynthesis and oxidative pentose phosphate pathway. Sucrose and glucose either act as substrates for cellular respiration or as osmolytes to maintain cell homeostasis, while fructose is not related to osmoprotection and seems related to secondary metabolites synthesis (Rosa et al. 2009). Soluble sugars, which are commonly termed as mono- and disaccharides, play a lead role in the structure and function of all living cells. Osmolytes and compatible solutes are overproduced under osmotic stress aiming to facilitate osmotic adjustment (Shao et al. 2005). Soluble sugars, e.g., glucose, fructose, and sucrose, are the major compounds accumulating and proving osmoprotection under various abiotic stresses (Kameli and Lösel 1995). The plant's ability to ferment available sugars for proper metabolic function, such as in rice (Mustroph and Albrecht 2003), green gram (Kumutha et al. 2008), pigeon pea (Manoharlallal 2014; Savita 2016), helps plants to tolerate a waterlogged environment.

Betaines are quaternary ammonium compounds which contain a carboxylic acid group. They may be generally regarded as fully N-methylated amino or imino acids. Different forms of betaines exist depending upon plant species including (GB), proline betaine, β-alanine betaine, choline-O-sulfate, and 3-dimethylsulfoniopropionate (Rhodes and Samaras 1994; McNeil et al. 1999). Glycine betaine is the most common among different kinds of betaine. Glycine betaine, also called as original betaine (N, N, N-trimethylglycine), was first discovered from sugar beet (*Beta vulgaris*) which is later found to be distributed in microorganisms, plants, and animals. It is one of the most abundant quaternary ammonium compounds that occur during dehydration stress in plants (Ashraf and Foolad 2007). Glycine betaine is synthesized from choline in a two-step oxidation by a ferredoxin (Fd)-dependent choline monooxygenase (CMO) and a betaine aldehyde dehydrogenase (BADH) with a strong preference for NAD^+. In almost all biological systems, conversion of choline to GB is through a two-step dehydrogenation/oxygenation via the unstable intermediate betaine aldehyde. In higher plants, GB biosynthesis occurs at chloroplasts. In the first step, choline is converted to an unstable compound betaine aldehyde by CMO. In the second step, betaine aldehyde is converted to GB by BADH (Rathinasabapathi et al. 1997). Besides this, other pathways of GB synthesis such as direct N-methylation of glycine also exist (Weretilnyk et al. 1989). Unlike higher plants, bacteria produce GB from choline using choline oxidase (COX) instead of CMO. The enzyme (CMO) is much responsible to synthesize GB, and in many plant studies, overexpression of BADH is found to be highly associated with GB synthesis and better tolerance to osmotic stress (Sakamoto and Murata 2001; Sulpice et al. 2003). It is mostly found in chloroplast where it takes part in protecting the thylakoid membrane and osmotic adjustment. Khan et al. (2014) showed that salicylic acid induced GB accumulation in *Vigna radiata* under salinity, with a consequent increase of glutathione, reduction of ethylene and oxidative stress, and improvement of photosynthesis. GB also protects photosynthesis by modifying the lipid composition of the thylakoid membranes in *Triticum aestivum* (Zhao et al. 2007). It increases soluble sugars and free amino acid accumulation to protect plant cells from salinity- and drought-induced osmotic stress in *Vigna unguiculata* (Manaf 2016), *Phaseolus vulgaris* (Osman and Salim 2016), and *Pisum sativum* (Osman 2015). GB specifically increases its own content and improves the activity of antioxidant enzymes and metabolites, such as SOD, CAT, APX, proline, and γ-amino butyric acid (GABA), reducing H_2O_2 and malondialdehyde (MDA) in *Prunus persica* (Shan et al. 2016), *Lolium perenne* (Hu et al. 2012), *Glycine max* (Malekzadeh 2015), and *Oryza sativa* (Yao et al. 2016). Rahman et al. (2002) reported that GB has the capability for the production of many vacuoles in the root cells to act as storage of Na^+ which was supposed to prevent Na^+ accumulation in the shoots and its subsequent damage effects. Foliar spraying of GB in salt-sensitive rice plants exposed to salt stress showed improved performance by maintaining better water use efficiency (WUE) and stabilizing photosynthetic pigment which led higher CO_2 assimilation and higher photosynthesis rate. Glycine betaine also had its effects on overall growth performance of rice plants (Cha-Um and Kirdmanee 2010).

Trehalose (Tre) is a nonreducing disaccharide composed of two glucose residues (α-D-glucopyranosyl-1 and 1-α-D-glucopyranoside) which are bonded by α-α-(1 1) linkage. Among the disaccharides, trehalose has particular properties, both of the reducing ends of the molecule form glycosidic bond. Therefore, trehalose is resistant to acidic hydrolysis and is stable in solution at high temperatures, even under acidic conditions (Richards et al. 2002). Trehalose is a membrane and molecule stabilizer; it bears water replacement mechanism where trehalose replaces water by establishing hydrogen bonds with membranes or macromolecules during dehydration or freezing (Crowe 2007). Trehalose plays multiple roles in confirming stress tolerance in plants. Trehalose biosynthesis in higher plants only occurs through trehalose phosphate synthase (TPS) and trehalose phosphate phosphatase (TPP) pathway (which is also known as OtsA-OtsB pathway). In the first step, binding of glucose-6-phosphate to uridine diphosphoglucose (UDP-glucose) is catalyzed by TPS to produce trehalose-6-phosphate (T6P). In the second step, trehalose-6-phosphate phosphatase (TPP) catalyzes the dephosphorylation of T6P to trehalose (Avonce et al. 2006; Paul et al. 2008).

Trehalose functions as a compatible solute which is well known. Exogenous Tre treatment improved antioxidant capacity and reduced Na^+/K^+ ratio and endogenous Pro level in salt stress-affected rice seedlings. Transcription of P5CS and P5CR was upregulated, and the activities of SOD and POX were decreased, whereas Tre increases the activity of APX. Tre upregulated the transcription of all antioxidant enzyme genes. Trehalose could not affect the growth inhibition during salt stress (Nounjan et al. 2012). Metabolic activity of maize seedlings was altered by Tre application which helped to alleviate damage effects of salt stress. Presoaking maize seeds with Tre (10 mM) increased photosynthetic pigment and nucleic acid content in salt-affected seedlings. Trehalose treatment reduced oxidative stress, which was indicated by stabilization of the plasma membranes, reduced rate of ion leakage, and reduced lipid peroxidation of maize root cells. Tre application increased the ratio of K^+/Na^+ in the leaves of maize seedlings (Zeid 2009). Exogenous application of Tre reduced Na^+ accumulation and growth inhibition of rice plant exposed to salt stress. Higher concentrations (10 mM) prevented chl loss in leaf blades and preserved root integrity (Garcia et al. 1997). Sorbitol (5 and 10 mM) and Tre (5 and 10 mM) showed protective roles in salt-sensitive cultivar of rice (*Oryza sativa* L. cv. KDML105) when subjected to salt stress (170 mM). Salinity markedly reduced growth of this cultivar. Salinity also resulted in oxidative stress as indicated by significant increase of H_2O_2, lipid peroxidation, and membrane electrolyte leakage. However, sorbitol and Tre supplementation ameliorated those adverse effects and improved the growth performance (Theerakulpisut and Gunnula 2012). Effects of osmolytes on abiotic stress tolerance mechanism of the crop under Fabaceae have been shown in Table 1.

3.2 *Plant Hormones*

Abscisic acid (ABA) is commonly known as the "stress hormone" that responds to variety of environmental stresses including both biotic and abiotic stress (Zhang 2014). The application of ABA to plant mimics the effect of a stress condition. Main function of ABA seems to be the regulation of plant water balance and osmotic stress tolerance. ABA is synthesized from β-carotene through several enzymatic steps. Conversion of β-carotene to ABA is mediated via number of enzyme-catalyzed steps.

Table 1 Improvement in growth and different physio-biochemical attributes by application of osmolytes in different species of peas under stress conditions

Osmolytes	Stress	Crop	Activities	References
Proline	Drought	Lentil (*Lens culinaris*)	– Increase in the content of free endogenous proline, total soluble protein and decrease MDA – Enhance CAT, APX, and GPX activities	Bekka et al. (2018)
Proline and betaine	Drought	Lentil (*Lens culinaris*)	– Both Bet or Pro separately increase the GSH content, maintain high activities of GST and Gly-I compared to the control and mostly also drought-stressed plants, with a concomitant decrease in GSSG content and H_2O_2 level – Bet and Pro together also decrease GSSG and H_2O_2 content	Molla et al. (2014)
Proline	Drought	Soybean (*Glycine max*)	– Genotypic differences exist in proline accumulation among soybean cultivars of different sensitivities to drought stress – Proline accumulation was higher in tolerant plants	Mwenye et al. (2016)
Proline/Glycine betaine (GB)	Salinity	Soybean (*Glycine max*)	– Increased accumulation of proline and antioxidant responses	Sabagh et al. (2015a, b)
Proline and glycine betaine (GB)	Water stress	Soybean (*Glycine max*)	– Protein and oil content was increased	Sabagh et al. (2015a, b)
Proline	Salinity	Common bean (*Phaseolus vulgaris* L.)	– Enhance SOD, CAT, and POD activities – Increase in the content of ascorbic acid and endogenous proline	Abdelhamid et al. (2013)
Proline	Heavy metal (Selenium)	Common bean (*Phaseolus vulgaris* L.)	– Enhance SOD, CAT, APX, and GR, GSH, and AsA activities as well as increase in the content of endogenous proline in higher toxicity level	Aggarwal et al. (2011)

(continued)

Table 1 (continued)

Osmolytes	Stress	Crop	Activities	References
Proline	Heat	Chickpea (*Cicer arietinum* L.)	– MDA content in shoot was decreased, but SOD, CAT, APX, GR activity were increased and – Nonenzymatic antioxidants ascorbate and glutathione content were also increased	Kaushal et al. (2011)
Proline	Cold	Chickpea (*Cicer arietinum* L.)	– Proline content level was increased	Saghfi and Eivazi (2014)
Proline	Salinity	Groundnut (*Arachis hypogaea* L.)	– At low concentrations increased plant growth by reducing salinity and also reduced MDA content	Hayat et al. (2012)
Proline and *Lolium perenne* (LP)	Nickel and/or salinity	Pea (*Pisum sativum*)	– Detoxified effects of Ni and salinity on growth, RWC, and various photosynthetic attributes of pea	Shahid et al. (2014)
Proline	Salinity	Sainfoin (*Onobrychis viciaefolia*)	– Increased in the content of free endogenous proline in both shoot and root, whereas decreased in the content of MDA only in shoot	Wu et al. (2017)
Proline and abscisic acid	Drought (water stress)	Faba Bean (*Vicia faba* L.)	– ABA and Pro alone as well as in combination increased in the content of endogenous proline, total soluble carbohydrate and enhanced CAT and POD activities	Ali et al. (2013)
Proline/Glycine betaine (GB)	Salinity	Mung bean (*Vigna radiata* L.)	– Increased AsA, GSH content, GSH/GSSG ratio and activities of APX, DHAR, GR, GST, GPX, CAT, Gly-I and Gly-II	Hossain et al. (2011)
Proline and glycine betaine (GB)	Salinity	Faba bean (*Viciafaba* L.)	– Reduced membrane injury, improved K^+ uptake and increased chlorophyll contents	Gadallah (1999)

(continued)

Table 1 (continued)

Osmolytes	Stress	Crop	Activities	References
Glycine betaine (GB)	Salinity	Green bean (*Phaseolus vulgaris* L.)	– Enhanced CAT, POD, and SOD activities – Increase in the content of proline	Abdelmotlb et al. (2019)
Glycine betaine (GB)	Salinity	Soybean (*Glycine max*)	– Proline and MDA contents were decreased and CAT and SOD activity were increased	Malekzadeh (2015)
Glycine betaine (GB)	Salinity	Soybean (*Glycine max*)	– Endogenous glycine betaine was increased	Rezaei et al. (2012)
Glycine betaine (Foliar application)	Chromium	Mung bean (*Vigna radiata* L.)	– Enhanced the activities of SOD, POD, and CAT in both shoots and roots	Jabeen et al. (2016)
Glycine betaine (GB) or Proline	Drought	Pea (*Pisum sativum*)	– Enhanced the activity of SOD, APX and CAT in leaves; – SOD activity increased but decreased APX activity in seeds	Osman (2015)
Glycine betaine (GB)	Cold	Chickpea (*Cicer arietinum* L.)	– Increased in floral retention, pod set and most of the yield parameters in cold-stressed plants – Chilling injury was reduced and elevation of relative leaf water content, chlorophyll and sucrose, and decrease in abscisic acid as well as active oxygen species (malondialdehyde and hydrogen peroxide)	Nayyar et al. (2015)
Glycine betaine (GB)	Drought	Chickpea (*Cicer arietinum* L.)	– Notably improved photosynthesis rate, transpiration rate, stomatal conductance, intercellular CO_2 concentration, photosynthetic water use efficiency, and carboxylation efficiency	Hosseinian et al. (2019)
Trehalose (Tre)	Drought	Cowpea (*Vigna sinensis*)	– Enhanced SOD, CAT, POX, and APX activities – Increased in the content of IAA, trehalose, total soluble sugar, carbohydrate, proline, free amino acids, and phenolic compound	Khater et al. (2018)

The abiotic stress which entails triggering of assorted ABA biosynthetic genes corresponding to zeaxanthin oxidase (ZEP), 9-cis-epoxycarotenoid dioxygenase (NCED), ABA-aldehyde oxidase (AAO), and molybdenum cofactor sulfurase (MCSU) might be because of calcium-dependent phosphorylation pathway (Tuteja 2007). ABA can enhance plant adaptation to various abiotic stresses. Increased level of endogenous ABA has also been observed in drought-stressed barley (Thameur et al. 2011), salt-stressed tobacco cells and alfalfa seedlings (Singh et al. 1987; Luo et al. 1992). External application of ABA has been revealed to increase drought tolerance in tomato and wheat species (Yadegari et al. 2014; Wei et al. 2015). ABA treatments could increase cold resistance in cucumber (Flores et al. 1988).

Indoleacetic acid (IAA) is one of the naturally occurring growth hormones that enhance cell division and elongation. The behavior of IAA in drought resistance is still rather contradictory. For a long time, it was generally assumed that water deficiency results in a decrease of IAA content (Pustovoitova and Zholkevich 1992). However, it has become more evident that adaptation to drought is accompanied by an increase in the IAA levels (Zholkevich and Pustovoitova 1993). Several reports have demonstrated that adaptation to drought is accompanied by an increase in the IAA levels (Zholkevich and Pustovoitova 1993). Indoleacetic acid (IAA) producing halotolerant bacteria to promote tomato (*Solanum lycopersicum* L.) growth and tolerance against salinity stress (Kang et al. 2019).

Brassinosteroids (BRs) are ubiquitous plant steroid hormones, playing diverse roles in plant growth, development, and stress responses (Ahammed et al. 2015). Defects in BRs biosynthesis or signaling result in physiologic and phenotypic abnormalities. To date, numerous studies have highlighted BR-induced stress tolerance to various environmental extremes such as high temperature, chilling, drought, salinity, and heavy metals in a range of plant species. In addition to this, BRs have the ability to confer resistance to plants against various biotic and abiotic stresses, such as salinity (Hayat et al. 2010), water stress (Fariduddin et al. 2009a), temperature extremes (Fariduddin et al. 2011; Gomes 2011), and heavy metals (Bajguz and Hayat 2005; Fariduddin et al. 2009b; Yusuf et al. 2011, 2012).

Salicylic acid (SA) is another molecule present in most plants (Shi et al. 2005) and has various functions, especially in the inhibition of germination, growth, and in the absorption of solutes through the roots (Ashraf et al. 2010). It plays an important role in the regulation of a number of physiological processes in the plant (Shi et al. 2005) and is an important component in the signals of translation of the main pathways for systemic acquired resistance (Song et al. 2004). Research has confirm that salicylic acid is a plant hormone and growth regulator that, in small amounts, promotes, inhibits, or modifies the physiological processes of the plants subjected to osmotic deficiency (Gonçalves 2013) and induces tolerance against biotic and abiotic stress factors (Kang et al. 2014). Salicylic acid has been studied extensively to evaluate its capability to provide protection to plants when they are subjected to water stress. Recent studies have shown that SA also confers plant resistance to various abiotic stresses. SA and methyl salicylate (MeSA) treatments increased resistance to chilling injury in maize (Janda et al. 1999), tomato (Ding et al. 2002), peaches

(Wang et al. 2006), and rice (Kang and Saltveit 2002). Potato microplants (Lopez-Delgado et al. 1998) are also shown to be more tolerant to heat stress when treated with exogenous SA. Additionally, SA is involved in regulating plant responses to heavy metal-induced toxicity (Wang et al. 2004a, b; Drazic and Mihailovic 2005; Pál et al. 2005). Interestingly, the majority of SA-regulated abiotic stresses in plants are involved in antioxidative responses, suggesting that SA as an internal signal molecule interacts with ROS signal pathway. For instance, treatment with exogenous SA caused H_2O accumulation in Arabidopsis, tobacco, and mustard (Rao et al. 1997). Both exogenous SA and H_2O_2 induced a high tolerance of potato microplants to high temperature (Lopez-Delgado et al. 1998).

Jasmonic acid (JA) is one of the most important plant growth regulators which is showing good potential in protecting plants from various types of stress conditions (Sirhindi et al. 2015). Jasmonic acid [3oxo-2-(2-cis-pentylcyclopentane 1-acetic acid)] acts as modulator by suppressing or enhancing the stress responses of plants (Popova et al. 1995). Promotion of leaf senescence estimated by a decrease in chlorophyll content and depression of photosynthesis-related gene was one of the first reported physiological effects of jasmonates group (He et al. 2002). Cell division, plant growth, stomatal conductance, and photosynthetic processes are all diversely affected by JA. It is reported from the studies carried out in previous years that application of JA could modulate the plant physiological processes toward abiotic stress tolerance (Walia et al. 2007) like salt stress (Kaur et al. 2013) and drought stress (Alam et al. 2014). Exogenous JA is effective in protecting plants from Cu stress by modulating photosynthetic pigments (Sharma et al. 2013).

Coronatine (COR) is a chlorosis-inducing nonhost-specific phytotoxin produced by several members of the pseudomonas syringae group of pathovars (Bender et al. 1999) and induces a wide array of effects in plants (Li et al. 2010). It leads to diffuse chlorosis of leaf, anthocyanin production, ethylene emission, auxin synthesis, tendril coiling, inhibition of root elongation, and hypertrophy (Feys et al. 1994; Uppalapati et al. 2008). Coronatine can increase defense-related protease inhibitors and secondary metabolites, such as volatiles, nicotine, and alkaloid as well as may play an important role in resistance to abiotic stress, such as salinity stress (Schuler et al. 2004; Xie et al. 2008). Ai et al. (2008) showed that COR increased the activities of SOD, CAT, POD, and GR in two rice cultivars under drought. Some studies indicated that COR acted as a mimic of jasmonates (JAs) and was more active than JAs with regard to production of secondary metabolites (Tamogami and Kodama 2000). Yan et al. (1999) reported that COR enhanced the resistance of millet seedlings to water stress. Relative water content (RWC) was used to identify drought tolerance in spring wheat cultivars (Strauss and Agenbag 2000). Wang et al. (2008) reported that COR alleviated drought stress in maize seedlings and low concentrations of COR increased the RWC in leaves of maize seedlings. Effects of plant hormones on abiotic stress tolerance mechanism of the crop under Fabaceae have been shown in Table 2.

Table 2 Improvement in growth and different physio-biochemical attributes by application of plant hormones in different species of peas under stress conditions

Osmoprotectants	Stress	Crop	Activities	References
Abscisic acid	Salinity	Common bean (*Phaseolus vulgaris* L.)	– Declined the activity of – 1,3-diaminopropan, malondialdehyde and superoxide dismutase – Declined the activities of proline and endogenous abscisic acid – Increased in the contents of Put, Spd, and Spe	Shevyakova et al. (2013)
Abscisic acid	Salinity	Common bean (*Phaseolus vulgaris* L.)	– Proline content in nodules decreased – Decrease in the content of amino acids and protein in nodule during 6th-day harvest – Endogenous ABA increased	Khadri et al. (2006)
Abscisic acid	Heat	Chickpea (*Cicer arietinum*)	– ABA increased in the contents of osmoprotectants like proline, glycine betaine, and trehalose, accompanied by improved plant tolerance	Kumar et al. (2012)
Abscisic acid	Drought	Grass pea (*Lathyrus sativus* L.)	– Increased in the content of ODAP and spermine	Xiong et al. (2006)
ABA (Abscisic acid)	Drought	Pea (*Pisum sativum*)	– Reduces the reduction in both fresh and dry weights shoot and root, proline content, total chlorophyll, SOD, ACP was increased, Partially overcame the decrease in IAA, GA contents and the adverse effect of water-deficit stress	Latif (2014)
Abscisic acid	High temperature and drought	Alfalfa (*Medicago sativa*)	– Proline increased in the leaf of tolerant genotype and endogenous abscisic acid increased both in tolerant and sensitive genotype	An et al. (2014)
Benzyladenine or putrescine	High temperature	Common bean (*Phaseolus vulgaris* L.)	– Increase in the content of protein, carbohydrate, and pod fiber in highest concentration application at medium and late sowing	Zaki et al. (2014)
β-Estradiol	Heavy metal (cadmium or copper)	Lentil (*Lens culinaris* Medik.)	– Enhance protease and amylase activities – Increase in the content of free amino acids and glucose	Chaoui and Ferjani (2013)

(continued)

Table 2 (continued)

Osmoprotectants	Stress	Crop	Activities	References
Brassinolide	Salinity	Cowpea (*Vigna sinensis*)	– Enhance SOD, POD, and PPO activities – Decreased lipid peroxidation – Increase in the content of ascorbic acid, tocopherol, and glutathione	El-Mashad and Mohamed (2012)
EBL (24-epibrassinolide)	Salt and zinc	Mung bean (*Vigna radiata* L.)	– Increased gas-exchange parameters and decreased in the electrolyte leakage. Increased carbonic anhydrase (CA) and nitrate reductase (NR) activity, proline content, and antioxidative enzymes, namely CAT, POX, SOD activity, were also increased	Mir et al. (2015)
EBL (24-epibrassinolide)-Foliar spray	Salinity	Pea (*Pisum sativum*)	– Significantly alleviated the deleterious effects of salinity by improving the plant biomass, water relations, and concentration of organic/inorganic osmolytes	Shahid et al. (2015)
EBL (24-epibrassinolide)	Cadmium (Cd) and/or salinity	Chickpea (*Cicer arietinum* L.)	– Significantly reduced the inhibitory effects on photosynthesis and increased total chlorophyll content, net photosynthetic rate, activities of nitrogenase, glutamate synthase, glutamine synthetase and glutamate dehydrogenase enzymes, antioxidant enzyme activities, and the proline content in both stress	Wani et al. (2017)
EBL (24-epibrassinolide) and Silicon	Cadmium (Cd)	Pea (*Pisum sativum*)	– Decreased the accumulation of H_2O_2, MDA content, electrolyte leakage, and methylglyoxal and maintained antioxidant levels and Gly-I, Gly-II levels were increased. Finally reduced Cd accumulation by enhancing the uptake of macronutrients and micronutrients in shoots and roots	Jan et al. (2018)

(continued)

Table 2 (continued)

Osmoprotectants	Stress	Crop	Activities	References
EBL (24-epibrassinolide)	Mercury	Chickpea (*Cicer arietinum* L.)	– Proline and glycine betaine contents were increased and reduced the production of hydrogen peroxide. Increased the activities of SOD, CAT, GST, and GPX, AsA, and GSH	Ahmad et al. (2018)
EBL (24-epibrassinolide)	Drought (water stress)	Grass pea (*Lathyrus sativus* L.)	– Increased in the content of β-N-oxalyl-L-α, β-diaminopropionic acid (β-ODAP) during moderate water stress	Xiong et al. (2016)
EBL (24-epibrassinolide)	Iron	Groundnut (*Arachis hypogaea* L.)	– Activities of antioxidant enzymes, i.e., SOD, POD, CAT, were increased	Song et al. (2016)
Etheopan (donor source of ethylene)	Waterlogged	Soybean (*Glycine max*)	– Significantly increased the photosynthesis pigment and GA_s content. The amino acid contents were increased and induced adventitious root initiation. Root surface area and expressions of glutathione transferases and relative glutathione activity, protein content and glutathione S-transferase (GSTs) were increased	Kim et al. (2018)
Gibberellic acid (GA_3)	Salinity	Alfalfa (*Medicago sativa* L.)	– Enhanced SOD, CAT, GPX, APX, and GR activities – Decreased in the content of MDA and total ascorbate level was higher in both cases	Younesi and Moradi (2014)
Gibberellic acid (GA_3)	Salinity	Mung bean (*Vigna radiata* L.)	– At low concentrations increased biomass production, chlorophyll content, and significant lowering SOD, CAT activities as well as proline, MDA and H_2O_2 contents in comparison to salt-treated test seedlings. Actually, at low concentrations of either gibberellic acid will be able to overcome the toxic effects of NaCl stress in mung bean seedlings	Ghosh et al. (2015)

(continued)

Table 2 (continued)

Osmoprotectants	Stress	Crop	Activities	References
Gibberellic acid (GA_3)	Salinity	Soybean (*Glycine max*)	– Bioactive gibberellins (GA_1 and GA_4) and jasmonic acid were increased but abscisic acid and salicylic acid were decreased. Adverse effect of salt stress was mitigated by regulating the level of phytohormones and restores normal growth	Hamayun et al. (2010)
Gibberellic acid (GA_3)	Chromium phytotoxicity	Pea (*Pisum sativum*)	– Decreased root and shoot nitrate reductase (NR), glutamine synthetase (GS), and glutamine 2-oxoglutarate amino transferase (GOGAT) activities, while glutamate dehydrogenase (GDH) activity and NH_4^+ content increased – The root and shoot activities of SOD and APX increased except APX activity, while CAT, GR, and DHAR activities were decreased – Total ascorbate and glutathione content in root and shoot decreased	Gangwar et al. (2011)
Gibberellic acid, indoleacetic acid (IAA)	Salinity	Lentil (*Lens culinaris*)	– Enhanced the activities of SOD, CAT, and POX – Increased in the content of proline, carbohydrate, and protein	Khattab et al. (2018)
Kinetin (Kn)	Salinity	Groundnut (*Arachis hypogaea* L.)	– Amino acid content was increased but reduced sugars and proline content	Elhamahmy Mohamed and Kotb (2014)
Kinetin (Kn)	Salinity	Soybean (*Glycine max*)	– Bioactive GA_1 and GA_4 contents were significantly higher in Kn-treated plants, endogenous JA, SA, and ABA contents were increased. Kn application plays an important role to reduce the adverse effects of salt stress	Hamayun et al. (2015)
Jasmonic acid (JA) or ascobin	Salinity	Soybean (*Glycine max*)	– Accumulation of nontoxic metabolites, i.e., sugars, amino acid proline, and proteins were enhanced	Sheteawi (2007)

(continued)

Table 2 (continued)

Osmoprotectants	Stress	Crop	Activities	References
Jasmonic acid (JA)	Heavy metal (Cd)	Faba bean (*Vicia faba*)	– JA restored growth, biomass yield, LRWC and pigment systems to appreciable levels and reduced levels of H_2O_2, MDA, and electrolyte leakage – Increased osmolytes and antioxidants	Ahmad et al. (2017a, b)
Jasmonic acid	Heavy metal (Copper)	Pigeon pea (*Cajanus cajan*)	– Reduced the oxidative stress and improved the physiological tolerance mechanism altering photosynthetic pigments, osmolytes, and antioxidants	Poonam et al. (2013)
Jasmonic acid	Heavy metal (Copper)	Pigeon pea (*Cajanus cajan*)	– Decreased in the content of MDA, proline. Glycine betaine decreased and reducing sugar increased in only high concentration of JA application – Enhanced POD activity in higher concentration of JA – Declined SOD and CAT activities	Sirhindi et al. (2015)
Methyl jasmonate	Drought	Cowpea (*Vigna sinensis*)	– Increased in the content of proline and total soluble sugars whereas decreased MDA	Sadeghipour (2018)
Methyl jasmonate	Salinity	Soybean (*Glycine max*)	– Reduced endogenous hormones and increased ABA levels, but GA4 content was decreased. Finally reduced the adverse effects of salinity on plant growth, chlorophyll content, leaf photosynthetic rate, leaf transpiration rate, and proline content	Yoon et al. (2009)
Methyl jasmonate	Drought	Soybean (*Glycine max*)	– Increased in activities of SOD, POD, CAT, proline but decreased MDA content	Anjum et al. (2011)
Coronatine (COR)	Heat	Chickpea (*Cicer arietinum* L.)	– MDA, proline content and SOD activity was decreased – But CAT, POX activity was increased; APX and GR activity showed a remarkable decrease but isoenzyme APX4 was increased	Ceylan et al. (2013)

(continued)

Table 2 (continued)

Osmoprotectants	Stress	Crop	Activities	References
Coronatine (COR)	Heat + PEG	Chickpea (*Cicer arietinum* L.)	– SOD activity was same, but proline content was decreased and increased the activities of CAT, POX; APX, GR activity and isoenzyme activity APX1, APX3 were decreased	Ceylan et al. (2013)
Coronatine (COR)	PEG osmotic stress	Chickpea (*Cicer arietinum* L.)	– MDA, proline content and SOD activity was decreased, but CAT, POX, APX, and GR activity were increased. An isoenzyme APX_1 and APX_4 were also increased	Ceylan et al. (2013)
Coronatine (COR; a structural and functional analog of jasmonates)	Drought	Groundnut (*Arachis hypogaea* L.)	– Osmolytes of peanut seedling leaves were increased, and MDA content was reduced. – The best concentration of COR was 0.01 μmol/L	Qin et al. (2009)
Salicylic acid (SA)	Heavy metal (Mercury)	Alfalfa (*Medicago sativa*)	– Enhanced NADH oxidase, APX and POD activities in the roots whereas slightly decline SOD activity – Increased in the content of ascorbate, glutathione, and proline in the roots	Zhou et al. (2009)
Salicylic acid (SA)	Salinity	Black gram (*Vigna mungo* L.)	– Increased in the content of total soluble sugar, reducing sugar, proline, and glycine betaine	Solanki et al. (2018)
Salicylic acid (SA)	Drought (water stress)	Cowpea (*Vigna sinensis*)	– Decreased proline content	Afshari et al. (2013)
Salicylic acid (SA)	Salinity	Lentil (*Lens culinaris*)	– Enhanced the activity of proline biosynthetic enzymes, viz. glutamyl kinase and pyrroline-5-carboxylate – Increased in the content of glycine betaine	Misra and Saxena (2009)
Salicylic acid (SA)	Salinity	Pea (*Pisum sativum*)	– Increased in the content of proline, total protein and sugar content as well as enhance SOD, CAT and POD activities at medium stress	Ahmad et al. (2017a)

(continued)

Table 2 (continued)

Osmoprotectants	Stress	Crop	Activities	References
Salicylic acid (SA)	Salinity	Pea (*Pisum sativum*)	– Reduction in soluble protein; – Increased of proline contents and SOD activity	Sajid et al. (2016)
Salicylic acid (SA)	Heavy Metals (As)	Soybean (*Glycine max*)	– Remarkable enhancement in proline levels was observed	Chandrakar et al. (2016)
Salicylic acid and *Moringa oleifera* leaf extract	Salinity	Common bean (*Phaseolus vulgaris* L.)	– Seed soaking with salicylic acid and foliar spray with *Moringa oleifera* leaf extract together increased total soluble sugars, free proline, and ascorbic acid	Rady et al. (2015)

3.3 Antioxidants

Ascorbic acid (AA) is a small antioxidant molecule as well as it is water soluble, reductone sugar acid with antioxidant properties, and acts as a primary substrate in the cyclic pathway for enzymatic detoxification of a number of reactive oxygen species (ROS) such as H_2O_2 and many other, harmful to normal functioning of plant metabolism. In addition, it acts directly to neutralize superoxide radicals ($O_2^{\cdot -}$), singlet oxygen ($O^{\cdot -}$), or hydroxyl radical ($OH^{\cdot -}$) simply by acting as a secondary antioxidant during reductive recycling of the oxidized form of a-tocopherol (Noctor and Foyer 1998). Several reports have demonstrated that its capability to provide protection to plants, when they are subjected to abiotic stress. Exogenous application of AsA enhanced plant height, leaf number, leaf area, and lateral shoot number in olive (Mayi et al. 2014), chlorophyll a and b, total soluble proteins, carbohydrates and carotenoids in wheat (Hussein et al. 2014), and leaf area and number of leaves in pearl millet (Hussein and Alva 2014) under drought stress. It is reported from the study carried out in canola application of AsA could modulate stearic acid, linoleic acid, palmitic acid percentage, and oil yield under drought conditions (Ahmed et al. 2013).

The tripeptide glutathione (GSH; γ-glutamate-cysteine-glycine) is sulfur containing nonprotein thiol. Reactivity of GSH depends on the thiol group (-SH). It can act as an effective electron accept or and donor for many biological reactions (Xiang et al. 2001). The nucleophilic nature of the thiol group is also important in the formation of mercaptide bonds with metals and for reacting with select electrophiles (Xiang et al. 2001). The reactivity, stability, and high water solubility of GSH make it an ideal biomolecule to plants against abiotic stresses. GSH is involved in many cellular processes including detoxification of xenobiotics (Dixon et al. 1998), sequestration of heavy metals (Cobbett and Goldsbrough 2002), and defense against ROS (Foyer and Noctor 2005). In addition, GSH is a substrate for glutathione S-transferase (GST) and glutathione peroxidase (GPX) which are also involved in the removal of

ROS (Noctor et al. 2002). GSH is synthesized from two-consecutive ATP-dependent reactions. In the first step, γ-glutamylcysteine (EC) is formed from L-glutamate and L-cysteine by γ-glutamylcysteine synthetase (ECS). The second step is catalyzed by glutathione synthetase which adds glycine to C-terminal of EC forming GSH. ECS is a major regulatory enzyme in GSH biosynthesis. The plastid is the compartment for GSH biosynthesis (Ching 2015).

Tocopherol (Toc) is a countable nonenzymatic antioxidant that is synthesized in the plastid of photosynthesizing organisms. In plants, four different types of Toc are found, including alpha (α), beta (β), gamma (γ), and delta (δ)-Toc. However, α-Toc is the most common form found in photosynthetic tissue. The accumulation of Toc varies greatly in different plant species and different plant parts as well. Tocopherols have diversified roles in plant growth and physiological processes that can also affect yield. Due to its antioxidant activity, Toc plays a vital role in conferring tolerance to several abiotic stresses (e.g., salinity, drought, metal toxicity, ozone, UV radiation). Several reports indicate that stress-tolerant plants exhibit an enhanced level of Toc, whereas sensitive ones show a decreased level of Toc under stressful conditions leading to oxidative damage. The most important role of Toc in mitigating abiotic stress-induced damage is its ability to scavenge or quench lipid peroxides, oxygen radicals, or singlet oxygen, resulting in detoxification of reactive oxygen species. Tocopherol also works in coordination with other antioxidants (e.g., ascorbate) and interacts with phytohormones such as ethylene, abscisic acid, salicylic acid, and jasmonic acid. Although a considerable number of studies have been done to explore the role of Toc in abiotic stress tolerance, there are still many gaps in grasping the basic mechanisms through which Toc confers abiotic stress tolerance (Hasanuzzaman et al. 2014b). Effects of antioxidants on abiotic stress tolerance mechanism of the crop under Fabaceae have been shown in Table 3.

3.4 Signaling Molecules

Nitric oxide (NO) is a simple molecule whose production is regulated by complex mechanisms, given the large number of synthesis and scavenging pathways that influence NO homeostasis. Apart from being the most abundant reactive nitrogen species (RNS) in plants, NO is considered a gasotransmitter with a pivotal role in a plethora of physiological processes throughout the plant life cycle, from the regulation of growth and development to biotic and abiotic stress tolerance (Sánchez-Vicente et al. 2019). Effects of signaling molecules on abiotic stress tolerance mechanism of the crop under Fabaceae have been shown in Table 4.

Table 3 Improvement in growth and different physio-biochemical attributes by application of antioxidants in different species of peas under stress conditions

Osmoprotectants	Stress	Crop	Activities	References
Ascorbic acid and gibberellin (GA_3)	Salinity	Common bean (*Phaseolus vulgaris* L.)	– The interaction of AsA with GA3 decreased the amounts of MDA and H_2O_2 whereas increased protein content – Enhance guaiacol peroxidase (GPOX) activity	Saeidi-Sar et al. (2013)
Glutathione (GSH)	Drought	Mung bean (*Vigna radiata* L)	– AsA content level did not increase significantly; – Proline, GSH, GSSG, GSH/GSSG ratio was increased	Nahar et al. (2015a)
Ascorbic acid, α -tocopherol and chitosan (foliar application)	Drought	Soybean (*Glycine max*)	– Effect of ascorbic acid increased the chlorophyll total, total of leaf area, number of filled pods and relative water content, while chitosan treatment increased stomatal density – All the antioxidants application showed the important of the role in reducing the effect of water stress on production and physiological characters of soybean	Hasanah et al. (2017)
Glutathione (GSH)	Salinity	Soybean (*Glycine max* L.)	– Increased in the content of MDA	Akram et al. (2017)
Glutathione (GSH)	High temperature (HT, 42 °C)	Mung bean (*Vigna radiata* L.)	– Increased APX, MDHAR, DHAR, GR, GPX, GST, CAT, Gly-I and Gly-II activities, GSH content and the GSH/GSSG ratio but decreased GSSG, MDA, H_2O_2 and MG content, $O_2\cdot$ generation rate and LOX activity were decreased	Nahar et al. (2015b)

(continued)

Table 3 (continued)

Osmoprotectants	Stress	Crop	Activities	References
Glutathione (GSH)	Salinity	Mung bean (*Vigna radiata* L)	– AsA content and activity of APX, MDHAR, DHAR, GR, SOD, CAT, GPX, GST, Gly-I and Gly-II were increased but – MDA and MG content was decreased	Nahar et al. (2015c)
Strigolactones (SL)	Dark chilling	Pea (*Pisum sativum*)	– Significantly greater shoot branching with higher leaf chlorophyll a/b ratios and carotenoid contents and decreased shoot fresh weights but increased leaf numbers; biomass (dry weight) accumulation; decreased photosynthetic carbon assimilation was observed	Cooper et al. (2018)

3.5 Polyamines

Polyamines (PAs) are low-molecular-weight aliphatic nitrogenous bases containing two or more amino groups, and they have potent biological activity (Xu et al. 2009; Vuosku et al. 2018). Putrescine (Put), spermidine (Spd), and spermine (Spm) are the main PAs in plants, and they are involved in the regulation of diverse physiological processes (Xu et al. 2014b; Mustafavi et al. 2018). They are also involved in responses to biotic and abiotic stresses (Vuosku et al. 2012; de Oliveira et al. 2016; Reis et al. 2016; Mustafavi et al. 2018). Putrescine is the central product of the common PA biosynthetic pathway. It contains two amino groups and is a synthetic precursor of Spd and Spm (Xu et al. 2009). Application of Spd and Spm improves grain filling and drought tolerant in wheat through relieving the inhibition caused by drought stress (Yang et al. 2016). Polyamines (Spm, Spd, and Put) can regulate the size of the potassium channel and the size of pores in the plasma membrane of guard cells, thereby strongly regulating pore opening and closing. In this way, PAs can control water loss in plants (Liu et al. 2000). In *Thymus vulgaris* L., Put treatment was shown to improve leaf water content, dry matter, antioxidant activities and decrease cell injury indices under water stress on plants (Mohammadi et al. 2018). Effects of PAs on abiotic stress tolerance mechanism of the crop under Fabaceae have been shown in Table 5.

Table 4 Improvement in growth and different physio-biochemical attributes by application of signaling molecules in different species of peas under stress conditions

Osmoprotectants	Stress	Crop	Activities	References
Nitric oxide (NO)	Heavy metal (Arsenic)	Common bean (*Phaseolus vulgaris* L.)	– Exogenous addition of NO significantly reversed the As-induced oxidative stress, maintaining H_2O_2 in a certain level through balanced alterations SOD, CAT, APX, DHAR, and GR activities and reduction of membrane damage	Talukdar (2013)
Nitric oxide (NO)	Salinity	Chickpea (*Cicer arietinum* L.)	– Increased the activity of SOD, CAT, APX, total soluble proteins, proline, and glycine betaine content	Ahmad et al. (2016)
Nitric oxide (NO)	Cadmium (Cd)	Groundnut (*Arachis hypogaea* L.)	– Increased the biomass and improved chlorophyll content, photosynthesis, and transpiration – Decreased accumulation of O_2, H_2O_2, and MDA content – Increased activities of SOD, POD, and CAT – Soluble protein and proline contents increased at sowing and seedling stages and improved the uptake of nutrient elements and reduced Cd accumulation in the roots, shoots, and kernels under Cd stress	Xu et al. (2014a, b)
Sodium nitroprusside (SNP)	Cadmium (Cd)	Groundnut (*Arachis hypogaea* L.)	– Effectively decrease O_2^- generation rate and MDA accumulation in leaves and roots. SOD, POD, and CAT were decreased, and nonenzymatic antioxidants (AsA and proline) was prevented	Dong et al. (2016)
Sodium nitroprusside (SNP)	Iron	Groundnut (*Arachis hypogaea* L.)	– Increased the activities of SOD, POD, and CAT – Increased the available Fe and chlorophyll concentrations in leaves, and promoted plant growth in Fe deficiency	Song et al. (2017)

(continued)

Table 4 (continued)

Osmoprotectants	Stress	Crop	Activities	References
SNP and polyamine (putrescine)	Salinity	Chickpea (*Cicer arietinum* L.)	– Positive effect on antioxidant enzymes – Putrescine increased the SOD activity, whereas nitric oxide increased the activities of CAT, POX, and APX	Sheokand et al. (2008)
NO and SA	Cadmium (Cd)	Groundnut (*Arachis hypogaea* L.)	– Improved growth, chlorophyll content, photosynthesis, and mineral nutrition – Decreased oxidative stress by increasing activities of antioxidant enzymes and content of nonenzymatic antioxidants – Increased Cd accumulation in the cell wall and decreased Cd distribution to organelles	Xu et al. (2014a, b)

3.6 Trace Elements

Silicon (Si) and selenium (Se) are widely studied trace elements, and their roles in plant growth and physiology are well documented. Recently, these trace elements have been found to be protective under abiotic stress conditions, and both were reported to play roles in conferring oxidative stress tolerance by enhancement of the antioxidant defense system in plants (Hasanuzzaman and Fujita 2011). A number of studies have showed that plants growing under heavy metal stress (cadmium) had reduced ROS contents when supplemented with Si compared to control plants (Hasanuzzaman et al. 2017). Soundararajan et al. observed an improved stomatal development in tissue-cultured carnation plants supplemented with Si. Similarly, Rios et al. proposed a model explaining how Si could improve stomatal functioning and enhance root hydraulic conductance through the regulation of aquaporins. Selenium's role in combating abiotic stresses caused by UV light, drought, heavy metals, cold, and salts is well documented in earlier literature. Previously, it helped promoting growth of rye grass, delaying drought, and inducing tolerance to UV-induced stress (Cartes et al. 2010; Hasanuzzaman and Fujita 2011; Hasanuzzaman et al. 2014a, b, c). Similarly, under drought stress, Se pretreatment of rapeseed enhanced antioxidant enzyme activities and increased concentration of AsA and glutathione (Hasanuzzaman and Fujita 2011). Selenium treatment also protected sunflower seedlings from cadmium-induced oxidative stress by boosting CAT, APX, and GR activities (Saidi et al. 2014). Effects of trace elements on abiotic stress tolerance mechanism of the crop under Fabaceae have been shown in Table 6.

Table 5 Improvement in growth and different physio-biochemical attributes by application of polyamines in different species of peas under stress conditions

Osmoprotectants	Stress	Crop	Activities	References
Putrescine	Drought	Alfalfa (*Medicago sativa*)	– Enhanced α-and β-amylases, invertase, and decline protease activities during PEG imposed drought – Declined α-and β-amylases, invertase, and protease activities and increase the polysaccharide and protein content	Zeid and Shedeed (2006)
Putrescine and arbuscular mycorrhiza	Salinity	Pigeon pea (*Cajanus cajan*)	– Modulating anabolic and catabolic enzyme activities responsible for putrescine biosynthesis	Garg and Sharma (2019)
Putrescine, spermidine, and spermine	Drought	Mung bean (*Vigna radiata* L.)	– Increased soluble protein content, relative water content, chlorophyll value, net photosynthetic rate, and more increased TSS and proline contents but decreased stomatal conductance and MDA level	Sadeghipour (2019)
Putrescine, spermidine, and spermine	Salinity	Mung bean (*Vigna radiata* L.)	– LOX activity, MDA content was reduced – Decreased DHA, GSSG content and increased content of AsA, GSH and the ratio of AsA/DHA, GSH/GSSG – CAT, DHAR, GPX, and GR was increased; APX and MDHAR did not show increases activity and GST activity remain same	Nahar et al. (2016)

(continued)

Table 5 (continued)

Osmoprotectants	Stress	Crop	Activities	References
Putrescine and arbuscular mycorrhiza	Salinity	Pigeon pea (*Cajanus cajan*)	– Modulating anabolic and catabolic enzyme activities responsible for putrescine biosynthesis	Garg and Sharma (2019)
Putrescine, spermidine, and spermine	Drought	Mung bean (*Vigna radiata* L.)	– Increased soluble protein content, relative water content, chlorophyll value, net photosynthetic rate, and more increased TSS and proline contents but decreased stomatal – conductance and MDA level	Sadeghipour (2019)

4 Summary and Conclusion

1. Osmolytes play an important and significant role in plant body. Osmolytes protect the plant from different stress condition and help the plants to overcome and recover from abiotic stress environment more rapidly.
2. Exogenous application of osmolytes under abiotic stress enhanced growth and development with desirable physiological characteristics of plants.
3. Exogenous osmolytes scavenge the ROS generated in plants under various abiotic stresses and result in reduction of oxidative damage to plant cells.
4. Exogenous application of osmoprotectant affects plant–water relations by maintaining turgidity of cells under stress and also increases the rate of photosynthesis with abiotic stress tolerance and improves crop yield.
5. Exogenous application of osmolytes to plants protects them from harmful effect of toxic elements like Cd and As.
6. Low concentrations of exogenous osmolytes protect plants from salinity, drought, and temperature stress. However, higher doses will cause toxic effects.
7. Keeping in view the potential of exogenous osmolytes and osmoprotectants may become a useful tool for successful crop production to counter the adverse effect of abiotic stressful environments.

Table 6 Improvement in growth and different physio-biochemical attributes by application of trace elements in different species of peas under stress conditions

Osmoprotectants	Stress	Crop	Activities	References
Silicon (Si)	Drought	Lentil (*Lens culinaris*)	– Decreased in the content of glycine betaine, proline, and total soluble sugar – Enhanced the activities of hydrolytic enzymes (α-amylase, β-amylase, and α-glucosidase) – Enhanced the activities of the antioxidant enzymes (APX, POX, CAT, and SOD)	Biju et al. (2017)
Silicon (Si)	Cadmium (Cd)	Pea (*Pisum sativum*)	– Restored the plant growth and membrane stability. Significantly increased the CAT, POD, SOD, and GR activities along with S-metabolites (cysteine, methionine, glutathione)	Rahman et al. (2017)
Silicon (Si)	Salinity	Alfalfa (*Medicago sativa* L.)	– Enhance APX activity in root, shoot, and leaves, CAT activity in leaves, POD activity in shoots – Decline SOD activity in shoots and roots	Wang et al. (2011)
Silicon (Si)	Drought	Chickpea (*Cicer arietinum* L.)	– H_2O_2, proline, MDA content, and LOX activity were decreased – Increased the SOD activity; CAT activity varied greatly in response to Si treatment. Three cultivars showed decrease in activity and four showed increasing activity, but APX activity significantly reduced	Gunes et al. (2007)

(continued)

Table 6 (continued)

Osmoprotectants	Stress	Crop	Activities	References
Silicon (Si)	Aluminum	Groundnut (*Arachis hypogaea* L.)	– Increased the activities of CAT, SOD, and POD but decreased MDA content in leaves and roots	Shen et al. (2014)
Silicon (Si)	Cadmium (Cd)	Groundnut (*Arachis hypogaea* L.)	– Reduced Cd toxicity in seedlings and enhancement of the SOD, POD, and CAT activities in roots	Shi et al. (2010)
Calcium (Ca)	Heavy metal (Cd)	Lentil (*Lens culinaris*)	– Modulate the activities of SOD, APX, CAT, DHAR, and GR	Talukdar (2012)
Calcium chloride ($CaCl_2$)	Salinity	Soybean (*Glycine max*)	– Increased content of glycine betaine but decreased proline content	Girija et al. (2002)
Potassium (K^+)	Salinity	Groundnut (*Arachis hypogaea* L.)	– Improves plant water status and overall physiology of plant	Chakraborty et al. (2016)
Potassium nitrate	Osmotic stress	Cowpea (*Vigna unguiculata)*	– Increased activity SOD, CAT, and APX in tolerant genotype	Araújo et al. (2017)
Lanthanum	Salinity	Mung bean (*Vigna radiate*)	– Enhanced the activities of APX, GR, MDHAR, DHAR, γ-ECS, and GalLDH – Increased in the content of AsA, GSH, and the ratios of AsA/DHA and GSH/GSSG reduce the MDA	Shan and Zhao (2014)

References

Abdelhamid MT, Rady MM, Osman AS, Abdalla MA (2013) Exogenous application of proline alleviates salt-induced oxidative stress in *Phaseolus vulgaris* L. plants. J Hortic Sci Biotech 88(4):439–446

Abdelmotlb NA, Abdel-All FS, Abd EL-Hady SA, EL-Miniawy SM, Ghoname AA (2019) Glycine betaine and sugar beet extract ameliorated salt stress adverse effect on green bean irrigated with saline water. Middle East J Agric Res 9:142–154

Acquaah G (2007) Principles of plant genetics and breeding. Blackwell, Oxford, p 385

Afshari M, Shekari F, Azimkhani R, Habibi H, Fotokian MH (2013) Effects of foliar application of salicylic acid on growth and physiological attributes of cowpea under water stress conditions. Iran Agric Res 32(1):55–70

Aggarwal M, Sharma S, Kaur N, Pathania D, Bhandhari K, Kaushal N, Kaur R, Singh K, Srivastava A, Nayyar H (2011) Exogenous proline application reduces phytotoxic effects of selenium by minimisingoxidative stress and improves growth in bean (*Phaseolus vulgaris* L.) seedlings. Biol Trace Element Res 140(3):354–367

Ahammed GJ, Xia XJ, Li X, Shi K, Yu JQ, Zhou YH (2015) Role of brassinosteroid in plant adaptation to abiotic stresses and its interplay with other hormones. Curr Protein Pept Sci 16(5):462–473

Ahmad P, Abdel Latef AA, Hashem A, Abd-Allah EF, Gucel S, Tran LS (2016) Nitric oxide mitigates salt stress by regulating levels of osmolytes and antioxidant enzymes in chickpea. Front Plant Sci 7:347

Ahmad F, Singh A, Kamal A (2017a) Ameliorative effect of salicylic acid in salinity stressed *Pisum sativum* by improving growth parameters, activating photosynthesis and enhancing antioxidant defense system. Biosci Biotech Res Commun 10:481–489

Ahmad P, Alyemeni MN, Wijaya L, Alam P, Ahanger MA, Alamri SA (2017b) Jasmonic acid alleviates negative impacts of cadmium stress by modifying osmolytes and antioxidants in *Faba bean* (*Vicia faba* L.). Archives of Agronomy and Soil Science 63(13):1889–1899

Ahmad P, Ahanger MA, Egamberdieva D, Alam P, Alyemeni MN, Ashraf M (2018) Modification of osmolytes and antioxidant enzymes by 24-epibrassinolide in chickpea seedlings under mercury (Hg) toxicity. J Plant Growth Regul 37(1):309–322

Ahmed F, Baloch DM, Hassan MJ, Ahmed N (2013) Role of plant growth regulators in improving oil quantity and quality of sunflower hybrids in drought stress. Biologia 59:315–322

Ai L, Li ZH, Xie ZX, Tian XL, Eneji AE, Duan LS (2008) Coronatine alleviates polyethylene glycol-induced water stress in two rice (*Oryza sativa* L.) cultivars. J Agron Crop Sci 194:360–368

Akram S, Siddiqui MN, Hussain BN, Al Bari MA, Mostofa MG, Hossain MA, Tran LS (2017) Exogenous glutathione modulates salinity tolerance of soybean [*Glycine max* (L.) Merrill] at reproductive stage. J Plant Growth Regul 36(4):877–888

Alam MM, Hasanuzzaman M, Nahar K, Fujita M (2013) Exogenous salicylic acid ameliorates short-term drought stress in mustard (Brassica juncea L.) seedlings by up-regulating the antioxidant defense and glyoxalase system. Austr J Crop Sc 7(7):1053–1063

Alam MM, Nahar K, Hasanuzzaman M, Fujita M (2014) Exogenous jasmonic acid modulates the physiology, antioxidant defense and glyoxalase systems in imparting drought stress tolerance in different *Brassica* species. Plant BiotechRep 8:279–293

Ali HM, Siddiqui MH, Al-Whaibi MH, Basalah MO, Sakran AM, El-Zaidy MO (2013) Effect of proline and abscisic acid on the growth and physiological performance of faba bean under water stress. Pak J Bot 45(3):933–940

An Y, Zhou P, Liang J (2014) Effects of exogenous application of abscisic acid on membrane stability, osmotic adjustment, photosynthesis and hormonal status of two lucerne (*Medicago sativa* L.) genotypes under high temperature stress and drought stress. Crop Pasture Sci 65(3):274–286

Anjum SA, Wang L, Farooq M, Khan I, Xue L (2011) Methyl jasmonate-induced alteration in lipid peroxidation, antioxidative defence system and yield in soybean under drought. J Agron Crop Sci 197(4):296–301

Araújo ED, Melo AS, Rocha M, Socorro D, Carneiro RF, Rocha MD (2017) Genotypic variation on the antioxidative response of cowpea cultivars exposed to osmotic stress. Revista Caatinga 30(4):928–937

Ashraf MF, Foolad M (2007) Roles of glycine betaine and proline in improving plant abiotic stress resistance. Environ Exp Bot 59(2):206–216

Ashraf M, Akram NA, Arteca RN, Foolad MR (2010) The physiological, biochemical and molecular roles of brassinosteroids and salicylic acid in plant processes and salt tolerance. Crit Rev Plant Sci 29(3):162–190

Avonce N, Mendoza-Vargas A, Morett E, Iturriaga G (2006) Insights on the evolution of trehalose biosynthesis. BMC Evol Biol 6(1):109–113

Bajguz A, Hayat S (2005) Effects of brassinosteroids on the plant responses to environmental stresses. Plant Physiol Biochem 47(1):1–8

Bekka S, Abrous-Belbachir O, Djebbar R (2018) Effects of exogenous proline on the physiological characteristics of *Triticumaestivum* L. and *Lens culinaris* Medik. under drought stress. Acta Agriculturae Slovenica 111(2):477–491

Bender CL, Alarcon CF, Gross DC (1999) *Pseudomonas syringae* phytotoxins: mode of action, regulation and biosynthesis by peptide and polypeptide synthetases. Microbiol Mol Biol Rev 63:266–292

Biju S, Fuentes S, Gupta D (2017) Silicon improves seed germination and alleviates drought stress in lentil crops by regulating osmolytes, hydrolytic enzymes and antioxidant defense system. Plant Physiol Biochem 119:250–264

Cartes P, Jara A, Pinilla L, Rosas A, Mora ML (2010) Selenium improves the antioxidant ability against aluminium-induced oxidative stress in ryegrass roots. Ann Appl Biol 156:297–307

Cevic S, Unyayar S (2015) The effects of exogenous application of ascorbate and glutathione on antioxidant system in cultivated *Cicer arietinum* and wild type *C. reticulatum* under drought stress. J Nat Appl Sci 19(1):91–97

Ceylan HA, Türkan I, Sekmen AH (2013) Effect of coronatine on antioxidant enzyme response of chickpea roots to combination of PEG-induced osmotic stress and heat stress. J Plant Growth Regul 32(1):72–82

Chakraborty K, Bhaduri D, Meena HN, Kalariya K (2016) External potassium (K^+) application improves salinity tolerance by promoting Na^+ -exclusion, K^+ -accumulation and osmotic adjustment in contrasting peanut cultivars. Plant Physiol Biochem 103:143–153

Chandrakar V, Dubey A, Keshavkant S (2016) Modulation of antioxidant enzymes by salicylic acid in arsenic exposed *Glycine max* L. J Soil Sci Plant Nutr 16(3):662–676

Chaoui A, Ferjani EE (2013) β-estradiol protects embryo growth from heavy-metal toxicity in germinating lentil seeds. J Plant Growth Regul 32(3):636–645

Cha-um S, Kirdmanee C (2010) Effect of glycinebetaine on proline, water use, and photosynthetic efficiencies, and growth of rice seedlings under salt stress. Turkish J Agric Forest 34(6):517–527

Ching HK (2015) Role of glutathione in abiotic stress tolerance of rice plants. J Taiwan Agric Res 64(3):167–176

Cobbett C, Goldsbrough P (2002) Phytochelatins and metallothioneins: roles in heavy metal detoxification and homeostasis. Annu Rev Plant Biol 53:159–182

Cooper JW, Beveridge CA, Hu Y, Beyyoudh L, Dasgan HY, Kunert K, Foyer CH (2018) Strigolactones positively regulate chilling tolerance in pea and in *Arabidopsis*. Plant, Cell Environ 41:1298–1310

Crowe JH (2007) Trehalose as a "chemical chaperone": fact and fantasy. Adv Exp Med Biol 594:143–158

de Oliveira LF, Elbl P, Navarro BV, Al E (2016) Elucidation of the polyamine biosynthesis pathway during Brazilian pine (Araucaria angustifolia) seed development. Tree Physiol 37:116–130

Demirevska K, Simova-Stoilova L, Fedina I, Georgieva K, Kunert K (2010) Response of oryzacystatin I transformed tobacco plants to drought, heat and light stress. J Agron Crop Sci 196(2):90–99

Dien DC, Mochizuki T, Yamakawa T (2019) Effect of various drought stresses and subsequent recovery on proline, total soluble sugar and starch metabolisms in rice (*Oryza sativa* L.) varieties. Plant Prod Sci 22(4):530–545

Ding CK, Wang CY, Gross KC, Smith DL (2002) Jasmonate and salicylate induce the expression of pathogenesis-related-protein genes and increase resistance to chilling injury in tomato fruit. Planta 214:895–901

Dixon DP, Cummins I, Cole DJ, Edwards R (1998) Glutathione-mediated detoxification systems in plants. Curr Opin Plant Biol 1:258–266

Dixon DP, Skipsey M, Edwards R (2010) Roles for glutathione transferases in plant secondary metabolism. Phytochemistry 71(4):338–350

Dong Y, Chen W, Xu L, Kong J, Liu S, He Z (2016) Nitric oxide can induce tolerance to oxidative stress of peanut seedlings under cadmium toxicity. Plant Growth Regul 79(1):19–28

Drazic G, Mihailovic N (2005) Modification of cadmium toxicity in soybean seedlings by salicylic acid. Plant Sci 168:511–517

Elhamahmy Mohamed A, Kotb MA (2014) Inducing beneficial effects on peanut growth, yield and quality by foliar spraying of kinetin under saline conditions. Zagazig J Agric Res 41(6):1145–1164

El-Mashad AA, Mohamed HI (2012) Brassinolide alleviates salt stress and increases antioxidant activity of cowpea plants (*Vigna sinensis*). Protoplasma 249(3):625–635

Fariduddin Q, Khanam S, Hasan SA, Ali B, Hayat S, Ahmad A (2009a) Effect of 28-homobrassinolide on drought stress induced changes in photosynthesis and antioxidant system of *Brassica juncea* L. Acta Physiol Plant 31:889–897

Fariduddin Q, Yusuf M, Hayat S, Ahmad A (2009b) Effect of 28-homobrassinolide on antioxidant capacity and photosynthesis in *Brassica juncea* plants exposed to different levels of copper. Environ Exper Botany 66:418–424

Fariduddin Q, Yusuf M, Chalkoo S, Hayat S, Ahmad A (2011) 28-homobrassinolide improves growth and photosynthesis in *Cucumis sativus* L. through an enhanced antioxidant system in the presence of chilling stress. Photosynthetica 49:55–64

Feys BJF, Benedetti CE, Penfold CN, Turner JG (1994) *Arabidopsis* mutants selected for resistance to the phytotoxin coronatine are male sterile, insensitive to methyl jasmonate, and resistant to a bacterial pathogen. Plant Cell 6:751–759

Flores A, Grau A, Laurich F, Dorffling K (1988) Effect of new terpenoid analogues of abscisic acid on chilling and freezing resistance. J Plant Physiol 132:362–369

Foyer CH, Noctor G (2005) Oxidant and antioxidant signaling in plants: a re-evaluation of the concept of oxidative stress in a physiological context. Plant, Cell Environ 28:1056–1071

Funck D, Winter G, Baumgarten L, Forlani G (2012) Requirement of proline synthesis during *Arabidopsis* reproductive development. BMC Plant Biol 12:191–203

Gadallah MA (1999) Effects of proline and glycinebetaine on *Vicia faba* responses to salt stress. Biol Plant 42(2):249–257

Gangwar S, Singh VP, Srivastava PK, Maurya JN (2011) Modification of chromium (VI) phytotoxicity by exogenous gibberellic acid application in *Pisum sativum* (L.) seedlings. Acta Physiologiae Plantarum 33(4):1385–1397

Garcia AB, Engler JD, Iyer S, Gerats T, Van Montagu M, Caplan AB (1997) Effects of osmoprotectants upon NaCl stress in rice. Plant Physiol 115(1):159–169

Garg N, Sharma A (2019) Role of putrescine (Put) in imparting salt tolerance through modulation of put metabolism, mycorrhizal and rhizobial symbioses in *Cajanus cajan* (L.). Symbiosis 79:1–16

Ghosh S, Mitra S, Paul A (2015) Physiochemical studies of sodium chloride on mungbean (*Vigna radiata* L. Wilczek) and its possible recovery with spermine and gibberellic acid. Sci World J 2015:1–8

Girija C, Smith BN, Swamy PM (2002) Interactive effects of sodium chloride and calcium chloride on the accumulation of proline and glycinebetaine in peanut (*Arachis hypogae*a L.). Environmental and Experimental Botany 47(1):1–10

Gomes MMA (2011) Physiological effects related to brassinosteroid application in plants. In: Hayat S, Ahmad A (eds) Brassinosteroids: a class of plant hormones. Springer, Dordrecht, Heidelberg, London, New York, pp 193–242

Gonçalves KS (2013) Aplicação de reguladores vegetais e de fosfito de potássio em mudas de eucalipto submetidas à deficiência hídrica. Tese (Doutorado em Agronomia: Área de Concentração em Irrigação e Drenagem)-Universidade Estadual Paulista, Botucatu

Gunes A, Pilbeam DJ, Inal A, Bagci EG, Coban S (2007) Influence of silicon on antioxidant mechanisms and lipid peroxidation in chickpea (*Cicer arietinum* L.) cultivars under drought stress. J Plant Interact 2(2):105–113

Hamayun M, Khan SA, Khan AL, Shin JH, Ahmad B, Shin DH, Lee IJ (2010) Exogenous gibberellic acid reprograms soybean to higher growth and salt stress tolerance. J Agric Food Chem 58(12):7226–7232

Hamayun M, Hussain A, Khan SA, Irshad M, Khan AL, Waqas M, Shahzad R, Iqbal A, Ullah N, Rehman G, Kim HY (2015) Kinetin modulates physio-hormonal attributes and isoflavone contents of soybean grown under salinity stress. Front Plant Sci 6:377

Hamilton EW, Heckathorn SA (2001) Mitochondrial adaptations to NaCl, complex I is protected by anti-oxidants and small heat shock proteins, whereas complex II is protected by proline and betaine. Plant Physiol 126(3):1266–1274

Hare PD, Cress WA, van Staden J (1999) Proline synthesis and degradation: a model system for elucidating stress-related signal transduction. J Exp Bot 50:413–434

Hasanah Y, Mawarni L, Irmansyah T (2017) Production and physiological characters of soybean under drought stress with foliar application of exogenous antioxidant. Conference on international research on food security, natural resource management and rural development, University of Bonn, Bonn, Germany, Sept 20–22

Hasanuzzaman M, Fujita M (2011) Selenium pretreatment upregulates the antioxidant defense and methylglyoxal detoxification system and confers enhanced tolerance to drought stress in rapeseed seedlings. Biol Trace Elem Res 143:1758–1776

Hasanuzzaman M, Nahar K, Fujita M (2013a) Plant response to salt stress and role of exogenous protectants to mitigate salt-induced damages. Ecophysiology and responses of plants under salt stress. Springer, New York, NY, pp 25–87

Hasanuzzaman M, Nahar K, Fujita M, Ahmad P, Chandna R, Prasad MN, Ozturk M (2013b) Enhancing plant productivity under salt stress: relevance of poly-omics. Salt stress in plants. Springer, New York, NY, pp 113–156

Hasanuzzaman M, Nahar K, Fujita M (2014a) Role of tocopherol (vitamin E) in plants: abiotic stress tolerance and beyond. In: Ahmed P, Rasool S (eds) Emerging technologies and management of crop stress tolerance, vol 2. A sustainable approach. Academic Press, Singapore, pp 267–289

Hasanuzzaman M, Nahar K, Fujita M (2014b) Exogenous proline and glycine betaine mediated upregulation of antioxidant defense and glyoxalase systems provides better protection against salt-induced oxidative stress in two rice (*Oryza sativa* L.) varieties. BioMed Res Int 757219

Hasanuzzaman M, Nahar K, Gill SS, Fujita M (2014c) Drought stress responses in plants, oxidative stress, and antioxidant defense. Clim Change Plant Abiot Stress Toler 18:209–250

Hasanuzzaman M, Nahar K, Anee TI, Fujita M (2017) Exogenous silicon attenuates cadmium-induced oxidative stress in *Brassica napus* L. by modulating AsA-GSH pathway and glyoxalase system. Front Plant Sci 8:1061

Hayat S, Hasan SA, Yusuf M, Hayat Q, Ahmad A (2010) Effect of 28-homobrassinolide on photosynthesis, fluorescence and antioxidant system in the presence or absence of salinity and temperature in *Vigna radiata*. Environ Exp Bot 69:105–112

Hayat S, Hayat Q, Alyemeni MN, Wani AS, Pichtel J, Ahmad A (2012) Role of proline under changing environments: a review. Plant Signal Behav 7(11):1456–1466

He Y, Fukushige H, Hildebrand DF, Gan S (2002) Evidence supporting a role of jasmonic acid in *arabidopsis* leaf senescence. Plant Physiol 128:876–884

Heuer B (2003) Influence of exogenous application of proline and glycinebetaine on growth of salt-stressed tomato plants. Plant Sci 165(4):693–699

Hoque MA, Banu MN, Nakamura Y, Shimoishi Y, Murata Y (2008) Proline and glycinebetaine enhance antioxidant defense and methylglyoxal detoxification systems and reduce NaCl-induced damage in cultured tobacco cells. J Plant Physiol 165(8):813–824

Hossain MA, Hasanuzzaman M, Fujita M (2011) Coordinate induction of antioxidant defense and glyoxalase system by exogenous proline and glycinebetaine is correlated with salt tolerance in mung bean. Front Agric China 5(1):1–4

Hosseinian SH, Akbari N, Eisvand HR, Akbarpour O, Saeedinia M (2019) Effect of drought stress and glycine betaine as foliar application on photosynthesis parameters of chickpea. J Water Irrig Manag 8(2):227–235

Hu L, Hu T, Zhang X, Pang H, Fu J (2012) Exogenous glycine betaine ameliorates the adverse effect of salt stress on perennial ryegrass. J Am Soc Hortic Sci 137:38–46

Hussein MM, Alva AK (2014) Effects of zinc and ascorbic acid application on the growth and photosynthetic pigments of millet plants grown under different salinity. J Agric Sci 5:1253–1260

Hussein NM, Hussein MI, Gadel Hak SH, Hammad MA (2014) Effect of two plant extracts and four aromatic oils on tuta absoluta population and productivity of tomato cultivar gold stone. Nat Sci 12:108–118

Jabeen N, Abbas Z, Iqbal M, Rizwan M, Jabbar A, Farid M, Ali S, Ibrahim M, Abbas F (2016) Glycinebetaine mediates chromium tolerance in mung bean through lowering of Cr uptake and improved antioxidant system. Arch Agron Soil Sci 62(5):648–662

Jan S, Alyemeni MN, Wijaya L, Alam P, Siddique KH, Ahmad P (2018) Interactive effect of 24-epibrassinolide and silicon alleviates cadmium stress via the modulation of antioxidant defense and glyoxalase systems and macronutrient content in *Pisum sativum* L. seedlings. BMC plant Biol 18(1):146

Janda T, Szalai G, Tari I, Páldi E (1999) Hydroponic treatment with salicylic acid decreases the effects of chilling injury in maize (*Zea mays* L.) plants. Planta 208:175–180

Jungklang J, Saengnil K, Uthaibutra J (2015) Efects of water-defcit stress and paclobutrazol on growth, relative water content, electrolyte leakage, proline content and some antioxidant changes *in Curcuma alismatifolia* Gagnep. cv. Chiang Mai Pink. Saudi J Biol Sci 24(7):1505–1512

Kameli A, Lösel V (1995) Contribution of carbohydrates and other solutes to osmotic adjustment in wheat leaves under water stress. J Plant Physiol 145:363–366

Kang HM, Saltveit ME (2002) Chilling tolerance of maize, cucumber and rice seedlings leaves and roots are differentially affected by salicylic acid. Physiol Plant 115:571–576

Kang G, Li G, Guo T (2014) Molecular mechanism of salicylic acid-induced abiotic stress tolerance in righer plants. Acta Physiol Plant 36(9):2287–2297

Kang SM, Shahzad R, Bilal S, Khan AL, Park YG, Lee KE, Asaf S, Khan MA, Lee IJ (2019) Indole-3-acetic-acid and ACC deaminase producing *Leclercia adecarboxylata* MO1 improves *Solanum lycopersicum* L. growth and salinity stress tolerance by endogenous secondary metabolites regulation. BMC Microbiol 19:80

Kaur H, Sharma P, Sirhindi G (2013) Sugar accumulation and its regulation by jasmonic acid in *Brassica napus* L. under salt stress. J Stress Physiol Biochem 9:53–64

Kaushal N, Gupta K, Bhandhari K, Kumar S, Thakur P, Nayyar H (2011) Proline induces heat tolerance in chickpea (*Cicer arietinum* L.) plants by protecting vital enzymes of carbon and antioxidative metabolism. Physiol Molec Biol Plants 17(3):203–213

Khadri M, Tejera NA, Lluch C (2006) Alleviation of salt stress in common bean (*Phaseolus vulgaris*) by exogenous abscisic acid supply. J Plant Growth Regul 25(2):110–119

Khan MI, Asgher M, Khan NA (2014) Alleviation of salt-induced photosynthesis and growth inhibition by salicylic acid involves glycinebetaine and ethylene in mungbean (*Vigna radiata* L.). Plant Physiol Biochem 80:67–74

Khater MA, Dawood MG, Sadak MS, Shalaby MA, El-Awadi ME, El-Din KG (2018) Enhancement the performance of cowpea plants grown under drought conditions via trehalose application. Middle East J Agric Res 7:782–800

Khattab EA, Badr EA, Afifi MH (2018) Improving the productivity of some varieties of lentil growing under salt stress. Asian J Res Crop Sci 2(3):1–10

Khedr AH, Abbas MA, Wahid AA, Quick WP, Abogadallah GM (2003) Proline induces the expression of salt-stress-responsive proteins and may improve the adaptation of *Pancratium maritimum* L. to salt-stress. J Experim Botany 54(392):2553–2562

Kim Y, Seo CW, Khan AL, Mun BG, Shahzad R, Ko JW, Yun BW, Park SK, Lee IJ (2018) Exo-ethylene application mitigates waterlogging stress in soybean (*Glycine max* L.). BMC Plant Biology 18(1):254

Kumar S, Kaushal N, Nayyar H, Gaur P (2012) Abscisic acid induces heat tolerance in chickpea (*Cicer arietinum* L.) seedlings by facilitated accumulation of osmoprotectants. Acta Physio Plantarum 34(5):1651–1658

Kumutha D, Sairam RK, Meena RC (2008) Role of root carbohydrate reserves and their mobilization in imparting waterlogging tolerance in greengraem [*Vigna radiate* (L.) Wilczek] genotypes. Indian Journal of Plant Physiology 33:735–744
Latif HH (2014) Physiological responses of *Pisum sativum* plant to exogenous ABA application under drought conditions. Pak J Bot 46(3):973–982
Li X, Shen X, Li J, Eneji AE, Li Z, Tian X, Duan L (2010) Coronatine alleviates water deficiency stress on winter wheat seedlings. J Integr Plant Biol 52(7):616–625
Liu K, Fu H, Bei Q, Al E (2000) Inward potassium channel in guard cells as a target for polyamine regulation of stomatal movements. Plant Physiol 124:1315–1325
Lopez-Delgado H, Dat JF, Foyer CH, Scott IM (1998) Induction of thermotol-erance in potato microplants by acetylsalicylic acid and H_2O. J Exper Botany 49:713–730
Luo M, Liu JH, Mohapatra S, Hill RD, Mohapatra SS (1992) Characterization of a gene family encoding abscisic acid- and environmental stress-inducible proteins of alfalfa. J Biol Chem 267(22):15367–15374
Lv WT, Lin B, Zhang M, Hua XJ (2011) Proline accumulation is inhibitory to Arabidopsis seedlings during heat stress. Plant Physiol 156(4):1921–1933
Mahajan S, Tuteja N (2005) Cold, salinity and drought stresses: an overview. Arch Biochem Biophys 444(2):139–158
Malekzadeh P (2015) Influence of exogenous application of glycinebetaine on antioxidative system and growth of salt-stressed soybean seedlings (*Glycine max* L.). Phys Molec Biol Plants 21:225–232
Manaf HH (2016) Beneficial effects of exogenous selenium, glycine betaine and seaweed extract on salt stressed cowpea plant. Anna Agric Sci 61:41–48
Manoharlallal (2014) Physiological evaluation of pigeonpea [*Cajanus cajans* (L.) Millsp.] genotypes for waterlogging and salinity tolerance. MSc Thesis, CCS Haryana Agricultural University, Hisar, India
Matysik J, Alia Bhalu B, Mohanty P (2002) Molecular mechanisms of quenching of reactive oxygen species by proline under stress in plants. Curr Sci 82(5):525–532
Mayi AA, Ibrahim ZR, Abdurrahman AS (2014) Effect of foliar spray of humic acid, ascorbic acid, cultivars and their interactions on growth of olive (*Olea european* L.) transplants cvs. Khithairy and Sorany. Agric Veter Sci 7:18–30
McNeil SD, Nuccio ML, Hanson AD (1999) Betaines and related osmoprotectants: targets for metabolic engineering of stress resistance. Plant Phys 120(4):945–949
Minhas PS, Rane J, Pasala RK (2017) Abiotic stresses in agriculture: An overview. In: Minhas P, Rane J, Pasala R (eds) Abiotic stress management for resilient agriculture. Springer, Singapore, pp 3–8
Mir BA, Khan TA, Fariduddin Q (2015) 24-epibrassinolide and spermidine modulate photosynthesis and antioxidant systems in *Vigna radiata* under salt and zinc stress. Int J Adv Res 3(5):592–608
Misra N, Saxena P (2009) Effect of salicylic acid on proline metabolism in lentil grown under salinity stress. Plant Sci 177(3):181–189
Mohammadi H, Ghorbanpour M, Brestic M (2018) Exogenous putrescine changes redox regulations and essential oil constituents in field-grown *Thymus vulgaris* L. under well-watered and drought stress conditions. Ind Crops Prod 122:119–132
Molla MR, Ali MR, Hasanuzzaman M, Al-Mamun MH, Ahmed A, Nazim-ud-Dowla MA, Rohman MM (2014) Exogenousproline and betaine-induced upregulation of glutathione transferase and glyoxalase I in lentil (*Lens culinaris*) under drought stress. Notulae Botanicae Horti Agrobotanici Cluj-Napoca 42(1):73–80
Munns R, Tester M (2008) Mechanisms of salinity tolerance. Annu Rev Plant Biol 59(651):681
Mustafavi SH, Badi HN, Sekara A, Al E (2018) Polyamines and their possible mechanisms involved in plant physiological processes and elicitation of secondary metabolites. Acta Physiol Plant 40:102

Mustroph A, Albrecht G (2003) Tolerance of crop plants to oxygen deficiency stress: fermentative activity and photosynthetic capacity of entire seedling under hypoxia and anoxia. Physiol Plant 117:508–520

Mwenye OJ, Van Rensburg L, Van Biljon A, Van der Merwe R (2016) The role of proline and root traits on selection for drought-stress tolerance in soybeans: a review. South Afr J Plant Soil 33(4):245–256

Nahar K, Hasanuzzaman M, Alam M, Fujita M (2015a) Glutathione-induced drought stress tolerance in mung bean: coordinated roles of the antioxidant defence and methylglyoxal detoxification systems. AoB PLANTS 7:plv069

Nahar K, Hasanuzzaman M, Alam MM, Fujita M (2015b) Exogenous glutathione confers high temperature stress tolerance in mung bean (*Vigna radiata* L.) by modulating antioxidant defense and methylglyoxal detoxification system. Environ Exp Bot 112:44–54

Nahar K, Hasanuzzaman M, Alam MM, Fujita M (2015c) Roles of exogenous glutathione in antioxidant defense system and methylglyoxal detoxification during salt stress in mung bean. Biol Plant 59(4):745–756

Nahar K, Hasanuzzaman M, Rahman A, Alam M, Mahmud JA, Suzuki T, Fujita M (2016) Polyamines confer salt tolerance in mung bean (Vigna radiata L.) by reducing sodium uptake, improving nutrient homeostasis, antioxidant defense, and methylglyoxal detoxification systems. Front Plant Sci 7:1104

Nayyar H, Chander K, Kumar S, Bains T (2015) Glycine betaine mitigates cold stress damage in Chickpea. Agron Sustain Dev 25(3):381–388

Noctor G, Foyer CH (1998) Ascorbate and glutathione; keeping active oxygen control. Annu Rev Plant Physiol Plant Mol Biol 49:249–279

Noctor G, Gomez L, Vanacker H, Foyer CH (2002) Interactions between biosynthesis, compartmentation and transport in the control of glutathione homeostasis and signaling. J Exp Bot 53:1283–1304

Nounjan N, Nghia PT, Theerakulpisut P (2012) Exogenous proline and trehalose promote recovery of rice seedlings from salt-stress and differentially modulate antioxidant enzymes and expression of related genes. J Plant Physiol 169(6):596–604

Osman HS (2015) Enhancing antioxidant-yield relationship of pea plant under drought at different growth stages by exogenously applied glycine betaine and proline. Ann Agric Sci 60:389–402

Osman HS, Salim BBM (2016) Influence of exogenous application of some phytoprotectants on growth, yield and pod quality of snap bean under NaCl salinity. Ann Agric Sci 61:1–13

Ozkur O, Ozdemir F, Bor M, Turkan I (2009) Physiochemical and antioxidant responses of the perennial xerophyte Capparis ovata Desf. to drought. Environment Exp Bot 66(3):487–492

Pál M, Horváthe E, Janda T, Páldi E, Szalai Szalai G (2005) Cadmium stimulates the accumulation of salicylic acid and its putative precursors in maize (*Zea mays*) plants. Physiol Plant 125:356–364

Paul MJ, Primavesi LF, Jhurreea D, Zhang Y (2008) Trehalose metabolism and signaling. Annu Rev Plant Biol 59:417–441

Pitman MG, Läuchli A (2002) Global impact of salinity and agricultural ecosystems. In: Läuchli A, Lüttge U (eds) Salinity: environment—plants—molecules. Springer, Dordrecht, pp 3–20

Poonam S, Kaur H, Geetika S (2013) Effect of jasmonic acid on photosynthetic pigments and stress markers *Cajanus cajan* (L.) Millsp. seedlings under copper stress. Am J Plant Sci 4(4):817–823

Popova LP, Stoinova ZG, Maslenkova LT (1995) Involvement of abscisic acid in photosynthetic process in *Hordeum vulgare* L. during salinity stress. J Plant Growth Regul 14:211–218

Pustovoitova T, Zholkevich V (1992) Basic trends in the investigation of drought effects on physiological processes in plants. Fiziol Biokhim Kul't Rast 24:14–27

Qin X, Sun SL, Wang ML, Lu CK (2009) Wang H (2009) Effect of COR on osmolytes and lipid peroxidation in peanut seedling leaves under drought stress. J Peanut Sci 1:18–21

Rady MM, Mohamed GF, Abdalla AM, Ahmed YH (2015) Integrated application of salicylic acid and *Moringa oleifera* leaf extract alleviates the salt-induced adverse effects in common bean plants. Int J Agric Technol 11(7):1595–1614

Rahman MS, Miyake H, Takeoka Y (2002) Effects of exogenous glycinebetaine on growth and ultrastructure of salt-stressed rice seedlings (*Oryza sativa* L.). Plant Prod Sci 5(1):33–44

Rahman MF, Ghosal A, Alam MF, Kabir AH (2017) Remediation of cadmium toxicity in field peas (Pisum sativum L.) through exogenous silicon. Ecotox Environ Saf 135:165–172

Rao MV, Paliyath G, Ormrod DP, Murr DP, Watkins CB (1997) Influence of salicylic acid on H_2O_2 production, oxidative stress, and H_2O_2—metabolizing enzymes. Plant Physiol 115:137–149

Rapacz M, Kościelniak J, Jurczyk B, Adamska A, Wójcik M (2010) Different patterns of physiological and molecular response to drought in seedlings of malt-and feed-type barleys (Hordeum vulgare). J Agron Crop Sci 196(1):9–19

Rathinasabapathi B, Burnet M, Russell BL, Gage DA, Liao PC, Nye GJ, Scott P, Golbeck JH, Hanson AD (1997) Choline monooxygenase, an unusual iron-sulfur enzyme catalyzing the first step of glycine betaine synthesis in plants: prosthetic group characterization and cDNA cloning. Proc Natl Acad Sci 94(7):3454–3458

Reddy PS, Jogeswar G, Rasineni GK, Maheswari M, Reddy AR, Varshney RK, Kavi Kishor PB (2015) Proline over-accumulation alleviates salt stress and protects photosynthetic and antioxidant enzyme activities in transgenic sorghum [*Sorghum bicolor* (L.) Moench]. Plant Physiol Biochem 94:104–113

Reis RS, Vale EM, Heringer AS, Al E (2016) Putrescine induces somatic embryo development and proteomic changes in embryogenic callus of sugarcane. J Proteom 130:170–179

Rezaei MA, Kaviani B, Masouleh AK (2012) The effect of exogenous glycine betaine on yield of soybean [*Glycine max* (L.) Merr.] in two contrasting cultivars pershing and DPX under soil salinity stress. Plant Omics 5(2):87–93

Rhodes D, Samaras Y (1994) Genetic control of osmoregulation in plants: cellular and molecular physiology of cell, vol regulation. CRC Press, Boca Raton, pp 347–361

Richards AB, Krakowka S, Dexter LB, Schmid H, Wolterbeek AP, Waalkens-Berendsen DH, Shigoyuki A, Kurimoto M (2002) Trehalose: a review of properties, history of use and human tolerance, and results of multiple safety studies. Food Chem Toxicol 40(7):871–898

Rohman MM, Begum S, Akhi AH, Ahsan AF, Uddin MS, Amiruzzaman M, Banik BR (2015) Protective role of antioxidants in maize seedlings under saline stress: exogenous proline provided better tolerance than betaine. Bothalia J 45(4):17–35

Rosa E, Prado C, Podazza G, Interdonato R, González JA, Hilal M, Prado FE (2009) Soluble sugars-Metabolism, sensing and abiotic stress, A complex network in the life of plants. Plant Sign Behav 4(5):388–393

Sabagh AE, Sorour S, Omar ABDE, Ragab A, Islam MS, Barutçular C, Ueda A, Saneoka H (2015a) Alleviation of adverse effects of salt stress on soybean (*Glycine max.* L.) by using osmoprotectants and organic nutrients. Int J Biol Biomol Agric Food Biotech Eng 9(9):921–925

Sabagh AE, Sorour S, Omar AE, Islam MS, Ueda A, Saneoka H, Barutçular C (2015b) Soybean (*Glycine max* L.) Growth enhancement under water stress conditions. In: International conference on chemical, agricultural and biological sciences (CABS-2015) 4–5 Sept 2015 Istanbul (Turkey), pp 144–148

Sadeghipour O (2018) Drought tolerance of cowpea enhanced by exogenous application of methyl jasmonate. Int J Mod Agric 7(4):51–57

Sadeghipour O (2019) Polyamines protect mung bean [*Vigna radiata* (L.) Wilczek] plants against drought stress. Biologia Futura 70(1):71–78

Saeidi-Sar S, Abbaspour H, Afshari H, Yaghoobi SR (2013) Effects of ascorbic acid and gibberellin A 3 on alleviation of salt stress in common bean (*Phaseolus vulgaris* L.) seedlings. Acta Physiol Plant 35(3):667–677

Saghfi S, Eivazi AR (2014) Effects of cold stress on proline and soluble carbohydrates in two chickpea cultivars. Int J Curr Microbiol Appl Sci 3(2):591–595

Saidi I, Chtourou Y, Djebali W (2014) Selenium alleviates cadmium toxicity by preventing oxidative stress in sunflower (*Helianthus annuus*) seedlings. J Plant Physiol 171:85–91

Sajid ZA, Safdar MU, Khilji SA (2016) Amelioration of salinity stress tolerance in pea (*Pisum sativum* L.) by exogenous application of salicylic acid. Biologia 62(1):69–78

Sakamoto A, Murata N (2001) The use of bacterial choline oxidase, a glycinebetaine-synthesizing enzyme, to create stress-resistant transgenic plants. Plant Physiol 125(1):180–188
Sánchez-Vicente I, Fernández-Espinosa MG, Lorenzo O (2019) Nitric oxide molecular targets: reprogramming plant development upon stress. J Exper Botany 70(17):4441–4460
Savita (2016) Effect of waterlogging, salinity and their interaction on growth, oxidative and carbohydrate metabolism in pigeonpea (*Cajanus cajan* L. Millsp.) genotypes. PhD thesis, Dept. of Botany and Plant Physiology, Chaudhary Charan Singh Haryana Agricultural University, Hisar, India
Schuler G, Mithofer A, Baldwin T, Berger S, Ebel J, Santos JG, Herrmann G, Holscher D, Kramell R, Kutchan TM, Mauche H, Schneider B, Stenzel I, Wasternack C, Boland W (2004) Coronalon: a powerful, tool in plant stress physiology. EFBS Lett 563:17–22
Shahid MA, Balal RM, Pervez MA, Abbas T, Aqeel MA, Javaid MM, Garcia-Sanchez F (2014) Exogenous proline and proline-enriched *Lolium perenne* leaf extract protects against phytotoxic effects of nickel and salinity in *Pisum sativum* by altering polyamine metabolism in leaves. Turkish J Botany 38(5):914–926
Shahid MA, Balal RM, Pervez MA, Abbas T, Aqeel MA, Riaz A, Mattson NS (2015) Exogenous 24-epibrassinolide elevates the salt tolerance potential of pea (*Pisum sativum* L.) by improving osmotic adjustment capacity and leaf water relations. J Plant Nutr 38(7):1050–1072
Shan C, Zhao X (2014) Effects of lanthanum on the ascorbate and glutathione metabolism of *Vigna radiata* seedlings under salt stress. Biol Plant 58(3):595–599
Shan T, Jin P, Zhang Y, Huang Y, Wang X, Zheng Y (2016) Exogenous glycine betaine treatment enhances chilling tolerance of peach fruit during cold storage. Postharvest Biol Technol 114:104–110
Shao HB, Liang ZS, Shao MA, Sun QZ, Hu M (2005) Investigation on dynamic changes of photosynthetic characteristics of 10 wheat *Triticum aestivum* L. genotypes during two vegetative growth stages at water deficits. Colloids Surf B 43:221–227
Sharma P, Kaur H, Sirhindi G (2013) Effect of jasmonic acid on photosynthetic pigments and stress markers in *Cajanus cajan* (L.) Millsp. seedlings under copper stress. Am J Plant Sci 4:817–823
Shen X, Xiao X, Dong Z, Chen Y (2014) Silicon effects on antioxidative enzymes and lipid peroxidation in leaves and roots of peanut under aluminum stress. Acta Physiol Plant 36(11):3063–3069
Sheokand S, Kumari A, Sawhney V (2008) Effect of nitric oxide and putrescine on antioxidative responses under NaCl stress in chickpea plants. Phys Molec Biol Plants 14(4):355–362
Sheteawi SA (2007) Improving growth and yield of salt-stressed soybean by exogenous application of *jasmonic acid* and *ascobin*. Int J Agric Biol (Pakistan) 9(3):473–478
Shevyakova NI, Musatenko LI, Stetsenko LA, Vedenicheva NP, Voitenko LP, Sytnik KM, Kuznetsov VV (2013) Effects of abscisic acid on the contents of polyamines and proline in common bean plants under salt stress. Russ J Plant Physiol 60(2):200–211
Shi O, Bao Z, Zhu Z, Ying Q, Qian Q (2005) Effects of different treatments of salicylic acid on heat tolerance, chlorophyll fluorescence, and antioxidant enzyme activity in seedlings of *Cucumis sativa* L. Russ J Plant Physiol 52(2):793–800
Shi G, Cai Q, Liu C, Wu L (2010) Silicon alleviates cadmium toxicity in peanut plants in relation to cadmium distribution and stimulation of antioxidative enzymes. Plant Growth Regul 61(1):45–52
Singh NK, Larosa PC, Handa AK, Hasegawa PM, Bressan RA (1987) Hormonal regulation of protein synthesis associated with salt tolerance in plant cells. Proc Natl Acad Sci USA 84(3):739
Sirhindi G, Sharma P, Singh A, Kaur H, Mir M (2015) Alteration in photosynthetic pigments, osmolytes and antioxidants in imparting copper stress tolerance by exogenous jasmonic acid treatment in *Cajanus cajan*. Int J Plant Phys Biochem 7(3):30–39
Slama I, Ghnaya T, Savouré A, Abdelly C (2008) Combined effects of long-term salinity and soil drying on growth, water relations, nutrient status and proline accumulation of *Sesuvium portulacastrum*. CR Biol 331:442–451

Solanki MV, Trivedi SK, Kandoliya UK, Golakiya BA (2018) Effect of exogenous application of salicylic acid on biochemical constituent in black gram (*Vigna mungo*L.) hepper irrigated with saline water. Eur J Biotech Biosci 6(5):28–34

Song JT, Lu H, McDowell JM, Greenberg JT (2004) A key role for ALD1 in activation of local and systemic defenses in *Arabidopsis*. Plant J 40(10):200–212

Song YL, Dong YJ, Tian XY, Kong J, Bai XY, Xu LL, He ZL (2016) Role of foliar application of 24-epibrassinolide in response of peanut seedlings to iron deficiency. Biol Plant 60(2):329–342

Song Y, Dong Y, Kong J, Tian X, Bai X, Xu L (2017) Effects of root addition and foliar application of nitric oxide and salicylic acid in alleviating iron deficiency induced chlorosis of peanut seedlings. J Plant Nutr 40(1):63–81

Strauss JA, Agenbag GA (2000) The use of physiological parameters to identify drought tolerance in spring wheat cultivars. South African Journal of Plant and Soil 17:20–29

Sulpice R, Tsukaya H, Nonaka H, Mustardy L, Chen TH, Murata N (2003) Enhanced formation of flowers in salt-stressed Arabidopsis after genetic engineering of the synthesis of glycine betaine. Plant J 36(2):165–176

Talukdar D (2012) Exogenous calcium alleviates the impact of cadmium-induced oxidative stress in *Lens culinaris* Medic. seedlings through modulation of antioxidant enzyme activities. J Crop Sci Biotech 15(4):325–334

Talukdar D (2013) Arsenic-induced oxidative stress in the common bean legume, *Phaseolus vulgaris* L. seedlings and its amelioration by exogenous nitric oxide. Phys Mol Biol Plants 19(1):69–79

Tamogami S, Kodama O (2000) Coronatine elicits phytoalexin production in rice leaves (*Oryza sativa* L.) in the same manner as jasmonic acid. Phytochemistry 54:689–694

Thameur A, Ferchichi A, López-Carbonell M (2011) Quantification of free and conjugated abscisic acid in five genotypes of barley (*Hordeum vulgare* L.) under water stress conditions. South Afr J Botany 77:222–228

Theerakulpisut P, Gunnula W (2012) Exogenous sorbitol and trehalose mitigated salt stress damage in salt-sensitive but not salt-tolerant rice seedlings. Asian J Crop Sci 4:165–170

Tiwari JK, Munshi AD, Kumar R, Pandey RN, Arora A, Bhat JS, Sureja AK (2010) Effect of salt stress on cucumber: Na^{+}–K^{+} ratio, osmolyte concentration, phenols and chlorophyll content. Physiol Plant 32(1):103–114

Tuteja N (2007) Abscisic acid and abiotic stress signaling. Plant Sig Behav 2(3):135–138

Uppalapati SR, Ayoubi P, Weng H, Palmer DA, Mitchell RE, Jones W, Bender CL (2008) The phytotoxin coronatine and methyl jasmonate impact multiple phytohormone pathways in tomato. Plant J 42:201–217

Vuosku J, Suorsa M, Ruottinen M, Sutela S, Muilu-Mäkelä R, Julkunen-Tiitto R (2012) Polyamine metabolism during exponential growth transition in scots pine embryogenic cell culture. Tree Physiol 32:1274–1287

Vuosku J, Karppinen K, Muilu-Mäkelä R, Kusano T, Sagor GHM, Avia K (2018) Scots pine aminopropyltransferases shed new light on evolution of the polyamine biosynthesis pathway in seed plants. Ann Bot 121:1243–1256

Walia H, Wilson C, Condamine P, Liu X, Ismail AM, Close TJ (2007) Large-scale expression profiling and physiological characterization of jasmonic acid-mediated adaptation of barley to salinity stress. Plant, Cell Environ 30:410–421

Wang SH, Yang ZM, Lu B, Li SQ, Lu YP (2004a) Copper-induced stress and antioxidative responses in roots of *Brassica juncea* L. Botan Bull Acad Sinica 45:203–212

Wang YS, Wang J, Yang ZM, Wang QY, Lü B, Li SQ, Lu YP, Wang SH, Sun X (2004b) Salicylic acid modulates aluminum-induced oxidative stress in roots of *Cassia tora*. Acta Botanica Sinica 46(7):819–828

Wang LJ, Chen SJ, Kong WF, Li SH, Archbold DD (2006) Salicylic acid pretreatment alleviates chilling injury and affects the antioxidant system and heat shock proteins of peaches during cold storage. Postharvest Biol Technol 41:244–251

Wang JP, Mao XG, Jing RL, Li RZ, Chang XP (2008) Single nucleotide polymorphism of TaCRT-D gene associated with drought resistance in wheat. Germplasm Sci Agric Sinica 41:3983–3990

Wang X, Wei Z, Liu D, Zhao G (2011) Effects of NaCl and silicon on activities of antioxidative enzymes in roots, shoots and leaves of alfalfa. Afr J Biotech 10(4):545–549

Wani AS, Tahir I, Ahmad SS, Dar RA, Nisar S (2017) Efficacy of 24-epibrassinolide in improving the nitrogen metabolism and antioxidant system in chickpea cultivars under cadmium and/or NaCl stress. Sci Hortic 225:48–55

Wei L, Wang L, Yang Y, Wang P, Guo T, Kang G (2015) Abscisic acid enhances tolerance of wheat seedlings to drought and regulates transcript levels of genes encoding ascorbate-glutathione biosynthesis. Front Plant Sci 6:458

Weretilnyk EA, Bednarek S, McCue KF, Rhodes D, Hanson AD (1989) Comparative biochemical and immunological studies of the glycine betaine synthesis pathway in diverse families of dicotyledons. Planta 178:342–352

Witt S, Galicia L, Lisec J, Cairns J, Tiessen A, Araus JL, Palacios-Rojas N, Fernie AR (2012) Metabolic and phenotypic responses of greenhouse-grown maize hybrids to experimentally controlled drought stress. Mol Plant 5(2):401–417

Wu GQ, Feng RJ, Li SJ, Du YY (2017) Exogenous application of proline alleviates salt-induced toxicity in cainfoin seedlings. J Animal Plant Sci 27:246–251

Xiang C, Wemer BL, Christensen EM, Oliver DJ (2001) The biological functions of glutathione revisited in Arabidopsis transgenic plants with altered glutathione levels. Plant Physiol 126:564–574

Xie ZX, Duan LS, Tian XL, Wang BQ, Eneji AE, Li ZH (2008) Coronatine alleviates salinity stress in cotton by improving the antioxidative defense system and radical-scavenging activity. J Plant Physiol 165:375–384

Xiong YC, Xing GM, Li FM, Wang SM, Fan XW, Li ZX, Wang YF (2006) Abscisic acid promotes accumulation of toxin ODAP in relation to free spermine level in grass pea seedlings (*Lathyrussativus* L.). Plant Physiol Biochem 44(2–3):161–169

Xiong JL, Kong HY, Akram NA, Bai X, Ashraf M, Tan RY, Zhu H, Siddique KH, Xiong YC, Turner NC (2016) 24-epibrassinolide increases growth, grain yield and β-ODAP production in seeds of well-watered and moderately water-stressed grass pea. Plant Growth Regul 78(2):217–231

Xu C, Wu X, Zhang H (2009) Impact of D-Arg on drought resistance and endogenous polyamines in mycorrhizal *Pinus massoniana*. J Nanjing Forest Univ 33:19–23

Xu L, Dong Y, Fan Z, Kong J, Liu S, Bai X (2014a) Effects of the application of exogenous NO at different growth stage on the physiological characteristics of peanut grown in Cd-contaminated soil. J Plant Inter 9(1):285

Xu L, Xing ST, Sun X (2014b) Effects of polyamines on hormones contents and the relationship with the flower bud differentiation in chrysanthemum. Plant Physiol 50:1195–1202

Yadav SK (2010) Heavy metals toxicity in plants: an overview on the role of glutathione and phytochelatins in heavy metal stress tolerance of plants. South Afr J Botany 76(2):167–179

Yadav SK, Singla-Pareek SL, Ray M, Reddy MK, Sopory SK (2005a) Methylglyoxal levels in plants under salinity stress are dependent on glyoxalase I and glutathione. Biochem Biophys Res Commun 337(1):61–67

Yadav SK, Singla-Pareek SL, Reddy MK, Sopory SK (2005b) Transgenic tobacco plants overexpressing glyoxalase enzymes resist an increase in methylglyoxal and maintain higher reduced glutathione levels under salinity stress. FEBS Lett 579(27):6265–6271

Yadegari LZ, Heidari R, Rahmani F, Khara J (2014) Drought tolerance induced by foliar application of abscisic acid and sulfonamide compounds in tomato. J Stress Physiol Biochem 10:327–334

Yan ZF, Wei JK, Zhou X (1999) Effects on the resistance water stress for India millet seedling by coronatine treatment. Chin Agric Sci Bull 15:11–14

Yancey PH (1994) Compatible and counteracting solutes. In: Strange SK (ed) Cellular and molecular physiology of cell regulation. CRC Press, Boca Raton, pp 81–109

Yang L, Liang H, Lv X, Liu D, Wen X, Liao Y (2016) Effect of polyamines on the grain filling of wheat under drought stress. Plant Physiol Biochem 100:113–129

Yao T, Shaharuddin A, Chai H, Mahmood M, Shaharuddin N, Chai Ling H (2016) Application of glycine betaine alleviates salt induced damages more efficiently than ascorbic acid in *In vitro* rice shoots. Aust J Basic Appl Sci 10:58–65

Yoon JY, Hamayan M, Lee SK, Lee IJ (2009) Methyl jasmonate alleviated salinity stress in soybean. J Crop Sci Biotech 12(2):63–68

Younesi O, Moradi A (2014) Effect of priming of seeds of *Medicago sativa* 'bami' with gibberellic acid on germination, seedlings growth and antioxidant enzymes activity under salinity stress. J Hortic Res 22(2):167–174

Yusuf M, Fariduddin Q, Hayat S, Hasan SA, Ahmad A (2011) Protective responses of 28 homobrssinolide in cultivars of *Triticum aestivum* with different levels of nickel. Arch Environ Contam Toxicol 60:68–76

Yusuf M, Fariduddin Q, Ahmad A (2012) 24-Epibrassinolide modulates growth, nodulation, antioxidant system, and osmolyte in tolerant and sensitive varieties of *Vigna radiate* under different levels of nickel: a shotgun approach. Plant Physiol Biochem 57:143–153

Zaki ME, Shalaby MA, Sohair Khalil I, Abou-Sedera FA, Abd MS (2014) Protective role of benzyladenine and putrescine on snap beans (*Phaseolus vulgaris* L.) productivity grown under high temperature. Middle East J Appl Sci 4(4):905–910

Zeid IM (2009) Trehalose as osmoprotectant for maize under salinity-induced stress. Res J Agric Biol Sci 5:613–622

Zeid IM, Shedeed ZA (2006) Response of alfalfa to putrescine treatment under drought stress. Biol Plant 50(4):635–640

Zhang J (2011) China's success in increasing per capita food production. J Exp Bot 62(11):3707–3711

Zhang D (2014) Abscisic acid: metabolism, transport and signaling. Springer, New York, NY

Zhao XX, Ma QQ, Liang C, Fang Y, Wang YQ, Wang W (2007) Effect of glycinebetaine on function of thylakoid membranes in wheat flag leaves under drought stress. Biol Plant 51:584–588

Zholkevich V, Pustovoitova T (1993) Growth and phytohormone content in *Cucumis sativus* L. leaves under water deficiency. Russ J Plant Physiol 40:595–599

Zhou ZS, Guo K, Elbaz AA, Yang ZM (2009) Salicylic acid alleviates mercury toxicity by preventing oxidative stress in roots of *Medicago sativa*. Environ Exp Bot 65(1):27–34

Role of Biostimulants for Enhancing Abiotic Stress Tolerance in Fabaceae Plants

Anamika Dubey, Ashwani Kumar, and Mohammad Latif Khan

Abstract Plants belong to the family *Fabaceae* play an important role in sustainable agriculture due to their high protein content and also in maintaining the N_2 economy of the soil in semi-arid and arid regions of the world. Presently, the impact of different abiotic and biotic stresses on the plants' growth and productivity is further accelerated by climate change and has resulted in reduced crop yield. Therefore, identification of physiological and biochemical traits that contribute to enhancing legumes growth and yield under stress conditions is a major challenge for the plant breeders and geneticists. Although the application of biostimulants/biofertilizers have emerged as eco-friendly solutions for tackling these problems. Biostimulant includes diverse substances and microorganisms (Plant Growth-Promoting Rhizobacteria (PGPR), bacterial and fungal endophytes or endosymbionts) which enhance plant growth and productivity, nutrition use efficiency and plant performance under stress conditions. Thus, these microbial bioinoculants play a dual role as biostimulant as well as biocontrol agent when applied to the plants. The global market for biostimulants reached up to $2,200 million by 2018. Extensive use of these biostimulants in agriculture filed, with particular reference to Fabaceae family, still lack peer-reviewed scientific evaluation. This chapter provides latest advances in the application of plant biostimulants and scientific information related to the nature, mechanism of action and effects of biostimulants on Fabaceae plants under abiotic stress condition.

Keywords Biostimulants · Fabaceae plants · PGPR · Mycorrhiza · Biofertilizers · Abiotic stress

1 Introduction

Maintaining global food production to feed rapidly increasing world population is considered as the greatest challenge nowadays. Rapid intensification in agriculture has put negative impacts on environmental health, which leads to a loss in countless

---

A. Dubey · A. Kumar (✉) · M. L. Khan
Metagenomics and Secretomics Research Lab., Department of Botany, Dr. Harisingh Gour University (A Central University), Sagar 470003, MP, India
e-mail: ashwaniiitd@hotmail.com

M. Hasanuzzaman et al. (eds.), *The Plant Family Fabaceae*,
https://doi.org/10.1007/978-981-15-4752-2_8

ecosystem services and global climate change. Therefore, there is an urgent need to promote organic farming to maintain soil health and plant productivity in a sustainable manner. One possible potential solution for this rising problem is to use plant biostimulants. Plant biostimulants, also known as agricultural biostimulators, are substances that can be added directly to the nearby plant environment to improve plant growth, nutrition uptake and plant performance under biotic and abiotic stress conditions (Dubey et al. 2018, 2019; Malla et al. 2018, 2019). There are different types of abiotic stress known to reduce crop production worldwide among them drought stress is considered as the most destructive abiotic stress that augment its intensity over the past few decades disturbing world's food production. The first legalidiom about plant biostimulants was provided in the Farm Bill, in December 2018 (https://www.congress.gov/115/bills/hr2/BILLS-115hr2enr.pdf). This describes a biostimulant as "a microbe or that substance, when applied to seeds, plants, or near the rhizosphere, stimulates usual processes to benefit or augment nutrient efficiency, nutrient uptake, stress tolerance in plants, crop yield." Global market for biostimulants reached $2.19 billion in 2018 and it is estimated to reach an annual growth rate of 12.5% from 2019 to 2024. Europe has the biggest market for biostimulants that is approximately 40% of the total market share; the North American market is estimated to reach $605.1 million in 2019.

Therefore, many biostimulants are directly added to the rhizosphere of the plant in order to facilitate nutrient uptake. Most of them show positive responses against severe environmental stresses like water scarcity, and soil salinity (Kumar et al. 2010; du Jardin 2015; Hashem et al. 2017; Ahmad et al. 2018; Dubey et al. 2018). Effects of abiotic stress tolerance in Fabaceae plants are shown in Fig. 1.

Newly emerged paradigms emphasize that plants are not separate entities instead, they are partners and host to other microbes. Plants harbour many microorganisms both on the surface as well as in their tissues, interact with them and allows them to tolerate different kinds of stresses faced by plants (Hassani et al. 2018). Therefore, if we functionally manipulate these interactions, we might fortify their role in plant stress tolerance (Sánchez-Cañizares et al. 2017).

Additionally, many reviews highlighted the role of various biostimulants for plant growth and performance, such as the use of seaweed extracts (Battacharyya et al. 2015; Crouch and Staden 2016; Shukla et al. 2018a, b), silicon (Ruppenthal et al. 2016), protein hydrolysates (Colla et al. 2015), humic and fulvic acids (Canellas et al. 2015), chitosan (Pichyangkura and Chadchawan 2015), the role of phosphate (Gómez-Merino and Trejo-Téllez 2015), arbuscular mycorrhizal fungi (Kumar et al. 2015; Rouphael et al. 2015; Hashem et al. 2019b), *trichoderma* (López-Bucio et al. 2015), plant growth-promoting rhizobacteria (Ruzzi and Aroca 2015; Vyas et al. 2018; Barnawal et al. 2019). Table 1 shows the summary of leguminous plant species, biostimulant and effects of stress.

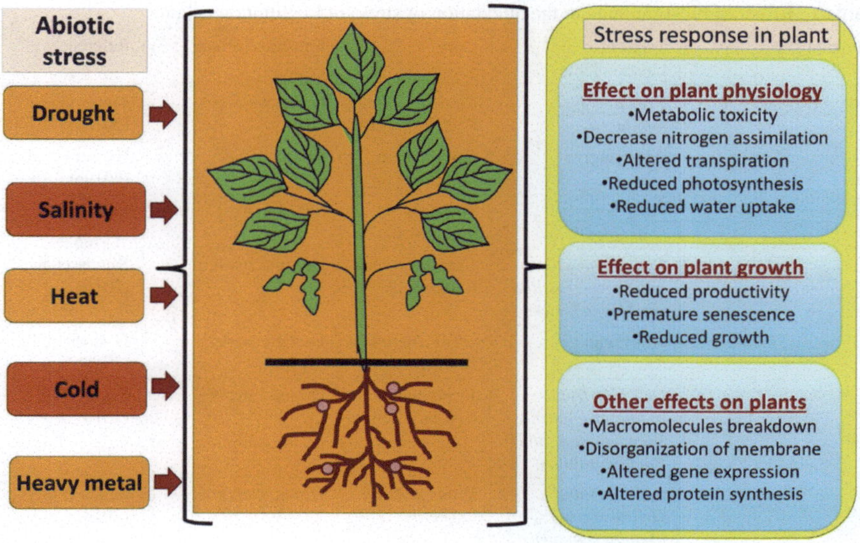

Fig. 1 Effect of abiotic stress tolerance in Fabaceae plants

2 Algal Extracts

Seaweed extracts (SWE) are used as biostimulants nowadays as commercial formulations for improving plant growth and abiotic stress tolerance in plants like to drought, salinity and heat, and. Algal extracts target a large number of pathways in order to increase tolerance under stress conditions. Seaweeds are green, brown and red macroalgae represent about 10% of the total productivity of marine ecosystem (Battacharyya et al. 2015; Crouch and Staden 2016). Macroalgae has been used as potential biofertilizers for many years and are still in use.

2.1 Seaweed Extracts

Extracts of seaweed are widely used on a variety of horticultural crops, and these biostimulants are available in powdered as well as in liquid form (Battacharyya et al. 2015). Liquid extracts of seaweeds are generally applied near the rhizospheric zone of the plant, however, this can be done by mixing the liquid extracts with irrigation water and then applied as drip irrigation to crops. Different scientists have used these seaweed extracts as foliar sprays on a variety of tree, vegetable, flowering and leguminous crops like in mango, almonds, plum, cherry, tomato, potato and soybean (Rathore et al. 2009; Colavita et al. 2011; Haider et al. 2012; Almohammedi et al. 2014; Backes et al. 2017; Kuisma 2018; Tomar et al. 2018). Several studies concluded

Table 1 Effect of biostimulants in the alleviation of stress in Leguminous plants

Type of BS	Types of biostimulants	Crop	Stress and effect	References
Microbial-derived	*Pseudomonas putida*	*Cicer arietinum* L. (Chick pea)	Drought stress	Tiwari et al. (2016)
	AM Fungi	*Glycine max* (Soybean)	Salt tress	Hashem et al. (2019a)
	Rhizobium leguminosarum	*Viciafaba* (Faba bean)	Salinity stress	Benidire et al. (2017)
	Pseudomonas putida and *Bacillus amyloliquefaciens*	*Cicer arietinum* L. (Chickpea)	Drought stress	Kumar et al. (2016)
	AM Fungi	*Cicer arietinum* L. (Chickpea)	Drought stress	Hashem et al. (2019b)
	Ochrobactrum pseudogrignonense, Pseudomonas and *Bacillus subtilis*	*Vigna mungo* L. (black gram)	Drought stress	Saikia et al. (2018)
	Bacillus subtilis, Ochrobactrum pseudogrignonense, and *Pseudomonas*	*Pisum sativum* L. (Pea)	Drought stress	Saikia et al. (2018)
	Pseudomonas aeruginosa	*Vigna radiata* (L.) (Mung bean)	Drought stress	Sarma and Saikia (2014)
	Pseudomonas simiae	*Vigna radiata* (L.) (Mung bean)	Drought stress	Kumari et al. (2016)
	Bacillus aryabhattai		Oxidative and nitrosative stress	Park et al. (2017)
Hormones	Salicylic and ascorbic acid	*Cicer arietinum* L. (chickpea)	Drought stress	Farjam et al. (2015)
	Salicylic acid and potassium	*Vigna radiata* L. (mungbean)	Drought stress	Majeed et al. (2016)
Amino acids	Proline	*Glycine max* (Soybean)	Drought stress	Mwenye et al. (2016)
Hormone	Melatonin	*Glycine max* (Soybean)	Drought stress	Wei et al. (2015)
	Melatonin	*Glycine max* (Soybean)	Salt stress	Wei et al. (2015)
Antioxidant	Glutathione	*Vigna radiata* L. (mungbean)	High-temperature stress	Nahar et al. (2015)
	Glycinebetaine	*Glycine max* (Soybean)	Salt stress	Malekzadeh (2015)
Plant growth regulator	Methyl jasmonate	*Glycine max* (Soybean)	Drought stress	Mohamed and Latif (2017)
	Silicon	*Glycine max* (Soybean)	Drought stress	Ruppenthal et al. (2016)
Seaweeds	*Ascophyllum nodosum* (ANE)	*Glycine max* (Soybean)	Drought stress	Shukla et al. (2018b)

(continued)

Table 1 (continued)

Type of BS	Types of biostimulants	Crop	Stress and effect	References
	Ascophyllum nodosum (ANE)	*Glycine max* (Soybean)	Salinity stress	Shukla et al. (2018a)
	Kappaphycus alvarezii	*Glycine max* (Soybean)	Flooding stress	Rathore et al. (2009)
Humic acid fulvic acid	Humic acid, organic fertilizer and Nitroxin Biofertilizer	*Cicer arietinum* L. (chickpea)	Drought stress	Hoseyn and Mahdi (2016)
	Humic acid, mycorrhiza and Rhizobium	*Cicer arietinum* L. (chickpea)	Drought stress	Morton (2013)
	Fulvic acid	*Glycine max* (Soybean)	Heat stress	Dinler et al. (2016)
	Fulvic acid	*Glycine max* (Soybean)	Salt stress	Dinler et al. (2016)

the fact that the foliar application of seaweed extracts is more effective when applied in the morning because at that time leaf stomata are generally open.

2.2 *Plant Biostimulants Activity of Seaweed Extract*

Plants generally absorb nutrients directly through the leaf surface or with the help of the roots. Extracts of seaweed alter biochemical, biological and physical properties of the soil and may also affect the roots facilitating nutrient uptake by plants. Seaweeds derived from brown algae composed of polyuronides like fucoidans and alginates. Alginic acids having metal chelating properties form high molecular weight polymers and also have soil-conditioning properties. Therefore, due to the presence of many cross-linked polymers the water holding capacity of the soil is improved (Dhaker et al. 2015). Various extracts of seaweed affect the regulation of those genes that are playing very crucial role in the uptake of nutrients. The study conducted by Castaings et al. (2011) and Krouk et al. (2010) showed that extract of seaweed, *Ascophyllum nodosum* upregulates the expression of a nitrate transporter gene (NRT1.1.) that helps to improve nitrogen sensing and auxin transport. Therefore, it results in enhanced nitrogen assimilation and improves the development of lateral roots. Mostly, the soil colonizing bacteria, rhizobia, forms root nodules with the roots of many leguminous plants and fix atmospheric nitrogen. A study conducted by Khan et al. (2009) found that rhizobia strain *Sinorhizobium melitoli* forms more number of nodules in alfalfa plant when supplied with extracts of *A. nodosum* extract (ANE). ANE induces the expression of NodC gene of bacteria that plays a key role in bacteria–plant signalling, by mimicking the effect of the luteolin. Plant growth-promoting properties of seaweed extracts have been well documented by various scientists in already published literature (Khan et al. 2009; Battacharyya et al. 2015; Crouch and Staden 2016; Shukla et al. 2018a) but very less is known about their mechanisms of action. Many commercial products are derivatives of brown algae like *Ascophyllum nodosum*, *Durvillae apotatorum*, red (ex *Lithothamnium calcareum*) macroalgae (Shukla et al. 2018a, b). At present, more than 47 companies are producing and selling a variety of algal extracts for increasing agricultural productivity therefore, the majority of the formulations are derived from macroalgae, *Ascophyllum nodosum* (brown algae).

Additionally, chemical-derived components of brown seaweed extract induce colonization and growth of positive soil fungi in the root. Alginic acid, a chief component of brown seaweed extracts promotes growth AM fungi which often improves phosphorus uptake (Ishii et al. 2009). Brown seaweeds such as *Ascophyllum nodosum, Alaria esculenta, Fucus serratus, F. vesiculosus, F. spiralis, Ectocarpus siliculosus, Laminaria digitata, L. saccharina, L. hyperborea, Pilayella littoralis and Halidrys siliquosa* contains many bioactive compounds, osmolytes like mannitol which is considered as an important shielding compound in response to different kinds of abiotic stresses. Spraying applications of IPA liquid seaweed extracts, enhance plant tolerance towards water stress (Mancuso et al. 2006).

2.3 Humic and Fulvic Acids

Humic acid and fulvic acid are the substances that are the main organic components of peat, lignites and soil. These substances are generally formed by biodegradation of organic matter resulting in a blend of acids that contains carboxyl groups and phenolate. Humic and Fulvic acids have low molecular weight and high oxygen content (Bulgari et al. 2015). Ample number of studies have been reported, the potential use of these substances for improving abiotic stress tolerance in different plant species like tomato, maize, chickpea, pea and sunflower (Morton 2013; Hoseyn and Mahdi 2016; Sharifi 2017; Khorasaninejad et al. 2018; Hatami 2019) (Fig. 2).

Dinler et al. (2016) in their study used fulvic acid for alleviating the effect of heat and salt stress in soybean plants. The study conducted by Hoseyn and Mahdi (2016) demonstrated the effect of humic acid, nitroxin biofertilizer and organic fertilizer in response to drought stress in the pea plant. Morton (2013) in his study used humic acid with *Rhizobium* and mycorrhiza for reducing the effect of drought stress in chickpea. Aydin (2012) in his study used humic acids for alleviating the effect of salinity stress in common bean (*Phaseolus vulgaris* L.) result in increase of endogenous levels of proline and reduced membrane leakage. Foliar application of salicylic acid, paclobutrazol and humic acid put a positive impact on anatomical, morphological, biochemical and seed yield characters of soybean (*Glycine max* L.) (El-Aal and Eid 2017). Some researchers used both humic acid and fulvic acid for reducing the impact of drought stress on various crop plants. Besides dicot plants humic acid extracts are also beneficial for monocot plants. When vermicompost extracts were applied to rice (*Oryza sativa* L.) leads to the production of ROS scavenging enzymes

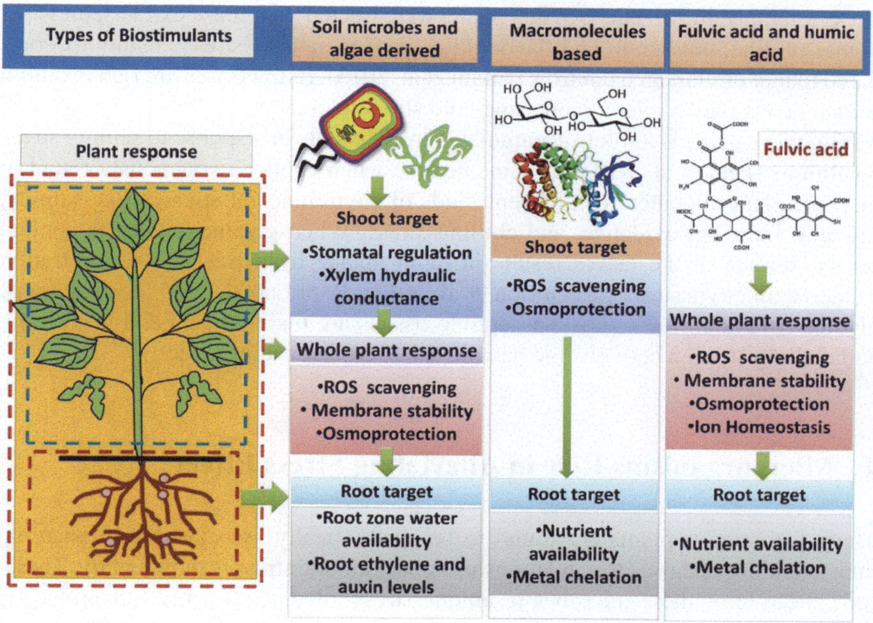

Fig. 2 Key mechanism targeted by biostimulants and plant response

as well as activating anti-oxidative enzymatic function. These enzymes are generally requisite to inactivate toxic-free oxygen radicals produced in plants under abiotic stress conditions (García et al. 2012).

3 Carbohydrates, Proteins, Amino Acids and Lipids

Protein hydrolysates (PH) are a blend of oligopeptides, polypeptides and amino acids formed after partial hydrolysis of agricultural by-product obtained from plants and animals (Colla et al. 2015). PH is generally sold as a formulation that consist of plant growth-promoting regulators. Most of the PH products, more than 90%, are derived from animal by-products formed by the chemical hydrolysis while plant-derived products are recently gaining more attention (Colla et al. 2015; Van Oosten et al. 2017).

Megafol is a commercial biostimulant used as anti-stress and growth activator in field consists of vitamins, proteins, betaines and amino acids obtained from extracts of algae and plants. Many of these extracts are obtained from Fabaceae plants and used for the growth of other plants. The study conducted by Ertani et al. (2013) used hydrolysate-based biostimulants extracted from *Medicago sativa* L., plants containing indole-3-acetic acid (IAA) and triacontanol (TRIA) which are known to improve stress tolerance in *Zea mays*. When plants were treated with these extracts, they had higher proline, potassium and flavonoid content in salinity stress conditions compared to that of untreated controls (Ertani et al. 2013). Extracts that are rich in amino acid contents play a role in increasing cold stress tolerance.

Glycine betaine is a solute accumulated inside plants in response to salinity stress conditions (Peel et al. 2010). Glycine betaine when applied exogenously found to increase plant tolerance for different kinds of environmental stresses like salinity, drought, freezing, oxidative, and chilling stress. When glycine betaine is applied on the surface of the leaves results in rapid uptake by leaves and concentration in meristematic tissues (Van Oosten et al. 2017). Rapid uptake and localization of glycine betaine in these most susceptible tissues are mainly beneficial in freezing and chilling stress where glycine betaine can exert a shielding effect.

4 Microorganisms Role in Alleviating Stress Effect

Many leguminous plants are found to be symbiotically associated with bacteria; understanding of those close relationships under abiotic stress is rudimentary. Therefore, the role of these microbes in abiotic stress tolerance has been identified by different scientist working in leguminous crop plants (Qiao et al. 2011; Sarma and Saikia 2014; Kumar et al. 2016; Kumari et al. 2016; Tiwari et al. 2016; Benidire et al. 2017; Park et al. 2017; Hashem et al. 2019b) (Table 1). Microbes having plant growth-promoting potential and that can cope up with severe environmental stress conditions can be used as biostimulants. These microbes have been isolated from different ecosystems with alkaline, saline, acidic, and arid soils. These bacteria belong to these genera such as *Rhizobium*, *Azotobacter*, *Azospirillum*, *Bradyrhizobium*, *Bacillus* and *Pseudomonas* have evolved different strategies to thrive under stress conditions (Kaushal and Wani 2016; Enebe and Babalola 2018; Krishna et al. 2019; Patil et al. 2020). Studies conducted by several researches reported the role of these microbes as biostimulants for alleviating the effect of abiotic stress on different leguminous plants.

Tiwari et al. (2016) used *Pseudomonas putida* for reducing the effect of drought stress chickpea (*Cicer arietinum* L.). A similar study conducted by Kumar et al. (2016) used *Pseudomonas putida* and *Bacillus amyloliquefaciens* for alleviating the effect of drought stress in chickpea (*Cicer arietinum* L.). Saikia et al. (2018) in their study uses a consortium of *Bacillus subtilis, Ochrobactrum pseudogrignonense,* and *Pseudomonas* for enhancing the yield of Peaplant (*Pisum sativum* L.) and black gram (*Vigna mungo* L.) under drought stress condition. Hashem et al. (2019a, b) uses AM fungi for abiotic stress tolerance in soybean (*Glycine max*) and chickpea (*Cicer arietinum* L.). Benidire et al. (2017), used *Rhizobium leguminosarum,* Sarma and Saikia (2014) used *Pseudomonas aeruginosa*. Kumari et al. (2016) used *Pseudomonas simiae* for increasing abiotic stress tolerance in Faba bean (*Vicia faba*) (Table 1).

5 Conclusion and Future Perspectives

Use of biostimulant for various agronomical crops has the potential to enhance plant resilience to different environmental perturbations. In order to adjust application rates, plant biostimulant techniques and specificities are recognized that may put the highest impact on stress tolerance, therefore, more concern should be given for better understanding of the functional mechanism of action of these biostimulants. Once a better understanding of these complex mechanisms has been reached; then we will be able to go ahead to the next generation of biostimulants where complementary mechanisms and synergies can be designed functionally. A systematic and comprehensive approach has been anticipated for discovering and characterizing novel biostimulants and thereby understanding their mechanism of action by using chemistry, biology and omics approaches. Meta-analysis of various effects of biostimulants on plants has been projected and an extensive meta-level assessment of the effect of humic substances on plant growth has been conducted. Analysis revealed that humic substances increase the overall dry weight of roots and shoots by at least 20%. Therefore, the identification of complementary/synergistic properties of these plant biostimulants are important for developing specific formulations that are targeted to improve plant response towards various types of abiotic stresses. For example, biostimulants for enhancing plant resilience in water-deficient environments should stimulate shoot versus root growth that will allow plants to explore the deeper layers of the soil during drought conditions and increasing the production of compatible solutes for reinstating positive water potential gradients and water uptake of diminishing soil water. Similarly, microbial-derived biostimulants also put positive effects by creating absorption surfaces around the roots of the plants and which helps plants to absorb more water from the soil. Therefore, the future of plant biostimulants should be driven by the following lines of force. With the development of plant biostimulants follow a traditional 'pharmacological' approach, where microorganisms or bioactive compounds produced by microbes are screened under controlled conditions and a sequential procedure is followed in order to select promising microbes, moving from the lab to more realistic natural conditions. This may be an efficient but severe and stepwise selection of active substances or microbes resulted in high production costs which are hardly justified by market-creating limited added value, like in agriculture and plant nutrition. Another way would start from field observations and lead back to the laboratory for the systematization of the scientific questions raised. For example, many ecologists and soil microbiologists are pointing out the inconsistency in the way individual plant cultivars interact with bacteria colonizing the rhizosphere and modulating the bacterial composition. Whether these genotype-dependent changes in the rhizospheric microbiome impact plant health and growth is an open question. These observations can be an early point for understanding the important interactions between plants PGPRs. From a more practical viewpoint, novel commercial approaches are being developed which aim at amplifying local beneficial microbiota instead of inoculating standardized microbial products. This approach is aggravated by the experimental fact that a limiting factor when using microbe based biostimulants is the capacity of the bioinoculant to survive and maintain adequate activity in the rhizosphere. This study can be parallel applied in the intestinal microbiome in human medicine: taking probiotics in the diet is one thing, but feeding the beneficial bacteria with probiotics seems even more important. This inspires new avenues to sustainable crop management, by developing breeding, plants varieties with enhanced ability to 'manage' their endospheric and rhizospheric microbiota and new fertilizing materials. With the technical breakthroughs and scientific knowledge now we have a complete picture of plant physiology. But research is focused on the limited number of the model plant under controlled conditions. A major challenge is to use this available knowledge and modern tools for the characterization of biostimulants and their effects on a broad range of cultivated plant species. For example, high-throughput plant phenotyping platforms have been developed for the characterization of mutants produced in functional genomics related studies, but they should (and do) inspire studies for understanding the mechanism of action of these biostimulants and their interactions with environmental stressors and with plant genotypes. Bridging this gap between and field data on mixtures and laboratory data on single biostimulants is equally important and challenging.

References

Aydin A (2012) Humic acid application alleviate salinity stress of bean (*Phaseolus vulgaris* L.) plants decreasing membrane leakage. Afr J Agric Res 7. https://doi.org/10.5897/ajar10.274

Ahmad P, Alyemeni MN, Ahanger MA et al (2018) Upregulation of antioxidant and glyoxalase systems mitigates nacl stress in *Brassica juncea* by supplementation of zinc and calcium. J Plant Interact 13:151–162. https://doi.org/10.1080/17429145.2018.1441452

Almohammedi AN, Almehemdi AF, Al-Ajeelee RK (2014) Impact of bat guano Otonycteris hemprichii Camd and seaweed extract on some growth and yield traits of barakaseed *Nigella sativa* L. J Biol Agric Healthc 4:57–65

Backes C, Villas Boas RL, Marques Santos AJ et al (2017) Foliar application of seaweed extract in potato culture. Rev De Agric Neotropical 4:53–57

Barnawal D, Singh R, Singh RP (2019) Role of plant growth promoting Rhizobacteria in drought tolerance: regulating growth hormones and osmolytes. PGPR Amelioration Sustain Agric, 107–128. https://doi.org/10.1016/B978-0-12-815879-1.00006-9

Battacharyya D, Babgohari MZ, Rathor P, Prithiviraj B (2015) Seaweed extracts as biostimulants in horticulture. Sci Hortic 196:39–48

Benidire L, Lahrouni M, El Khalloufi F et al (2017) Effects of *Rhizobium leguminosarum* inoculation on growth, nitrogen uptake and mineral assimilation in *Vicia faba* plants under salinity stress. J Agric Sci Technol 19:889–901

Bulgari R, Cocetta G, Trivellini A et al (2015) Biostimulants and crop responses: a review. Biol Agric Hortic 31:1–17

Canellas LP, Olivares FL, Aguiar NO et al (2015) Humic and fulvic acids as biostimulants in horticulture. Sci Hortic 196:15–27. https://doi.org/10.1016/j.scienta.2015.09.013

Castaings L, Marchive C, Meyer C, Krapp A (2011) Nitrogen signalling in Arabidopsis: how to obtain insights into a complex signalling network. J Exp Bot 62:1391–1397

Colavita GM, Spera N, Blackhall V, Sepulveda GM (2011) Effect of seaweed extract on pear fruit quality and yield. In: Acta Horticulturae, pp 601–608

Colla G, Nardi S, Cardarelli M, Ertani A, Lucini L, Canaguier R, and Rouphael Y (2015) Protein hydrolysates as biostimulants in horticulture. Sci Hortic 28–38. https://doi.org/10.1016/j.scienta.2015.08.037
Crouch IJ, Van Staden J (2016) Commercial seaweed products as biostimulants in horticulture commercial seaweed products as biostimulants in horticulture. J Home Consu Hortic 4682:19–76. https://doi.org/10.1300/J280v01n01
Dhaker SC, Mundra SL, Dhaker RC, Sumeriya HK (2015) Effect of weed management and sulphur on nutrient content and uptake by weeds and soybean. Legume Res 38:411–414. https://doi.org/10.5958/0976-0571.2015.00098.3
Dinler BS, Gunduzer E, Tekinay T (2016) Pre-treatment of fulvic acid plays a stimulant role in protection of soybean (*Glycine max* L.) leaves against heat and salt stress. Acta Biologica Cracov Ser Bot 58:29–41. https://doi.org/10.1515/abcsb-2016-0002
du Jardin P (2015) Plant biostimulants: definition, concept, main categories and regulation. Sci Hortic 196:3–14
Dubey A, Kumar A, Abd Allah EF, et al (2018) Growing more with less: breeding and developing drought resilient soybean to improve food security. Ecol Ind 105:0–1. https://doi.org/10.1016/j.ecolind.2018.03.003
Dubey A, Malla MA, Khan F et al (2019) Soil microbiome: a key player for conservation of soil health under changing climate. Biodivers Conserv. https://doi.org/10.1007/s10531-019-01760-5
El-Aal M, Eid R (2017) Optimizing growth, seed yield and quality of soybean (*Glycine max* L.) plant using growth substances. Asian Res J Agric 6:1–19. https://doi.org/10.9734/arja/2017/36034
Enebe MC, Babalola OO (2018) The influence of plant growth-promoting rhizobacteria in plant tolerance to abiotic stress: a survival strategy. Appl Microbiol Biotechnol 102:7821–7835
Ertani A, Schiavon M, Muscolo A, Nardi S (2013) Alfalfa plant-derived biostimulant stimulate short-term growth of salt stressed *Zea mays* L. plants. Plant Soil 364:145–158. https://doi.org/10.1007/s11104-012-1335-z
Farjam S, Kazemi-Arbat H, Siosemardeh A et al (2015) Effects of salicylic and ascorbic acid applications on growth, yield, water use efficiency and some physiological traits of chickpea (*Cicer arietinum* L.) under reduced irrigation. Legume Res 38:66–71. https://doi.org/10.5958/0976-0571.2015.00011.9
García AC, Santos LA, Izquierdo FG et al (2012) Vermicompost humic acids as an ecological pathway to protect rice plant against oxidative stress. Ecol Eng 47:203–208. https://doi.org/10.1016/j.ecoleng.2012.06.011
Gómez-Merino FC, Trejo-Téllez LI (2015) Biostimulant activity of phosphite in horticulture. Sci Hortic 196:82–90. https://doi.org/10.1016/j.scienta.2015.09.035
Haider MW, Ayyub CM, Pervez MA et al (2012) Impact of foliar application of seaweed extract on growth, yield and quality of potato (*Solanum tuberosum* L.). Soil Environ 31:157–162
Hashem A, Abd Allah EF, Alqarawi AA, et al (2017) Plant defense approach of *Bacillus subtilis* (Bera 71) against *Macrophomina phaseolina* (tassi) goid in mung bean. J Plant Interact 12:390–401. https://doi.org/10.1080/17429145.2017.1373871
Hashem A, Abd Allah EF, Alqarawi AA, et al (2019a) Comparing symbiotic performance and physiological responses of two soybean cultivars to arbuscular mycorrhizal fungi under salt stress. Saudi J Biol Sci 26:38–48. https://doi.org/10.1016/j.sjbs.2016.11.015
Hashem A, Kumar A, Al-Dbass AM et al (2019b) Arbuscular mycorrhizal fungi and biochar improves drought tolerance in chickpea. Saudi J Biol Sci 26:614–624. https://doi.org/10.1016/j.sjbs.2018.11.005
Hassani MA, Durán P, Hacquard S (2018) Microbial interactions within the plant holobiont. Microbiome 6:58
Hatami H (2019) The effect of zinc and humic acid applications on yield and yield components of sunflower in drought stress. J Adv Agric Technol 4:36–39. https://doi.org/10.18178/joaat.4.1.36-39
Hoseyn A, Mahdi AM (2016) Pea response to drought stress, humic acid, organic fertilizer and nitroxin biofertilizer. Ecol Environ Conserv 22:473–477

Ishii T, Aikawa J, Nakamura N et al (2009) Effect of citrus juice pomace extracts on in vitro hyphal growth of vesicular-arbuscular mycorrhizal fungi and their in vivo infections of citrus roots. Engei Gakkai Zasshi 69:9–14. https://doi.org/10.2503/jjshs.69.9

Kaushal M, Wani SP (2016) Plant-growth-promoting rhizobacteria: drought stress alleviators to ameliorate crop production in drylands. Ann Microbiol 66:35–42. https://doi.org/10.1007/s13213-015-1112-3

Khan W, Rayirath UP, Subramanian S, Jithesh MN, Rayorath P, Hodges DM, et al (2009) Seaweed extracts as biostimulants of plant growth and development. J. Plant Growth Regul 386–399. https://doi.org/10.1007/s00344-009-9103-x

Khorasaninejad S, Alizadeh Ahmadabadi A, Hemmati K (2018) The effect of humic acid on leaf morphophysiological and phytochemical properties of *Echinacea purpurea* L. under water deficit stress. Sci Hortic 239:314–323. https://doi.org/10.1016/j.scienta.2018.03.015

Krishna SBN, Dubey A, Malla MA et al (2019) Integrating microbiome network: establishing linkages between plants, microbes and human health. Open Microbiol J 13:330–342. https://doi.org/10.2174/1874285801913020330

Krouk G, Lacombe B, Bielach A et al (2010) Nitrate-regulated auxin transport by NRT1.1 defines a mechanism for nutrient sensing in plants. Dev Cell 18:927–937. https://doi.org/10.1016/j.devcel.2010.05.008

Kuisma P (2018) The effect of foliar application of seaweed extract on potato. Agric Food Sci 61:371–377. https://doi.org/10.23986/afsci.72367

Kumar A, Dames JF, Gupta A et al (2015) Current developments in arbuscular mycorrhizal fungi research and its role in salinity stress alleviation: a biotechnological perspective. Crit Rev Biotechnol 35:461–474. https://doi.org/10.3109/07388551.2014.899964

Kumar A, Sharma S, Mishra S (2010) Influence of arbuscular mycorrhizal (AM) fungi and salinity on seedling growth, solute accumulation, and mycorrhizal dependency of *Jatropha curcas* L. J Plant Growth Regul 29:297–306. https://doi.org/10.1007/s00344-009-9136-1

Kumar M, Mishra S, Dixit V, et al (2016) Synergistic effect of *Pseudomonas putida* and *Bacillus amyloliquefaciens* ameliorates drought stress in chickpea (*Cicer arietinum* L.). Plant Signal Behav 11. https://doi.org/10.1080/15592324.2015.1071004

Kumari S, Vaishnav A, Jain S et al (2016) Induced drought tolerance through wild and mutant bacterial strain *Pseudomonas simiae* in mung bean (*Vigna radiata* L.). World J Microbiol Biotechnol 32:1–10. https://doi.org/10.1007/s11274-015-1974-3

López-Bucio J, Pelagio-Flores R, Herrera-Estrella A (2015) Trichoderma as biostimulant: exploiting the multilevel properties of a plant beneficial fungus. Sci Hortic 196:109–123. https://doi.org/10.1016/j.scienta.2015.08.043

Majeed S, Akram M, Latif M et al (2016) Mitigation of drought stress by foliar application of salicylic acid and potassium in mungbean (*Vigna radiata* L.). Legume Res 39:208–214. https://doi.org/10.18805/lr.v39i2.9532

Malekzadeh P (2015) Influence of exogenous application of glycinebetaine on antioxidative system and growth of salt-stressed soybean seedlings (*Glycine max* L.). Physiol Mol Biol Plants 21:225–232. https://doi.org/10.1007/s12298-015-0292-4

Malla MA, Dubey A, Yadav S et al (2019) Exploring the human microbiome: the potential future role of next-generation sequencing in disease diagnosis and treatment. Front Immunol 9:1–23. https://doi.org/10.3389/fimmu.2018.02868

Malla MA, Dubey A, Yadav S, et al (2018) Understanding and designing the strategies for the microbe-mediated remediation of environmental contaminants using Omics approaches. Front Microbiol 9. https://doi.org/10.3389/fmicb.2018.01132

Mancuso S, Azzarello E, Mugnai S, Briand X (2006) Marine bioactive substances (IPA extract) improve foliar ion uptake and water stress tolerance in potted Vitis vinifera plants. Adv Hortic Sci 20:156–161

Mohamed HI, Latif HH (2017) Improvement of drought tolerance of soybean plants by using methyl jasmonate. Physiol Mol Biol Plants 23:545–556. https://doi.org/10.1007/s12298-017-0451-x

Morton (2013) Reducing effects of drought stress by application of humic acid, mycorrhiza and Rhizobium on chickpea. Int J, 1775–1778

Mwenye OJ, van Rensburg L, van Biljon A, van der Merwe R (2016) The role of proline and root traits on selection for drought-stress tolerance in soybeans: a review. South Afr J Plant Soil 33:245–256

Nahar K, Hasanuzzaman M, Alam MM, Fujita M (2015) Exogenous glutathione confers high temperature stress tolerance in mung bean (*Vigna radiata* L.) by modulating antioxidant defense and methylglyoxal detoxification system. Environ Exp Bot 112:44–54. https://doi.org/10.1016/j.envexpbot.2014.12.001

Park YG, Mun BG, Kang SM et al (2017) Bacillus aryabhattai SRB02 tolerates oxidative and nitrosative stress and promotes the growth of soybean by modulating the production of phytohormones. PLoS ONE 12:1–28. https://doi.org/10.1371/journal.pone.0173203

Patil A, Dubey A, Malla MA, et al (2020) Complete genome sequence of *Lactobacillus plantarum* strain JDARSH, isolated from sheep milk. Microbiol Resour Announcements. https://doi.org/10.1128/mra.01199-19

Peel GJ, Mickelbart MV, Rhodes D (2010) Choline metabolism in glycinebetaine accumulating and non-accumulating near-isogenic lines of *Zea mays* and *Sorghum bicolor*. Phytochemistry 71:404–414. https://doi.org/10.1016/j.phytochem.2009.11.002

Pichyangkura R, Chadchawan S (2015) Biostimulant activity of chitosan in horticulture. Sci Hortic 196:49–65. https://doi.org/10.1016/j.scienta.2015.09.031

Qiao G, Wen XP, Yu LF, Ji XB (2011) The enhancement of drought tolerance for pigeon pea inoculated by arbuscular mycorrhizae fungi. Plant Soil Environ 57:541–546

Rathore SS, Chaudhary DR, Boricha GN et al (2009) Effect of seaweed extract on the growth, yield and nutrient uptake of soybean (*Glycine max*) under rainfed conditions. South Afr J Botany 75:351–355. https://doi.org/10.1016/j.sajb.2008.10.009

Rouphael Y, Franken P, Schneider C et al (2015) Arbuscular mycorrhizal fungi act as biostimulants in horticultural crops. Sci Hortic 196:91–108. https://doi.org/10.1016/j.scienta.2015.09.002

Ruppenthal V, Zoz T, Steiner F, et al (2016) Silicon does not alleviate the adverse effects of drought stress in soybean plants. Semina: Ciencias Agrarias 37:3941–3954. https://doi.org/10.5433/1679-0359.2016v37n6p3941

Ruzzi M, Aroca R (2015) Plant growth-promoting rhizobacteria act as biostimulants in horticulture. Sci Hortic 196:124–134. https://doi.org/10.1016/j.scienta.2015.08.042

Saikia J, Sarma RK, Dhandia R, et al (2018) Alleviation of drought stress in pulse crops with ACC deaminase producing rhizobacteria isolated from acidic soil of Northeast India. Sci Reports 8. https://doi.org/10.1038/s41598-018-21921-w

Sánchez-Cañizares C, Jorrín B, Poole PS, Tkacz A (2017) Understanding the holobiont: the interdependence of plants and their microbiome. Curr Opin Microbiol 38:188–196

Sarma RK, Saikia R (2014) Alleviation of drought stress in mung bean by strain *Pseudomonas aeruginosa* GGRJ21. Plant and Soil 377. https://doi.org/10.1007/s11104-013-1981-9

Sharifi P (2017) Studying maize growth indices in different water stress conditions and the use of humic acid. Biomed Pharmacol J 10:303–310. https://doi.org/10.13005/bpj/1111

Shukla PS, Prithiviraj B, Shotton K, et al (2018a) Seaweed extract improve drought tolerance of soybean by regulating stress-response genes. AoB Plants

Shukla PS, Shotton K, Norman E, et al (2018b) Seaweed extract improve drought tolerance of soybean by regulating stress-response genes. AoB Plants 10. https://doi.org/10.1093/aobpla/plx051

Tiwari S, Lata C, Chauhan PS, Nautiyal CS (2016) *Pseudomonas putida* attunes morphophysiological, biochemical and molecular responses in *Cicer arietinum* L. during drought stress and recovery. Plant Physiol Biochem 99:108–117. https://doi.org/10.1016/j.plaphy.2015.11.001

Tomar IS, Yadav RK, Tripathi RK, et al (2018) Effect of nutrient management on yields, economics and nutrient content and their uptake of soybean (*Glyscine max* L.) under rainfed conditions. J Agri Search 5. https://doi.org/10.21921/jas.5.2.3

Van Oosten MJ, Pepe O, De Pascale S, et al (2017) The role of biostimulants and bioeffectors as alleviators of abiotic stress in crop plants. Chem Biol Technol Agric 4

Vyas P, Kumar D, Dubey A, Kumar A (2018) Screening and characterization of *Achromobacter xylosoxidans* isolated from rhizosphere of *Jatropha curcas* L. (Energy Crop) for plant-growth-promoting traits. J Adv Res Biotechnol

Wei W, Li QT, Chu YN et al (2015) Melatonin enhances plant growth and abiotic stress tolerance in soybean plants. J Exp Bot 66:695–707. https://doi.org/10.1093/jxb/eru392

Abiotic and Biotic Stresses Interaction in *Fabaceae* Plants. Contributions from the Grain Legumes/Soilborne Vascular Diseases/Drought Stress Triangle

Susana T. Leitão, Susana Araújo, Diego Rubiales, and Maria Carlota Vaz Patto

Abstract As sessile organisms, plants are constantly exposed to simultaneously abiotic and biotic stresses that impact growth thus resulting in significant yield losses. An example is drought and root infecting pathogens, which combined cause greater damage to plants than the stresses individually. Substantial information is available on the physiological, molecular, and metabolic changes in *Fabaceae* plants exposed to individual stresses, but little is known about how plants respond to multiple stresses. This is of primary importance for the development of breeding approaches based on the trade-off between plant defense response mechanisms, and high and consistent yield under field conditions. A better knowledge of the mechanisms by which legume plants perceive and transduce simultaneous or sequential combination of stress signals to initiate diverse adaptive responses is essential for breeding multiple stress-tolerant crop cultivars. In this chapter, we assess the relevance of understanding legume combined responses to abiotic and biotic stresses for production and breeding, focusing on soilborne vascular diseases and drought interaction in grain legumes. Particular attention is given to the crosstalk between signaling pathways of the "stress triangle" pathogen/host/environment interactions and to the application of integrated breeding methods aiming at multiple stress-resistant legume crops better adapted to climate change.

Keywords Abiotic/biotic interaction · Grain legumes · Drought · Soilborne vascular diseases · Climate change

---

S. T. Leitão · S. Araújo · M. C. Vaz Patto (✉)
Instituto de Tecnologia Química e Biológica António Xavier, Universidade Nova de Lisboa (ITQB NOVA), Oeiras, Portugal
e-mail: cpatto@itqb.unl.pt

D. Rubiales
Institute for Sustainable Agriculture, CSIC, Córdoba, Spain

M. Hasanuzzaman et al. (eds.), *The Plant Family Fabaceae*,
https://doi.org/10.1007/978-981-15-4752-2_9

1 Introduction

In nature, plants are exposed to various and complex types of stresses involving numerous environmental factors, both abiotic and biotic. They include drought, heat, cold, salinity, high light intensity, high CO_2 concentrations, diseases, and pests. In most cases, these stresses are studied individually, although in the field they occur in combination. Recent studies have revealed that the response of plants to a simultaneous or sequential combination of stresses cannot be simply inferred from the response to each one applied individually (Atkinson et al. 2013; Bai et al. 2018; Kissoudis et al. 2014, 2015; Zhang and Sonnewald 2017). Nevertheless, there are common features shared by all stress conditions, like a reduced plant growth, reflecting what is commonly known as the growth/defense trade-off (Huot et al. 2014).

To facilitate adaptation to a given stress or stress combination at minimal costs, plants require an energy-efficient and flexible system. A first stress factor preceded by another stress factor may either endure (due to priming effect) or predispose the plants to the subsequent stress (Mauch-Mani et al. 2017; Pandey et al. 2017). In general, abiotic stresses, such as drought or nutrient deprivation, result in the weakening of plant defenses and increased susceptibility to necrotrophic pathogens, including otherwise weakly virulent facultative pathogens. However, the same abiotic stresses seem to reduce plant susceptibility to biotrophic pathogens (Saijo and Loo 2019).

Pathogenic infections can disrupt the normal stomatal regulation of transpiration, reverting abscisic acid (ABA)-mediated stomatal closure, and reducing photosynthesis and water use efficiency (Grimmer et al. 2012; Bilgin et al. 2010). Furthermore, the combination of extreme temperature with viral infection might decrease the plant heat stress response efficiency, via down-regulation of heat shock transcription factors (Anfoka et al. 2016). On the contrary, fungal endophyte colonization can cause epigenetic modifications linked to an improvement of drought tolerance in wheat (Hubbard et al. 2014).

Genome-wide association studies in the model plant *Arabidopsis thaliana* suggested that tolerance against single abiotic and combined biotic/abiotic stresses is mediated by different genetic loci (Davila Olivas et al. 2017; Thoen et al. 2017). Consistent with this, transcriptome responses in the same species also differ between combined stresses and individual stresses (Atkinson et al. 2013; Cohen and Leach 2019; Prasch and Sonnewald 2013; Rasmussen et al. 2013). However, the genetic basis for combined stress responses remains largely unknown (Saijo and Loo 2019). Recent studies suggested that conflicting stress responses are balanced/prioritized within the plant body in an age- or stage-specific and spatial/temporal manner to maintain growth and reproduction under multiple stresses (Berens et al. 2019; Betsuyaku et al. 2017; Karasov et al. 2017; Smakowska et al. 2016).

Humanity is facing a major challenge in producing enough food for an additional three billion people, i.e., about 50% more than nowadays, within the next 50 years. This will take place in a context of arable land shrinking and degradation, nutrient deficiencies, increased water scarcity, and uncertainty due to predicted climatic changes (FAO 2017). On the other hand, forecasted global climate change effects increase the chances of plants' multi-stress interactions (Kissoudis et al. 2014), challenging breeding programs to develop crops that thrive and maintain a high yield at constantly varying field conditions.

This chapter aims to provide a general overview of the main molecular and physiological responses triggered by both biotic and abiotic stress in plants, with an emphasis on legumes. We also provide a critical discussion of the main challenges and achievements in this field.

2 Plant Defense Responses Against Biotic and Abiotic Stresses

As a part of the plant defense toward the pathogen attack, metabolic changes are induced in plants, such as the secretion of pathogenesis-related proteins, ascorbate and glutathione peroxidases (H_2O_2-scavengers), proteases, xyloglucans, phenols, phytoalexins, and lignin-like compounds (Yadeta and Thomma 2013). These metabolic changes influence morpho-physiological traits such as stomatal movement, cell membrane stability, cell wall reinforcement, leaf water potential, plant growth and root morphology which were shown to be distinctly altered under combined stress (Irulappan and Senthil-Kumar 2018). The phenylpropanoid pathway has a particularly important function in the production of lignin, flavonoids, and phytoalexins in response to wounding and pathogen attack (Fraire-Velázquez et al. 2011). For instance, in *A. thaliana*, secondary cell wall-associated genes were suggested to be involved in plant's response toward combined drought and wilt pathogen infection, but not in combined drought and foliar pathogen infection (Gupta et al. 2016). Therefore, it appears that the outcome of the biotic/abiotic interaction is, on most occasions, organ, plant species and genotype, pathogen, and abiotic stress intensity dependent.

2.1 Phytohormones Signaling Pathways

Plants have evolved complex molecular systems to adapt to rapidly evolving pathogens and changing climates. Plants defense responses against both abiotic and biotic stresses are prioritized and fine-tuned in a context-dependent manner, to facilitate adaptation at the minimal costs (Nobori and Tsuda 2019). This is enabled by a complex crosstalk between various phytohormones signaling pathways, including synergistic and antagonistic interactions between salicylic acid (SA), jasmonic acid (JA), ethylene (ET), or abscisic acid (ABA) (Fujita et al. 2006; Mauch-Mani and Mauch 2005; Pieterse et al. 2012). Ramegowda and Senthil-Kumar (2015) hypothesized that under combined stresses, the first signaling effects operate through ABA and JA pathways, in a non-specific manner. However, the same authors suggested that the modulation of stress signal, crosstalk, and subsequent downstream events are specific and unique to the combination of stresses faced by the plant. For instance, during the response against biotrophic and hemibiotrophic pathogens, which activate the SA signaling pathways, plants become more susceptible to attacks by necrotrophic pathogens and insect herbivory, which involves the activation of JA- and ET-mediated responses (Fraire-Velázquez et al. 2011; Thaler et al. 2012). In rice, abiotic stress up-regulated genes responsive to ABA, auxin, JA, and SA, while biotic stress up-regulated genes responsive to the same hormones plus cytokinin and ET (Cohen and Leach 2019). Interestingly, in this study, ABA, JA, and SA-mediated pathways were the most significantly up-regulated hormone pathways in both abiotic and biotic stress. Furthermore, ABA signaling suppresses SA to prioritize abiotic stress responses over biotic stress responses (Yasuda et al. 2008). Altogether, these studies indicate that specific hormone signatures are triggered depending on the type of stress involved, either as if it occurs single or combined.

Auxin is generally thought to be a "master hormone" because it influences every aspect of plant growth and development (Balzan et al. 2014; Grossmann 2010). This phytohormone acts on the transcriptional regulation of target genes, mainly through auxin response factors (ARFs) transcription factors, which were found to be mediators of the auxin action in biotic and abiotic stress responses in tomato (Bouzroud et al. 2018). However, auxin homeostasis is also relevant for stress adaptation mainly by interacting with SA and ABA signaling pathways. For example, during pathogen infection, auxin negatively affects SA-mediated defense responses in Arabidopsis (Park et al. 2007).

Besides phytohormone balance, the trade-off between development and defense under abiotic and biotic stress conditions relies also on other interconnected signaling pathways, including Ca^{2+} sensing, production of reactive oxygen species (ROS), production of secondary metabolites, and activation of kinase cascades (Kudla et al. 2018; Liang and Zhou 2018; Prasch and Sonnewald 2015; Zhang et al. 2018). As an example, ROS generation by respiratory burst oxidase homolog (RBoh) proteins, after pathogen elicitors recognition, leads to ABA-induced stomatal closure and to hypersensitive cell death (Fujita et al. 2006). ROS and Ca^{2+} also indirectly modulate the expression of genes like kinases and transcription factors involved in the crosstalk between signaling cascades response to abiotic/biotic stress interaction (Fraire-Velázquez et al. 2011). In particular, the role of WRKY transcription factors in mediating plant biotic and abiotic stress response through SA, JA, or ET signaling has been well reported. For example, *AtWRKY30* is induced under ROS-mediated oxidative stress and pathogen attack and its overexpression in *A.thaliana* improved abiotic stress tolerance during seed germination (Besseau et al. 2012).

Although transcriptional regulation is important for stress adaptations, the post-transcriptional regulation mediated by microRNAs (miRNAs) adds another layer in the regulatory network, which might be crucial for the coordination of developmental and environmental responses (Shriram et al. 2016). The miRNAs are endogenous small non-coding RNAs (~21–22 nucleotides) that negatively regulate gene expression, by either cleaving their complementary messenger RNA (mRNA) targets or repressing their translation (Jones-Rhoades et al. 2006). Manipulation of mRNA transcript abundance via miRNA control provides a unique strategy for modulating differential plant gene expression. As a consequence, miRNAs are emerging as the next-generation targets for genetic engineering to improve the agronomic properties of crops (Djami-Tchatchou et al. 2017). Recently, Sanz-Carbonell et al. (2019) identified 24 miRNAs expressed in response to abiotic (cold, drought, salinity, and short day) and biotic (viroid infection, *Agrobacterium* infiltration, and ascomycete root infection) stress conditions imposed individually in melon (*Cucumis melo* L.). Most of these miRNAs showed a decrease in their abundances after stress imposition and were predicted to target well-known transcriptions factors as miR156/SPL, miR164/NAC, and miR167/ARF6. This suggests that these genes need to be expressed during stress imposition. Since the miRNA-mediated response triggered by biotic stresses is distinguishable from the triggered by abiotic ones, according to their total expression values (Sanz-Carbonell et al. 2019), it would be interesting to investigate the effect of combined abiotic/biotic stresses on these stress-specific miRNA profiles.

Many of the genes with key roles in abiotic/biotic interactions are included in the same functional categories to those identified in single stress studies, i.e., stress hormone responses, ROS, mitogen-activated protein kinase (MAPK) signaling and regulation of hypersensitivity response (Kissoudis et al. 2014). However, the action of these genes is neutralized or prioritized under stress combination in comparison with the individual stress, revealing antagonistic interactions with detrimental effects on plant adaptation under multifactorial stress. In addition to the integrative signaling pathway regulation, the deployment of genes that have a protective function on proteins and cellular components under abiotic stress, such as dehydrins (DHN), and other late embryogenesis abundant (LEA) proteins or RNA chaperones was also described to minimize the antagonist interaction with biotic stress signaling (Kang et al. 2013). DHN overexpression was suggested to be involved in ABA and JA signaling pathways in *Arachis duranensis*, the wild ancestor of peanut (Mota et al. 2019). In this study, the gene *AdDHN1* showed an opposite response to abiotic and biotic stress, with a positive regulation under water deficit and a negative regulation upon nematode infection. In rice, the transcriptomic response to a variety of stresses, namely bacterial, fungal and virus infection, and drought, salt, cold and heat, identified several genes significantly expressed only under a unique stress or, in opposition, "universally" expressed (Narsai et al. 2013).

2.2 Functional Proteins

Besides phytohormones, signaling components and their regulators, other functional genes are involved in plant antagonist responses to combined abiotic and biotic stresses. Aung et al. (2018) described the effect of water in modulating plant/pathogens interactions. Indeed, many pathogens translocate virulence proteins (effectors) into host cells to target different components of plant pattern-triggered immunity and redirect sugars and water into the apoplast, where many pathogens live.

Also, the photosynthetic pathway was considered a hub of crosstalk in growth and defense trade-off during plant/pathogen/environment interactions, with bZIP transcription factors as good candidate regulators of stress response (Cohen and Leach 2019).

Moreover, in simultaneous abiotic stress and pathogen infection, one of the earliest responses occurs at the cell wall with stress-induced regulation of transcripts encoding cell wall proteins and cellulose synthesis. The so-called cell wall degrading enzymes (CWDE), like cellulases, have been shown to contribute directly to virulence and to an increased susceptibility to drought by disturbing the cellular cytoskeletal and microtubules dynamics (Kesten et al. 2017). Furthermore, acidic endochitinase and chitinase, which have an essential role in plant defense not only by functioning against fungal pathogens through degrading chitin but also in improving tolerance against temperature, salt and wounding stresses, were up-regulated under biotic infections, and down-regulated in response to abiotic stress in rice (Narsai et al. 2013). Additionally, the peroxisomal catalase *Sc*CAT1, one ROS scavenger enzyme, has been suggested to be a positive responsive component of abiotic and biotic stresses, avoiding oxidative damage and promoting hypersensitive cell death response in sugarcane (Su et al. 2014).

3 Climate Change Effect on Abiotic/Biotic Interactions: Future Constraints to Legume Production in the Mediterranean Region

Climate change, with increased frequencies of severe droughts and flooding, is predicted to have a progressively negative effect on the yield of food crops, particularly in the absence of efforts to mitigate global CO_2 emissions (Luck et al. 2011). The aggravation of existing environmental and agronomic problems due to climate change will be particularly important for the most vulnerable Mediterranean basin (Cramer et al. 2018). This region will face the big challenge of meeting higher water demands with less available water resources. Future warming in the Mediterranean region is expected to exceed worldwide rates by 25%, with summer warming surpassing the global rates mean by 40% (Lionello and Scarascia 2018). A global atmospheric temperature increase of 2 °C will probably be accompanied by a reduction in summer precipitation of around 10–15% in Southern France, Northwestern Spain, and the Balkans, and up to 30% in Turkey and Portugal (Vautard et al. 2014). On the other hand, flash floods, caused by short and locally heavy rains that regularly affect the Mediterranean region, will be aggravated and more frequent (Cramer et al. 2018).

Global changes in climatic seasonality are also affecting the emergence of pests and diseases and, consequently, plant/pathogen interactions. Besides the pathosystem-specific effects, there is a growing consensus that elevated temperatures with longer growing seasons will result in pathogen geographic expansion and enhanced reproductive potential. Both the life-cycle dynamics of the pathogenic organisms and the geographic distribution and importance of each disease and crop are sensitive to weather conditions (Garrett et al. 2006; Miraglia et al. 2009). This will increase the chances for host range expansion and the rise of a more rapid evolution of virulent strains, threatening durability of resistance (Engering et al. 2013; Garrett et al. 2006). In addition, environmental conditions may modulate host susceptibility/resistance responses to pathogens. Their influence on plant disease severity and spread might be positive, negative, or neutral, depending on the host/pathogen interaction, with some diseases becoming less important and others more important for particular crops, in particular regions (Elad and Pertot 2014; Velásquez et al. 2018).

The projected increase in global temperature will most likely change the regional distribution in which a crop is susceptible to a particular pathogen. It is expected that the ones with cold, heat, or desiccation resistance structures, which inocula overwinter for the following crop-growing season, will cause more severe and frequent epidemics (Ma et al. 2015). For the same crop, the life cycle of some pathogens might be favored by increasing temperatures, like the one of stem rust of wheat (*Puccinia graminis*), while for others, such as yellow rust (*P. striiformis*), the increase in temperature will be detrimental. However, an already occurring outcome of warming temperatures is the selection of new pathogen strains better adapted to these temperatures. This is the case of new races of *P. striiformis*, more aggressive at higher temperatures that have become more prevalent worldwide in only a few years (Milus et al. 2009). An increase in temperature can also inactivate temperature-sensitive resistances in plants (Huot et al. 2017). Elevated temperature breaks down effector-triggered immunity and promotes disease in many plant-pathosystems (Alcázar and Parker 2011). As an example, Landa et al. (2006) found that, with an increase in 3 °C, a fusarium wilt-resistant chickpea variety becomes susceptible. Moreover, different races of *Fusarium oxysporum* f. sp. *ciceri* showed distinct virulence with temperature. Nevertheless, in crops like wheat and pea, some genes were already identified conferring resistance to fungal diseases at higher temperatures (Chen 2013; Fondevilla et al. 2006).

Other climatic factors, such as increasing atmospheric CO_2, may provide greater biomass in C3 plants, which combined with increased canopy humidity might promote foliar diseases such as rusts, powdery mildews, leaf spots and blights (Manning and von Tiedemann 1995). Many plant diseases are less severe under low moisture conditions. The effects of drought might vary depending on the pathogen life-cycle stage it occurs. If drought occurs during pathogen germination, sporulation, and dispersal, the potential for disease development might be reduced. However, drought occurring after host infection might favor the disease through reduced host vigor or by impairing the production of plant defenses (Wakelin et al. 2018).

Climate change effects on the plant/pathogen interaction may be different, and depending on the degree of physiological interaction, pathogens have with their plant hosts (Eastburn et al. 2011). For instance, biotrophic and hemibiotrophic pathogens have extended periods of physiological interaction with their hosts, as they derive nutrients from living cells. Therefore, the factors that affect plant growth, such as water deficit, may also affect the colonization of host tissues.

As described above, climate change effects are already acting on the so-called disease triangle interplay between pathogens, hosts, and the environment (Zhang and Sonnewald 2017). Still, exposure to combined abiotic and biotic stresses is predicted to become more frequent (Kissoudis et al. 2014), with a negative influence on crop productivity and quality (Scheelbeek et al. 2018). To improve crop yield, broad efforts are needed to understand mechanisms underlying plant responses to simultaneous or sequential exposure to different abiotic and biotic stresses. The next section will focus on the particularities of the drought/soilborne vascular disease plant interaction responses in legumes.

4 The Challenges of Drought and Soilborne Vascular Diseases

Grain legumes are key components of the Mediterranean diet (Vaz Patto et al. 2015). Proteins are major components, but grain legume seeds are also rich in slowly digestible starch, soluble sugars, fiber, minerals, and vitamins, as well as secondary metabolites such as isoflavonoids, polyphenols, and phytosterols with potential effect in preventing cardiovascular disease (Arnoldi et al. 2015). The unique ability of legumes to fix atmospheric nitrogen, thus with little or no requirement for N-fertilizers inputs, makes grain legumes very appropriate partners in intercropping systems aiming at a more sustainable agriculture. In the case of Europe, the paradigm of grain legume production has changed dramatically in the last 60 years. Despite the nutritional value that grain legumes provide for both humans and livestock, their cultivation in Europe has been constantly decreasing over the last 50 years (Zander et al. 2016). In 1961, around 10 million ha were cropped with various species of grain legumes, namely common bean (*Phaseolus vulgaris* L.), pea (*Pisum sativum* L.), faba bean (*Vicia faba* L.), chickpea (*Cicer arietinum* L.), lentil (*Lens culinaris* Medik.), cowpea (*Vigna unguiculata* L.), grass pea (*Lathyrus sativus* L.), vetches (*Vicia* sp.), and lupins (*Lupinus* sp.). More than half of these crops were for direct human consumption. Diet change, with increased consumption of meat, and a lack in modernization and mechanization of the legume farming system in Europe, rendered legumes less economically attractive to farmers, leading to a drop to less than 5 million ha of area currently cultivated with the same legume crops (FAOSTAT 2017). Indeed, the sensitivity of grain legumes to various pests and diseases and the limited adaptability of available cultivars to a broad range of environmental conditions, with consequences in yield potential, have contributed to a lower investment in legume cultivation compared to competing cereal crops (Rubiales et al. 2015; Siddique et al. 2012). Diseases and pests are frequent in temperate and tropic regions, being important factors compromising legume production (Assefa et al. 2019; De Ron 2015). Drought is currently the most widespread and yield-limiting abiotic stress in many crops, and particularly in legumes (Vadez et al. 2012), being considered a bigger threat to legume survival than pathogen infection (Fujita et al. 2006). Drought at flowering and reproductive growth stages is estimated to cause 50–70% yield losses, depending on the legume crop (Farooq et al. 2017).

The sensitivity of grain legumes to drought is associated with leaf senescence, oxidative damage of photosynthetic apparatus, reduced carbon fixation, inhibited flowering and reproductive development, sterility of pollen grains and reduced grain development (reviewed by Farooq et al. 2017). During the last decade, some molecular players and physiological pathways underlying water deficit responses of a broad range of legume species have been elucidated (for a comprehensive review, see Araújo et al. 2015). As an example, water channel aquaporins have been described to be involved in drought tolerance in faba bean and common bean (Cui et al. 2008; Montalvo-Hernández et al. 2008). In chickpea, several transcription factors, such as DREB2A, AP2/ERF, zinc finger family, MYB domain, WRKY factor, and auxin response factor, were implicated in water deficit response, with a role in major metabolic pathways such as ABA, proline, and flavonoid biosynthesis (Hiremath et al. 2011; Mahdavi Mashaki et al. 2018; Nayak et al. 2009). In common bean, DREB2 and LEA3 were also identified as candidate targets responsive to drought (Barrera-Figueroa et al. 2007; Cortés et al. 2012). In cowpea (*Vigna unguiculata* (L.) Walp.), several drought stress-inducible genes involved in ABA biosynthesis were identified, namely cowpea responsive to dehydration (CPRD) genes like dehydrin (CPRD22), alcohol dehydrogenase (CPRD12), NADPH-dependent aldehyde reductase (CPRD14), and lipoxygenase (CPRD46), phospholipase D (VuPLD1) or 9-cis-epoxycarotenoid dioxygenase (Muchero et al. 2010).

Besides drought, fusarium wilts and root rots are major constraints to the production of legume crops worldwide (De Ron 2015; Rubiales et al. 2015). The most important pathogens causing wilt in legumes are various host-specific forms of *Fusarium oxysporum*, whereas causal agents of root rots are mainly *F. solani*, *Rhizoctonia* spp., *Verticillium* spp., and *Aphanomyces euteiches* (Bodah 2017; Kraft et al. 1988). Naturally found in fields, all of these soilborne pathogens directly penetrate roots, invade the cortex, and colonize the vascular tissue of the host plant. *F. oxysporum* and *F. solani* are responsible for vascular disease in chickpea, common bean, cowpea, pea, and lentil in most growing regions (Afouda et al. 2009; Chaudhary et al. 2006; Chen et al. 2016). Their development is favored by warm and dry conditions and visual symptoms include rotting of the root system, wilting, stunting, chlorosis, and lesions on the stem and/or leaves of infected plants. *Fusarium* spp. can survive in the soil for extended periods in the absence of the host, mainly in the form of thick-walled chlamydospores (Coleman 2016; Di Pietro et al. 2003). On the other hand, *Rhizoctonia* spp., *Verticillium* spp., and *Aphanomyces euteiches* are favored by humid and cooler climate as found in northern Europe and North America and are major constraints in the production of cool-season legumes such as pea, chickpea, lentil, or faba bean (Kraft and Pfleger 2001; Wu et al. 2018). The symptoms are similar to those of fusarium wilt, with generalized wilting, dark brown lesions above the collar region on the main stem and lower branches, and drooping of petioles and leaflets (Nene et al. 2012).

To understand the molecular mechanisms and genetic control of legume resistance to some of these pathogens, quantitative trait loci (QTL) mapping studies have been described. Regarding root rot, genomic regions associated with *F. solani* f. sp. *phaseoli*, *R. solani*, and *A. euteiches* partial resistance were detected in common bean (Hagerty et al. 2015; Kamfwa et al. 2013; Oladzad et al. 2019; Schneider et al. 2001). In pea, several QTLs associated with resistance to *F. solani* f. sp. *pisi* and to *A. euteiches* were reported (Coyne et al. 2019; Desgroux et al. 2016; Hamon et al. 2011, 2013; Pilet-Nayel et al. 2005). Moreover, a major QTL conferring partial resistance to *Verticillium* wilt was identified in the model legume *Medicago truncatula* (Ben et al. 2013). In the case of fusarium wilt, QTLs were identified in chickpea associated with *F. oxysporum* f. sp. *ciceris*, in cowpea associated with *F. oxysporum* f. sp. *tracheiphilum*, and in lentil associated with *F. oxysporum* f. sp. *lentis* partial resistance (Hamwieh et al. 2005; Pottorff et al. 2012; Sabbavarapu et al. 2013; Sharma et al. 2004).

5 Interaction of Drought and Soilborne Vascular Diseases in Grain Legumes. What Is Known so Far?

Little is known concerning the complex details underlying the abiotic/biotic interactions in legumes. Among the few reports available, there is one on the model legume *M. truncatula* (ecotype Jemalong A17), where simultaneous drought and spider mite attack (*Tetranychus urticae*) stresses were applied and biochemistry assays and gene expression analysis performed to understand the chemical defense responses involved. An increase in the production of peroxide (representative of ROS) was observed, which enhanced signaling processes and accumulation of antioxidants and osmoprotective compounds, such as proline (Antoniou et al. 2018).

In common bean, Ambachew et al. (2015) described the negative effect of combined drought stress and *Ophiomyia* spp. (bean fly) infestation on the mean seed yield and leaf chlorophyll content. This study examined trait interrelations among two diverse recombined inbred line populations resulting from crosses of parental lines of different gene pools of origin, for the combined effect of drought and bean fly infestation. The authors concluded that vertical root pulling resistance, stem diameter, and pod harvest index should be included as selection criteria for improving seed yield in common beans under drought and bean fly infestation conditions.

Enhanced resistance to powdery mildew (*Erysiphe pisi*) induced by ozone exposure has been demonstrated using two pea cultivars (Rush and Laurence 1993). Although plant defense mechanisms induced by the ozone exposure were not unveiled, the authors concluded that it was not correlated with pisatin content, a pea phytoalexin.

Plant/water relations influence plant defense, with stomatal regulation having a critical role (Beattie 2011). Under drought environments, plant/pathogen interaction can inhibit the normal ABA-mediated stomatal closure response of plants to drought, with a clear impact on crop performance (Grimmer et al. 2012). Both pathogen infection and plant host resistance mechanisms can result in alterations in guard cell regulation of stomatal aperture. A wide range of foliar fungi has been described to disrupt the cuticular or stomatal regulation of transpiration, thus affecting legume/water relations. For example, *Erysiphe pisi* in pea, *Uromyces appendiculatus* and *Colletotrichum lindemuthianum* in common bean, and *Botrytis fabae* in faba bean were described to cause a decrease in the stomatal aperture during the day or impairment of stomatal closure during the night (reviewed by Grimmer et al. 2012). The negative influence of foliar pathogens decreasing net photosynthesis and water use efficiency, defined as the ratio of CO_2 assimilation rate to transpiration rate at the leaf level, was also described in different legume-pathogen systems: *Bean yellow mosaic virus* in faba bean (Radwan et al. 2008), *Phaeoisariopsis griseola* and *U. appendiculatus* in common bean (Jesus-Junior et al. 2001; Meyer et al. 2001), or *E. pisi* in pea (Ayres 1976).

Drought stress has a known agronomic interaction with soilborne pathogens that infect roots causing vascular diseases in grain legumes. For instance, drought stress was reported to increase the incidence of *R. bataticola* and *F. solani* in chickpea (Sinha et al. 2019). In this work, the combination of drought and two fungal diseases reduced the primary root length, crop growth, and ultimately yield, as compared to the situation in which drought was combined with a single pathogen stress. In common bean, the simultaneous exposure of plants to drought stress and to the fungal pathogen *Macrophomina phaseolina* (causal agent of charcoal rot and seedling blight) resulted in a higher leaf transpiration rate and temperature when compared to plants subjected only to drought stress (Mayek-Pérez et al. 2002). Furthermore, QTLs have been reported in pea governing both drought resistance and resistance to the root parasitic weed *Orobanche crenata*, suggesting that they might be controlling water and solute losses due to drought or *O. crenata* presence (Iglesias-García et al. 2015). In this context, it is worth to mention that often drought stress results in root exudation of amino acids and carbohydrates, which become available for priming plant and soil microbial activity after rewetting (Karlowsky et al. 2018). Actually, these exudates have been reported as a source of nutrients for the growth of soilborne pathogens (Duniway 1977). More recently, Bani et al. (2018) described how pea root exudates can modulate, either stimulating or inhibiting *F. oxysporum* spore germination. In this study, pisatin was identified as having an important role in the constitutive antifungal defense of pea, and its amount in the root exudates was negatively correlated with the extent of *F. oxysporum* germination. On the other hand, pathogen infection also modulates the root exudates profile, leading to the recruiting of beneficial rhizosphere communities, with an effect on the resistance of subsequent plant populations growing in the same soil (Yuan et al. 2018).

The identification and location of genes controlling the interaction of multiple traits in grain legume species is still an important missing information in the present context of climate change. The molecular mechanisms behind multiple stress interactions have been studied until now using mainly model plants like *Arabidopsis*, tomato, and tobacco (Atkinson et al. 2013; Kissoudis et al. 2015; Prasch and Sonnewald 2015) (see Sect. 2 in chapter "Fabaceous Plants Under Abiotic Stresses and Biochemical Functions of Micronutrients" for more details).

In *A. thaliana* model plant, a genome-wide association study (GWAS) analyzed contrasting and similar responses to a set of different abiotic and biotic stresses, applied as single or double stresses, including the combination of drought with the necrotrophic fungal pathogen *Botrytis cinerea* (Thoen et al. 2017). Many of the candidate genes identified in this study were involved in phytohormone-mediated processes, highlighting the role that phytohormones play on biotic/abiotic stress responses. Another GWAS identified key genes related to the *A. thaliana* adaptive responses to *B. cinerea* under sequential drought stress (Coolen et al. 2019). As a major outcome, this study showed that genes involved in JA biosynthesis and plant invertase/pectin methylesterase inhibitor were playing a role in resistance to both stresses. The interaction of drought with herbivory by caterpillars of *Pieris rapae* in *Arabidopsis* was also analyzed by GWAS (Davila Olivas et al. 2017). This study demonstrated the role played by bHLH transcription factors (e.g., MYC family), which were considered major regulators of JA- and ABA-mediated responses, insect resistance, and drought responses.

Attempts to understand the molecular mechanism underlying the interaction between soilborne pathogens causing vascular diseases and drought in legumes are now giving the first steps. Sinha et al. (2017) investigated the effects of the combination of water deficit and a bacterial wilt disease (caused by *Ralstonia solanacearum*) in chickpea using a transcriptomic approach. Genes implicated in osmoprotectant and phytohormone biosynthesis, signaling, xylem differentiation, and cellulose and lignin deposition were found differentially expressed. In another work, an increase in osmoprotectants such as sugars, with a concurrent decrease in many organic acids such as citric acid and malic acid were described for *M. truncatula* plants subjected to drought and *F. oxysporum* (Dickinson et al. 2018).

Plant responses to a combination of abiotic stress (e.g., drought) and pathogen infection strongly depend on the pathogen and crop involved, plant developmental stage, and intensity and duration of each stress (Rejeb et al. 2014). The limited data available on plant responses in general, and on grain legumes in particular, under abiotic and biotic stress combinations, makes essential and urgent a better understanding of the molecular mechanisms used by these particular plant species to balance growth and defense against diverse combinations of environmental constraints to maximize grain legumes fitness through plant breeding (Huot et al. 2014). The insufficient understanding of the complex molecular mechanisms of plant response to simultaneous stresses supports the need for an integrated approach combining functional genomics to accurately identify key genes and biomarkers regulating multi-tolerances in legume breeding programs.

6 Future Perspectives in Legume Abiotic/Biotic Stress Integrated Resistance Breeding

The availability of multi-stress-resistant legume varieties would decrease pesticide application and increase legume production under harsh environments, making them pillars for the development of sustainable agriculture.

Available data and projections on the effect of climate change highlight an increase in the chances of plants encountering abiotic and/or biotic stress, with more frequent stress interactions (Kissoudis et al. 2014). Under this stressed environment scenario, breeding in legumes should be guided, primarily by biotic and abiotic stress resistance, to provide stability to high yield potential. However, breeding efforts must be aligned/combined with the need to maintain particular quality traits and market class characteristics, which are essential for meeting consumer preference in various markets (Assefa et al. 2019; Vaz Patto et al. 2015).

Goals regarding biotic and abiotic stress resistance tend to be specific to a particular location, varying, for example, according to the temperature, soil fertility, and water availability. However, some constraints are becoming common to all legume production regions. Examples are various root rot and wilt diseases that are worldwide problems, intensified by the more frequent water scarcity (Elad and Pertot 2014).

Additionally, biological control using antagonistic fungal and bacterial agents against pathogens has received increasing attention and might be a valuable ally to plant defense and an integrated resistance breeding. The ability of several *Trichoderma* spp. to control *Fusarium solani*, *Pythium aphanidermatum*, *Rhizoctonia solani*, *Macrophomina phaseolina*, and *Sclerotium rolfsii* in mungbean (*Vigna radiata* L.) and faba bean has been tested with promising results (El-Mougy and Abdel-Kader 2008; Ramzan et al 2016). Also, the combination of soil treatment with *Trichoderma* spp. with seed coating with thiophanate-methyl reduced the incidence of *F. solani* and *F. oxysporum* in common bean (Abd-El-Khair et al. 2019). The authors demonstrated a positive correlation between the biocontrol activation of *Trichoderma* isolates and the enhancement of peroxidase, polyphenoloxidase, and chitinase enzymes in bean plants to resist infection with soilborne pathogens. Furthermore, the use of plant growth-promoting rhizobacteria may support the plant's ability to tolerate abiotic stresses and boost crop yield (Enebe and Babalola 2018). Some drought-resistant rhizosphere competent bacteria are known to improve plant health and promote growth during abiotic stresses. For example, *Rhizobium* was able to promote plant growth, nutrient content, and increased yield in drought-stressed common bean (Yanni et al. 2016), whereas *Pseudomonas putida* was shown to ameliorate drought stress effects and assist recovery in chickpea by enhancing osmolytes and ROS production, and up-regulation of various stress-responsive genes (Tiwari et al. 2016).

The success of plant breeding over the past century has been associated with a narrowing of the available genetic diversity within elite germplasm, particularly for some species such as peanut and soybean. To overcome the yield potential and stability gap in stressed environments, new sources of variation are needed and these include landraces and wild relatives of crop species (Tester and Langridge 2010).

Pea, faba bean, lentil, grass pea, and chickpea are some of the world's oldest domesticated crops that arose in the Fertile Crescent of Mesopotamian agriculture (Smýkal et al. 2015). Common bean, on the other hand, has a particular evolutionary history, with two geographically isolated and genetically differentiated gene pools independently domesticated in the Mesoamerican and in the Andes (Bitocchi et al. 2013). Since the domestication of these legume species, many landraces have been developed encompassing much of the original diversity, plus accumulated mutations and genetic recombination. Traditional landraces that are well adapted to local environmental conditions may represent important sources of combined biotic and abiotic resistance.

Legumes (grain and forage) constitute the second largest group of crop world accessions conserved in gene banks (1,041,345 accessions, 15% of all) after cereals (FAO 2010). However, there is an urgency to ensure that the diversity in legume landraces is sampled and conserved in ex situ gene banks, especially as farming becomes dominated by modern uniform cultivars, resulting in the erosion of crop genetic diversity. A similar caution also applies to the in situ populations of wild relatives (Smýkal et al. 2015). For an efficient use of these underexplored plant genetic resources in the context of multi-stress resistance, we need to expand the phenotyping and genotyping of these germplasm collections, increase our understanding of the molecular basis for key resistance/tolerance traits, and develop new breeding strategies that allow introgression of multiple traits (Feuillet et al. 2008).

7 Conclusion

Breeding for resistance against multiple stresses is a challenge. Each abiotic stress/pathogen/host combination should be treated independently as, despite the potential universal applicability of some interactions, many unique interactions may be crucial for the phenotypic response (Kissoudis et al. 2014). Since genotype-by-environment interactions indicative of adaptive plasticity cannot be assessed in non-stress conditions, to uncover beneficial alleles in key genes, the multi-trait screening of these genetic resources should be performed in combined stress environments (Kissoudis et al. 2016). According to Berens et al. (2019), special attention should also be given to the effect of different plant development stages in the outcome of the stress interaction.

The success in developing legume varieties resistant to abiotic/biotic stresses depends not only on the availability of good sources of resistance but also on the inheritance of the resistance. The identification and location of genes controlling these traits may be accomplished by association mapping studies based on the existing natural variation, validated by functional analysis using transcriptomic, proteomic, and metabolomics data of contrasting accessions (Davila Olivas et al. 2017; Thoen et al. 2017). As already mentioned, it is also essential that these genetic studies are performed under combined stress conditions (Bai et al. 2018). By analyzing several traits simultaneously, these studies enabled a comprehensive and integrative view of the genetic architecture and molecular pathways underlying combined abiotic and biotic stress resistance in legumes (Dickinson et al. 2018; Sinha et al. 2017).

The search for the molecular mechanisms behind the abiotic/biotic stresses interactions has been the main driving force of some research consortium established in the past. One example is the ABSTRESS European FP7-KBBE project, which aimed to explore the gene networks implicated in the interaction of drought stress and fusarium infection in legumes by combining integrated systems biology and comparative genomics approaches (ABSTRESS).

The omics combination will narrow down the number of candidate genes for combined stress tolerance, identifying major regulatory hubs, such as transcription factors (Kissoudis et al. 2014). Besides the deployment of these individual common regulators in novel varieties, an educated pyramiding of non-(negatively) interacting resistant components appears as an alternative breeding strategy to achieve tolerance to combined stress in crops. These might be disease resistance genes (R-genes) and inducible abiotic stress resistance genes involved in physical barriers against pathogen penetration (Bai et al. 2018; Kissoudis et al. 2014). Resistance to pathogens seems to be genetically simpler than tolerance to abiotic stresses, such as drought, as it is often conferred by single major genes (Kissoudis et al. 2016). In this way, a marker-assisted backcross could be the first step in the development of multi-stress-tolerant varieties with the prerequisite of testing the stability of R-genes under different stress scenarios. The subsequent introgression of quantitative abiotic stress resistance governed by several minor genes/QTLs will be more challenging. However, with the advances in genomic selection research for complex traits in many crops, as well as in legumes (Annicchiarico et al. 2019; Roorkiwal et al. 2018; Varshney et al. 2019), this might become an alternative to achieve balanced signaling pathways in novel varieties, contributing to a sustained performance of crops under diverse environmental conditions.

Acknowledgements Financial support by Fundação para a Ciência e Tecnologia (FCT), Portugal, is acknowledged through grant SFRH/BD/92160/2013 (STL), DL57 PhD holder contract (SA), IF/01337/2014 FCT Investigator contract (MCVP), research project BeGeQA (PTDC/AGR-TEC/3555/2012) and research unit GREEN-IT "Bioresources for Sustainability" (UID/Multi/04551/2019).

References

Abd-El-Khair H, Elshahawy IE, Haggag HEK (2019) Field application of Trichoderma spp. combined with thiophanate-methyl for controlling *Fusarium solani* and *Fusarium oxysporum* in dry bean. Bull Natl Res Centre 43:19. https://doi.org/10.1186/s42269-019-0062-5

ABSTRESS European FP7-KBBE.2011.1.1-02 project (2012–2016) Improving the resistance of legume crops to combined abiotic and biotic stress. Grant agreement ID: 289562. https://ec.europa.eu/eip/agriculture/en/find-connect/projects/abstress-improving-resistance-legume-crops

Afouda L, Wolf G, Wydra K (2009) Development of a sensitive serological method for specific detection of latent infection of *Macrophomina phaseolina* in cowpea. J Phytopathol 157:15–23. https://doi.org/10.1111/j.1439-0434.2008.01453.x

Alcázar R, Parker JE (2011) The impact of temperature on balancing immune responsiveness and growth in *Arabidopsis*. Trends Plant Science 16:666–675. https://doi.org/10.1016/j.tplants.2011.09.001

Ambachew D, Mekbib F, Asfaw A, Beebe SE, Blair MW (2015) Trait associations in common bean genotypes grown under drought stress and field infestation by BSM bean fly. Crop J 3:305–316. https://doi.org/10.1016/j.cj.2015.01.006

Anfoka G, Moshe A, Fridman L, Amrani L, Rotem O, Kolot M, Zeidan M, Czosnek H, Gorovits R (2016) Tomato yellow leaf curl virus infection mitigates the heat stress response of plants grown at high temperatures. Sci Rep. 6:19715

Annicchiarico P, Nazzicari N, Pecetti L, Romani M, Russi L (2019) Pea genomic selection for Italian environments. BMC Genom 20:603. https://doi.org/10.1186/s12864-019-5920-x

Antoniou C, Fragkoudi I, Martinou A, Stavrinides MC, Fotopoulos V (2018) Spatial response of *Medicago truncatula* plants to drought and spider mite attack. Plant Physiol Biochem 130:658–662. https://doi.org/10.1016/j.plaphy.2018.08.018

Araújo SS, Beebe S, Crespi M, Delbreil B, González EM, Gruber V, Lejeune-Henaut I, Link W, Monteros MJ, Prats E, Rao I, Vadez V, Vaz Patto MC (2015) Abiotic stress responses in Legumes: strategies used to cope with environmental challenges. Crit Rev Plant Sci 34:237–280. https://doi.org/10.1080/07352689.2014.898450

Arnoldi A, Zanoni C, Lammi C, Boschin G (2015) The role of grain legumes in the prevention of hypercholesterolemia and hypertension. Crit Rev Plant Sci 34:144–168. https://doi.org/10.1080/07352689.2014.897908

Assefa T, Assibi Mahama A, Brown AV, Cannon EKS, Rubyogo JC, Rao IM, Blair MW, Cannon SB (2019) A review of breeding objectives, genomic resources, and marker-assisted methods in common bean (*Phaseolus vulgaris* L.). Mol Breed 39:20. https://doi.org/10.1186/s13059-016-0883-6

Atkinson NJ, Lilley CJ, Urwin PE (2013) Identification of genes involved in the response of *Arabidopsis* to simultaneous biotic and abiotic stresses. Plant Physiol 162:2028–2041. https://doi.org/10.1104/pp.113.222372

Aung K, Jiang Y, He SY (2018) The role of water in plant–microbe interactions. Plant J 93:771–780. https://doi.org/10.1111/tpj.13795

Ayres PG (1976) Patterns of stomatal behaviour, transpiration, and CO_2 exchange in pea following infection by powdery mildew (*Erysiphe pisi*). J Exp Bot 27:1196–1205. https://doi.org/10.1093/jxb/27.6.1196

Bai Y, Kissoudis C, Yan Z, Visser RGF, van der Linden G (2018) Plant behaviour under combined stress: tomato responses to combined salinity and pathogen stress. Plant J 93(4):781–793. https://doi.org/10.1111/tpj.13800

Balzan S, Johal GS, Carraro N (2014) The role of auxin transporters in monocots development. Front Plant Sci 5:393. https://doi.org/10.3389/fpls.2014.00393

Bani M, Cimmino A, Evidente A, Rubiales D, Rispail N (2018) Pisatin involvement in the variation of inhibition of *Fusarium oxysporum* f. sp. *pisi* spore germination by root exudates of *Pisum* spp. germplasm. Plant Pathol 67:1046–1054. https://doi.org/10.1111/ppa.12813

Barrera-Figueroa BE, Peña-Castro JM, Acosta-Gallegos JA, Ruiz-Medrano R, Xoconostle-Cázares B (2007) Isolation of dehydration-responsive genes in a drought tolerant common bean cultivar and expression of a group 3 late embryogenesis abundant mRNA in tolerant and susceptible bean cultivars. Funct Plant Biol 34(4):368–381. https://doi.org/10.1071/FP06224

Beattie GA (2011) Water relations in the interaction of foliar bacterial pathogens with plants. Annu Rev Phytopathol 49:533–555. https://doi.org/10.1146/annurev-phyto-073009-114436

Ben C, Toueni M, Montanari S, Tardin M-C, Fervel M, Negahi A, Saint-Pierre L, Mathieu G, Gras M-C, Noël D, Prospéri J-M, Pilet-Nayel M-L, Baranger A, Huguet T, Julier B, Rickauer M, Gentzbittel L (2013) Natural diversity in the model legume *Medicago truncatula* allows identifying distinct genetic mechanisms conferring partial resistance to *Verticillium* wilt. J Exp Bot 64:317–332. https://doi.org/10.1093/jxb/ers337

Berens ML, Wolinska KW, Spaepen S, Ziegler J, Nobori T, Nair A, Krüler V, Winkelmüller TM, Wang Y, Mine A, Becker D, Garrido-Oter R, Schulze-Lefert P, Tsuda K (2019) Balancing trade-offs between biotic and abiotic stress responses through leaf age-dependent variation in stress hormone cross-talk. Proc Natl Acad Sci 116:2364–2373. https://doi.org/10.1073/pnas.1817233116

Besseau S, Li J, Palva ET (2012) WRKY54 and WRKY70 co-operate as negative regulators of leaf senescence in *Arabidopsis thaliana*. J Exp Bot 63:2667–2679. https://doi.org/10.1093/jxb/err450

Betsuyaku S, Katou S, Takebayashi Y, Sakakibara H, Nomura N, Fukuda H (2017) Salicylic acid and jasmonic acid pathways are activated in spatially different domains around the infection site during effector-triggered immunity in *Arabidopsis thaliana*. Plant Cell Physiol 59:8–16. https://doi.org/10.1093/pcp/pcx181

Bilgin DD, Zavala JA, Zhu JIN, Clough SJ, Ort DR, DeLucia EH (2010) Biotic stress globally downregulates photosynthesis genes. Plant Cell Environ 33:1597–1613. https://doi.org/10.1111/j.1365-3040.2010.02167.x

Bitocchi E, Bellucci E, Giardini A, Rau D, Rodriguez M, Biagetti E, Santilocchi R, Spagnoletti Zeuli P, Gioia T, Logozzo G, Attene G, Nanni L, Papa R (2013) Molecular analysis of the parallel domestication of the common bean (*Phaseolus vulgaris*) in Mesoamerica and the Andes. New Phytol 197:300–313. https://doi.org/10.1111/j.1469-8137.2012.04377.x

Bodah E (2017) Root rot diseases in plants: a review of common causal agents and management strategies. Agric Res Technol 5:3. https://doi.org/10.19080/ARTOAJ.2017.05.555661

Bouzroud S, Gouiaa S, Hu N, Bernadac A, Mila I, Bendaou N, et al. (2018) Auxin Response Factors (ARFs) are potential mediators of auxin action in tomato response to biotic and abiotic stress (Solanum lycopersicum). PLoS ONE 13(2): e0193517. https://doi.org/10.1371/journal.pone.0193517

Chaudhary S, Anderson TR, Park SJ, Yu K (2006) Comparison of screening methods for resistance to *Fusarium* root rot in common beans (*Phaseolus vulgaris* L.). J Phytopathol 154:303–308. https://doi.org/10.1111/j.1439-0434.2006.01097.x

Chen X (2013) High-temperature adult-plant resistance, key for sustainable control of stripe rust. Am J Plant Sci 04:608–627. https://doi.org/10.4236/ajps.2013.43080

Chen W, Sharma HC, Muehlbauer FJ (2016) Compendium of chickpea and lentil diseases and pests. APS Press, St. Paul, Minnesota, USA

Cohen SP, Leach JE (2019) Abiotic and biotic stresses induce a core transcriptome response in rice. Sci Rep 9:6273. https://doi.org/10.1038/s41598-019-42731-8

Coleman JJ (2016) The *Fusarium solani* species complex: ubiquitous pathogens of agricultural importance. Mol Plant Pathol 17:146–158. https://doi.org/10.1111/mpp.12289

Coolen S, Van Pelt JA, Van Wees SCM, Pieterse CMJ (2019) Mining the natural genetic variation in *Arabidopsis thaliana* for adaptation to sequential abiotic and biotic stresses. Planta 249:1087–1105. https://doi.org/10.1007/s00425-018-3065-9

Cortés AJ, This D, Chavarro C, Madriñán S, Blair MW (2012) Nucleotide diversity patterns at the drought-related *DREB2* encoding genes in wild and cultivated common bean (*Phaseolus vulgaris* L.). Theor Appl Genet 125:1069–1085. https://doi.org/10.1007/s00122-012-1896-5

Coyne CJ, Porter LD, Boutet G, Ma Y, McGee RJ, Lesné A, Baranger A, Pilet-Nayel M-L (2019) Confirmation of *Fusarium* root rot resistance QTL *Fsp-Ps 2.1* of pea under controlled conditions. BMC Plant Biol 19:98. https://doi.org/10.1080/07060661.2018.1429494

Cramer W, Guiot J, Fader M, Garrabou J, Gattuso J-P, Iglesias A, Lange MA, Lionello P, Llasat MC, Paz S, Peñuelas J, Snoussi M, Toreti A, Tsimplis MN, Xoplaki E (2018) Climate change and interconnected risks to sustainable development in the Mediterranean. Nat Clim Change 8:972–980. https://doi.org/10.1038/s41558-018-0299-2

Cui X-H, Hao F-S, Chen H, Chen J, Wang X-C (2008) Expression of the *Vicia faba VfPIP1* gene in *Arabidopsis thaliana* plants improves their drought resistance. J Plant Res 121:207–214. https://doi.org/10.1007/s10265-007-0130-z

Davila Olivas NH, Kruijer W, Gort G, Wijnen CL, van Loon JJA, Dicke M (2017) Genome-wide association analysis reveals distinct genetic architectures for single and combined stress responses in *Arabidopsis thaliana*. New Phytol 213:838–851. https://doi.org/10.1111/nph.14165

De Ron AM (2015) Grain Legumes. In: De Ron AM (ed) Handbook of Plant Breeding. Springer, New York. ISBN 978-1-4939-2796-8

Desgroux A, L'Anthoëne V, Roux-Duparque M, Rivière J-P, Aubert G, Tayeh N, Moussart A, Mangin P, Vetel P, Piriou C, McGee RJ, Coyne CJ, Burstin J, Baranger A, Manzanares-Dauleux M, Bourion V, Pilet-Nayel M-L (2016) Genome-wide association mapping of partial resistance to *Aphanomyces euteiches* in pea. BMC Genom 17:124. https://doi.org/10.1186/s12864-016-2429-4

Di Pietro A, Madrid MP, Caracuel Z, Delgado-Jarana J, Roncero MIG (2003) *Fusarium oxysporum*: exploring the molecular arsenal of a vascular wilt fungus. Mol Plant Pathol 4:315–325. https://doi.org/10.1046/j.1364-3703.2003.00180.x

Dickinson E, Rusilowicz MJ, Dickinson M, Charlton AJ, Bechtold U, Mullineaux PM, Wilson J (2018) Integrating transcriptomic techniques and k-means clustering in metabolomics to identify markers of abiotic and biotic stress in *Medicago truncatula*. Metabolomics 14:126. https://doi.org/10.1007/s11306-018-1424-y

Djami-Tchatchou AT, Sanan-Mishra N, Ntushelo K, Dubery IA (2017) Functional roles of microRNAs in agronomically important plants-potential as targets for crop improvement and protection. Front Plant Sci 8:378. https://doi.org/10.3389/fpls.2017.00378

Duniway JM (1977) Predisposing effect of water stress on the severity of Phytophthora root rot in safflower. Phytopathol 67:884–889. https://doi.org/10.1094/Phyto-67-884

Eastburn DM, McElrone AJ, Bilgin DD (2011) Influence of atmospheric and climatic change on plant–pathogen interactions. Plant Pathol 60:54–69. https://doi.org/10.1111/j.1365-3059.2010.02402.x

Elad Y, Pertot I (2014) Climate change impacts on plant pathogens and plant diseases. J Crop Improv 28:99–139. https://doi.org/10.1080/15427528.2014.865412

El-Mougy NS, Abdel-Kader MM (2008) Long-term activity of bio-priming seed treatment for biological control of faba bean root rot pathogens. Australas Plant Pathol 37:464–471. https://doi.org/10.1071/AP08043

Enebe MC, Babalola OO (2018) The influence of plant growth-promoting rhizobacteria in plant tolerance to abiotic stress: a survival strategy. Appl Microbiol Biotechnol 102:7821–7835. https://doi.org/10.1007/s00253-018-9214-z

Engering A, Hogerwerf L, Slingenbergh J (2013) Pathogen-host-environment interplay and disease emergence. Emerg Microbes Infect 2(2):e5. https://doi.org/10.1038/emi.2013.5

FAO (2010) The second report on the state of the world's Plant Genetic Resources for food and agriculture, Rome

FAO (2017) The future of food and agriculture—Trends and challenges. Rome

FAOSTAT (2017) Food and Agriculture Organization of the United Nations, FAOSTAT. Rome, Italy. Available at http://fao.org/faostat/en/#data/QC. Accessed 13 Oct 2019

Farooq M, Gogoi N, Barthakur S, Baroowa B, Bharadwaj N, Alghamdi SS, Siddique KHM (2017) Drought stress in grain legumes during reproduction and grain filling. J Agron Crop Sci 203:81–102. https://doi.org/10.1111/jac.12169

Feuillet C, Langridge P, Waugh R (2008) Cereal breeding takes a walk on the wild side. Trends Genet 24:24–32. https://doi.org/10.1016/j.tig.2007.11.001

Fondevilla S, Carver TLW, Moreno MT, Rubiales D (2006) Macroscopic and histological characterisation of genes *er1* and *er2* for powdery mildew resistance in pea. Eur J Plant Pathol 115:309–321. https://doi.org/10.1007/s10658-006-9015-6

Fraire-Velázquez Sl, Rodríguez-Guerra Rl, Sánchez-Calderón L (2011) Abiotic and biotic stress response crosstalk in plants. In: Shanker A, Venkateswarlu B (eds) Abiotic stress response in plants—physiological, biochemical and genetic perspectives. IntechOpen. https://doi.org/10.5772/23217

Fujita M, Fujita Y, Noutoshi Y, Takahashi F, Narusaka Y, Yamaguchi-Shinozaki K, Shinozaki K (2006) Crosstalk between abiotic and biotic stress responses: a current view from the points of convergence in the stress signaling networks. Curr Opin Plant Biol 9:436–442. https://doi.org/10.1016/j.pbi.2006.05.014

Garrett KA, Dendy SP, Frank EE, Rouse MN, Travers SE (2006) Climate change effects on plant disease: genomes to ecosystems. Annu Rev Phytopathol 44:489–509. https://doi.org/10.1146/annurev.phyto.44.070505.143420

Grimmer MK, John Foulkes M, Paveley ND (2012) Foliar pathogenesis and plant water relations: a review. J Exp Bot 63:4321–4331. https://doi.org/10.1093/jxb/ers143

Grossmann K (2010) Auxin herbicides: current status of mechanism and mode of action. Pest Manag Sci 66:113–120. https://doi.org/10.1002/ps.1860

Gupta A, Sarkar AK, Senthil-Kumar M (2016) Global transcriptional analysis reveals unique and shared responses in *Arabidopsis thaliana* exposed to combined drought and pathogen stress. Front Plant Sci 7:686. https://doi.org/10.3389/fpls.2016.00686

Hagerty CH, Cuesta-Marcos A, Cregan PB, Song Q, McClean P, Noffsinger S, Myers JR (2015) Mapping *Fusarium solani* and *Aphanomyces euteiches* root rot resistance and root architecture quantitative trait loci in common bean. Crop Sci 55:1969–1977. https://doi.org/10.2135/cropsci2014.11.0805

Hamon C, Baranger A, Coyne CJ, McGee RJ, Le Goff I, L'Anthoëne V, Esnault R, Rivière J-P, Klein A, Mangin P, McPhee KE, Roux-Duparque M, Porter L, Miteul H, Lesné A, Morin G, Onfroy C, Moussart A, Tivoli B, Delourme R, Pilet-Nayel M-L (2011) New consistent QTL in pea associated with partial resistance to *Aphanomyces euteiches* in multiple French and American environments. Theor Appl Genet 123:261–281. https://doi.org/10.1007/s00122-011-1582-z

Hamon C, Coyne CJ, McGee RJ, Lesné A, Esnault R, Mangin P, Hervé M, Le Goff I, Deniot G, Roux-Duparque M, Morin G, McPhee KE, Delourme R, Baranger A, Pilet-Nayel M-L (2013) QTL meta-analysis provides a comprehensive view of loci controlling partial resistance to *Aphanomyces euteiches* in four sources of resistance in pea. BMC Plant Biol 13:45. https://doi.org/10.1186/1471-2229-13-45

Hamwieh A, Udupa SM, Choumane W, Sarker A, Dreyer F, Jung C, Baum M (2005) A genetic linkage map of *Lens* sp. based on microsatellite and AFLP markers and the localization of fusarium vascular wilt resistance. Theor Appl Genet 110:669–677. https://doi.org/10.1007/s00122-004-1892-5

Hiremath PJ, Farmer A, Cannon SB, Woodward J, Kudapa H, Tuteja R, Kumar A, Bhanuprakash A, Mulaosmanovic B, Gujaria N, Krishnamurthy L, Gaur PM, Kavikishor PB, Shah T, Srinivasan R, Lohse M, Xiao Y, Town CD, Cook DR, May GD, Varshney RK (2011) Large-scale transcriptome analysis in chickpea (*Cicer arietinum* L.), an orphan legume crop of the semi-arid tropics of Asia and Africa. Plant Biotechnol J 9:922–931. https://doi.org/10.1111/j.1467-7652.2011.00625.x

Hubbard M, Germida JJ, Vujanovic V (2014) Fungal endophytes enhance wheat heat and drought tolerance in terms of grain yield and second-generation seed viability. J Appl Microbiol 116:109–122. https://doi.org/10.1111/jam.12311

Huot B, Yao J, Montgomery BL, He SY (2014) Growth–defense tradeoffs in plants: a balancing act to optimize fitness. Mol Plant 7:1267–1287. https://doi.org/10.1093/mp/ssu049

Huot B, Castroverde CDM, Velásquez AC, Hubbard E, Pulman JA, Yao J, Childs KL, Tsuda K, Montgomery BL, He SY (2017) Dual impact of elevated tmperature on plant defence and baterial virulence in *Arabidopsis*. Nat Commun 8:1808. https://doi.org/10.1038/s41467-017-01674-2

Iglesias-García R, Prats E, Fondevilla S, Satovic Z, Rubiales D (2015) Quantitative trait loci associated to drought adaptation in pea (*Pisum sativum* L.). Plant Mol Biol Rep 33:1768–1778. https://doi.org/10.1007/s11105-015-0872-z

Irulappan V, Senthil-Kumar M (2018) Morpho-physiological traits and molecular intricacies associated with tolerance to combined drought and pathogen stress in plants. In: Gosal SS, Wani SH (eds) Biotechnologies of crop improvement, vol 3. Genomic Approaches. Springer International Publishing, Cham, pp 59–74

Jesus-Junior WC, Vale FXR, Martinez CA, Coelho RR, Costa LC, Zambolim BH (2001) Effects of angular leaf spot and rust on leaf gas exchange and yield of common bean (*Phaseolus vulgaris*). Photosynthetica 39:603–606. https://doi.org/10.1023/A:1015620532467

Jones-Rhoades MW, Bartel DP, Bartel B (2006) MicroRNAs and their regulatory roles in plants. Annual Rev Plant Biol 57:1, 19–53. https://doi.org/10.1146/annurev.arplant.57.032905.105218

Kamfwa K, Mwala M, Okori P, Gibson P, Mukankusi C (2013) Identification of QTL for *Fusarium* root rot resistance in common bean. J Crop Improv 27:406–418. https://doi.org/10.1080/15427528.2013.786775

Kang H, Park SJ, Kwak KJ (2013) Plant RNA chaperones in stress response. Trends Plant Sci 18:100–106. https://doi.org/10.1016/j.tplants.2012.08.004

Karasov TL, Chae E, Herman JJ, Bergelson J (2017) Mechanisms to mitigate the trade-off between growth and defense. Plant Cell 29:666–680. https://doi.org/10.1105/tpc.16.00931

Karlowsky S, Augusti A, Ingrisch J, Akanda MKU, Bahn M, Gleixner G (2018) Drought-induced accumulation of root exudates supports post-drought recovery of microbes in mountain grassland. Front Plant Sci 9:1593. https://doi.org/10.3389/fpls.2018.01593

Kesten C, Menna A, Sánchez-Rodríguez C (2017) Regulation of cellulose synthesis in response to stress. Curr Opin Plant Biol 40:106–113. https://doi.org/10.1016/j.pbi.2017.08.010

Kissoudis C, van de Wiel C, Visser RGF, van der Linden G (2014) Enhancing crop resilience to combined abiotic and biotic stress through the dissection of physiological and molecular crosstalk. Front Plant Sci 5:207. https://doi.org/10.3389/fpls.2014.00207

Kissoudis C, Chowdhury R, van Heusden S, van de Wiel C, Finkers R, Visser RGF, Bai Y, van der Linden G (2015) Combined biotic and abiotic stress resistance in tomato. Euphytica 202:317–332. https://doi.org/10.1007/s10681-015-1363-x

Kissoudis C, Sunarti S, van de Wiel C, Visser RGF, van der Linden CG, Bai Y (2016) Responses to combined abiotic and biotic stress in tomato are governed by stress intensity and resistance mechanism. J Exp Bot 67:5119–5132. https://doi.org/10.1093/jxb/erw285

Kraft JM, Pfleger FL (2001) Compendium of pea diseases and pests. APS Press, St Paul, Minnesota, USA

Kraft JM, Haware MP, Hussein MM (1988) Root rot and wilt diseases of food legumes. In: Summerfield RJ (ed) World crops: Cool season food legumes: a global perspective of the problems and prospects for crop improvement in pea, lentil, faba bean and chickpea. Springer, Netherlands, Dordrecht, pp 565–575. ISBN 978-94-009-2764-3

Kudla J, Becker D, Grill E, Hedrich R, Hippler M, Kummer U, Parniske M, Romeis T, Schumacher K (2018) Advances and current challenges in calcium signaling. New Phytol 218:414–431. https://doi.org/10.1111/nph.14966

Landa BB, Navas-Cortés JA, del Mar Jiménez-Gasco M, Katan J, Retig B, Jiménez-Díaz RM (2006) Temperature response of chickpea cultivars to races of *Fusarium oxysporum* f. sp. *ciceris*, causal agent of fusarium wilt. Plant Dis 90:365–374. https://doi.org/10.1094/PD-90-0365

Liang X, Zhou J-M (2018) Receptor-like cytoplasmic kinases: central players in plant receptor kinase–mediated signaling. Annu Rev Plant Biol 69:267–299. https://doi.org/10.1146/annurev-arplant-042817-040540

Lionello P, Scarascia L (2018) The relation between climate change in the Mediterranean region and global warming. Reg Environ Change 18:1481–1493. https://doi.org/10.1007/s10113-018-1290-1

Luck J, Spackman M, Freeman A, Trebicki P, Griffiths W, Finlay K, Chakraborty S (2011) Climate change and diseases of food crops. Plant Pathol 60:113–121. https://doi.org/10.1111/j.1365-3059.2010.02414.x

Ma L, Qiao J, Kong X, Zou Y, Xu X, Chen X, Hu X (2015) Effect of low temperature and wheat winter-hardiness on survival of *Puccinia striiformis* f. sp *tritici* under controlled conditions. PLOS One 10(6): e0130691. https://doi.org/10.1371/journal.pone.0130691

Mahdavi Mashaki K, Garg V, Nasrollahnezhad Ghomi AA, Kudapa H, Chitikineni A, Zaynali Nezhad K, Yamchi A, Soltanloo H, Varshney RK, Thudi M (2018) RNA-seq analysis revealed genes associated with drought stress response in kabuli chickpea (*Cicer arietinum* L.). PLoS ONE 13:e0199774. https://doi.org/10.1371/journal.pone.0199774

Manning WJ, von Tiedemann A (1995) Climate change: potential effects of increased atmospheric carbon dioxide (CO_2), ozone (O_3), and ultraviolet-B (UV-B) radiation on plant diseases. Environ Pollut 88(2):219–245. https://doi.org/10.1016/0269-7491(95)91446-r

Mauch-Mani B, Mauch F (2005) The role of abscisic acid in plant–pathogen interactions. Curr Opin Plant Biol 8:409–414. https://doi.org/10.1016/j.pbi.2005.05.015

Mauch-Mani B, Baccelli I, Luna E, Flors V (2017) Defense priming: an adaptive part of induced resistance. Annu Rev Plant Biol 68:485–512. https://doi.org/10.1146/annurev-arplant-042916-041132

Mayek-Pérez N, GarcÍa-Espinosa R, López-Castañeda C, Acosta-Gallegos JA, Simpson J (2002) Water relations, histopathology and growth of common bean (*Phaseolus vulgaris* L.) during pathogenesis of *Macrophomina phaseolina* under drought stress. Physiol Mol Plant Pathol 60:185–195. https://doi.org/10.1006/pmpp.2001.0388

Meyer S, Saccardy-Adji K, Rizza F, Genty B (2001) Inhibition of photosynthesis by *Colletotrichum lindemuthianum* in bean leaves determined by chlorophyll fluorescence imaging. Plant, Cell Environ 24:947–955. https://doi.org/10.1046/j.0016-8025.2001.00737.x

Milus EA, Kristensen K, Hovmøller MS (2009) Evidence for increased aggressiveness in a recent widespread strain of *Puccinia striiformis* f. sp. *tritici* causing stripe rust of wheat. Phytopathology 99:89–94. https://doi.org/10.1094/PHYTO-99-1-0089

Miraglia M, Marvin HJP, Kleter GA, Battilani P, Brera C, Coni E, Cubadda F, Croci L, De Santis B, Dekkers S, Filippi L, Hutjes RWA, Noordam MY, Pisante M, Piva G, Prandini A, Toti L, van den Born GJ, Vespermann A (2009) Climate change and food safety: an emerging issue with special focus on Europe. Food Chem Toxicol 47:1009–1021. https://doi.org/10.1016/j.fct.2009.02.005

Montalvo-Hernández L, Piedra-Ibarra E, Gómez-Silva L, Lira-Carmona R, Acosta-Gallegos JA, Vazquez-Medrano J, Xoconostle-Cázares B, Ruíz-Medrano R (2008) Differential accumulation of mRNAs in drought-tolerant and susceptible common bean cultivars in response to water deficit. New Phytol 177:102–113. https://doi.org/10.1111/j.1469-8137.2007.02247.x

Mota APZ, Oliveira TN, Vinson CC, Williams TCR, Costa MMdC, Araujo ACG, Danchin EGJ, Grossi-de-Sá MF, Guimaraes PM, Brasileiro ACM (2019) Contrasting effects of wild *Arachis* dehydrin under abiotic and biotic Stresses. Front Plant Sci 10:497. https://doi.org/10.3389/fpls.2019.00497

Muchero W, Ehlers JD, Roberts PA (2010) Restriction site polymorphism-based candidate gene mapping for seedling drought tolerance in cowpea [*Vigna unguiculata* (L.) Walp]. Theor Appl Gen 120(3):509–518. https://doi.org/10.1007/s00122-009-1171-6

Narsai R, Wang C, Chen J, Wu J, Shou H, Whelan J (2013) Antagonistic, overlapping and distinct responses to biotic stress in rice (*Oryza sativa*) and interactions with abiotic stress. BMC Genom 14:93. https://doi.org/10.1186/1471-2164-14-93

Nayak SN, Balaji J, Upadhyaya HD, Hash CT, Kishor PBK, Chattopadhyay D, Rodriquez LM, Blair MW, Baum M, McNally K, This D, Hoisington DA, Varshney RK (2009) Isolation and sequence analysis of DREB2A homologues in three cereal and two legume species. Plant Sci 177:460–467. https://doi.org/10.1016/j.plantsci.2009.07.009

Nene YL, Reddy MV, Haware MP, Ghanekar AM, Amin KS, Pande S, Sharma M (2012) Field diagnosis of chickpea diseases and their control. Information Bulletin No. 28 (revised), International Crops Research Institute for the Semi-Arid Tropics, Patancheru, Andhra Pradesh, India

Nobori T, Tsuda K (2019) The plant immune system in heterogeneous environments. Curr Opin Plant Biol 50:58–66. https://doi.org/10.1016/j.pbi.2019.02.003

Oladzad A, Zitnick-Anderson K, Jain S, Simons K, Osorno JM, McClean PE, Pasche JS (2019) Genotypes and genomic regions associated with *Rhizoctonia solani* resistance in common bean. Front Plant Sci 10:956. https://doi.org/10.3389/fpls.2019.00956

Pandey P, Irulappan V, Bagavathiannan MV, Senthil-Kumar M (2017) Impact of combined abiotic and biotic stresses on plant growth and avenues for crop improvement by exploiting physiomorphological traits. Front Plant Sci 8:537. https://doi.org/10.3389/fpls.2017.00537

Park E, Park J-Y, Kim Y-S, Staswick PE, Jeon J, Yun J, Kim S-Y, Kim J, Lee Y-H, Park C-M (2007) GH3-mediated auxin homeostasis links growth regulation with stress adaptation response in *Arabidopsis*. J Biol Chem 282:10036–10046. https://doi.org/10.1074/jbc

Pieterse CMJ, van der Does D, Zamioudis C, Leon-Reyes A, Wees SCMV (2012) Hormonal modulation of plant immunity. Annu Rev Cell Dev Biol 28:489–521. https://doi.org/10.1146/annurev-cellbio-092910-154055

Pilet-Nayel ML, Muehlbauer FJ, McGee RJ, Kraft JM, Baranger A, Coyne CJ (2005) Consistent quantitative trait loci in pea for partial resistance to *Aphanomyces euteiches* isolates from the United States and France. Phytopathology 95:1287–1293. https://doi.org/10.1094/PHYTO-95-1287

Pottorff M, Wanamaker S, Ma YQ, Ehlers JD, Roberts PA, Close TJ (2012) Genetic and physical mapping of candidate genes for resistance to *Fusarium oxysporum* f.sp. *tracheiphilum* race 3 in cowpea [*Vigna unguiculata* (L.) Walp]. PLoS ONE 7:e41600. https://doi.org/10.1371/journal.pone.0041600

Prasch CM, Sonnewald U (2013) Simultaneous application of heat, drought, and virus to *Arabidopsis* plants reveals significant shifts in signaling networks. Plant Physiol 162:1849–1866. https://doi.org/10.1104/pp.113.221044

Prasch CM, Sonnewald U (2015) Signaling events in plants: Stress factors in combination change the picture. Environ Exp Bot 114:4–14. https://doi.org/10.1016/j.envexpbot.2014.06.020

Radwan DEM, Lu G, Fayez KA, Mahmoud SY (2008) Protective action of salicylic acid against bean yellow mosaic virus infection in *Vicia faba* leaves. J Plant Physiol 165:845–857. https://doi.org/10.1016/j.jplph.2007.07.012

Ramegowda V, Senthil-Kumar M (2015) The interactive effects of simultaneous biotic and abiotic stresses on plants: mechanistic understanding from drought and pathogen combination. J Plant Physiol 176:47–54. https://doi.org/10.1016/j.jplph.2014.11.008

Ramzan N, Noreen N, Perveen Z, Shahzad S (2016) Effect of seed pelleting with biocontrol agents on growth and colonisation of roots of mungbean by root-infecting fungi. J Sci Food Agric 96:3694–3700. https://doi.org/10.1002/jsfa.7553

Rasmussen S, Barah P, Suarez-Rodriguez MC, Bressendorff S, Friis P, Costantino P, Bones AM, Nielsen HB, Mundy J (2013) Transcriptome responses to combinations of stresses in *Arabidopsis*. Plant Physiol 161:1783–1794. https://doi.org/10.1104/pp.112.210773

Rejeb IB, Pastor V, Mauch-Mani B (2014) Plant responses to simultaneous biotic and abiotic stress: molecular mechanisms. Plants (Basel) 3:458–475. https://doi.org/10.3390/plants3040458

Roorkiwal M, Jarquin D, Singh MK, Gaur PM, Bharadwaj C, Rathore A, Howard R, Srinivasan S, Jain A, Garg V, Kale S, Chitikineni A, Tripathi S, Jones E, Robbins KR, Crossa J, Varshney RK (2018) Genomic-enabled prediction models using multi-environment trials to estimate the effect of genotype × environment interaction on prediction accuracy in chickpea. Sci Rep 8:11701. https://doi.org/10.1038/s41598-018-30027-2

Rubiales D, Fondevilla S, Chen W, Gentzbittel L, Higgins TJV, Castillejo MA, Singh KB, Rispail N (2015) Achievements and challenges in legume breeding for pest and disease resistance. Crit Rev Plant Sci 34:195–236. https://doi.org/10.1080/07352689.2014.898445

Rush H, Laurence JA (1993) Interactive effects of ozone and powdery mildew on pea seedlings. Phytopathology 83:1258–1263. https://doi.org/10.1094/Phyto-83-1258

Sabbavarapu MM, Sharma M, Chamarthi SK, Swapna N, Rathore A, Thudi M, Gaur PM, Pande S, Singh S, Kaur L, Varshney RK (2013) Molecular mapping of QTLs for resistance to *Fusarium* wilt (race 1) and *Ascochyta* blight in chickpea (*Cicer arietinum* L.). Euphytica 193:121–133. https://doi.org/10.1007/s10681-013-0959-2

Saijo Y, Loo EPI (2019) Plant immunity in signal integration between biotic and abiotic stress responses. New Phytol. https://doi.org/10.1111/nph.15989

Sanz-Carbonell A, Marques MC, Bustamante A, Fares MA, Rodrigo G, Gomez G (2019) Inferring the regulatory network of the miRNA-mediated response to biotic and abiotic stress in melon. BMC Plant Biol 19:78. https://doi.org/10.1186/s12870-019-1679-0

Scheelbeek PFD, Bird FA, Tuomisto HL, Green R, Harris FB, Joy EJM, Chalabi Z, Allen E, Haines A, Dangour AD (2018) Effect of environmental changes on vegetable and legume yields and nutritional quality. Proc Natl Acad Sci 115:6804–6809. https://doi.org/10.1073/pnas.1800442115

Schneider KA, Grafton KF, Kelly JD (2001) QTL analysis of resistance to *Fusarium* root rot in bean. Crop Sci 41:535–542. https://doi.org/10.2135/cropsci2001.412535x

Sharma KD, Winter P, Kahl G, Muehlbauer FJ (2004) Molecular mapping of *Fusarium oxysporum* f. sp. *ciceris* race 3 resistance gene in chickpea. Theor Appl Genet 108:1243–1248. https://doi.org/10.1007/s00122-003-1561-0

Shriram V, Kumar V, Devarumath RM, Khare TS, Wani SH (2016) MicroRNAs as potential targets for abiotic stress tolerance in plants. Front Plant Sci 7:817. https://doi.org/10.3389/fpls.2016.00817

Siddique KHM, Johansen C, Turner NC, Jeuffroy M-H, Hashem A, Sakar D, Gan Y, Alghamdi SS (2012) Innovations in agronomy for food legumes. A review. Agron Sustain Dev 32:45–64. https://doi.org/10.1007/s13593-011-0021-5

Sinha R, Gupta A, Senthil-Kumar M (2017) Concurrent drought stress and vascular pathogen infection induce common and distinct transcriptomic responses in chickpea. Front Plant Sci 8:333. https://doi.org/10.3389/fpls.2017.00333

Sinha R, Irulappan V, Mohan-Raju B, Suganthi A, Senthil-Kumar M (2019) Impact of drought stress on simultaneously occurring pathogen infection in field-grown chickpea. Sci Rep 9:5577. https://doi.org/10.1038/s41598-019-41463-z

Smakowska E, Kong J, Busch W, Belkhadir Y (2016) Organ-specific regulation of growth-defense tradeoffs by plants. Curr Opin Plant Biol 29:129–137. https://doi.org/10.1016/j.pbi.2015.12.005

Smýkal P, Coyne CJ, Ambrose MJ, Maxted N, Schaefer H, Blair MW, Berger J, Greene SL, Nelson MN, Besharat N, Vymyslický T, Toker C, Saxena RK, Roorkiwal M, Pandey MK, Hu J, Li YH, Wang LX, Guo Y, Qiu LJ, Redden RJ, Varshney RK (2015) Legume crops phylogeny and genetic diversity for science and breeding. Crit Rev Plant Sci 34:43–104. https://doi.org/10.1080/07352689.2014.897904

Su Y, Guo J, Ling H, Chen S, Wang S, Xu L, Allan AC, Que Y (2014) Isolation of a novel peroxisomal catalase gene from sugarcane, which is responsive to biotic and abiotic stresses. PLoS ONE 9:e84426. https://doi.org/doi.org/10.1371/journal.pone.0084426

Tester M, Langridge P (2010) Breeding technologies to increase crop production in a changing world. Science 327:818–822. https://doi.org/10.1126/science.1183700

Thaler JS, Humphrey PT, Whiteman NK (2012) Evolution of jasmonate and salicylate signal crosstalk. Trends Plant Sci 17:260–270. https://doi.org/10.1016/j.tplants.2012.02.010

Thoen MPM, Davila Olivas NH, Kloth KJ, Coolen S, Huang P-P, Aarts MGM, Bac-Molenaar JA, Bakker J, Bouwmeester HJ, Broekgaarden C, Bucher J, Busscher-Lange J, Cheng X, Fradin EF, Jongsma MA, Julkowska MM, Keurentjes JJB, Ligterink W, Pieterse CMJ, Ruyter-Spira C, Smant G, Testerink C, Usadel B, van Loon JJA, van Pelt JA, van Schaik CC, van Wees SCM, Visser RGF, Voorrips R, Vosman B, Vreugdenhil D, Warmerdam S, Wiegers GL, van Heerwaarden J, Kruijer W, van Eeuwijk FA, Dicke M (2017) Genetic architecture of plant stress resistance: multi-trait genome-wide association mapping. New Phytol 213:1346–1362. https://doi.org/10.1111/nph.14220

Tiwari S, Lata C, Chauhan PS, Nautiyal CS (2016) Pseudomonas putida attunes morphophysiological, biochemical and molecular responses in *Cicer arietinum* L. during drought stress and recovery. Plant Physiol Biochem 99:108–117. https://doi.org/10.1016/j.plaphy.2015.11.001

Vadez V, Berger JD, Warkentin T, Asseng S, Ratnakumar P, Rao KPC, Gaur PM, Munier-Jolain N, Larmure A, Voisin A-S, Sharma HC, Pande S, Sharma M, Krishnamurthy L, Zaman MA (2012) Adaptation of grain legumes to climate change: a review. Agron Sustain Dev 32:31–44. https://doi.org/10.1007/s13593-011-0020-6

Varshney RK, Pandey MK, Bohra A, Singh VK, Thudi M, Saxena RK (2019) Toward the sequence-based breeding in legumes in the post-genome sequencing era. Theor Appl Genet 132:797–816. https://doi.org/10.1007/s00122-018-3252-x

Vautard R, Gobiet A, Sobolowski S, Kjellström E, Stegehuis A, Watkiss P, Mendlik T, Landgren O, Nikulin G, Teichmann C, Jacob D (2014) The European climate under a 2 °C global warming. Environ Res Lett 9:3

Vaz Patto MC, Amarowicz R, Aryee ANA, Boye JI, Chung H-J, Martín-Cabrejas MA, Domoney C (2015) Achievements and challenges in improving the nutritional quality of food legumes. Crit Rev Plant Sci 34:105–143. https://doi.org/10.1080/07352689.2014.897907

Velásquez AC, Castroverde CDM, He SY (2018) Plant-Pathogen warfare under changing climate conditions. Curr Biol 28:R619–R634. https://doi.org/10.1016/j.cub.2018.03.054

Wakelin SA, Gomez-Gallego M, Jones E, Smaill S, Lear G, Lambie S (2018) Climate change induced drought impacts on plant diseases in New Zealand. Australas Plant Pathol 2018(47):101–114. https://doi.org/10.1007/s13313-018-0541-4

Wu L, Chang K-F, Conner RL, Strelkov S, Fredua-Agyeman R, Hwang S-F, Feindel D (2018) *Aphanomyces euteiches*: a threat to Canadian field pea production. Engineering 4:542–551

Yadeta KA, Thomma BPHJ (2013) The xylem as battleground for plant hosts and vascular wilt pathogens. Front Plant Sci 4:97. https://doi.org/10.3389/fpls.2013.00097

Yanni Y, Zidan M, Dazzo F, Rizk R, Mehesen A et al. (2016) Enhanced symbiotic performance and productivity of drought stressed common bean after inoculation with tolerant native rhizobia in extensive fields. Agric, Ecosyst Environ 232:119–128. https://doi.org/10.1016/j.agee.2016.07.006

Yasuda M, Ishikawa A, Jikumaru Y, Seki M, Umezawa T, Asami T, Maruyama-Nakashita A, Kudo T, Shinozaki K, Yoshida S, Nakashita H (2008) Antagonistic interaction between systemic acquired resistance and the abscisic acid–mediated abiotic stress response in *Arabidopsis*. Plant Cell 20:1678–1692. https://doi.org/10.1105/tpc.107.054296

Yuan J, Zhao J, Wen T et al (2018) Root exudates drive the soil-borne legacy of aboveground pathogen infection. Microbiome 6(1):156. https://doi.org/10.1186/s40168-018-0537-x

Zander P, Amjath-Babu TS, Preissel S, Reckling M, Bues A, Schläfke N, Kuhlman T, Bachinger J, Uthes S, Stoddard F, Murphy-Bokern D, Watson C (2016) Grain legume decline and potential recovery in European agriculture: a review. Agron Sustain Dev 36:26. https://doi.org/10.1007/s13593-016-0365-y

Zhang H, Sonnewald U (2017) Differences and commonalities of plant responses to single and combined stresses. Plant J 90:839–855. https://doi.org/10.1111/tpj.13557

Zhang M, Su J, Zhang Y, Xu J, Zhang S (2018) Conveying endogenous and exogenous signals: MAPK cascades in plant growth and defense. Curr Opin Plant Biol 45:1–10. https://doi.org/10.1016/j.pbi.2018.04.012

Leguminosae (*nom. alt.* Fabaceae)—Its Diversity, Use and Role in Environmental Conservation in the Harsh Environs of the Cold Deserts of North-West India

Vaneet Jishtu and Gurinderjit Singh Goraya

Abstract Leguminosae, known as the legume, pea or bean family, form one of the most diverse, climatically resilient, and socio-economically important group of flowering plants in the trans-Himalayan cold deserts. Whereas many of the native plants belonging to the genera *Caragana* and *Astragalus*, form vast expanses of low thickets in the region holding on to the dry soil and forming food source both to the wild fauna and the domestic livestock. Many of the cultivated legumes like pea, chick pea, clovers and a variety of beans have adapted well to the harsh climes of the region adding to the socio-economy of local communities. The Leguminosae also form one of the important sources of fodder and fuel wood, while many of its *taxa* are also used in religious rituals and as popular folk medicine. This work on Leguminosae (*nom. alt.* Fabaceae) from the trans-Himalayan cold desert areas of the north-west and western part of India is based on the published floristic and ethnobotanical accounts from the region, supplemented with information gathered by the authors during their visits to the area over a period of more than twenty years. The work has resulted in upgrading the floristic and ethnobotanical documentation of the Leguminosae from the region.

Keywords Economic botany · Biodiversity · Fuelwood and fodder · Cold desert · Desertification

1 Introduction

The Leguminosae (*nom. alt.* Fabaceae) family is economically interesting in the Indian cold deserts and many plants are valued as food, fodder, fuelwood, and as folk medicine. Hardy shrub species viz. *Caragana* spp., occupying high alpine dry slopes, in addition to checking desertification in the cold deserts, are the mainstay food of

V. Jishtu (✉)
Himalayan Forest Research Institute, Conifer Campus, Panthaghati, Shimla 171013, HP, India
e-mail: vjishtuv@gmail.com; coolbotanist@yahoo.com

G. S. Goraya
Himachal Pradesh Forest Department, Former PCCF & HoFF, Himachal Pradesh, Talland, Shimla 171002, HP, India

M. Hasanuzzaman et al. (eds.), *The Plant Family Fabaceae*,
https://doi.org/10.1007/978-981-15-4752-2_10

herbivore fauna. These *Caragana* shrubs have also been collected for hundreds of years as fuelwood in this treeless region. The roots, leaves, and fruits of some *taxa* are medicinally important, while some *taxa* like *Cicer microphyllum* are valuable fodder stored for winters.

Indian north-west and western Himalaya has a good representation of Leguminosae, especially in the cold desert regions covering Ladakh, northern parts of Himachal Pradesh, and north-western Uttarakhand. Leguminosae is a large family and stands tall among the dominant families of flowering plants across the globe. Varying numbers of genera and species ranging from 550 to 690 genera and 12,000 to 19,500 species have been cited by various workers (Airy-Shaw 1966; Hutchinson 1964 and Melchior 1964; Judd et al. 1999; LPWG 2017; Anonymous[1]). In India, 1252 legume species distributed in 199 genera have been documented from across the tropical and the temperate regions. Srivastava and Shukla (2015) have documented 93 species and two varieties of legumes distributed in 24 genera from the trans-Himalayan cold deserts of the north-western India.

Most diverse genera in Leguminosae is *Astragalus* L., a large genus of over 3000 species of herbs and small shrubs and is probably the largest genus of flowering plants across the globe (Lock and Simpson 1991; Mabberley 1997; Maassoumi 1998; Ranjbar and Karamian 2002; Chaudhary et al. 2008). In India, about 80 species of *Astragalus* L. have been documented (Chaudhary et al. 2008); of which 75 species occur in the north-western and western Himalayas, one of the main centers of diversity for this genus (Hooker 1872–1897). This statement is strengthened by the fossil and phylogenetic evidence, which suggests that the legumes originally evolved in arid and/or semi-arid regions along the Tethys seaway during the early Tertiary (Schrire et al. 2005). In India, 40 species of *Astragalus* have been reported from the cold deserts (Srivastava and Shukla 2015), 22 species of this genus have been reported from Lahaul and Spiti (Aswal and Mehrotra 1994), and 17 species from the Pin Valley National Park (PVNP) in Spiti (Chandra Sekar and Srivastava 2009). The genus *Astragalus* is also well represented in the neighboring countries, with 135 species reported from Pakistan (Anonymous[2]) and 167 genera and 1673 species reported from China (Anonymous[3]). Of late, some species from these genera have been relocated to some other genera like *Ophiocarpus* (Bunge) Ikonn., *Podolotus* Benth., and *Podlechiella* Maassoumi and Kaz. Osaloo; for example, *Astragalus marschallianus* is now *Astracantha denudata* (Steven) Podlech., *Astragalus strobiliferus* now *Astracantha strobilifera* (Benth.) Podlech. (Ranjbar and Karamian 2002; Chaudhary et al. 2008; Anonymous[4]).

Another, important genus from this family in the cold deserts is *Caragana* L., which is represented by more than 100 species worldwide of which 66 are recorded from China alone (Kumar et al. 2016). In India, we have ten species reported with

---

[1]http://lib.unipune.ac.in:8080/xmlui/bitstream/handle/123456789/7452/04_chapter%201.pdf. Accessed on 19 Nov 2019.

[2]http://www.efloras.org/florataxon.aspx?flora_id=5&taxon_id=102978. Accessed on 18 Nov 2019.

[3]http://www.efloras.org/florataxon.aspx?flora_id=2&taxon_id=10335. Accessed on 18 Nov 2019.

[4]http://www.theplantlist.org/.

three species from the cold desert regions of NW India (Srivastava and Shukla 2015) of which *C. versicolor* being the most dominant among them. However, today some species from the genus are re-allotted into other genera, e.g., *Caragana cuneata* (Benth) Baker is now *Chesneya cuneata* (Benth.) Ali, *Caragana moorcroftiana* Benth., now treated as *Sophora moorcroftiana* (Benth.) Baker, *Caragana nubigena* (D. Don) Bunge as *Spongiocarpella nubigena* (D. Don) Yakovlev. *Oxytropis* is the other genera with around ten species reported from the region, while the other genera have limited number of species. The other major genera in the region include *Chesneya*, *Cicer*, *Colutea*, *Lotus*, *Medicago*, *Melilotus*, *Pisum*, *Sophora*, *Thermopsis*, *Trifolium*, *Trigonella* and *Vicia*.

The facts provided above highlight the importance of working out the Leguminosae diversity, its use, and the vital role it plays for environmental conservation in the harsh environs of the cold deserts of North-West India, also often referred to as the Trans-Himalayan cold deserts.

2 Geography and Vegetation

The Indian cold deserts in the north-western and western part of the country encompass the trans-Himalayan regions of the Union Territory of Ladakh, Lahaul and Spiti district and Pooh Sub-Division of Kinnaur district in Himachal Pradesh, and a small region of Uttarakhand (include Nilang, Mana and Niti valleys in Garhwal and Johar, inner Darma and Byans valleys in Kumaun). This region occupies a position of great bio-geographical and geostrategic importance, bordered by Pakistan, China, and the western margin of the Tibetan Plateau. The cold deserts in India lie in between 31°44′ to 36°0′ N latitude and 75°15′ to 80°15′ E longitude with a total geographical area which comes under the cold desert in this region is approximately 74,809 km^2 (Jishtu et al. 2003) (Table 1).

The geology of the cold deserts has great variations; as a result, the soil found here undergoes significant changes, depending mainly upon the topography, terrain, parent rock, and the nature and type of vegetation. The principal soil types encountered in the cold desert regions are red and black soils (light to dark brown), ferruginous red soils (red), brown forest soils, mountain and hill soils, the high altitude meadow soils, and the alpine soils. Physically, these soils are coarse (gravelly loamy sand) to fine (silty clay loam), shallow to deep and moderate to poor in mineral nutrients (Sharma et al. 2006). The pH of these soils generally ranges between 6 and 10, with a high degree of variability in soil properties is however evident from one valley to the other.

The region is apparently devoid of any monsoonal precipitation, and most precipitation otherwise is in the form of snow during the prolonged winter months; annual precipitation in general is normally less than 200 mm. Mean temperatures are below 0 °C from November to March (going up to −70 °C in Dras, while summer temperatures are high (peaking up to 40 °C). These physiographic peculiarities and ensuing

Table 1 Statewise distribution of cold deserts in India

State	Cold desert region	Lat. (N)	Long. (E)	Approx. area (km^2)
Jammu and Kashmir	Leh and Kargil district of Ladakh	32° 15′–36°	75° 15′–80°	68,321 km^2 (With 27,555 km^2 Chinese and Pakistan occupation)
Himachal Pradesh	Lahaul and Spiti district. A small pocket of Bharmour in Chamba district	31° 44′–32° 59′	76° 46′–78° 44′	6488 km^2
Uttaranchal[a]	Small pocket in the Janvi valley of Uttarkashi district			
Sikkim[a]	Barren northern tip			

[a]Information about the approximate area of cold deserts in these states is not available

harsh climatic conditions have led to the emergence of a unique cold desert ecosystem as well as culture of the community, which is unlike any other in the world (Anonymous[5]). This ecosystem is also home to highly adaptive, rare endangered fauna such as Asiatic Ibex, Tibetan Argali, Ladakh Urial, Bharal, Tibetan Antelope, Tibetan Gazelle, Wild Yak, Snow Leopard, Brown Bear, Tibetan Wolf, Wild Dog, Tibetan Wild Ass and the Black Necked Crane, along with the flora which is rich in endemic and economically important species (Anonymous[6]; Kala 2000).

These cold deserts encase deep incised valleys offering inhospitable terrain and climate, but providing life sustaining ecosystem services to local communities such as timber, fuelwood, fodder, medicine, and food. Because, of the remoteness and difficult terrain, the basic developmental services like transport, health, and education are hindered. Furthermore, the local environment too is very harsh, with unexpected fluctuations in seasonal temperature, light intensity, and ultraviolet (UV) radiations, along with thin rarefied air. Due to this remoteness of the region, the welfare programmes and facilities provided by government take time to reach them or at times left out. Therefore, majority of the communities residing in the far flung remote valleys of the region are considered among the poorest. Because of this, in such circumstances, local communities have relied on the available native plant-based traditional means for their day-to-day livelihood sustenance. Their tried and tested

[5]https://whc.unesco.org/en/tentativelists/6055/. Accessed on 15 Nov 2019.

[6]http://environmentportal.in/files/India_Fourth_National_Report-FINAL.pdf. Accessed on 15 Nov 2019.

knowledge of the nearby natural resources utilization to meet their day-to-day living has always been more or less sustainable.

Due to prolonged and extreme cold winters, the agriculture season is very short which starts from May and ends by September depending upon different altitudes. The cold desert thus, represents a diverse and characteristics vegetation distributed over a wide range of topographical variation from dry scree slopes to moist river valleys. The high endemism of its flora and fauna justifies the diversity and uniqueness of the Indian trans-Himalaya. The great variation in elevation, temperature, precipitation and harsh physical parameters have resulted in a diversity of biotic communities with a relatively unique flora of flowering plants, well adapted to the region (823 species including grasses and sedges (Stewart 1917); 662 species from Ladakh (Kachroo et al. 1977); 700 species (Chaurasia et al. 2011); 647 species from Ladakh (Kala 2011); 1195 species (plus 30 sub species, 42 varieties, and 1 forma) from Ladakh (Srivastava and Shukla 2015); 985 species from Lahaul and Spiti (Aswal and Mehrotra 1994); and 513 species from PVNP, Spiti (Chandra Sekar and Srivastava 2009).

Vegetation is overwhelmingly herbaceous—comprising of a few tree species and some shrub species. Trees and shrubs, particularly *Juniperus* and *Betula utilis* which have aromatic as well as medicinal properties seem to have been over harvested. Additionally, manmade forests of *Populus*, *Salix*, *Hippophae* and *Myricaria* can also be seen along river banks, rivulets, and smaller streams. The region in general supports very scanty and highly specialized vegetation, characteristic to the cold desert ecosystem.

The flora of the cold deserts is dominated by xerophytes (Chaurasia and Singh 1996), well adaptive to extreme cold weather. There are three main elements in the flora of the region: oasitic, desert steppe and alpine arid rangelands, which in itself are easily recognizable and separable (Stewart 1917). Oasitic vegetation is restricted to small patches in moist niches, more abundant at intermediate and lower altitudes, especially along streams and rivulets. Desertic vegetation is suited to dry, a sandy, and rocky arid slopes that have adaptations to store water in its tissues and dense hairs, waxy, or modified leaves to reduce transpiration. Alpine vegetation is adapted to the harsh environment with low temperatures, dryness, ultraviolet radiation and a short growing season. It includes perennial grasses, sedges, forbs, cushion plants, mosses and lichens (Fig. 3).

The life-form spectrum in the cold deserts under reference reflects desert steppe and steppe vegetation (Klimes 2003). It has been reported by earlier workers that the members of Leguminosae do favor steppe vegetation; in regions where trees are rare the shrubby and herbaceous members become important features of vegetation. The vegetation consists mainly of hemicryptophytes—herbaceous perennials, including sedges and grasses, which produce perennating buds at the soil surface, buds being protected by leaf or stem bases. They are followed by therophytes—annual plants that complete their life cycle in a short favorable period and survive the major part of the harsh environmental conditions as seed. The rest of the life forms like hamaephytes, geophytes, phanerophytes, and hydrophytes are limited in number (Dickoré and Miehe 2002; Rawat 2008; Kumar et al. 2016).

3 Representation of the Legumes in the Cold Desert Regions: Some Analytic Studies

The Leguminosae form one of the largest groups of plants in the trans-Himalayan cold deserts and play a very significant role in the socio-economy of the region. Some *taxa* of this family play a vital role for energy production in this region, whereas many others form a crucial source of fodder (Fig. 4).

The work incorporates information from various valid publications especially floras as standard references including Hooker (1872–1897), Stewart (1916, 1917), Ali (1977), Kachroo et al. (1977), Dhar and Kachroo (1983), Polunin and Stainton (1984), Aswal and Mehrotra (1994), Chaurasia and Singh (1996a, b 1998, 2001), Klimes and Dickoré (2005), Chandra Sekar and Srivastava (2009) and Srivastava and Shukla (2015), besides other valid publications.

General ethnobotanical information is based on the studies on the subject carried out by various authors (Buth and Irshad 1988; Sood et al. 2001; Samant et al. 2011; Ballabh et al. 2007; Chaurasia et al. 2008; Samant et al. 2011; Ahmad et al. 2015; Saroya 2017; Nautiyal et al. 2018).

In the latest and the most comprehensive work on the plants of cold deserts, Srivastava and Shukla (2015) have documented 95 taxa of legumes from the region. The present study enlists a total of 114 plants in 29 genera as a consolidated list from the cold desert region (Table 2).

From the list given above as Table 2 and its graphical representation in Fig. 1, it is observed that the major genera are *Astragalus* (57%) with 39 species and *Oxytropis* (18%) with 12 species, together accounting for almost 75% representation of the legumes in the cold desert regions of the Indian trans-Himalaya. These observations are in agreement with the earlier studies on floristics from the region as mentioned ahead. Stewart (1916–1917), documented 59 species from Leguminosae of which 28 belonged to the genus *Astragalus* alone, accounting for almost 48%. *Oxytropis* followed distant second with only seven species (12%). These two genera are more numerous and common in the Russian parts of Central Asia too (Grubov 2003). Flora of Lahaul Spiti (Aswal and Mehrotra 1994), documents 53 species from Leguminosae of which 22 belonged to the genus *Astragalus* alone, accounting for almost 42%. *Oxytropis* followed distant second with only six species (12%). The flora of the cold desert by Srivastava and Shukla (2015) documents 93 species with two varieties in 24 genera of legumes from the cold deserts. The genus *Astragalus* accounts for 40 of these species (43%), whereas the genus *Oxytropis* is represented by 11 species and the genus *Caragana* is represented by three species. It is also clear from the graphical representation that 11 genera are represented only by a single species while nine genera have just two species each.

The genera *Astragalus* is the most predominant followed by *Oxytropis*, even though two species from *Astragalus* are now placed in the new formed genera of *Astracantha* while two members of *Caragana* are now shifted to the genus *Spongiocarpella* (Anonymous[4]).

Table 2 Consolidated list of Leguminosae from the cold desert region under review

Sl. No	Plant name	Habit
1	*Astracantha denudata* (Steven) Podlech	Herb
2	*Astracantha strobilifera* (Benth.) Podlech	Under shrub
3	*Astragalus amherstianus* Benth.	Herb
4	*Astragalus arnoldii* Hemsl. & H Pearson	Herb
5	*Astragalus bicuspis* Fisch.	Under shrub
6	*Astragalus chlorostachys* Lindl.	Under shrub
7	*Astragalus coluteocarpus* Boiss.	Herb
8	*Astragalus confertus* Bunge	Herb
9	*Astragalus densiflorus* Kar. & Kir.	Herb
10	*Astragalus drasianus* H.J. Chowdhery, Uniyal & Balodi	Herb
11	*Astragalus falconeri* Bunge	Herb
12	*Astragalus frigidus* (L) A. Gray	Herb
13	*Astragalus gracilipes* Bunge	Herb
14	*Astragalus grahamianus* Benth.	Under Shrub
15	*Astragalus graveolens* Benth.	Under Shrub
16	*Astragalus hendersonii* Baker	Herb
17	*Astragalus heydei* Baker	Herb
18	*Astragalus himalayanus* Klotzsch	Herb
19	*Astragalus hoffmeisteri* (Klotzsch) Ali	Herb
20	*Astragalus ladakhensis* R.R. Rao & Balodi	Herb
21	*Astragalus leucocephalus* Bunge	Under Shrub
22	*Astragalus macropterus* DC.	Under Shrub
23	*Astragalus malacophyllus* Hort. ex G. Don	Herb
24	*Astragalus maxwellii* Benth.	Herb
25	*Astragalus melanostachys* Bunge	Herb
26	*Astragalus multiceps* Benth.	Herb
27	*Astragalus munroi* Bunge	Herb
28	*Astragalus nivalis* Kar. & Kir.	Herb
29	*Astragalus ophiocarpus* Boiss.	Herb
30	*Astragalus oplites* Benth. ex R. Parker	Herb
31	*Astragalus oxyodon* Baker	Herb
32	*Astragalus peduncularis* Royle	Herb
33	*Astragalus polyacanthus* Benth.	Under Shrub
34	*Astragalus rhizanthus* Benth.	Herb
35	*Astragalus subuliformis* DC.	Herb
36	*Astragalus subumbellatus* Klotzsch	Herb

(continued)

Table 2 (continued)

Sl. No	Plant name	Habit
37	*Astragalus tibetanus* Bunge	Herb
38	*Astragalus tribulifolius* Bunge	Herb
39	*Astragalus webbianus* Benth.	Herb
40	*Astragalus zanskarensis* Bunge	Under Shrub
41	*Astragalus rhizanthus* subsp. *candolleanus* (Benth.) Podlech	Shrub
42	*Astragalus strictus* Benth.	Herb
43	*Caragana brevifolia* Kom.	Shrub
44	*Caragana brevispina* Benth.	Shrub
45	*Caragana conferta* Baker.	Shrub
46	*Caragana gerardiana* Benth.	Shrub
47	*Caragana polyacantha* Royle	Shrub
48	*Caragana versicolor* Benth.	Shrub
49	*Chesneya cuneata* (Benth.) Ali	Herb
50	*Chesneya popovii* Kamelin & Yakovlev	Herb
51	*Cicer arietinum* L.	Herb
52	*Cicer microphyllum* Benth.	Herb
53	*Colutea nepalensis* Sims	Shrub
54	*Desmodium elegans* DC.	Shrub
55	*Desmodium multiflorum* DC.	Shrub
56	*Hedysarum astragaloides* Benth.	Herb
57	*Hedysarum microcalyx* Baker	Under Shrub
58	*Hylodesmum podocarpum* (DC.) H. Ohashi & R.R. Mill	Herb
59	*Indigofera heterantha* Brandis	Shrub
60	*Lathyrus aphaca* L.	Herb
61	*Lathyrus emodi* (Fritsch) Ali	Herb
62	*Lathyrus sativus* L.	Herb
63	*Lens culinaris* Medik.	Herb
64	*Lespedeza juncea* (L.f.) Pers.	Herb
65	*Lotus corniculatus* L.	Herb
66	*Lotus japonicus* (Regel) K. Larsen	Herb
67	*Medicago edgeworthii* Sirj	Herb
68	*Medicago falcata* L.	Herb
69	*Medicago lupulina* L.	Herb
70	*Medicago polyceratia* (L.) Trautv.	Herb
71	*Medicago polymorpha* L.	Herb
72	*Medicago sativa* L.	Herb

(continued)

Table 2 (continued)

Sl. No	Plant name	Habit
73	*Melilotus indicus* (L.) All.	Herb
74	*Melilotus officinalis* (L.) Pall.	Herb
75	*Melilotus albus* Medik.	Herb
76	*Oxytropis cachemiriana* Cambess.	Herb
77	*Oxytropis chiliophylla* Benth.	Herb
78	*Oxytropis densa* Bunge	Herb
79	*Oxytropis humifusa* Kar. & Kir.	Herb
80	*Oxytropis hypoglottoides* (Baker) Ali	Herb
81	*Oxytropis immersa* (Baker) B. Fedtsch.	Herb
82	*Oxytropis lapponica* (Wahlenb.) Gay	Herb
83	*Oxytropis microphylla* (Pall.) DC.	Herb
84	*Oxytropis mollis* Benth.	Herb
85	*Oxytropis pusilla* Bunge	Herb
86	*Oxytropis shivai* Aswal, Goel & Mehrotra	Herb
87	*Oxytropis tatarica* Baker	Herb
88	*Parochetus communis* D. Don	Herb
89	*Phaseolus coccineus* L.	Herb
90	*Phaseolus vulgaris* L.	Herb
91	*Piptanthus nepalensis* (Hook.) D. Don	Herb
92	*Pisum sativum* L.	Herb
93	*Robinia pseudoacacia* L.	Tree
94	*Sophora alopecuroides* L.	Shrub
95	*Sophora moorcroftiana* (Benth.) Baker	Shrub
96	*Spongiocarpella nubigena* (D. Don) Yakovlev	Shrub
97	*Stracheya tibetica* Benth.	Herb
98	*Thermopsis barbata* Benth.	Herb
99	*Thermopsis inflata* Cambess.	Herb
100	*Trifolium alexandrinum* L.	Herb
101	*Trifolium dubium* Sibth.	Herb
102	*Trifolium pratense* L.	Herb
103	*Trifolium repens* L.	Herb
104	*Trigonella cachemiriana* Cambess.	Herb
105	*Trigonella corniculata* Sibth & Sm	Herb
106	*Trigonella emodi* Benth.	Herb
107	*Trigonella foenum-graecum* L.	Herb
108	*Trigonella gracilis* Benth.	Herb

(continued)

Table 2 (continued)

Sl. No	Plant name	Habit
109	*Vicia bakeri* Ali	Herb
110	*Vicia faba* L.	Herb
111	*Vicia pallida* Hook. & Arn.	Herb
112	*Vicia sativa* L.	Herb
113	*Vicia tenuifolia* Roth	Herb
114	*Vigna vexillata* (L.) A. Rich.	Herb

Fig. 1 Genera wise species distribution

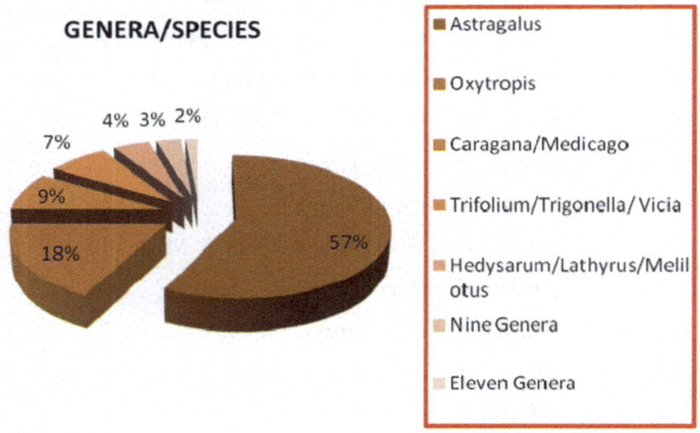

As per the graphical representation above in Fig. 2, it is evident that the native vegetation is herbaceous with a few shrubs. The table shows a majority are herbs in habitat (87%) including the under shrubs, with the arboreal elements limited to just 13%. A single tree is represented by the non-native *Robinia pseudoacacia*, which has been planted by the government agencies across the cold desert landscape, initially under the Desert Development Programme (DDP) of the government during the 1990s, implemented mainly in the cold desert regions of the state of Himachal Pradesh.

Fig. 2 Habitwise species distribution

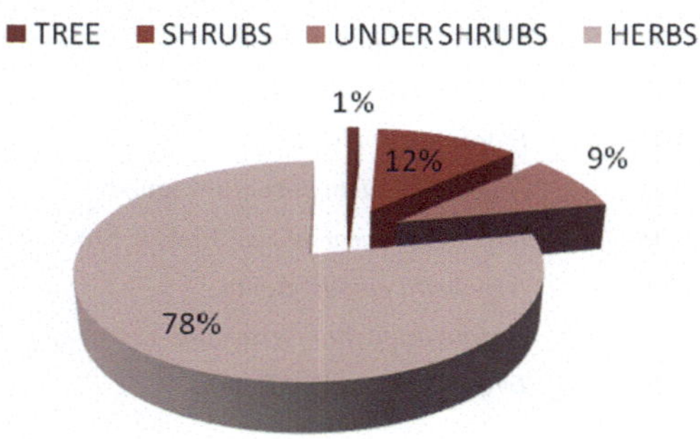

Fig. 3 Diverse vegetational elements adapted to the harsh cold desert environment

(A) *Parochetus communis* D. Don (B) *Colutea nepalensis* Sims (C) *Oxytropis microphylla* (Pall.) DC. (D) *O. microphylla* Flower close up (E) *Piptanthus nepalensis* (Hook.) D. Don (F) *Lotus corniculatus* L. (G) *Medicago sativa* L. (H) *Melilotus officinalis* (L.) Pall.

Fig. 4 Striking bright flowers of the Leguminosae in the stark landscape of the cold deserts

The leguminous trees in general are absent from this region and the diversity of shrubs and herbs is being used by the local communities for meeting the demands of their fuel wood, fodder, medicine, basket making, agricultural implements, along with several other traditional usages. Our observations and interaction with people in the study area suggest that some genera like *Astragalus* and *Caragana* can certainly be considered as keystone species of high conservation concern in the arid cold desert landscape. In addition to their use by local communities and their role in binding the loose soil strata, these plants also provide habitat to a variety of birds, insects and small mammals. Species like *Caragana gerardiana* and *C. versicolor* have been considered as species of high conservation concern as studied under various research programmes (Rawat 2008; Kumar et al. 2016). The Indian Council of Forestry Research and Education (ICFRE), through its regional Himalayan Forest Research Institute (HFRI) at Shimla, has an integrated research programme for the genera, *Caragana*, *Astragalus* and *Sophora* being undertaken at their field research station at Tabo (Spiti).

A vivid glimpse of the varied uses, the members of Leguminosae are put to in the western Himalayan cold deserts is presented below.

3.1 Traditional Medicine

The cold desert region has been traditionally rich in ethnic folklore with distinct culture as yet undisturbed by external influences. It follows the Tibetan system of medicine which is also known as the '*Amchi*' system of medicine that has been passed on from generations within the village. This indigenous healthcare system plays a major part in the overall health care of local communities. A large number of local plants are being used in medicine and general health care and wellness; below are described in brief the members of the Leguminosae, being used for health care by the local communities (Buth and Irshad 1988; Bhattacharyya 1991; Gurmet et al. 1998; Jishtu et al. 2003; Kala 2006; Chaurasia et al. 2008; Ballabh et al. 2008; Singh et al. 2009; Rawat et al. 2009; Chaurasia et al. 2011; Angmo 2012; Rana et al. 2014; Abbas et al. 2017; Namtak and Sharma 2018). In modern research, alkaloids from *Sophora moorcroftiana* are thought to be a potential drug to treat human *echinococcosis*—a parasitic disease caused by tapeworms (Fabian et al. 2018).The more prominent plants which find use in traditional medicine are listed below in Table 3.

3.2 Fuelwood

Owing to extreme winters, the fuelwood collected from the wild is the major source of domestic energy in the cold deserts. Use of shrubs for fuel is a way of life for the people living in harsh desertic environment (Young et al. 1989). The people in

Table 3 Legume plants used in local medicine in the cold desert region

Plant species	Vernacular name	Ethno-medicinal uses
Astragalus rhizanthus subsp. *candolleanus* (Benth.) Podlech	Zomoshing, Sarmang	Root as heart stimulant and also for various skin related problems Source of a well known drug 'Rudanti' or 'Rudravanti' used for tuberculosis, coughs and blood purifier; also used for enhancing the general metabolism
Astragalus annularis Forssk. (= *A. subulatus*)	Rungentso	Used for a number of skin-related problems
Astragalus multiceps Benth.	Sarmang	The entire plant is used as tonic for treatment of diabetes and nephritis and also for colic-related problems in the abdomen caused by wind or obstruction of the intestines
A. ladakensis Balak	Serpang	All parts are used as diuretic
A. himalayanus Klotzsch	Kayaba Chhutup	Flowers and seed are used in strangury (symptom characterized by painful, frequent urination) and also in leprosy
A. nivalis Kar. & Kir.	Keechu	Root is used as blood purifier
A. subuliformis DC.	Zomoshing	A gum like substance obtained, is used as diuretic
A. grahamianus Royle ex Benth.	Rangchawag	Root and leaf used for treatment of many skin related problems
Astragalus zanaskariensis Benth.	Chisigma.	Roots are sun-dried and powdered. It is then dissolved in luke warm milk and given to expel intestinal worms in children. Also, the extract of fresh roots is claimed to be effective against ring worm infection
Astragalus chlorostachys Royle ex Benth.	Zomoshing	Tonic, febrifuge and also used in the cure of tuberculosis
Astragalus rhizanthus Benth.	Srad-sa, Sermang, Krelseng	Women supposedly chew the roots to trigger better lactation. The roots are also used for treating various kinds of tumors relating to the liver, throat, chest and eyes; besides, treating pain in the chest and back

(continued)

Table 3 (continued)

Plant species	Vernacular name	Ethno-medicinal uses
Astragalus strictus Benth.	Serpang	Diuretic (increases the flow of urine)
Astragalus tribulifolius Bunge	Yanglo	Diuretic (increases the flow of urine)
Sophora moorcroftiana (Benth.) Baker	Takay vonpo	Claimed to act as an excellent blood purifier. Leaves are collected fresh and then boiled in milk, thereafter cooled. It is taken in the morning for a week to purify the blood. The leaves are also dried and kept in powdered form for use as an antiseptic agent as and when required
Cicer microphyllum Benth.	Sari, Serri, Baniyarts, Vananyarcha, Chiri	All parts used to combat mountain sickness like high altitude mountain sickness, jaundice and stress. Also used for sore mouth disease in cattle and tongue infection
Caragana versicolor Benth.	Tsaon, Trama, Dam, Bramswak, Zomoshing	Root is used for curing fever and throat infections, besides relief from food poisoning. Also used locally for menstrual cramps (Dysmenorrhea)
Chesneya cuneata (Benth.) Ali	Bi-gang-bo	The roots are antiseptic
Hedysarum falconeri Baker		The fresh roots are taken orally for loss of appetite
Lotus corniculatus L.	Mentok	All parts anti-inflammatory
Medicago lupulina L.	Bu-su-hang-pho, Gunyok	All plant parts used in Liver and lung diseases, kidney disorder, pneumonia and also for cure of cough
Medicago sativa L.	Yarkandi Ole	Seeds used in formulation of remedies for gynaecological disorders by Amchi in Ladakh. Tablets of 1–2 grams are prepared; administered orally for Menorrhagia and other pathogenic diseases in the uterus

(continued)

Table 3 (continued)

Plant species	Vernacular name	Ethno-medicinal uses
Oxytropis lapponica (Wahl).Gay.	Sumtuk	All plant parts used in case of joint aches.
O. tatarica Baker	Shusha	All parts of the plant used as diuretic
Sophora alopecuroides L.		Fresh leaves are applied directly (Topical) for ear ache and wound healing.
Trifolium pratense L.	Bhukshum	All parts of the plant used in cure of cough and other bronchial problems besides, venereal diseases
T. repens L.	Bhukshum	The fresh leaves being astringent, are applied directly (Topical) for ear ache and wounds
Trigonella emodi Benth	Bugsug, Kuchona, Ool buk	The leaves and stem are used for fever, anemia and peptic ulcer; the paste of leaves is applied on the wounds
Trigonella polycerata L. ·	Tongzil	Dried seeds are powdered and 1/2 tablespoon given thrice a day for three days with hot water in fever and also for cough and cold. Used in diarrhea and as vegetable and green fodder
Thermopsis inflata Camb.	Dugsrad, Bu-su-hang-pho	Used for remedy from various kind of swellings
Melilotus officinalis (L.) Pall.	rGya-sPos- dMan-pa, Bhukshuk	The entire plant is used during menstruation, for insomnia, palpitations and also for wounds, cut and bruises

these parts spend about seven to eight months literary inside their homes and require energy to keep themselves warm and comfortable in the harsh environment. Stewart has reported the use of *Caragana pygmea* as fuel, way back in 1869 (Bhattacharyya 1991). With little tree flora, the collection of fuelwood is limited to cutting of sturdy bushes like *Caragana* and *Astragalus* species. In Ladakh and Lahaul and Spiti, people do use shrubs of *Hippophae* species as fuelwood; however, even it is not enough to meet the domestic fuelwood demand in the region. Plantations of some non-native tree species like *Robinia pseudoacacia* have been raised in homesteads and field margins to meet fuelwood demand. Major part of the demand is, however, being met through purchase of fuelwood from the government subsidised Forest Department

Depots, who get their supply from the lower parts of the state. The consumption of fuelwood increases during the long winters and gradually decreases during the short summer season. Local communities collect/extract the shrubs at the end of growing season and stack them on their mud roofs for winter use. This, stockade of shrubs on the roof top serve dual purpose, as it also keeps their homes warm from the continuous snow that accumulates during the long harsh winter. Major species exploited for their deep roots and branches as fuel wood includes *Astragalus rhizanthus*, *Caragana brevifolia*, *C. gerardiana*, *C. versicolor*, *Colutea nepalensis*, *Robinia pseudoacacia*, *Sophora mollis* and *S. moorcroftiana* (Fig. 5).

3.3 Fodder

In the cold deserts, livestock forms an important and integral part of their livelihood. Due to a prolonged winter season, livestock fodder is mostly collected during the short summers and stored. Here, the diversity, utilization pattern, and season of availability/collection are important for the prioritization of selected wild species for fodder use. Also, there are limited fodder yielding herbaceous plants and shrubs in the dry alpine scrubland and moist slopes in the cold deserts. Traditional knowledge system plays a key role in the use of any plant species as fodder with species preference (Nautiyal et al. 2018). The plant that is mostly preferred and thus extensively collected from the wild as winter fodder is *Cicer microphyllum*. In the highland pastures, normally above 4000 m, *Caragana* (mainly *C. gerardiana* and *C. versicolor*) is the most important and highly preferred fodder and fuelwood species. *Caragana* is among the few plant species grazed by livestock as well as the wild ungulates during winter season when the resources are scarce (Kumar et al. 2016). Other plant species from this family include *Astragalus candolleanus*, *A. chlorostachys*, *A. rhizanthus* (roots also in this case), *Caragana brevispina*, *Oxytropis humifusa*, *O. microphylla*, *Robinia pseudoacacia*, *Thermopsisinflata*, *T. barbata*, *Trigonella emodi*, *Vicia faba* (escape from cultivation), *V. tenuifolia* and *V. sativa.* Roots of certain species like *Astragalus marschallianus* Fisch., are being used as nutritious fodder for livestock. During winter, the roots of certain plants like *A. grahamianus* are dug out and used as fodder. Roots of *A. rhizanthus* are in popular use as nutritious fodder for 'Churu'/'Zomo' (Yak), especially during the lean period, when all is cold and dry.

Besides the above-mentioned wild plants, a number of other species are cultivated as fodder crops; *Trifolium repens*, *T. pratense*, *Medicago sativa*, *M. falcata*, *M. lupulina*, *Melilotus albus*, *M. indica* and *M. officinalis* (Fig. 6).

Fig. 5 Shrubs as fuel is a way of life for the communities inhabiting the cold deserts

Fig. 6 Some important herbaceous fodder species from the Desert Steppe

3.4 Wild Edibles

The leguminous flora of this harsh region is of great importance to the local communities, particularly for edible purpose. The harsh climate, prolonged cold winter prevailing in this zone always enforced the far-flung inhabitants to search for their food, shelter, medicine, fodder, and fuel (Ballabh et al. 2007; Chaurasia et al. 2011). Given in Table 4 are some members of Leguminosae which are edible and more readily consumed by the local communities.

Table 4 Some Leguminoceae species which are edible and more readily consumed by the local communities

Plant name	Vernacular name	Resource use
Caragana brevispina Royle	Thama/Thangchhar	Fresh buds are consumed raw
Cicer microphyllum Benth.	Seri or Srad-Kar	Unripe seeds consumed by shepherds and children; the seeds also cooked as a vegetable by Spitians
Lathyrus sativus L.	Khesar, Sanma	Seeds and pods are eaten raw and relished
Lathyrus humilis (Ser.) Spreng.	Kaown	Seeds and pods eaten raw
Trifolium repens L.	Shaftal/Kut	Leaves are consumed raw
Trigonella emodi Benth.	Kuchona	Tender shoots used as pot herb (vegetable)
Trigonella foenum-graecum L.		Cultivated in kitchen gardens; tender shoots/leaves used as pot herb

Besides the above-mentioned wild plants, a number of other species are cultivated as crops. The major cultivated species from the family include *Phaseolus coccineus*, *P. vulgaris*, *Pisum sativum*, *P. s* var. *pratense*, *Trifolium repens*, *T. pratense*, *Vicia faba*, *Medicago sativa*, *M. falcata*, *M. lupulina*, *Melilotus albus*, *M. indica*, *M. officinalis* and *Vicia faba*.

3.5 Other Ethno Botanical Uses

Besides, the above-mentioned uses, the other plants of Leguminosae that find use in the day to day life of the communities are given in Table 5.

The use of plants described above showcase how native communities have intrinsic value on all living organisms and their natural environment; reflected by the way they deal with nature and further supplemented by their attitude toward this natural environment.

3.6 Ecology and Desertification

Various plant communities have been an integral part of the cold deserts for millennia and thus the ecology of this unique biome remains an important part. Biodiversity plays an important role in this ecosystem as most species that live and thrive in these environments do not exist in other biomes. However, this arid biome is facing threat from various factors, including climatic variations and human activities leading to desertification. This desertification is exacerbated by human exploitation of ecosystems, causing land degradation, soil erosion and loss of productive ecosystems and their biodiversity. As regards the ecology and desertification of the cold desert, the genus *Caragana* plays a pivotal role in the highlands for checking soil erosion and

Table 5 Uses of some Fabaceae plants

Usage	Plant names
Flowers in rituals	*Astragalus candolleanus* Royle ex. Benth., *Robinia pseudoacacia* L., *Sophora moorcroftiana* (Benth.) Baker
Agricultural implements/basketry	*Colutea nepalensis* Sims, *Robinia pseudoacacia* L., *Sophora moorcroftiana* (Benth.) Baker
Cattle care (veterinary)	*Caragana brevispina* Royle (increase lactation); *Cicer microphyllum* L. (mouth sores)
Soap	Fine paste made from leaves of *Astragalus grahamianus* Royle ex Benth.
Miscellaneous	*Astragalus rhizanthus* Royle ex Benth., for the manufacture of paper; *Trigonella emodi* Benth., used in Lahaul as an aromatic and insect repellent

Fig. 7 *Caragana* spp.

at the same time a vital source of fuelwood for winter use storage. It is a common sight to see patchy dark green mosaics in the highlands and high passes of Spiti and the Rupshu highlands in Ladakh. The species *C. versicolor* (*C. pygmaea* Baker) supports typical scrub steppe vegetation on high pasturelands with altitudes ranging between 4000 and 5000 m (Rawat and Adhikari 2002), alongside species of *Eurotia* and *Oxytropis* aided by sedges (*Carex* and *Kobresia*). However, this plant community is facing threat from various anthropogenic pressures like fuelwood extraction, rising tourism and the vagaries of nature in the form of climate change and global warming. These plant communities are also the backbone of the nomadic Changpas who herd their livestock in the Rupshu and Changthang highlands of Ladakh and share it with the wild herbivores like Ibex, Wild Ass, Blue Sheep, Yaks, etc. Thus, the increasing livestock population and fast degrading *Caragana* plant communities have resulted into a key conservation issue under desertification (Fig. 7).

Caragana plant species can easily be considered as a keystone species of high conservation concern in the arid cold desert landscape. Besides, its role in binding the loose soil strata, it provides habitat to a number of birds, insects, and small mammals. *C. versicolor* as a species of high conservation concern has also been studied by various researches (Rawat 2008; Kumar et al. 2016). Apart from animal fodder and their role in soil stabilization, the whole plant is extracted by local inhabitants and transhumant pastoralists for fuel primarily because of the lack of other energy sources in the region (Rawat 2008). Some studies on certain species (like *Caragana nubigena* (now *Chesneya nubigena* (D. Don) Ali), have been reported to moderate the physical environment, thus providing a micro-habitat suitable for colonization by herbaceous communities of grasses and forbs (Jacquez and Patten 1996).

However, of late, the sudden influx of tourists to remote locations of this barren landscape has put pressures on its natural resources, including the clearing of *Caragana* shrubs for tenting and more use of fuelwood for cooking. This destructive and excessive removal of these particular plant *taxa* has led to site degradation, resulting in soil erosion, the major cause for desertification. It is well documented that, native plants play an important role in slope stabilization and checking soil erosion. Herbaceous members from Leguminosae that play a major role in soil scree stabilization include *Chesneya nubigena, C. cuneata, Cicer microphyllum, Hedysarum microcalyx, Oxytropis cachemiriana, O. densa, O. humifusa, O. hypoglottoides, O. lapponica, O. microphylla, O. tatarica, Thermopsis inflata* besides members of the genera *Caragana* and *Astragalus* as discussed earlier.

4 Final Discussion and Conclusions

Even as fairly comprehensive documentation of the species diversity of the plants of the north-west Himalayan cold deserts is available, the population density of various species and their conservation status remains poorly assessed even today. Wild populations of many plant species of the region are on a continuous decline due to various pressures including those on account of unsustainable exploitation and development of infrastructure. Most of the shrubby vegetation in the region, including members of Leguminosae, forms a ready browse for the domestic as well as wild animals, which relish their leaves and tender shoots; whereas their underground parts are extracted by the local communities, nomadic graziers, and tourist campers for fuel. Both these factors severely impact their regeneration, which is so crucial for sustenance of the wild fauna as well as domestic herbivores and for local pastoral communities. Along with the increasing biotic interference, the fragile ecosystem of this region is also highly vulnerable to degradation due to climate change and global warming.

Considering the role of native flora, especially the Leguminosae, in the local socio-economy and in maintaining the fragile ecological stability in the region, it becomes prudent to look into their conservation aspects and balance it with community needs. Sustainable utilization of the natural resources in the cold deserts is a complex issue that encompasses local community needs, their cultural values, the ever-rising anthropogenic pressures in the form of rising population, defense expansion, other developmental activities, including tourism and also the change in local climate. Fodder and fuelwood being the mainstay of local communities as well as for the native fauna is renewable and its utilization can only be sustained if the rate of use does not exceed its growth/regeneration rate.

Ecosystem services and the potential of indigenous traditional knowledge of cold desert plants in general and those of Leguminosae in particular are not valued as it should be. The contribution of Leguminosae in the Himalayan cold desert regions of Ladakh and Lahaul and Spiti should be studied for various parameters to put a value to their contribution. The aim here is to review the entire work carried out on Leguminosae in the cold deserts of India and to portray at one place the scattered information on the subject for better fruitful understanding.

Such studies involving flora of a region are indispensable for all botanical progress in any part of the globe. More importantly, such documentations provide a ready reference on the plant wealth of a defined geographical area. It acts as a mine of information on its constituent *taxa*, describing their nomenclature, economic utility, conservation status, as well as the overall ecosystem services rendered. Such revisions and studies on a particular group or family of plants are an important means of understanding their detailed information from a particular region and also aids in understanding floristics and other botanical studies.

References

Abbas Z, Khan S, Alam J et al (2017) Medicinal plants used by inhabitants of the Shigar Valley, Baltistan region of Karakorum range-Pakistan. J Ethnobiol Ethnomed 13:53. https://doi.org/10.1186/s13002-017-0172-9

Ahmad S, Singh J, Verma D (2015) Inventory of important fodder plant species of Ladakh Himalaya, India. 23rd International Grassland Congress at New Delhi

Airy-Shaw HK (1966) A dictionary of the flowering plants and ferns In: Willis JC (revised 7th ed. by Airy-Shaw). Cambridge

Ali SI (1977) Papilionaceae. In: Nasir YJ, Ali SI (eds) Flora of (West) Pakistan 100—Karachi

Angmo K (2012) A study on ethno-flora with special reference to traditional health care system in Western Ladakh, Jammu and Kashmir. Ph D. thesis, Forest Research Institute University (Deemed), Dehradun, Uttarakhand, p 238

Aswal BS, Mehrotra BN (1994) Flora of Lahaul-Spiti (A cold desert in north west Himalaya). Bishen Singh Mahendra Pal Singh, Dehradun, India, pp 10–15

Ballabh B, Chaurasia OP (2011) Herbal formulations from cold desert plants used for gynecological disorders. Ethnobotany Res Appl 9:059–066

Ballabh B, Chaurasia O, Pande PC (2007) Raw edible plants of cold desert Ladakh. Ind J Tradit knowl 6

Ballabh B, Chaurasia OP, Ahmed Z, Singh SB (2008) Traditional medicinal plants of cold desert Ladakh-used against kidney and urinary disorders. J Ethnopharmacol 118(2):331–9

Bhattacharyya A (1991) Ethnobotanical observations in the Ladakh Region of Northern Jammu and Kashmir State, India. Econ Bot 45(3):305–308. https://doi.org/10.1007/BF02887067

Buth GM, Irshad A Navchoo (1988) Ethnobotany of Ladakh (India) plants used in health care. J Ethnobiol 8(2):185–194

Chandra Sekar K, Srivastava SK (2009) Flora of the Pin Valley National Park, Himachal Pradesh. Botanical Survey of India, Calcutta

Chaudhary L, Rana T, Anand K (2008) Current status of the systematics of *Astragalus* L. (Fabaceae) with special reference to the Himalayan Species in India. Taiwania 53(4):338. https://doi.org/10.6165/tai

Chaurasia OP, Singh B (1996a–b, 1997, 1998, 2001) Cold desert plants 1–5. Leh

Chaurasia OP, Khatoon N, Singh SB (2008) Field guide—Floral Diversity of Ladakh; World Wide Fund for Nature-India. Jammu and Kashmir. Department of Wildlife Protection; Defence Institute of High Altitude Research, DRDO

Chaurasia OP, Singh B (1996–2001) Cold desert plants. Field Research Laboratory, DRDO, Leh, pp 1–4

Chaurasia OP, Ballabh, Basant (2011) Herbal formulations from cold desert plants used for gynaecological disorders. Ethnobot Res Appl 9. https://doi.org/10.17348/era.9.0.59-66

Dhar U, Kachroo P (1983) Alpine flora of Kashmir Himalaya. Jodhpur

Dickoré WB, Miehe G (2002) Cold spots in the highest mountains of the world. In: Körner C, Spehn EM (eds) Diversity patterns and gradients in the flora of the Karakorum. Mountain Biodiversity: A global assessment, London, pp 129–147

Fabian D, Garschall K, Klepsatel P, Santos-Matos G, Sucena E, Kapun M, Lemaitre B, Schlötterer C, Arking R, Flatt T (2018) Evolution of longevity improves immunity in Drosophila. Evolut Lett 2. https://doi.org/10.1002/evl3.89

Grubov VI (2003) Plants of Central Asia-plant collection from China and Mongolia vol 8a, Leguminosae. CRC Press, p 184

Gurmet P, Chaurasia OP, Singh B, Singh DV (1998) Some traditional medicinal plants of Khardungla and Changla (Ladakh) used in Amchi medicine. Bull Med-Ethnobot Res 163

Hooker JD (1872–1897) Flora of British India. L. Reeve & Co., London, pp 1–7

Hooker JD (1879) The Flora of British India. Salicaceae to Cornaceae, Published under the Authority of the Secretary of State for India in Council, London, vol 2, pp 115–117

Hutchinson J (1964) The genera of flowering plants. Dicotyledones. vol 1. Oxford University Press, London, p xi+516 pp

Jacquez GM, Patten DT (1996) Mt Res Develop 16(3):265–273

Jishtu V, Subramani SP, Kapoor KS, Goraya GS (2003) Medicinal plants from the cold deserts of north-West India. Non Timber Forest Products of India (Nautiyal, S. and Kaul, A. K. eds.). Jyoti Publishers and Distributors, Dehradun, pp 59–94

Jishtu V (2005) Studies on the Floristics and Associated Mycorrhiza of Baspa Valley of Kinnaur, Himachal Pradesh. PhD Thesis, 393 pp

Judd WS, Campbell CS, Kellogg EA, Stevens PF (1999) Plant systematics: a phylogenetic approach. Sinauer Associates, Sunderland, p 464

Kachroo P, Bansilal S, Dhar U (1977) Flora of Ladakh: an ecological and taxonomical appraisal. International Book Distributors, Dehradun

Kala CP (2000) Status and conservation of rare and endangered medicinal plant in the Indian trans-Himalaya. Biol Conserv 93:371–379

Kala CP (2006). Medicinal plants of high altitude cold deserts in India: diversity, distribution and traditional uses. Int J Biodivers Sci, Ecosyst Serv & Manage 2:43–56

Kala CP (2011) Floral diversity and distribution in the high altitude cold desert of ladakh, India. J Sustain Forest 30:360–369. https://doi.org/10.1080/10549811.2011.534036

Klimes L (2003) Life-forms and clonality of vascular plants along an altitudinal gradient in E Ladakh (NW Himalayas). Basic App Ecol 4:317– 328. https://doi.org/10.1078/1439-1791-00163

Klimes L, Dickoré WB (2005) A contribution to the vascular plant flora of Lower Ladakh (Jammu & Kashmir, India). Willdenowia 35:125–153. ISSN 0511-9618

Kumar A, Adhikari BS, Rawat GS (2016) Caragana versicolor Benth. (Fabaceae), a keystone species of high conservation concern in the Hindu Kush Himalayan region. Cur Sci 111(6):985–987

Lock JM, Simpson K (1991) Legumes of west Asia, a checklist. Royal Botanic Gardens, Kew, India, pp 96–102

LPWG (2017) A new subfamily classification of the Leguminosae based on a taxonomically comprehensive phylogeny. Taxon 66:44–77. https://doi.org/10.12705/661.3

Maassoumi AA (1998) *Astragalus* L. in the World, check list. Research Institute of Forests and Rangelands, Tehran, Iran, 618 pp

Mabberley DJ (1997) The Plant-Book, a portable dictionary of the vascular palnts, edn 2. Cambridge University Press, Cambridge, UK, 858 pp

Melchior H (1964) A Engler's syllabus der pflanzenfamilien, vol II. Gebruder Borntraeger, Berlin

Namtak S, Sharma RC (2018) Medicinal plant resources in Skuru watershed of Karakoram wildlife sanctuary and their uses in traditional medicines system of Ladakh, *India*. Int J Complim Altern Med 11(5):294–302

Nautiyal M, Tiwari P, Tiwari JK, Rawa DS (2018) Fodder diversity, availability and utilization pattern in Garhwal Himalaya, Uttarakhand. Plant Archives 18(1):279–287

Polunin O, Stainton A (1984) Flowers of the Himalaya. Oxford University Press, Delhi

Rana PK, Kumar P, Singhal VK, Rana JC (2014) Uses of local plant biodiversity among the tribal communities of Pangi Valley of District Chamba in cold desert Himalaya. Sci World J, India, p 753289

Ranjbar M, Karamian R (2002) Astragalus sect. Astragalus (Fabaceae) in Iran, complementary notes with a key to the species. Nord J Bot 22:177–181

Rawat GS (2008) ENVIS bulletin: wildlife and protected areas, vol 11, no 1. Wildlife Institute of India, Dehradun, India, pp 1–7

Rawat GS, Adhikari BS (2002) Vegetation characteristics and patterns of livestock grazing in Changthang plateau, eastern Ladakh. Report submitted to International Centre for Integrated Mountain Development, Kathmandu, Nepal, and Wildlife Institute of India, Dehra Dun, Uttaranchal, India, p 26

Rawat RS, Jishtu V, Kapoor KS (2009) Medicinal and aromatic plant diversity of Himalayan cold desert with reference to Spiti valley of North-West Himalayas. Ind Forester 135(7):891–904

Samant SS, Singh Ashok, Lal Manohar, Sharma Pankaj (2011) Diversity, distribution and conservation prioritization of economically important species in Lahaul valley, North-Western Himalaya, India. Global change. Biodiversity and livlihoods in cold desert region of Asia. BSMPS, Dehradun, India, pp 31–42

Saroya Singh A (2017) Ethnobotany. Indian Council of Agricultural Research, New Delhi. ISBN: 978–81-7164-168-0

Schrire BD, Lavin M, Lewis GP (2005) Global distribution patterns of the Leguminosae: insights from recent phylogenies. Biol Skr 55:375–422

Sharma VK, Dwivedi KS, Tripathi D, Ahmed Z (2006) Status of available major and micro-nutrients in the soils of different blocks of Leh district of cold arid region of Ladakh in relation to soil characteristics. J Ind Soc Soil Sci 54:248–250

Singh A, Lal M, Samant SS (2009) Diversity indigenous uses and conservation prioritization of medicinal plants in Lahaul Valley, proposed Cold Desert biosphere Reserve. Ind Intl J Biodiv Sci Mgmt 5(3):132

Sood SK, Nath R, Kalia DC (2001) Ethnobotany of cold desert tribes of Lahoul-Spiti (North west Himalaya). Deep Publication, New Delhi, pp 4, 123

Srivastava SK, Shukla AN (2015) Flora of cold desert, Western Himalaya, India 2. Botanical Survey of India

Stewart RR (1972) An annotated catalogue of the vascular plants of West Pakistan and Kashmir. Fakhri Press, Karachi, Pakistan

Stewart RR (1916) The flora of Ladakh, Western Tibet I. Discussion of the flora. Bull Torrey Bot Club 43:571–590

Stewart RR (1917) The flora of Ladakh, Western Tibet II. List of Ladakh plants. Bull Torrey Bot Club 43:625–650

Young James A, Jerry D, Buddy, Evans AR (1989) Use of shrubs for fuel. In: McKell CM (eds) The biology and utilisation of shrubs. Academic Press, San Diego, p 656

Abiotic Stress Responses and Tolerance

Morphological, Physiobiochemical and Molecular Adaptability of Legumes of Fabaceae to Drought Stress, with Special Reference to *Medicago Sativa* L.

Akbar Hossain, Muhammad Farooq, Ayman EL Sabagh, Mirza Hasanuzzaman, Murat Erman, and Tofazzal Islam

Abstract Drought stress (DS) is one of the most hostile limitations for sustainable crop production. Developing DS-tolerant crop cultivars and the use of better crop management practices may help improve crop performance under drought. In this chapter, the adverse effect of drought on the growth and development of legumes and the morphological, physiobiochemical, and molecular basis of adaptability to drought are described. Under drought, overproduction of reactive oxygen species causes oxidative damage. The role of osmolytes and antioxidants in countering the oxidative damages has been widely described. Moreover, "omics-based approaches," such as proteomics, metabolomics–transcriptomics, and genomics are promissory approaches to identify drought-tolerant genes, decode complex gene networks, and numerous signaling cascades involved in drought tolerance in legumes. The recently

A. Hossain (✉)
Bangladesh Wheat and Maize Research Institute (BWMRI), Dinajpur 5200, Bangladesh
e-mail: akbarhossainwrc@gmail.com; tanjimar2003@yahoo.com

M. Farooq
Department of Crop Sciences, College of Agricultural and Marine Sciences, Sultan Qaboos University, Al-Khoud-123, Muscat, Oman

Department of Agronomy, University of Agriculture, Faisalabad 38040, Pakistan

The UWA Institute of Agriculture and School of Agriculture, & Environment, the University of Western Australia, Perth, WA 6001, Australia

A. EL Sabagh
Department of Agronomy, Faculty of Agriculture, Kafrelsheikh University, Kafr El-Shaikh 33516, Egypt

M. Hasanuzzaman
Department of Agronomy, Faculty of Agriculture, Sher-e-Bangla Agricultural University, Dhaka, Bangladesh

A. EL Sabagh · M. Erman
Department of Field Crops, Faculty of Agriculture, Siirt University, Siirt, Turkey

T. Islam
Institute of Biotechnology and Genetic Engineering (IBGE), Bangabandhu Sheikh Mujibur Rahman Agricultural University, Gazipur 1706, Bangladesh

M. Hasanuzzaman et al. (eds.), *The Plant Family Fabaceae*,
https://doi.org/10.1007/978-981-15-4752-2_11

developed CRISPR-Cas technology has already been used in precision breeding of many plants including the members of Fabaceae such as alfalfa is also discussed in the chapter.

Keywords Alfalfa · Drought stress · Omics approaches · Antioxidative protection · Ascorbate-glutathione cycle

Abbreviations

ABA	Abscisic acid
APX	Ascorbate peroxidase
AsA	Ascorbic acid
CAT	Catalase
CS	Compatible solutes
DHA	Dehydroascorbate (oxidized ascorbate)
DHAR	Dehydroascorbate reductase
DE	Drought escaping
DS	Drought stress
GB	Glycine betaine
GPX	Glutathione peroxidase
GR	Glutathione reductase
GPX	Glutathione peroxidase
GST	Glutathione-S-transferase
H_2O_2	Hydrogen peroxide
JA	Jasmonic acid
MDA	Malondialdehyde
MDAR	Monodehydroascorbate reductase
MDHAR	Monodehydroascorbate reductase
$^1O_2$	Single oxygen
$O^{2\bullet-}$	Uperoxide anion free radical
OH•	Hydroxyl-free radical
PEG	Polyethylene glycol
POD	Peroxidase
RLD	Root length and density
RDW	Root dry weight
RO	Alkoxy radical
ROS	Reactive oxygen species
RSR	Root shoot ratio
RT-qPCR	Real-time quantitative polymerase chain reaction
RWC	Relative water content
SOD	Superoxide dismutase
TCA	Trichloroacetic acid

1 Introduction

The fast-growing global population is estimated to rise by 9.7 billion by the year 2050 (Varshney et al. 2011). Therefore to meet the food demand of the increasing population, it is anticipated that food production, including animal-based proteins, should be increased by about 70% by the year 2050 (Fischer 2009; Searchinger et al. 2019). Furthermore, the global agricultural land area needs to be 593 million hectares (double to the area of the country India) to produce the desired crop and pasture yields (Blum 2013; Searchinger et al. 2019). In fact, ten most populated countries in the world have already faced a decreasing trend in per capita arable land (Gomiero 2016; Smith et al. 2016). At the same time abiotic stresses particularly drought is one of the greatest significant limitations which severely affect the crops' productivity leading to substantial yield losses globally (Li et al. 2014). The IPCC (2014) already reported that the declining rate of agricultural productivity has occurred as a result of the hostile effects of DS, which postures a thoughtful threat to agricultural production (Anjum et al. 2011a, b; Zandalinas et al. 2018).

The antagonistic effects of DS on grain yield and quality of crops are complex and depending on the inherent physiobiochemical activities of plants including the assimilation and accumulation of plants' nutrition, and their remobilization to various plants' organs (Farooq et al. 2017a, b; Prasad et al. 2017; Sehgal et al. 2018). Under stressful conditions, plants build physiobiochemical mechanisms to subsist against stress (Kathuria et al. 2009; Hasanuzzaman et al. 2012a, b; 2013a, b, c). Generally, most plants accrue water-soluble compounds known as "low molecular weight compatible solutes (CS)" to survive against the adverse effect of reactive oxygen species (oxidative stress; ROS) (Hasanuzzaman et al. 2013a, c). Accumulation of these compounds under harsh environments protect plants through adjusting cellular osmotic pressure, detoxifying ROS, stabilizing enzymatic activity, and also protecting membrane integrity (Gill and Tuteja 2010; Singh et al. 2015). Among CS, glycine betaine (GB) is the most important one which protectants the adverse effect of ROS (Fariduddin et al. 2013; Gill et al. 2014). This chapter highlights and deliberates the morphological, physiological, biochemical, and molecular adaptability of legumes, particularly the legume alfalfa (*Medicago sativa* L.) to DS.

2 Plant Alfalfa and Its Importance

The legume alfalfa belongs to the family of Fabaceae (Campanelli et al. 2013). It is originating in Transcaucasia, Turkmenistan highlands, Iran and Asia Minor. The "alfalfa" refers to the main species *M. sativa* subsp. *sativa* as it is the hybrid between subsp. *sativa* and subsp. *Falcate.* Since, alfalfa is closely related to *M. sativa* in morphologically. Alfalfa is a perennial predominantly cross-pollinated plant (Radovic et al. 2009). It is called "the queen of forages" due to its high yield perspective with decent herbage quality and also resistance to recurrent cutting (Radovic et al. 2009;

Shi et al. 2017a, b). As it is a perennial forage and a fairly hardy species, it has numerous advantages including wide adaptability, a relatively high level of abiotic stresses including cold resistance, heat, drought, heavy metal, and salinity. Drought affects almost all phases of the plant ranging from germination of the seed to the maturity stage of plants (Xu et al. 2014; Nadeem et al. 2018, 2019). Since, the alfalfa has a deep roots system, which helps the plant to survive under long-term drought without any damage to its re-growing process (Safarmejad 2008).

3 The Adverse Effect of Drought on the Growth and Development of Alfalfa

Drought is the most erratic restriction with significant hostile effects on crop production worldwide (Golldack et al. 2014; Anjum et al. 2017; Hussain et al. 2018), particularly in the arid and semiarid regions. The sensitivity of plant to DS is a composite singularity that depends on the growth stage of the affected plants, genetic makeup, level, duration, and extent of the harshness of DS (Zhu 2002; Hamidi and Safarnejad 2010; Yousfi et al. 2016; Nadeem et al. 2019). The adverse effects of drought are linked to the increased the oxidative damage, changes in leaf gas exchange and decreasing the carbon assimilation rate and turgor, affect the leaf development, enzymatic activity, ion balance and thus leading to a decrease in the yield of plants (Nadeem et al. 2019). During drought conditions, root and stem length of alfalfa were decreased (Castroluna et al. 2014). While Castroluna et al. (2014) observed that stem mid-node length and leaf size of alfalfa at the seedling stage were decreased drastically under water-deficit condition (drought stress), which lead to hamper the growth and development, the final yield of alfalfa. Since, the alfalfa has a deep root system and has the ability to osmotic regulation, which helps the plant to survive under long-term drought (Safarmejad 2008; Hamidi and Safarnejad 2010; Castroluna et al. 2014; Yousfi et al. 2016).

4 Physiobiochemical Adaptability of Alfalfa Against Drought Stress

Stress-tolerant plants exhibit several adaptation mechanisms (morphological, physiobiochemical, and inherent genetic makeup) to persist against water-deficit conditions. Among them, the most important mechanisms are discussed below.

4.1 Inherent Characters to Escape or Avoid Drought

The primary adaptation mechanism of plants under drought stress (DS) is escaping which consists of faster growth for enabling a quick ended the life span before the beginning of water scarcity. Legumes including alfalfa can escape the DS by shortening their life span to evade DS and also can retain higher water in their tissue through enlightening water-holding capacity (Siddique et al. 1993). The area where terminal DS predominates but the crops growing season is shorter, plants could escape drought if phenological stages of growing crops are positively matched with periods of available soil moisture status (Farooq et al. 2014). For example, (Hall 2012) found that traditional legumes including cowpea and alfalfa flowering time in the Sahara desert of Africa is matched with the time of termination of the monsoon (rainy season), as a survival tactic to survive under drought. Early flowering and seed set prior to DS events are also important strategies to escape DS (Shavrukov et al. 2017). Calvache et al. (1997) observed that alfalfa, common bean, and cowpea can alleviate the negative impact of DS through generating new plants' structures during the phase of DS. For example, legume alfalfa inherently is a recurrent growth habit with a deep-root system, and thus, it has better phenotypic structure or competence to tolerate water-deficit stress than the plants which have shallow-root systems (Chowdhury et al. 2016). However, Calvache et al. (1997) found that when DS occurs at the earlier growth stages of crops, plants escape the DS through producing osmolytes and high water use efficiency.

4.2 Screening Potential Traits for Drought Tolerance

A potential trait of legumes which is positively correlated with DS tolerance is root architecture that might be considered in breeding programs to improve DS-tolerant cultivars, as this character is supposed to be controlled by a single gene (Vadez et al. 2008; Sofi et al. 2018). In another study, roots and root nodules are found key features and essential sensors to adapt legumes to drought (Kunert et al. 2016). Therefore, during screening for DS-tolerant legumes, the architecture of roots and root nodules along with the documentation of genes, proteins, and metabolites that are answerable for DS tolerance should be considered.

Generally, perennial herbs like alfalfa have healthier root: shoot ratios and a wider-lateral and deeper-root systems than annual crops (Schenk and Jackson 2002). Since, plants with deeper-root systems are capable to uptake water and nutrients from deep soil horizons, which ultimately balance nutrients water and nutrients balance in different soil horizons, particularly under dryland farming systems (Cocks 2001; Dear et al. 2003; McCulley et al. 2004; Del Pozo et al. 2017). Similarly, as a perennial and deep-rooted crop, alfalfa can able to persist against soils water deficit and can improve soil aggregation and water infiltration (McCallum et al. 2004; Masri and John Ryan 2006; Dear and Ewing 2008; Li et al. 2008, 2010) and also reinstates soil nitrogen, carbon, and microbial activity (Ward et al. 2006; Li et al. 2012). Since, the alfalfa has deep root system and ability to perform osmotic regulation, alfalfa is able to achieve high yield and persists long dry periods under the Mediterranean environments (Humphries and Auricht 2001; Dear et al. 2003; Benabderrahim et al. 2015) and more resilient to the changing climate conditions (Ghahramani and Moore 2013). For example, Hamidi and Safarnejad (2010) tested few years ago six alfalfa genotypes for measuring the percentage of germination, root, and shoot growth under various osmotic potential (DS). They observed a significant decrease in % germination, radicle length and also decreased seed vigor, while the ratio of radicle and plumule was improved under higher DS. Among six genotypes of alfalfa, genotypes "Yazdi" and "Maupa" had a maximum % germination under severe DS which indicates that these cultivars were tolerance to DS at the seed germination stage than other genotypes (Hamidi and Safarnejad 2010). Another research conducted by Ramamoorthy et al. (2017) with 12 genotypes of the legume chickpea aimed to find out the genotypes which are tolerant to drought through observing the root characters such as root length and density (RLD), root dry weight (RDW), and root:shoot ratio (RSR) and found a significant variation between RLD, RDW, deep RDW, and RSR under drought conditions. Finally, the study revealed that genotypes with maximum RLD and RDW may be the best selection criteria for drought tolerance genotype (Ramamoorthy et al. 2017). Additionally, early maturing and deep-root legumes such as chickpeas and pigeon peas were found drought tolerant (Upadhyaya et al. 2012). Therefore, potential characters accompanying with DS tolerance in legumes could be the greatest encouraging way for using in breeding programs for improvement of DS tolerance cultivars.

4.3 Accumulation Compatible Solutes (CS) Under Drought Stress (DS)

Stress-tolerant plants accumulate compatible solutes under DS for osmoprotection which is an important tactic to survive under DS (Nadeem et al. 2019). The osmoprotection has a closer connection with non-toxic elements with abundant constituents of the cell, while osmotic regulation contributes to retaining cell turgor pressure by maintaining the water contents in the affected cells (Slama et al. 2015). Initially, compatible solutes accumulate in the cells that are affected by DS without inquisitive with the macromolecules as well as hydroxyl compounds or nitrogen-containing compounds (Majumdar et al. 2016). Among CS, non-reducing sugars (di-, tri-, and tetra-saccharides) contribute to upholding the cell membranes integrity under DS (Ramanjulu and Bartels 2002), where mannitol promote the stabilization of the structure of macromolecules including phosphoribulokinase, glutathione, thioredoxin, ferredoxin, and also scavenge-free hydroxyl radicals (Bhauso et al. 2014). Membrane lipids (trehalose), protein, and cell-biological structures lead to progress the activity of photosynthetic (Ibrahim and Abdellatif 2016; Khater et al. 2018). Besides these CS, proline acts as a signaling compound to adjust mitochondria function, retains integrity of cell membrane through reducing lipids' oxidation, scavenge-free hydroxyl radicals and also activates the particular genes for stress recovery (Amede and Schubert 2003; Solanki and Sarangi 2015; Shinde et al. 2016).

4.4 Antioxidant Defense Against Oxidative Stress

Stress-induced oxidative stress means the production of reactive oxygen species (ROS) such as alkoxy radical (RO), hydroxyl radical ($^{\bullet}OH$), hydrogen peroxide (H_2O_2), superoxide radical ($O_2^{\bullet -}$), and single oxygen ($^1O_2$) which damage macromolecules and cell membrane (Farnese et al. 2016); since ROS acts as a signaling apparatus to stimulate defense mechanisms against stress (Choudhury et al. 2017). Too much production of ROS in plant cells under DS causes oxidative stress, leads to the destruction of the cell's fundamental components such as cell lipids, proteins, and nucleic acids, finally death of all affected cells (Kurutas 2016). Chakrabarty et al. (2016) observed that enzymatic antioxidants, i.e., catalase (CAT), superoxide dismutase (SOD), ascorbate peroxide (APX), glutathione peroxidase (GPX), monodehydroascorbate reductase (MDHAR), dehydroascorbate reductase (DHAR), and glutathione reductase (GR), and non-enzymatic antioxidants include ascorbate, glutathione, tocopherols, ascorbic acid, phenolics and carotenoids regulate the defensive mechanism against the excessive production of ROS in the plant cells. Al Hassan et al. (2017); Sahitya et al. (2018) revealed that stress-tolerant plants can stabilize or production of a greater concentration of antioxidants or antioxidant enzymes to make a defensive mechanism against DS. Nadeem et al. (2019) found that ABA, JA (jasmonic acid), and Ca^{2+} were triggered in the plants' cell under the water-deficit environment (Fig. 1). They also observed that DS prompted the biosynthesis of JA and ABA which leads to the transcription of the ion carrier genes. Overexpression of transcription factors such as GmNACs, WRKY, AP2/ERF, DREB, MYB, and ZIP was also recorded in the plant cell under drought (Nadeem et al. 2019).

It is also observed that (Fig. 1) under stressful conditions enzymatic antioxidant SOD recovers the adverse effect of superoxide radicals and H_2O_2 (Nadeem et al. 2019), while APX assistances to create $NADP^+$ and also helps to alter H_2O_2 to H_2O (Zoz and Castagnara 2013), whereas other enzymatic antioxidants such as GR and DHAR assist to provide a substrate for mitigating the adverse effect of stress-induced oxidative stress. During stress-induced oxidative stress, the concentration of antioxidants was increased in the cells of plants tolerant to DS, as observed in alfalfa (Hamidi and Safarnejad 2010), green bean (Yasar et al. 2013), pea (Osman 2015), soybean (Guler and Pehlivan 2016), and chickpea (Patel et al. 2011). The enzymatic antioxidants such as SOD, APX, GR, GST, GPX, and POD activities were increased under DS in horse gram and common bean also observed by Saglam et al. (2011), Bhardwaj and Yadav (2012).

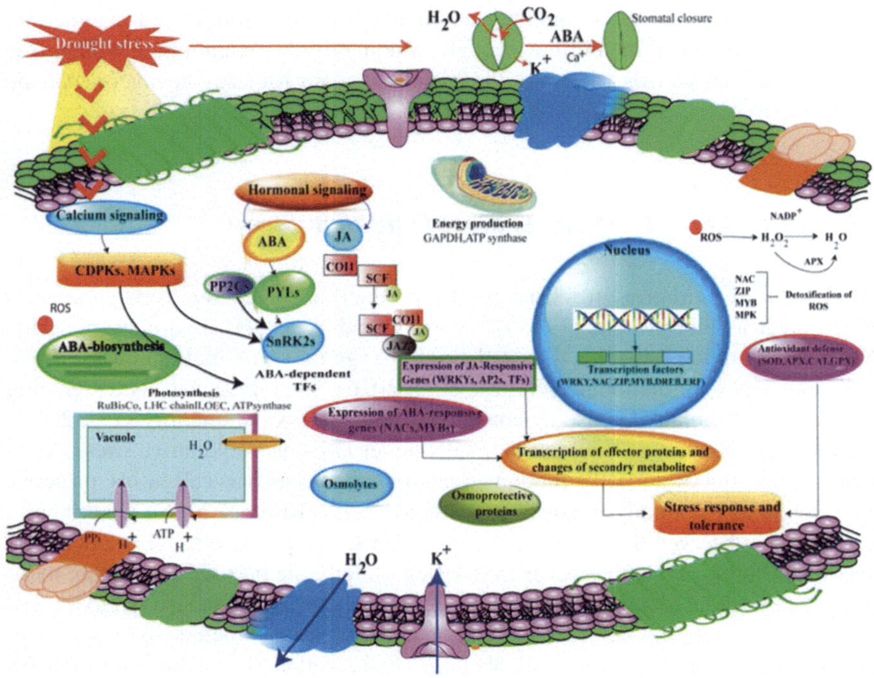

Fig. 1 Drought tolerance mechanism in legumes under the family Fabaceae. ABA, abscisic acid; JA, jasmonic acid; and ROS, reactive oxygen species (Adapted from Nadeem et al. 2019)

4.5 Regulation of Phytohormone During Drought Stress

All parts of plant growth (under normal and stressful conditions) are fully controlled by several hormones such as gibberellins, cytokinins, auxins, ABA, and ethylene (Ullah et al. 2018). For example, Bielach et al. (2017) observed that under DS the phytohormone cytokinin level was risen in xylem sap for stimulating stomatal opening through lessening its sensitivity to ABA (Bielach et al. 2017). Weyers and Paterson (2001) observed that ethylene and ABA hormones were increased in plants under water deficit, while the concentration of cytokinins, gibberellins, and auxin was decreased. The increasing level of ABA concentrations in plant cells is due to the decreasing trend of ABA catabolism that hampers the entrance of ABA from the phloem and rhizosphere (Hartung et al. 2002). For example, Miyashita et al. (2005) found that an increase in ABA concentration prompted the declined trend of the stomatal conductance in kidney bean under re-watering (Miyashita et al. 2005). Hydraulic conductivity in the roots of the plant is controlled by ABA hormone which is positively linked to water uptake and transport (Merilo et al. 2015; Park et al. 2017). The ABA also controls the formation of ROS such as $O_2^{\bullet-}$ and H_2O_2, through promoting the activities of GR (enzymatic antioxidant) (Park et al. 2017). Plant hormone

JA also has a crucial role to alleviate DS (Fig. 1; Nadeem et al. 2019). The JA is linked with other hormones to improve drought tolerance ability of plants through improving root architecture improvement, scavenging of free radicals of ROS, and also controlling stomatal conductance (Mohamed and Hanan 2017).

5 Molecular and Biotechnological Approaches for Drought Tolerant

To enhance the productivity of legume under water-deficit conditions is crucial for the improvement of DS-tolerant genotypes. The combined use of improved up-to-date tools with traditional breeding procedures may be also useful for the improvement of DS-tolerant crop cultivars in the scenario of climate change. Therefore, to fulfill the food demand of increasing population under the changing climate, breeding, molecular, and transgenic approaches may help to improve drought resistance legume cultivars. A description of breeding, molecular, and transgenic biotechnological approaches are discussed in the following subsections.

5.1 Development of Drought-Tolerant Cultivars Through Conventional Breeding

Development of DS tolerance in plants by conventional breeding is a convenient tactic and a foremost strategic for crop improvement (Farooq et al. 2014). While development of variety through the conventional breeding is a more time-consuming procedure than biotechnological and genetic modification strategies (Zheng et al. 2017). In the breeding process, the quantitative genetic basis is frequently inadequate and also lacking physiological knowledge underlying the response to drought (Torres et al. 2010). It also needs a substantial heritable variation, which is sometimes absent due to high genotype and environment interactions. Besides, improved categorization of the environment is a precondition to improve the usefulness of target traits (Mir et al. 2012). Some legumes have a root architecture adequate to survive under water-deficit conditions. For example, alfalfa, cowpea, chickpea, common bean, and soybean have fibrous, densely and depth root systems for avoiding/escape drought (Khan et al. 2010; Duc et al. 2015).

Traits as cooler canopies, high stomatal conductance (Duc et al. 2015), early flowering and maturity (Duc et al. 2015) have been linked with escaping the DS. Approaches as canopy spectral reflectance as a high-throughput phenotyping technique (Montes et al. 2007; Chapman 2008) enable quick and easy measurements for assortment of the cultivars against DS (Duc et al. 2015). Plant canopy temperature is also an extensively relevant selection criterion that is connected with canopy conductance at the vegetative stage and also linked with plant water status (Hou et al.

2018). Besides these, infrared thermography (thermal infrared imaging) may also be engaged to select legumes for DS resistance as the trait is linked with leaf canopy temperature (Martynenko et al. 2016). Alongside these criteria, late leaf senescence may be helpful for better grain and biomass yields and indirect selection criteria prior to drought tolerance (Muchero et al. 2013). Furthermore, significant physiological features such as faster and depth root architecture (Ribeiro et al. 2019), leaf canopy temperature (Khan et al. 2010; Martynenko et al. 2016), cooler canopies, high stomatal conductance (Duc et al. 2015), higher water use efficiency (Amede et al. 1999), and carbon isotope discrimination ($\Delta^{13}C$) may be constructive for selection of legume genotypes including alfalfa, especially in the dry regions where drought is frequently occurs. Therefore, foremost priority should be taken for the above traits during selection or screening genotypes for cultivation under drought area through the conventional breeding process.

5.2 Genomic and Biotechnological Approaches to Survive Against Drought Stress

The development of stress-tolerant varieties only by conventional breeding is a long-term strategy than biotechnological and genetic modification strategies (Zheng et al. 2017). Therefore, traditional breeding with the combination of transgenic approaches may be highly efficient to improve drought-tolerant crop cultivars including legumes (Zhang et al. 2005; Vasconcelos et al. 2008; Suárez et al. 2009; Li et al. 2010). Transgenic, genetic modification and other gene-based approaches such as gene editing have been found as the most valuable methodology for understanding the drought resistance mechanisms of desirable plants (Zhang et al. 2015). Legumes that were genetically manipulated with single-gene approaches for improving drought tolerance are listed in Table 1.

Additionally, the modern biotechnological tools are providing new knowledge that allows to categorize genes or group of genes which help the plant to survive against extreme DS. The transgenic or genetic modification based approaches such as biolistic or *Agrobacterium*-mediated transformation are now greatly considerable to develop legume cultivar against drought. Alfalfa drought tolerance ability is a significant feature for improving the yield potential under the water-deficit environment. The research on the molecular and genetic engineering for DS responses in alfalfa was first originated in the 1990s (Luo et al. 1991, 1992; Laberge et al. 1993), but the available knowledge about the genetic and physiological mechanisms of alfalfa under DS is still scarce (Zhang et al. 2015). A description of some genomic and biotechnological approaches for improving drought tolerance in alfalfa and other legumes is discussed in the following subheading.

Table 1 Single gene responsible for the drought tolerance in Fabaceae

Legumes	Latin name	Gene	Specific function	Reference
Alfalfa	*Medicago sativa*	*AtEDT1*	Confers drought tolerance	Zheng et al. (2017)
		CsLEA	Confers drought tolerance	Zhang et al. (2016)
		codA	Confers drought tolerance	Li et al. (2014)
		HaHB11	Confers drought tolerance	Cabello et al. (2017)
		SPL13	Improve drought tolerance	Arshad et al. (2017)
		GsZFP1	Confers drought tolerance	Tang et al. (2013)
		AVP1	Enhance drought tolerance	Bao et al. (2009)
Soybean	*Glycine max*	*PgTIP1*	Confers drought tolerance	An et al. (2018)
		GmDREB2	Confers drought tolerance	(Savitri and Fauziah et al. 2018)
		GmRACK1	Confers drought tolerance	Li et al. (2018)
		AtABF3	Confers drought tolerance	Kim et al. (2018)
		GmFDL19	Confers drought tolerance	Li et al. (2017)
		GmSK1	Confers drought tolerance	Chen et al. (2018)
		GmNAC, GmDREB, GmZIP, ERF089	Transcription factors	(Manavalan et al. 2009)
		DREB1A, rd29A	Transcription factors	(Bhatnagar-Mathur et al. 2007)
		GmBIN2	Enhance tolerance to drought	Wang et al. (2018)
		GmCaM4	Confers drought tolerance	Yoo et al. (2005)
		CDPK	Confers drought tolerance	Guenther et al. (2003)

(continued)

Table 1 (continued)

Legumes	Latin name	Gene	Specific function	Reference
		GmHK, GmCLV1A, GmCLV1B, GmRLK1, GmRLK2, GmRLK3, GmRLK4	Osmosensor	Yamamoto et al. (2000)
Mungbean	*Vigna radiata*	*VrbZIP*	Confers drought tolerance	Wang et al. (2018)
		codA	Confers drought tolerance	Baloda et al. (2017)
		VrWRKY	Confers drought tolerance	Srivastava et al. (2018)
Pigeon pea	*Cajanus cajan*	*C.cajan_29830, C.cajan_33874*	Improve drought tolerance	Saxena et al. (2011)
		WRKY, MyB, NF-Y	Transcription factors	Yang et al. (2006)
Cowpea	*Vigna unguiculata*	*VuPLD1, VuNCED1,*	ABA-biosynthesis during drought stress	Muchero et al. (2010)
Chickpea	*Cicer arietinum*	*Aquaporins*	Drought stress tolerance	Azeem et al. (2019)
		DREB2A	Transcription factors	Nayak et al. (2009)
		MYB, WRKY, bZIP	Transcription factors	Hiremath et al. (2011)
		MyB, AP2/ERF, XPB1	Transcription factors	Deokar et al. (2011)
Broad bean	*Vicia faba*	*VfPIP1*	Aquaporin or water transport under DS	Cui et al. (2008)
Common bean	*Phaseolus vulgari*	*Asr1, Asr2*	ABA signaling pathway	Cortés et al. (2012)
		PvLEA3	Protein stabilization	Barrera-Figueroa et al. (2007)

5.2.1 Improvement of Alfalfa Through Genetic Engineering

In dry areas (including arid and semiarid regions), due to the unpredictability and scarcity of rainfall as well as insufficient irrigation water, the development of drought resistance alfalfa cultivars is a key breeding aim for the sustainability of alfalfa production. Target drought tolerance genes may be assisted to develop drought tolerant and water use efficiency in alfalfa (Zhang et al. 2015). Genetic engineering has been established as a modern strategy to produce transgenic plants by transferring a (or few) target gene associated with the ability to withstand various abiotic stresses (Ashraf 2010; Turan et al. 2012). The first report of genetic engineered (transgenic) lines of alfalfa was reported by Deak et al. (1986). Then, many other transgenic alfalfa plants through gene transfer methods were also generated (Zhang et al. 2005; Ramon et al. 2009; Tang et al. 2013). These methodologies have been used to allow the expression of genes that have been linked with significant metabolic pathways, allowing the improvement of nutrient uptake and water use efficiency, quality forage yield, and also the capability to fight against abiotic stress including drought.

For the genetic improvement of alfalfa, generally, three main approaches are normally followed such as direct and indirect transformation, and germplasm line transformation. The technique transformation through germplasm line is generally made with own plants' pollen, and also ovary, and other germ cells through the ovary injection and pollen tube method for familiarizing exogenous genes (Shi et al. 2017a, b). To obtain transgenic plants through the direct transformation approach, chemical or physical methods such as ultrasonic, biolistic, microinjection, and polyethylene glycol (PEG) mediated are generally used for introducing of exogenous DNA into plant cells. While genes transformation through *Agrobacterium* is used in the indirect transformation methods for introducing DNA into the plant cell (Shi et al. 2017a, b).

5.2.2 Structural and Transcription Factors (TFs) Genes for Drought Resistance in Alfalfa

Both structural and TFs genes play a key role in drought response plants via multifaceted and systematic cross-talk (Zheng et al. 2017). Several categories of TFs (i.e., *MYBs*, *DREBs*, *bZIPs*, *NACs*, and *bHLHs*) were found suitable for improving DS tolerance in several plant species via ectopic expression (Yamaguchi-Shinozaki and Shinozaki 2006; Agarwal et al. 2018). As one example, Liu et al. (2013) found that transgenic *Arabidopsis* plants overexpressing the *DREB2.7* gene from maize have increased drought tolerance. Both DS and salinity stress were improved in transgenic *Arabidopsis* expressing the *CmMYB2* from *Chrysanthemum* (Shan et al. 2012), *CabZIP1* from pepper and *ZmbZIP72* from maize (Lee et al. 2006; Ying et al. 2012). Zhai et al. (2016) found that the overexpression of *TabHLH39* from wheat enhanced salt, drought, and freezing tolerance ability in transgenic *Arabidopsis* during the seedling stage. Similarly, TFs GhNAC2 from cotton, *ZmNAC55* from maize, *MlNAC9* from *Miscanthus*, and *OsNAC6* from rice improved the drought tolerance ability in different types of plants (Mao et al. 2016; Gunapati et al. 2016; Zhao et al. 2016; Lee et al. 2017).

The transcription factors (TF) genes could enhance the tolerance ability of plants against abiotic stress, which is generally introduced in alfalfa from numerous plant species, including the model legumes *M. truncatula* and soybean to increase survival ability against abiotic stress (Zheng et al. 2017). Jiang et al. (2009) reported that transgenic alfalfa plants expressing an ethyl-responsive *MtWXP*1 gene from *M. truncatula* showed a better drought tolerance ability as a result of the enhanced cuticular wax buildup on leaves, decreased water loss, and improved photosynthesis. In another study, Tang et al. (2013) found that a *Cys2/His2*-type zinc-finger protein gene *GsZFP1* from *Glycine soja*, when inserted into alfalfa, enhanced the DS fighting ability by triggering the expression of numerous stress-responsive genes in alfalfa. Furthermore, other functional genes such as the *CodA* gene from bacteria, the orange gene from sweet potato, the *ALDH* and *ZFP* genes from *Cleistogenes songorica*, and the *NHX* and *VP1* genes from *Zygophyllum xanthoxylum* have also been introduced into alfalfa for improving the DS response of alfalfa (Tang et al. 2013; Duan et al. 2015; Wang et al. 2015; Bao et al. 2016; Zhang et al. 2016).

The mitochondrial alternative oxidase (Aox) gene family is composed by 3–5 genes dispersed in two subfamilies (*Aox1* and *Aox2*) (Cavalcanti et al. 2013). Several kinds of research have been focused on explaining the function of Aox in non-thermogenic plants. The *Aox* plays a significant role in the prevention of ROS and upkeeps the TCA cycle in the circumstances of high-cellular ATP concentration (Arnholdt-Schmitt et al. 2006). The *Aox* mediates the physiological and molecular processes of plants (Sircar et al. 2012). The Aox genes such as *Aox1* (*Aox2a* and an *Aox2b*—named here *Aox2b1* and *Aox2b2*) were described in two leguminous species such as *Medicago sativa* and *M. truncatula* (Cavalcanti et al. 2013). In plants, *Aox* is encrypted by a small nuclear family with two sub-families such as *Aox1* (all angiosperms) and *Aox2* (only in eudicot species) (Borecký et al. 2006). Earlier study observed the *Aox1* gene is a stress-responsive gene and available in plant *Arabidopsis thaliana*, while *Aox2* (*Aox2a* and *Aox2b*) was observed other plant species which belongs to the order Fabales including the plants alfalfa, soybean, *Vigna unguiculata*, among others (Considine et al. 2002; Cavalcanti et al 2013). Among *Aox2a* and *Aox2b*, *Aox2b* is also induced by stress leguminous *Vigna unguiculata* (Cavalcanti et al. 2013).

Zheng et al. (2017) also established transgenic alfalfa line overexpressing *EDT1* from *Arabidopsis* (*AtEDT1*) and observed better performance of these lines under both laboratory and field drought conditions. All the transgenic lines performed healthier and exposed to less leaf wilting than wild-type plants when watered after 20 days (Zheng et al. 2017; Fig. 2). The developed transgenic lines recovered quickly when re-watered for 5 days than wild-type plants. When re-watered for 10 days, more than 67% transgenic plants were survived, while wild-type plants survived only 17% (Zheng et al. 2017; Fig. 2). Therefore, this supports the assumption that *AtEDT1* could play a major role in DS tolerance in plant species under the Mediterranean regions.

5.3 Quantitative Trait Loci (QTL) as a Potential Genome-Based Approaches for Drought Tolerance

Numerous breeding approaches have been used in the development of DS tolerance in legumes based on marker-assisted selection (MAS) (Nadeem et al. 2019).

Fig. 2 Enhanced the drought tolerance ability of alfalfa as a result of *AtEDT1* transgenic approach (Adapted from Zheng et al. 2017)

Since several findings revealed that desired alleles in different quantitative trait loci (QTLs) related to DS tolerance in alfalfa is one of the most useful MAS genome-based methodologies. For example, for DS tolerance in the seedling stage of plants, Hamwieh et al. (2013) recognized 12 QTLs such as H6C-07, NCPGR-50, H5E-02, TR-50, H1B-04, SCEA19, H5G-01, H6C-07, H1F-21, TAA-58, TA-113, and

H6C-07, that were found the most promising in chickpea, while QTL Qncl.Sw1 was linked with the grain yield in the same species (Radhika et al. 2007). Furthermore, seven markers such as CPRD8-1, VuPAT1-2, ACC-3, CPRD14-2, CPRD14-3, CPRD22-4, and CPRD22-2 were found to be linked to the DS tolerance in cowpea (Muchero et al. 2009). Similarly, QTLs significant traits for surviving under DS, such as root architecture, leaf ash, and WUE were also described (Manavalan et al. 2009; Pipolo et al. 2011). Abdel-Haleem et al. (2011) recognized five QTLs, namely Gm01, Gm02, Gm03, Gm04, and Gm20, associated with fibrous root system of soybean, an important feature for coping against DS. QTLs for stomatal characteristics on chromosome II of faba bean were identified by applying SNPs derived from *M. truncatula* L. (Khazaei et al. (2014). Mukeshimana et al. (2014) used SNPs from BARCBean6K_3 Beadchip for the identification of 14 QTLs to DS tolerance in common bean. Also in common bean, Briñez et al. (2017) mapped QTLs for DS in a cross between AND-277 x SEA-5 with SSRs and SNP markers.

Omics-based approaches, such as transcriptome, genome, microRNAome, proteome, and metabolome data (Fig. 3), can be used to identify candidate genes, thereby assisting in QTL mapping. The integration of "omics" approaches would be a leading approach for sustainable legumes production under DS (Nadeem et al. 2019).

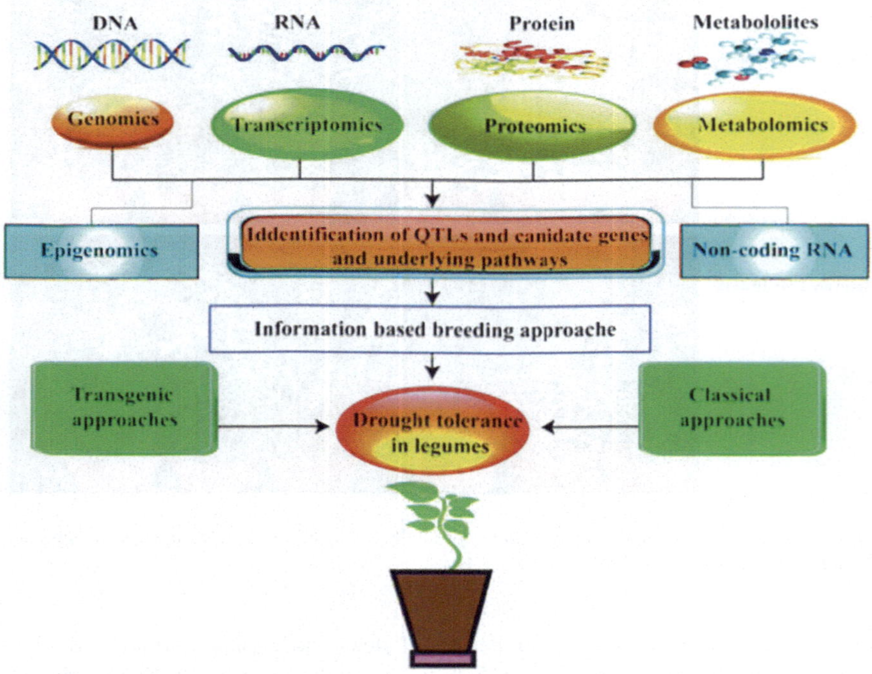

Fig. 3 Omics-based approach for the improvement of drought tolerance in legumes (Adapted from Nadeem et al. 2019)

5.4 Development of Drought Tolerance in Plants Through Genome Editing: CRISPR/Cas9

The most innovative and accurate genome editing (GE) tool is CRISPR-Cas9. The adaptive-immune system CRISPR (clustered regularly interspaced short palindromic repeats)-Cas (CRISPR-associated) in prokaryotes has led to a revolution in targeted genome editing of plants with high precision and accuracy. This technology allows the manipulation or modification of crop plant genome in several ways—(i) by simply incorporating random mutation (insertion or deletion) through non-homologous end joining to disrupt gene(s); (ii) by generating targeted point mutations in genes using precise base editors; and (iii) by a whole-gene insertion employing the cell's homology-directed repair pathway. Use of CRISPR-Cas in deletion/adding the crop plant desired genome is emerging rapidly for the increased yield, quality, domestication and stress tolerance including DS (Haque et al. 2018; Islam 2019; Nadeem et al. 2019).

Since its development, the CRISPR/Cas9 gene-editing approach has been used successfully used for improving different traits in wheat (Sanchez-Leon et al. 2018), maize (Zhu et al. 2016; Agarwal et al. 2018), barley (Lawrenson et al. 2015; Kapusi et al. 2017), rice (Huang et al. 2018), potato (Nakayasu et al. 2018), and soybean (Cai et al. 2015, 2018). Hence, CRISPR/Cas9 GE has still a huge application for the development of elite legumes cultivars grown under the diverse environmental conditions.

A schematic representation of the development of DS tolerance in plants through genome editing (GE) with Cas9/sgRNA is depicted in Fig. 4. After the selection of the target gene (in Fig. 3), target gene sgRNAs are designed following the online tools. For the edition of the target gene, the sgRNA and Cas9 casette (or with its variant) is cloned into a plant binary vector to be used for agrobacterium-mediated transformation. Then, edited plants are then screened for the expression of the Cas9 and sgRNA. Finally, by using PCR/RE genotyping and sequencing approaches, plants with the preferred edition are selected (Nadeem et al. 2019).

The CRISPR-Cas genome editing shows high promise to address environmental problems associated with global climate change and develop resistant plants (rice, maize, soybean) to DS (Tang et al. 2016; Shi et al. 2017a, b; Shim et al. 2018; Islam 2019). To what concerns legumes, CRISPR-Cas technology has been used for modulating DS and salinity tolerance in soybean (Li et al. 2015). Also in the same species, the disruption of the *Drb2a* and *Drb2b* genes (double-stranded RNA-binding proteins 2 genes) for DS and salt tolerance in soybean resulted in a transgene-free variety that received USDA green pass for cultivation in the USA in 2017 (Islam 2019). In the future, it is expected that the revolutionary CRISPR technology would be used in improvement of DS tolerance in many members of the family Fabaceae and by this way contribute to mitigate the crop losses due to global climate change in the dry regions of the world.

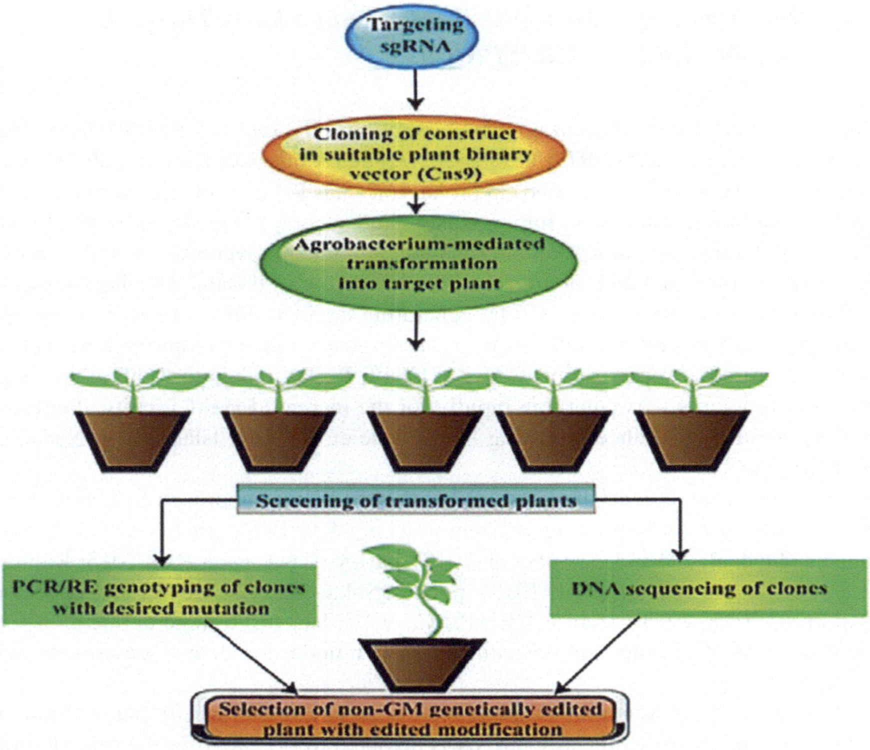

Fig. 4 Development of drought tolerance in plants through genome editing (GE) with Cas9/sgRNA (Adapted from Nadeem et al. 2019)

5.5 MicroRNAs Are Universal Gene Regulators for Drought Stress Tolerance in Alfalfa

MicroRNAs (miRNAs) are negative post-transcriptional regulators of gene expression (either transcript cleavage or translation repression) in plants, with potential to be used in the genetic engineering approaches (Jones-Rhoades et al. 2006; Macovei et al. 2012; Zhou and Luo 2013; Aung et al. 2015a, b). Numerous physiological processes such as plant phenology, apical dominance, biomass, root systems, and seed development and also environmental stress response are controlled by numerous miRNAs (Zhang and Wang 2015). Kantar et al. (2010) found that 28 new miRNAs, belonging to 18 families, were expressed in response to dehydration stress in barley. In another study conducted by Sunkar and Zhu (2004), 26 miRNAs from 34 loci were up- or down-regulated in response to abiotic stress in *Arabidopsis*. Arshad et al. (2017) recognized that miR156 can be a suitable approach for modulating many traits in plants, including DS tolerance. The overexpression of miR156 (miR156OE) in alfalfa improved DS tolerance, by comparison to the wild-type plants (Arshad et al. 2017). Beside higher survival and reduced water loss during drought stress, miR156OE genotypes also maintained higher stomatal conductance compared to WT. The miR156 overexpression also boosted the higher accumulation of proline, ABA, and antioxidants (Arshad et al. 2017).

6 Conclusions and Future Research Perspectives

Due to the cumulative incidences of abiotic stresses under the changing climate context, the sustainable agricultural productivity threatened globally. Among the various abiotic stresses, drought stress stood out since it causes a significant yield loss in crops including legumes. Under DS key plant processes such as morphological physiological, biochemical, and metabolic paths are disrupted, which ultimately impacts on crop productivity and yield. Drought often triggers an overproduction of reactive oxygen species which causes oxidative damage. The role of osmolytes and antioxidants in countering the oxidative damages have been widely studied and described. The recently developed CRISPR-Cas technology has already been used for precision breeding of many plants including the members of Fabaceae and are discussed in this chapter. Moreover, to meet the food security of ever-increasing population of the world, "omics-based approaches," such as proteomics, metabolomics transcriptomics, and genomics could contribute to update our knowledge on drought-tolerant genes, unveiling complex gene networks, by unveiling drought-tolerant candidate genes and intricate gene networks, and various signaling cascades involved in DS tolerance in legumes.

References

Abdel-Haleem H, Lee GJ, Boerma RH (2011) Identification of QTL for increased fibrous roots in soybean. Theor Appl Genet 122:935–946

Agarwal A, Yadava P, Kumar K, Singh I (2018) Insights into maize genome editing via CRISPR/Cas9. Physiol Mol Biol Plants 24:175–183

Al Hassan M, Chaura J, Donat-torres P, Boscaiu M (2017) Antioxidant responses under salinity and drought in three closely related wild monocots with different ecological optima. AoB Plants 9, plx009. https://doi.org/10.1093/aobpla/plx009

Amede T, Kittlitz EV, Schubert S (1999) Differential drought responses of faba bean (*Vicia faba L.*) inbred lines. J Agron Crop Sci 183:35–45

Amede T, Schubert S (2003) Mechanisms of drought resistance in grain legumes i: osmotic adjustment. Ethiop J Sci 26:37–46

An J, Cheng C, Hu Z, Chen H, Cai W, Yu B (2018) The Panax ginseng PgTIP1 gene confers enhanced salt and drought tolerance to transgenic soybean plants by maintaining homeostasis of water, salt ions and ROS. Environ Exp Bot 155:45–55

Anjum SA, Ashraf U, Zohaib A, Tanveer M, Naeem M, Ali I, Tabassum T, Nazir U (2017) Growth and developmental responses of crop plants under drought stress: a review. Zemdirbyste Agric 104:267–276

Anjum SA, Wang LC, Farooq M, Hussain M, Xue LL, Zou CM (2011a) Brassinolide application improves the drought tolerance in maize through modulation of enzymatic antioxidants and leaf gas exchange. J Agron Crop Sci 197:177–185

Anjum SA, Xie XY, Wang LC, Saleem MF, Man C, Lei W (2011b) Morphological, physiological and biochemical responses of plants to drought stress. Afr J Agric Res 6:2026–2032

Arnholdt-Schmitt B, Costa JH, de Melo DF (2006) AOX—a functional marker for efficient cell reprogramming under stress? Trends Plant Sci 11(6):281–287

Arshad M, Feyissa BA, Amyot L, Aung B, Hannoufa A (2017) MicroRNA156 improves drought stress tolerance in alfalfa (Medicago sativa L.) by silencing SPL13. Plant Sci 258:122–136

Ashraf M (2010) Inducing drought tolerance in plants: recent advances. Biotechnol Adv 28(1):169–183

Aung B, Gruber MY, Amyot L, Omari K, Bertrand A, Hannoufa A (2015a) MicroRNA156 as a promising tool for alfalfa improvement. Plant Biotech J 13:779–790

Aung B, Gruber MY, Hannoufa A (2015b) The MicroRNA156 system: a tool in plant biotechnology. Biocatal Agric Biotechnol 4:432–442

Azeem F, Bilal A, Rana MA, Muhammad AA, Habibullah N, Sabir H, Sumaira R, Hamid M, Usama A, Muhammad A (2019) Drought a_ects aquaporins gene expression in important pulse legume chickpea (*Cicer arietinum L.*). Pak J Bot 51:81–88

Baloda A, Madanpotra S, Aiwal PK (2017) Transformation of mungbean plants for salt and drought tolerance by introducing a gene for an osmoprotectant glycine betaine. J Plant Stress Physiol 3:5. https://doi.org/10.19071/jpsp.2017.v3.3148

Bao AK, Du BQ, Touil L, Kang P, Wang QL, Wang SM (2016) Co-expression of tonoplast Cation/H + antiporter and H+ -pyrophosphatase from xerophyte Zygophyllum xanthoxylum improves alfalfa plant growth under salinity, drought and field conditions. Plant Biotech J 14:964–975

Bao AK, Wang SM, Wu GQ, Xi JJ, Zhang JL, Wang CM (2009) Overexpression of the Arabidopsis H+ -PPase enhanced resistance to salt and drought stress in transgenic alfalfa (*Medicago sativa L.*). Plant Sci 176:232–240

Barrera-Figueroa BE, Peña-Castro JM, Acosta-Gallegos JA, Ruiz-Medrano R, Beatriz XC (2007) Isolation of dehydration-responsive genes in a drought tolerant common bean cultivar and expression of a group 3 late embryogenesis abundant mRNA in tolerant and susceptible bean cultivars. Funct Plant Biol 34:368–381

Benabderrahim MA, Hamza H, Haddad M, Ferchichi A (2015) Assessing the drought tolerance variability in Mediterranean alfalfa (*Medicago sativa L.*) genotypes under arid conditions. Plant Biosyst 149:395–403

Bhardwaj J, Yadav SK (2012) Comparative study on biochemical parameters and antioxidant enzymes in a drought tolerant and a sensitive variety of horsegram under drought stress. Am J Plant Physiol 7:17–29

Bhatnagar-Mathur P, Devi MJ, Reddy DS, Lavanya M, Vadez V, Serraj R, Yamaguchi-Shinozaki K, Sharma KK (2007) Stress-inducible expression of At DREB1A in transgenic peanut (*Arachis hypogaea* L.) increases transpiration efficiency under water-limiting conditions. Plant Cell Rep 26:2071–2082

Bhauso TD, Radhakrishnan T, Kumar A, Mishra GP, Dobaria JR (2014) Overexpression of bacterial mtlD gene in peanut improves drought tolerance through accumulation of mannitol. Sci World J 10, Article ID 125967. https://doi.org/10.1155/2014/125967

Bielach A, Hrtyan M, Tognetti VB (2017) Plants under stress: involvement of auxin and cytokinin. Int J Mol Sci 18:1427. https://doi.org/10.3390/ijms18071427

Blum WE (2013) Soil and land resources for agricultural production: general trends and future scenarios—a worldwide perspective. Int Soil Water Conserv Res 1(3):1–14

Borecký J, Nogueira FT, De Oliveira KA, Maia IG, Vercesi AE, Arruda P (2006) The plant energy-dissipating mitochondrial systems: depicting the genomic structure and the expression profiles of the gene families of uncoupling protein and alternative oxidase in monocots and dicots. J Exp Bot 57(4):849–864

Briñez B, Morini J, Cardoso K, Rosa JS, Bassi D, Gonçalves GR, Almeida C, Fausto J, Paulino DC, Blair MW et al (2017) Mapping QTLs for drought tolerance in a SEA 5_AND 277 common bean cross with SSRs and SNP markers. Genet Mol Biol 823:813–823

Cabello JV, Giacomelli JI, Gómez MC, Chan RL (2017) The sunflower transcription factor HaHB11 confers tolerance to water deficit and salinity to transgenic Arabidopsis and alfalfa plants. J Biotechnol 257:35–46

Cai Y, Chen L, Liu X, Guo C, Sun S, Wu C, Jiang B, Han T, Hou W (2018) CRISPR/Cas9-mediated targeted mutagenesis of GmFT2a delays flowering time in soybean. Plant Biotechnol J 16:176–185

Cai Y, Chen L, Liu X, Sun S, Wu C, Jiang B, Han T, Hou W (2015) CRISPR/Cas9-mediated genome editing in soybean hairy roots. PLoS ONE 10:e0136064

Calvache M, Reichardt K, Bacchp OOS (1997) Deficit irrigation at different growth stages of the common bean. Sci Agric 54:1–16

Campanelli A, Ruta C, Morone-Fortunato I, De Mastro G (2013) Alfalfa (Medicago sativa L.) clones tolerant to salt stress: in vitro selection. Cent Eur J Biol 8(8):765–76

Castroluna A, Ruiz OM, Quiroga AM, Pedranzani HE (2014) Effects of salinity and drought stress on germination, biomass and growth in three varieties of *Medicago sativa* L. Avances en Investigación Agropecuaria 18(1):39–50

Cavalcanti JH, Oliveira GM, da Cruz Saraiva KD, Torquato JP, Maia IG, de Melo DF, Costa JH (2013) Identification of duplicated and stress-inducible Aox2b gene co-expressed with Aox1 in species of the *Medicago* genus reveals a regulation linked to gene rearrangement in leguminous genomes. J Plant Physiol 170(18):1609–1619

Chakrabarty A, Aditya M, Dey N, Banik N (2016) Antioxidant signaling and redox regulation in drought- and salinity-stressed plants. In: Drought stress tolerance in plants, vol 1. Springer, Berlin. ISBN 9783319288994

Chapman SC (2008) Use of crop models to understand genotype by environment interactions for drought in real-world and simulated plant breeding trials. Euphytica 161:195–208

Chen Y, Chi Y, Meng Q, Wang X, Yu D (2018) GmSK1, an SKP1 homologue in soybean, is involved in the tolerance to salt and drought. Plant Physiol Biochem 127:25–31

Choudhury FK, Rivero RM, Blumwald E, Mittler R (2017) Reactive oxygen species, abiotic stress and stress combination. Plant J 90:856–867

Chowdhury JA, Karim MA, Khaliq QA, Ahmed AU, Khan MSA (2016) Effect of drought stress on gas exchange characteristics of four soybean genotypes. Bangladesh J Agric Res 41:195–205

Cocks PS (2001) Ecology of herbaceous perennial legumes: a review of characteristics that may provide management options for the control of salinity and waterlogging in dryland cropping systems. Aust J Agric Res 52:137–151

Considine MJ, Holtzapffel RC, Day DA, Whelan J, Millar AH (2002) Molecular distinction between alternative oxidase from monocots and dicots. Plant Physiol 129:949–953

Cortés AJ, This D, Carolina C, Madriñán S, Blair MW (2012) Nucleotide diversity patterns at the drought-related DREB2 encoding genes in wild and cultivated common bean (*Phaseolus vulgaris L.*). Theor Appl Genet 125:1069–1085

Cui XH, Hao FS, Chen H, Chen J, Wang XC (2008) Expression of the Vicia faba VfPIP1 gene in Arabidopsis thaliana plants improves their drought resistance. J Plant Res 121:207–214

Deak M, Kiss GB, Koncz C, Dudits D (1986) Transformation of *Medicago* by Agrobacterium mediated gene transfer. Plant Cell Rep 5(2):97–100

Dear BS, Ewing MA (2008) The search for new pasture plants to achieve more sustainable production systems in southern Australia. Aust J Exp Agric 48:387–396

Dear BS, Moore GA, Hughes SJ (2003) Adaptation and potential contribution of temperate perennial legumes to the southern Australian wheat belt: a review. Aust J Exp Agric 43:1–18

Del Pozo A, Ovalle C, Espinoza S, Barahona V, Gerding M, Humphries A (2017) Water relations and use-efficiency, plant survival and productivity of nine alfalfa (*Medicago sativa L.*) cultivars in dryland Mediterranean conditions. Eur J of Agron 84:16–22

Deokar AA, Kondawar V, Jain PK, Karuppayil SM, Raju NL, Vadez V (2011) Comparative analysis of expressed sequence tags (ESTs) between drought-tolerant and -susceptible genotypes of chickpea under terminal drought stress. BMC Plant Biol 11:70. https://doi.org/10.1186/1471-2229-11-70

Duan Z, Zhang D, Zhang J, Di H, Wu F, Hu X, Meng X, Luo K, Zhang J, Wang Y (2015) Co-transforming bar and CsALDH genes enhanced resistance to herbicide and drought and salt stress in transgenic alfalfa (*Medicago sativa L.*). Front Plant Sci 6:1115. https://doi.org/10.3389/fpls.2015.01115

Duc G, Agrama H, Bao S, Berger J, Bourion V, De Ron AM, Gowda CL, Mikic A, Millot D, Singh KB, Tullu A (2015) Breeding annual grain legumes for sustainable agriculture: new methods to approach complex traits and target new cultivar ideotypes. Crit Rev Plant Sci 34:381–411

Fariduddin Q, Varshney P, Yusuf M, Ali A, Ahmad A (2013) Dissecting the role of glycine betaine in plants under abiotic stress. Plant Stress 7(1):8–18

Farnese FS, Menezes-silva PE, Gusman GS, Oliveira JA (2016) When bad guys become good ones: The key role of reactive oxygen species and nitric oxide in the plant responses to abiotic stress. Front Plant Sci 7:471. https://doi.org/10.3389/fpls.2016.00471

Farooq M, Farooq M, Hussain M, Siddique KHM (2014) drought stress in wheat during flowering and grain-filling periods. Crit Rev Plant Sci 33:331–349

Farooq M, Gogoi N, Barthakur S, Baroowa B, Bharadwaj N, Alghamdi SS, Siddique KHM (2017a) Drought stress in grain legumes during reproduction and grain filling. J Agron Crop Sci 203:81–102

Farooq M, Nadeem F, Gogoi N, Ullah A, Alghamdi SS, Nayyar H, Siddique KH (2017b) Heat stress in grain legumes during reproductive and grain-filling phases. Crop Pasture Sci 68:985–1005

Fischer G (2009) World food and agriculture to 2030/50. In: Technical paper from the expert meeting on how to feed the world in 2050, pp 24–26

Ghahramani A, Moore AD (2013) Climate change and broadacre livestock production across southern Australia. 2. Adaptation options via grassland management. Crop Pasture Sci 64:615–630

Gill SS, Gill R, Anjum NA (2014) Target osmoprotectants for abiotic stress tolerance in crop plants—glycine betaine and proline. In: Plant adaptation to environmental change: significance of amino acids and their derivatives, vol 10, pp 97–108

Gill SS, Tuteja N (2010) Reactive oxygen species and antioxidant machinery in abiotic stress tolerance in crop plants. Plant Physiol Biochem 48(12):909–930

Golldack D, Li C, Mohan H, Probst N (2014) Tolerance to drought and salt stress in plants: unraveling the signaling networks. Front Plant Sci 5:151. https://doi.org/10.3389/fpls.2014.00151

Gomiero T (2016) Soil degradation, land scarcity and food security: reviewing a complex challenge. Sustainability 8(3):281. https://doi.org/10.3390/su8030281

Guenther JF, Chanmanivone N, Galetovic MP, Wallace IS, Cobb JA, Roberts DM (2003) Phosphorylation of soybean nodulin 26 on serine 262 enhances water permeability and is regulated developmentally and by osmotic signals. Plant Cell 15:981–991

Guler NS, Pehlivan N (2016) Exogenous low-dose hydrogen peroxide enhances drought tolerance of soybean (*Glycine max L.*) through inducing antioxidant system. Acta Biol Hung 67:169–183

Gunapati S, Naresh R, Ranjan S, Nigam D, Hans A, Verma PC, Gadre R, Pathre UV, Sane AP, Sane VA (2016) Expression of GhNAC2 from *G. herbaceum*, improves root growth and imparts tolerance to drought in transgenic cotton and Arabidopsis. Sci Rep 6:24978. https://doi.org/10.1038/srep24978

Hall AE (2012) Phenotyping cowpeas for adaptation to drought. Front Physiol 3:155. https://doi.org/10.3389/fphys.2012.00155

Hamidi H, Safarnejad A (2010) Effect of drought stress on alfalfa cultivars (*Medicago sativa* L.) in germination stage. Am Eur J Agric Environ Sci 8(6):705–709

Hamwieh A, Imtiaz M, Maize I, Malhotra RS (2013) Multi-environment QTL analyses for drought-related traits in a recombinant inbred population of chickpea (*Cicer arietinum* L.). Theor Appl Genet 126:1025–1103

Haque E, Taniguchi H, Hassan MM, Bhowmik P, Karim MR, Smiech M, Zhao K, Rahman M, Islam T (2018) Application of CRISPR/Cas9 genome editing technology for the improvement of crops cultivated in tropical climates: recent progress, prospects, and challenges. Front Plant Sci 9:617. https://doi.org/10.3389/fpls.2018.00617
Hartung W, Sauter A, Hose E (2002) Abscisic acid in the xylem: Where does it come from, where does it go to? J Exp Bot 53:27–32
Hasanuzzaman M, Hossain MA, Silva JAT, Fujita M (2012a) Plant responses and tolerance to abiotic oxidative stress: antioxidant defenses is a key factor. In: Shanker AK, Shanker C, Mandapaka M (eds) Crop stress and its management: perspectives and strategies: bandi V. Springer, Berlin, pp 261–316
Hasanuzzaman M, Nahar K, Alam MM, Fujita M (2012b) Exogenous nitric oxide alleviates high temperature induced oxidative stress in wheat (*Triticum aestivum L.*) seedlings by modulating the antioxidant defense and glyoxalase system. Aust J Crop Sci 6:1314–1323
Hasanuzzaman M, Nahar K, Fujita M (2013a) Plant response to salt stress and role of exogenous protectants to mitigate salt-induced damages. In: Azooz MM, Prasad MNV, Ahmad P (eds) Ecophysiology and responses of plants under salt stress. Springer, New York, pp 25–87
Hasanuzzaman M, Gill SS, Fujita M (2013b) Physiological role of nitric oxide in plants grown under adverse environmental conditions. In: Tuteja N, Gill SS (eds) Plant acclimation to environmental stress. Springer: New York, pp 269–322
Hasanuzzaman M, Nahar K, Fujita M (2013c) Extreme temperatures, oxidative stress and antioxidant defense in plants. In: Vahdati K, Leslie C (eds) Abiotic stress—plant responses and applications in agriculture. Tech: Rijeka Croatia, pp 169–205
Hiremath PJ, Farmer A, Cannon SB, Woodward J, Kudapa H, Tuteja R, Kumar A, BhanuPrakash A, Mulaosmanovic B, Gujaria N, Krishnamurthy L (2011) Large-scale transcriptome analysis in chickpea (*Cicer arietinum L.*), an orphan legume crop of the semi-arid tropics of Asia and Africa. Plant Biotechnol J 9:922–931
Hou M, Tian F, Zhang L, Li S, Du T, Huang M, Yuan Y (2018) Estimating crop transpiration of soybean under different irrigation treatments using thermal infrared remote sensing imagery. Agronomy 9:8. https://doi.org/10.3390/agronomy9010008
Huang L, Zhang R, Huang G, Li Y, Melaku G, Zhang S, Chen H, Zhao Y, Zhang J, Zhang Y, Hu F (2018) Developing superior alleles of yield genes in rice by artificial mutagenesis using the CRISPR/Cas9 system. Crop J 6:475–481
Humphries AW, Auricht GC (2001) Breeding lucerne for Australia's southern dryland cropping environments. Aust J Agric Res 52(2):153–169
Hussain M, Farooq S, Hasan W, Ul-allah S, Tanveer M (2018) Drought stress in sunflower: physiological effects and its management through breeding and agronomic alternatives. Agric Water Manag 201:152–166. https://doi.org/10.1016/j.agwat.2018.01.028
Ibrahim HA, Abdellatif YMR (2016) Effect of maltose and trehalose on growth, yield and some biochemical components of wheat plant under water stress. Ann Agric Sci 61:267–274
IPCC (2014) Climate change synthesis report contribution of working groups I. II and III to the fifth assessment report of the intergovernmental panel on climate change. IPCC, Geneva, p 151
Islam T (2019) CRISPR-cas technology in modifying food crops. In: CAB reviews (in press)
Jiang QZ, Zhang JY, Guo XL, Monteros MJ, Wang ZY (2009) Physiological characterization of transgenic alfalfa (*Medicago sativa*) plants for improved drought tolerance. Int J Plant Sci 170:969–978
Jones-Rhoades MW, Bartel DP, Bartel BB (2006) MicroRNAs and their regulatory roles in plants. Annu Rev Plant Biol 57:19–53
Kantar M, Unver T, Budak H (2010) Regulation of barley miRNAs upon dehydration stress correlated with target gene expression. Funct Integr Genom 10:493–507
Kapusi E, Corcuera-Gómez M, Melnik S, Stoger E (2017) Heritable genomic fragment deletions and small indels in the putative ENGase gene induced by CRISPR/Cas9 in barley. Front Plant Sci 8:540. https://doi.org/10.3389/fpls.2017.00540

Kathuria H, Giri J, Nataraja KN, Murata N, Udayakumar M, Tyagi AK (2009) Glycinebetaine-induced water-stress tolerance in codA-expressing transgenic indicarice is associated with up-regulation of several stress responsive genes. Plant Biotechnol J 7:512–526

Khan HR, Paull JG, Siddique KHM, Stoddard FL (2010) Faba bean breeding for drought-affected environments: a physiological and agronomic perspective. Field Crops Res 115:279–286

Khater MA, Dawood MG, Sadak MS, Shalaby MAF, El-Din KG (2018) Enhancement the performance of cowpea plants grown under drought conditions via trehalose application. Middle-East J Agric Res 7:782–800

Khazaei H, Sullivan DMO, Sillanpää MJ, Stoddard FL (2014) Use of synteny to identify candidate genes underlying QTL controlling stomatal traits in faba bean (*Vicia faba* L.). Theor Appl Genet 127:2371–2385

Kim HJ, Cho HS, Pak JH, Kwon T, Lee J, Kim D, Lee DH, Kim C, Chung Y (2018) Molecules and cells confirmation of drought tolerance of ectopically expressed AtABF3 gene in soybean. Mol Cells 41:413–422

Kunert KJ, Vorster BJ, Fenta BA, Kibido T, Dionisio G, Foyer CH (2016) Drought stress responses in soybean roots and nodules. Front Plant Sci 7:1015. https://doi.org/10.3389/fpls.2016.01015

Kurutas EB (2016) The importance of antioxidants which play the role in cellular response against oxidative/nitrosative stress: current state. Nutr J 15:1–22

Laberge S, Castonguay Y, Vezina LP (1993) New coldand drought-regulated gene from Medicago sativa. Plant Physiol 101:1411–1412

Lawrenson T, Shorinola O, Stacey N, Li C, Østergaard L, Patron N, Uauy C, Harwood W (2015) Induction of targeted, heritable mutations in barley and Brassica oleracea using RNA-guided Cas9 nuclease. Genome Biol 16:258. https://doi.org/10.1186/s13059-015-0826-7

Lee DK, Chung PJ, Jeong JS, Jang G, Bang SW, Jung H, Kim YS, Ha SH, Choi YD, Kim JK (2017) The rice OsNAC6 transcription factor orchestrates multiple molecular mechanisms involving root structural adaptions and nicotianamine biosynthesis for drought tolerance. Plant Biotechnol J 15:754–764

Lee SC, Choi HW, Hwang IS, Choi DS, Hwang BK (2006) Functional roles of the pepper pathogen-induced bZIP transcription factor, CAbZIP1, in enhanced resistance to pathogen infection and environmental stresses. Planta 224:1209–1225

Li DH, Li W, Li HY, Guo JJ, Chen FJ (2018) The soybean GmRACK1 gene plays a role in drought tolerance at vegetative stages. Russ J Plant Physiol 65:541–552

Li GD, Nie ZN, Boschma SP, Dear BS, Lodge GM, Hayes RC, Clark B, Hughes SJ, Humphries AW (2010) Persistence and productivity of *Medicago sativa* subspecies sativa, caerulea, falcata and varia accessions at three intermittently dry sites in south-eastern Australia. Crop Pasture Sci 61:645–658

Li H, Wang Z, Ke Q, Ji CY, Jeong JC, Lee HS, Lim YP, Xu B, Deng XP, Kwak SS (2014) Overexpression of codA gene confers enhanced tolerance to abiotic stresses in alfalfa. Plant Physiol Biochem 85:31–40

Li JH, Jiao SM, Qing Gao RQ, Richard DB (2012) Differential effects of legume species on the recovery of soil microbial communities, and carbon and nitrogen contents in abandoned fields of the loess plateau. China Environ Manage 50:1193–1203

Li S, Yang G, Yi X, Ma G, Ma XZ, Li Y, Cao Y (2008) Transformation of Alfalfa (*Medicago sativa* L.) by agrobacterium-mediated process with DsNRT2 gene from *Dunaliella salina*. J Sichuan Univ 45:409–412

Li XM, Chao DY, Wu Y, Huang X, Chen K, Cui LG, Su L, Ye WW, Chen H, Chen HC, Dong NQ (2015) Natural alleles of a proteasome $\alpha 2$ subunit gene contribute to thermos-tolerance and adaptation of African rice. Nat Gen 47:827–833. https://doi.org/10.1038/ng.3305

Li Y, Chen Q, Nan H, Li X, Lu S, Zhao X, Liu B, Guo C, Kong F, Cao D (2017) Overexpression of GmFDL19 enhances tolerance to drought and salt stresses in soybean. PLoS ONE 12:e0179554. https://doi.org/10.1371/journal.pone.0179554

Liu S, Wang X, Wang H, Xin H, Yang X, Yan J, Li J, Tran LS, Shinozaki K, Yamaguchi-Shinozaki K, Qin F (2013) Genome-wide analysis of ZmDREB genes and their association with natural

variation in drought tolerance at seedling stage of *Zea mays* L. PLoS Genet 9:e1003790. https://doi.org/10.1371/journal.pgen.1003790
Luo M, Lin L, Hill RD, Mohapatra SS (1991) Primary structure of an environmental stress and abscisic acid inducible alfalfa protein. Plant Mol Biol 17:1267–1269
Luo M, Liu JH, Mohapatra S, Hill RD, Mohapatra SS (1992) Characterization of a gene family encoding abscisic acid- and environmental stress-inducible proteins of alfalfa. J Biol Chem 267:15367–15374
Macovei A, Gill SS, Tuteja N (2012) MicroRNAs as promising tools for improving stress tolerance in rice. Plant Signal Behav 7:1296–1301
Majumdar R, Barchi B, Turlapati SA, Gagne M (2016) Glutamate, ornithine, arginine, proline, and polyamine metabolic interactions: the pathway is regulated at the post-transcriptional level. Front Plant Sci 7:78. https://doi.org/10.3389/fpls.2016.00078
Manavalan LP, Guttikonda SK, Tran LP, Nguyen HT (2009) Physiological and molecular approaches to improve drought resistance in soybean. Plant Cell Physiol 50:1260–1276
Mao H, Yu L, Han R, Li Z, Liu H (2016) ZmNAC55, a maize stress-responsive NAC transcription factor, confers drought resistance in transgenic Arabidopsis. Plant Physiol Biochem 105:55–66
Martynenko A, Shotton K, Astatkie T, Petrash G, Fowler C, Neily W, Critchley AT (2016) Thermal imaging of soybean response to drought stress: the effect of *Ascophyllum nodosum* seaweed extract. Springer plus 5:1393. https://doi.org/10.1186/s40064-016-3019-2
Masri Z, John Ryan J (2006) Soil organic matter and related physical properties in a Mediterranean wheat-based rotation trial. Soil Till Res 87:146–154
McCallum MH, Kirkegaard JA, Green TW, Cresswell HP, Davies SL, Angus JF, Peoples MB (2004) Improved subsoil macroporosity following perennial pastures. Aust J Exp Agric 44:299–307
McCulley RL, Jobbágy EG, Pockman WT, Jackson RB (2004) Nutrient uptake as a contributing explanation for deep rooting in arid and semi-arid ecosystems. Oecologia 141:620–628
Merilo E, Pirko J, Kristiina L, Omid M, Hanna H, Hannes K, Mikael B (2015) Abscisic acid transport and homeostasis in the context of stomatal regulation. Mol Plant 8:1321–1333
Mir RR, Zaman-Allah M, Sreenivasulu N, Trethowan R, Varshney RK (2012) Integrated genomics, physiology and breeding approaches for improving drought tolerance in crops. Theor Appl Genet 125:625–645
Miyashita K, Tanakamaru S, Maitani T, Kimura K (2005) Recovery responses of photosynthesis, transpiration, and stomatal conductance in kidney bean following drought stress. Environ Exp Bot 53:205–214
Mohamed IH, Hanan HL (2017) Improvement of drought tolerance of soybean plants by using methyl jasmonate. Physiol Mol Biol Plants 23:545–556
Montes JM, Melchinger EA, Reif JC (2007) Novel throughput phenotyping platforms in plant genetic studies. Trends Plant Sci 12:433–436
Muchero W, Je V, Close TJ, Roberts PA (2009) Mapping QTL for drought stress-induced premature senescence and maturity in cowpea (*Vigna unguiculata* L.). Theor Appl Genet 118:849–863
Muchero W, Je V, Roberts PA (2010) Restriction site polymorphism-based candidate gene mapping for seedling drought tolerance in cowpea (*Vigna unguiculata* (L.)Walp.). Theor Appl Genet 120:509–518
Muchero W, Roberts PA, Diop NN, Drabo I, Cisse N, Close TJ, Muranaka S, Boukar O, Ehlers JD (2013) Genetic architecture of delayed senescence, biomass, and grain yield under drought stress in cowpea. PLoS ONE 8:e70041
Nadeem M, Li J, Wang M, Shah L, Lu S, Wang X, Ma C (2018) Unraveling field crops sensitivity to heat stress: mechanisms, approaches, and future prospects. Agronomy 8:128. https://doi.org/10.3390/agronomy8070128
Nadeem M, Li J, Yahya M, Sher A, Ma C, Wang X, Qiu L (2019) Research progress and perspective on drought stress in legumes: a review. Int J Mol Sci 20(10):254. https://doi.org/10.3390/ijms20102541

Nakayasu M, Akiyama R, Lee HJ, Osakabe K, Osakabe Y, Watanabe B, Sugimoto Y, Umemoto N, Saito K, Muranaka T et al (2018) Generation of solanine-free hairy roots of potato by CRISPR/Cas9 mediated genome editing of the St16DOX gene. Plant Physiol Biochem 131:70–77

Nayak SN, Balaji J, Upadhyaya HD, Hash CT, Kishor PBK, Blair MW, Baum M, Chattopadhyay D, Marı L, Mcnally K et al (2009) Plant science isolation and sequence analysis of DREB2A homologues in three cereal and two legume species. Plant Sci 177:460–467

Osman HS (2015) Enhancing antioxidant yield relationship of pea plant under drought at different growth stages by exogenously applied glycine betaine and proline. Ann Agric Sci 60:389–402

Park J, Lee Y, Martinoia E, Geisler M (2017) Plant hormone transporters: What we know and what we would like to know. BMC Biol 15:93. https://doi.org/10.1186/s12915-017-0443-x

Patel PK, Hemantaranjan A, Sarma BK, Singh R (2011) Growth and antioxidant system under drought stress in chickpea (*Cicer arietinum L.*) as sustained by salicylic acid. J Stress Physiol Biochem 7:130–144

Pipolo AE, Boerma HR, Sinclair TR (2011) Identification of QTLs associated with limited leaf hydraulic conductance in soybean. Euphytica 186:679–686

Prasad PVV, Bheemanahalli R, Jagdish SK (2017) Field crops and the fear of heat stress-opportunities, challenges and future directions. Field Crops Res 200:114–121

Radhika P, Gowda SJM, Kadoo NY, Mhase LB, Jamadagni BM, Sainani MN, Chandra S, Gupta VS (2007) Development of an integrated intraspecific map of chickpea (*Cicer arietinum L.*) using two recombinant inbred line populations. Theor Appl Genet 115:209–216

Radović J, Sokolović D, Marković J (2009) Alfalfa-most important perennial forage legume in animal husbandry. Biotechnol Anim Husb 25:465–475

Ramamoorthy P, Lakshmanan K, Upadhyaya HD, Vadez V, Varshney RK (2017) Root traits confer grain yield advantages under terminal drought in chickpea (*Cicer arietinum L.*). F Crop Res 201:146–161

Ramanjulu S, Bartels D (2002) Drought and desiccation induced modulation of gene. Plant Cell Environ 25:141–151

Ramon S, Cecilia C, Gabriel I (2009) Enhanced tolerance to multiple abiotic stresses in transgenic alfalfa accumulating trehalose. Crop Sci 49:1791–1799

Ribeiro T, Da Silva DA, Esteves JADF, Azevedo CVG, Gonçalves JGR, Carbonell SAM, Chiorato AF (2019) Evaluation of common bean genotypes for drought tolerance. Bragantia 78:1–11

Safarmejad A (2008) Morphological and biochemical response to osmotic stress in alfalfa (*Medicago sativa* L.). Pak J Bot 40(2):735–746

Saglam A, Saruhan N, Terzi R, Kadioglu A (2011) The relations between antioxidant enzymes and chlorophyll fluorescence parameters in common bean cultivars differing in sensitivity to drought stress. Russ J Plant Physiol 58:60–68

Sahitya UL, Krishna MSR, Prasad GS, Kasim DP, Deepthi RS (2018) Seed antioxidants interplay with drought stress tolerance indices in chilli (*Capsicum annuum* L.) seedlings. Biomed Res Int 14, Article ID 1605096, 14 p. https://doi.org/10.1155/2018/1605096

Sanchez-Leon S, Gil-humanes J, Ozuna CV, Sousa C, Voytas DF, Barro F (2018) Low-gluten, nontransgenic wheat engineered with CRISPR/Cas9. Plant Biotechnol J 16:902–910

Savitri ES, Fauziah SM (2019) Characterization of drought tolerance of GmDREB2 soybean mutants (Glycine max (L.) Merr) by ethyl methane sulfonate induction. In: AIP Conference Proceedings 2018 Oct 10, vol 1, p 020017. AIP Publishing LLC. https://doi.org/10.1063/1.5061853

Saxena KB, Singh G, Gupta HS, Mahajan V, Kumar RV, Singh B, Vales MI (2011) Enhancing the livelihoods of Uttarakhand farmers by introducing pigeonpea cultivation in hilly areas. J Food Legum 24:128–132

Schenk HJ, Jackson RB (2002) Rooting depths, lateral root spreads and below-ground/above-ground allometries of plants in water-limited ecosystems. J Ecol 90:480–494

Searchinger T, Waite R, Hanson C, Ranganathan J, Dumas P, Matthews F (2019) World resources final report 2019: Creating a sustainable food future—a menu of solutions to feed nearly 10 billion people by 2050. World Resource Institute, Final Report 2019, vol 564. https://wrr-food.wri.org/sites/default/files/2019-07/WRR_Food_Full_Report_0.pdf. Accessed 29 Aug 2019

Sehgal A, Sita K, Siddique KHM, Kumar R, Bhogireddy S, Varshney RK, Hanumantha Rao B, Nair RM, Prasad PVV, Nayyar H (2018) Drought or/and heat-stress effects on seed filling in food crops: impacts on functional biochemistry, seed yields, and nutritional quality. Front Plant Sci 9:1705. https://doi.org/10.3389/fpls.2018.01705

Shan H, Chen S, Jiang J, Chen F, Chen Y, Gu C et al (2012) Heterologous expression of the chrysanthemum R2R3-MYB transcription factor CmMYB2 enhances drought and salinity tolerance, increases hypersensitivity to ABA and delays flowering in Arabidopsis thaliana. Mol Biotechnol 51:160–173

Shavrukov Y, Kurishbayev A, Jatayev S, Shvidchenko V, Zotova L, Koekemoer F, De Groot S, Soole K, Langridge P (2017) Early flowering as a drought escape mechanism in plants: How can it aid wheat production? Front Plant Sci 8:1950. https://doi.org/10.3389/fpls.2017.01950

Shi J, Gao H, Wang H, Lafitte HR, Archibald RL, Yang M, Hakimi SM, Mo H, Habben JE (2017a) ARGOS8 variants generated by CRISPR-Cas9 improve maize grain yield under field drought stress conditions. Plant Biotechnol J 15(2):207–216

Shi S, Nan L, Smith KF (2017b) The current status, problems, and prospects of alfalfa (*Medicago sativa L.*) breeding in China. Agronomy 7(1):1. https://doi.org/10.3390/agronomy7010001

Shim JS, Oh N, Chung PJ, Kim YS, Choi YD, Kim JK (2018) Overexpression of osnac14 improves drought tolerance in rice. Front Plant Sci 9:310. https://doi.org/10.3389/fpls.2018.00310

Shinde S, Villamor JG, Lin W, Sharma S, Verslues PE (2016) Proline co-ordination with fatty acid synthesis and redox metabolism of chloroplast and mitochondria. Plant Physiol 172:1074–1088

Siddique KHM, Walton GH, Seymour M (1993) A comparison of seed yields of winter grain legumes in Western Australia. Aust J Exp Agric 33:15–22

Singh M, Kumar J, Singh S, Singh VP, Prasad SM (2015) Roles of osmoprotectants in improving salinity and drought tolerance in plants: a review. Rev Environ Sci Biotechnol 14(3):407–426

Sircar D, Cardoso HG, Mukherjee C, Mitra A, Arnholdt-Schmitt B (2012) Alternative oxidase (AOX) and phenolic metabolism in methyl jasmonate-treated hairy root cultures of *Daucus carota* L. J Plant Physiol 169(7):657–663

Slama I, Abdelly C, Bouchereau A, Flowers T, Savouré A (2015) Diversity, distribution and roles of osmoprotective compounds accumulated in halophytes under abiotic stress. Ann Bot 115:433–447

Smith P, House JI, Bustamante M, Sobocká J, Harper R, Pan G, West PC, Clark JM, Adhya T, Rumpel C, Paustian K (2016) Global change pressures on soils from land use and management. Global Change Biol 22(3):1008–1028

Sofi PA, Djanaguiraman M, Siddique KHM, Prasad PVV (2018) Reproductive fitness in common bean (*Phaseolus vulgaris L.)* under drought stress is associated with root length and volume. Indian J Plant Physiol 23:796–809

Solanki JK, Sarangi SK (2015) Effect of drought stress on proline accumulation in peanut genotypes. Int J Adv Res 2:301–309

Srivastava R, Kumar S, Kobayashi Y, Kusunoki K, Tripathi P, Kobayashi Y, Koyama H, Sahoo L (2018) Comparative genome-wide analysis of WRKY transcription factors in two Asian legume crops: Adzuki bean and Mung bean. Sci Report 8:16971. https://doi.org/10.1038/s41598-018-34920-8

Suárez R, Calderón C, Iturriaga G (2009) Enhanced tolerance to multiple abiotic stresses in transgenic alfalfa accumulating trehalose. Crop Sci 49:1791–1799

Sunkar R, Zhu JK (2004) Novel and stress-regulated microRNAs and other small RNAs from Arabidopsis. Plant Cell 16(8):2001–2019

Tang L, Cai H, Ji W, Luo X, Wang Z, Wu J, Wang X, Cui L, Wang Y, Zhu Y, Bai X (2013) Overexpression of GsZFP1 enhances salt and drought tolerance in transgenic alfalfa (*Medicago sativa L.*). Plant Physiol Biochem 71:22–30

Tang N, Ma S, Zong W, Yang N, Lv Y, Yan C, Guo Z, Li J, Li X, Xiang Y, Song H (2016) MODD mediates deactivation and degradation of OsbZIP46 to negatively regulate ABA signaling and drought resistance in rice. Plant Cell 28:2161–2177

Torres AM, Avila CM, Gutierrez N, Palomino C, Moreno MT, Cubero JI (2010) Marker-assisted selection in faba bean (*Vicia faba L.*). Field Crops Res 115:243–252

Turan S, Cornish K, Kumar S (2012) Salinity tolerance in plants: breeding and genetic engineering. Aust J Crop Sci 6(9):1337–1348

Ullah A, Manghwar H, Shaban M, Khan AH, Akbar A, Ali U, Ali E, Fahad S (2018) Phytohormones enhanced drought tolerance in plants: a coping strategy. Environ Sci Pollut Res 25:33103–33118

Upadhyaya HD, Kashiwagi J, Varshney RK, Gaur PM, Saxena KB, Krishnamurthy L, Gowda CL, Pundir RP, Chaturvedi SK, Basu PS, Singh IP (2012) Phenotyping chickpeas and pigeonpeas for adaptation to drought. Front Physiol 3:179. https://doi.org/10.3389/fphys.2012.00179

Vadez V, Rao S, Kholova J, Krishnamurthy L, Kashiwagi J, Ratnakumar P (2008) Root research for drought tolerance in legumes: Quo vadis? J Food Legum 21:77–85

Varshney RK, Bansal KC, Aggarwal PK, Datta SK, Craufurd PQ (2011) Agricultural biotechnology for crop improvement in a variable climate: hope or hype? Trends Plant Sci 16:363–371

Vasconcelos ESD, Barioni Junior W, Cruz CD, Ferreira RDP, Rassini JB et al (2008) Alfalfa genotype selection for adaptability and stability of dry matter production. Acta Sci Agron 30:339–343

Wang L, Zhu J, Li X, Wang S, Wu J (2018) Salt and drought stress and ABA responses related to bZIP genes from *V. radiata* and *V. angularis*. Gene 651:152–160

Wang Z, Ke Q, Kim MD, Kim SH, Ji CY, Jeong JC, Lee HS, Park WS, Ahn MJ, Li H, Xu B (2015) Transgenic alfalfa plants expressing the sweetpotato Orange gene exhibit enhanced abiotic stress tolerance. PLoS ONE 10:e0126050

Ward P, Micin M, Dunin F (2006) Using soil, climate and agronomy to predict soil water use by Lucerne compared with soil water use by annual crops or pastures. Aust J Soil Res 57:347–354

Weyers JDB, Paterson NW (2001) Plant hormones and the control of physiological processes. New Phytol 152:375–407

Xu J, Yuan Y, Xu Y, Zhang G, Guo X, Wu F, Wang Q, Rong T, Pan G, Cao M et al (2014) Identification of candidate genes for drought tolerance by whole-genome resequencing in maize. BMC Plant Biol 14:83. https://doi.org/10.1186/1471-2229-14-83

Yamaguchi-Shinozaki K, Shinozaki K (2006) Transcriptional regulatory networks in cellular responses and tolerance to dehydration and cold stresses. Annu Rev Plant Biol 2(57):781–803

Yamamoto E, Karakaya HC, Knap HT (2000) Molecular characterization of two soybean homologs of Arabidopsis thaliana CLAVATA1 from the wild type and fasciation. Biochim Biophys Acta 1491:333–340

Yang S, Pang EW, Ash EG, Huttner E, Zong EX, Kilian EA (2006) Low level of genetic diversity in cultivated Pigeonpea compared to its wild relatives is revealed by diversity arrays technology. Theor Appl Genet 113:585–595

Yasar F, Uzal O, Yasar TO (2013) Investigation of the relationship between the tolerance to drought stress levels and antioxidant enzyme activities in green bean (*Phaseolus Vulgaris* L.) genotypes. Afr J Agric Res 8:5759–5763

Ying S, Zhang DF, Fu J, Shi YS, Song YC, Wang TY, Li Y (2012) Cloning and characterization of a maize bZIP transcription factor, ZmbZIP72, confers drought and salt tolerance in transgenic Arabidopsis. Planta 235:253–266

Yoo JH, Park CY, Kim JC, Do Heo W, Cheong MS, Park HC, Kim MC, Moon BC, Choi MS, Kang YH, Lee JH (2005) Direct interaction of a divergent CaM isoform and the transcription factor, MYB2, enhances salt tolerance in Arabidopsis. J Biol Chem 280:3697–3706

Yousfi N, Sihem N, Ramzi A, Abdelly C (2016) Growth, photosynthesis and water relations as affected by different drought regimes and subsequent recovery in *Medicago laciniata* (L.) populations. J Plant Biol 59:33–43

Zandalinas SI, Mittler R, Balfagón D, Arbona V, Gómez-Cadenas A (2018) Plant adaptations to the combination of drought and high temperatures. Physiol Plant 162:2–12

Zhai Y, Zhang L, Xia C, Fu S, Zhao G, Jia J, Kong X (2016) The wheat transcription factor, TabHLH39, improves tolerance to multiple abiotic stressors in transgenic plants. Biochem Biophys Res Commun 473:1321–1327

Zhang B, Wang Q (2015) MicroRNA based biotechnology for plant improvement. J Cell Physiol 230:1–15

Zhang J, Duan Z, Zhang D, Zhang J, Di H, Wu F, Wang Y (2016) Co-transforming bar and CsLEA enhanced tolerance to drought and salt stress in transgenic alfalfa (*Medicago sativa L.*). Biochem Biophys Res Commun 472:75–82

Zhang JY, Broeckling CD, Blancaflor EB, Sledge MK, Sumner W, Wang ZY (2005) Overexpression of WXP1, a putative *Medicago truncatula* AP2 domain-containing transcription factor gene, increases cuticular wax accumulation and enhances drought tolerance in transgenic alfalfa (*Medicago sativa*). Plant J 42:689–707

Zhang T, Yu LX, Zheng P, Li Y, Rivera M, Main D, Greene SL (2015) Identification of loci associated with drought resistance traits in heterozygous auto tetraploid Alfalfa *(Medicago sativa L.)* Using genome-wide association studies with genotyping by sequencing. PLoS ONE 10(9):e0138931

Zhao X, Yang X, Pei S, He G, Wang X, Tang Q, Jia C, Lu Y, Hu R, Zhou G (2016) The Miscanthus NAC transcription factor MlNAC9 enhances abiotic stress tolerance in transgenic Arabidopsis. Gene 586:158–169

Zheng G, Fan C, Di S, Wang X, Xiang C, Pang Y (2017) Over-expression of arabidopsis EDT1 gene confers drought tolerance in Alfalfa (*Medicago sativa* L.). Front Plant Sci 8:2125. https://doi.org/10.3389/fpls.2017.02125

Zhou M, Luo H (2013) MicroRNA-mediated gene regulation: potential applications for plant genetic engineering. Plant Mol Biol 83:59–75

Zhu J, Song N, Sun S, Yang W, Zhao H, Song W, Lai J (2016) Efficiency and inheritance of targeted mutagenesis in maize using CRISPR-Cas9. J Genet Genom 43:25–36

Zhu JK (2002) Salt and drought stress signal transduction in plants. Annu Rev Plant Biol 53:247–273

Zoz T, Castagnara D (2013) Peroxidase activity as an indicator of water deficit tolerance in soybean cultivars. Biosci J 29:1664–1671

Phaseolus Species Responses and Tolerance to Drought

Jose A. Polania, Caspar C. C. Chater, Alejandra A. Covarrubias, and Idupulapati M. Rao

Abstract Drought is the main abiotic stress limiting bean crop yield in smallholder systems in developing countries. *Phaseolus* is one of the main genera among the cultivated grain legumes, with a wide range of adaptations to different environments, from humid tropics to deserts. Among the *Phaseolus* species, five are known to have been domesticated and cultivated; common bean (*P. vulgaris*) being the most economically important, while tepary bean (*P. acutifolius*) is better adapted to drought. Drought stress affects growth, development, and yield of beans. Terminal drought can be catastrophic for farmers' crop yields. Several studies have been carried out to: characterize the physiological responses to drought stress; identify developmental and physiological traits related to drought tolerance; and identify the possible genes involved. Crop responses to drought are very varied, depending on the species and cultivar; reflecting their unique environmental, evolutionary, and/or domestication history. Potential target traits include water use efficiency, effective use of water, rooting depth, fine root production, root hair length and density, stomatal traits, biomass accumulation, harvest index, mobilization of photoassimilates to pod and seed formation and filling, and carbon fixation by non-foliar photosynthesis. In this chapter, we discuss the main responses of the *Phaseolus* genus to drought stress.

Keywords CO_2 recycling · Harvest index · Pod epidermis · Shoot traits · Sink strength · Root traits

J. A. Polania (✉) · C. C. C. Chater · A. A. Covarrubias
Departamento de Biología Molecular de Plantas, Instituto de Biotecnología, Universidad Nacional Autónoma de México, 62210 Cuernavaca, Mexico
e-mail: josepolania03@gmail.com

C. C. C. Chater
Department of Molecular Biology and Biotechnology, University of Sheffield, Sheffield S10 2TN, UK

I. M. Rao
International Center for Tropical Agriculture (CIAT), Km 17, Recta Cali-Palmira, A.A. 6713, Cali, Colombia

Present Address:
Agricultural Research Service, Department of Agriculture, Plant Polymer Research Unit, National Center for Agricultural Utilization Research, 1815 North University Street, Peoria, IL 61604, USA

M. Hasanuzzaman et al. (eds.), *The Plant Family Fabaceae*,
https://doi.org/10.1007/978-981-15-4752-2_12

1 Introduction

Beans (*Phaseolus spp.*) and especially common bean (*Phaseolus vulgaris* L.) represent the most important legumes worldwide for direct consumption. Their nutritional qualities, as sources of high quality protein, micronutrients, carbohydrates, and fiber (Beebe et al. 2014; Rao 2014), make them valuable food sources for human health. As legume species that perform symbiotic nitrogen fixation (SNF) with rhizobial bacteria, bean crops are important components in agro-sustainability, particularly in rotation with cereals in production systems. There are around 70 species in the genus *Phaseolus*, most of which are located in Mesoamerica. Five species were domesticated: the common bean *P. vulgaris*; the year-long bean *P. dumosus*; the runner bean *P. coccineus*; the tepary bean *P. acutifolius*; and the Lima bean *P. lunatus*. Of these, *P. vulgaris* and *P. lunatus* underwent two independent and isolated domestications, one in Mesoamerica and the other in the Andes (Gepts 2001; Freytag and Debouck 2002; Delgado-Salinas et al. 2006).

Of the five cultivated *Phaseolus* species, four of them (*P. vulgaris*, *P. dumosus*, *P. coccineus*, and *P. acutifolius*) belong to the Vulgaris group. In particular, *P. vulgaris, P. dumosus*, and *P. coccineus* are very closely related (Bitocchi et al. 2017). However, they present some differences in their mating systems and life cycles, as *P. vulgaris* is predominantly autogamous and annual, *P. acutifolius* is autogamous and annual, *P. coccineus* is predominantly allogamous and perennial, and *P. dumosus* has intermediate characteristics between *P. coccineus* and *P. vulgaris* (Freytag and Debouck 2002; Bitocchi et al. 2017). The Lima bean belongs to the Lunatus group and is predominantly an autogamous species that includes both annuals and perennials (Delgado-Salinas et al. 2006; Bitocchi et al. 2017).

As a result of its geographical distribution, evolution, and domestication, the cultivated species of *Phaseolus* have adapted to and been selected for different environmental conditions, including extreme cold and heat, deficiency or excess of water, photoperiod, various soil conditions, and responses to pests and diseases (Bitocchi et al. 2017; De Ron et al. 2019). While *P. vulgaris* is generally adapted to warmer temperatures in semi-humid to semi-arid environments, *P. coccineus* is generally adapted to cooler temperatures in more humid environments at higher altitudes. *P. dumosus* is characterized by intermediate adaptation, and *P. acutifolius* has generally been selected for extreme drought tolerance, because of its origin from hotter and more arid environments of northern Mexico and the southeast USA. *P. lunatus* is adapted to humid and sub-humid climates, as well as to warm temperate zones (Butare et al. 2011; Beebe et al. 2013; Bitocchi et al. 2017).

Beans are usually cultivated by smallholder farmers in many regions of America and Africa, where they are often exposed to unfavorable growth conditions, including many biotic and abiotic stresses (Beebe et al. 2013, 2014; Polania et al. 2016a). Among the abiotic constraints, drought is the most limiting factor for bean growth and seed production, causing yield losses that can reach up to 100% in cases of severe terminal drought (Rao 2014; Polania et al. 2016c, 2017b). The adaptive responses to drought stress in *Phaseolus* species are diverse and related to the environment where they evolved; some species such as *P. coccineus* and *P. dumosus* have few adaptations to drought stress, while *P. acutifolius* is the best adapted (Rao et al. 2013; Polania et al. 2016a, 2017a). Similarly, plant characteristics that confer drought tolerance are diverse; some traits are species-specific, while others are shared across *Phaseolus* species due to the high degree of collinearity between genomes which indicates conserved gene function (Gujaria-Verma et al. 2016; Bitocchi et al. 2017; Lobaton et al. 2018).

Efforts have been made for the phenotypic characterization of bean drought response, as well as the molecular characterization of tolerance to this stress. There has been significant progress in defining phenotypic responses to drought and this knowledge is useful to identify potential target traits for developing drought-tolerant breeding lines. There are several target traits that are proving useful. These include, but are not limited to, rooting depth, fine root production, root hair length and density, water use efficiency, effective use of water, stomatal development and physiology, osmotic control, biomass accumulation, harvest index, mobilization of photoassimilates to pod and seed formation and filling, superior enzymatic activity related to sink strength, and carbon fixation by non-foliar photosynthesis (Cuellar-Ortiz et al. 2008; Polania et al. 2016a, 2017a; Chater et al. 2017; Rao et al. 2017; González et al. 2019). In this chapter, we discuss the role of these traits in the responses of the *Phaseolus* genus to drought stress, their phenotypic variation (particularly between *P. vulgaris* and *P. acutifolius*) and the potential of *Phaseolus* wild relatives and interspecies hybridization for future crop improvement to improve survival and yields under drought stress.

1.1 Phenotypic Characterization Under Drought Stress

Of the cultivated *Phaseolus* species, common beans (*P. vulgaris*) have been the subject of more phenotypic studies of drought stress responses than the other species, followed by tepary beans (*P. acutifolius*). These studies have focused on root system responses, leaf photosynthesis, plant growth and development, biomass accumulation, and partitioning of assimilates to the reproductive structures of the plant (Beebe et al. 2013; Rao et al. 2013; Araújo et al. 2015; Polania and Rao 2019).

1.1.1 Leaf Photosynthetic Responses to Drought Stress

Leaf photosynthesis is a widely studied process which is severely affected by drought stress, as reduced water availability limits gas exchange and therefore CO_2 fixation (Farooq et al. 2016; Mathobo et al. 2017; Nadeem et al. 2019). However, there is limited evidence of differences in leaf photosynthesis between tolerant and drought-sensitive common bean genotypes (Lizana et al. 2006; Rosales et al. 2012). Under drought, chlorophyll content, measured using both destructive and nondestructive techniques, is shown to increase in the leaf. However, this observed higher chlorophyll concentration per unit leaf area is primarily a consequence of the inhibition of leaf expansion under drought (Rosales-Serna et al. 2004; Polania et al. 2012), rather than a compensatory upregulation of photosynthesis. This interpretation is also supported by the poor correlation between leaf chlorophyll content and seed yield under drought stress in different field and greenhouse studies (Rosales-Serna et al. 2004; Polania et al. 2012). Chlorophyll content, therefore, allows discrimination between non-stress and stress treatments, but is not a useful trait as a selection tool for drought tolerance. A similar conclusion can be reached with respect to differences observed in the photosynthetic efficiency of photosystem II (PSII) in irrigated and drought stressed common beans, which suggest an effect of water limitation on quantum yield. However, no clear relations have been found between photosynthetic efficiency, genotype, and seed yield within and between treatments. Again, this trait appears to be of limited use for the selection of bean genotypes adapted to drought in breeding programs (Polania and Rao 2019).

1.1.2 Water Use Under Drought Stress: Spenders and Savers

Characteristics related to water use show a high correlation with seed production under drought stress, as well as broad differences between bean genotypes under these conditions. Some traits related with water status, water use, and transpiration include stomatal conductance, stomatal density, canopy temperature depression, and carbon isotope discrimination (Polania et al. 2012, 2016b, c, 2017b; Beebe et al. 2013; Rao et al. 2017). Several reports indicate that these traits could serve as selection criteria for drought tolerance in breeding programs. Water use efficiency (WUE), the amount of carbon gain relative to water loss, or 'more crop per drop,' is considered an important component of drought tolerance in different crops (Blum 2009; Sinclair 2012; Vadez et al. 2014). WUE is a complex trait and difficult to phenotype, preventing many breeding programs from using it directly (Araus et al. 2002; Easlon et al. 2014; Leakey et al. 2019). It has been reported that traits related to conserving water at the vegetative stage (e.g., lower leaf conductance, lower stomata density, smaller leaf canopy), slow down the depletion of available water in the soil profile and save water for reproductive growth and seed filling, resulting in better seed yield under terminal drought stress conditions (Zaman-Allah et al. 2011; Araújo et al. 2015; Chater et al. 2019). However, increases in WUE may reduce transpiration rate at the expense of carbon uptake, biomass accumulation, and, ultimately, yield potential (Blum 2005; Polania et al. 2016a). Nevertheless, the selection of genotypes for higher WUE is not only an appropriate approach for breeding for severe droughts or terminal drought in arid environments. It may also provide large yield gains with little carbon cost in legumes compared to non-legume crops, particularly in resource-limited smallholder systems where nitrogen fertilizer application is limited (Adams et al. 2016). In contrast to WUE, effective use of water (EUW) implies breeding for maximal soil moisture capture for transpiration and decreased non-stomatal transpiration (Blum 2009). EUW is relevant when there is still soil water available, when deep-rooted genotypes access water deep in the soil profile that is not normally available, or under moderate or intermittent droughts (Araus et al. 2002; Polania et al. 2016a, 2017a).

Bean genotypes can be classified into two phenotypic groups based on water use traits: water savers (the isohydric plant model) and water spenders (the anisohydric plant model). This grouping facilitates targeting genotypes to specific agro-ecological niches (Polania et al. 2016a, 2017a, b; Polania and Rao 2019). In the field, phenotypes with the water spending strategy under drought stress generate deeper roots and can, therefore, access more water from deep soil layers, which results in an improved plant water status. This increase in water use is associated with greater carbon assimilation (increased stomatal conductance) and plant growth (higher EUW). If combined with better photosynthate remobilization to pod and seed production, this can improve seed yield under certain drought conditions. Water spending genotypes could prove useful for cultivation in agro-ecological regions where seasonal rainfall is intermittent and where there is available water deep in the soil profile (Polania et al. 2017a; Polania and Rao 2019).

Drought-tolerant genotypes that are classified in the 'water saver' ideotype generate fewer roots and have access to less water, resulting in reduced stomatal conductance, and moderate gas exchange and plant growth (higher WUE). Combined with better photosynthate partitioning, this results in a moderate level of seed yield under severe drought stress. Water-saving genotypes can be more suitable to semi-arid and arid environments dominated by terminal drought stress. Approaches to improve WUE and the water-saving strategy include the optimization of stomatal development and stomatal physiology. Both approaches aim to minimize total plant water loss by reducing the available pore area for transpiration, and these molecular mechanisms appear to be highly conserved across land plants (Chater et al. 2016, 2017, 2019). Inhibition of stomatal development contributes to the reduction of leaf stomatal density (the number of stomata per unit area), thereby enhancing WUE. As with leaf chlorophyll content, stomatal densities tend to increase and stomatal size decreases during drought as cell and leaf expansion rates are inhibited, so downregulation of stomatal development plays an important role to counteract this effect. This compensatory response may also be observed in tepary bean pods compared to common bean pods in Fig. 1a, b. Bean pre-breeding efforts to target stomatal physiology, in particular for the water saver ideotype, could include the identification of variation in stomatal conductance responses. Differences in stomatal kinetics (the rates of opening and closing in response to environmental factors such as sunflecks, VPD, etc.), as well as maximal pore aperture, and guard cell size, across genotypes are not only inheritable but also selectable traits that can be translated into improved WUE. Natural variation in stomatal densities, patterning, and responses across crop genotypes and wild relatives represents a valuable resource to be exploited for improving drought tolerance in common bean (Blair et al. 2016; Chater et al. 2019).

The classification of bean genotypes into water savers or spenders seems to be related to their place of origin and domestication. Common bean races (e.g., Durango), and teparies that evolved in arid parts of Mexico, tend to be water savers, while Andean races and those that evolved in more temperate and semi-arid environments, behave as water spenders. Most domesticated tepary beans were bred for and

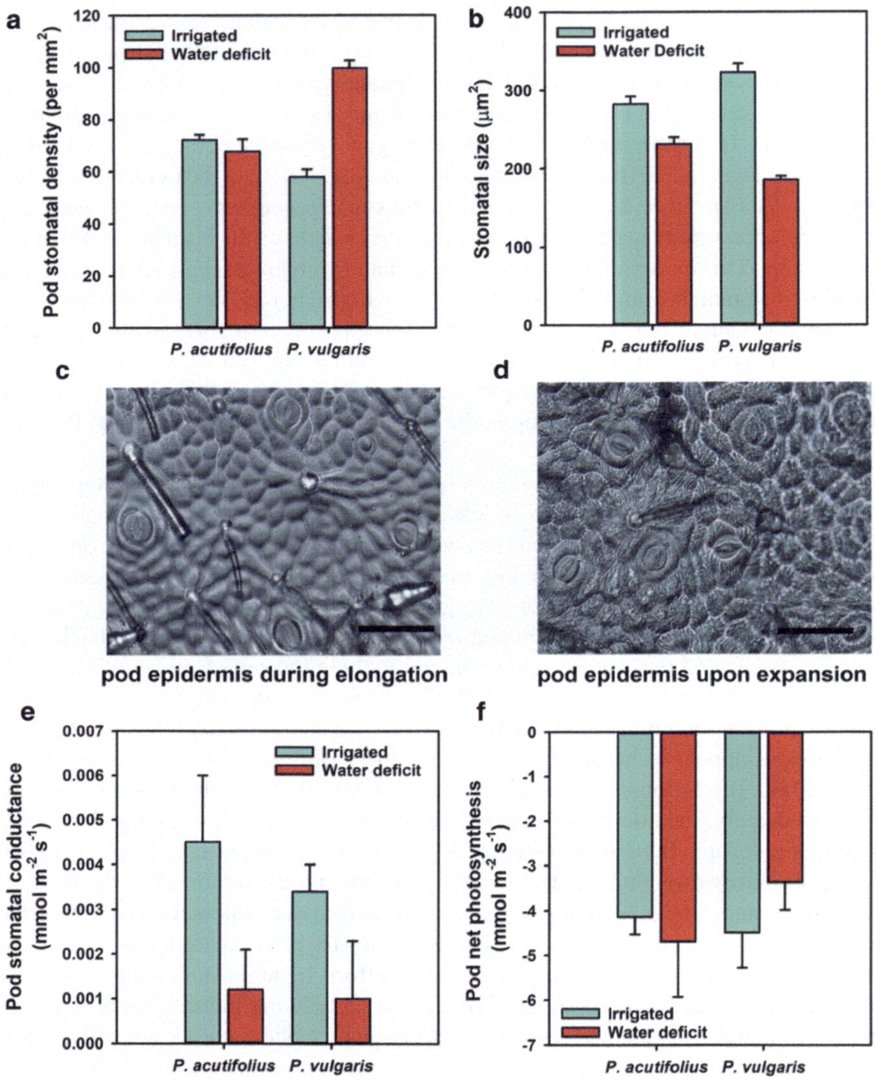

Fig. 1 Common bean and tepary bean pod responses to water deficit Comparison of pod stomatal density (**a**) and size (**b**) responses to water deficit between *Phaseolus vulgaris* (Pinto Saltillo) and *P. acutifolius* (Tepary VSU #10). Impressions of *P. acutifolius* (Tepary VSU #10) pod epidermis during pod elongation (**c**) and upon expansion (pre-midpod filling) (**d**), showing enhanced cuticle patterning, epicuticular wax production, and possible occlusions of the stomatal pores. Scale bars = 50 μm. Conductance (**e**) and net photosynthesis (**f**) at midpod filling stage in *Phaseolus vulgaris* (Pinto Saltillo) and *P. acutifolius* (Tepary TARS 32) under irrigated and water deficit conditions

are grown in the deserts of northern Mexico and the south west USA and are well suited to semi-arid-to-arid conditions (Beebe et al. 2014), traditionally relying on fleeting floodplain waters for rapid growth. The tepary bean presents traits related to the water saver model and is different to common bean, with small and thin leaves, fine roots, and reduced stomatal conductance and water loss. Tepary also shows more inhibitory stomatal density responses (Fig. 1a) by counteracting drought-associated reductions in cell size (Fig. 1b), better osmotic control, and a greater ability for carbon remobilization compared to common bean (Mohamed et al. 2005; Butare et al. 2011; Rao et al. 2013; Beebe et al. 2014).

1.1.3 Mobilization of Photosynthates in Response to Drought Stress

A superior mobilization of photosynthates to seed production, often measured as harvest index (HI), is central to drought tolerance mechanisms and genetic improvement across bean species. Several studies in common bean and tepary bean have shown that photosynthate mobilization to pods and seeds contributes to better yield under drought (Beebe et al. 2008, 2013; Devi et al. 2013; Rao 2014; Polania et al. 2016a, 2017b; Rao et al. 2017; Chaves et al. 2018; Mukankusi et al. 2018). This has been demonstrated phenotypically through the use of dry matter partitioning indices, sugar determination, and carbon labeling techniques. Dry matter partitioning indices include: the pod partitioning index (PPI) which indicates the extent of mobilization of assimilates from vegetative structures to pods, and the pod harvest index (PHI) that indicates the degree of mobilization of assimilates from the pod wall to seed formation and filling (Assefa et al. 2013; Beebe et al. 2013; Rao et al. 2013, 2017; Polania et al. 2016a, c, 2017b; Chaves et al. 2018).

Genotypes tolerant to drought, independent of their water use strategy (savers or spenders) show higher PPI and PHI; in other words, a greater mobilization and contribution of photosynthates to the pod and final yield. Whereas drought-susceptible genotypes, such as *P. coccineus* and *P. dumosus*, have a limited capacity for photosynthate mobilization to reproductive organs, with lower rates of carbon translocation (low HI and PHI values), highly tolerant beans, such as *P. acutifolius*, show the highest rates of carbon translocation to pod and seed formation (high HI and PHI values) (Rao et al. 2013; Polania et al. 2016a, 2017a). Low HI and PHI also correlate with more vegetative growth, poor pod formation, seed set, and seed quality in drought-susceptible genotypes (Butare et al. 2011, 2012).

Photosynthate mobilization can also explain differences in drought response between tolerant Mesoamerican and susceptible Andean *P. vulgaris* gene pools (Polania et al. 2016c). The ability to mobilize photosynthates (PPI and PHI) is greater in Mesoamerican genotypes than in Andean genotypes, and Andean drought sensitivity may be a direct result of a limited genetic capacity to redirect plant reserves to pod and seed production under water limitation. This stark difference may be a result of the divergent evolutionary events undergone before and during the domestication of these races; historically, the Mesoamerican gene pool was exposed to more extreme drought stresses than the Andean lineages (Polania et al. 2016c; Polania and Rao 2019).

Superior photosynthate mobilization to pod and seed formation under drought stress has not only been demonstrated through use of partitioning indices but also by biochemical analyses. Techniques such as ^{14}C-labelling, used to quantify sugar accumulation and partitioning, show that high seed yield under terminal drought stress is associated with an efficient carbon mobilization from leaves to pods and to seeds (Cuellar-Ortiz et al. 2008; Rosales et al. 2012). These lines of evidence indicate that efficient photosynthate mobilization is often a key mechanism for bean drought tolerance and that it involves total non-structural carbohydrate allocation from stems to pods and seed filling. As sucrose is the main plant long-distance transport carbohydrate from the source through the phloem, sucrose production, mobilization, and unloading are considered to be major factors for improving drought tolerance (Li et al. 2017). Under drought stress, significant changes observed in the expression of sucrose transporter genes (SUTs) suggest a critical role for sucrose transport for drought tolerance (Xu et al. 2018). However, sink strength limitations to sucrose unloading under drought may prove more important than phloem loading and transport, as observed under heat stress in common beans (Soltani et al. 2019).

1.1.4 Reproductive Sink Strength and Sink Carbon Recycling Under Drought

The flowering and reproductive phases are highly vulnerable and affected by drought stress, the ability of the plant to maintain flowers and form pods (avoid abortion) is fundamental in drought tolerance in beans (Nadeem et al. 2019). A major effect of drought stress on *Phaseolus* species and grain legumes in general is the reduction in yield components, such as pod number per area (PNA) and seed number per area (SNA) (Assefa et al. 2015; Polania et al. 2017b; Rao et al. 2017). Seed number per pod is determined by the potential of the ovules formed by the plant, pollination processes, and the capacity of the plant to transport carbohydrates toward the developing embryos (Farooq et al. 2016). *Phaseolus* plants, even under optimal water supply, sacrifice one or two ovules to optimize seed formation within a pod, and under drought conditions, a plant may abort half or more of its ovules depending on the species. However, embryo and seed abortion in drought-tolerant genotypes is minimized, and this, combined with efficient photoassimilate mobilization from vegetative structures and pod wall, ensures adequate seed formation and filling per pod. This is evident in *P. acutifolius*, which experiences less embryo abortion and displays superior seed formation compared to other *Phaseolus* species under drought. *P. acutifolius*' superior sink strength could be linked to its small seed size (Rao et al. 2013); a trait that has parallels with the small-seeded drought-tolerant Mesoamerican *P. vulgaris* genotypes. In contrast, species with large seeds, such as *P. coccineus,* and *P. dumosus*, and Andean *P. vulgaris* genotypes, display high levels of embryo abortion and lower seed numbers per pod under drought, indicating poor sink strength. By selecting bean genotypes with higher sink strength, resulting in a greater pod and seed number per area, we may increase seed yield under drought conditions (Rao et al. 2017).

Under water limitation, not only are pod and seed formation highly sensitive to the supply of source leaf photoassimilates but they may also depend on the remobilization of stored assimilates in stem tissues (Blum 2009; Rao et al. 2017). The processes of embryo development and seed filling have high carbon demands that are usually satisfied by phloem unloading (Smith et al. 2018). Although, like leaves, pod tissue is green and photosynthetic until senescence, some studies have shown that mature pod gas exchange is very low by comparison (Crookston et al. 1974). Constraints to mature pod gas exchange include low stomatal densities (Fig. 1a), reduced stomatal opening, and the development of thick waxy cuticle that may occlude the mature stomatal pore (Fig. 1c, d), which contribute to the very low stomatal conductance (Fig. 1e) and null values of net photosynthesis in mature pods (Fig. 1f). These effects are evident from determination of photosynthetic rates in the pod using a portable gas exchange system that showed negative values of pod net photosynthesis due to, not only low capacity of the pod to perform gas exchange but also from respiration of developing embryos, as shown in Fig. 1f. These observations suggest that, whereas stomata may play an active role during pod differentiation and expansion, their contribution to mature pod carbon assimilation and seed filling may be negligible. However, results from chickpea (*Cicer arietinum*) demonstrate that non-foliar podwall photosynthesis plays a fundamental role in the refixation of CO_2 released from respiring embryos (Furbank et al. 2004). Furthermore, under drought, this internal CO_2 recycling could be an important trait for plant carbon partitioning efficiency, providing a source of assimilates for seed filling at a time when externally-derived CO_2 is severely limited due to stomatal closure (Ma et al. 2001; Furbank et al. 2004; Brazel and Ó'Maoiléidigh 2019; González et al. 2019).

1.1.5 Reprioritizing Sinks to Reproduction: Nodule Function Under Drought Stress

Phaseolus species have a lower symbiotic nitrogen fixation (SNF) capacity compared to many legumes (Hardarson et al. 1993; Hardarson 2004; Peoples et al. 2009), but SNF remains crucial for bean and soil nutrition in low-input rainfed agri-systems that are most at risk of drought. As SNF has a high carbon cost (12–17 g of ATP for every gram of N_2 fixed) and is very sensitive to water deficit (Devi et al. 2013; Ramaekers et al. 2013; Polania et al. 2016b; Barbosa et al. 2018), carbon limitation under drought rapidly reduces nodule photosynthate supply, and thus downregulates nitrogen fixation (Gonzalez et al. 1995; Sassi et al. 2008). There is wide variation in the level of SNF between *Phaseolus* species and across cultivars within species (Wilker et al. 2019), particularly in their response to drought, again reflecting divergent evolutionary and domestication processes. *P. acutifolius* and Mesoamerican *P.*

vulgaris 'water saver' genotypes drastically reduce SNF under drought, suggesting that strong inhibitory signaling mechanisms interfering with SNF have been selected in tolerant beans (Polania et al. 2016b). This strategy leads to a rapid decrease of nodule carbon supply, favoring the mobilization of this carbon to pod and seed formation, while simultaneously increasing WUE and nitrogen use efficiency (Polania et al. 2016b). Nevertheless, this strategy may only be optimal in arid-to-semi-arid environments. In drought-tolerant Andean and Mesoamerican *P. vulgaris* water spender genotypes, a combination of superior seed yield and better SNF under drought stress have been reported. This spender response may be due to better source/sink relationships, using the acquired nitrogen (N) more efficiently to produce seed, improved remobilization of stored C and N to seed, and more abundant thick roots for accessing water and nutrients in the soil profile (Polania et al. 2017a).

1.1.6 Root Responses Under Drought Stress

Root system traits—like shoot traits—associated with drought tolerance follow a similar trend that reflects the ancestral and domestication environment of the germplasm (Strock et al. 2019) and their water saving or spending strategy. The *P. vulgaris* domestication process has selected for bean plants with greater rooting depth than their wild counterparts under drought stress (Berny Mier y Teran et al. 2018). Evaluation of root traits under field and greenhouse conditions have demonstrated the contribution of deep rooting to improving drought tolerance through increased water acquisition (Sponchiado et al. 1989; White and Castillo 1992; Polania et al. 2009, 2012, Polania et al. 2017a, b; Rao 2014; Rao et al. 2016; Strock et al. 2019).

A range of genotypes with differential root system architecture should be utilized for the development of drought-tolerant lines for specific agro-ecological conditions (Heng et al. 2018). There is extensive genotypic diversity in root system characteristics and drought responses between *Phaseolus* species. Whereas *P. acutifolius* has a fine root system and *P. coccineus* has a thick root system; *P. vulgaris* tends to have an intermediate root system by comparison (Butare et al. 2011, 2012; Polania et al. 2017a). In *P. vulgaris*, Andean genotypes develop a long seedling taproot and abundant basal roots, while Mesoamerican genotypes show the shortest seedling taproot length and few basal roots (Strock et al. 2019). As indicated before, drought-tolerant *P. vulgaris* water spenders have vigorous and deeper root systems (Polania et al. 2017a), and this rooting strategy combined with better photosynthate remobilization results in higher seed yield under drought. Some studies also suggest that higher root hair densities help the root to penetrate the soil and help provide essential functions in water uptake and nodule formation (Rivera et al. 2019). Higher basal root abundance and greater adventitious root development are also associated with improved water uptake and bean drought tolerance (Strock et al. 2019).

Water-saving genotypes combine higher seed yield under more extreme drought stress with a slower growing and shallower root system (Polania et al. 2017a). This strategy of water conservation and higher WUE is also complemented by more efficient photosynthate remobilization to pod and seed sinks (Polania et al. 2017a). As with rhizobial nodules, roots represent a large carbon sink. One model, to increase carbon use efficiency per area and improve drought tolerance, proposes to increase *Phaseolus* root porosity. This anatomical engineering could be achieved in several ways: by limiting cortical parenchyma production through the reduction of cortical cell files; through loss of cortical parenchyma and formation of root cortical aerenchyma and root cortical senescence; and by increasing cortical cell size (Lynch 2018). By limiting the carbon demands of the root system, mobilization of photosynthates to reproduction could increase yields whether the plant is under water limitation or not.

P. acutifolius has a water saver root system, with fine and shallow roots and efficient carbon use per unit root length; in other words, a higher specific root length or more root length per gram of biomass (Polania et al. 2017a). These root traits contribute to the enhanced photosynthate remobilization and early maturity characteristic of tepary beans. This root structure, combined with small leaves and greater stomatal control for reduced water use, results in extreme drought tolerance (Mohamed et al. 2005; Butare et al. 2011; Rao et al. 2013; Beebe et al. 2014; Polania et al. 2016a).

However, the relationship between root system characteristics and shoot growth can be complex; a genotype with a vigorous and deeper root system could fail in its performance under drought stress due to excessive vegetative growth and poor seed production. A vigorous and deeper root system, with rapid growth is useful, but not sufficient, for improved drought tolerance in bean. Instead, the strategic breeding of combinations of traits, including better root systems, photosynthate remobilization from vegetative structures, and increased pod and seed sink strength, are needed to reduce yield losses caused by drought (Beebe et al. 2014; Polania et al. 2016a, 2017a; Rao et al. 2017).

1.2 Molecular Approaches to Address Drought Tolerance Mechanisms

Molecular biology offers a powerful tool for accelerating the identification of superior genotypes under drought stress, through the generation of molecular markers of key processes related to drought tolerance. In *Phaseolus* species, molecular biology has already successfully identified and generated molecular markers related to disease resistance (Assefa et al. 2019), but their use for improvement of abiotic stress resistance and specifically of drought resistance has been more complex (Chater et al. 2019). This limitation exhibits the complexities of drought resistance as a multifactorial trait involving multiple plant processes and gene responses. Adjustment responses are highly influenced by the severity and duration of stress conditions, the developmental stage of the plant, tissue identity, genetic background, and the combination of these factors. An additional factor to be considered is that abiotic stresses usually occurring in a combined form (i.e., water deficit and high temperatures, water deficit, low temperatures, etc.).

Different genotyping technologies have been developed, utilizing single nucleotide polymorphisms (SNPs) that have become available through next-generation sequencing (NGS). These SNP-based markers have been applied to study diversity, genome mapping, and detection of quantitative trait loci (QTLs) for biotic and abiotic stresses in common beans (Diaz et al. 2018). Furthermore, the availability of reference genomes for common bean and other *Phaseolus* species, the reduction in sequencing costs, and the development of different methods for bioinformatic analyses are contributing to the hastening of knowledge acquisition. SNP and QTL-based methods will soon contribute to efficient, feasible, and pertinent molecular-based technologies for drought resistance marker assisted breeding (Lobaton et al. 2018).

By unraveling the genetic underpinnings of the drought tolerance traits and phenotypes described above, we may be able to accelerate bean molecular breeding. Efforts in this direction are the QTL analyses to find genome and chromosomal regions related to key traits under well-watered and drought conditions. These traits include, among others, seed yield, yield components (PHI, PPI, HI, number of pods per plant, seeds per pod), biomass partitioning, canopy biomass, and days-to-flowering. In this regard, a recent study found QTLs for PHI, some of which co-localized with three out of eight QTLs for seed yield, emphasizing the relevance of an efficient carbon allocation for seed production. As expected, for complex traits such as stress tolerance, this study presented evidence suggesting epistatic networks in traits such as yield and PHI (Berny Mier y Teran et al. 2019).

Other reports have focused on the characterization of the molecular basis of deep rooting, using a recombinant inbred line (RIL) population of DOR 364 × BAT 477 and QTL study of root traits under drought stress. Several QTLs on linkage groups b01 or b11 were identified, which explained up to 41% of the genetic variance (Asfaw and Blair 2012). Furthermore, two major and stable QTLs were mapped to Pv01 and Pv02 for grain yield under multiple stresses. The Pv01 QTL explained up to 37% of the phenotypic variance for seed yield, whereas the Pv02 QTL was detected under drought stress ($R^2 = 33\%$) and multiple stresses ($R^2 = 17–23\%$). Some of the SNPs linked to these QTLs were associated to genes whose transcript levels were correlated with water deficit responses. This study also provided evidence of the significance of phenological plasticity in adaptation to drought: although late maturity is associated with increased yield under optimal conditions, early maturity is associated with higher yield under terminal drought (Trapp et al. 2015).

Using a drought-tolerant Mesoamerican x drought-susceptible Andean population of *P. vulgaris*, twenty-two QTLs were detected for chlorophyll, leaf and stem fresh biomass, leaf dry biomass, leaf temperature, number of pods and seeds per plant, seed weight, days to flowering, pod biomass and yield under well-watered and drought conditions (Briñez et al. 2017). All QTLs detected under drought conditions came from the resistant parent, among the most important being leaf area, fresh mass and pod dry weight. A SNP characterization using drought-tolerant Durango genotypes detected 83 SNPs with particularly significant association to phenotypes such as days-to-flowering and seed biomass (Villordo-Pineda et al. 2015). Some of the SNPs were associated to genes involved in the plant response to drought. For example, the authors identified genes implicated in ABA mediated processes, starch biosynthesis, proline biosynthesis, reactive oxygen species generation, kinases, late embryogenesis abundant (LEA) proteins, and putative transcription factors, among others.

Overall, these different marker studies emphasize that drought tolerance is a complex quantitative trait, where many diverse genes/proteins participate to allow plant adjustment to water scarcity. Nevertheless, these works among others contribute to the knowledge required for the development of molecular marker assisted approaches for the selection of drought tolerance in common bean. These approaches will have to be tailored according to particular genotypes, environments, and the plant developmental stages of interest.

2 Future Perspectives

Results of phenotypic characterization of beans under drought stress indicate that tolerance is not related to a single trait or process, and, instead, it is the strategic combination of a few characteristics that results in higher seed productivity under drought conditions. Some traits, such as water use efficiency related to stomatal density and physiology, and greater carbon mobilization to pod and grain formation, may already be common characteristics in tolerant *Phaseolus* genotypes. However, the control and molecular bases of these processes have not been completely described. Studies on the molecular control of stomatal density and physiology, as well as its relationship with other plant processes could open up a wealth of novel breeding targets. Similarly, carbon mobilization is a key trait in drought tolerance across *Phaseolus* crops, but little is known about the biochemical and molecular processes that control this characteristic. In particular, the roles of transporters in photoassimilate loading and unloading to and from the phloem, and processes related to greater sink strength and capacity, could lead to novel candidates for yield improvement under drought. Research on the understanding of the molecular control of these key processes will allow for the development of molecular markers that will increase the efficiency of drought breeding programs in beans and other legumes.

Acknowledgements C. Chater is grateful for funding from the European Union's Horizon 2020 research and innovation program under the Marie Skłodowska-Curie grant agreement No. 700867. J. A. Polania was supported by DGAPA-UNAM postdoctoral fellowship. This work was partially supported by a CONABIO grant to AAC. We thank the Covarrubias-Reyes laboratory at IBT-UNAM and the bean team at CIAT for their contribution to this work. C. Chater would like to thank Dr. Harbans Bhardwaj (Virginia State University) for the gift of tepary seed VSU #10, and Dr. Tim Porch (USDA TARS) for the gift of tepary seed TARS 32.

References

Adams MA, Turnbull TL, Sprent JI, Buchmann N (2016) Legumes are different: leaf nitrogen, photosynthesis, and water use efficiency. Proc Natl Acad Sci 113:4098–4103. https://doi.org/10.1073/pnas.1523936113

Araújo SS, Beebe S, Crespi M et al (2015) Abiotic stress responses in legumes: strategies used to cope with environmental challenges. CRC Crit Rev Plant Sci 34:237–280. https://doi.org/10.1080/07352689.2014.898450

Araus JL, Slafer GA, Reynolds MP, Royo C (2002) Plant breeding and drought in C3 cereals: What should we breed for? Ann Bot 89:925–940. https://doi.org/10.1093/aob/mcf049

Asfaw A, Blair MW (2012) Quantitative trait loci for rooting pattern traits of common beans grown under drought stress versus non-stress conditions. Mol Breed 30:681–695. https://doi.org/10.1007/s11032-011-9654-y

Assefa T, Assibi Mahama A, Brown AV et al (2019) A review of breeding objectives, genomic resources, and marker-assisted methods in common bean (Phaseolus vulgaris L.). Mol Breed 39:1–23. https://doi.org/10.1007/s11032-018-0920-0

Assefa T, Beebe S, Rao IM et al (2013) Pod harvest index as a selection criterion to improve drought resistance in white pea bean. F Crop Res 148:24–33. https://doi.org/10.1016/j.fcr.2013.04.008

Assefa T, Wu J, Beebe S et al (2015) Improving adaptation to drought stress in small red common bean: phenotypic differences and predicted genotypic effects on grain yield, yield components and harvest index. Euphytica 303:477–489

Barbosa N, Portilla E, Buendia HF et al (2018) Genotypic differences in symbiotic nitrogen fixation ability and seed yield of climbing bean. Plant Soil 428:223–239. https://doi.org/10.1007/s11104-018-3665-y

Beebe S, Rao IM, Blair MW, Acosta-Gallegos JA (2013) Phenotyping common beans for adaptation to drought. Front Physiol 4:1–20. https://doi.org/10.3389/fphys.2013.00035

Beebe S, Rao IM, Cajiao C, Grajales M (2008) Selection for drought resistance in common bean also improves yield in phosphorus limited and favorable environments. Crop Sci 48:582–592. https://doi.org/10.2135/cropsci2007.07.0404

Beebe S, Rao IM, Devi M, Polania J (2014) Common beans, biodiversity, and multiple stress: challenges of drought resistance in tropical soils. Crop Pasture Sci 65:667–675. https://doi.org/10.1071/CP13303

Berny Mier y Teran JC, Konzen ER, Medina V et al (2018) Root and shoot variation in relation to potential intermittent drought adaptation of Mesoamerican wild common bean (Phaseolus vulgaris L.). Ann Bot 1–16. https://doi.org/10.1093/aob/mcy221

Berny Mier y Teran JC, Konzen ER, Palkovic A et al (2019) Effect of drought stress on the genetic architecture of photosynthate allocation and remobilization in pods of common bean (Phaseolus vulgaris L.), a key species for food security. BMC Plant Biol 19:1–15

Bitocchi E, Rau D, Bellucci E et al (2017) Beans (Phaseolus ssp.) as a model for understanding crop evolution. Front Plant Sci 8:21. https://doi.org/10.3389/fpls.2017.00722

Blair MW, Cortés AJ, This D (2016) Identification of an ERECTA gene and its drought adaptation associations with wild and cultivated common bean. Plant Sci 242. https://doi.org/10.1016/j.plantsci.2015.08.004

Blum A (2005) Drought resistance, water-use efficiency, and yield potential—are they compatible, dissonant, or mutually exclusive? Aust J Agric Res 56:1159–1168. https://doi.org/10.1071/AR05069

Blum A (2009) Effective use of water (EUW) and not water-use efficiency (WUE) is the target of crop yield improvement under drought stress. F Crop Res 112:119–123. https://doi.org/10.1016/j.fcr.2009.03.009

Brazel AJ, Ó'Maoiléidigh DS (2019) Photosynthetic activity of reproductive organs. J Exp Bot 70:1737–1754

Briñez B, Perseguini JMKC, Rosa JS et al (2017) Mapping QTLs for drought tolerance in a SEA 5 x AND 277 common bean cross with SSRs and SNP markers. Genet Mol Biol 40:813–823. https://doi.org/10.1590/1678-4685-gmb-2016-0222

Butare L, Rao IM, Lepoivre P et al (2011) New genetic sources of resistance in the genus Phaseolus to individual and combined aluminium toxicity and progressive soil drying stresses. Euphytica 181:385–404

Butare L, Rao IM, Lepoivre P et al (2012) Phenotypic evaluation of interspecific recombinant inbred lines (RILs) of Phaseolus species for aluminium resistance and shoot and root growth response to aluminium-toxic acid soil. Euphytica 186:715–730

Chater CCC, Caine RS, Fleming AJ, Gray JE (2017) Origins and evolution of stomatal development. Plant Physiol 174:624–638. https://doi.org/10.1104/pp.17.00183

Chater CCC, Caine RS, Tomek M, et al (2016) Origin and function of stomata in the moss Physcomitrella patens. Nat Plants 2:. https://doi.org/10.1038/nplants.2016.179

Chater CCC, Covarrubias AA, Acosta-Maspons A (2019) Crop biotechnology for improving drought tolerance: targets, approaches, and outcomes. Annu Plant Rev online 2:1–39. https://doi.org/10.1002/9781119312994.apr0669

Chaves N, Polania J, Muñoz C et al (2018) Caracterización fenotípica por resistencia a sequía terminal de germoplasma de frijol común. Agron Mesoam 29:1–17

Crookston RK, O'Toole J, Ozbun JL (1974) Characterization of the bean pod as a photosynthetic organ. Crop Sci 14:708–712. https://doi.org/10.2135/cropsci1974.0011183x001400050030x

Cuellar-Ortiz SM, De La Paz Arrieta-Montiel M, Acosta-Gallegos JA, Covarrubias AA (2008) Relationship between carbohydrate partitioning and drought resistance in common bean. Plant Cell Environ 31:1399–1409. https://doi.org/10.1111/j.1365-3040.2008.01853.x

De Ron AM, Álvarez-García S, Casquero PA, et al (2019) Common bean genetics, breeding, and genomics for adaptation to changing to new agri-environmental conditions. In: Kole C (ed) Genomic designing of climate-smart pulse crops. Springer, pp 1–106

Delgado-Salinas A, Bibler R, Lavin M (2006) Phylogeny of the genus Phaseolus (Leguminosae): a recent diversification in an ancient landscape. Syst Bot 31:779–791

Devi M, Sinclair TR, Beebe S, Rao IM (2013) Comparison of common bean (Phaseolus vulgaris L.) genotypes for nitrogen fixation tolerance to soil drying. Plant Soil 364:29–37. https://doi.org/10.1007/s11104-012-1330-4

Diaz LM, Ricaurte J, Tovar E et al (2018) QTL analyses for tolerance to abiotic stresses in a common bean (Phaseolus vulgaris L.) population. PLoS One 1–26. https://doi.org/10.1371/journal.pone.0202342

Easlon HM, Nemali KS, Richards JH et al (2014) The physiological basis for genetic variation in water use efficiency and carbon isotope composition in Arabidopsis thaliana. Photosynth Res 119:119–129. https://doi.org/10.1007/s11120-013-9891-5

Farooq M, Gogoi N, Barthakur S et al (2016) Drought stress in grain legumes during reproduction and grain filling. J Agron Crop Sci 203:81–102. https://doi.org/10.1111/jac.12169

Freytag G, Debouck DG (2002) Taxonomy, distribution, and ecology of the genus Phaseolus (Leguminosae-Papilionodeae) in North America, Mexico and Central America. Botanical Research Institute of Texas (BRIT), Forth Worth, TX, USA

Furbank RT, White R, Palta JA, Turner NC (2004) Internal recycling of respiratory CO_2 in pods of chickpea (Cicer arietinum L.): the role of pod wall, seed coat, and embryo. J Exp Bot 55:1687–1696

Gepts P (2001) Phaseolus vulgaris (Beans). In: Encyclopedia of Genetics. Academic Press, pp 1444–1445

Gonzalez EM, Gordon AJ, James CL, Arrese-Igor C (1995) The role of sucrose synthase in the response of soybean nodules to drought. J Exp Bot 46:1515–1523

González I, Acosta A, Cetz J, et al (2019) Relevance of carbon distribution towards reproductive organs in the resistance to terminal drought of common bean cultivars. In: 9th international conference on legume genetics and genomics (ICLGG). Dijon, Francia, p 1

Gujaria-Verma N, Ramsay L, Sharpe AG et al (2016) Gene-based SNP discovery in tepary bean (Phaseolus acutifolius) and common bean (P. vulgaris) for diversity analysis and comparative mapping. BMC Genom 17:1–16. https://doi.org/10.1186/s12864-016-2499-3

Hardarson G (2004) Enhancement of symbiotic nitrogen fixation in grain legumes: selected results from the FAO/IAEA program. In: Serraj R (ed) Symbiotic nitrogen fixation: prospects for enhanced application in tropical agriculture. Oxford & IBH Publishing, New Delhi, pp 163–171

Hardarson G, Bliss FA, Cigales-Rivero MR et al (1993) Genotypic variation in biological nitrogen fixation by common bean. Plant Soil 152:59–70. https://doi.org/10.1007/BF00016333

Heng Y, Manish R, Babu V et al (2018) Genetic diversity of root system architecture in response to drought stress in grain legumes. J Exp Bot 69:3267–3277

Leakey ADB, Ferguson JN, Pignon CP et al (2019) Water use efficiency as a constraint and target for improving the resilience and productivity of C 3 and C 4 crops. Annu Rev Plant Biol 70:781–808. https://doi.org/10.1146/annurev-arplant-042817-040305

Li G, Pan J, Cui K et al (2017) Limitation of unloading in the developing grains is a possible cause responsible for low stem non-structural carbohydrate translocation and poor grain yield formation in rice through verification of recombinant inbred lines. Front Plant Sci 8:1–16. https://doi.org/10.3389/fpls.2017.01369

Lizana C, Wentworth M, Martinez JP et al (2006) Differential adaptation of two varieties of common bean to abiotic stress: I. Effects of drought on yield and photosynthesis. J Exp Bot 57:685–697

Lobaton J, Miller T, Gil J et al (2018) Resequencing of common bean identifies regions of inter-gene pool introgression and provides comprehensive resources for molecular breeding. Plant Genome 11:1–21

Lynch J (2018) Rightsizing root phenotypes for drought resistance. J Exp Bot 69:3279–3292. https://doi.org/10.1093/jxb/ery048

Ma Q, Behboudian MH, Turner NC, Palta JA (2001) Gas exchange by pods and subtending leaves and internal recycling of CO_2 by pods of chickpea (Cicer arietinum L.) subjected to water deficits. J Exp Bot 52:123–131. https://doi.org/10.1093/jxb/52.354.123

Mathobo R, Marais D, Steyn JM (2017) The effect of drought stress on yield, leaf gaseous exchange and chlorophyll fluorescence of dry beans (Phaseolus vulgaris L.). Agric Water Manag 180:118–125. https://doi.org/10.1016/j.agwat.2016.11.005

Mohamed F, Mohamed M, Schmitz-Eiberger N et al (2005) Comparative drought postponing and tolerance potentials of two tepary bean lines in relation to seed yield. African Crop Sci J 13:49–60

Mukankusi C, Raatz B, Stanley Nkalubo, et al (2018) Genomics, genetics and breeding of common bean in Africa: a review of tropical legume project. Plant Breed 1–14. https://doi.org/10.1111/pbr.12573

Nadeem M, Li J, Yahya M et al (2019) Research progress and perspective on drought stress in legumes: a review. Int J Mol Sci 20:1–32. https://doi.org/10.3390/ijms20102541

Peoples MB, Brockwell J, Herridge DF et al (2009) The contributions of nitrogen-fixing crop legumes to the productivity of agricultural systems. Symbiosis 48:1–17. https://doi.org/10.1007/BF03179980

Polania J, Poschenrieder C, Beebe S, Rao IM (2016a) Effective use of water and increased dry matter partitioned to grain contribute to yield of common bean improved for drought resistance. Front Plant Sci 7:1–10. https://doi.org/10.3389/fpls.2016.00660

Polania J, Poschenrieder C, Beebe S, Rao IM (2016b) Estimation of phenotypic variability in symbiotic nitrogen fixation ability of common bean under drought stress using 15 N natural abundance in grain. Eur J Agron 79:66–73. https://doi.org/10.1016/j.eja.2016.05.014

Polania J, Poschenrieder C, Rao IM, Beebe S (2017a) Root traits and their potential links to plant ideotypes to improve drought resistance in common bean. Theor Exp Plant Physiol 29:143–154. https://doi.org/10.1007/s40626-017-0090-1

Polania J, Rao IM, Beebe S, Garcia R (2009) Root development and distribution under drought stress in common bean (Phaseolus vulgaris L.) in a soil tube system. Agron Colomb 27:25–32

Polania J, Rao IM, Cajiao C et al (2016c) Physiological traits associated with drought resistance in Andean and MesoAmerican genotypes of common bean (Phaseolus vulgaris L.). Euphytica 210:17–29. https://doi.org/10.1007/s10681-016-1691-5

Polania J, Rao IM, Cajiao C et al (2017b) Shoot and root traits contribute to drought resistance in recombinant inbred lines of MD 23–24 × SEA 5 of common bean. Front Plant Sci 8:1–18. https://doi.org/10.3389/fpls.2017.00296

Polania J, Rao IM, Mejía S et al (2012) Características morfo-fisiológicas de frijol común (Phaseolus vulgaris L.) relacionadas con la adaptación a sequía. Acta Agron 61:197–206

Polania JA, Rao IM (2019) Drought resistance of common bean. In: Pessarakli Mohammad (ed) Handbook of plant and crop stress, 4th edn. CRC Press, Boca Raton, FL., pp 805–812

Ramaekers L, Galeano CH, Garzón N et al (2013) Identifying quantitative trait loci for symbiotic nitrogen fixation capacity and related traits in common bean. Mol Breed 31:163–180. https://doi.org/10.1007/s11032-012-9780-1

Rao IM (2014) Advances in improving adaptation of common bean and Brachiaria forage grasses to abiotic stresses in the tropics. In: Pessarakli M (ed) Handbook of plant and crop physiology, 3rd edn. CRC Press, Taylor and Francis Group, Boca Raton, FL, pp 847–889

Rao IM, Beebe S, Polania J et al (2013) Can tepary bean be a model for improvement of drought resistance in common bean? African Crop Sci J 21:265–281

Rao IM, Beebe S, Polania J et al (2017) Evidence for genotypic differences among elite lines of common bean in the ability to remobilize photosynthate to increase yield under drought. J Agric Sci 155:857–875. https://doi.org/10.1017/S0021859616000915

Rao IM, Miles JW, Beebe S, Horst WJ (2016) Root adaptations to soils with low fertility and aluminium toxicity. Ann Bot 118:593–605. https://doi.org/10.1093/aob/mcw073

Rivera M, Polania J, Ricaurte J et al (2019) Soil compaction induced changes in morpho-physiological characteristics of common bean. J Soil Sci Plant Nutr 19:217–227

Rosales-Serna R, Kohashi-Shibata J, Acosta-Gallegos JA et al (2004) Biomass distribution, maturity acceleration and yield in drought-stressed common bean cultivars. F Crop Res 85:203–211. https://doi.org/10.1016/S0378-4290(03)00161-8

Rosales MA, Ocampo E, Rodríguez-Valentín R et al (2012) Physiological analysis of common bean (Phaseolus vulgaris L.) cultivars uncovers characteristics related to terminal drought resistance. Plant Physiol Biochem 56:24–34. https://doi.org/10.1016/j.plaphy.2012.04.007

Sassi S, Gonzalez EM, Aydi S et al (2008) Tolerance of common bean to long-term osmotic stress is related to nodule carbon flux and antioxidant defenses: evidence from two cultivars with contrasting tolerance. Plant Soil 312:39–48. https://doi.org/10.1007/s11104-008-9613-5

Sinclair TR (2012) Is transpiration efficiency a viable plant trait in breeding for crop improvement? Funct Plant Biol 39:359–365. https://doi.org/10.1071/FP11198

Smith MR, Rao IM, Merchant A (2018) Source-sink relationships in crop plants and their influence on yield development and nutritional quality. Front Plant Sci 9:1–10. https://doi.org/10.3389/fpls.2018.01889

Soltani A, Weraduwage SM, Sharkey TD, Lowry DB (2019) Elevated temperatures cause loss of seed set in common bean (Phaseolus vulgaris L.) potentially through the disruption of source-sink relationships. BMC Genom 20:1–18

Sponchiado BN, White JW, Castillo JA, Jones PG (1989) Root growth of four common bean cultivars in relation to drought tolerance in environments with contrasting soil types. Exp Agric 25:249–257. https://doi.org/10.1017/S0014479700016756

Strock CF, Burridge J, Massas ASF et al (2019) Seedling root architecture and its relationship with seed yield across diverse environments in Phaseolus vulgaris. F Crop Res 237:53–64

Trapp JJ, Urrea CA, Cregan PB, Miklas PN (2015) Quantitative trait loci for yield under multiple stress and drought conditions in a dry bean population. Crop Sci 55:1596–1607. https://doi.org/10.2135/cropsci2014.11.0792

Vadez V, Kholova J, Medina S et al (2014) Transpiration efficiency: new insights into an old story. J Exp Bot 65:6141–6153. https://doi.org/10.1093/jxb/eru040

Villordo-Pineda E, González-Chavira MM, Giraldo-Carbajo P et al (2015) Identification of novel drought-tolerant-associated SNPs in common bean (Phaseolus vulgaris). Front Plant Sci 6:1–9. https://doi.org/10.3389/fpls.2015.00546
White JW, Castillo JA (1992) Evaluation of diverse shoot genotypes on selected root genotypes of common bean under soil water deficits. Crop Sci 32:762–765
Wilker J, Navabi A, Rajcan I et al (2019) Agronomic performance and nitrogen fixation of heirloom and conventional dry bean varieties under low-nitrogen field conditions. Front Plant Sci 10:1–21. https://doi.org/10.3389/fpls.2019.00952
Xu Q, Chen S, Ren Y et al (2018) Regulation of sucrose transporters and phloem loading in response to environmental cues. Plant Physiol 176:930–945. https://doi.org/10.1104/pp.17.01088
Zaman-Allah M, Jenkinson DM, Vadez V (2011) Chickpea genotypes contrasting for seed yield under terminal drought stress in the field differ for traits related to the control of water use. Funct Plant Biol 38:270–281. https://doi.org/10.1071/FP10244

Fabaceae Plants Response and Tolerance to High Temperature Stress

Khursheda Parvin, Kamrun Nahar, Tasnim Farha Bhuiyan, and Mirza Hasanuzzaman

Abstract Fabaceae is the third largest flowering plant family, commonly known as legumes that are the major source of dietary protein; some are source of oil, fodder, timber, and medicine, and some are ornamental plants. Legume enriches soil health-adding atmospheric nitrogen. This potential crop is often under threat in terms of growth, physiology, developmental processes, and yield potential due to high temperature induced by climate change. There are various approaches to combat high temperature stress. Agronomic approaches and exogenous application of phytoprotectants including osmoprotectants, nutrients, antioxidants, polyamines, trace elements, phytohormone, and signaling molecule are potential ways to mitigate high temperature damages of Fabaceae plants. Few research findings also developed transgenic plants showing improved tolerance to high temperature stress. This chapter accommodates information regarding high temperature effects on Fabaceae plants and different approaches mitigating high temperature induced-damages which will draw the attention of the scientists and researchers and will help them in converting the Fabaceae plants to be more prominent against the threat of climate change.

Keywords High temperature stress · Oxidative stress · Antioxidant defense · Signaling molecule · Transgenic approach

K. Parvin
Laboratory of Plant Stress Responses, Department of Applied Biological Sciences, Faculty of Agriculture, Kagawa University, Miki-Cho, Kita-Gun, Kagawa 761-0795, Japan

Department of Horticulture, Sher-e-Bangla Agricultural University, Sher-e-Bangla Nagar, Dhaka 1207, Bangladesh

K. Nahar · T. F. Bhuiyan
Department of Agricultural Botany, Sher-e-Bangla Agricultural University, Sher-e-Bangla Nagar, Dhaka 1207, Bangladesh

M. Hasanuzzaman (✉)
Department of Agronomy, Sher-e-Bangla Agricultural University, Sher-e-Bangla Nagar, Dhaka 1207, Bangladesh
e-mail: mhzsauag@yahoo.com

M. Hasanuzzaman et al. (eds.), *The Plant Family Fabaceae*,
https://doi.org/10.1007/978-981-15-4752-2_13

1 Introduction

Among the flowering plant families, Fabaceae commonly known as legumes are the third largest family comprising of about 20,000 species. Based on nutrition and production, legumes occupy the second position after cereal (Kumar et al. 2018). Legume seeds are mostly eaten/uptaken for protein supply. Some legumes provide us with edible oils. A broad range of legumes are cultivated for fodder or/and forage. Some Fabaceae family plants are cultivated for getting wood and timber. Legume plants are also grown in a wide range for their flower or as ornamental plants. A group of Fabaceae plants have medicinal value. Almost all the Fabaceae plants enrich soil-fixing atmospheric nitrogen and also adding biofertilizer after incorporation. Without these, new by-products are also adding, which are also from Fabaceae plant derivatives (Voisin et al. 2014). Thus, Fabaceae plants occupy broad space in our everyday life.

Rising temperature is one of the most primary effects of global climate change which is resulted from number of reasons. The Intergovernmental Panel on Climate Change 2014 (IPCC 2014) revealed that the number of warm days and nights and the frequency of heat waves have been greater than before in many regions of this earth. The global surface temperature change is projected to exceed 2 °C by the end of the twenty-first century, compared to 1850–1900 (IPCC 2014). This increasing pattern is more severe in tropical/subtropical and dry regions. High temperature (HT) often does not come alone; it can associate with increased drought stress, and it can be combined with high salinity; it causes the melting of glaciers causing sea level rise. So, temperature rise is considered as the most severe environmental concern (Mittler and Blumwald 2010). The rise of temperature reduces the production of food crops, timber production, and medicinal plants. High temperature can cause total crop failure, and even extinction of species can be consequence of HT at extreme level. Similar to other plants, the Fabaceae plants are also prone to diminished growth and yield potential under HT stress. High temperature alters soil physical, biochemical, and biophysical properties that affecting crop growth. The preliminary plant growth processes are also affected like germination, seedlings growth, and vital physiological processes which then cause crop loss. Knowing the effects of any stresses including the HT stress is the prerequisite for developing the tolerant variety. Various approaches to combat HT stress tolerance have been discovered/invented with the development of science. So, the present chapter collects and accommodates the information regarding the HT effects on Fabaceae plants and differential approaches to sustain crop productivity and to improve HT tolerance of plants within the same frame.

2 Importance of Fabaceae Plants

Based on worldwide distribution and nutritional value, Fabaceae plants are the most demanded plants group after cereal and thus occupied the second position (Lewis et al. 2005; Kumar et al. 2018). It includes important crop species including herbs, like soybean (*Glycine max*), Lentil (*Lens culinaris*), groundnut (*Arachis hypogaea*), common bean (*Phaseolus vulgaris*), mung bean (*Vigna radiata*), chickpea (*Cicer arietinum*), garden pea (*Pisum sativum)* pigeon pea (*Cajanus cajan*), and alfalfa (*Medicago sativa*) playing an important role as human and animals dietary components (Considine et al. 2017). In addition, legume also includes shrubs: *Sesbania sesban*, *S. grandiflora*, *Gliricidia sepium*, *Calliandra calothyrsus*, and trees *Leucaena leucocephala*, *S. grandiflora*, *Moringa oleifera*, etc.

Legume, bean, or pea family known as Fabaceae are very important both from economical and ecological point of view. Fabaceae plants are considered extraordinarily important because of their abundance and exceptional diversity together with their multiple uses as food, feed, fodder, medicines, etc. (Voisin et al. 2014). Grain legumes also known as pulses are the enriched and potential sources of plant proteins, calories, vitamins, minerals, amino acids, crude fibers, etc. Due to rapid global population growth, millions of people especially from underdeveloped and developing countries are suffering from protein malnutrition (Hasanuzzaman et al. 2015). Pulses are known as poor people's meat, when supplemented in diets can play vital role as the alternative source of expensive animal protein along with the improvement of nutritional status. Thus, pulses are essential food crops for ensuring global food security for ever-increasing population (Iqbal et al. 2006; Nasr Esfahani et al. 2017). In the present world, groundnut and soybean contribute about 33% of total dietary nitrogen (N) and 40% of cooking oil, respectively, where about 70% foods and 25% feed demands are covered by pulses with serving more 5% for other uses in many countries including Europe, Australia, etc. (Zhang et al 2017).

In the present era of global climate change, legume crops can act as potential tools those can ensure sustainable agricultural system through their ability of biological nitrogen fixation, increasing retention of nutrients, improving soil fertility, reducing soil erosion as well as reducing global warming (Fig. 1). For fulfilling food demand of ever-increasing populations, the continuous cultivation making the soil exhausted, and in this condition, legumes are the only key those can meet both environmental sustainability and food demand. The year 2016 has already been declared by the United Nations as the International Year of Pulses (grain legumes) focusing on a motto of 'Nutritious Seeds for a Sustainable Future.' They marked as the most important cultivated crops and those can enrich soil fertility and organic matter through forming symbiotic relationships with certain bacteria (i.e., *Rhizobium* and *Bradyrhizobium*). About 30–200 kg N ha^{-1} can be fixed yearly by pulse crops. Thus, each pulse crop is called 'mini N factory' as they can convert atmospheric free nitrogen to usable nitrogen compounds which are further used by themselves or the subsequent plants along with soil stabilization, restoration, and improvement of soil properties (Khedar et al. 2008; Nulik et al. 2013; Foyer et al. 2016; Cao et al. 2017).

A number of pulses also can play role in freeing the soil-bound phosphorus (P) and help in providing optimum soil as well as plant nutrition. Legumes are also known to improve soil organic carbon stocks which further improve the potential yield of crops (Rose et al. 2010; Dhakal et al. 2016). So, cultivating pulse crop makes the soil ecologically more friendly and financially more sound through sustaining the soil productivity.

Fabaceae plants have much more diversified uses. In many areas, after harvesting the main crops, Fabaceae plants are grown as cover crops due to their soil fertility enrichment capacity by fixing N, weed suppression, controlling erosion, enhanced infiltration rate, and having some economic benefits. As a result, poor farmers often get the scope to reduce the cost of buying N fertilizers, and when due to continuous farming, depletion of soil N occurs; legumes can be used in crop rotation such as catch crop, intercrop, alley crop, and green manure to replenish the soil. Therefore, the risk of crop failure and summer-fallowing due to sole cropping of cereal and other crops are also replaced by extended crop rotation with pulses. Many legumes are also used in the projects of reforestation and revegetation, globally. In developing countries, due to their short life cycle, they are observed as necessary crops to be included as intercrop in the cropping pattern. Increased crop yield, enhanced nitrogen-use efficiency, diversification of crops, and decreased appearance of plant diseases are the benefits of intercropping or rotating legumes with cereals or other crops. So, enhanced cropping intensity can also be gained. Legumes also possess the characteristics of restorative crops, forage, and fodder crops. In pasture alfalfa, clover, etc., are gazed by animals, whereas *Leucaena and Albizia* are regularly used as fodder (Hasanuzzaman et al. 2019). Besides providing human food, legumes serve various other purposes like forage, pharmaceuticals, timber, industrial, and aesthetic. *L. leucocephala* or *G. sepium* trees sp. serve multipurpose as livestock forage, timber, shade plant, and reforestation cover. In addition, they have the capacity to withstand different environmental stresses under changing climate. It was revealed that symbiotic associations by-product play roles to tolerate drought, heavy metal, and oxidative stresses. Furthermore, associated bacteria also involved in the bioremediation of metal, pollutants, organic pollutants degradation, and photostabilization (Sultana et al. 2014; Hasanuzzaman et al. 2015). It is already reported that, legumes have the capacity to mitigate the adverse climate change through reducing greenhouse gases (GHG) emission compared to other crops (Stagnari et al. 2017), because they release 5–7 times less GHG per unit area and also permit soil carbon (C) sequestration; thus, influence saving of more fossil energy is in the cropping system.

Legumes are traditionally popular for their medicinal values. Folk medicines are composed of some legumes. Isoflavones from *Glycine max* have been reported as anticarcinogenic compound along with reducing the cancer risk and serum cholesterol reduction. For the postmenopausal women, another compound provided by soy food and soybean called phytoestrogens has been documented to act as the replacement of hormone therapy (Kennedy 1995; Molteni et al. 1995). Legume seeds are suggested beneficial for human health because they contain lower saturated fatty acids, total oil, glycaemic index (blood sugar indicator) and higher unsaturated fatty acids, protein, and fibers. Higher intake of *L. culinaris* showed lowered blood pressure, lowered glycemic load, anticarcinogenic, and hypocholesterolemic effects. It also helped in preventing the cardiovascular and chronic diseases, cancer, and diabetes due to high contents of phytochemicals and fibers (Shahwar et al. 2017). Lupin seeds play a significant role in controlling diabetes, overweight, lowering hypertension, balancing energy, and improving blood lipid (Ryan et al. 2007; Capraro et al. 2010). Celiac patients can also be benefitted by pulses since they are devoid of gluten. Cardiovascular diseases, diabetes, breast cancer risk, and obesity can also be prevented by the composition of pulses (Campos-Vega et al. 2010).

Moreover, biodegradable plastics, dyes, inks, gums, oils, etc., industrial products can be prepared from legumes. In sezile textiles and papers gums preparation, *Cyamopsis* sp. and *Sesbania* sp. are used. Nowadays, bio-diesel fuels from soybean are powering the vehicles in the USA and other countries (Morris 1997).

3 Effects of High Temperature Stress on Fabaceae Plants

As the factor of abiotic stresses, HT causes the negative changes in plant growth and physiology resulting in a great yield loss of Fabaceae plants. Heat stress restricts plant phenological development through imbalancing cell water, inhibiting cell growth, disruption in light capture, inactivating enzymatic functions, limiting photosynthesis, hampering reproductive functions. In this section, we will briefly summarize the plant sufferings from HT focusing on the growth, physiological, and yield aspects (Table 1).

3.1 Growth

High temperature hampers the plants normal growth procedure by affecting differential stages from seed germination to senescence. High temperature-induced stunted growth might be associated with inhibition in cell division and cell elongation pointed out by lower plant height and leaf area (Siddiqui et al. 2015). Different genotypes of *Vicia faba* from different geographical location were exposed to 31 and 37 °C, respectively, to evaluate the growth responses (Siddiqui et al. 2015). About 60-day-old seedlings of Zafar 1, Zafar 2, Shebam1, Makamora, Espan, Giza Blanka, Giza 3, C4, C5, and G853 genotypes of *V. faba* were treated with respective HT for 48 h. Therefore, the growth of all the genotypes was reduced assessed by plant height, shoot fresh weight (FW), shoot dry weight (DW), and leaf area. But C5 showed the comparative better growth than others due to its highest values in all studied parameters and thus considered tolerant, while Epsan showed the lowest values due to its sensitivity to HT. On the basis of the screening study of Siddiqui et al. (2015), *V. faba* cv. C5 was further treated with HT by providing 38 °C, where stressed plants

Table 1 Responses of Fabaceae plants under high temperature stress

Crops	Stress	Plant responses	References
Vicia faba L. cv. Zafar 1, Zafar 2, Shebam1, Zafar 1, Zafar 2, Shebam1, Giza Blanka, Giza 3, C4, C5 and G853	31 and 37 °C, 48 h	• Decreased plant height, shoot biomass accumulation and leaf area • Reduced leaf relative water content (LRWC) with higher proline (pro) content • Genotypes C5 showed higher tolerance due to its maximum value of all studied parameters under both stresses	Siddiqui et al. (2015)
V. faba L. cv. C5	38 °C, 48 h	• Reduced plant height, shoot fresh weight (FW), shoot dry weight (DW), leaf area • Enhanced osmoprotectant accumulation such as pro and glycine betaine (GB) • Increased leaf electrolyte leakage (EL) • Increased chlorophyll (Chl) degradation and thus reduced total Chl content • Decreased carbonic anhydrase (CA) and RuBisCO activities	Siddiqui et al. (2018)
V. faba L. cv. C5	37 °C, 48 h	• Decreased membrane integrity indicated by higher EL • Reduced total Chl content with the increase of Chl degradation • Enhanced P5CS activity while PDH activity reduced • Reduced LRWC	Alamri et al. (2019)
V. faba L. cv. C5	42 °C, 48 h	• Decreased membrane integrity indicated by higher EL • Reduced total Chl content with the increase of Chl degradation • Enhanced P5CS activity while PDH activity reduced • Decreased LRWC • Increased endogenous pro and NO content	Alamri et al. (2019)

(continued)

Table 1 (continued)

Crops	Stress	Plant responses	References
Vigna radiata L.	>40 °C, reproductive phase	• Increased leaf chlorosis, scorching and wilting and reduced leaf area with lowered Chl accumulation • Reduced above ground total plant biomass up to 76% • Inhibited photochemical efficiency (PS II, Fv/Fm) • Caused early flowering with decrease in flowering and podding duration • Reduced total flower number up to 46% • Decreased pollen and ovule viability • Lowered pod number along with increased pod abortion • Decreased pod size, pod and seed yield	Sharma et al. (2016)
V. radiata L. cv. SML 832 (tolerant) and SML 668 (sensitive)	40/28 °C, during reproductive stage	• Increased leaf pro content by 3.4-fold with lower LRWC • Enhanced both Pyrroline-5-carboxylate synthase (P5CS) and pyrroline-5-carboxylate reductase (P5CR) enzymes activities • Decreased Pro dehydrogenase (PDH) activity • Increased membrane damage up to 27% with 33% increase of cellular viability (CV) • Showed 29% reduction in Chl content while reduced photochemical efficiency and increased stomatal conductance (gS), increased RuBisCO activity • Declined reproductive function by decreasing pollen and ovule viability • Lowered yield up to 23% by decreasing pod number and pod weight	Priya et al. (2019a)
V. radiata L. cv. SML 832 and SML 668	45/33 °C, during reproductive stage	• Showed up to 2.2-fold lowered leaf pro accumulation with lower LRWC • Increased P5CS and P5CR activities in tolerant but decreased in sensitive genotypes • Decreased PDH activity up to 5.1-fold • Increased membrane damage up to 30% with decrease of CV up to 55% • Reduced Chl content up to 53%, while reduced the photochemical efficiency and increased gS, decreased RuBisCO activity • Declined reproductive function by decreasing pollen and ovule viability • Lowered yield up to 44% by decreasing pod number and pod weight	Priya et al. (2019a)

(continued)

Table 1 (continued)

Crops	Stress	Plant responses	References
V. radiata L. cv. Binamoog-1	42 °C, 24 h	• Decreased LRWC by 20% • Reduced Chl *a* content by 11% • Increased Pro level	Nahar et al. (2015a)
V. radiata L. cv. Binamoog-1	42 °C, 48 h	• Decreased LRWC by 26% • Reduced Chl *a* content by 14% • Increased Pro level	Nahar et al. (2015a)
V. radiata L. cv. BARI Mung-2	40 °C, 48 h	• Increased putrescine (Put) content by 423% • Decreased LRWC by 19% with 21% pro level • Decreased Chl ($a + b$) content by 24%, while reduced leaf area • Affected seedling vigor negatively, while shoot and root length decreased by 18 and 22%, respectively • Reduced seedling DW	Nahar et al. (2016)
V. radiata L. cv. SML 668	40/30, and 45/35 °C, 10 day	• Decreased seed germination by 36% • Lowered seedling shoot and root length by 34 and 23%, respectively • Possessed 12 and 47% less LRWC and Chl content, respectively	Kumar et al. (2011)
V. radiata L. cv. SML832 and SML668	43/30 °C 45/32 °C	• Reduced LRWC and gS • Decreased Chl content by 37 and 39% at flowering and podding stage respectively • Lowered Chl florescence of PS II • Decreased biomass content and pod setting • Decreased filled pod up to 39%, while seed number and seed weight decreased by 48 and 40%, respectively • Hampered reproduction function	Kaur et al. (2015)
V. radiata L.	45/30 °C, 7 d	• Reduced plant dry matter and grain yield by 24 and 59%, respectively • Inhibited bud, flower, and pod number drastically • Decreased pollen viability • Reduced seed germination by 12%	Patriyawaty et al. (2018)

(continued)

Table 1 (continued)

Crops	Stress	Plant responses	References
Lablab purpureus L. cv. VRBSEM-15	40–42 °C, 21 d	• Decreased membrane stability by increasing EL • Reduced leaf Chl and carotenoid (Car) contents • Decrease plant water relation and photosynthesis • Affected Fv/Fm of PS II • Decreased plant growth and yield	Rai et al. (2018a)
Cicer arietinum L.	30/25 °C, 3 d	• Reduced Fv/Fm • Decreased total biomass content, pod number • Declined grain yield	Makonya et al. (2019)
C. arietinum L.	35/30 °C, 3 d	• Caused reduction in Fv/Fm • Reduced total biomass content, pod number • Declined grain yield drastically	Makonya et al. (2019)
C. arietinum L.	40/30 and 45/35 °C	• Lowered Chl content up to 32% while photochemical efficiency reduced by 53% • Decreased total plant biomass up to 22% • Decreased pollen viability, pollen germination, pollen tube growth, and stigma receptibility • Inhibited pod setting • Produced up to 37% lower pod under 40/30 °C • Maximum 3–5 pods were produced under 45/35 °C • Reduced seed yields up to 45%	Kumar et al. (2013)
C. arietinum L. cv. GPF2	40/35 and 45/40 °C, 10 d	• Decreased shoot length by 64% • Declined survival rate by 80% • Increased pro, GB and trehalose content under 40/35 °C, but decreased all these at 45/40 °C temperature • Decreased Chl level by 41%, while EL extremely increased	Kumar et al. (2012)
Phaseolus vulgaris L.	40 °C, 20 d	• Decreased photosynthesis (Pn), gS and Fv/Fm significantly • Reduced total Chl and Car contents with the reduction of pro accumulation	Chavez-Arias et al. (2018)

showed poor performance in plant height, shoot FW and DW with their reduced leaf area (Siddiqui et al. 2018).

Likely, *P. vulgaris* is suffered from HT by limiting growth performance as genotype dependently (Chavez-Arias et al. 2018). Four cultivars of *P. vulgaris* including 'Cerinza,' 'Bachue,' 'Bacata,' and 'Bianca' were tested with different temperatures such as 30, 35, and 40 °C for 20 days, respectively. Among them, Cerinza has been accepted more by farmers for its higher adaptability to HT by its production stability, which might be due to its best tolerant index upon 40 °C.

Six different varieties of *L. culinaris* such as Asha, Subrata, IPL 406, IPL 81, Lv, and Sehore were evaluated by exposing to 35–40 °C for 4 h (Chakraborty and Pradhan 2011). Here, higher temperature significantly reduced seed germination and caused stunted growth. Based on tolerance index, it was revealed that Sehore and Lv were susceptible to elevated temperatures while IPL 406, IPL 81, Asha, and Subrata were tolerant.

Two cultivars of *V. radiata* viz. SML 832 and SML 668 were studied under HT of 40/25 °C while its optimum levels for growth were 28–30 °C (Kaur et al. 2015). This HT caused the significant reduction in the dry biomass of shoots by 16 and 19% from SML 832 and SML 668, respectively. High temperature (40 °C) limits the seedling vigor of *V. radiata* cv. BARI mung 2 (Nahar et al. 2016). In this study, growth of *V. radiata* was also drastically reduced under heat stress confirmed by decreased shoot height and root length about 18 and 22%, respectively. However, heat stress also reduced the leaf area and seedling DW in *V. radiata*.

An experiment was conducted to evaluate the response of *V. aconitifolia* upon short-term exposure to heat stress by taking 37 genotypes under consideration (32 mutants and five varieties) (Harsh et al. 2016). Heat stress conditions were formed by exposing seven-day-old seedlings at 42 °C for 1 h in hot air oven. Among 37 genotypes, 6 mutants/genotypes, i.e., 40-300-17-14-6, SN 5-1, SN 3-3, 40-300-17-10-3, 423-8-9-10-1, and RMO-40 showed high survival percentage at 42 °C, while thirteen genotypes showed average survival and less survival was recorded from eighteen genotypes under heat stress. Hence, stress tolerance was evaluated from the number of plants those showed wilting symptoms and consequent damages, viz. leaf margin or tip burning, stem lodging, leaf drooping.

Therefore, HT stress restricts the plant normal growth, biomass accumulation, and total photosynthetic area by which yield loss accelerated and thus threatens the food security (Fig. 2).

3.2 Physiology

Due to global warming, plants are frequently experienced by HT which alters the normal plant physiology including, membrane disorganization, disturbance in cell water status, photosynthesis, respiration, etc. (Fig. 2). The severity of heat stress-induced damages is crop species-dependent, while the tolerant varieties show comparatively better performances.

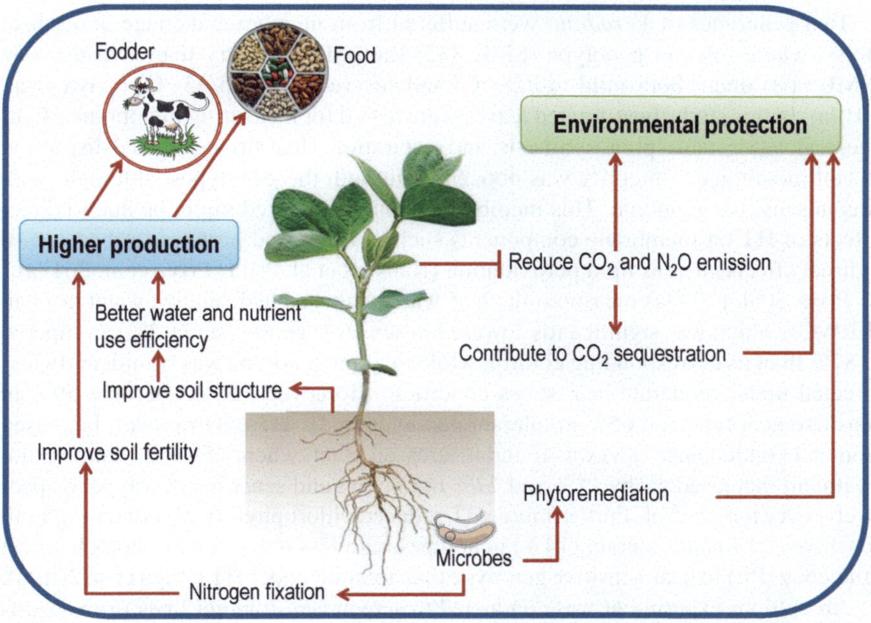

Fig. 1 Role of legumes in sustainable agriculture

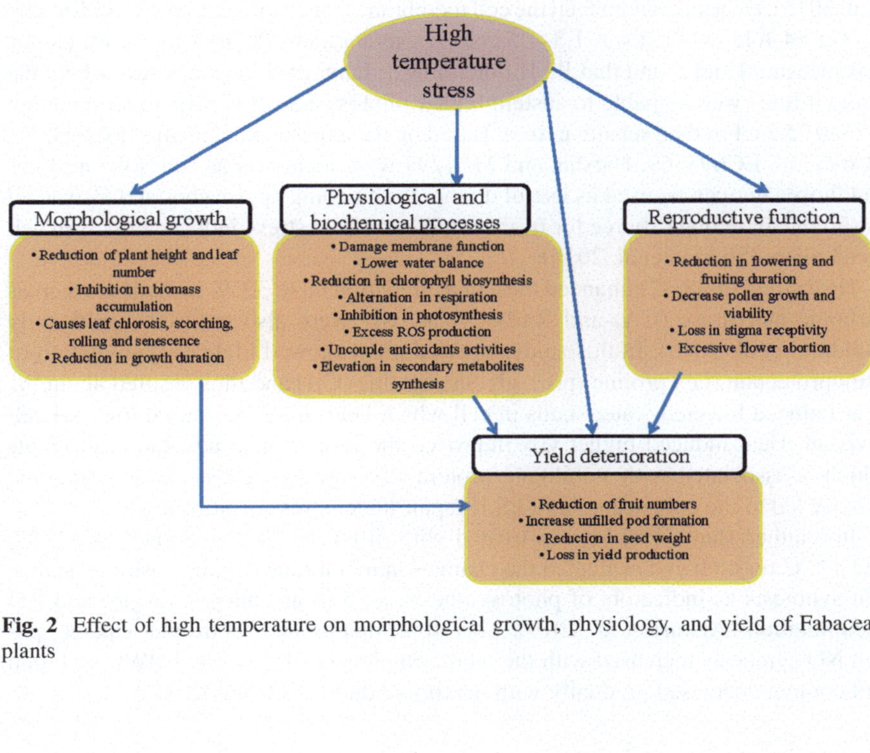

Fig. 2 Effect of high temperature on morphological growth, physiology, and yield of Fabaceae plants

Two genotypes of *V. radiata* were suffered from membrane damage under heat stress, where tolerant genotype (SML 832) showed less injury than sensitive one (SML 668) under both mild (40/28 °C) and severe stress (45/33 °C) (Priya et al. 2019a). In this study, heat-treated leaves were tested for measuring membranes damage, leaf water status, photosynthesis, and respiration. Heat stress-induced disruption in cell membranes' integrity was noticeably in both the genotypes, although more was in sensitive genotype. This membrane damage occurred might be due to direct effects of HT on membrane components such as lipids and proteins, or because of indirect effects such as lipid peroxidation (Kaushal et al. 2011; Priya et al. 2019a).

Priya et al. (2019a) measured the leaf water status as leaf relative water content (LRWC), which was significantly lowered in sensitive genotype (71.3%), compared to 87% than its corresponding control, while tolerant genotype was significantly less affected under moderate heat stress condition. Moreover, LRWC became 59% in sensitive genotype and 65% in tolerant genotype under HT environment. Increased stomatal conductance (gS) was found in stressed plant, where 45/33 °C temperature treatment increased gS by 11% and 22% in tolerant and sensitive genotype, respectively, over the control. Furthermore, HT reduced chlorophyll (Chl) content in both sensitive (29%) and tolerant (14%) genotypes with less reduction of photochemical efficiency (PE) in heat-sensitive genotype than tolerant under HT (Priya et al. 2019a).

An outdoor experiment was conducted to screen heat-tolerant lines from 41 different lines of *V. radiata* under full irrigation condition; seeds were late-sown (LS) in April to ensure higher temperature (>40/28 °C) during the reproductive stage (Sharma et al. 2016). Heat stress damaged the cell membranes much higher in sensitive lines by 1.53–1.84-fold, while it was 1.3–1.5-fold in tolerant lines. Photosynthetic efficiency was measured and found that PS II function was hampered by heat stress where the tolerant lines were capable to sustain higher photosystem II (PS II) function about 0.73–0.75 Fv/Fm than sensitive lines. Based on the experimental results, EC693357, EC693358, EC693369, Harsha, and ML1299 were identified as heat-tolerant lines and those later can be used as useful donors for breeding approaches and also acted as suitable base plant source for further studies on heat stress-induced effects in cell metabolism (Sharma et al. 2016).

Heat stress of 38 °C enhanced the Chl degradation up to 303% in *V. faba*, whereas carbonic anhydrase (CA) and RuBisCO activity were also reduced significantly (Siddiqui et al. 2018). In this study, stressed plant showed higher accumulation of osmoprotectant, i.e., proline (pro), glycine betaine (GB) and thus ensured about the heat-induced lowered water status in cell which later might be caused lowered cell division. Heat-induced higher GB improved the biochemical adaptation in plants which is associated with stabilizing protein structure as well as better enzymatic activity led to the enhancement in PS II repair under stress condition.

In another study, *V. faba* was treated with different HT levels such as 32, 37, and 42 °C for 48 h and evaluated the changes in membrane damage, osmotic status, Chl synthesis as indicators of photosynthesis, as well as endogenous pro and NO accumulation (Alamri et al. 2019). Therefore, membrane electrolyte leakage, pro and NO synthesis increased with the increasing level of HT, while LRWC and total Chl content decreased gradually with maximum damages from 42 °C.

P. vulgaris cv. 'Cerinza,' 'Bachue,' 'Bacata,' and 'Bianca' were studied under HT stress (40 °C) for 20 days (Chavez-Arias et al. 2018). Growth of *P. vulgaris* plant was reduced when subjected to heat stress along with physiological damages including imbalanced water status, gS, chlorophyll florescence (maximum quantum yield of PS II, Fv/Fm; photochemical quenching, qP; and electron transport rate ETR; and photosynthesis, Pn. Hence, heat stress caused around 80% reduction of gS and Pn while other parameters also inhibited in all genotypes compared to control treatment.

3.3 Yield

As a consequence of restriction in plant growth and physiological processes, heat stress limits the yield contributing characters as well as yield of Fabaceae plants. High temperature causes the inhibition of flower differentiation and developmental processes resulted in the reduction of ovules size in flowers (Johkan et al. 2011). Therefore, flower production, fruit setting, seed number, and seed yield are also restricted by HT in these plants. Exposure of heat stress during the early reproductive phase causes failure of seed setting in Fabaceae plants which become impossible to be rescued and later leading to fatal and irreversible yield loss (Liu et al. 2019).

Most of the cool and warm season crops are seriously affected by temperatures over 35/25 °C, especially in their reproductive phase, followed by pollen infertility to stimulate loss of flowers and potential pods (Priya et al. 2019a). It is well reported that, reproductive stage is very sensitive to heat stress which accelerates the yield loss in many Fabaceae crops such *C. arietinum* (Devasirvatham et al. 2012), *L. culinaris* (Sita et al. 2017), and *V. radiata* (Kaur et al. 2015). This heat stress-induced damage in the reproductive stage is also species-dependent. *V. radiata* genotypes including heat-tolerant SML 832 and heat-sensitive SML 668 were treated by mild stress, 40/28 °C and severe stress, 45/33 °C to evaluate how heat stress negatively affected the crop yield depending on genotypes (Priya et al. 2019a). It was found that, severe stress caused more losses in reproduction attributes compared to MS such as pollen viability became lower such as 43% in tolerant genotype and 18% in sensitive genotype under HS condition which was 83–85% in control treatment. High temperature (45 °C) decreased the pollen germination up to 41 and 15%, and stigma receptivity about 48 and 73%, and ovule fertility by 35 and 65% in tolerant and sensitive genotypes, respectively, over the control. Therefore, heat-sensitive one faced greater yield loss by decreasing pod number, pod yield, and seed yield compared to tolerant one under HT (Priya et al. 2019a).

V. radiata showed the significant reduction in pod setting, number of filled pods (32–38%), seed number (43–47%), and seed yield (35–40%) plant^{-1} under 40 °C (Kaur et al. 2015). This yield reduction was due to the heat stress-induced reduction of reproductive function, for example, pollen viability, pollen germination, pollen tube growth, pollen load, and stigma receptivity. Thereafter, Kaur et al. (2015) revealed that decrease in sucrose concentration with the lowered sucrose synthesizing enzymes activities (sucrose synthase, sucrose phosphate synthase) and hydrolyzing enzymes (acid invertase) which played vital role in inhibiting yields under HT.

High temperature (>40 °C) on reproductive stage of *V. radiata* was studied in a field study conducted by Sharma et al. (2016). Hence, reduction in leaf area and shortening the duration of flowering and podding were might be involved in decreasing the pod and seed yields. Sharma et al. (2016) suggested that, heat-induced inhibition in photosynthetic ability (Chl content, PS II function), carbon fixation (RuBisCO activity), and assimilation processes (sucrose synthesis and its hydrolysis) later resulted in lower yield production. Thus, HT adversely affected the potentiality of reproduction of plants.

On the basis of superior grain yield potentiality, four genotypes of *C. arietinum* genotypes such as Acc#RR-2, Acc#RR-3, Acc#7, and Acc#8 were chosen to conduct a field experiment to evaluate the heat stress-induced yield loss under 3 days of MS (30/25 °C) and HS (35/30 °C) (Makonya et al. 2019). Regardless of tolerance level, all genotypes showed the significant decline in grain yield under both MS and HS. Relative to control, the most and least decline in grain yield about 94 and 65% were observed from Acc#8 and Acc#7, respectively, under high heat stress. Among four genotypes, Acc#7 had the highest grain yield when exposed to the moderate heat stress although it also showed a comparative higher yield under HS (Makonya et al. 2019). The higher pod number from Acc#7 might be contributed in producing higher yield under stress condition.

High temperature (40–42 °C) significantly hampered the yield-related components of *Lablab purpureus* L. cv. VRBSEM-15 (Rai et al. 2018a). Exposure to HT, *L. purpureus* showed the delay in fruit setting as well as total yield. Different temperature treatments such as 30 and 34 °C were exposed to *V. faba* where heat stress reduced the yield potentiality ensured by the reduction in bean number, pod number, bean number per pods (Bishop et al. 2016).

4 Combined Effects of High Temperature and Other Stresses

Plants are mostly cultivated in the open field condition where plants often face variety of environmental conditions and sometimes combined stress effects have also been reported. High temperature stress together with drought is the most common phenomenon. In addition, HT and salinity, HT with heavy metal, or other stresses can distort the plant physiological processes (Mittler and Blumwald 2010). Short-term exposure of HT causes programmed cell death within couple of minutes. Extreme dryness drawn from drought stress may lead to irreversible cellular damage in very short term. Long-term exposure of different stresses alone or in combination gradually imposes stress to cell in such a way that their functioning is hindered to cause aberration in continuing normal physiological processes.

Drought and heat stress adversely affect the developmental processes of *P. vulgaris* plant. In a field experiment, common bean plants of 25 days old were subjected to normal and drought (50% reduction of irrigation) conditions. Drought and HT decreased number of pods plant^{-1}, seeds pod^{-1}, harvest index, and seed yield of different varieties of common bean (Kazai et al. 2019). Consequences of drought, heat, and combination of drought and heat stresses were evaluated in the leaves of two soybean varieties, Surge and Davison. Due to differentially expressed photosynthesis-related proteins, different varieties showed different responses. RuBisCO regulation, electron transport, Calvin cycle, and carbon fixation were affected in both varieties under drought and heat stresses. Higher level expressions of heat shock-related protein regulated heat tolerance characteristics. Over generation of reactive oxygen species (ROS) was also increased under those stresses including cytotoxic concentrations of hydrogen peroxide (H_2O_2) (Das et al. 2016). *P. vulgaris* genotypes were examined for phenology, partitioning, plant–water relations, photosynthetic parameters, and shoot growth. Leaf water content of tolerant genotypes was managed by the reductions in leaf water potential and shoot extension in response to heat and drought stresses. Tolerant genotypes showed higher biomass translocation to pods and higher pod set in branches (Omae et al. 2012). *C. arietinum* cv. ICC 4958 plants were subjected to osmotic stress and HT stress (35–44 °C) in combination. Stresses caused higher production of H_2O_2, malondialdehyde (MDA), pro and differential modulation of peroxidase and CAT activities. The relative growth rate of chickpea plants was negatively affected by combined osmotic and HT stresses (Ceylan et al. 2013). Mung bean (*V. radiata* L. cv. BARI Mung-2) plants were simultaneously exposed to HT stress of 40 °C and drought (applying 5% polyethylene glycol). Those stresses resulted in lipid peroxidation through creating oxidative stress by increasing H_2O_2 and $O_2^{\bullet -}$, lipoxygenase (LOX) activity which was also due to disturbance of antioxidant defense system. Among the antioxidant components ascorbate (AsA) and glutathione (GSH) pool, activities of superoxide dismutase (SOD), CAT, glutathione peroxidase (GPX), dehydroascorbate reductase (DHAR), glutathione reductase (GR) were perturbed which was vital to contribute oxidative stress. Both the HT and drought stresses increased the highly reactive methylglyoxal (MG) together with the modulation of glyoxalase I (Gly I) and glyoxalase II (Gly II) enzymes. The extremity of MG production increased when HT and drought stresses were imposed concommitantly (Nahar et al. 2017). *L. culinaris* plants were exposed to HT and drought (50% field capacity, 33/28 °C day/night temperature) resulted in noticeable decline in the rate and duration of seed filling to diminish the final seed size; hampered the content of seed starch, storage proteins and their fractions, minerals, and several amino acids (Sehgal et al. 2017). The forage crop guar (*Cyamopsis tetragonoloba*) was subjected to multiple stresses like HT (42 °C), drought stress (40% of field capacity), and salt stress (200 mM NaCl) separately and simultaneously. High temperature hindered root growth and water use efficiency extremely but moderately affected leaf area, stomatal conductance, and number of pods. Drought decreased leaf area, plant height, and prolonged days to flowering. It also showed slight effect on some other physiological parameters as it reduced leaf number, biomass, stomatal conductance, and number of pods. Salt exposure had a moderately negative impact on all studied traits except leaf number.

However, the simultaneous imposition of multiple stresses brutally exaggerated all the physiological and growth parameters (except water use efficiency) proving by the combined HT and other stresses, compared to HT stress alone.

5 Oxidative Stress and Antioxidant Defense Against High Temperature

Enzyme-dependent physiological and metabolic pathways of plant cells are very sensitive to HTs. Upon heat stress, these enzymatic and metabolic activities become disturbed and consequently produce excess ROS—singlet oxygen ($^1O_2$), superoxide anion ($O_2^{\bullet-}$), H_2O_2, and hydroxyl radical ($OH^\bullet$) which causes oxidative stress (Hasanuzzaman et al. 2013). As balanced and low concentrated ROS acts as secondary messenger and also involved in plant growth and developmental processes, but heat stress-induced overproduced ROS causes cellular damages which threats survival by enhancing oxidation of lipids, proteins, nucleic acid; inactivation of enzymes ultimately resulting in programmed cell death (PCD) (Nahar et al. 2015a). Therefore, heat-induced photosystem damages cause the less absorption of excess photon intensity beyond the requirement for the CO_2 assimilation and later serve as a source of excess ROS (Asada 2006; Nahar et al. 2015a).

A study was conducted hydroponically to evaluate the oxidative damages in *V. radiata* under 40 °C for 24 h (Nahar et al. 2016). Hence, higher accumulation of $O_2^{\bullet-}$ and H_2O_2 in leaves under heat stress was visualized by histochemical staining followed by biochemical estimation. In addition, HT caused higher LOX activities followed by elevated level of MDA content and thus ensured heat-induced oxidative stress (Fig. 3).

A group of scientist observed a greater DNA damage in heat shock stressed *V. faba* which was correlated with stress-mediated extreme accumulation of H_2O_2 where lipid peroxidation was higher as measured by acute amount of MDA (Siddiqui et al. 2018). In this study, *V. faba* cv. C5 plants were treated by 38 °C to create HT for 48 h. Twenty-one-day-old seedlings of *L. purpureus* were treated with HT up to 42 days old (Rai et al. 2018b). Hence, HT stress resulted in higher production of ROS such as $O_2^{\bullet-}$ and H_2O_2 as oxidative stress marker with greater lipid preoxidation. Later, HT enhanced DNA damages in *L. purpureus* confirmed by both increase and decrease in bands profiles of methylation and demethylation patterns, respectively.

Both heat-tolerant (ICCV07110, ICCV92944) and heat-sensitive (ICC14183, ICC5912) genotypes of *C. arietinum* were treated with various temperature of 30/20, 35/25, 40/30, and 45/35 °C during 50% flowering stage for next 18–7 days depending on genotypes (Kumar et al. 2013). Therefore, MDA and H_2O_2 contents were evaluated to detect the severity of heat-induced oxidative stress. Comparatively greater oxidative damages were observed in sensitive genotypes than tolerant genotypes upon varying degree of heat stress. Sensitive genotypes showed higher MDA content at 29–32% and 40–49% under 40/30 and 45/35 °C temperatures, respectively,

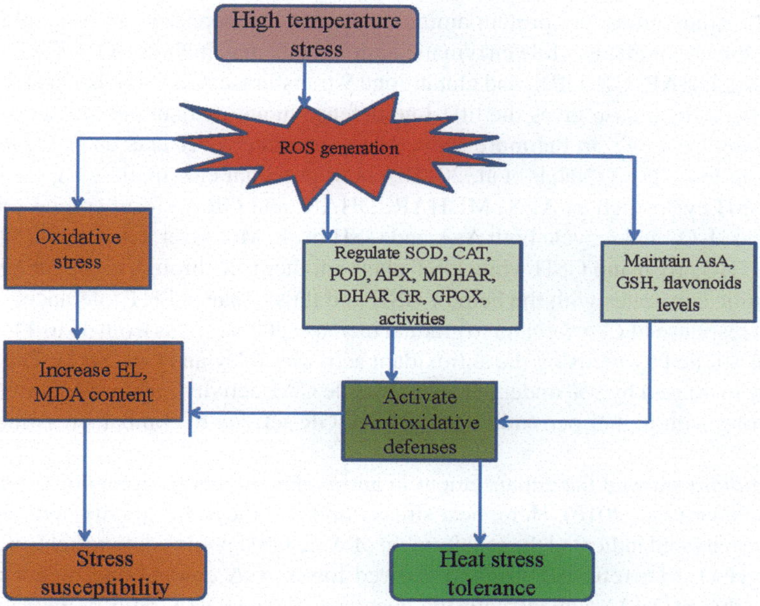

Fig. 3 High temperature-induced oxidative stress and antioxidant defense in Fabaceae plants. ROS-reactive oxygen species, MDA-malondialdehyde, EL- electrolyte leakage, AsA-ascorbate, GSH-glutathione, SOD-superoxide dismutase, CAT-catalase, POD-peroxidase, APX- ascorbate peroxidase, MDHAR- monodehydroascorbate reductase, DHAR-dehydroascorbate reductase, GR-glutathione reductase, GPOX-guaiacol peroxidase

compared to tolerant genotypes, while 45/35 °C temperature induced higher H_2O_2 content by two-fold in sensitive genotypes than tolerant one. Thus, tolerant *C. arietinum* genotypes have the capacity to less suffering in heat-induced oxidative stress than their sensitive genotypes.

The MDA and H_2O_2 accumulation were increased in *L. culinaris*, irrespective of varieties, upon exposure to HT (Chakraborty and Pradhan 2011). However, susceptible varieties of Sehore and Lv showed more than 50% higher MDA content than the other four tolerant varieties (IPL 406, IPL 81, Asha and Subrata) under 50 °C.

The increase of day temperature enhanced mainly MDA content in *P. vulgaris* cv. Cerinza, Bachue, Bacata, and Bianca regardless of the nature of genotypes (Chavez-Arias et al. 2018). Hence, 'Bianca' showed comparatively less MDA accumulation than other three cultivars when exposed to HT stress of 40 °C for 20 days.

Plant antioxidant defense systems are highly accompanied with better stress tolerance as well as to protect HT-induced ROS production and oxidative stress. Both enzymatic and non-enzymatic antioxidants are spontaneously involved in this antioxidant defense system to detoxify ROS as cyclic manner and ensure the plant survival (Hasanuzzaman et al. 2013). Ascorbic acid (AsA), glutathione (GSH), alkaloids, phenolic compounds, nonprotein amino acids, and α-tocopherol are principle non-enzymatic antioxidants while enzymatic antioxidants may include SOD, CAT, APX, MDHAR, DHAR, GR, GPX, and glutathione *S*-transferase (GST) (Nahar et al. 2016). Superoxide dismutase gives the first-line defense in antioxidant defense system by converting toxic $O_2^{\bullet -}$ to the more stable H_2O_2, and later CAT acts on H_2O_2 to produce non-toxic H_2O (Nahar et al. 2015b). Four potential enzymatic components of AsA-GSH cycle such as APX, MDHAR, DHAR, and GR perform crucial roles in reducing H_2O_2 and recycle both AsA and GSH back. Moreover, GPX and/GST also reduce H_2O_2 by using GSH while GST plays another role through conjugating with xenobiotic substrates with the help of GSH and thus enhanced HT tolerance.

The response of *C. arietinum* to gradual increase of heat stress from 35 to 44 °C for 1 h was studied by focusing the antioxidant activities (Ceylan et al. 2013). The SOD activity increased by 9% under heat stress while CAT activity increased about three-fold along with higher peroxidase (POD) and GR activity to combat stress-induced ROS.

V. radiata showed the enhancement in antioxidant activities upon 40 °C temperature (Nahar et al. 2016). Here, heat stress improved the SOD activity and reduced CAT activity and indicated the requirement of AsA-GSH pool to defend heat-induced higher H_2O_2. Therefore, *V. radiata* showed lower AsA content and higher dehydroascorbate (DHA) content with the decrease of AsA/DHA ratio by 64%, while GSH/GSSG ratio also decreased. Heat stress increased APX and GR activity by 42 and 50% respectively, with the reduction of both MDHAR and DHAR activities. The GPX and GST activities also increased in stressed *V. radiata*. In another study, 6-day-old *V. radiata* seedlings were treated with 42 °C temperature for 24 and 48 h (Nahar et al. 2015a). Regardless of treatment duration, stressed seedlings showed the lower content of AsA with the increased endogenous GSH and GSSG contents resulted in decreased GSH/GSSG ratio. It was observed that APX, GR, and GST activities increased, while MDHAR, DHAR, GPX, and CAT activities reduced in stressed conditions.

Genotypical variation in antioxidant activities was studied in *V. radiata* by taking five tolerant (EC693357, EC693358, EC693369, HARSHA, ML1299) and five sensitive (KPS1, EC693361, EC693363, EC693370, IPM02-3) lines under heat stress about >40 °C (Sharma et al. 2016). Non-enzymatic antioxidants such as AsA and GSH remarkably increased in all lines under heat stress while the increase trend in these components was comparatively higher in tolerant lines (AsA, 1.75–2-fold; and GSH, 11.81–1.99-fold), than sensitive lines (AsA, 1.26–1.52-fold and GSH, 1.38–1.65-fold). Moreover, SOD activity increased in both tolerant and sensitive lines by 1.69–1.85-fold and 1.79–2.18-fold, respectively, under heat stress. But, interestingly CAT activity increased as the same extent in all lines. Again, 1.48–1.77-fold and 1.27–1.37-fold increased APX were found in tolerant and sensitive lines, respectively, upon heat stress condition, where GR activity highly increased in tolerant than sensitive lines.

L. purpureus was treated with 40–42 °C to observe their antioxidant defense mechanisms (Rai et al. 2018b). Then, stressed plants responded to HT stress by increasing CAT, guaiacol peroxidase (GPOX), POD, APX, GR, and ferric-reducing antioxidant power (FRAP) activity significantly.

38 °C increased antioxidant enzymes activities such as SOD, CAT, and POD in *V. faba* plants (Siddiqui et al. 2018).

Seven-day-old seedlings of 37 genotypes of *V. aconitifolia* were treated with 42 °C temperature for 1 h (Harsh et al. 2016). This short-term heat exposure increased CAT, GPOX, and SOD activities in most of the genotypes.

6 Use of Exogenous Phytoprotectants to Enhance High Temperature Tolerance

Plants have the ability to protect the abiotic stresses including HT depending upon the types of plant species and their genotypes. In this section, we will discuss external phytoprotectant-induced improvement in HT tolerance of Fabaceae plants.

6.1 Phytohormones

Use of exogenous phytohormones to alleviate the heat stress-induced toxicity in Fabaceae plants is one of the major ways to protect plants from the harmful effects of HT (Hasanuzzaman et al. 2013). Plant hormones affect the plant growth and reproduction by modulating the plant antioxidants activities, physiological processes, and development under HT stress (Ozga et al. 2017).

Besides being a signaling molecule, abscisic acid (ABA) is well known as stress hormone due to its association in thermotolerance in heat-stressed plants. An investigation was carried out on the effect of exogenous ABA to mitigate the HT toxicity on heat-sensitive *C. arietinum* (Kumar et al. 2012). Very low concentrated ABA (2.5 μM) was found effective as co-treatment with various degrees of HT such as 40/35 and 45/40 °C. In this study, 4-day-old seedlings were exposed to HT stress with and without ABA for the next 10 days. After that, Kumar et al. (2012) reported that, exogenous ABA increased seedlings growth in terms of shoot and root length and Chl synthesis with the enhancement of endogenous ABA and osmolytes like pro, GB, and trehalose (Tre). The ABA-mediated higher survival and growth were associated with ABA-induced reduction in oxidative stress which was confirmed from lower content of MDA and H_2O_2 and lesser (EL). The ABA-induced heat tolerance in *C. arietinum* was checked by applying inhibitor of ABA biosynthesis, fluridone (FLU) under stress condition. Therefore, FLU application reverted the ABA-induced actions in seedling survival, growth, Chl synthesis as well as oxidative stress, thus showed an obvious role of ABA in enhancing heat tolerance in *C. arietinum*.

Brassinosteroids (BRs) are another group of plant hormone, having the potential influence on cell division and stem elongation by pursuing diverse physiological processes (Mitchell et al. 1970). Thereafter, an experiment was conducted by spraying different doses of BRs as β-sitosterol (25, 50, and 100 mg L^{-1}) on 30-day-old *P. vulgaris* under HT up to 35 °C (El-Bassiony et al. 2012). It was found that, BRs increased vegetative growth, yield, with higher bean quality by increasing the nutrient content of N, P, K, and free amino acid and phenolic acids. Later, this study suggested the role of BRs on decreasing the HT-induced oxidative stress which resulted in significant betterment of plant growth and yield.

L. purpureus plants were subjected to HT stress of 40–42 °C with the external application of various doses of salicylic acid (SA: 0.5, 1.0, 1.5, 2.0 mM) (Rai et al. 2018b). Salicylic acid regulated the growth and biochemical processes in heat-stressed plants by the modulation of antioxidants activities. Heat-stressed plants showed higher growth by increased plant height and better yield indicated by number of pods, pod size, and yield plant^{-1}. Exogenous SA enhanced Chl content with higher leaf area which might be directly involved in producing higher yield. Heat stress elevated the accumulation of MDA and H_2O_2 contents with extreme membrane damage (EL), which were minimized by SA treatments where SA-induced higher involvement of antioxidants activities were observed. In this study, SA application enhanced the contents of AsA and GSH with the higher activities of APX and GR in stressed plants. The higher activities of SOD, CAT, POD, GPOX, and FRAP in SA applied plants under stress could be involved in the reduction of ROS content. Finally, from the overall observation, 1.0 mM dose of SA showed the best performance in reducing heat-induced toxicity in *L. purpureus*.

A structural and functional mimic of jasmonic acid (JA) is coronatine (COR)—involved in the growth and developmental processes of plants. It was reported that, COR and JA are also similar in activities but not identical (Uppalapati et al. 2005), while COR enhanced the protease inhibitors and secondary metabolites production (Ceylan et al. 2013). An experiment was conducted to make correlation between exogenously applied COR and plant tolerance to abiotic stress including HT (44 °C) by taking *C. arietinum* as a test plant (Ceylan et al. 2013). In this study, the relative growth rate (RGR), content of MDA, pro, and H_2O_2 along with the activities of antioxidant enzymes in *C. arietinum* roots were observed with or without 0.01 μM COR application. The COR application reduced MDA, pro and H_2O_2 contents while RGR was increased in stressed seedlings compared to stressed alone. Moreover, COR application increased the activities of CAT, POD and decreased APX and GR activities in stressed plants for which oxidative stress was reduced.

Exogenous foliar application of auxin [4-chloroindole-3-acetic acid (4-Cl-IAA)] enhanced the seed yield of *P. sativum* under heat stress (Abeysingha 2015).

6.2 *Osmoprotectants*

Accumulation of osmolytes is the adaptive mechanisms of tolerance of any kind of stress condition including HT. The osmoprotectants, Pro, GB, and trehalose have been shown to enhance under heat stress. In the present days, for mitigation of

HT-induced damages in plants, exogenous use of osmoprotectants has been found beneficial. These molecules provide protection against heat stress through managing ROS by uplifting antioxidant capacities (Akhtar et al. 2015; Nahar et al. 2015b). As non-essential amino acid, Pro is proved as the most extensively studied and revealed thermoprotectant. Proline protects from stress-induced damages through performing various roles such as C and N storehouse, ROS scavenger, molecular chaperone, protein and membrane stabilizer. (Ashraf and Foolad 2007; Szabados and Savoure 2010). Through different physiological roles pro protect plants from HT damages. Thus, most of the studies showed the positive correlation between pro and heat stress, whereas, in few cases, pro also showed some negative correlation with heat stress. Two genotypes (heat tolerant and sensitive) of *V. radiata* subjected to HT stress (45/33 °C) until reproductive stage with or without 5 mM pro treatment. Heat stress resulted in considerable reduction in pollen germination, stigma receptivity, and ovule viability together high cell membranes damage, lowered water content, cellular respiration, stomatal conductance, photosynthesis, C fixation and assimilation enzymes activity, consequently affecting pod number, pod weight, seed weight plant^{-1}. Exogenous pro treatment significantly acts in thermotolerance through elevating endogenous pro and noticeable pollen fertility improvement by proper functioning of stigma and ovule. At the same time, plant water status activated C fixation, increased photosynthetic ability, and assimilated enzymes ultimately increased the pod number pod wt, filled pod , and seed weight plant^{-1}. Proline treatment showed significant effects on heat-sensitive genotype (Priya et al. 2019a). A study with chickpea demonstrated the beneficial roles of exogenously applied Pro under HT stress. *C. arietinum* root and shoot growth was inhibited at HT stress at 40/35 °C and 45/40 °C (day/night), respectively. Different physiological disorders were observed with the enhanced level of MDA, H_2O_2, membrane damages as well as increased oxidative injury. Plant water status was also reduced with reduced cellular respiration, and Chl contents. Enzymatic (SOD, CAT, APX, and GR) and non-enzymatic (AsA and GSH) antioxidants activity were decreased with the inhibition of C fixation enzyme (RuBisCO), sucrose phosphate synthase, and sucrose hydrolysis (invertase) activity. When the plants were exogenously supplied with 10 μM pro, inhibited shoot and root growth was improved with elevated endogenous Pro (63 μmol g^{-1}) level, recovered membrane damage, improved water status, RWC, Chl contents specially at 45/40 °C because of enhanced activity of vital antioxidants components as well as reduced level of oxidative damages under heat stress (Kaushal et al. 2011). Kumar et al. (2012) found that heat stress (45/40 °C) induced growth inhibition of *C. arietinum* was altered and promoted through the exogenous application of pro, GB, and trehalose (10 μM). The effect of heat stress and exogenous protectants on *L. purpureus* was investigated. It was demonstrated that heat stress reduced growth and biomass, damaged membranes, and generated ROS. Exogenous proline effectively controlled oxidative damages and regulated the bio-physiological growth and antioxidant defense system and thus developed heat stress tolerance (Rai et al. 2018a). Organic compatible solute GB is another important amphoteric quaternary amine that plays a significant role in protecting plants against stresses through reducing ROS, osmotic adjustment, photosynthetic apparatus, and protein stabilization (Ashraf and

Foolad 2007). Exogenous application of GB improved plant biomass and yield. It also improved the attributes related to reproductive biology. Similarly, in *M. sativa* exogenously applied GB showed protective effects against stress (Zhao et al. 1992). Heat stress induced damage in mitigation by exogenous GB in legumes is needed to be explored more.

Trehalose (Tre) is a reducing disaccharide that plays a role as a stress protectant. It acts as a molecule and membrane stabilizer, forms hydrogen bonds with biomolecules, membranes, replaces water, and thus, gives protection against stresses (Crowe 2007; Nahar 2015b). Gao et al. (2013) conducted an experiment with *V. faba* under heat stress. Pretreatment of *V. faba* leaf disks with Tre (1, 10, 25, 50, and 100 μM) protected thylakoid membrane proteins increased photosynthetic efficiency and photochemical quenching, decreased MDA, H_2O_2 contents, EL due to upregulated antioxidant defense system. Hence, the role of Tre for heat stress tolerance is yet to be explored.

6.3 Plant Nutrients

There are 17 elements marked important for the life cycle, growth, and the development of plants is known as essential plant nutrients. They are further subdivided into macronutrients (C, H, O, N, P, K Ca, Mg, S) and micronutrients (Fe, Mn, Zn, Cu, Mo, Cl, B, Ni) depending on their more and less necessity for the plant's life cycle, growth, and development, respectively. To attain the proper growth, metabolism, higher productivity, as well as sustainable agricultural system, proper nutrients management is very important (Noreen et al. 2018). Although, the main function of plant nutrients is to run the plant growth and development smoothly, plants tolerance against various abiotic stresses can be achieved through proper management of nutrients. Under heat stress condition, proper nutrients management can reduce the adverse effects of stress. High temperature stress causes higher generation of ROS, oxidative damages of chloroplast, membranes, etc. Addition of proper nutrients scavenge the toxic ROS by enhanced antioxidants, photosynthesis, properly maintained membrane and chloroplast integrity, and finally improved the tissue water status of plant under heat stress (Waraich et al. 2012). Through different mechanisms, exogenous use of different plant nutrients has shown to develop tolerance against HT-induced injuries by improving growth and uplifting antioxidant defense mechanism (Hasanuzzaman et al. 2015). Siddiqui et al. (2018) conducted an experiment to investigate the effect of heat stress on *V. faba* and mitigation of adverse heat stress through exogenous magnesium (Mg) application. High temperature decreased the total Chl, CA, and RuBisCO activities as well as growth attributes but elevated the level of osmolytes (Pro, GB) accumulation, antioxidant enzymes (SOD, POD, CAT) activities. Increased EL, MDA, H_2O_2 contents, DNA damage were also evident, whereas exogenously applied Mg caused significant enhancement of Pro, GB together with increased activities of SOD, POD, CAT. On contrary, reduced damage of DNA, electrolyte leakage, contents of MDA and H_2O_2 clearly indicated that exogenously supplied Mg, besides playing essential role in growth and development, improved the heat stress tolerance of *V. faba* plant by reducing cellular oxidative damages through enhanced accumulation of osmolytes and antioxidant enzymes activities. In a study done by Thalooth et al. (2006) demonstrated that Zn when foliarly applied gave better growth, yield components as well as better protection to dehydrated and stressed *V. radiata*. It was demonstrated that *Glycine max* when supplemented with boron (B) under stress condition had enhanced nitrogen metabolism, fatty acid oil, and seed protein content compared to plant did not supplied with B (Bellaloui 2011).

Adequate amount of Zn nutrition is very crucial to increase the growth of *C. arietinum* plants under HT (Ullah et al. 2019). Ullah et al. (2019) had been conducted a study by using two levels of Zn including low dose (0.3 mg kg^{-1} of soil) and high dose (3 mg kg^{-1} of soil) to mitigate the heat (35/30 °C day/night temperature) stress on *C. arietinum*. Here, higher Zn alleviates the adverse effects of HT by decreasing oxidative and osmotic stress along with improvement in photosynthesis rate, PS II efficiency, plant growth, biomass accumulation, and Zn uptake. Enhanced activity of SOD, APX, and higher content of AsA were observed in sufficient Zn applied heat-stressed plant which correlated the Zn-induced modulated antioxidant defense for suppressing oxidative stress. Hence, they suggested that Zn not only supply the plant nutrient but can also be act as HT stress inhibitor when applied sufficiently.

6.4 *Antioxidants*

Non-enzymatic antioxidants—AsA, GSH, pro are directly involved in scavenging ROS and enhancing heat tolerances in Fabaceae plants. Therefore, exogenous application of these compounds shows the very potential way to protect plant from heat toxicity. Among non-enzymatic antioxidant, AsA is very potential one, which protects plant cells from oxidative stress. Alleviation of HT toxicity by exogenous AsA was evaluated on *V. radiata* (Kumar et al. 2011). The AsA (50 μM) was used with and without varying degree of HT including 30/20, 35/25, 40/30, and 45/35 °C on hydroponically grown seedlings in growth chamber. Therefore, the effect of AsA was evaluated on growth, membrane damage, Chl loss, leaf water status, oxidative stress, and antioxidant components. Seed germination and seedlings growth were enhanced by AsA treatment in stressed condition. The severity of Chl, water loss, and oxidative damage (MDA and H_2O_2 content) were inhibited in AsA treated plants under heat condition which can be correlated with AsA-induced higher activities of SOD, CAT, APX, and GR. Exogenous AsA increased endogenous AsA, GSH, and pro accumulation in stressed plants and indicated the role of AsA to mitigate heat-induced chlorosis, growth retardation, and oxidative stress in *V. radiata*.

Again, exogenous GSH treated *V. radiata* showed the heat tolerance by regulating the antioxidant and glyoxalase systems significantly upon short-term exposure to HT stress (42 °C) (Nahar et al. 2015a). High temperature stress on 6-day-old seedling showed the increased MDA, MG, pro contents with higher generation of ROS and LOX activity. Therefore, seedlings suffered from Chl degradation and water crisis.

In this situation, endogenous AsA content reduced with higher GSH and GSSG with a result of lower GSH/GSSG. But, it was noticeable that 0.5 mM GSH application on stressed seedlings as pretreatment for 24 h duration altered above-mentioned heat-induced all parameters as well as enhanced better seedlings growth. Application of GSH also increased APX, MDHAR, DHAR, GR, CAT, GPX, GST activities significantly in *V. radiata* seedlings under heat stress. The activity of Gly I and Gly II also enhanced by GSH which resulted in the reduction of MG content under stressed condition. The result of this well-organized study suggested the potential of exogenous GSH to enhance heat tolerance and physiological adaptations of *V. radiata*. The role of pro was described above by keeping in the group of osmoprotectant that is why here we avoid it making repetition as a role of antioxidant.

6.5 Polyamines

Polyamines including spermidine (Spd), putrescine (Put), and spermine (Spm) are universally distributed in living organism as low molecular weight aliphatic polycations. It has been already established that, these three PAs are accumulated in plants upon exposure to abiotic stress including heat (Hussain et al. 2011; Hasanuzzaman et al. 2013). Polyamines protect plants from HT stress by involving in securing photosynthetic apparatus with the regulation of photosynthesis and maintaining membrane thermostability (Hasanuzzaman et al. 2013). Therefore, exogenous PAs-regulated antioxidant defense and glyoxalase systems to attain heat tolerance in *V. radiata* were extensively studied by applying Spm against 40 °C (Nahar et al. 2016). Heat caused extreme oxidative stress indicated by higher ROS generation, MDA content, membrane damage and LOX activity with suppression of seedling growth and Chl content. But the application of 0.2 mM Spm (as pretreatment for 24 h) to stressed seedlings showed the improvement in seedling vigor, growth and biomass accumulation with the increase of water and Chl contents. Thus, it was observed that, Spm reduced ROS generation, MDA content, LOX activity in stressed seedling. Nahar et al. (2016) reported the Spm induced upregulation in antioxidant and glyoxalase enzymes activities in heat-stressed seedlings which resulted in higher tolerance. Exogenous Spm increased AsA and GSH contents while enhanced the enzymatic activities of SOD, CAT, GPX, DHAR, and GR—conferred the oxidative stress tolerance to HT stress. Stressed seedlings also showed higher content of MG due to uncoupling the glyoxalase enzymes activities, but Spm reduced MG level by increasing the glyoxalase enzymes activities. Finally, Spm modulated higher endogenous PAs levels were responsible for attaining higher tolerance in *V. radiata* under HT.

Exogenous applications of PAs strengthen the enhancement of heat tolerance in *G. max* through modulation of the endogenous PAs content (Amooaghaie and Moghym 2011). Application of 1 mM of each Put, Spd, and Spm was used in germinating seeds of *G. max* as pretreatment (2 h) prior to 45 °C (2 h), then improved both roots and hypocotyls growth were observed (Amooaghaie and Moghym 2011). The effect of exogenous PAs also reduced the EL and MDA levels from root and hypocotyls under heat stress. Thus, it has been suggested that Put, Spd, and Spm protects membrane integrity under HT stress.

6.6 Signaling Molecules

Nitric oxide (NO) is vital signaling molecule due to its important roles in numerous physiological processes in plants and also acting in increasing plants tolerance to HT stress (Hasanuzzaman et al. 2013). It has been shown that exogenous NO (as sodium nitroprusside, SNP) application advances the plant developmental process with higher growth (Hu et al. 2017).

Application of SNP as NO donor enhanced the tolerance of *L. purpureus* plants to HT stress of 40–42 °C. Experimental result indicated in terms of growth and yield enhancement, reduction of oxidative stress through upregulating the antioxidants activities (Rai et al. 2018b). Here, SNP treatment significantly modulated the activities of FRAP, POD, APX, GR, GPOX with maintaining higher content of AsA and GSH, which were cumulatively involved in reducing ROS to tackle oxidative stress under HT. SNP also increased plant water content and Chl content which imposed higher photosynthesis and growth in stressed plants as well as improved plant tolerance. Rai et al. (2018b) also reported that, NO regulated DNA methylation pattern and stress-responsive genes which later conveyed the HT tolerance.

Hydrogen peroxide (H_2O_2) enhanced the alfalfa seed germination under HT stress such as 60, 70, and 80 °C up to 24 h (Hong and Kang 2016).

A nonprotein amino acid is γ-aminobutyric acid (GABA) which also acts as signaling molecule in response to stress condition. It has been reported that GABA modulates the tolerance of *V. radiata* against HT stress (45/28 °C) (Priya et al. 2019b). In this study, heat-stressed plants suffered from the declining GABA concentrations in leaf and anther by 49 and 60%, respectively, led to heat sensitivity of reproductive function. But 1 mM GABA treatment improved reproductive function in heat-stressed plants significantly in terms of pollen germination, pollen viability, stigma receptivity, and ovule viability. In addition, heat-stressed plants had lesser damage to cell membranes, Chl content, Chl fluorescence, RuBisCO activity, carbon assimilation, and C utilization upon GABA application. Exogenous GABA also increased the osmolytes accumulation indicated by Pro and Tre leading to improved leaf water status. Consequently, GABA-treated stressed plants produced more pods number and seed weight by 28 and 27%, respectively. Priya et al. (2019b) also observed GABA mediated reduction in oxidative stress confirmed by lower H_2O_2 production and MDA content. Application of GABA enhanced the activities of SOD, CAT, APX, and GR with higher contents of both AsA and GSH in heat-stressed *V. radiata*.

7 Agronomic Approaches to Combat High Temperature Stress

In the present climate change scenario, crop plants are regularly facing the ever-fluctuating temperature extremes. In an open field, even though it is not an easy task to regulate the growth temperature required for the crops, some options are still there to combat the extreme heat stress. To alleviate the negative effects of HT stress, manipulation of some agronomic crop management practices, like time and method of sowing, tillage, irrigation practices, mulching, intercropping, crop rotation, choice of tolerant varieties can result in great success and escape the crops critical growth period from extreme heat stress. High temperature causes altered evaporation and infiltration of soil water, heat exchange between soil and environment. Thus, ultimately resulted in the changed temperature and water balance of the soil as well as altered soil surface (Ferrero et al. 2005; Sekhon et al. 2010). The soil surface alteration can be controlled by the operation of tillage and mulching. As the stored soil water and plants water uptake from soil is greatly hampered by the heat-induced roughness of soil surface, evaporation, etc., deep tillage can be an effective way to improve the rooting depth of plants in a compact soil (Lipiec et al. 2012; Martínez et al. 2012). Organic mulches have low thermal conductivity. So, the use of organic mulches can be another protective way from extreme evaporation by reducing surface temperature and ultimately, improving water content of soil (Khan et al. 2000; Mulumba and Lal 2008). Conservation tillage practices are very appropriate for dry-land agriculture, for soil structure conservation through reduced water infiltration, runoff, etc. In *C. arietinum*, conservation tillage practices along with mulching have showed higher productivity in the seasonal periods of higher heat and minimum precipitation in Middle Indo-Gangetic Plains (Mishra et al. 2012b). In a study conducted with *L. culinaris*, mulching effect on soil, canopy, and leaf temperature under stress condition was investigated. It was revealed that the highest growth and productivity of lentil were achieved when irrigated after 70 mm evaporation and 2 ton ha^{-1} mulch was applied together (Amini and Alami-Milani 2013). Different varieties also behave differently under high heat and scanty water conditions. For example, 'JG 16' genotype of *C. arietinum* could perform better under adequate moisture level, whereas 'KWR 108' genotype provided good productivity under scanty moisture level. Especially, the early maturing crop varieties under stress condition could escape terminal water shortages by their ability to establish early crop stand and reduced heat stress (Sekhon et al. 2010). So, the selection of appropriate varieties of pulses under various weather extremes can also play role in reducing the risk of crop failure and yield loss (Mishra et al 2012b; Singh et al 2015). Varietal diversification under stress condition can also reduce the risk of yield loss compared to growing a single crop. To optimize the production, in adverse environmental situation, a contingency planning for setting planting date can reduce the loss (Praharaj et al. 2015). Rain-fed regions are usually chosen for pulse production. Hence, water-saving irrigation method employment, scientific irrigation scheduling, and proper application method can ensure higher irrigation area coverage, nutrients and water use efficiency and good crop stand (DAC GoI 2012). Under stressful condition, like water shortages condition adoption of some techniques, those are efficiently water saving are very effective. Some modern irrigation practices save more than 50% water, like sprinkler, drip, micro, and film hole irrigation. Compared to surface irrigation, they have more water use efficiency as well as provide more yields. But the problem associated is the high-cost and energy requirements for the system (Jensen 2013). Zhang et al. (2015) conducted an experiment to see the impact of scheduled irrigation in *G. max* and found that irrigation decreased the extreme heat effects and played beneficial role to achieve higher yield under climate change risk. In *C. arietinum* and *P. sativum*, raising furrow irrigated bed can conserve soil moisture together with the increased productivity. Moreover, to overcome the terminal moisture stress, for increasing water use efficiency and getting higher yield, single irrigation at branching stage is effective (Mishra et al. 2012a). Alteration of sowing time would be beneficial to escape the heat stress at crops' developmental stages sensitive to HT stress. Cooler weather at early spring and progressively hotter at the mid-summer are the characteristics of the subtropical region. Concurrent production of agroforestry trees and agricultural crops in the same field can reduce heat stress by balancing the temperature in the region. Similarly, using shade above the agricultural field zone can be another protective measure to mitigate HT stress (Ismail et al. 1999; Khan et al. 2000). Water shortages and alteration of temperature due to climate change can also remarkably affect the geographical distribution and infestation of diseases and pests, and those can significantly limit the production potential of the existing crops (Vadez et al. 2011; Sharma 2014). In such condition, crop rotation with the short duration species can be an effective measure to control the outbreak of disease and pest. In addition, new control measures should also be adopted. In *P. sativum* plant, use of several biofertilizers, arbuscular mycorrhizal fungi showed increased water, and nutrient use efficiencies under stress condition. However, to combat climate change-induced heat stress, gene mining for improving heat stress tolerance, efficient water and nutrients management, changing existing cropping pattern, using alternate legume gene bank, avoiding stress vulnerable regions, shed management, introducing microirrigation, manipulating sowing time, etc., are some important measures to increase pulse production. By adopting the above-mentioned management practices, in spite of water and soil conservation, increased organic C sequestration in soil and reduced GHGs emission will occur those will finally reduce the adverse climate change effects (Basu et al. 2016).

8 Transgenic Approaches to Enhance High Temperature Stress Tolerance

Development of plants tolerant to HT stress is of colossal worth in light of global warming. Re-programming the genetic machinery plants often survives and perpetuates their life process under HT stress. Transgenic approaches aim to develop HT stress-tolerant plant species through over-expressing heat shock protein genes or by altering levels of heat shock factors. Over-expression of some other genes/proteins has proven to improve HT stress tolerance which may regulate osmoprotecting function, antioxidant defense system and reactive oxygen species, membrane properties, photosynthetic reactions, production of metabolites having capacities to modulate plant physiological processes (Grover et al. 2013). Among the presently existing/published articles, only few are experimented with transgenic approaches of HT stress tolerance in Fabaceae plants (Table 2).

The *VfHsp17.9-CII* from *G. max* is a HT transcript. It was induced into *V. faba*. The transgenic *V. faba* demonstrated a clear difference for HT stress tolerance in

Table 2 Transgenic approaches to develop high temperature stress tolerant legume plants

Host plants/source	Genes	Transgenic plants	Tolerance responses	References
Glycine max L.	*VfHsp17.9-CII*	*Vicia faba* L.	Expression of VfHsp17.9-CII transcript was higher and is related to HT tolerance in seedling and reproductive stage. Increased pollen viability with the increase of transcript level of gene	Kumar et al. (2015)
Medicago sativa L.	*MsHsp23*	*M. sativa* L.	Caused less wilting, and rendered greenish leaves, increased APX activity levels	Lee et al. (2017)
Saccharomyces cerevisae	*ScTPS1–ScTPS2*	*M. sativa* L.	Increased survival percentage with no visible necrosis damage. Higher Tre content, relative water content and fresh weight	Suárez et al. (2009)

leaves of young seedlings and flowering plants. Expression of *VfHsp17.9-CII* transcript was 350-fold higher under 38 °C for 1 h, whereas the extent was 620-fold after 2 h of heat shock. The loss of pollen viability was higher in 2 h of HT stress, compared to 4 h of HT indicating the function of this gene in maintaining pollen viability. The *VfHsp17.9-CII* transcript confirmed an obvious HT induction pattern in leaves of young seedlings and flowering plants (Kumar et al. 2015). Trehalose-6-phosphate synthase (*TPS1*) and trehalose-6-phosphate phosphatase (*TPS2*) gene expression encouraged tolerance to multiple stresses. Transgenic *M. sativa* encoded *ScTPS1-ScTPS2* gene which was transformed from *Saccharomyces cerevisiae. M. sativa* was exposed to extreme temperatures, drought, or salt stress. Plants exposed to preadaptation at 37 °C for 4 h, before a heat shock at 40, 45, 50, and 55 °C for 1 h and then were subjected to 10 d of recovery. Compared to wild type, transgenic plants are biosynthesized higher amount of Tre. This osmoprotectant confirmed higher relative water content. Transgenic plants also had higher percentage of survival with higher foliage fresh weight (Suárez et al. 2009). Heat shock proteins (Hsps) are molecular chaperone participating in protein folding and signal transduction, maintaining cellular homeostasis. These characteristics of Hsps contributed to developmental processes and abiotic stress protection. *MsHsp23* (heat shock protein) of wild *M. sativa* exploited to develop transgenic *M. sativa* which showed enhanced tolerance to HT (42 °C) and oxidative stress. The transgenic plants reduced wilting and increased greenness of leaves (Lee et al. 2017).

9 Conclusion

High temperature is imposing a great threat to growth and the development of Fabaceae plants either alone or together with other stress. Upon stresses, plants faced suppression in their normal growth and physiology followed by a devastating obstacle in reproductive development and yield loss. In addition, HT accelerates the excessive production of reactive oxygen species and disruption in plant antioxidant defense systems. Here, plants fail to make balance in non-enzymatic antioxidants accumulation and activate many enzymatic antioxidants activities. As an outcome, plants suffer from oxidative damage including cellular injury and even death. Due to higher nutritional value and demand, Fabaceae plants have a large position in global food production. Thus, it is important to protect Fabaceae plants from HT stress for assuring the world food security. There are some researches to improve the tolerance of Fabaceae plants to heat stress by adopting exogenous phytoprotectants and managing agronomic cultural practices which are not sufficient. On the other hand, there are still lacking in introducing transgenic technique to unlocking the potentiality of Fabaceae for increasing HT tolerance along with higher yield. Henceforth, both laboratory and field experiments are extensively demanded to explore different biochemical and molecular manners incorporation with agronomic practices to know the actual HT responses and their tolerances tactics.

References

Abeysingha GLDN (2015) The effects of auxins on seed yield parameters in wheat, pea and canola grown under controlled environmental and Western Canadian field conditions. MSc thesis. University of Alberta, Edmonton (AB)

Akhtar S, Naik A, Hazra P (2015) Harnessing heat stress in vegetable crops towards mitigating impacts of climate change. In: Choudhary ML, Patel VB, Siddiqui MW, Mahdl SS (eds) Climate dynamics in horticultural science. CRC Press, Boca Raton, FL, pp 173–200

Alamri SA, Siddiqui MH, Al-Khaishany MY, Khan MN, Ali HM, Alakeel KA (2019) Nitric oxide-mediated cross-talk of proline and heat shock proteins induce thermotolerance in *Vicia faba* L. Environ Exp Bot 161:290–302

Amini R, Alami-Milani M (2013) Effect of mulching on soil, canopy and leaf temperature of lentil (*Lens culinaris* Medick.). Int J Farm Appl Sci 2:797–802

Amooaghaie R, Moghym S (2011) Effect of polyamines on thermotolerance and membrane stability of soybean seedling. Afr J Biotechnol 10:9677–9682

Asada K (2006) Production and scavenging of reactive oxygen species in chloroplasts and their functions. Plant Physiol 141:391–396

Ashraf M, Foolad MR (2007) Roles of glycine betaine and proline in improving plant abiotic stress tolerance. Environ Exp Bot 59:206–216

Basu PS, Singh UMMED, Kumar ANIL, Praharaj CS, Shivran RK (2016) Climate change and its mitigation strategies in pulses production. Indian J Agron 61:71–82

Bellaloui N (2011) Effect of water stress and foliar boron application on seed protein, oil, fatty acids, and nitrogen metabolism in soybean. Am J Plant Sci 2:692–701

Bishop J, Jones HE, Lukac M, Potts SG (2016) Insect pollination reduces yield loss following heat stress in faba bean (*Vicia faba* L.). Agric Ecosyst Environ 220:89–96

Campos-Vega R, Loarca-Piña G, Oomah BD (2010) Minor components of pulses and their potential impact on human health. Food Res Int 43:461–482

Cao D, Takeshima R, Zhao C, Liu B, Jun A, Kong F (2017) Molecular mechanisms of flowering under long days and stem growth habit in soybean. J Expt Bot 68:1873–1884

Capraro J, Clemente A, Rubio LA, Magni C, Scarafoni A, Duranti M (2010) Assessment of the lupin seed glucose-lowering protein intestinal absorption by using in vitro and ex vitro models. Food Chem 125:1279–1283

Ceylan HA, Turkan I, Sekmen AH (2013) Effect of coronatine on antioxidant enzyme response of chickpea roots to combination of PEG-induced osmotic stress and heat stress. J Plant Growth Regul 32:72–82

Chakraborty U, Pradhan D (2011) High temperature-induced oxidative stress in *Lens culinaris*, role of antioxidants and amelioration of stress by chemical pre-treatments. J Plant Interact 6:43–52

Chavez-Arias CC, Ligarreto-Moreno GA, Restrepo-Díaz H (2018) Evaluation of heat stress period duration and the interaction of daytime temperature and cultivar on common bean. Environ Exp Bot 155:600–608

Considine MJ, Siddique KHM, Foyer CH (2017) Nature's pulse power: legumes, food security and climate change. J Expt Bot 68:1815–1818

Crowe JH (2007) Trehalose as a "chemical chaperone": fact and fantasy. Adv Exp Med Biol 594:143–158

DAC GoI (2012) Report of expert group on pulses. Department of Agriculture and Cooperation, Government of India, Ministry of Agriculture, New Delhi, pp 1–148

Das A, Eldakak M, Paudel B, Kim D-W, Hemmati H, Basu C, Rohila JS (2016) Leaf proteome analysis reveals prospective drought and heat stress response mechanisms in soybean. Biomed Res Int. https://doi.org/10.1155/2016/6021047

Devasirvatham V, Gaur PM, Mallikarjuna N, Tokachichu RN, Trethowan RM, Tan DK (2012) Effect of high temperature on the reproductive development of chickpea genotypes under controlled environments. Funct Plant Biol 39:1009–1018

Dhakal Y, Meena RS, Kumar S (2016) Effect of INM on nodulation, yield, quality and available nutrient status in soil after harvest of green gram. Legume Res 39:590–594

El-Bassiony AM, Ghoname AA, El-Awadi ME, Fawzy ZF, Gruda N (2012) Ameliorative effects of brassinosteroids on growth and productivity of snap beans grown under high temperature. Gesunde Pflanzen 64:175–182

Ferrero A, Usowicz B, Lipiec J (2005) Effects of tractor traffic on spatial variability of soil strength and water contentin grass covered and cultivated sloping vineyard. Soil Tillage Res 84:127–138

Foyer CH, Lam HM, Nguyen HT, Siddique KH, Varshney RK, Colmer TD, Cowling W, Bramley H, Mori TA, Hodgson JM, Cooper JW (2016) Neglecting legumes has compromised human health and sustainable food production. Nat Plants 2:16112

Gao J, Wang N, Xu SS, Li Y, Wang Y, Wang GS (2013) Exogenous application of trehalose induced H_2O_2 production and stomatal closure in *Vicia faba*. Biol Plant 57:380–384

Grover A, Mittal D, Negi M, Lavania D (2013) Generating high temperature tolerant transgenic plants: achievements and challenges. Plant Sci 205:38–47

Harsh A, Sharma YK, Joshi U, Rampuria S, Singh G, Kumar S, Sharma R (2016) Effect of short-term heat stress on total sugars, proline and some antioxidant enzymes in moth bean (*Vigna aconitifolia*). Ann Agric Sci 61(1):57–64

Hasanuzzaman M, Nahar K, Alam M, Roychowdhury R, Fujita M (2013) Physiological, biochemical, and molecular mechanisms of heat stress tolerance in plants. Int J Mol Sci 14:9643–9684

Hasanuzzaman M, Nahar K, Alam MM, Ahmad S, Fujita M (2015) Exogenous application of phytoprotectants in legumes against environmental stress. In: Azooz MM, Ahmad P (eds) Legumes under environmental stress: yield, improvement and adaptations. Wiley, Chichester, pp 161–197

Hasanuzzaman M, Masud AAC, Bhuiyan TF, Anee TI (2019) Legumes for soil fertility management and sustainable agriculture. SATSA Mukhapatra—Annu Tech Issue 23:29–46

Hong EJ, Kang DH (2016) Effect of sequential dry heat and hydrogen peroxide treatment on inactivation of *Salmonella Typhimurium* on alfalfa seeds and seeds germination. Food Microbiol 53:9–14

Hu J, Yang H, Mu J, Lu T, Peng J, Deng X, Kong Z, Bao S, Cao X, Zuo J (2017) Nitric oxide regulates protein methylation during stress responses in plants. Mol Cell 67:702–710

Hussain SS, Ali M, Ahmad M, Siddique KHM (2011) Polyamines: natural and engineered abiotic and biotic stress tolerance in plants. Biotechnol Adv 29:300–311

IPCC (2014) Climate change 2014: Synthesis report. Contribution of working groups I, II and III to the fifth assessment report of the intergovernmental panel on climate change. In: Core Writing Team, Pachauri RK, Meyer LA (eds) IPCC, Geneva, Switzerland, p 151

Iqbal A, Khalil IA, Ateeq N, Khan MS (2006) Nutritional quality of important food legumes. Food Chem 97:331–335

Ismail AM, Hall AE, Close TJ (1999) Purification and partial characterization of adehydrin involved in chilling tolerance during seedling emergence of cowpea. Plant Physiol 96:13566–13570

Jensen R (2013) Some options for securing water resources for agricultural production. In: Grzesiak M, Rzepka A, Hura T, Grzesiak S (eds) Plant functioning under environmental stress. Cracow-Plant-Stress Conference, 2013, pp 9–25

Johkan M, Oda M, Maruo T, Shinohara Y (2011) Crop production and global warming. In: Casalegno S (ed) Global warming impacts—case studies on the economy, human health, and on urban and natural environments. InTech, Rijeka, Croatia, pp 139–152

Kaur R, Bains TS, Bindumadhava H, Nayyar H (2015) Responses of mungbean (*Vigna radiata* L.) genotypes to heat stress: effects on reproductive biology, leaf function and yield traits. Sci Hortic 197:527–541

Kaushal N, Gupta K, Bhandhari K, Kumar S, Thakur P, Nayyar H (2011) Proline induces heat tolerance in chickpea (*Cicer arietinum* L.) plants by protecting vital enzymes of carbon and antioxidative metabolism. Plants Physiol Mol Biol 17:203–213

Kazai P, Noulas C, Khah E, Vlachostergios D (2019) Yield and seed quality parameters of common bean cultivars grown under water and heat stress field conditions. AIMS Agric Food 4(2):285–302

Kennedy AR (1995) The evidence for soybean products as cancer preventative. J Nutr 125:733–743

Khan AR, Chandra D, Quraishiand S, Sinha RK (2000) Soilaeration under different soil surface conditions. J Agron Crop Sci 185:105–112

Khedar OP, Singh RV, Shrimali M, Singh NP (2008) Pulses: status and cultivation technology. Aavishkar Publishers, Jaipur

Kumar S, Kaur R, Kaur N, Bhandhari K, Kaushal N, Gupta K, Bains TS, Nayyar H (2011) Heat-stress induced inhibition in growth and chlorosis in mungbean (*Phaseolus aureus* Roxb.) is partly mitigated by ascorbic acid application and is related to reduction in oxidative stress. Acta Physiol Plant 33:2091. https://doi.org/10.1007/s11738-011-0748-2

Kumar S, Kaushal N, Nayyar H, Gaur P (2012) Abscisic acid induces heat tolerance in chickpea (*Cicer arietinum* L.) seedlings by facilitated accumulation of osmoprotectants. Acta Physiol Plant 34:1651–1658

Kumar S, Thakur P, Kaushal N, Malik JA, Gaur P, Nayyar H (2013) Effect of varying high temperatures during reproductive growth on reproductive function, oxidative stress and seed yield in chickpea genotypes differing in heat sensitivity. Arch Agron Soil Sci 59:823–843

Kumar R, Lavania D, Singh AK, Negi M, Siddiqui MH, Al-Whaibi MH, Grover A (2015) Identification and characterization of a small heat shock protein 17.9-CII gene from faba bean (*Vicia faba* L.). Acta Physiol Plant 37(9):190. https://doi.org/10.1007/s11738-015-1943-3

Kumar N, Hazra KK, Nath CP, Praharaj CS, Singh U (2018) Grain legumes for resource conservation and agricultural sustainability in south Asia. In: Meena RS, Das A, Yadav GS, Lal R (eds) Legumes for soil health and sustainable management. Springer, New York, pp 77–108

Lee K-W, Rahman M, Choi GJ, Kim K-Y, Ji HC, Hwang TY, Lee SH (2017) Expression of small heat shock protein23 enhanced heat stress tolerance in transgenic alfaalfa plants. J Anim Plant Sci 27(4):1238–1244

Lewis G, Schrire B, Mackinder B, Lock M (2005) Legumes of the World. Royal Botanic Gardens, Kew

Lipiec J, Horn R, Pietrusiewicz J, Siczek A (2012) Effectsof soil compaction on root elongation and anatomy of different cereal plant species. Soil Tillage Res 121:74–81

Liu Y, Li J, Zhu Y, Jones A, Rose RJ, Song Y (2019) Heat stress in legume seed setting: effects, causes, and future prospects. Front Plant Sci 10:938. https://doi.org/10.3389/fpls.2019.00938

Makonya GM, Ogola JB, Muasya AM, Crespo O, Maseko S, Valentine AJ, Ottosen CO, Rosenqvist E, Chimphango SB (2019) Chlorophyll fluorescence and carbohydrate concentration as field selection traits for heat tolerant chickpea genotypes. Plant Physiol Biochem 141:172–182

Martínez IG, Prat C, Ovalle C, delPozo A, Stolpe N, Zagal E (2012) Subsoiling improves conservation tillage in cereal production of severely degraded Alfisols under Mediterranean climate. Geoderma 18:10–17

Mishra JP, Praharaj CS, Singh KK (2012a) Enhancing water use efficiency and production potential of chickpea and fieldpea through seed bed configurations and irrigation regimesin North Indian Plains. J Food Legumes 25:310–313

Mishra JP, Praharaj CS, Singh KK, Kumar N (2012b) Impact of conservation practices on crop water use and productivity in chickpea under middle Indo-Gangetic plains. J Food Legumes 25:41–44

Mitchell JW, Mandava NB, Worley JF, Plimmer JR, Smith MV (1970) Brassins: a new family of plant hormones from rape pollen. Nature 225:1065–1066

Mittler R, Blumwald E (2010) Genetic engineering for modern agriculture: challenges and perspectives. Annu Rev Plant Biol 61:443–462

Molteni A, Brizio-Molteni L, Persky V (1995) In vitro hormonal effects of soybean isoflavones. J Nutr 125:S751–S756

Morris JB (1997) Special purpose legume genetic resources conserved for agricultural, industrial and pharmaceutical use. Econ Bot 51:251–263

Mulumba LN, Lal R (2008) Mulching effects on selected soil physical properties. Soil Tillage Res 98:106–111

Nahar K, Hasanuzzaman M, Ahamed KU, Hakeem KR, Ozturk M, Fujita M (2015a) Plant responses and tolerance to high temperature stress: role of exogenous phytoprotectants. In: Hakeem K (ed) Crop production and global environmental issues. Springer, Cham, pp 385–435

Nahar K, Hasanuzzaman M, Alam MM, Fujita M (2015b) Exogenous glutathione confers high temperature stress tolerance in mung bean (*Vigna radiata* L.) by modulating antioxidant defense and methylglyoxal detoxification system. Environ Exp Bot 112:44–54

Nahar K, Hasanuzzaman M, Alam MM, Rahman A, Mahmud JA, Suzuki T, Fujita M (2016) Insights into spermine-induced combined high temperature and drought tolerance in mung bean: osmoregulation and roles of antioxidant and glyoxalase system. Protoplasma 254(1):445–460

Nahar K, Hasanuzzaman M, Alam MM, Rahman A, Mahmud JA, Suzuki T, Fujita M (2017) Insights into spermine-induced combined high temperature and drought tolerance in mung bean: osmoregulation and roles of antioxidant and glyoxalase system. Protoplasma 254(1):445–460

Nasr Esfahani M, Inoue K, Chu HD, Nguyen KH, Van Ha C, Watanabe Y, Burritt DJ, Herrera-Estrella L, Mochida K, Tran LSP (2017) Comparative transcriptome analysis of nodules of two Mesorhizobium–chickpea associations with differential symbiotic efficiency under phosphate deficiency. Plant J 91:911–926

Noreen S, Fatima Z, Ahmad S, Ashraf M (2018) Foliar application of micronutrients in mitigating abiotic stress in crop plants. In: Fujita M, Oku H, Nahar K, Hawrylak-Nowak B (eds) Hasanuzzaman M. Plant nutrients and abiotic stress tolerance Springer, Singapore, pp 95–117

Nulik J, Dalgliesh N, Cox K, Gabb S (2013) Integrating herbaceous legumes into crop and livestock systems in eastern Indonesia. Australian Centre for International Agricultural Research (ACIAR)

Omae H, Kumar A, Shono M (2012) Adaptation to high temperature and water deficit in the common bean (*Phaseolus vulgaris* L.) during the reproductive period. J Bot 2012:803413. https://doi.org/10.1155/2012/803413

Ozga JA, Kaur H, Savada RP, Reinecke DM (2017) Hormonal regulation of reproductive growth under normal and heat-stress conditions in legume and other model crop species. J Exp Bot 68:1885–1894

Patriyawaty NR, Rachaputi RC, George D (2018) Physiological mechanisms underpinning tolerance to high temperature stress during reproductive phase in mungbean (*Vigna radiata* (L.) Wilczek). Environ Exp Bot 150:188–197

Praharaj CS, Kumar N, Singh U, Singh SS, Singh J (2015) Transplanting in pigeonpea—a contingency measure for realizing higher productivity in Eastern Plains. J Food Legumes 28:34–39

Priya M, Sharma L, Kaur R, Bindumadhava H, Nair RM, Siddique KHM, Nayyar H (2019a) GABA (γ-aminobutyric acid), as a thermo-protectant, to improve the reproductive function of heat-stressed mungbean plants. Sci Rep 9(1):7788. https://doi.org/10.1038/s41598-019-44163-w

Priya M, Sharma L, Singh I, Bains TS, Siddique KH, Bindumadhava H, Nair RM, Nayyar H (2019b) Securing reproductive function in mungbean grown under high temperature environment with exogenous application of proline. Plant Physiol Biochem 140:136–150

Rai KK, Rai N, Rai SP (2018a) Response of *Lablab purpureus* L. to high temperature stress and role of exogenous protectants in mitigating high temperature induced oxidative damages. Mol Biol Rep 45:1375–1395

Rai KK, Rai N, Rai SP (2018b) Salicylic acid and nitric oxide alleviate high temperature induced oxidative damage in *Lablab purpureus* L. plants by regulating bio-physical processes and DNA methylation. Plant Physiol Biochem 128:72–88

Rose TJ, Hardiputra B, Rengel Z (2010) Wheat, canola and grain legume access to soil phosphorus fractions differs in soils with contrasting phosphorus dynamics. Plant Soil 326:159–170

Ryan E, Galvin K, O'Connor TP, Maguire AR, O'Brien NM (2007) Phytosterol, squalene, tocopherol content and fatty acid profile of selected seeds, grains and legumes. Plant Foods Hum Nut 62:85–91

Sehgal A, Sita K, Bhandari K, Kumar S, Kumar J, Prasad PVV, Siddique KHM, Nayyar H (2017) Influence of drought and heat stress, applied independently or in combination during seed development, on qualitative and quantitative aspects of seeds of lentil (*Lens culinaris* Medikus) genotypes, differing in drought sensitivity. Plant Cell Environ 42:198–211

Sekhon HS, Singh G, Sharma P, Bains TS (2010) Water use efficiency under stress environments. In: Yadav SS, McNeil DL, Redden R, Patil SA (eds) Climate change and management of cool season grain legume crops. Springer Press, New York

Shahwar D, Bhat TM, Ansari MYK, Chaudhary S, Aslam R (2017) Health functional compounds of lentil (*Lens culinaris* Medik): a review. Int J Food Prop. https://doi.org/10.1080/10942912.2017.1287192

Sharma HC (2014) Climate change effects on insects: implications for crop protection and food security. J Crop Improv 28:229–259

Sharma L, Priya M, Bindumadhava H, Nair RM, Nayyar H (2016) Influence of high temperature stress on growth, phenology and yield performance of mungbean [*Vigna radiata* (L.) Wilczek] under managed growth conditions. Sci Hortic 213:379–391

Siddiqui MH, Al-Khaishany MY, Al-Qutami MA, Al-Whaibi MH, Grover A, Ali HM, Al-Wahibi MS (2015) Morphological and physiological characterization of different genotypes of faba bean under heat stress. Saudi J Biol Sci 22(5):656–663

Siddiqui MH, Alamri SA, Al-Khaishany MY, Al-Qutami MA, Ali HM, Al-Whaibi MH, Al-Wahibi MS, Alharby HF (2018) Mitigation of adverse effects of heat stress on *Vicia faba* by exogenous application of magnesium. Saudi J Biol Sci 25(7):1393–1401

Singh U, Praharaj CS, Singh SS, Kumar N (2015) Influence of crop establishment practices and genotypes in pigeonpea—wheat system under IGP of India. J Food Legumes 28:315–319

Sita K, Sehgal A, Hanumantha Rao B, Nair RM, Vara Prasad PV, Kumar S, Nayyar H (2017) Food legumes and rising temperatures: Effects, adaptive functional mechanisms specific to reproductive growth stage and strategies to improve heat tolerance. Front Plant Sci 8:1–30. https://doi.org/10.3389/fpls.2017.01658

Stagnari F, Maggio A, Galieni A, Pisante M (2017) Multiple benefits of legumes for agriculture sustainability: an overview. Chem Biol Technol Agric 4:2017. https://doi.org/10.1186/s40538-016-0085-1

Suárez R, Calderón C, Iturriaga G (2009) Enhanced tolerance to multiple abiotic stresses in transgenic alfalfa accumulating trehalose. Crop Sci 49(5):1791–1799

Sultana R, Choudhary AK, Pal AK, Saxena KB, Prasad BD, Singh R (2014) Abiotic stresses in major pulses: current status and strategies. In: Gaur RK, Sharma P (eds) Approaches to plant stress and their management. Springer, New Delhi, pp 173–190

Szabados L, Savoure A (2010) Pro: a multifunctional amino acid. Trends Plant Sci 15:89–97

Thalooth AT, Tawfik MM, Mohamed HM (2006) A comparative study on the effect of foliar application of zinc, potassium and magnesium on growth, yield and some chemical constituents of mungbean plants grown under water stress conditions. World J Agric Sci 2:37–46

Ullah A, Romdhane L, Rehman A, Farooq M (2019) Adequate zinc nutrition improves the tolerance against drought and heat stresses in chickpea. Plant Physiol Biochem 143:11–18

Uppalapati SR, Ayoubi P, Weng H, Palmer DA, Mitchell RE, Jones W, Bender CL (2005) The phytotoxin coronatine and methyl jasmonate impact multiple phytohormone pathways in tomato. Plant J 42:201–217

Vadez V, Kholova J, Choudhary S, Zindy P, Terrier M, Krishnamurthy L, Kumar PR, Turner NC (2011) Whole plant response to drought under climate change. In: Yadav SS, Redden R, Hatfield JL, Lotze-Campen H, Hall AE (eds) Crop adaptation to climate change. Wiley, Chichester

Voisin AS, Guéguen J, Huyghe C, Jeuffroy MH, Magrini MB, Meynard JM, Mougel C, Pellerin S, Pelzer E (2014) Legumes for feed, food, biomaterials and bioenergy in Europe: a review. Agron Sustainable Dev 34:361–380

Waraich EA, Ahmad R, Halim A, Aziz T (2012) Alleviation of temperature stress by nutrient management in crop plants: a review. J Soil Sci Plant Nutr 12:221–244

Zhang T, Lin X, Sassenrath GF (2015) Current irrigation practices in the central United States reduce drought and extreme heat impacts for maize and soybean, but not for wheat. Sci Total Environ 508:331–342

Zhang H, Chu S, Zhang D (2017) Transcriptome dataset of soybean (*Glycine max*) grown under phosphorus-deficient and -sufficient conditions. Data 2:17. https://doi.org/10.3390/data2020017
Zhao Y, Aspinall D, Paleg LG (1992) Protection of membrane integrity in *Medicago sativa* L. by glycine betaine against the effects of freezing. J Plant Physiol 140:541–543

Legume Responses and Adaptations to Nutrient Deficiencies

Rafael D. C. Duarte, Carla S. Santos, and Marta W. Vasconcelos

Abstract Legumes have unique mechanisms to respond to nutrient deficiencies that can be considered as important advantages for agricultural purposes. The preponderance of plant-based protein is on the rise, and the market value of protein crops is expected to be worth billions by 2025. To match the global demand for plant-based products, crops productivity must be ensured; however, this might be impaired either by environmental or anthropogenic pressures that lead to soil nutrient disturbance. The responses activated by legumes to nutrient deficiencies and the mechanisms they utilize to adapt to such conditions will be discussed in this chapter. The study of these factors enables breeding programs specific for legumes and crop improvement. Understanding legumes responses also allows for a better management of agricultural practices and the adoption of more sustainable methods. It is important to reflect on the impact of climate change and intensive farming on food quality and on the future of agriculture, and this chapter contributes with important facts about the role of legumes in our current scenario.

Keywords Plant-based protein · Sustainability · Climate change · Environmental factors

1 Introduction

Plants are responsible for the preservation of the atmosphere, supply of raw materials, and provision of all the food we eat, directly or indirectly. Among the different and various families in the Plantae kingdom, a few stand out and are valued worldwide as a sustainable food source, as is the case for legumes (Shafique et al. 2014; Martin et al. 2013).

R. D. C. Duarte · C. S. Santos · M. W. Vasconcelos (✉)
CBQF—Centro de Biotecnologia E Química Fina—Laboratório Associado, Universidade Católica Portuguesa, Escola Superior de Biotecnologia, Rua Diogo Botelho 1327, 4169-005 Porto, Portugal
e-mail: mvasconcelos@porto.ucp.pt

M. Hasanuzzaman et al. (eds.), *The Plant Family Fabaceae*,
https://doi.org/10.1007/978-981-15-4752-2_14

Legumes family, *Fabaceae*, comprises over 20,000 species and includes peas, lentils, soybeans, beans, and peanuts among others. If we were to consider their economic value and importance, it would be only eclipsed by *Gramineae*, a family that includes all three major crops (wheat, maize, and rice) (Michaels 2016; Maphosa and Jideani 2017). This prominence of legumes to world crop productions is comprehensible, as they can be key contributors to healthier human diets, while having great shelf-times and still play a crucial role in the improvement of soil quality (Sathe 2016).

Starting by nutrition, legumes provide not only proteins, with numerous essential amino acids, but also complex carbohydrates, vitamins, minerals, and dietary fiber (Maphosa and Jideani 2017). Recent studies, also strongly suggest that more predominantly plant-based pattern regimens, like the Mediterranean Diet, are linked to lower risk of chronic diseases (type 2 diabetes mellitus) and other metabolic syndromes (Murphy et al. 2018). Considering all the factors, it is clear that legumes can be a first-choice for health-conscious consumers, which opens a window of opportunity to the growth of legumes market. A shift in the mindset of consumers towards more plant-based diets has become perceivable (Ellet 2019; Morgan 2019b).

In addition to their nutritional value, legumes are also accountable for several other important roles in modern agriculture. For instance, and thanks to their unique symbiotic relationship with Rhizobium bacteria, they are renowned for their ability of carrying out nitrogen (N) fixation, improving soil properties. As a consequence, in the cropping systems of many countries, legumes are widely used as part of an intercropping strategy based on the rotation of crops (Michaels 2016).

Nevertheless, as much as legumes stand as a beneficial and important crop, they can be extremely sensible to environmental stresses (Scheelbeek 2018). Thus, in a world of global climate change, it is important to understand the serious effects that these atmosphere and weather fluctuations may have on plants, especially deviations on the availability of certain nutrients (Morgan 2019b.). These changes include unpredictable and extreme temperatures, decreased water availability, and elevated concentrations of atmospheric carbon dioxide that strongly affect plant yield and nutritional quality (Andjelkovic 2018). For example, since the beginning of the Industrial Revolution, in the eighteenth century, atmospheric concentration of CO_2 has practically doubled from 280 to 408 μmol mol^{-1}. Given that CO_2 is essential for photosynthesis, this condition may improve the overall yield, while having a negative influence in the nutrient content of edible parts of plants, either through promoting sugar accumulation or limiting the uptake of N (Dong et al. 2018; Myers et al. 2014). Therefore, in order to fully comprehend these effects and how legumes respond to nutrient deficiencies, it is important to understand plant nutrition and the mechanisms beneath.

Legumes, as it is the case for the majority of plants, mainly obtain their nutrients from the soil and require both macronutrients, the "building blocks" and micronutrients, cofactors for enzyme activity (Afzal et al. 2015; White and Brown 2010). These nutrients can be oftentimes unavailable to plants for different reasons, and throughout time, plants have adapted to absorb and utilize them very efficiently. As previously mentioned, while N uptake can be achieved through rhizobial symbionts, phosphorus (P) assimilation may require an increase in root-hair production or production of rootlets, for example, (Sulieman and Tran 2017). The efficient use of these nutrients is a limiting factor of crops productivity. Although sometimes misjudged as pests or diseases, either due to low nutrient phytoavailability or high concentrations of toxic elements, nutrient deficiencies affected crops can exhibit clear signs of deterioration, as death of plant tissue, stunted growth, and others (White and Brown 2010).

There are many methods and strategies developed to enhance the uptake of different nutrients, and these may vary from plant to plant and depend on plants' native soils (Reid and Hayes 2003; Hernández et al. 2007), as will be further discussed during this chapter.

2 Legumes

Legumes, one of the first domesticated plant families in the Mediterranean, are found in most areas of the world and have a vast number of species which include peas, alfalfa, peanuts, soybeans, beans, and many others. Usually, cultivated legumes fall into two classes: grain and forage. Forage legumes, as alfalfa, are grown in pasture for grazing by livestock or to be harvested as hay, frequently for industrial purposes (Beltran and Canas 2018). Grain legumes, on the other hand, are cultivated for their seeds and are mainly used for human and animal consumptions (González-Pérez and Arellano 2009). These edible seeds of legume plants are known as pulses, which grow in pods that come in different shapes, sizes, and colors. The main distinction to other vegetable crops is that pulses are harvested exclusively as dry grains. Recognizing the importance of these high protein seeds, the Food and Agriculture Organization (FAO) of the United Nations designated the year of 2016 as the international year of pulses.

These crops, that can be either annual, biennial or perennial, are grouped into the C3 category of plants (Archimède et al. 2011). Comparing C3 and C4 plants, both rely on photosynthesis (Calvin cycle) to make sugars from CO_2 but differentiate in how plants respond to CO_2, which leads to certain advantages and disadvantages for different habitats or conditions (Sage and Zhu 2011). C3 legumes like beans, alfalfa, and lentils, are more sensitive to higher CO_2 concentrations than typical C4 plants (for example, maize and sorghum). This sensitivity can affect differently the nutrients under analysis, but evidence suggests a negative impact in some minerals, zinc (Zn) and iron (Fe), and protein concentrations (Perring et al. 2010; Myers et al. 2014). These nutritional losses, in a modern world where dietary micronutrient deficiencies prevail, contribute to food insecurity and low quality agri-food cycles (in terms of nutrition). However, legumes, especially pulses, can be an inexpensive source of plant-based protein that can help to achieve numerous of the sustainable development goals proposed by the UN (Calles et al. 2019).

2.1 Benefits to the Environment

The unique ability of legumes to form symbiotic relations with N-fixing microorganisms, enables legumes to thrive in soils with low N content, where others cannot, and therefore, contributing to soil improvement and improving ecosystem sustainability (Goh et al. 2013; Michaels 2016). These *Rhizobium* bacteria are responsible for employing the solar energy seized by the plant to break the inert atmospheric N into reactive N species, such as ammonium (Liu et al. 2018). Moreover, legumes also contribute to soil carbon sequestration and overall reduced fossil energy inputs in farming systems (Stagnari et al. 2017).

The agronomic impact that legumes can achieve through intercropping strategies is significant, with higher protein yields per hectare, drop harsher pesticide practices thanks to crop diversification, and reduced fertilizer usage rate (Nemecek et al. 2008). Due to the latter, legume cultivation can have as a consequence the reduction of greenhouse gas (GHG) emissions in arable systems in about 5–7 times less per unit area compared with other crops (Jeuffroy et al. 2013).

2.2 Benefits to Human Health

Humans require essential amino acids, carbohydrates, essential fatty acids, and basically, 28 vitamins and minerals to sustain life and health. Usually, the sources for alimentation are either plant or animal based, with the last having more prominence in the last century (Williams et al. 2012). Nevertheless, in the case of plant-based diets, they can meet or suppress the recommended intakes of most nutrients while having the advantage of providing several health benefits when incorporated in the common Western diet, supported by metabolic measures in health and disease (Martin et al. 2013). Nowadays, western consumers, that were being overwhelmed with an excess macronutrient intake and insufficient micronutrient consumption, are gaining more awareness into integrating plant-based meals to their daily life. This growth in public popularity is evident and, likewise, the scientific community is going along as it is noticeable, in the last 10 years, almost a 300% increase in the frequency of scientific publications including the terms "vegan," "vegetarian," and "plant-based" (Medawar et al. 2019).

Besides being highly nutritious, legumes are low in fat, cholesterol free, and excellent sources of protein, fiber, carbohydrates, and a wide range of minerals and vitamins (Polak et al. 2015). Also, several bioactive legume compounds hold antioxidant properties with a number of beneficial physiological properties against chronic diseases in general. Therefore, considering the overall nutritional legume properties and the current world food security scenario, they become increasingly attractive both in developed countries where weight management problems and decurrent health issues prevail, and in developing countries, posing as an important protein/energy source to help to eradicate malnutrition (Fao 2019).

2.3 Legumes Contribution for a Sustainable Future

As in current knowledge there are 17 sustainable development goals (SDGs) proposed by the UN to achieve a more sustainable future for all. Oftentimes overlooked, legumes may be a key factor in the much-needed change as their (re)introduction in humans' diets can be a building block to several of the SDGs.

In the case of the SDGs "no poverty" and "decent work and economic growth," legumes can contribute with increased employment, income-earning opportunities, and fair trade. As European policies are beginning to promote short supply chains to increase regional sustainability (Deppermann et al. 2018), legumes can gain competitiveness, since they have very high protein production levels per hectare (Pilorgé and Muel 2016). These shorter supply chains also promote the "sustainable cities and communities" SDG. Moreover, legumes can be consumed, at low expense, for their green pods, immature seeds, tubers, leaves and mature seeds, being a flexible food and commodity, and assuring "zero hunger" (Vasconcelos and Gomes 2016). Additionally, when considering legume dried grains, their long-term shelf-life constitutes an obvious advantage.

Legumes are also recognized for their health promoting characteristics, thus contributing for the SDG "good health and well-being." Through the improvement of body weight, oxidative stress and inflammatory status (Mirmiran et al. 2018), this food can help to reduce the risk of obesity, diabetes, cardiovascular diseases, and even cancer (Santos et al. 2017). Additionally, studies demonstrate that women's knowledge and education on nutrition and gender empowerment ("quality education" and "gender equality SDGs") in rural regions improve health status and diet quality (Olney et al. 2015).

Legume inclusion in agricultural practices can drive environmental, economic, and human development (Mcdermott and Wyatt 2017). The impacts of changing the current agricultural scenario to a more legume-inclusive one have been well described. Specifically, legumes require low amounts of water, being tolerant to drought; they are N fixers, reducing the need for fertilizers, improving soil quality and being well suited for crop rotation; and help to reduce greenhouse gas emissions (Lienhardt et al. 2019). In this way, they address global climate constraints and the SDGs "clean water and sanitation," "responsible consumption and production," "climate action," and "life on land."

Finally, the involvement of the academia and industry in legumes capitalization has led to the development of a wide range of new legume-based products ("industry, innovation, and infrastructure"). For example, simultaneously responding to the "affordable and clean energy," new types of renewable energy sources based on legumes like a biogas are being developed, helping in energy saving (Stinner 2015). Within the innovation context, food biotechnology and market trends are strongly responding to the new trends and plant-based food alternatives availability is increasing. Legume protein is being often used in protein powders, veggie burgers, and bakery. Novel approaches for the use of legume protein include the development of alcoholic beverages and formulas for babies and toddlers. The type of products made with legumes is becoming more refined and complex, and their nutritional specifications are getting more precise and prepared by demand.

2.4 Legume Nutrition

To achieve their optimal growth and dietary quality plants require some elements, like C, H, and O, that can be absorbed from the atmosphere or water, while the majority of the other elements, commonly known as mineral elements, are found in the soil (López-Arredondo et al. 2017). In order to sustain plant growth, legumes require 14 essential elements, with several others that are considered beneficial but do not stand as essential (for example, silicon and selenium). These essential elements can be divided into macronutrients and micronutrients, with the split being based on the actual amount of nutrient required for adequate plant growth (White and Brown 2010). Macronutrients, required in larger amounts, like P, N and potassium (K), are among the most limiting factor for crop yield. On the other hand, and even though plants require considerably smaller amounts of micronutrients, a deficiency in these minerals may prejudice them as much as a lack of N, P, or K. For instance, Fe deficiency conditions impact chlorophyll production and may lead to serious adverse effects like interveinal chlorosis, reduced growth, and yellowing of leaves (Santos et al. 2016) (Fig. 1).

Regarding the mechanisms of uptake, for most of the macronutrients, the pathways of absorption and transport within the plant are reasonably well understood, while the situation is not as clear with micronutrients, maybe except Cl, Fe, and Zn (Reid and Hayes 2003).

Fig. 1 Visual representation of Fe deficiency symptoms (yellowing of the leaves, interveinal chlorosis) in a *Glycine max* plant

When the appropriate mineral supplementation cannot be achieved and plant response is unable to ensure the proper uptake, nutrient deficiencies are reflected in numerous physiological, biochemical, and molecular changes (Santos et al. 2015, 2019b).

3 Current Scenario and Obstacles for Legume Nutrition

Several factors can impact the efficiency of nutrient acquisition by legumes, principally abiotic stresses that are now even more unpredictable and harsh to crops, given the status of global climate change (Smith 2001; Soares et al. 2019b).

Where we stand today, even if we disregard GHG, the disruptive anthropogenic activities lead, inevitability, to an increase in the average temperature, more likelihood of strong, and persistent heat waves, as well as the occurrence of erratically distributed precipitation in many areas (Andjelkovic 2018). As a result, plants will have to cope with drought, heat, salinity, and soil constrains with low nutrient availabilities. On this matter, legumes will be one of the most affected crops thanks to their higher sensibility when compared to other staple crops, cereals for example (Scheelbeek 2018). An overview of the predominant obstacles for legume nutrition and further solutions is demonstrated in Table 1.

Starting by the global impact of climate change, this is maybe the most relevant threat to plant growth and productivity, with concerning effects on either plant quality or quantity (Soares et al. 2019b). Several climate models predict that the extended and randomized periods of rainfall and drought, will increase the risk of soil erosion. Consequently, the topsoil layer which is typically richer in nutrients may lose a great part of its nutrients, affecting the overall process of nutrient cycling (Nearing et al.

Table 1 Factors impacting legume nutrition and potential answers

Obstacles	Negative impact	Possible solutions	References
Monocropping	Nutrient depletion, loss of genetic diversity	Crop rotation involving legumes	Jacques and Jacques (2012), Sharma (2009)
Tilling	Organic matter loss	No-tillage crop production	Curci et al. (1997), Triplett and Dick (2008)
Weather adversities	Erosion, distressed nutrient cycling	Increased water-use efficiency	Nearing et al. (2004), Hatfield and Dold (2019)
Synthetic fertilizers	Reduced microbiological diversity	Precision fertilization	Tian and Niu (2015)
Pesticides	Diminished nutrient availability	Inclusion of non-toxic alternatives	Machekano et al. (2019)

2004). Also, higher temperatures can lead to an increased decomposition of organic matter which could reduce the carbon available for plants. Although the majority of carbon is obtained through photosynthesis, soil carbon is also important as it helps to retain water and other nutrients (Conant et al. 2011, 2008). Given the multifactorial stresses associated with climate change, it is also relevant the effect of elevated CO_2 (eCO_2) on the decrease of protein grain concentration (harshly impacting human nutrition) and disturbance of nutrient accumulation in important crops (Beach et al. 2019; Soares et al. 2019a). Furthermore, studies that evaluated longer treatments with eCO_2 reported acclimation of photosynthesis (Kaplan et al. 2012; Ainsworth et al. 2004), accelerated leaf senescence (Ludewig and Sonnewald 2000), and stunted growth (Ainsworth et al. 2004).

In addition to these concerning climate change effects, bad practices related to industrial farming is another powerful contributor for soil health and respective nutrient content depletion (Yang et al. 2018). Not only human population is increasing, but also the average calorie intake per person is rising rapidly (Popkin et al. 2012) and, in order to achieve the production amount necessary to meet consumers demand, the agricultural business is currently compelling farmers to further engage into these bad practices (Hazell and Wood 2008).

For example, monocropping, the practice of growing in consecutive years, the same crop into the same plot of land, has led to an intense depletion of the nutrients in the soil, negatively impacting productivity, and promoting the growth of pathogenic microorganisms that suppress plant development (Aparicio and Costa 2007). The tillage operation is a groundwork-type element of cropping which alters the physicochemical properties of soil often speeding up the mineralization rate, at the cost of losing a portion of soil organic matter (Curci et al. 1997). Despite their role in weed control, to avoid the consequent negative impact, a no-tillage crop management is now being more often promoted with significantly improvement of crop-soil association, either by accumulation of organic matter, improved water retention and infiltration with higher vegetative growth (Daryanto et al. 2017).

Additionally, even though synthetic fertilizers played a key role in the demographic explosion, helping farmers to increase their productivities and to feed billions of new people (Erisman et al. 2008), their unbalanced and abusive utilization has resulted in decreased soil microbiological diversity (specially harmful for legumes), acidification and increased salt and heavy metal contamination of soils (Zhou et al. 2017; Tian and Niu 2015). Moreover, N fertilizers could potentially lead to the aggravation of the climate change problematic since it alters global N cycle by increasing emissions of nitrous oxide, not only a GHG but also especially harmful to ozone (Fagodiya et al. 2017).

Finally, the use of pesticides to either control the growth of weeds or the propagation of insects and fungi may affect microbial diversity and potentially suppress N-fixing bacteria, common to leguminous plants, from replenishing natural N fertilizer in soil (Potera 2007). Such is the case of glyphosate (commonly known as RoundUp), that when used outside, the recommended limits has a negative effect in soil microorganisms and binds with essential macro/micronutrients; therefore, reducing their availability and preventing the uptake of those required components to plant processes (Mertens et al. 2018).

To reply to these potential antagonists, legumes have a wide range of responses. These can be dependent on the species or genotype in question, or on the type and severity of stress they face. In fact, these plants have evolved to overcome nutrient limitations by several pathways, either through nodules formation and other root adaptations, the production of certain compounds to attract *Rhizobium* soil bacteria, increase allocation of resources and several others, as further detailed below.

4 Legumes Responses and Adaptations to Nutrient Deficiencies

Given the examples of either environmental or anthropogenic causes to soil nutrient disturbance, it is now important to address the responses and adaptations that legumes employ to overcome this deficit. Since these stress factors majorly impact crop yield, in this section, it will be prioritized nutrients that limit plant growth and development, as is the case for phosphate, Fe, and Zn. The difference between yield potential and actual yield (e.g., yield gap) is predicted to increase in the harsh forthcoming environmental changes (Slattery et al. 2013).

The responses and adaptions that these plants have to undergo to thrive under nutrient limiting conditions are various; nonetheless, they can exhibit similar mechanisms when different nutrients are lacking (Fig. 2).

Alterations in root system architecture, in genetic regulation of certain transporters, mycorrhizal symbiosis adjustments, and exudates release are among the most important features that will be discussed.

4.1 Root System Architecture

As the availability of nutrients in soil is heterogenous and easily impactable, the majority of plants has evolved in a way that their root system architecture can be altered to efficiently forage and absorb these essential minerals. Although some morphological changes can be, to a certain extent, nutrient specific, several adaptions in root architectural patterns are more generalistic and follow the same premises.

For example, in case of P deficiency, that is perhaps the most limiting macronutrient to legume crops' growth, root architecture adjustments are encouraged to further forage the surface soil layers that are richer in P, due to its low mobility (Liao et al. 2004). In that way, signaling pathways will engage into an enhanced lateral root formation, which implies a shallower and broader root system that boosts plant capacity to explore the desired upper layers (Liao et al. 2004; Ho et al. 2005). Increased root-hair abundance and length is also one of the typical responses since it can be

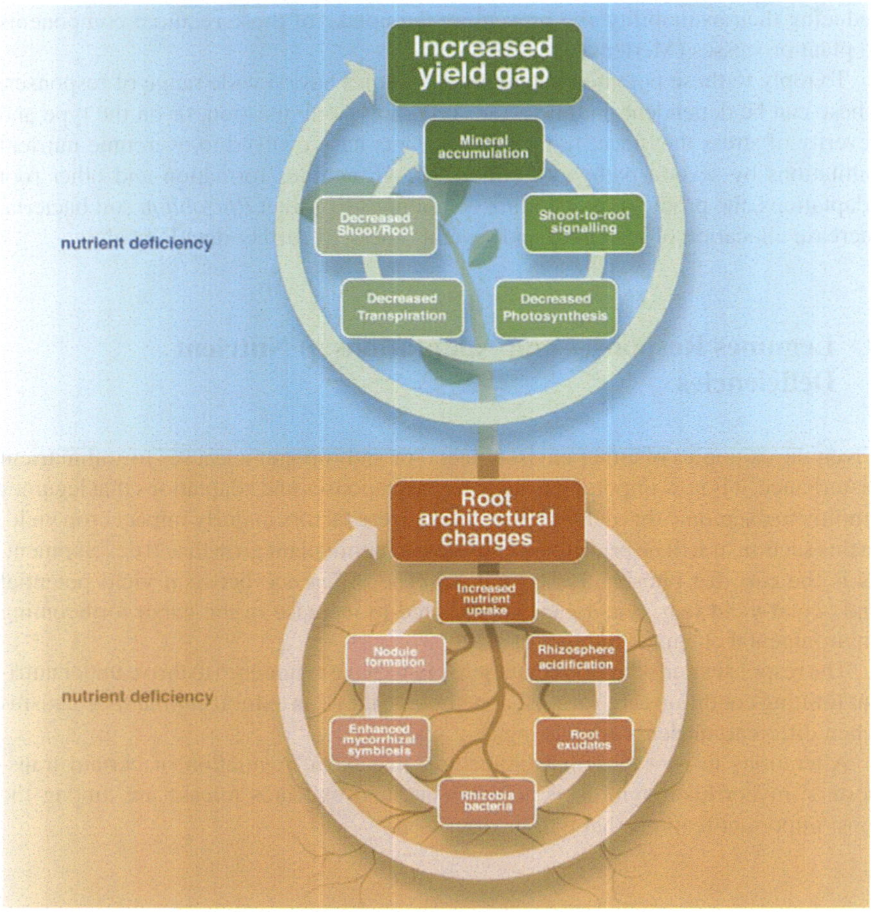

Fig. 2 Root and shoot responses to nutrient deficiency. In order to increase nutrient uptake, the roots undergo architectural changes that enable the increase of absorption area; legume plants release protons to acidify the soil and organic compounds (exudates) to solubilize and better absorb the nutrients; bacteria are attracted to these compounds and mycorrhizal symbiosis is enhanced, leading to nodule formation. In the shoots, when nutrients are deficient, the final yield of the plant does not reach the plants' potential yield, resulting in a yield gap; in order to accumulate nutrients in the aerial organs, legumes activate a shoot-to-root signaling pathway and decrease physiological processes (like photosynthesis and transpiration); shoot-to-root ratio is decreased, since the assimilates are directed to root growth

responsible for 90% of P acquisition (Lambers et al. 2006; Sánchez-Calderón et al. 2006). Root hairs may also be induced in response to many other nutrient deficiencies such as K, magnesium, Fe, and manganese (Jung et al. 2009; Niu et al. 2014; Yang et al. 2008).

One other common response is the formation of cluster roots, densely clustered secondary roots, to increase root area and exudation of enzymes and organic acids. This engages legume species into a more prolific nutrient acquisition as studied in white lupin models (Vance 2001). Cluster development is controlled by a systemic signal from the shoot with sugar and phloem transport being essential to the regulation of the process (Liu et al. 2005).

Similarly to P, Fe, and Zn exhibit limited solubility in soils, and angiosperms usually rely on two different mechanisms to acquire them from the soil, reduction-based strategy (Strategy I), and chelation-based strategy (Strategy II) (Walker and Connolly 2008). Legume species utilize Strategy I, and it is based on enzymatic activities in the plasma membrane of cells in the outer layer of the root that are induced in nutrient limiting conditions (Saini et al. 2016). Initially, H^+-ATPases release protons in the rhizosphere, acidifying the surroundings which will then solubilize Fe and Zn promoting their uptake (Marschner and Römheld 1994; Sinclair and Kramer 2012).

4.2 *Physiological Changes*

In order to achieve mineral accumulation and compensate for nutrient deficiencies, in general, some physiological statuses are adjusted to meet the greater shoot demand. One of the impacted processes is photosynthesis, which seems to decrease in response to nutrient limitations, as is the case of N (Antal et al. 2010) and P (Abdelrahman et al. 2018). In consequence, the conversion efficiency of intercepted radiation into biomass decreases (Slattery et al. 2013). As resources for respiration decrease, resource allocation changes and legumes direct them toward nutrient uptake, by increasing root growth, thus decreasing shoot-to-root ratio (Kleinert et al. 2014). Since under these stress conditions, roots are the major sinks, transpiration rates decrease, modulating the acquisition of several nutrients (Matimati et al. 2014).

Another strategy is the nutrient remobilization of short or mid-term storage, through phloem transport, which may help to compensate a transient lack of mineral uptake by roots. Usually, this occurs during vegetative growth (low soil nutrient availability) as mature leaves become sources to support the growth of news organs (Maillard et al. 2015).

4.3 *Legumes Related Symbionts*

One unique way that legumes utilize to enhance nutrient acquisition is through mycorrhizal symbiosis (Santos et al. 2019a). For example, the fungi colonization of plant cells that act as an extension of the root system, facilitating the uptake and contributing to a more effective nutrient translocation in the host plant (Gutjahr and Paszkowski 2013; Thuynsma et al. 2014). Among the most important fungi, the arbuscular mycorrhizal fungi (AMF), belonging to *Glomeromycota* phylum, stand as a crucial partner for P acquisition (Scheublin et al. 2004). Furthermore, rhizobia bacteria are responsible for producing N-fixing nodules on legumes where, in case of nutrient deficiency, the uptake and recycling are enhanced (Udvardi and Poole 2013). From the rhizobia, *Rhizobium* genus is the most known since these gram-negative and motile bacteria are deeply associated with the leguminous plants (Parker 2001).

Particularly to rhizobia-legume symbiosis and nodule formation, these well-organized and differentiated structures are molded after rhizobia attachment to root-hair cells, with nodule organogenesis being coordinated by a series of molecular interactions between receptor-like kinases among bacteria and their legume host (Popp and Ott 2011). Afterward, the bacteria end up differentiating into bacteroids (symbiotic N-fixing form) that rely on legumes as a source of carbon and other nutrients, while ensuring N fixation (Roriz et al. 2020). Therefore, it is easy to underline legumes natural advantage in N depleted soils, as this exclusive plant–microbe interaction can ensure the uptake of C and other critical nutrients from the soil outside the rhizosphere and translocate them to the host plants (Yoneyama et al. 2012). Root nodulation also impact P uptake, for example, through increased recycling and phosphate conservation mechanisms that prevent complete depletion through nutrient reallocation from the root (Sulieman and Tran 2017; Nasr Esfahani et al. 2016). Some legumes even have the ability to mobilize P from nucleic acids and phospholipids, with recent reports suggesting that in some cases, these alternate routes to conserve P are preferred to the direct uptake from soil (Vardien et al. 2016).

On the other hand, nodules have a strong Fe requirement, as higher rates of N fixation seem to depend on increased concentrations of this micronutrient (Rotaru, 2009). Moreover, under Fe limitation, it was shown that the allocation of Fe ions to nodules increase to support nodule growth, as well as the activity of ferric chelate reductase that is a key enzyme in Fe reduction and uptake processes (Slatni et al. 2008).

4.4 Exudates Release

The increase in the release of exudates to enhance nutrient soil availability depends not only on the damaging agricultural practices previously mentioned but also on the inherent characteristics of the environment, as it is common that a vast amount of nutrients may be present in chemical forms inaccessible to plants (Tate 2005).

One of the most common responses developed by plants is the production of certain types of enzymes and organic acids, as citrate, malate, and oxalate. These are usually induced under nutrient deficient conditions (Yadav and Tarafdar 2001) or, when present, when nutrients are unavailable for absorption (Dakora and Phillips 2002). For example, in order to solubilize the P provided by fertilizers, some extracellular enzymes produced by legumes, for example, phosphatases, release this P into usable forms for soil biota and plants through the cleavage of ester linkage sites (Dotaniya et al. 2019). In fact, it has been reported higher phosphatase activity in leguminous crops when compared to maize and wheat species which reinforces this strong tendency of legumes response to variations in soil P availabilities (Yadav and Tarafdar 2001).

Root exudates have also an important role in micronutrient availability. Organic compounds, like phenolics, organic acids, sugars, and flavins, have been studied in the last years for their important role in the Fe solubilization and mobilization (Sisó-Terraza et al. 2015; Fourcroy et al. 2014; Schmid et al. 2014), as well in metal detoxification (Chen et al. 2017). These compounds have also the ability to behave as signaling molecules, binding to receptors in rhizobial cell surfaces enhancing the number of root nodules through an increased expression of nodule-synthesis related genes (*nod*) (Sugiyama and Yazaki 2012). Hence, there is also a link between root exudates and rhizobia attraction that will consequently increase functional nodules development, whose importance was already discussed.

4.5 Gene Expression Modulation

The expression of transporters and enzymes important in nutrient acquisition is highly impacted by nutrient deficiencies (Santos et al. 2019a). For example, in the response to low P, the expression of the high-affinity phosphate transporter GmPT5 has been shown to increase, which impacts nodulation and plant growth, since it controls P transport from roots to nodules (Ho et al. 2005). Nonetheless, among the P transporters, the ones belonging to the Pht1 family are the most commonly induced by P starvation, under the control of the phosphate starvation response transcription factors (Gu et al. 2016). In fact, for *Glycine max* species, from the 15 Pht1 known paralogs, all are up-regulated in low P conditions (Fan et al. 2013).

For instance, ferritin which is essential for oxidative stress prevention and adaptation to adverse environmental situations has been shown to be up-regulated in plants that endure better under Fe-deficient conditions. Especially, the induction of certain ferritin genes, *fer3*, is believed to be an efficient response mechanism to improve Fe acquisition (Boamponsem et al. 2017). Also, in Fe limiting soils, legumes will tend to enhance the production of coumarins through the bHLH transcription factor, one of the major regulators of Fe deficiency responses FIT (FER-like Iron deficiency-induced Transcription Factor) (Schmidt et al. 2014). FIT is considered to be a network of transcriptional regulators which allow the plant to cope with not only Fe limiting conditions (Chutia et al. 2019) but also to interact with P an Cu uptake machinery under decrement of these nutrients (Filiz and Kurt 2019). Concerning other mechanisms of uptake, it has been shown that after Fe^{3+} reduction, Fe^{2+} is transported to the root by Fe-regulated transporters (IRT), that are express only under Fe-deficient conditions (Connolly et al. 2002). Besides Fe absorption, these transporters may also be responsible for the transport of other divalent metals (Barberon et al. 2011) and Fe translocation (Xiong et al. 2014).

Phytohormones can also positively regulate Fe deficiency responses, as it was demonstrated the abscisic acid (ABA) promotion of Fe transportation from root-to-shoot, alongside with phenolic compound secretion from the roots (Lei et al.

2014). Although the molecular mechanisms behind ABA action in this context are still unknown, evidences suggest the involvement of ABI5 in the ABA-mediated alleviation of chlorosis under Fe deficiency (Liu et al. 2017) and regulation of Fe levels by inducing ferritin expression (Santos et al. 2019a).

This kind of research is still expanding since recently, and only for common bean species, over a hundred genes regulated by P availability have been identified in roots and nodules with several transcription factors, microRNA and photosynthates acknowledged as part of the deprivation signaling pathway (Liu et al. 2018). Many and different elements are controlled at the transcriptional level with legumes remodeling their transcriptome and proteome in different ways to coordinate the required metabolic and morphological adaptations (Tran et al. 2010).

5 Conclusion and Future Perspectives

For more than 5000 years, legumes have been one of the first-choice crops to feed the world's population (Ahmed and Hasan 2014), being recognized for their nutritious content, health benefits, and positive impact on ecosystem sustainability. However, decades of bad farming practices and exacerbated effects of climate change severely impacts legume nutrition and plant overall productivity. Some of the effects caused by these deficiencies are easily observable while others require a more cautious approach as it is crucial a correct diagnose.

Nonetheless, legumes have evolved to cope with some of adversities and the dissection of the physiological and molecular adaptations is an important field of study. As advances in genomics may allow the breeding of nutrient-deficiency tolerant legumes, a correct understanding of the inherent strategies and mechanisms that deal with these limitations is of utmost importance. Either through enhancement of nutrient acquisition, improved remobilization, and optimized legume–rhizobium associations, enabling or improving these processes will allow the future population to meet the agriculture demands for higher yields and endure climate change negative effects on food quality.

Acknowledgements This work was funded by National Funds from FCT—Fundação para a Ciência e a Tecnologia through project PTDC/AGRPRO/3972/2014. We would also like to thank the scientific collaboration under the FCT project UID/Multi/50016/2019.

References

Abdelrahman M, El-Sayed MA, Hashem A, Abd_Allah EF EF, Alqarawi AA, Burritt DJ, Tran LSP (2018) Metabolomics and transcriptomics in legumes under phosphate deficiency in relation to nitrogen fixation by root nodules. Front Plant Sci 9:922

Afzal F, Khan T, Khan A, Khan S, Raza H, Ihsan Ahanger MA, Kazi AG (2014) Nutrient deficiencies under stress in legumes. In: Azooz MM, Ahmad P (eds) Legumes underenvironmental stress: yield, improvement and adaptations. Wiley, Hoboken, pp 53–65

Ahmed S, Hasan M (2014) Legumes: an overview. J Pharm Pharm Sci 2311–4673(2):34–38

Ainsworth EA, Rogers A, Nelson R, Long SP (2004) Testing the "source–sink" hypothesis of down-regulation of photosynthesis in elevated [CO_2] in the field with single gene substitutions in Glycine max. Agric For Meteorol 122:85–94

Andjelkovic V (2018) Introductory chapter: climate changes and abiotic stress in plants

Antal T, Mattila H, Hakala-Yatkin M, Tyystjärvi T, Tyystjärvi E (2010) Acclimation of photosynthesis to nitrogen deficiency in Phaseolus vulgaris. Planta 232:887–898

Aparicio V, Costa JL (2007) Soil quality indicators under continuous cropping systems in the Argentinean Pampas. Soil Tillage Res 96:155–165

Archimède H, Eugène M, Marie Magdeleine C, Boval M, Martin C, Morgavi DP, Lecomte P, Doreau M (2011) Comparison of methane production between C3 and C4 grasses and legumes. Anim Feed Sci Technol 166–167:59–64

Barberon M, Zelazny E, Robert S, Conejero G, Curie C, Friml J, Vert G (2011) Monoubiquitin-dependent endocytosis of the iron-regulated transporter 1 (IRT1) transporter controls iron uptake in plants. Proc Natl Acad Sci U S A 108:E450–E458

Beach RH, Sulser TB, Crimmins A, Cenacchi N, Cole J, Fukagawa NK, Mason-D'croz D, Myers S, Sarofim MC, Smith M, Ziska LH (2019) Combining the effects of increased atmospheric carbon dioxide on protein, iron, and zinc availability and projected climate change on global diets: a modelling study. Lancet Planet Health 3:e307–e317

Beltran JP, Canas LA (2018) Grain and forage legumes: nutritional value and agriculture sustainability. Methods Mol Biol 1822:1–10

Boamponsem GA, Leung DWM, Lister C (2017) Insights into resistance to Fe deficiency stress from a comparative study of invitro-selected novel Fe-efficient and Fe-inefficient potato plants. Front Plant Sci 8:1581

Calles T, Xipsiti M, Del Castello R (2019) Legacy of the International Year of Pulses. Environ Earth Sci 78:124

Chen YT, Wang Y, Yeh KC (2017) Role of root exudates in metal acquisition and tolerance. Curr Opin Plant Biol 39:66–72

Chutia R, Abel S, Ziegler J (2019) Iron and phosphate deficiency regulators concertedly control coumarin profiles in arabidopsis thaliana roots during iron, phosphate, and combined deficiencies. Front Plant Sci 10:113

Conant RT, Drijber RA, Haddix ML, Parton WJ, Paul EA, Plante AF, Six J, Steinweg JM (2008) Sensitivity of organic matter decomposition to warming varies with its quality. Glob Change Biol 14:868–877

Conant R, Ryan M, Ågren GI, Birgé H, Davidson E, Eliasson P, Evans S, Frey S, Giardina C, Hopkins F, Hyvönen R, Kirschbaum M, Lavallee J, Leifeld J, Parton W, Steinweg JM, Wallenstein M, Wetterstedt M, Bradford M (2011) Temperature and soil organic matter decomposition rates—synthesis of current knowledge and a way forward. Glob Change Biol 17:3392–3404

Connolly EL, Fett JP, Guerinot ML (2002) Expression of the IRT1 metal transporter is controlled by metals at the levels of transcript and protein accumulation. Plant Cell 14:1347–1357

Curci M, Pizzigallo MDR, Crecchio C, Mininni R, Ruggiero P (1997) Effects of conventional tillage on biochemical properties of soils. Biol Fertil Soils 25:1–6

Dakora F, Phillips D (2002) Root exudates as mediators of mineral acquisition in low-nutrient environments. Plant Soil 245:35–47

Daryanto S, Wang L, Jacinthe P-A (2017) Impacts of no-tillage management on nitrate loss from corn, soybean and wheat cultivation: a meta-analysis. Scientific Reports 7:12117

Deppermann A, Havlík P, Valin H, Boere E, Herrero M, Vervoort J, Mathijs E (2018) The market impacts of shortening feed supply chains in Europe. Food Security 10:1401–1410

Dong J, Gruda N, Lam SK, Li X, Duan Z (2018) Effects of elevated CO_2 on nutritional quality of vegetables: a review. Frontiers in Plant Science 9:924

Dotaniya ML, Aparna K, Dotaniya CK, Singh M, Regar KL (2019) Role of soil enzymes in sustainable crop production (Chap. 33). In: Kuddus M (ed) Enzymes in food biotechnology. Springer, NY

Ellet LJ (2019) Rise of the health conscious consumer continue to redefefine consumer markets (Online). Available https://home.kpmg/uk/en/home/media/press-releases/2019/03/rise-of-the-health-conscious-consumer-continue-to-redefine-consumer-markets.html. Accessed 4 Nov 2019

Erisman JW, Sutton MA, Galloway J, Klimont Z, Winiwarter W (2008) How a century of ammonia synthesis changed the world. Nat Geosci 1:636–639

Fagodiya RK, Pathak H, Kumar A, Bhatia A, Jain N (2017) Global temperature change potential of nitrogen use in agriculture: A 50-year assessment. Sci Reports 7:44928

Fan C, Wang X, Hu R, Wang Y, Xiao C, Jiang Y, Zhang X, Zheng C, Fu Y-F (2013) The pattern of Phosphate transporter 1 genes evolutionary divergence in Glycine maxL. BMC Plant Biol 13:48

Fao I, UNICEF, WFP and WHO (2019) The state of food security and nutrition in the world 2019. Safeguarding against economic slowdowns and downturns (Online). Rome, FAO [Accessed]

Filiz E, Kurt F (2019) FIT (Fer-like iron deficiency-induced transcription factor) in plant iron homeostasis: genome-wide identification and bioinformatics analyses. J Plant Biochem Biotech

Fourcroy P, Siso-Terraza P, Sudre D, Saviron M, Reyt G, Gaymard F, Abadia A, Abadia J, Alvarez-Fernandez A, Briat JF (2014) Involvement of the ABCG37 transporter in secretion of scopoletin and derivatives by Arabidopsis roots in response to iron deficiency. New Phytol 201:155–167

Goh CH, Veliz Vallejos DF, Nicotra AB, Mathesius U (2013) The impact of beneficial plant-associated microbes on plant phenotypic plasticity. J Chem Ecol 39:826–839

González-Pérez S, Arellano JB (2009) 15—vegetable protein isolates. In: Phillips GO, Williams PA (eds) Handbook of hydrocolloids, 2nd edn. Woodhead Publishing, Cambridge

Gu M, Chen A, Sun S, Xu G (2016) Complex regulation of plant phosphate transporters and the gap between molecular mechanisms and practical application: what is missing? Mol Plant 9:396–416

Gutjahr C, Paszkowski U (2013) Multiple control levels of root system remodeling in arbuscular mycorrhizal symbiosis. Front Plant Sci 4:204

Hatfield JL, Dold C (2019) Water-use efficiency: advances and challenges in a changing climate. Front Plant Sci 10

Hazell P, Wood S (2008) Drivers of change in global agriculture. Philosophical transactions of the Royal Society of London. Series B, Biol Sci 363:495–515

Hernández G, Ramírez M, Valdés-López O, Tesfaye M, Graham MA, Czechowski T, Schlereth A, Wandrey M, Erban A, Cheung F, Wu HC, Lara M, Town CD, Kopka J, Udvardi MK, Vance CP (2007) Phosphorus stress in common bean: root transcript and metabolic responses. Plant Physiol 144:752–767

Ho M, Sotomayor J, Brown K, Lynch J (2005) Root architectural tradeoffs for water and phosphorus acquisition. Funct Plant Biol 32

Jacques P, Jacques J (2012) Monocropping cultures into ruin: the loss of food varieties and cultural diversity. Sustainability 4:2970–2997

Jeuffroy MH, Baranger E, Carrouée B, De Chezelles E, Gosme M, Hénault C, Schneider A, Cellier P (2013) Nitrous oxide emissions from crop rotations including wheat, oilseed rape and dry peas. Biogeosciences 10:1787–1797

Jung J-Y, Shin R, Schachtman DP (2009) Ethylene mediates response and tolerance to potassium deprivation in Arabidopsis. Plant Cell 21:607–621

Kaplan F, Zhao W, Richards JT, Wheeler RM, Guy CL, Levine LH (2012) Transcriptional and metabolic insights into the differential physiological responses of arabidopsis to optimal and supraoptimal atmospheric CO2. PLoS ONE 7:e43583

Kleinert A, Venter M, Kossmann J, Valentine A (2014) The reallocation of carbon in P deficient lupins affects biological nitrogen fixation. J Plant Physiol 171:1619–1624

Lambers H, Shane MW, Cramer MD, Pearse SJ, Veneklaas EJ (2006) Root structure and functioning for efficient acquisition of phosphorus: matching morphological and physiological traits. Ann Bot 98:693–713

Lei GJ, Zhu XF, Wang ZW, Dong F, Dong NY, Zheng SJ (2014) Abscisic acid alleviates iron deficiency by promoting root iron reutilization and transport from root to shoot in Arabidopsis. Plant, Cell Environ 37:852–863

Liao H, Yan X, Rubio G, Beebe S, Blair M, Lynch J (2004) Basal root gravitropism and phosphorus acquisition efficiency in common bean. Funct Plant Biol 31:959

Lienhardt T, Black K, Saget S, Costa MP, Chadwick D, Rees RM, Williams M, Spillane C, Iannetta PM, Walker G, Styles D (2019) Just the tonic! Legume biorefining for alcohol has the potential to reduce Europe's protein deficit and mitigate climate change. Environ Int 130:104870

Liu J, Samac DA, Bucciarelli B, Allan DL, Vance CP (2005) Signaling of phosphorus deficiency-induced gene expression in white lupin requires sugar and phloem transport. Plant J 41:257–268

Liu J, Fan Y, Zou J, Fang Y, Wang L, Wang M, Jiang X, Liu Y, Gao J, Zhang C (2017) A RhABF2/Ferritin module affects rose (Rosa hybrida) petal dehydration tolerance and senescence by modulating iron levels. Plant J 92:1157–1169

Liu A, Contador CA, Fan K, Lam HM (2018) Interaction and regulation of carbon, nitrogen, and phosphorus metabolisms in root nodules of legumes. Front Plant Sci 9:1860

López-Arredondo DL, Sánchez-Calderón L, Yong-Villalobos L (2017) Molecular and genetic basis of plant macronutrient use efficiency: concepts, opportunities, and challenges. In: Hossain MA, Kamiya T, Burritt DJ, Tran LSP, Fujieara T (eds) plant macronutrient use efficiency. Academic Press, Cambridge

Ludewig F, Sonnewald U (2000) High CO_2-mediated down-regulation of photosynthetic gene transcripts is caused by accelerated leaf senescence rather than sugar accumulation. FEBS Lett 479:19–24

Machekano H, Mvumi BM, Chinwada P, Kageler SJ, Rwafa R (2019) Evaluation of alternatives to synthetic pesticides under small-scale farmer-managed grain storage conditions. Crop Protect 126:104941

Maillard A, Diquélou S, Billard V, Laîné P, Garnica M, Prudent M, Garcia-Mina JM, Yvin JC, Ourry A (2015) Leaf mineral nutrient remobilization during leaf senescence and modulation by nutrient deficiency. Front Plant Sci 6:317

Maphosa Y, Jideani VA (2017) The role of legumes in human nutrition, functional food. In: Hueda MC (ed) Improve health through adequate food. Intech Open. https://doi.org/10.5772/intechopen.69127

Marschner H, Römheld V (1994) Strategies of plants for acquisition of iron. Plant Soil 165:261–274

Martin C, Zhang Y, Tonelli C, Petroni K (2013) Plants, diet, and health. Annu Rev Plant Biol 64:19–46

Matimati I, Verboom GA, Cramer MD (2014) Do hydraulic redistribution and nocturnal transpiration facilitate nutrient acquisition in Aspalathus linearis? Oecologia 175:1129–1142

Mcdermott J, Wyatt AJ (2017) The role of pulses in sustainable and healthy food systems. Ann N Y Acad Sci 1392:30–42

Medawar E, Huhn S, Villringer A, Veronica Witte A (2019) The effects of plant-based diets on the body and the brain: a systematic review. Translat Psych 9:226

Mertens M, Höss S, Neumann G, Afzal J, Reichenbecher W (2018) Glyphosate, a chelating agent-relevant for ecological risk assessment? Environ Sci Pollut Res Int 25:5298–5317

Michaels TE (2016) Grain legumes and their dietary impact: overview. Reference Module in Food Science. Elsevier

Mirmiran P, Hosseinpour-Niazi S, Azizi F (2018) Therapeutic lifestyle change diet enriched in legumes reduces oxidative stress in overweight type 2 diabetic patients: a crossover randomised clinical trial. Eur J Clin Nutr 72:174–176

Morgan JB (2019b) Nature news (Online). Available https://www.nature.com/scitable/knowledge/library/plant-soil-interactions-nutrient-uptake-105289112/. Accessed 4 Nov 2019

Murphy KJ, Marques-Lopes I, Sánchez-Tainta A (2018) Cereals and legumes (Chap 7). In: Sánchez-Villegas A, Sánchez-Tainta A (eds) The prevention of cardiovascular disease through the mediterranean diet. Academic Press, Cambridge

Myers SS, Zanobetti A, Kloog I, Huybers P, Leakey ADB, Bloom AJ, Carlisle E, Dietterich LH, Fitzgerald G, Hasegawa T, Holbrook NM, Nelson RL, Ottman MJ, Raboy V, Sakai H, Sartor KA, Schwartz J, Seneweera S, Tausz M, Usui Y (2014) Increasing CO2 threatens human nutrition. Nature 510:139–142

Nasr Esfahani M, Kusano M, Nguyen KH, Watanabe Y, Ha CV, Saito K, Sulieman S, Herrera-Estrella L, Tran LS (2016) Adaptation of the symbiotic Mesorhizobium-chickpea relationship to phosphate deficiency relies on reprogramming of whole-plant metabolism. Proc Natl Acad Sci USA 113:E4610–E4619

Nearing M, Pruski F, O'neal MR (2004) Expected climate change impacts on soil erosion rates: a review. J Soil Water Conserv 59:43

Nemecek T, Richthofen J-S, Dubois G, Casta P, Charles R, Pahl H (2008) Environmental impacts of introducing grain legumes into European crop rotations. Eur J Agron 28:380–393

Niu Y, Chai R, Liu L, Jin G, Liu M, Tang C, Zhang Y (2014) Magnesium availability regulates the development of root hairs in Arabidopsis thaliana (L.) Heynh. Plant, Cell Environ 37:2795–2813

Olney DK, Pedehombga A, Ruel MT, Dillon A (2015) A 2-year integrated agriculture and nutrition and health behavior change communication program targeted to women in Burkina Faso reduces anemia, wasting, and diarrhea in children 3—12.9 months of age at baseline: a cluster-randomized controlled trial. J Nutr 145:1317–1324

Parker J (2001) Rhizobium. In: Brenner S, Miller JH (eds) Encyclopedia of genetics. Academic Press, New York

Perring M, Cullen B, Johnson I, Hovenden M (2010) Modelled effects of rising CO2 concentration and climate change on native perennial grass and sown grass-legume pastures. Clim Res 42:65–78

Pilorgé E, Muel F (2016) What vegetable oils and proteins for 2030? Would the protein fraction be the future of oil and protein crops?⋆. OCL 23:D402

Polak R, Phillips EM, Campbell A (2015) Legumes: health benefits and culinary approaches to increase intake. Clin Diab Publ Am Diab Assoc 33:198–205

Popkin BM, Adair LS, Ng SW (2012) Global nutrition transition and the pandemic of obesity in developing countries. Nutr Rev 70:3–21

Popp C, Ott T (2011) Regulation of signal transduction and bacterial infection during root nodule symbiosis. Curr Opin Plant Biol 14:458–467

Potera C (2007) Agriculture: pesticides disrupt nitrogen fixation. Environ Health Perspect 115:A579–A579

Reid R, Hayes J (2003) Mechanisms and control of nutrient uptake in plants. Int Rev Cytol 229:73–114

Roriz M, Barros M, Castro PML, Carvalho S, Vasconcelos MW (2020) Improving iron nutrition in plant foods. In: Noureddine Benkeblia (ed) The Role of Legumes and Soil Microbes. https://doi.org/10.1002/9781119511144.ch6

Rotaru V (2009) Interactive influence of phosphorus and iron on nitrogen fixation by soybean. Environ Experim Botany 66:94–99

Sage RF, Zhu X-G (2011) Exploiting the engine of C4 photosynthesis. J Exp Bot 62:2989–3000

Saini RK, Nile S, Keum YS (2016) Food science and technology for management of iron deficiency in humans: a review. Trends in Food Science & Technology 53:13

Sánchez-Calderón L, López-Bucio J, Chacón-López A, Gutiérrez-Ortega A, Hernández-Abreu E, Herrera-Estrella L (2006) Characterization of <em> low phosphorus insensitive </em> mutants reveals a crosstalk between low phosphorus-induced determinate root development and the activation of genes involved in the adaptation of arabidopsis to phosphorus deficiency. Plant Physiol 140:879–889

Santos CS, Roriz M, Carvalho SMP, Vasconcelos MW (2015) Iron partitioning at an early growth stage impacts iron deficiency responses in soybean plants (Glycine max L.). Front Plant Sci 6:325

Santos C, Serrão I, Vasconcelos M (2016) Comparative analysis of iron deficiency chlorosis responses in soybean (Glycine max) and barrel medic (Medicago truncatula). Revista de Ciencias Agrarias 39:70–81

Santos CS, Benkeblia N, Vasconcelos MW (2017) Strategies for enhancing phytonutrient content in plant based foods. In: Benkeblia N (ed) Phytonutritional improvement of crops. Wiley, Sussex, pp 203–232

Santos CS, Deuchande T, Vasconcelos MW (2019a) Molecular aspects of iron nutrition in plants. In: Cánovas FM, Lüttge U, Leuschner C, Risueño M-C (eds) Progress in Botany, vol 81. Springer, Cham, pp 125–56

Santos CS, Ozgur R, Uzilday B, Turkan I, Roriz M, Rangel A, Carvalho SMP, Vasconcelos MW (2019b) Understanding the role of the antioxidant system and the tetrapyrrole cycle in iron deficiency chlorosis. Plants (Basel) 8:348

Sathe SK (2016) Beans, overview. Reference module in food science. Elsevier

Scheelbeek P (2018) Effect of environmental changes on vegetable and legume yields and nutritional quality. Proc Nat Acad Sci 115:6804

Scheublin TR, Ridgway KP, Young JPW, Van Der Heijden MGA (2004) Nonlegumes, legumes, and root nodules harbor different arbuscular mycorrhizal fungal communities. Appl Environ Microbiol 70:6240

Schmid NB, Giehl RFH, Döll S, Mock H-P, Strehmel N, Scheel D, Kong X, Hider RC, Von Wirén N (2014) Feruloyl-CoA 6′-Hydroxylase1-dependent coumarins mediate iron acquisition from alkaline substrates in arabidopsis. Plant Physiol 164:160–172

Schmidt H, Günther C, Weber M, Spörlein C, Loscher S, Böttcher C, Schobert R, Clemens S (2014) Metabolome analysis of arabidopsis thaliana roots identifies a key metabolic pathway for iron acquisition. PLoS ONE 9:e102444

Shafique A, Rehman S, Khan A, Kazi AG (2014) Improvement of legume crop production under environmental stresses through biotechnological intervention (Chap 1). In: Ahmad P, Rasool S (eds) Emerging technologies and management of crop stress tolerance. Academic Press, San Diego

Sharma RC (2009) Cereal-based cropping systems in Asia: nutrition and disease management (Chap. 5). In: Sadras V, Calderini D (eds) Crop physiology. Academic Press, San Diego

Sinclair SA, Kramer U (2012) The zinc homeostasis network of land plants. Biochim Biophys Acta 1823:1553–1567

Sisó-Terraza P, Rios J, Abadía J, Abadía A, Alvarez-Fernández A (2015) Flavins secreted by roots of iron-deficient Beta vulgaris enable mining of ferric oxide via reductive mechanisms. New Phytol 209:733

Slatni T, Krouma A, Aydi S, Chaiffi C, Gouia H, Abdelly C (2008) Growth, nitrogen fixation and ammonium assimilation in common bean (Phaseolus vulgaris L) subjected to iron deficiency. Plant Soil 312:49–57

Slattery RA, Ainsworth EA, Ort DR (2013) A meta-analysis of responses of canopy photosynthetic conversion efficiency to environmental factors reveals major causes of yield gap. J Exp Bot 64:3723–3733

Smith FW (2001) Plant responses to nutritional stresses. In: molecular analysis of plant adaptation to the environment. Springer, Dordrecht, pp 249–269

Soares J, Deuchande T, Valente LMP, Pintado M, Vasconcelos MW (2019a) Growth and nutritional responses of bean and soybean genotypes to elevated CO_2 in a controlled environment. Plants 8:465

Soares JC, Santos CS, Carvalho SMP, Pintado MM, Vasconcelos MW (2019b) Preserving the nutritional quality of crop plants under a changing climate: importance and strategies. Plant Soil 443:1–26

Stagnari F, Maggio A, Galieni A, Pisante M (2017) Multiple benefits of legumes for agriculture sustainability: an overview. Chem Biol Technol Agric 4:2

Stinner PW (2015) The use of legumes as a biogas substrate—potentials for saving energy and reducing greenhouse gas emissions through symbiotic nitrogen fixation. Energy, Sustain Soc 5:4

Sugiyama A, Yazaki K (2012) Root exudates of legume plants and their involvement in interactions with soil microbes

Sulieman S, Tran LS (2017) Legume nitrogen fixation in soils with low phosphorus availability: adaptation and regulatory implication

Tate RLI (2005) Encyclopedia of soils in the environment: volume 1–4. Soil Sci 170:669

Thuynsma R, Valentine A, Kleinert A (2014) Phosphorus deficiency affects the allocation of below-ground resources to combined cluster roots and nodules in Lupinus albus. J Plant Physiol 171:285–291

Tian D, Niu S (2015) A global analysis of soil acidification caused by nitrogen addition. Environ Res Lett 10:024019

Tran HT, Hurley BA, Plaxton WC (2010) Feeding hungry plants: the role of purple acid phosphatases in phosphate nutrition. Plant Sci 179:14–27

Triplett G, Dick W (2008) No-tillage crop production: a revolution in agriculture. Agron J 100:153

Udvardi M, Poole PS (2013) Transport and metabolism in legume-rhizobia symbioses. Annu Rev Plant Biol 64:781–805

Vance CP (2001) Symbiotic nitrogen fixation and phosphorus acquisition. Plant nutrition in a world of declining renewable resources. Plant Physiol 127:390–397

Vardien W, Steenkamp ET, Valentine AJ (2016) Legume nodules from nutrient-poor soils exhibit high plasticity of cellular phosphorus recycling and conservation during variable phosphorus supply. J Plant Physiol 191:73–81

Vasconcelos MW, Gomes AM (2016) The legume grains: when tradition goes hand in hand with nutrition. In: Kristbergsson K, Oliveria J (eds) Traditional foods: general and consumer aspects. Springer, US, Boston, MA

Walker EL, Connolly EL (2008) Time to pump iron: iron-deficiency-signaling mechanisms of higher plants. Curr Opin Plant Biol 11:530–535

White P, Brown P (2010) Plant nutrition for sustainable development and global health. Ann Bot 105:1073–1080

Williams M, Lean M, Pogson B, Martin C (2012) Plants, food and human health: you are what you eat? The plant cell

Xiong H, Guo X, Kobayashi T, Kakei Y, Nakanishi H, Nozoye T, Zhang L, Shen H, Qiu W, Nishizawa NK, Zuo Y (2014) Expression of peanut iron regulated transporter 1 in tobacco and rice plants confers improved iron nutrition. Plant Physiol Biochem 80:83–89

Yadav RS, Tarafdar J (2001) Influence of organic and inorganic phosphorus supply on the maximum secretion of acid phosphatase by plants. Biol Fertil Soils 34:140–143

Yang TJW, Perry PJ, Ciani S, Pandian S, Schmidt W (2008) Manganese deficiency alters the patterning and development of root hairs in Arabidopsis. J Exp Bot 59:3453–3464

Yang Q, Li Z, Lu X, Duan Q, Huang L, Bi J (2018) A review of soil heavy metal pollution from industrial and agricultural regions in China: pollution and risk assessment. Sci Total Environ 642:690–700

Yoneyama K, Xie X, Kim HI, Kisugi T, Nomura T, Sekimoto H, Yokota T, Yoneyama K (2012) How do nitrogen and phosphorus deficiencies affect strigolactone production and exudation? Planta 235:1197–1207

Zhou J, Jiang X, Wei D, Zhao B, Ma M, Chen S, Cao F, Shen D, Guan D, Li J (2017) Consistent effects of nitrogen fertilization on soil bacterial communities in black soils for two crop seasons in China. Sci Rep 7:3267

Nutrient Management for Improving Abiotic Stress Tolerance in Legumes of the Family Fabaceae

Akbar Hossain, Ayman EL Sabagh, Murat Erman, Shah Fahad, Tofazzal Islam, Rajan Bhatt, and Mirza Hasanuzzaman

Abstract Grain legumes are rich in carbohydrate, protein, vitamin, mineral, fiber, and essential amino acids. Besides, legumes play a vital role in fixing atmospheric nitrogen (N) which ultimately improves soil fertility through the symbiotic process. However, in the changing climate, the sustainability of grain legumes production is vulnerable due to the extreme events of abiotic stresses such as drought, salinity, heat stress, and heavy metals. These abiotic stresses are linked with the physiological, biochemical, and morphological changes that prevent the full genetic potential productivity of the legume crops. Plants need an ample amount of mineral nutrients (micro-and macronutrients) in each stage of the development to achieve maximum yield. Among these mineral nutrients, macronutrients, particularly N, phosphorus (P), and potassium (K) and micronutrients particularly iron (Fe), zinc (Zn), silicon (Si), and selenium (Se) involve in several physiological, biochemical, and morphological processes in plants. These nutrient elements also play a vital role in increasing

A. Hossain (✉)
Bangladesh Wheat and Maize Research Institute (BWMRI), Dinajpur 5200, Bangladesh
e-mail: akbarhossainwrc@gmail.com; tanjimar2003@yahoo.com

A. EL Sabagh
Department of Agronomy, Faculty of Agriculture, Kafrelsheikh University, Kafr el-Sheikh 33516, Egypt

A. EL Sabagh · M. Erman
Department of Field Crops, Faculty of Agriculture, Siirt University, Siirt, Turkey

S. Fahad
Department of Agriculture, University of Swabi, Khyber Pakhtunkhwa, Pakistan

T. Islam
Institute of Biotechnology and Genetic Engineering (IBGE), Bangabandhu Sheikh Mujibur Rahman Agricultural University, Gazipur 1706, Bangladesh

R. Bhatt
Regional Research Station, Kapurthala, India

Punjab Agricultural University, Ludhiana, Punjab 144601, India

M. Hasanuzzaman
Department of Agronomy, Faculty of Agriculture, Sher-e-Bangla Agricultural University, Dhaka, Bangladesh

M. Hasanuzzaman et al. (eds.), *The Plant Family Fabaceae*,
https://doi.org/10.1007/978-981-15-4752-2_15

plant resistance to environmental stresses. An adequate balance of nutrient supply is needed for each stage of the development and to achieve maximum yield potential of legume crops. While imbalanced use of mineral nutrients may result in negative impacts on environmental and also increase the cost of production. Therefore, adequate balanced mineral nutrient management is necessary for the sustainable production of legume crops under the changing climate. This chapter described the negative impacts of abiotic stresses on legumes under the family Fabaceae and also highlighted the essential roles of balanced mineral nutrients in each developmental phage of plants for attaining maximum yield even under the abiotic stressful environment.

Keywords Abiotic stresses · Legumes · Fabaceae family · Nutrient management

1 Introduction

After the family *Poaceae* or grass family, legumes (family *Leguminosae* or *Fabaceae*) are considered the second most important plant family rich in carbohydrates, proteins, vitamins, minerals, fiber, and essential amino acids (Ara et al. 2019). Therefore, grain legumes or pulse crops are a key source of carbohydrates, proteins, vitamins, minerals, fiber, and essential amino acids in the human diets, particularly in the developing countries (Mishra et al. 2014). Besides, legumes play a vital role in fixing atmospheric nitrogen (N) which ultimately improves soil fertility through symbiosis, and thus reduces the dependence on chemical fertilizers (Rubiales and Mikic 2015). Hence, legumes are used as an intercrop with cereals as the concept of climate-smart agriculture (FAO 2013) for improving soil fertility and crop productivity through biological N-fixation (Qureshi et al. 2010; Rao et al. 2013; Farooq et al. 2017).

All plants including legumes are sessile organisms and thus face adverse abiotic stresses such as soil salinity, drought, heat, cold, flooding, and heavy metal contamination in their whole life span ranging from germination to the maturity that leads to decline the crop productivity (Xu et al. 2014; Negrão et al. 2017; Nadeem et al. 2018, 2019). Similarly, due to climate change, the extreme events of abiotic stresses also affect on the N-fixation in legumes through altering the physiological, biochemical, and morphological processes of plants that limit the full genetic potential for higher yield (Beebe et al. 2013).

Plants need an ample amount of mineral nutrients (micro- and macronutrients) in each stage of the development to achieve maximum yield (López-Arredondo et al. 2013). Particularly, macronutrients such as N, P, and K (Edmeades 2003) and micronutrients (Fe, Zn, and Mn) are involved in several physiological, biochemical, and morphological processes in plants for improving plant resistance to abiotic stresses (Fageria et al. 2002) although the requirement of the amount and types of nutrients are varied with species to species and genotype to genotypes according to their biological processes (Rengel and Damon 2008; Shukla et al. 2014). However, an adequate balance of nutrients is needed for each stage of development and to achieve maximum yield potential (López-Arredondo et al. 2013). While the imbalance use of mineral nutrients may result in negative impacts on the environment and also increase the cost of production (Hawkesford 2012, 2014), therefore, adequate balanced mineral nutrient management is necessary for sustainable legumes production under the changing climate. In this chapter, we reviewed the adverse effect of abiotic stresses on legumes under the family of Fabaceae and also highlighted that balanced nutrient can improve the abiotic stress tolerance in plants.

2 The Adverse Effect of Abiotic Stresses on the Growth and Development of Legumes

2.1 Drought

Drought stress (DS) poses a constant challenge for agricultural crops. It is considered as a severe constraint for global agricultural productivity. Its intensity and severity are predicted to increase in the near future due to global changing climate (Baroowa and Gogoi 2013; Ghassemi-golezani et al. 2013). Legumes are mostly grown in dry areas since they are moderately tolerant to drought stress (Nadeem et al. 2019). However, severe DS leads to remarkable yield losses of crops due to their high sensitivity against drought during the seedling and reproductive stages (Farooq et al. 2016; Nadeem et al. 2019). The most prominent effects of DS are limited germination, shrinkage plant growth through decrease in net photosynthesis as a result of abolishing the photosynthetic apparatus (Chowdhury et al. 2016), reduced carbon and N-fixation (Beebe et al. 2013; Nadeem et al. 2018, 2019), hampered assimilate translocation (Mondal et al. 2011; Zlatev and Lidon 2012), and hindered the development of reproductive organs such as pollen grain sterility (Sehgal et al. 2018). The DS leads to reduce the number of pods and grains (Liu et al. 2004; Vadez et al. 2012), shortens or fastens the time of flowering (Samarah et al. 2009), weakens sink movement (Andersen et al. 2015), and also condenses nutrients uptake (Nadeem et al. 2019).

For enhancing water use efficiency of plants under DS, stomatal control is considered as one of the major physiological responses in plants (Makbul et al. 2011). Since the rate of stomatal conductance is linked with the rate of transpiration (De Souza et al. 1997), DS ultimately controls the rate of photosynthesis. Stomatal conductance in soybean was decreased significantly as compared to the control under drought stress (Hao et al. 2013; Mak et al. 2014; Mutava et al. 2014). Similarly, the germination rate in soybean (Heatherly 1993), chickpea (Awari and Mate 2015), and in fababean (Li et al. 2018) was decreased significantly under drought stress. Atti et al. (2004) reported that during DS, the declining rate of photosynthetic, stomatal conductance, and transpiration was 78.4, 92, and 85.4%, respectively. Under drought stress, the transpiration rate of legumes was lessened by more than 50% (Hao et al. 2013; Mak et al. 2014).

2.2 Salinity

Due to the global climate change, areas and severity of soil salinity are increasing globally which is threatening to the global food security of the increasing population. FAO (2008) estimated that more than 800 million hectares of land across the globe are affected by soil salinity. During salt stress, plants are normally facing osmotic stress, hormones, ionic, and nutritional imbalances that lead to occur oxidative stress in the plant cells (El Sayed 2011; Deinlein et al. 2014). During salt stress, two types of challenges are tackled by the plant: Firstly, high concentrations of solute interrupt the osmotic balance, resulting in the physiological drought which restricts water and nutrients uptake from root zone soil solution (Munns 2011; Farooq et al. 2015); and secondly, ionic imbalances from transformed K^+/Na^+ ratios and Na^+, Cl^- ion concentrations lead to occur oxidative stress in plant cells (Rejili et al. 2007; Haro et al. 2010). High concentrations of Na^+ interrupt the osmotic balance, resulting in the physiological drought which restricts water and nutrients uptake from root zone soil solution of the affected plants (Farooq et al. 2015), although some plant species are also sensitive to Cl^-, due to the presence of major anion in salt-affected soils (Nadeem et al. 2019). Under the extreme salt stress, the excessive concentrations of Na^+ and Cl^- ions in the root zone of legumes hamper the availability of other micro- and microelements including B, Zn, Ca, Cu, Mg, Fe, N, P, and K (Yadav et al. 1989; El Sayed 2011).

Similar to other plants, the seedling of grain legumes establishment and reproductive stages are also sensitive to high salinity stress (Al-Mutata 2003). Overall plant growth in legumes under salinity has been occurred due to hormonal, ionic, and nutritional imbalances and also a lack of cell wall acidification (Pitann et al. 2011). Sometimes leaf senescence in legumes is observed as a result of chlorosis and necrosis (Sehrawat et al. 2013a, b). Salt stress also hampered the growth and development of major grain legumes such as *Phaseolus vulgaris* L. (Ferri et al. 2000), *Cicer arietinum* L. (Sadiki and Rabih 2001), *Vigna radiata* L. Wilczek (Kabir et al. 2004), *Glycine max* L. (Luo et al. 2006), *Vicia faba* L. (Pitann et al. 2011), and *Lens culinaris* L. (Bandeoglu et al. 2004).

If salt stress is affected by the reproductive stages of legumes, flower formation is delayed and also reduced the number of flowers and pods which lead to reduce the grain yield and quality (Khan et al. 2016a, b). Limitation of nitrate (NO_3^-) absorption from the soil solution under salt stress reduces the grain protein and distressed N metabolism in the grain of legumes (Ghassemi-Golezani et al. 2010). Similar to grain protein, starch mobilization in grain legumes is also highly sensitive to salt stress (Flowers 2004).

Salt stress also restricts N uptake and metabolism in the grain of legumes through hampering the biological N-fixation (BNF) (Lopez et al. 2008; Nadeem et al. 2019). It is well-known that BNF generally occurs in the nodules of plant roots, and this process is also highly sensitive to salinity stress. Salt stress significantly reduces the number and movement of nodules in grain legumes including *Vicia faba* L. (Cordovilla et al.

1994; Rabie and Almadini 2005) and *Cajanus cajan* L. (Matamoros et al. 1999; Garg and Manchanda 2008) which ultimately limit the BNF (Cordovilla et al. 1994; Delgado et al. 1994).

2.3 Heat Stress

Heat stress (HS) is one of the foremost abiotic stresses causing an enormous loss in the productivity of crops (IPCC 2014; Asseng et al. 2015; Fahad et al. 2017). The increasing temperature has already changed the growing periods in many regions of the world (Porter 2005). Even though growing temperatures have a positive effect on crop production particularly in temperate regions, generally the impact of increasing temperature is still undesirable (Challinor et al. 2014).

Wahid et al. (2007) revealed that under HS, growth, and development of plants are significantly influenced by a series of morpho- and physio-biochemical changes. Fahad et al. (2017) noticed that HS reduces the rate of germination and also seedling stand establishment (Fahad et al. 2017). Under HS, growth, density, and length of the roots are also reduced which leads to limit the uptake of water and nutrients from soil solution to the aboveground parts of the plant (Wahid et al. 2007; Huang et al. 2012). Vollenweider and Gunthardt-Goerg (2005) observed that HS causes leaf senescence through scorching of the leaf twigs ultimately shrinkage growth and development of plants through decreasing in net photosynthesis as a result of abolishing the photosynthetic apparatus. HS causes serious destruction of proteins in plant cells through interrupting their synthesis, deactivate the enzymatic activity, destruction of the cell membranes, and also hinder the cell division process (Smertenko et al. 1997) since the adverse effects of HS on crops including legumes fluctuate with the duration, timing, and severity of the HS (Fahad et al. 2016). For example, HS triggered the considerable yield loss in *Arachis hypogea* L. and *Phaseolus vulgaris* L. (Vara Parasad et al. 1999; Rainey and Griffiths 2005). The HS affects the nutrient cycle, uptake, and availability and also obstructing different physiological functions of plants. HS hinders the activity of enzyme-like nitrate reductase since this enzyme is directly linked with N metabolism (Basirirad 2000; Klimenko et al. 2006; Rennenberg et al. 2006).

2.4 Heavy Metal Stress

Generally, heavy metals (HMs) such as Fe, Zn, Mn, Cu, Hg, Ni, Cd, Co, and arsenic are accumulated in soils through imbalanced use of pesticides, chemical fertilizers, and also from waste industrial and disposal sewage water (Aydinalp and Marinova 2009). These HMs contaminate groundwater through leaching and also store in soil surface (Gupta and Ali 2012; Dağhan and Ozturk 2015; Hakeem et al. 2015; Ozturk et al. 2015a, b; Basheer 2018). Plants are bio-accumulators of various heavy metals naturally from the HMs contaminated soil and water (Celik et al. 2010; Haribabu and Sudha 2011; Ozturk et al. 2017). Although some of HMs with a specific concentration is essential as micronutrients for different physio-biochemical processes in plants, whereas their excessive concentration has damaging impacts on the physio-biochemical process of the plants (Ghori et al. 2019).

Leaf chlorosis, poor growth, burning of the root, and death of the plants are the major toxicity symptoms of HMs (Ozturk et al. 2008, 2015b) although HM-tolerant plants could accumulate HMs to some extent. However, the level of their tolerances varies from species to species and genotypes to genotypes (Abolghassem et al. 2018). The high concentration of HMs accumulation in plants hinders many physio-biochemical and morphological process which leads to the decline of crop productivity (Shahid et al. 2015). Grain legumes such as *Pisum sativum* L. (Wani et al. 2008a), *Cicer arietinum* L. (Wani et al. 2008b; Wani and Khan 2010), and *Vigna radiata* (L.) Wilczek (Wani et al. 2007) faced nutrients deficiency or toxic effects when grown in HMs-enriched soil.

The HMs accumulation in soils reduced the microbial activities (Paudyal et al. 2007; Wani et al. 2008a) leading to a subsequent reduction of soil fertility (Khan et al. 2009; Krujatz et al. 2011). Likewise, several earlier reports revealed that higher levels of HMs accumulation in soil reduced the rhizobial activity in host legumes (Heckman et al. 1987; Broos et al. 2007) and concurrently decreased the yields of affected legumes (Moftah 2000).

2.5 Abiotic Stress-Induced Excessive Production of ROS

Under stressful conditions (including heavy metal stress, HMs), molecular oxygen is reduced in plant cells, while produce extremely oversensitive intermediary products such as H_2O_2, hydroxyl radicals, and superoxide radicals that are recognized as ROS. The production of ROS is one of the most common and most studied response of plants toward HMs (Jalmi et al. 2018), which affects various enzymatic activities, denatures proteins and lipid through chain reactions and ultimately damages cell membranes and DNA (Rascio and Navari-Izzo 2011; Barconi et al. 2011; Ahmad et al. 2012). These reactions also disturb the integrity of membranes, resulting in the alteration of basic plant metabolic reactions such as photosynthesis, respiration, and homeostasis (Hossain et al. 2012). The ROS also heading to lipid peroxidation, particularly the cellular membranes affecting them to leak, damaging biomolecules, and also cleavage of DNA strands collectively known as oxidative stress (Barconi et al. 2011; Ahmad et al. 2012).

3 Nutrient Management for Improving Abiotic Stress Tolerance in Plants

3.1 Drought

Drought stress (DS) is a severe constraint for global agricultural productivity (Baroowa and Gogoi 2013; Ghassemi-golezani et al. 2013). The DS decreases seed germination, seedling growth, and development of plants through inhibiting net photosynthesis (Chowdhury et al. 2016), hinders the development of reproductive organs such as pods/plant and grains/plant (Vadez et al. 2012; Sehgal et al. 2018), reduces the time of flowering (Samarah et al. 2009; Andersen et al. 2015), and also reduces the nutrients uptake (Nadeem et al. 2019).

The DS can be mitigated in two ways: (i) selection of crop cultivars that are tolerant to DS and (ii) improved agronomic approaches. Since breeding approaches are time-consuming and need skill technical knowledge, the DS can be mitigated through enhancing water use efficiency (WUE), which is linked with the rate of stomatal conductance and the transpiration (Makbul et al. 2011; Hao et al. 2013; Mak et al. 2014). Agronomic approaches such as the adjustment in time of sowing, plant geometry, and fertilizer management can help to improve drought tolerance in plants through enhancing WUE (Siddique et al. 2012; Jha et al. 2015). Besides these approaches, zero tillage, mulching, and deep plowing could also enhance the WUE for alleviating the adverse effects of DS in plants (Mupangwa et al. 2007; Johnson et al. 2018). Since fertilizer management is the best option for increasing drought tolerance in plants to mitigate the harmful effects of DS (Hansel et al. 2017), Jin et al. (2006) found that both P and N fertilizers can increase WUE in crops by decreasing the rate of evaporation. They also observed that P nourishment enhanced the rate of photosynthesis in soybean through controlling the leaf water potential, membrane stability, stomatal conductance, and development of root numbers and density under DS (Jin et al. 2006). Similarly, N fertilization enhanced the grain protein concentration, yield and yield-related traits in chickpea (Palta et al. 2005). Sangakkara et al. (1999) observed that the application of balanced K improved the photosynthetic efficiency in legume by keeping higher tissue water potential during drought. Besides chemical fertilizers, the combined application of organic and inorganic fertilizers enhances the WUE and can also increase the final grain yield of soybean (Hati et al. 2006; Bandyopadhyay et al. 2010).

During the DS, the productivity of legume crops was enhanced by the addition of selenium (Se) through the improvement of WUE in the root system of legumes (Leport et al. 2006). Likewise, Se can enhance the growth of aging seedlings and delay leaf senescence (Grams et al. 2007). Mohammadi et al. (2011) also described that Se contracted the production of lipid peroxidation through increasing enzymatic antioxidants including GPX and SOD in soybean and chickpea under DS. Furthermore, under drought stress, the silicon (Si) application also improved the tonoplast and plasma membrane structure in terms of integrity and function and chlorophyll fluorescence (Xu et al. 2015). Hattori et al. (2005) observed that Si significantly enhanced the relative water content (RWC) in plants by enlightening the levels of proline and glycine betaine (GB), whereas Mali and Aery (2008) found that N-fixation in cowpea was increased by Si nutrition. The addition of Si in combination with K significantly improved the shoot dry weight of in chickpea under DS (Kurdali et al. 2016, 2019). Zn application increased the grain yield and yield attributes of chickpea under drought stress (Shaban et al. 2012). Grain yield in mung bean was also increased due to the foliar application of Zn (Thalooth et al. 2006). Similarly, an adequate amount of both Fe and Zn increased the RWC that leads to increased micronutrients concentration and protein content of grains in legumes (Yadavi et al. 2014; Farooq et al. 2017). Boron is an important micronutrient, which is helpful in N-fixation through improving the number of nodules in soybean during DS (Bellaloui et al. 2013; Bellaloui and Mengistu 2015).

3.2 Salinity

Similar to other abiotic stress, soil salinity (SS) is considered as one of the furthermost significant abiotic stresses which disturb the morphological and physio-biochemical process of the plants, which leads to expressively shrinkages the yield of the crops including legumes (Bano et al. 2012; Yildiz and Terz 2013).

The SS reduces the ability of plants to take up water (Munns 2002, 2005). However, plants can overcome the adverse effect of SS through the accumulation and production of proline, GB, soluble sugars (Sakamoto and Murata 2002; Murakeözy et al. 2003) and also the absorption of mineral ions (Meloni et al. 2004; Slama et al. 2008), which in turn gives the possibility to absorb water and restore plant turgor (Xiong and Zhu 2002). Without physiological approaches, a number of management strategies have been considered for alleviating the negative effect of SS in plants (Khan et al. 2013). Among them, adequate balance mineral nutrition is found one of the best strategies to alleviate the SS in crop plants (Rengel and Marschner 2005). It is due to the availability of essential plant nutrients under SS that could alleviate the adverse effect of SS through declining permeability of plasma membranes that subsequently conserve membrane integrity and utility (Singh et al. 2014).

It is well-known that N is the foremost crucial macronutrients for proper growth and development in plants. However, it is the key constituent of all amino acids and various N comprising compounds such as proline, GB, amides, amino acids and polyamine, which are the most important compounds in plants exposed to SS (Siddiqui et al. 2010). Under SS, usually, N availability and uptake are poor (Siddiqui et al. 2010), due to high Cl^- content in the soil. However, foliar application of N may reduce the effect of SS and enhanced the growth of the plants through increasing the availability and uptake of N lead to enhance the growth and productivity in chickpea (Garg and Chandel 2011).

Similar to N, K^+ availability in mild SS conditions enhanced the water relations, growth as well as grain yield of mung bean also reported by Kabir et al. (2004). Kafi et al. (2012) found that both sufficient presence of available K^+ and Ca^{2+} could alleviate the injurious effects of SS in lentil. Superficially provided Ca^{2+} reduced the contaminated effects of Na^+ and Cl^-, apparently by facilitating a higher K^+ to Na^+ selectivity (Latef 2011). However, the application of Ca^{2+} also maintains membrane integrity and selectivity, thus reducing Na^+ and Cl^- toxicity in the plants (Grattan and Grieve 1999).

Although legumes are highly sensitive to SS and contain fairly low Si contents, exogenous application of Si reduces Na^+ and Cl^- contents in cowpea, common bean, and faba bean and improves yield components in faba bean under SS conditions (Kardoni et al. 2013; Farooq et al. 2017). Moreover, the application of both nitrogenous and potassic fertilizers, in combination with the application of Si could help legumes to survive against SS. Garg and Singla (2016) recently found that *Arbuscular mycorrhiza* (AM) combined with Si availability improve the nutrient uptake which leads to improve the growth and yield attributes in chickpea under SS. They also revealed that AM improves Si uptake, which limits uptake and injurious effect of Na^+ lead to the increased productivity of crops (Garg and Bhandari 2016).

3.3 Heat Stress

Plants nutrients are not only required for better plant growth and development but also help to improve survival ability against various abiotic stresses including heat stress (HS) (Waraich et al. 2011; Tahir et al. 2011). Among the abiotic stresses, HS affects the uptake of mineral nutrients in plants ultimately limits the plant growth negatively (Waraich et al. 2011, 2012). Among the mineral nutrient, N plays a vital role in the utilization of absorbed light energy and photosynthetic carbon metabolism (Huang et al. 2004) under HS. Similarly, N in the form of NO is extremely reactive and membrane-permeable free radical which has a broad spectrum of regulatory functions in many physiological processes of plants under various abiotic and biotic stresses including HS (Yang et al. 2006; Zhang et al. 2006; Zhao et al. 2007). Like N, K nutrition also plays a vital role in the survival of crop plants under environmentally stressful conditions. Since K is a vital element for numerous physio-biochemical processes of plants (Mengel and Kirkby 2001), thus it minimizes the excessive production of ROS (Waraich et al. 2011, 2012).

Calcium nutrition also plays a vital role in regulating a number of physiological processes at the tissue, cellular, and molecular levels that stimulate both growths and abiotic stresses including HS (Waraich et al. 2011, 2012). Kolupaev et al. (2005) revealed that the exogenous application of Ca inducing the HS resistance in plants by enhancing the antioxidant defense. Likewise, Jiang and Huang (2001) reported that the constructive effect of Ca under HS may be due to maintaining the antioxidant activity.

It is well-known that under HS conditions plants accumulate HS-induced excessive ROS, which leads to cell death, and finally death of the affected plants (Molassiotis et al. 2006). The tolerant plants generally survive by accumulating numerous enzymatic and non-enzymatic antioxidants to scavenge HS-induced ROS. Application of available boron (B) could improve the antioxidant activities of plants for mitigating the negative effects ROS generated through HS through maintaining numerous physio-biochemical processes (Tavallali et al. 2018).

Comparatively, low level of Se also was documented as an indispensable component in plants, which can enhance plants to survive against several abiotic stress-induced oxidatives damage in plant tissues including HS (Hawrylak-Nowak et al. 2010; Valadabadi et al. 2010; Cartes et al. 2010). Se plays roles in the synthesis of a major antioxidant enzyme and glutathione peroxidase enzyme.

Magnesium (Mg) surplus and insufficiency have harmful effects on plant photosynthesis through significant deterioration of leaf chlorophyll pigments (Shabala and Hari 2005; Hermans and Verbruggen 2005). The production of excessive HS-induced ROS in plant cells (del Rio et al. 2006; Navrot et al. 2007) leads to the destruction of proteins, lipids, carbohydrates, and DNA in plants' cells (Khan and Singh 2008; Tuteja 2007, 2010; Gill et al. 2011). Waraich et al. (2011, 2012) observed that Mg has been involved in many physio-biochemical processes in plants for proper growth and development during HS.

Under extreme events of abiotic stresses, plants cannot uptake nutrients and prompt many morpho-physiological disorders in plants. Mn is an essential trace element particularly involved in photosynthesis, nitrogen metabolism, and to form other compounds required for plant metabolism. Recently, scientists found that Mn has a decisive role in the reduction of the production of ROS through increasing the enzymatic and non-enzymatic antioxidant compounds against abiotic stresses including HS (Aloni et al. 2008; Waraich et al. 2011, 2012).

3.4 *Heavy Metal Stress*

Naturally occurring elements with a density >6 kg dm^{-3}, 53 are considered as heavy metals (HMs) as they have (Hooda 2010). The HMs include Cd, Ni, Pb, Fe, Zn, Co, As, Cr, silver (Ag), and platinum (Pt) (Bhat et al. 2019). Depending on the substances and physicochemical conditions of environment, HMs are classified into two categories such as nonessential elements (Cd, Pd, Hg, Cr, As, and Ag), which do not have any biological activity and, even at low concentrations, may have a harmful effect on plants (Lasat 2002; Kabata-Pendias 2011); and essential micronutrients (Cu, Zn, Fe, Mn, Mo, Ni, and Co) (Kabata-Pendias 2011) which are involved in many important biochemical and physiological processes of plants (Kabata-Pendias 2011; Kalaivanan and Ganeshamurthy 2016).

However, plants follow numerous molecular and physiological mechanisms through the production of enzymatic and non-enzymatic antioxidants as well as other compounds such as amino acids, glutathione, phytochelatins, metallothioneins, and enzymes such as superoxide dismutase and peroxide (Hossain et al. 2012) to overcome HMs stress (Krzesłowska 2011).

Besides these physiological mechanisms, balance and adequate amount of nutrients management could overcome the harmful/toxic effect of HMs. Zhang et al. (2014) observed that N efficiently relieved the Cd toxicity in plants through unclogging the pathway of Cd inhibition on chlorophyll synthesis. The foliar application of nitric oxide sinking the unhelpful action of HMs (Kopyra and Gwózdz 2003); nitric oxide encourages the stomatal closure and enhances the adaptive mechanism against stress including HMs also (Garcia-Mata and Lamattina 2001). Likewise, Li et al. (2007) reported that addition of balance amount of N (NO_3^- or NH_4^+) in nutrient solution could enhance the development of root systems ultimately enhanced the adaptive mechanism against Cd and Zn (Yang et al. 2004).

Nzihou and Sharrock (2010) revealed that soluble phosphate could be used as a means of chemical treatment for the conversion of HMs into insoluble compounds. They also found that phosphate and apatite M10(PO4)6(OH)2, could be removed HMs as the immobilize form from the soil solution, wastewater, groundwater. Several hydrogen or dihydrogen phosphates have been considered for scavenging soluble HMs (Nzihou and Sharrock 2010).

Hydrargyrite, Pb5(PO4)3Cl, is a typical example of a lead phosphate with very low solubility in a wide pH range (Nriagu 1984). Another frequently encountered example is hydroxylapatite, Ca5(PO4)3(OH), (HA) as the lead member of a large class of substituted compounds of similar structures. The HA has a low solubility product (Jaynes et al. 1999) in Pb contaminated soils since Pb could be effectively removed after reaction with HA (Ma et al. 1993) as the form of hydroxypyromorphite [Pb10(PO4)6(OH)2] (Ma et al. 1993).

K is an essential plant nutrient for growth and development of plants (Cakmak et al. 2010) since more than 10% biomass of plants are connected with K (Siddique et al. 2012). It is also involved in the defensive mechanism of plants under numerous stress conditions including HMs also through the enhancing stomatal movement and gas-exchange capacity (Mengel and Kirkby 2001; Bybordi 2015). Addition of K reduced the Cd-induced oxidative stress in plants through increasing the synthesis of carbohydrates, amino acids, nitrogenous compounds (Zorb et al. 2014) and also improving the accumulation of secondary metabolites, antioxidants, and nutrients in plants which helped in mitigation of Cd stress (Yasin et al. 2017).

Similar to other essential elements, Zn also takes part in physiological functions of plants including alleviating the HMs stress in various plants (Hassan et al. 2017). It is well-documented that HMs-induced oxidative stress limits the growth and development of plants by hampering the photosynthesis as well as inhabits metabolic functions. However, Zn could alleviate the toxicity of HMs by stimulating the antioxidant defense (Hassan et al. 2017).

Similar to other stresses, Si could also alleviate the antagonistic effects of HMs stress (Liang et al. 2007; Kim et al. 2014) through stimulating the enzymatic and non-enzymatic antioxidants defense system (Bhat et al. 2019). Abolghassem et al. (2018) observed that under HMs, Si facilitates for altering pH value, the formation of Si-HMs, and stimulating the enzymatic and non-enzymatic antioxidants.

Se is also an essential trace element, which plays an innovative role in plant biology under abiotic stresses including HMs (Hatfield et al. 2014; Wu et al. 2015). Earlier research findings documented that foliar application of Se with fertilizer proliferate the Se concentration in plants' tissue which led to alleviate the injurious effects of different abiotic stresses including HMs stress (Ardebili et al. 2014; Ahmad et al. 2016; Kaur et al. 2016; Wu et al. 2016a). Wu et al. (2016b) observed that exogenous application of Se significantly reduced the detrimental effect of Cd and Pb.

3.5 *Abiotic Stress-Induced Excessive Production of ROS*

Although ROS is considered as harmful species, however, recently scientists found that they could also play a vital role under different stress conditions and also assist as pointers for certain stages of plant development including germination, senescence, and stomatal closure (Yuan et al. 2013). They also act stress responses in plants through working as signaling molecules and also activate the gene expression process for the production of stress-defensive enzymatic and non-enzymatic antioxidants (i.e., ascorbic acid, glutathione, tocopherol, and enzymes including glutathione reductase, peroxidase, catalase, ascorbate peroxidase, and superoxide dismutase) and hormones (Sytar et al. 2013; Baxter et al. 2013).

Of all the ROS, H_2O_2 involves in cell-to-cell signaling acclamatory for tolerance to various stresses through controlling the germination, directive the leaf senescence, stomatal aperture, growth, and development and also programmed cell death (Pitzschke et al. 2006; Mittler et al. 2004) and also links with various steps in the pathway of the mitogen-activated protein kinase by reacting with transcriptional factors, phosphatases and kinases (Opdenakker et al. 2012). The plant utilizes ROS species for the formation of lignin in the cell walls to protect against pathogen attack and water-loss (Underwood 2012).

4 Summary and Conclusion

Legumes play a vital role in agriculture, as grain legumes are rich in carbohydrate, protein, vitamin, mineral, fiber, and essential amino acids and also could fix atmospheric N. Therefore, crop rotations with legume crops are an effective strategy to improve the soil fertility and also to reclaim eroded or barren areas, making them crucial for agricultural and environmental sustainability. However, in the changing climate, the sustainability of grain legumes production is vulnerable due to the extreme events of abiotic stresses such as drought, salinity, heat stress, and heavy metals. Since plants follow numerous physiological mechanisms, particularly the production of enzymatic and non-enzymatic antioxidants is to overcome the adverse effect of abiotic stress. While a balanced and adequate amount of macro- and micronutrients are involved in the physiological, biochemical, and morphological mechanisms for achieving the potential yield as well as overcome the harmful effect of abiotic stresses. While imbalanced use of mineral nutrients has a negative impact on the environment which also increases the production cost of crops. Therefore, it is important to apply balanced nutrients for sustainable legume production under the changing climate. In the following subsections, we discussed the nutrient management for improving abiotic stress tolerance in legumes crops elaborately. This chapter overviewed the adverse effect of abiotic stresses on legumes under the family of Fabaceae and also highlighted the requirements for a balanced nutrient supply for improvement of the abiotic stress tolerance in plants.

Conflict of Interest Statement The authors declare that the research was conducted in the absence of any commercial or financial relationships that could be construed as a potential conflict of interest.

References

Abolghassem E, Yulong D, Yinfeng X, Sirous S (2018) Silicon Mechanisms to Ameliorate Heavy Metal Stress in Plants. BioMed Res Int Article ID 8492898, 10 p. https://doi.org/10.1155/2018/8492898

Ahmad K, Bhatti IA, Muneer M, Iqbal M, Iqbal Z (2012) Removal of heavy metals (Zn, Cr, Pb, Cd, Cu and Fe) in aqueous media by calcium carbonate as an adsorbent. Int J Chem Biochem Sci 2:48–53

Ahmad R, Waraich EA, Nawaz F, Ashraf MY, Khalid M (2016) Selenium (Se) improves drought tolerance in crop plants–a myth or fact? J Sci Food Agric 96:372–380

Al-Mutata M (2003) Effect of salinity on germination and seedling growth of chickpea (*Cicer arietinum*) genotypes. Int J Agric Biol 5:226–229

Aloni B, Karni L, Deventurero G, Turhan E, Aktas H (2008) Changes in ascorbic acid concentration, ascorbate oxidase activity, and apoplastic pH in relation to fruit development in pepper (*Capsicum annuum* L.) and the occurrence of blossomend rot. J Hort Sci Biotechnol 83:100–105

Andersen MN, Asch F, Wu Y, Jensen CR, Næsted H, Mogensen VO, Koch KE (2015) Soluble invertase expression is an early target of drought stress during the critical, abortion-sensitive phase of young ovary development in maize. Plant Physiol 130:591–604

Ara A, Sofi PA, Rather MA, Rashid M, Gull M (2019) Abiotic Stress Tolerance in Legumes – Critical Approaches. Int J Curr Microbiol App Sci 8:1991–2000

Ardebili NO, Saadatmand S, Niknam V, Khavari-Nejad RA (2014) The alleviating effects of selenium and salicylic acid in salinity exposed soybean. Acta Physiol Plant 36:3199–3205

Asseng S, Ewert F, Martre P, Rötter RP, Lobell DB, Cammarano D, Kimball BA, Ottman MJ, Wall GW, White JW, Reynolds MP (2015) Rising temperatures reduce global wheat production. Nat Clim Chang 5:143–147

Atti S, Bonnell R, Smith D, Prasher S (2004) Response of an indeterminate soybean (Glycine Max (L.) Merr) to chronic water deficit during reproductive development under greenhouse conditions. Can Water Resour J 29:209–222

Awari VR, Mate SN (2015) Effect of drought stress on early seedling growth of chickpea (*Cicer arietinum* L.) genotypes. Life Sci Int Res J 2:356–361

Aydinalp C, Marinova S (2009) The effects of heavy metals on seed germination and plant growth on alfalfa plant (*Medicago sativa*). Bulg J Agric Sci 15(4):347–350

Bandeoglu E, Eyidogan F, Yücel M, Öktem HA (2004) Antioxidant responses of shoots and roots of lentil to NaCl-salinity stress. Plant Growth Regul 42:69–77

Bandyopadhyay KK, Misra AK, Ghosh PK, Hati KM (2010) Effect of integrated use of farmyard manure and chemical fertilizers on soil physical properties and productivity of soybean. Soil Tillage Res 110:115–125

Bano S, Ashraf M, Akram NA, Al-Qurainy F (2012) Regulation in some vital physiological attributes and antioxidative defense system in carrot (*Daucus carota* L.) under saline stress. J Appl Bot Food Qual 85:105–115

Barconi D, Bernardini G, Santucci A (2011) Linking protein oxidation to environmental pollutants: redox proteome approaches. J Proteom 74:2324–2337

Baroowa B, Gogoi N (2013) Biochemical changes in two *Vigna* spp. during drought and subsequent recovery. Indian J Plant Physiol 18:319–325

Basheer AA (2018) Chemical chiral pollution: impact on the society and science and need of the regulations in the 21st century. Chirality 30(4):402–406

Basirirad H (2000) Kinetics of nutrient uptake by roots: responses to global change. New Phytol 147:155–169

Baxter A, Mittler R, Suzuki N (2013) ROS as key players in plant stress signalling. J Exp Bot 65(5):1229–1240

Beebe S, Rao IM, Mukankusi C, Buruchara R (2013) Improving resource use efficiency and reducing risk of common bean production in Africa, Latin America and the Caribbean. In: Eco-efficiency: from vision to reality, p 117

Bellaloui N, Mengistu A (2015) Effects of Boron Nutrition and Water Stress on Nitrogen Fixation, Seed δ^{15}N and δ^{13}C Dynamics, and Seed Composition in Soybean Cultivars Differing in Maturities. Sci World J Article ID 407872, 11 p. http://dx.doi.org/10.1155/2015/407872

Bellaloui N, Hu Y, Mengistu A, Kassem MA, Abel CA (2013) Effects of foliar boron application on seed composition, cell wall boron, and seed δ(15)N and δ(13)C isotopes in water-stressed soybean plants. Front Plant Sci 4:270. https://doi.org/10.3389/fpls.2013.00270

Bhat JA, Shivaraj SM, Singh P, Navadagi DB, Tripathi DK, Dash PK, Solanke AU, Sonah H, Deshmukh R (2019) Role of silicon in mitigation of heavy metal stresses in crop plants. Plants 8(3):71. https://doi.org/10.3390/plants8030071

Broos K, Macdonald LM, Warne MSJ, Heemsbergen DA, Barnes MB, Bell M, Mc Laughlin MJ (2007) Limitations of soil microbial biomass carbon as an indicator of soil pollution in the field. Soil Biol Biochem 39:2693–2695

Bybordi A (2015) Influence of exogenous application of silicon and potassium on physiological responses, yield, and yield components of saltstressed wheat. Commun Soil Sci Plant Anal 46:109–122

Cakmak I, Pfeiffer WH, McClafferty B (2010) Biofortification of durum wheat with zinc and iron. Cereal Chem 87:10–20

Cartes P, Jara A, Pinilla L, Rosas A, Mora M (2010) Selenium improves the antioxidant ability against aluminium-induced oxidative stress in ryegrass roots. Ann Appl Biol 156:297–307

Celik S, Yucel E, Celik S, Gucel S, Ozturk M (2010) Carolina poplar (Populus × canadensis Moench) as a biomonitor of trace elements in the West Black Sea region of Turkey. J Environ Biol 31(Special Issue):225–232

Challinor AJ, Watson J, Lobell DB, Smith DR, Howden SM, Chhetri N (2014) A meta-analysis of crop yield under climate change and adaptation. Nat Clim Change 4:287–291

Chowdhury JA, Karim MA, Khaliq QA, Ahmed AU, Khan MSA (2016) Effect of drought stress on gas exchange characteristics of four soybean genotypes. Bangladesh J Agric Res 41:195–205

Cordovilla MP, Ocana A, Ligero F, Lluch C (1994) The effect of salinity on N fixation and assimilation in *Vicia faba*. J Exp Bot 45:1483–1488

Dağhan H, Ozturk M (2015) Soil pollution in Turkey and remediation methods. In: Hakeem K, Sabir M, Ozturk M, Mermut A (eds) Soil remediation and plants: prospects and challenges. Elsevier, New York, pp 287–312

De Souza PI, Egli DB, Bruening WP (1997) Water stress during seed filling and leaf senescence in soybean. Agron J 89:807–812

Deinlein U, Stephan AB, Horie T, Luo W, Xu G, Schroeder JI (2014) Plant salt-tolerance mechanisms. Trends Plant Sci 19(6):371–379 del Rio LA, Sandalio LM, Corpas FJ, Palma JM, Barroso JB (2006) Reactive oxygen species and reactive nitrogen species in peroxisomes. Production, scavenging, and role in cell signaling. Plant Physiol 14:330–335

Delgado MJ, Ligero F, Lluch C (1994) Effects of salt stress on growth and nitrogen fixation by pea, faba bean, common bean, and soybean plants. Soil Biol Biochem 26:371–376

Edmeades DC (2003) The long-term effects of manures and fertilizers on soil productivity and quality: a review. Nutr Cycl Agroecosyst 66:165–180

El Sayed HESA (2011) Influence of NaCl and Na_2SO_4 treatments on growth development of broad bean (*Vicia faba* L.) plant. J Life Sci 5:513–523

Fageria NK, Baligar VC, Clark RB (2002) Micronutrients in crop production. Adv Agron 77:185–268

Fahad S, Hussain S, Saud S, Khan F, Hassan S, Nasim W, Arif M, Wang F, Huang J (2016) Exogenously applied plant growth regulators affect heat-stressed rice pollens. J Agron Crop Sci 202:139–150

Fahad S, Bajwa AA, Nazir U, Anjum SA, Farooq A, Zohaib A, Sadia S, Nasim W, Adkins S, Saud S, Ihsan MZ, Alharby H, Wu C, Wang D, Huang J (2017) Crop production under drought and heat stress: plant responses and management options. Front Plant Sci 8:1147. https://doi.org/10.3389/fpls.2017.01147

FAO (2008) FAO land and plant nutrition management service. Available online link http://www.fao.org/ag/agl/agll/spush/

FAO (2013) Climate-smart agriculture sourcebook. Available online link http://www.fao.org/3/a-i3325e.pdf

Farooq M, Hussain M, Wakeel A, Siddique KH (2015) Salt stress in maize: effects, resistance mechanisms, and management. A review. Agron Sustain Develop 35:461–481

Farooq M, Gogoi N, Barthakur S, Baroowa B, Bharadwaj N, Alghamdi SS (2016) Drought stress in grain legumes during reproduction and grain filling. J Agron Crop Sci 203:81–102

Farooq M, Gogoi N, Hussain M, Barthakur S, Paul S, Bharadwaj N, Migdadi HM, Alghamdi SS, Siddique KH (2017) Effects, tolerance mechanisms and management of salt stress in grain legumes. Plant Physiol Biochem 118:199–217

Ferri A, Lluch C, Caana A (2000) Effect of salt stress on carbon metabolism and bacteriod respiration in root nodules of common bean (*Phaseolus vulgaries* L.). Plant Biol 2:396–402. https://doi.org/10.1055/s-2000-5956

Flowers TJ (2004) Improving crop salt tolerance. J Exp Bot 55:307–319. https://doi.org/10.1093/jxb/erh003

Frehner M, Sangakkara UR, Nosberger J (1999) Effect of soil moisture and potassium fertilizer on shoot water potential, photosynthesis and partitioning of carbon in mungbean and cowpea. J Agron Crop Sci 185:201–207

Garcia-Mata C, Lamattina L (2001) Nitric oxide induces stomatal closure and enhances the adaptive plant responses against drought stress. Plant Physiol 126:1196–1204. https://doi.org/10.1104/pp.126.3.1196

Garg N, Chandel S (2011) Effect of mycorrhizal inoculation on growth, nitrogen fixation, and nutrient uptake in Cicer arietinum (L.) under salt stress. Turkish J Agric For 35:205–214

Garg N, Manchanda G (2008) Effect of arbuscular mycorrhizal inoculation of salt-induced nodule senescence in *Cajanus cajan* L. (pigeonpea). J Plant Growth Regul 27:115–124

Garg N, Singla P (2016) Stimulation of nitrogen fixation and trehalose biosynthesis by naringenin (Nar) and arbuscular mycorrhiza (AM) in chickpea under salinity stress. J Plant Growth Regul 80(1):5–22

Ghassemi-Golezani K, Taifeh-Noori M, Oustan S, Moghaddam M, Seyyed-Rahmani S (2010) Oil and protein accumulation in soybean grains under salinity stress. Not Sci Biol 2:64–67. https://doi.org/10.15835/nsb.2.2.4590

Ghassemi-golezani K, Ghassemi S, Bandehhagh A (2013) Effects of water supply onfield performance of chickpea (*Cicer arietinum* L.) cultivars. Int J Agron Plant Prod 4:94–97

Ghori NH, Ghori T, Hayat MQ, Imadi SR, Gul A, Altay V, Ozturk M (2019) Heavy metal stress and responses in plants. Int J Environ Sci Technol 16(3):1807–1828

Gill SS, Khan NA, Anjum NA, Tuteja N (2011) Amelioration of cadmium stress in crop plants by nutrients management: morphological, physiological and biochemical aspects. Plant Stress 5:1–23

Grams TEE, Koziolek C, Lautner S, Matyssek R (2007) Distinct roles of electric and hydraulic signals on the reaction of leaf gas exchange upon re-irrigation in *Zea mays* L. Plant, Cell Environ 30:79–84

Grattan SR, Grieve CM (1999) Salinity-mineral nutrient relations in horticultural crops. Sci Hort 78:127–157

Gupta VK, Ali I (2012) Environmental water: advances in treatment, remediation and recycling. Elsevier, Dordrecht

Hakeem KR, Sabir M, Ozturk M, Mermut A (2015) Soil remediation and plants: prospects and challenges. Elsevier, London, p 724

Hansel FD, Amado TJC, Diaz DAR, Rosso LHM, Nicoloso FT, Schorr M (2017) Phosphorus fertilizer placement and tillage affect soybean root growth and drought tolerance. Agron J109:2936–2944

Hao X, Li P, Feng Y, Han X, Gao J, Lin E, Han Y (2013) Effects of fully open-air [CO2] elevation on leaf photosynthesis and ultrastructure of *Isatis indigotica*. Fort PLoS ONE 8:e74600

Haribabu TE, Sudha PN (2011) Effects of heavy metals copper and cadmium exposure on the antioxidants properties of the plant cleome gynandra. Int J Plant Anim Environ Sci 1:80–87

Haro R, Bañuelos MA, Rodríguez-Navarro A (2010) High-affinity sodium uptake in land plants. Plant Cell Physiol 51(1):68–79

Hassan, ZU, Ali S, Rizwan M, Hussain A, Akbar Z, Rasool N, Abbas F (2017) Role of zinc in alleviating heavy metal stress. In: Essential plant nutrients 2017. Springer, Cham, pp 351–366. https://doi.org/10.1007/978-3-319-58841-4_14

Hatfield DL, Tsuji PA, Carlson BA, Gladyshev VN (2014) Selenium and selenocysteine: roles in cancer, health, and development. Trends Biochem Sci 39:112–120. https://doi.org/10.1016/j.tibs.2013.12.007

Hati KM, Mandal KG, Misra AK, Ghosh PK, Bandyopadhyay KK (2006) Effect of inorganic fertilizer and farmyard manure on soil physical properties, root distribution, and water-use efficiency of soybean in Vertisols of central India. Bioresour Technol 97:2182–2188

Hattori T, Inanaga S, Araki H, An P, Morita S, Luxová M, Lux A (2005) Application of silicon enhanced drought tolerance in Sorghum bicolor. Physiol Plant 123:459–466

Hawkesford MJ (2012) Improving nutrient use efficiency in crop. In eLS. Wiley, Chichester, UK

Hawkesford MJ (2014) Reducing the reliance on nitrogen fertilizer for wheat production. J Cereal Sci 59:276–283

Hawrylak-Nowak B, Matraszek R, Szymanska M (2010) Selenium modifies the effect of short-term chilling stress on cucumber plants. Biol Trace Elem Res 138:307. https://doi.org/10.1007/s12011-010-8613-5

Heatherly LG (1993) Drought stress and irrigation effects on germination of harvested soybean seed. Crop Sci 33:777–781

Heckman JR, Angle JS, Chaney RL (1987) Residual effects of sewage sludge on soybean: II. Accumulation of soil and symbiotically fixed nitrogen. J Environ Qual 16:117–124

Hermans C, Verbruggen N (2005) Physiological characterization of Mg deficiency in Arabidopsis thaliana. J Exp Bot 418:2153–2161

Hooda PS (2010) Trace elements in soils. Wiley, Chichester

Hossain MA, Piyatida P, Silva JAT, Fujita M (2012) Molecular mechanism of heavy metal toxicity and tolerance in plants: central role of glutathione in detoxification of reactive oxygen species and methylglyoxal and in heavy metal chelation. J Bot 2012:1–37

Huang ZA, Jiang DA, Yang Y, Sun JW, Jin SH (2004) Effects of nitrogen deficiency on gas exchange, chlorophyll fluorescence, and antioxidant enzymes in leaves of rice plants. Photosynthetica 42:357–364

Huang B, Rachmilevitch S, Xu J (2012) Root carbon and protein metabolism associated with heat tolerance. J Exp Bot 63:3455–3465. https://doi.org/10.1093/jxb/ers003

IPCC (2014) Climate change 2014: synthesis report. In: Proceedings of the contribution of working groups I, II and III to the fifth assessment report of the intergovernmental panel on climate change, eds core writing team, pachauri RK., and Meyer LA, Geneva, IPCC, p 151

Jalmi SK, Bhagat PK, Verma D, Noryang S, Tayyeba S, Singh K, Sharma D, Sinha AK (2018) Traversing the links between heavy metal stress and plant signaling. Front Plant Sci 9:12. https://doi.org/10.3389/fpls.2018.00012

Jaynes WF, JrPA Moore, Miller DM (1999) Solubility and ion activity products of calcium phosphate minerals. J Environ Qual 28(2):530–536

Jha S, Sehgal VK, Rao YVS (2015) Effect of sowing direction and crop geometry on water use efficiency and productivity of Indian mustard (*Brassica juncea* L.) in semi-arid region of India. J Oilseed Brassica 6:257–264

Jiang Y, Huang B (2001) Plants and the environment. Effects of calcium on antioxidant activities and water relations associated with heat tolerance in two cool-season grasses. J Exp Bot 52:341–349. https://doi.org/10.1093/jexbot/52.355.341

Jin J, Wang G, Liu X, Pan X, Herbert SJ, Tang C (2006) Interaction between phosphorus nutrition and drought on grain yield, and assimilation of phosphorus and nitrogen in two soybean cultivars differing in protein concentration in grains. J Plant Nutr 29:1433–1449

Johnson YK, Ayuke FO, Kinama JM, Sijali IV (2018) Effects of tillage practices on water use efficiency and yield of different effects of tillage practices on water use efficiency and yield of different drought tolerant common bean varieties in machakos county, Eastern Kenya. Am Sci Res J Eng Technol Sci 40:217–234

Kabata-Pendias A (2011) Trace elements in soils and plants, 4th edn. Chemical Rubber Company Press, Boca Raton

Kabir ME, Karim MA, Azad MAK (2004) Effect of potassium on salinity tolerance of mungbean (*Vigna radiata* L. Wilczek). J Biol Sci 4:103–110

Kafi MO, Goldani MO, Jafari MH (2012) Effectiveness of nutrient management in managing saline agro-ecosystems: a case study of Lens culinaris Medik. Pak J Bot 44:269–274

Kalaivanan D, Ganeshamurthy AN (2016) Mechanisms of heavy metal toxicity in plants. In: Abiotic stress physiology of horticultural crops. Springer, Berlin, Germany, pp 85–102

Kardoni F, Mosavi SS, Parande S, Torbaghan ME (2013) Effect of salinity stress and silicon application on yield and component yield of faba bean (*Vicia faba*). Int J Agric Crop Sci (IJACS) 6:814–818

Kaur S, Kaur N, Siddique KHM, Nayyar H (2016) Beneficial elements for agricultural crops and their functional relevance in defence against stresses. Arch Agron Soil Sci 62:905–920

Khan NA, Singh S (2008) Abiotic stress and plant responses. IK International, New Delhi

Khan MS, Zaidi A, Wani PA, Oves M (2009) Role of plant growth promoting rhizobacteria in the remediation of metal contaminated soils. Environ Chem Lett 7:1–19

Khan MM, Al-Mas'oudi RSM, Al-Said F, Khan I (2013) Salinity effects on growth, electrolyte leakage, chlorophyll content lipid peroxidation in cucumber (*Cucumis sativus* L.). Int Conf Food Agric Sci IPCBEE 55 (2013). https://doi.org/10.7763/ipcbee.2013.V55.6

Khan HA, Siddique KHM, Colmer TD (2016a) Vegetative and reproductive growth of salt-stressed chickpea are carbon-limited: sucrose infusion at the reproductive stage improves salt tolerance. J Exp Bot. https://doi.org/10.1093/jxb/erw177

Khan HA, Siddique KHM, Colmer TD (2016b) Salt sensitivity in chickpea is determined by sodium toxicity. Planta 244:623–637. https://doi.org/10.1007/s00425-016-2533-3

Kim YH, Khan AL, Kim DH, Lee SY, Kim KM, Waqas M, Jung HY, Shin JH, Kim JG, Lee IJ (2014) Silicon mitigates heavy metal stress by regulating P-type heavy metal ATPases, Oryza sativa low silicon genes, and endogenous phytohormones. BMC Plant Biol 14:13

Klimenko S, Peshkova A, Dorofeev N (2006) Nitrate reductase activity during heat shock in winter wheat. J Stress Physiol Biochem 2:50–55

Kolupaev Y, Akinina G, Mokrousov A (2005) Induction of heat tolerance in wheat coleoptiles by calcium ions and its relation to oxidative stress. Russian J Plant Physiol 52:199–204. https://doi.org/10.1007/s11183-005-0030-9

Kopyra M, Gwózdz EA (2003) Nitric oxide stimulates seed germination and counteracts the inhibitory effect of heavy metals and salinity on root growth of Lupinus luteus. Plant Physiol Biochem 41:1011–1017. https://doi.org/10.1016/j.plaphy.2003.09.003

Krujatz F, Haarstrick A, Nörtemann B, Greis T (2011) Assessing the toxic effects of nickel, cadmium and EDTA on growth of the plant growth-promoting rhizobacterium Pseudomonas brassicacearum. Water Air Soil Pollut. https://doi.org/10.1007/s11270-011-0944-0

Krzesłowska M (2011) The cell wall in plant cell response to trace metals: polysaccharide remodeling and its role in defense strategy. Acta Physiol Plant 33(1):35–51

Kurdali F, Al-chammaa M, Mouasess A (2016) Growth and nitrogen fixation in silicon and/or potassium fed chickpeas grown under drought and well-watered conditions. J Stress Physiol Biochem 9:385–406

Kurdali F, Al-Chammaa M, Al-Ain F (2019) Growth andN_2 fixation in saline and/or water stressed Sesbania aculeate plants in response to silicon application. Silicon 11:781–788

Lasat MM (2002) Phytoextraction of toxic metal: a review of biological mechanisms. J Environ Qual 31:109–120

Latef AAHA (2011) Ameliorative effect of calcium chloride on growth, antioxidant enzymes, protein patterns and some metabolic activities of canola (Brassica napus L.) under seawater stress. J Plant Nutr 34:1303–1320

Leport L, Turner NC, Davies SL, Siddique KHM (2006) Variation in pod production and abortion among chickpea cultivars under terminal drought. Eur J Agron 24:236–246

Li JG, Jin SL, Chen YQ, Lin GL, Han XR, Li TQ, Yang XE, Zhu E (2007) Effects of nitrogen fertilizer on the root morphology and cadmium accumulation in low cadmium treatment Sedum alfredii Hance. Chin Agric Sci Bull 23:260–265 (In Chinese)

Li P, Zhang Y, Wu X, Liu Y (2018) Drought stress impact on leaf proteome variations of faba bean in the Qinghai-Tibet Plateau of China. Biotech 8(3):110. https://doi.org/10.1007/s13205-018-1088-3

Liang Y, Sun W, Zhu Y-G, Christie P (2007) Mechanisms of silicon-mediated alleviation of abiotic stresses in higher plants: a review. Environ Pollut 147:422–428

Liu F, Jensen CR, Andersen MN (2004) Pod set related to photosynthetic rate and endogenous ABA in soybeans subjected to di_erent water regimes and exogenous ABA and BA at early reproductive stages. Ann Bot 94:405–411

Lopez M, Herrera-Cervera JA, Iribarne C, Tejera NA, Lluch C (2008) Growth and nitrogen fixation in *Lotus japonicus* and *Medicago truncatula* under NaCl stress: nodule carbon metabolism. J Plant Physiol 165:641–650

López-Arredondo DL, Leyva-González MA, Alatorre-Cobos F, Herrera-Estrella L (2013) Biotechnology of nutrient uptake and assimilation in plants. Int J Develop Biol 57:595–610

Luo GZ, Wang YJ, Xie ZM, Gai JY, Zhang JS, Chen SY (2006) The putative ser/thr protein kinase gene GmAAPK from soybean is regulated by abiotic stress. J Integrat Plant Biol 48:327–333. https://doi.org/10.1111/j.1744-7909.2006.00169.x

Ma QY, Traina SJ, Logan TJ, Ryan JA (1993) In situ lead immobilization by apatite. Environ Sci Technol 27(9):1803–1810

Mak M, Babla M, Xu SC, O'Carrigan A, Liu XH, Gong YM, Holford P, Chen ZH (2014) Leaf mesophyll K^+, H^+ and Ca^{2+}fluxes are involved in drought-induced decrease in photosynthesis and stomatal closure in soybean. Environ Exp Bot 98:1–12

Makbul S, Güler NS, Durmu SN, Güven S (2011) Changes in anatomical and physiological parameters of soybean under drought stress. Turk J Bot 35:369–377

Mali M, Aery NC (2008) Silicon effects on nodule growth, dry-matter production, and mineral nutrition of cowpea (*Vigna unguiculata* L.). J Plant Nutr Soil Sci 171:835–840

Matamoros MA, Baird LM, Escuredo PR, Dalton DA, Minchin FR, Iturbe-Ormaetxe I, Rubio MC, Moran JF, Gordon AJ, Becana M (1999) Stress induced legume root nodule senescence: physiological, biochemical and structural alterations. Plant Physiol 121:97–111

Meloni DA, Gulotta MR, Martinez CA, Oliva MA (2004) The effects of salt stress on growth, nitrate reduction and proline and glycinebetaine accumulation in Prosopis alba. Braz J Plant Physiol 16(1):39–46

Mengel K, Kirkby EA (2001) Principles of Plant Nutrition, 5th edn. Kluwer Academic Publishers, Dordrecht

Mishra S, Panda SK, Sahoo L (2014) Transgenic Asiatic grain legumes for salt tolerance and functional genomics. Rev Agric Sci 2:21–36. https://doi.org/10.7831/ras.2.21

Mittler R, Vanderauwera S, Gollery M, Van Breusegem F (2004) Reactive oxygen gene network of plants. Trends Plant Sci 9(10):490–498

Moftah JM (2000) Physiological response of lead polluted tomato and egg plant to the antioxidant ethylene diurea. Menufiya Agric Res 25:933–955

Mohammadi A, Habibi D, Rohami M, Mafakheri S (2011) Effect of drought stress on antioxidant enzymes activity of some chickpea cultivars. Am-Eurasian J Agric Environ Sci 11:782–785

Molassiotis A, Sotiropoulos T, Tanou G, Diamantidis G, Therios I (2006) Boron-induced oxidative damage and antioxidant and nucleolytic responses in shoot tips culture of the apple rootstock EM9 (*Malus domestica* Borkh). Environ Exp Bot 56:54–62

Mondal MM, Fakir MS, Juraimi AS, Hakim MA, Islam MM, Shamsuddoha AT (2011) Effects of flowering behavior and pod maturity synchrony on yield of mungbean (*Vigna radiata* (L.) Wilczek). Aust J Crop Sci 5:945–953

Munns R (2002) Comparative physiology of salt and water stress. Plant, Cell Environ 25:239–250

Munns R (2005) Genes and salt tolerance: bringing them together. New Phytol 167(3):645–663

Munns R (2011) Plant adaptations to salt and water stress: differences and commonalities. In: Advances in botanical research 2011 Jan 1 (vol 57, pp 1–32). Academic Press

Mupangwa W, Maize I, Twomlow S, Walker S, Africa S, Hove L (2007) Effect of minimum tillage and mulching on maize (*Zea mays* L.) yield and water content of clayey and sandy soils. Phys Chem Earth 32:1127–1134

Murakeözy ÉP, Nagy Z, Duhazé C, Bouchereau A, Tuba Z (2003) Seasonal changes in the levels of compatible osmolytes in three halophytic species of inland saline vegetation in Hungary. J Plant Physiol 160(4):395–401

Mutava RN, Prince SJ, Syed NH, Song L, Valliyodan B, Chen W, Nguyen HT (2014) Understanding abiotic stress tolerance mechanisms in soybean: a comparative evaluation of soybean response to drought and flooding stress. Plant Physiol Biochem 86:109–120. https://doi.org/10.1016/j.plaphy.2014.11.010

Nadeem M, Li J, Wang M, Shah L, Lu S, Wang X, Ma C (2018) Unraveling field crops sensitivity to heat stress: Mechanisms, approaches, and future prospects. Agron 8:128. https://doi.org/10.3390/agronomy8070128

Nadeem M, Li J, Yahya M, Wang M, Ali A, Cheng A, Wang X, Ma C (2019) Grain legumes and fear of salt stress: Focus on mechanisms and management strategies. Int J Mol Sci 20:799. https://doi.org/10.3390/ijms20040799

Navrot N, Rouhier N, Gelhaye E, Jaquot JP (2007) Reactive oxygen species generation and antioxidant systems in plant mitochondria. Physiol Plant 129:185–195

Negrão S, Schmöckel SM, Tester M (2017) Evaluating physiological responses of plants to salinity stress. Ann Bot 119:1–11

Nriagu JO (1984) Formation and stability of base metal phosphates in soils and sediments. In: Nriagu JO, Moore PB (eds) Phosphate minerals. Springer-Verlag, London, UK

Nzihou A, Sharrock P (2010) Role of phosphate in the remediation and reuse of heavy metal polluted wastes and sites. Waste and Biomass Valorization 1(1):163–174

Opdenakker K, Remans T, Vangronsveld J, Cuypers A (2012) Mitogen-activated protein (MAP) kinases in plant metal stress: regulation and responses in comparison to other biotic and abiotic stresses. IntJ Mol Sci 13(6):7828–7853

Ozturk M, Yucel E, Gucel S, Sakcali S, Aksoy A (2008) Plants as biomonitors of trace elements pollution in soil. In: Prasad MNV (ed) Trace elements: environmental contamination, nutritional benefits and health implications. Wiley, New York, pp 723–744

Ozturk M, Ashraf M, Aksoy A, Ahmad MSA (eds) (2015a) Phytoremediation for green energy. Springer, New York

Ozturk M, Ashraf M, Aksoy A, Ahmad MSA (eds) (2015b) Plants, pollutants and remediation. Springer, New York

Ozturk M, Altay V, Karahan F (2017) Studies on trace elements in Glycyrrhiza taxa distributed in Hatay-Turkey. Int J Plant Environ 3(2):01–07

Palta JA, Kumari S, Spring C, Turner NC (2005) Foliar nitrogen applications increase the seed yield and protein content in chickpea (*Cicer arietinum* L.) subject to terminal drought. Aust J Agric Res 56:105–112

Paudyal SP, Aryal RR, Chauhan SVS, Maheshwari DK (2007) Effect of heavy metals on growth of Rhizobium strains and symbiotic efficiency of two species of tropical legumes. Sci World 5:27–32

Pitann B, Kranz T, Zorb C, Walter A, Schurr U, Mühling KH (2011) Apoplastic pH and growth in expanding leaves of *Vicia faba* under salinity. Environ Exp Bot 74:31–36. https://doi.org/10.1016/j.envexpbot.2011.04.015

Pitzschke A, Forzani C, Hirt H (2006) Reactive oxygen species signaling in plants. Antioxid Redox Signal 8(9–10):1757–1764

Porter JR (2005) Rising temperatures are likely to reduce crop yields. Nature 436:174. https://doi.org/10.1038/436174b

Qureshi MI, Muneer S, Bashir H, Ahmad J, Iqbal M (2010) Nodule physiology and proteomics of stressed legumes. Adv Bot Res 56:1–48. https://doi.org/10.1016/B978-0-12-381518-7.00001-7

Rabie GH, Almadini AM (2005) Role of bioinoculants in development of salt tolerance of *Vicia faba* plants under salinity stress. Afr J Biotechnol 4:210–222

Rainey K, Griffiths P (2005) Evaluation of Phaseolus acutifolius A. Gray plant introductions under high temperatures in a controlled environment. Genet Resour Crop Evol 52:117–120. https://doi.org/10.1007/s10722-004-1811-2

Rao IM, Beebe SE, Polania J, Ricaurte J, Cajiao C, Garcia R, Rivera M (2013) Can tepary bean be a model for improvement of drought resistance in common bean? Afr Crop Sci J 21:265–281

Rascio N, Navari-Izzo F (2011) Heavy metal hyperaccumulating plants: how and why do they do it? And what makes them so interesting? Plant Sci 180(2):169–181

Rejili M, Vadel AM, Guetet A, Neffatti M (2007) Effect of NaCl on the growth and the ionic balance K^+/Na^+ of two populations of *Lotus reticus* (L.) (Papilionaceae). South Afri J Bot 73:623–631

Rengel Z, Damon PM (2008) Crops and genotypes differ in efficiency of potassium uptake and use. Physiol Plant 133:624–636

Rengel Z, Marschner P (2005) Nutrient availability and management in the rhizosphere: exploiting genotypic differences. New Phytol 168:305–312

Rennenberg H, Loreto F, Polle A, Brilli F, Fares S, Beniwal RS, Gessler A (2006) Physiological responses of forest trees to heat and drought. Plant Biol 8:556–571. https://doi.org/10.1055/s-2006-924084

Rubiales D, Mikic A (2015) Introduction: legumes in sustainable agriculture. CRC Crit Rev Plant Sci 34:2–3

Sadiki M, Rabih K (2001) Selection of chickpea (*Cicer arietinum* L) for yield and symbiotic nitrogen fixation ability under salt stress. Agronomie 21:659–666. https://doi.org/10.1051/agro:2001158

Sakamoto A, Murata N (2002) The role of glycine betaine in the protection of plants from stress: clues from transgenic plants. Plant, Cell Environ 25:163–171

Samarah NH, Haddad N, Alqudah AM (2009) Yield potential evaluation in chickpea genotypes under late terminal drought in relation to the length of reproductive stage. Ital J Agron 3:111–117

Sehgal A, Sita K, Siddique KH, Kumar R, Bhogireddy S, Varshney RK, HanumanthaRao B, Nair RM, Prasad PV, Nayyar H (2018) Drought or/and heat-stress effects on seed filling in food crops:

impacts on functional biochemistry, seed yields, and nutritional quality. Front Plant Sci 9:1705. https://doi.org/10.3389/fpls.2018.01705

Sehrawat N, Bhat KV, Sairam RK, Jaiwal PK (2013a) Screening of mungbean (*Vigna radiata* L. Wilczek) genotypes for salt tolerance. Int J Plant Anim Environ Sci 4:36–43

Sehrawat N, Bhat KV, Sairam RK, Toomoka N, Kaga A, Shu Y, Jaiwal PK (2013b) Diversity analysis and confirmation of intra-specific hybrids for salt tolerance in mungbean (*Vigna radiata* L. Wilczek). Int J Integrat Biol 14:65–73

Shabala S, Hari Y (2005) Effects of magnesium availability on the activity of plasma membrane ion transporters and light-induced responses from broad bean leaf mesophyll. Planta 221:56–65

Shaban M, Lak M, Hamidvand Y, Nabaty E, Khodaei F, Yarahmadi M, Azimi SM, Rahmat MG, Shaban M, Motlagh ZR (2012) Response of chickpea (*Cicer arietinum* L.) cultivars to integrated application of Zinc nutrient with water stress. Int J Agric Crop Sci 4:1074–1082

Shahid M, Khalid S, Abbas G, Shahid N, Nadeem M, Sabir M, Aslam M, Dumat C (2015) Heavy metal stress and crop productivity. In: Hakeem K et al (eds) Crop production and global environmental issues. Springer, Cham

Shukla R, Sharma YK, Shukla AK (2014) Molecular mechanism of nutrient uptake in plant. Int J Curr Res Acad Rev 2:142–154

Siddique KHM, Johansen C, Alghamdi SS (2012) Innovations in agronomy for food legumes, a review. Agron Sustain Dev 32:45–64

Siddiqui MH, Mohammad F, Khan MN, Al-Whaibi MH, Bahkali AHA (2010) Nitrogen in relation to photosynthetic capacity and accumulation of osmoprotectant and nutrients in Brassica genotypes grown under salt stress. Agric SciChina 5:671–680

Singh M, Kumar J, Singh VP, Prasad SM (2014) Plant tolerance mechanism against salt stress: the nutrient management approach. Biochem Pharmacol (Los Angel) 3:e165. https://doi.org/10.4172/2167-0501.1000e165

Slama I, Ghnaya T, Savouré A, Abdelly C (2008) Combined effects of long-term salinity and soil drying on growth, water relations, nutrient status and proline accumulation of *Sesuvium portulacastrum*. CR Biol 331:442–451

Smertenko A, Draber P, Viklicky V, Opatrny Z (1997) Heat stress affects the organization of microtubules and cell division in Nicotiana tabacum cells. Plant, Cell Environ 20:1534–1542. https://doi.org/10.1046/j.1365-3040.1997.d01-44.x

Sytar O, Kumar A, Latowski D, Kuczynska P, Strzałka K, Prasad MN (2013) Heavy metal-induced oxidative damage, defense reactions, and detoxification mechanisms in plants. Acta Physiol Plant 35(4):985–999

Tahir MA, Aziz T, Rahmatullah (2011) Silicon induced growth and yield enhancement in two wheat genotypes differing in salinity tolerance. CommunSoil Sci Plant Anal 42:395–407

Tavallali V, Karimi S, Espargham O (2018) Boron enhances antioxidative defense in the leaves of salt-affected Pistacia vera seedlings. Hort J 87(1):55–62

Thalooth AT, Tawfik MM, Mohamed HM (2006) A comparative study on the effect of foliar application of zinc, potassium and magnesium on growth, yield and some chemical constituents of mungbean plants grown under water stress conditions. World J Agric Sci 2:37–46

Tuteja N (2007) Mechanisms of high salinity tolerance in plants, Meth. Enzymol.: Osmosens. Osmosignal. 428:419–438

Tuteja N (2010) Cold, salt and drought stress. Plant stress biology: from genomics towards system biology. Wiley, Weinheim, Germany, pp 137–159

Underwood W (2012) The plant cell wall: a dynamic barrier against pathogen invasion. Front Plant Sci 3:85. https://doi.org/10.3389/fpls.2012.00085

Vadez V, Berger JD, Warkentin T, Asseng S, Ratnakumar P, Rao KP, Gaur PM, Munier-Jolain N, Larmure A, Voisin AS, Sharma HC (2012) Adaptation of grain legumes to climate change: a review. Agron Sustain Dev 32:31–44

Valadabadi SA, Shiranirad AH, Farahani HA (2010) Ecophysiological influences of zeolite and selenium on water deficit stress tolerance in different rapeseed cultivars. J Ecol Nat Environ 2:154–159

Vara Parasad PV, Craufurd PQ, Summerfield RJ (1999) Fruit number in relation to pollen production and viability in groundnut exposed to short episodes of heat stress. Ann Bot 84:381–386. https://doi.org/10.1006/anbo.1999.0926

Vollenweider P, Gunthardt-Goerg MS (2005) Diagnosis of abiotic and biotic stress factors using the visible symptoms in foliage. Environ Pollut 137:455–465. https://doi.org/10.1016/j.envpol.2005.01.032

Wahid A, Gelani S, Ashraf M, Foolad MR (2007) Heat tolerance in plants: an overview. Environ Exp Bot 61:199–223. https://doi.org/10.1016/j.envexpbot.2007.05.011

Wani PA, Khan MS (2010) Bacillus species enhance growth parameters of chickpea (*Cicer arietinum* L.) in chromium stressed soils. Food Chem Toxicol 48:3262–3267

Wani PA, Khan MS, Zaidi A (2007) Impact of heavy metal toxicity on plant growth, symbiosis, seed yield and nitrogen and metal uptake in chickpea. Aust J Exp Agric 47:712–720

Wani PA, Khan MS, Zaidi A (2008a) Impact of zinc-tolerant plant growth promoting rhizobacteria on lentil grown in zinc-amended soil. Agron Sustain Dev 28:449–455

Wani PA, Khan MS, Zaidi A (2008b) Chromium-reducing and plant growth-promoting Mesorhizobium improves chickpea growth in chromium-amended soil. Biotechnol Lett 30:159–163

Waraich EA, Ahmad R, Ashraf MY, Saifullah AM (2011) Improving agricultural water use efficiency by nutrient management in crop plants. Acta Agric Scandinavica, Section B—Plant Soil Sci 61(4):291–304

Waraich EA, Ahmad R, Halim A, Aziz T (2012) Alleviation of temperature stress by nutrient management in crop plants: a review. J Soil Sci Plant Nutr 12(2):221–244

Wu Z, Bañuelos GS, Lin ZQ, Liu Y, Yuan L, Yin X, Li M (2015) Biofortification and phytoremediation of selenium in China. Front Plant Sci 6:136. https://doi.org/10.3389/fpls.2015.00136

Wu Z, Yin X, Bañuelos GS, Lin ZQ, Liu Y, Li M, Yuan L (2016a) Indications of selenium protection against cadmium and lead toxicity in oilseed rape (*Brassica napus* L.). Front Plant Sci 7:1875. https://doi.org/10.3389/fpls.2016.01875

Wu Z, Yin X, Bañuelos GS, Lin ZQ, Zhu Z, Liu Y, Yuan L, Li M (2016b) Effect of selenium on control of postharvest gray mold of tomato fruit and the possible mechanisms involved. Front Microbiol 6:1441. https://doi.org/10.3389/fmicb.2015.01441

Xiong L, Zhu J-K (2002) Molecular and genetic aspects of plant responses to osmotic stress. Plant, Cell Environ 25:131–139

Xu J, Yuan Y, Xu Y, Zhang G, Guo X, Wu F, Wang Q, Rong T, Pan G, Cao M, Tang Q (2014) Identification of candidate genes for drought tolerance by whole-genome resequencing in maize. BMC Plant Biol 14:83. https://doi.org/10.1186/1471-2229-14-83

Xu CX, Ma YP, Liu YL (2015) Effects of silicon (Si) on growth, quality and ionic homeostasis of aloe under salt stress. South Afr J Bot 98:26–36

Yadav HD, Yadav OP, Dhankar OP, Oswal MC (1989) Effect of chloride salinity and boron on germination, growth and mineral composition of chickpea (*Cicer arietinum* L.). Ann Arid Zone 28:63–67

Yadavi A, Movahhedi-dehnavi M, Balouchi H (2014) Effect of micronutrients foliar application on grain qualitative characteristics and some physiological traits of bean (*Phaseolus vulgaris* L.) under drought stress. Indian J Fundam Appl Life Sci 4:124–131

Yang XE, Long XX, Ye HB, He ZL, Calvert DV, Stoffella PJ (2004) Cadmium tolerance and hyperaccumulation in a new Zn-hyperaccumulating plant species (*Sedum alfredii* Hance). Plant Soil 259(1–2):181–189

Yang JD, Yun JY, Zhang TH, Zhao HL (2006) Presoaking with nitric oxide donor SNP alleviates heat shock damages in mung bean leaf discs. Bot Stud 47:129–136

Yasin NA, Zaheer MM, Khan WU, Ahmad SR, Ahmad A, Ali A, Akram W (2017) The beneficial role of potassium in Cd-induced stress alleviation and growth improvement in *Gladiolus grandiflora* L. Int J Phytoremed 20:274–283

Yildiz M, Terz H (2013) Effect of NaCl stress on chlorophyll biosynthesis, proline, lipid peroxidation and antioxidative enzymes in leaves of salt-tolerant and salt-sensitive barley cultivars. J Agric Sci 19:79–88

Yuan L, Liu X, Luo M, Yang S, Wu K (2013) Involvement of histone modifications in plant abiotic stress responses. J Integr Plant Biol 55(10):892–901

Zhang Y, Wang L, Liu Y, Zhang Q, Wei Q, Zhang W (2006) Nitric oxide enhances salt tolerance in maize seedlings through increasing activities of proton-pump and Na1/H1 antiport in the tonoplast. Planta 224:545–555

Zhang F, Wan X, Zhong Y (2014) Nitrogen as an important detoxification factor to cadmium stress in poplar plants. J Plant Interact 9(1):249–258

Zhao MG, Tian QY, Zhang WH (2007) Nitric oxide synthase-dependent nitric oxide production is associated with salt tolerance in Arabidopsis. Plant Physiol 144:206–217

Zlatev Z, Lidon FC (2012) An overview on drought induced changes in plant growth, water relations and photosynthesis. Emirates J Food Agric 24:57–72

Zorb C, Senbayramb M, Peiter E (2014) Potassium in agriculture status and perspectives. J Plant Physiol 171:656–669. https://doi.org/10.1016/j.jplph.2013.08.008

Fabaceous Plants Under Abiotic Stresses and Biochemical Functions of Micronutrients

Shyam Narain Pandey

Abstract Fabaceous plants are mostly grown in tropical and subtropical areas of the world. These are rich sources of minerals and proteins to the human beings and animals. They have a specific function nitrogen fixation in their root nodules through symbiotic association with prokaryotic organisms. Thus, they provide staple protein-rich food to the world's increasing population on one hand and contribute soil fertility by providing available nitrogen on the other hand. These processes are vital for maintaining the ecosystem. Due to the rapid fluctuation in environmental variables, Fabaceous plants are facing various abiotic stress problems such as drought, minerals disorders, salinity, temperature, water and losing their growth reproductive yields, and other products of various importances emerge as a great problem and big challenge for their management. Micronutrients such as zinc, iron, copper, manganese, molybdenum, boron, chlorine are essential to normal growth and biomolecular functions of all the plants. Fabaceous plants also require these essential micronutrients for their growth, metabolism, and reproductive yield. The assessment and management of micronutrient status in soil–plants can protect the Fabaceous plants to grow normally and keep their specific function nitrogen fixation, even under abiotic stress conditions. Soils in adverse environmental conditions (drought, salinity, degradation through erosion, etc.) mostly show the deficiency of micronutrients which account for a wide gap in between the production level of the food crops including Fabaceous plants and also necessitate their prompt management.

Keywords Fabaceous plants · Micronutrients · Abiotic stresses · Nitrogen fixation · Biomolecular functions

1 Introduction

Fabaceous plants are mostly pulses also called as grain legumes. The grain legumes belong to the family Fabaceae which include about 670 genera and 20,000 species of herbs, shrubs, and trees. Pulses are growing mostly in tropical and tropical regions of

S. N. Pandey (✉)
Department of Botany, University of Lucknow, Lucknow 226007, India
e-mail: snpandey511@gmail.com

M. Hasanuzzaman et al. (eds.), *The Plant Family Fabaceae*,
https://doi.org/10.1007/978-981-15-4752-2_16

the world. Just after cereals, pulses ranked second as a human staple food; the grains from different Fabaceous plants provide rich protein source to human beings and animals. The major producer of grain legumes are China, India, Mexico, Brazil, Nigeria, and Turkey, and most of pulse producing countries are facing their low production year by year due to climate change and abiotic stresses (Gupta and Pandey 2019). The abiotic stress conditions like drought, salinity, nutrients deficiency are major factors for the low production of pulses (Muszynska and Labudda 2019). The evolved high yielding varieties of cereals over the past decade pushed out the production of pulses. The Fabaceous plants not only the rich source of protein but nodules in their root contain nitrogen-fixing bacteria. The nitrogen-fixing nodules supply sufficient nitrogen to the plants and also contribute it to the soil for good fertility (Marschner 1995; Pandey et al. 2018). Some common Fabaceous plants are pigeon pea (*Cajanus cajan*), chickpea (*Cicer arietinum*), lentil (*Lens esculenta*), broad bean (*Vicia faba*), cowpeas (*Vigna unguiculata*), pea (*Pisum sativum*), soybean (*Glycine max*), common bean (*Phaseolus species*), cluster bean (*Cyamopsis tetragonoloba*) (also used for green manure), groundnut (*Arachis hypogaea*) (also used as oilseeds), etc. Some Fabaceous plants are also used as vegetables. Fabaceous plants show many growths and biomolecular functions disorder under abiotic stress conditions (Pandey and Gautam 2009; Upadhyay et al. 2013). Some other common Fabaceous plants are *Acacia* sp., *Medicago sativa* (Alfalfa), Genus *Trifolium* (clover), *Lespedeza* sp. (bush clover), Cytisus sp. (broom), *Trigonella foenum-graecum* (Fenugreek), *Lablab purpureus* (Hyacinth bean), *Indigofera* sp. (indigo), *Laburnum* sp. (Laburnum), *Lathyrus* sp. Beach pea (*L. japonicas*) and sweet pea (*L. odoratus*), *Haematoxylum campechianum* (logwood), *Prosopis* sp. (Mesquite), *Mimosa* sp., *Delonix regia* (Royal Poinciana), *Senna* sp., Albizia sp. (Silk tree), *Crotalaria juncea* (Sun hemp), *Tamarindus indica* (Tamarind), etc., are useful in various ways. Although several minerals and favorable environmental conditions are required for optimum growth, metabolism, and reproductive yield of Fabaceous plants, micronutrients are very significant for protection under stresses, regulation of biomolecular functions and optimum production of grains (Sharma et al. 1996; Pandey and Gautam 2009; Kannaujia and Pandey 2013). Micronutrients such as iron, copper, zinc, molybdenum, boron, manganese, chlorine are well known for functioning against oxidative stresses (Alscher et al. 2002; Cakmak 2000; Sharma 2006), regulating biomolecular functions (Pandey 2018), in biological nitrogen fixation (Ma et al. 2017; Pandey et al. 2018) and to maintaining reproductive yield (Sharma et al. 1990; Nautiyal and Chatterjee 1999; Pandey et al. 2006). Most of the micronutrients are involved in redox reactions of many enzymatic functions due to their variable oxidation states. Least attention is given to the assessment and management of the micronutrients in soil–plant systems in agriculture. Fabaceous plants (*Vicia faba*, green gram) show reproductive yield due to ultrastructural changes in pollen grains (Pandey et al. 1995). For the achievement of optimum growth, cellular metabolism, reproductive yield, and protection against abiotic stresses adequate micronutrient status can play a significant role in maintaining Fabaceous plants to survive better in adverse conditions (Saikia et al. 2018; Muszynska and Labudda 2019). Mostly, soil conditions inhibited the availability of zinc in the soil for absorption by plants, foliar application of zinc can enhance reproductive efficiency of Fabaceous plants (Pathak et al. 2012). For sustainable production of pulse crops under abiotic stresses, some plant-growth-promoting rhizobacteria and normal critical levels of micronutrients such as iron, zinc, copper, manganese can be applied (Pandey et al. 2018; Saikia et al. 2018). The oxidative damage (Cakmak 2000) and production of ethylene due to abiotic stress in Fabaceous plants can be managed with adequate supply of micronutrients and with regulating enzyme activity in plants (Sharma 2006; Pandey 2018; Gupta and Pandey 2019). The application of micronutrients may alleviate abiotic stresses by protecting the plants through the regulation of biomolecular functions (Tewari et al. 2013; Muszynska and Labudda 2019).

2 Micronutrients

Micronutrients are essential elements along with macronutrients for normal plant growth and to complete their life cycle, and their function is specific and cannot be substituted with other elements. The plants are unable to grow normally in the absence of micro- and macronutrients therefore essential to growth, metabolism, and reproductive yield of plants. Micronutrients are required in traces (<1 ppm), but they are specific to particular cellular functions and vital to plants. The considered essential micronutrients are manganese (Mn), copper (Cu), zinc (Zn), boron (B), chlorine (Cl), molybdenum (Mo), and iron (Fe) which play a very significant role in plants growth and their biochemical activities (Arnon 1954; Sharma 2006). Micronutrients Fe, Mn, Cu, and Zn bear positive ions placed in the first transition series of Period IV. The Mo is of the second transition series of Period V. In addition, nickel and cobalt have been claimed as essential micronutrients (Ahmed and Evans 1960; Hewitt and Bond 1966; Gerendas et al. 1999). Cobalt is essential for nitrogen fixation in Fabaceous plants (Dilworth et al. 1979) and non-Fabaceous plants (Hewitt and Bond 1966). Cobalt plays a role for the activity of B_{12} Coenzyme-dependent ribonucleotide reductase has been observed in Rhizobium species (Cowles et al. 1969). In plants, nickel is present as free ionic form, and as chelate mostly located in phloem and xylem sap, its deficiency causes chlorosis in interveinal areas of monocotyledons (Brown et al. 1987). Micronutrients have variable oxidation states, form complexes with proteins and organic ligands, and form metalloenzymes function as catalyst in biological system (Marschner 1995; Pandey 2018). Some beneficial elements such as aluminum, cobalt, selenium, silicon also improve physiological functions in Fabaceous plants under abiotic stress conditions such as drought, salinity, and nutrient disorders (Pilon-Smits et al. 2009; Gupta and Pandey 2019).

3 Plant Micronutrients and Abiotic Stresses

Fabaceous crops grow in tropical and subtropical areas of the world facing various environmental problems. Under the present fluctuating environmental conditions, these crops are facing a variety of abiotic stresses such as high temperature, salinity, drought, heavy metal pollutants, and nutrients disorder. These stress conditions cause low productivity of Fabaceous plants throughout the world. The pulse grain production has been also decreased due to the biotic stresses induced by abiotic stresses in root zones (Pandey et al. 2018). Since essential nutrients play a very vital role in growth and cellular metabolism of plants (Muszynska and Labudda 2019), their suitable availability to the plants is affected by various soil factors. The low or elevated levels of these nutrients also cause nutrients disorders in the plants (Kannaujiya and Pandey 2013). Some nutrients are specifically important in pulse crops such as nitrogen, iron, copper, zinc to biosynthesis of more protein contents (Sharma 2006; Pandey 2018). Under abiotic stress conditions, Fabaceous plants show disturbed cellular metabolism (Pandey and Verma 2010; Ravichandran and Pathmanabhan2014). Along with nutrient stresses various stress conditions like water, light, temperature, heavy metals affect normal metabolic processes and produce a large number of harmful reactive oxygen species (ROS) such as hydrogen peroxide (H_2O_2), superoxide anion ($O_2^{\cdot -}$), and hydroxyl radical (OH) (Tewari et al. 2013). These highly ROS cause oxidative stress to cellular organelles through lipid peroxidation (Sharma 2006; Apel and Hirt 2004). The stress conditions caused due to micronutrient deficiencies intensify the adverse effects of ROS due to poor cellular defense mechanisms. A large number of micronutrient elements are involved as a constituent or activator of many enzymes involved in the cellular defense system (Pandey et al. 2002; Jaleel et al. 2009; Pandey 2018). Some beneficial elements can help in survival and promotion in the production of Fabaceous plants under abiotic stresses (Pilon-Smits et al. 2009). The nanoparticle size of the trace element application protects plants against abiotic stresses such as drought, salinity, flood, temperature variations (Khan and Upadhyay 2019; Muszynska and Labudda 2019). Fabaceous plants fix atmospheric dinitrogen into available form not only for their own use but also for other plants; at the same time, Fabaceous root nodules require many essential trace elements such as Zn, Mo, Mn, Ni for proper activity of enzymes which involve in nitrogen fixation (Sharma 2006; Pandey et al. 2018). The adequate trace metal content enhances cellular defense system though it acts as activator of many antioxidative enzymes (Pandey 2018; Gupta and Pandey 2019) and biosynthesis of antioxidant biomolecules (Pandey et al. 2002; Saikia et al. 2018). Under iron, stress conditions due to iron deficiency in leaves induce reactive oxygen species generation and cause oxidative stress (Tewari et al. 2005, 2013).

4 Functions of Micronutrients

Majority of micronutrient elements are cofactors or activators of many enzymes. They participate in oxidation–reduction reactions and metal–protein functions as electron carriers. Micronutrients play significant and specific role in electron transport chains in chloroplast and mitochondrial membranes for the generation of ATP (photo and oxidative photophosphorylation) (Sharma 2006; Pandey 2018). In chloroplast, iron and copper are coordinating the flow of electrons from pigment system II to pigment system I during photosynthesis. Copper and iron micronutrients are activators of electron transport protein at mitochondrial inner membrane. The most important role of micronutrients is catalytic and regulates enzyme-catalyzed reactions in cellular metabolism (Mendel and Hansch 2002; Alscher 2000). Copper, zinc, iron, manganese, and molybdenum are the constituents of oxidoreductase enzymes involved in various vital processes in plants such as photosynthesis, respiration, ion uptake, metabolism of minerals, nitrogen fixation, and various biosynthetic pathways (Nautiyal and Chatterjee 1999; Dordas and Brown 2000; Ma et al. 2017).

Zinc and manganese are involved in the activity of hydrolases (Berg and Shi, 1996). Iron, zinc, and manganese are involved in the activity of lyases. Variable oxidation states of manganese are involved in photooxidation of water in initial reaction of light reaction (PS II) during photosynthesis. Also, manganese catalyzes the enzymatic reduction of $N = N$ bond during the conversion of molecular nitrogen (N_2) to ammonia (Sharma 2006; Pandey 2018). Copper and zinc have played to a role in photosynthesis and transpiration observed in safflower (Pandey and Sharma 1989; Pandey and Sharma 1996).

Manganese (Mn^{2+}) activates NAD^+—isocitrate dehydrogenase enzyme which catalyzes the reduction of NAD^+ to NADH in matrix part of mitochondria. Micronutrients are allosteric effectors and influence the activity of many enzymes. The role of some micronutrients is known to regulate the plant–water relationship (Flowers 1988; Xu et al. 2000); particularly, chlorine has a direct role in osmoregulation in plant cells. Chlorine concentration also influences the turgor-dependent extension growth of cells. These processes regulate the closing and opening of stomata. Several micronutrients are involved in biosynthesis of signaling molecules, growth hormones, and secondary metabolites. Micronutrients are also involved in biosynthesis of secondary metabolites through being as cofactors of many enzymes catalyze the biosynthetic reactions of secondary metabolites (Hermann and Weaver, 1999; Gupta and Pandey 2019). Kaurene synthase enzyme catalyzes the reaction of synthesis of gibberellins precursor ent-kaurene activated by Co^{2+} and Mn^{2+}. Gibberellins are major signaling biochemicals in plants (Sun and Gubler 2004). Jasmonic acid biosynthetic enzymes contain iron as a cofactor. Micronutrients are also cofactors of enzymes involved in biosynthesis of abscisic acid and ethylene plant hormone (Prescott and John 1996; Chapple 1998; Saikia et al. 2018). Cobalt is a micronutrient in legumes during symbiotic nitrogen fixation. Boron affects bacteria-legume cell–surface interactions and induces nodulation in the root of leguminous plants (Bolanos et al. 1996). Leguminous plant shows better germination and growth responses with adequate boron in plant (Bell et al. 1989). Under the deficiency of iron, plants have ferric reducing activity as observed in *Phaseolus vulgaris* (Sharma 2006). Molybdenum reserve in sufficient amount in seeds of legumes (*Phaseolus vulgaris*) grown in tropical regions promotes nitrogen fixation and growth of plants. Micronutrients redistribute and accumulate in pods during pod filling in legumes (Caballero et al. 1996). Zinc concentration in Fabaceous plants affects the enzyme's activity for scavenging ROS and hydrogen peroxide oxidants, which affect growth, protein metabolism, and indole-3-acetic acid along with some other phytohormones in Fabaceous plants (Cakmak et al. 1989).

4.1 Photosynthesis

Several micronutrients such as copper, iron, manganese play an essential role in photosynthetic process of plants including Fabaceous plants. Copper is also bound to polypeptide functions in maintaining lipid environment which favors the movement of plastoquinone molecules in pigment system II electron (Maksymiec 1997). The deficiency of copper in safflower decreases photosynthesis (Pandey and Sharma 1996). Due to the important role of copper in chloroplast organization, photosynthetic pigments, and leaf carbon dioxide conductance, deficiency of copper decreases photosynthetic activity (Casimira 1990).

Zinc plays crucial role in photosynthetic process of plants; it is important in the activity of more than 200 enzymes that involve in various biochemical processes in plants (Pandey 2018). A decrease in photosynthetic activity in plants due to zinc deficiency has been observed (Shrotri et al 1981; Pandey and Sharma 1989). Zinc deficiency damages thylakoid membrane due to ROS (Marschner and Cakmak 1989) and abnormalities of chloroplast in bundle sheath in mesophyll cells (Shrotri et al 1978). Zinc-deficient plant shows the inhibition of carbonic anhydrase activity in C_3 and C_4 plants (Hatch and Burnell 1990; Cooper et al 1969). Carbonic anhydrase contains Zn as cofactor which is critical in photosynthesis of C_4 plants. Carbonic anhydrase is localized mainly in cytosol of mesophyll cells site for phosphoenolpyruvate carboxylase (uses bicarbonate as substrate). Carbonic anhydrase catalyzes the production of HCO_3 (Hatch and Burnell 1990) in C_4 plants. Carbonic anhydrase maintains the carbon dioxide concentration at the site of the activity of RuBP carboxylase through the conversion of bicarbonate to carbon dioxide in C_3 plants (Cooper et al. 1969). Sharma et al. (1996) reported the effect of zinc deficiency on carbonic anhydrase activity regulate stomatal opening in Fabaceous plants (*Vicia faba*). Inadequate supply of zinc in Fabaceous and other plants limits the photoassimilates and their availability to biosynthetic process of cellular molecules (Sharma 2006; Pandey 2018). Zinc plays its role in the synthesis of auxin (IAA) an important plant growth hormone (Cakmak et al. 1989).

Molybdenum (Mo) is a constituent of more than thirty enzymes in plants. It catalyzes the redox reactions. The enzyme nitrate reductase and xanthine oxidase contain molybdenum as cofactor involve in biosynthetic pathway of auxins and abscisic acid

(Taylor 1991; Koshiba et al. 1996). Molybdenum is a constituent of nitrogenase, plays a significant role in atmospheric nitrogen fixation in root nodules of Fabaceous plants. Therefore, the Fabaceous plants are very sensitive to molybdenum nutrition. The deficiency of molybdenum in plants adversely affects the growth and metabolism of plants (particularly Fabaceous plants), and also decreases the sugar levels in leaves (Agrawala and Hewitt 1955).

Boron is a constituent of several compounds in plants (Dembitsky et al. 2002). Its role in cross-linking of cell wall polysaccharides has been reported (O'Neill et al. 2004). Boron concentration in plants influences the pentose phosphate pathway (forms an alternative to glycolysis). An increase in the activity of acid phosphatase due to deficiency of boron has been observed (Sharma and Abidi 1986; Agarwala et al. 1991). Chlorine is a component in the structure of oxygen evolution complex, which also contains manganese (in photo system II). Low chlorine content in plants decreases oxygen evolution (Kelley and Izawa 1978). Chlorine deficiency also causes reduced leaf lamina, restricted cell divisions, and reduction of extension growth of cells (Terry 1977; Pandey 2018).

4.1.1 Protective Role Under Abiotic Stresses

Fabaceous plants are facing various stress conditions such as drought, high temperature, salinity, and essential nutrients disorder conditions (Pandey and Gautam 2009). The essential micronutrients zinc, copper, manganese, etc., protect Fabaceous plants during stress conditions through regulating the biomolecular functions (Pandey 2018). Highly reactive oxygen species (ROS) that generate during abiotic stresses cause oxidative stress, and micronutrients play an important role in their detoxifications. Most of the micronutrients behave as a cofactor of several enzymes in cellular antioxidant system that provides protection against ROS damage of cellular membranes. The ROS generates during electron transport system in chloroplast and mitochondria (Elstner 1991; Asada 1994). The first line defensive enzyme against ROS is superoxide dismutase that contains micronutrients as its constituents (as a cofactor of Fe-SOD, Cu Zn SOD, Mn-SOD) activate the action of SOD (Alscher et al. 2002). The main location of Mn-SOD activity is in the mitochondria, whereas Cu-Zn SOD and Fe SOD is localized in chloroplast for converting ROS into H_2O_2. The elevated H_2O_2 concentrations in cells oxidize the lipid membrane, and the H_2O_2 levels control by heme enzymes catalase and ascorbate peroxidase in plants. The ascorbate peroxidase uses ascorbate as an electron donor and reduces H_2O_2 to H_2O (Sharma 2006; Pandey 2018). Catalase rapid breaks down the H_2O_2 and catalase enzyme concentrated in peroxisomes where further H_2O_2 is formed through the conversion of glyoxylate. Therefore, the above functions indicate the significant role of micronutrients iron, zinc, copper, and Mn as a protectant of oxidative stress in cells during abiotic stress conditions (Asada 1997). The deficiency of micronutrients in plants weakens the cellular defense system, plants facing abiotic stress conditions such as high temperature, salinity, nutrient deficiency/excess damaged by ROS (Cakmak 2000). Some micronutrients also behave as osmoprotectant and maintain the structure of cellular membrane under temperature and salinity stresses (Gorham 1995). Iron enhances the production of ROS such as toxic hydroxyl radicals (Halliwell and Gutteridge 1989) and compounds ferryl ($Fe^{2+}O$) or perferryl ($Fe^{2+}O_2$) damage proteins, membrane lipids and nucleic acids in plants. The iron enzyme lipoxygenase causes hydroxylation of linoleic acid and also catalyzes the production of singlet oxygen species ($O_2^{\cdot}$). Thus, elevated iron boosts up the production of ROS in addition to production through abiotic stresses (Sharma 2006). But, also iron protects the plants against ROS and salt stresses through becoming a cofactor of antioxidative enzyme (Fe-SOD). Copper deficiency weakens antioxidant enzyme system due to decrease in copper-zinc SOD activity in plants. Excess heavy metals accumulation in plants induces to form phytochelatins, zinc bind to them from metalloproteins. Metallo-polypeptides contribute tolerance mechanisms against excess heavy metal accumulation in plants. As being a constituent of copper-zinc SOD, zinc plays its role in antioxidant defense mechanism in plants and catalyzes the detoxification of ROS by producing H_2O_2 which can be further detoxifies by micronutrient linked enzymes such as catalase and ascorbate peroxidase (Pandey 2018) during drought, salinity, nutrients disorder conditions, etc., Superoxide ions cause the production of hydroxyl radicals which are more reactive than superoxide ions severally degenerates cellular lipid membranes, various proteins and nucleic acids (Cakmak 2000; Beak et al. 2015). Molybdenum participates in oxidation–reduction reaction as being cofactor of enzyme involves in biological nitrogen fixation (Sigel and Sigel 2002). In the process of nitrate accumulation and dinitrogen fixation by microorganism, other nutrients sulfur, copper, and iron are also involved (Hille 1996; Sigel and Sigel 2002). These processes improve soil fertility and reduce mineral stresses in plants (Pandey et al. 2018; Mendal and Hansch 2002). Plants suffering from abiotic stresses show disturbances in turgor pressure and damages in cell walls and membranes due to high efflux of potassium ions, sucrose, phenolics, and amino acids (Cakmak et al. 1995; Yu et al. 2002). Boron maintains the integrity of cell wall and cellular membrane and maintains the turgidity of cells during abiotic stress conditions (Yu et al. 2003; Pandey 2018). Molybdenum increases fruiting and plants growth in acid soil by promoting pollen fertility (Gubler et al. 1982). Sufficient boron content in plant tissues supports the synthesis of lignin, flavonoids (Carpena et al. 1984), alkaloids, and polyphenols (Carpena et al. 1984; Srivastava et al. 1985). These compounds biosynthesized with boron involvement enhance plant defense against stresses (Sharma 2006). Plants facing abiotic stresses imposed redox imbalance generate ROS and change cell wall structure, and severity increases under boron deficiency. Cultured BY-2 tobacco cells show overexpression of salicylate-inducible glutathione-S-transferase (GST) and glucosyltransferase (ToGT) genes which involve in plant defense system to protect from oxidative stress and maintaining cell wall structure (Kobayashi et al. 2004). Boron deficiency lowers the relative water content in plants (Sharma and Sharma 1987), and proline accumulation is only due to water stress conditions in plants (Sharma and Sharma 1987; Marschner 1995).

4.2 Role in Reproductive Yield

Micronutrients play a crucial role in the reproductive yield of Fabaceous plants (Pandey and Gautam 2009). Several workers are established the involvement of micronutrients in the development of reproductive parts of plants (Sharma et al. 1990; Pandey et al. 2006). The involvement of TFIII A type zinc finger controls the expansion of specific cell in plants (Takatsuji et al. 1992) and plays role in the development of female gametophyte and seed development (Brive et al. 2001). The reproductive yield of plants affected by tissue concentration of micronutrients that controlled by morphological features and plant factors, and they affect the translocation of micronutrients (Brown et al. 2002; Takahashi et al. 2003). The plant's reproductive development such as seed's set and development supported by adequate manganese content (Pandey et al. 2018). Seeds of *Pisum sativum* and beans (Fabaceous plants) exhibit browning symptoms under moderate manganese deficiency. Manganese deficiency induces plants to go in the early reproductive phase (Sharma 2006). Cereals and cotton plants have been observed to show decrease in tillering, and lateral branching due to manganese deficiency (Pandey et al. 2002; Sharma 1992; Otegui et al. 2002) reported the accumulation of manganese-phytate at the chalazal end of the endosperm and its movement toward the endosperms toward the embryo at the late bent cotyledon stage in *Arabidopsis*. Micronutrient copper is also copper which is beneficial to the development of anther (Dell 1981), functional tapetum and prevents dehiscence of anther due to poor lignin synthesis for anther cell walls critical for dehiscence of pollen grains (Dell 1981). Copper affects the maturation of seeds and maintains seed quality through the activity of amylase and invertase during grain filling stage (Nautiyal and Chatterjee 1999; Nautiyal et al. 1999). Zinc in plants plays an important role in the reproductive yield of crops including Fabaceous plants (Pandey et al. 2000). The deficiency of zinc in plants affects the initiation of flowering, gametogenesis (in male and female), zygote formation, and seed development. For the process of microsporogenesis, pollen grain's production, pollen fertility, and seed setting adequate zinc content are required (Sharma et al. 1990). Pollen grains at their tip of the germ tube of Fabaceous plants (*Vicia faba*) contain higher zinc content than other parts of plant (Polar 1970). The reduction in pollen viability and alteration in morphology of exine in Fabaceous plants (*Phaseolus vulgaris* and *Lablab purpureus*) have been reported by Pandey et al. (1995). Zinc deficiency decreases RNA content with increasing activity of RNAase (Wetzel and Jensen 1992) lead changes in morphology of exine. Zinc-deficient Fabaceous plants (such as *Vicia faba*) show decrease in stigmatic size and pollen receptive area of the stigma tip observed under scanning electron microscope (Pandey et al. 1995; Pathak et al. 2012). Zinc-deficient Fabaceous plants take longer time for the maturation of seeds and show changes in seed coat topography possibly inhibit their germination (Sharma et al. 1995; Baylock 1995). Zinc-deficient Fabaceous plant (*Vicia faba*) shows ultrastructural changes on stigma may cause poor pollen fertility (Pandey et al. 1995). Molybdenum (Mo) status in soil and plants leads to low reproductive yield in plants (Pandey 2018). In acid soils, the application of Mo promotes fruiting (Gubler et al. 1982) through beneficial effects on pollen fertility. In Fabaceous plant (chickpea), the deficiency of Mo affects protein quality in seeds and reduces legumins and vicillins and increases albumins (Nautiyal et al. 2005). Low molybdenum in plant tissues reduces seed dormancy and causes pre-harvest sprouting of grains (Cairns and Kurtzinger 1992), as the role of molybdenum as a constituent of aldehyde oxidase catalyzes the abscisic acid (Seo and Koshiba 2002) which suppress premature germination of seeds. Molybdenum content-rich seeds benefit the seedlings growth and seed production of plants cultivated under molybdenum deficient soils (Gurley and Giddens 1969; Chatterjee and Nautiyal 2001). Plants are more sensitive to their reproductive development for boron content than their vegetative parts development. Relatively high concentration of boron in anthers, stigma, and ovary (Loomis and Durst 1992), and abnormalities in floral parts of plants under low boron status have been reported (Xu et al. 1993). Restricted and delayed flowering, pollen sterility, premature bud abscission, poor seed development under boron deficiency indicate boron requirement in the reproductive biology of plants (Adams et al. 1975; Chatterjee and Nautiyal 2001). Reduction in flower-bearing branches and premature bud abscission causes a decrease in number of flowers in plants with boron deficiency (Sharma 2006). Boron involves in the development of cytoskeletal structure preceding pollen germination and pollen tube growth (Li et al. 2001). Boron deficiency enhances the accumulation of phenolic compounds on stigmatic surface which inhibits the germination of pollen grains. The role of boron in seed yield, pollen fertility, seed maturation, and seed germination has been established in Fabaceous and other group of plants (Bell et al. 1989; Sharma 2006; Pandey et al. 2018). Abiotic stress conditions such as temperature, humidity, water, light, and salinity affect boron availability to Fabaceous plants at the critical stages of reproductive development (Bell et al. 1989; Sharma 2006).

5 Nitrogen Fixation in Fabaceous Plants

Essential micronutrients such as molybdenum, manganese, iron, nickel, cobalt, zinc play a very significant role in nitrogen fixation in leguminous plants. Iron is an important component in protein leghaemoglobin, which is oxygen-binding protein found in root nodules of Fabaceous plants (Eide et al. 1996). During the nitrogen fixation of molecular nitrogen, nitrogenase activity is very important; nitrogenase is inhibited by oxygen content in nodules. Thus, iron-containing leghaemoglobin binds the oxygen and prevents the activity of nitrogenase for normal nitrogen fixation process in root nodules of Fabaceous plants (Pandey et al. 2018). The nodulating Fabaceous plants need sufficient level of iron content for the nitrogen fixation processes (Udavardi and Day 1997). Because iron is an important constituent of the nitrogenase complex, ferredoxin and leghaemoglobin involve in molecular atmospheric nitrogen fixation. Abiotic stresses such as salinity and sodicity, low or very high temperature, with very low water content affect the availability of iron to the plants, and Fabaceous plants show poor nitrogen fixation under these abiotic stress conditions (Sharma 2006; Pandey 2018). Low iron content in Fabaceous plants shows poor nitrogen fixation and nodulation (Tang et al. 1992). A high accumulation of ureide due to decrease in the activity of allantoate amidohydrolase under manganese deficiency reduces nitrogen fixation in Fabaceous plants. Manganese content enhances the degradation of ureide and nitrogen fixation in soybean under low water stress conditions (Purcell et al. 2000). Zinc and copper protect Fabaceous plants from oxidative stresses such as being constituent of antioxidative enzymes and prevent nitrogen fixation and nodule formation (Sharma 2006; Pandey 2018). Molybdenum micronutrient plays an essential role in Fabaceous plants for dinitrogen fixation by microorganisms (Agarwala and Hewitt 1995). Nitrogenase is a prokaryotic enzyme which is a complex of enzymes dinitrogenase (Molybdenum-iron protein) and dinitrogenase reductase (Fe-protein). Symbiotic bacteria require FeMoCo (iron-molybdenum cofactor) for their nitrogenase activity. The single 4Fe–4S cluster in dinitrogenase reductase provides high reducing power to dinitrogen. Biological fixation of nitrogen needs an input of eight high potential electrons and eight protons for the reduction of one molecule of nitrogen to ammonia ($N_2 \rightarrow NH_3$) (Sharma 2006).

$$N_2 + 16\ ATP + 8e^- + 8H^+ \rightarrow 2\ NH_3 + H_2O + 16\ ADP + 16\ P$$

Some Fabaceous plants such as soybean and cowpea fix atmospheric nitrogen in root nodules and reach to the aerial parts as ureides. Boron is involved in ureide's catabolism via the allantoate amidohydrolase pathway. Boron-deficient plants show decrease in the activity of allantoate amidohydrolase which catalyzes the oxidative decarboxylation of allantoate to uredoglycine, with concomitant accumulation of allantoate (Sharma 2006). Therefore, the specific function of Fabaceous plants is nitrogen fixation that must require a large number of micronutrients to complete the process normally (Marschner 1995).

6 Conclusions

Pulse crops (grain legumes) belong to the family Fabaceae are rich sources of nitrogen and proteins and contribute in available form of nitrogen into the soil through atmospheric N_2 fixation (dinitrogen) in symbiotic associations with prokaryotes in their root nodules. Mostly, Fabaceous crops are grown in tropical and subtropical regions of the world facing a lot of abiotic stress problems. Drought, salinity temperature fluctuations, nutrient disorders, and several other abiotic stresses are lowering the growth, metabolism, and production of Fabaceous plants. Micronutrients, along with other essential nutrients, play a very significant role in maintaining and strengthening the growth, reproductive yields, and internal defense system of the Fabaceous plants during various abiotic stress conditions. Various enzymatic activities and biomolecular functions involve in the nitrogen fixation process, nodulation, and symbiotic microbes' activities are maintained and protected by adequate quantity of micronutrients (iron, manganese, zinc, molybdenum, copper, boron, etc.) in Fabaceous plants. During abiotic stress conditions, essential micronutrients are

Table 16.1 Some important enzymes having micronutrient element as cofactor (Sharma 2006)

Enzyme (EC)	Micronutrient (cofactor)
Catalase (EC 1.11.1.6)	Iron
Peroxidase (EC 1.11.1.7)	Iron
Superoxide dismutase (EC 1.15.1.1)	Copper-Zinc, Manganese, Iron
Cytochrome c oxidase (EC 1.9.3.1)	Iron, Copper
Nitrite reductase (EC 1.7.7.1)	Iron
Glutamate synthase (EC 1.4.7.1)	Iron
Nitrate reductase (EC 1.6.6.1)	Molybdenum
Laccase (EC 1.10.3.2)	Copper
Ascorbate oxidase (EC 1.10.3.3)	Copper
Copper amine oxidase (EC 1.4.3.6)	Copper
Alcohol dehydrogenase (EC 1.1.1.2)	Zinc
Glutamine dehydrogenase (EC 1.4.1.3)	Zinc
Allantoate amidohydrolase (EC 3.5.3.9)	Manganese
Carboxypeptidase A (EC 3.4.17.1)	Zinc
Aconitase (EC 4.2.1.11)	Iron
Enolase (EC 4.2.1.11)	Manganese, Magnesium
Phosphoenolpyruvate carboxylase (EC 4.1.1.31)	Manganese, Magnesium
Carbonic anhydrase (EC 4.2.1.1)	Zinc

less available to the plants root for the absorption; thus, their deficiencies cause low growth, poor cellular defense, and reproductive yield to the Fabaceous plants. The genetic variability of Fabaceous plants also major concerns to uptake, translocation, and distribution of micronutrients in the internal plant system. The assessment of micronutrients under various environmental conditions and their management in growth medium can protect and enhance the growth and production of Fabaceous plants growing under various abiotic stress conditions (Table 16.1).

Acknowledgments Author expresses his gratitude to the late Professor C. P. Sharma, former head of the Botany Department and in-charge of the Indian Council of Agricultural Research, the Department of Botany, University of Lucknow, Lucknow, and Dr. B. D. Nautiyal, the Department of Botany (Rtd.), University of Lucknow, Lucknow for his fruitful discussion and suggestions regarding Plant Nutritional Physiology under various stresses. He also acknowledges the help of Isha Verma during finalizing the paper.

References

Adams P, Graves CJ, Winson GW (1975) Some effects of copper and boron deficiencies on the growth and flowering of *Chrysanthemum morifolium* cv. Hurricane. J Sci Food Agric 26:899–901

Agarawala SC, Hewitt EJ (1955) Molybdenum as a plant nutrient V. The interrelationship of molybdenum and notrate supply in the concentration of sugars, nitrate and organic nitrogen in cauliflower plants grown in sand culture. J Hort Sci 30:151–162
Agarwala SC, Hewitt EJ (1995) Molybdenum as a plant nutrient V. The interrelationship of molybdenum and nitrate supply in the concentration of sugars, nitrate and organic nitrogen in cauliflower plants grown in sand culture. J Hort Sci 61:397–405
Ahmed S, Evans HJ (1960) Cobalt: a micronutrient for the growth of soybean plants under symbiotic conditions. Soil Sci 90:205–210
Alscher RG, Erturk N, Heath LS (2002) Role of superoxide dismutases (SODs) in controlling oxidative stress. J Exp Bot 53:1331–1341
Apel K, Hirt H (2004) Reactive oxygen species: metabolism, oxidative stress and signal transduction. Ann Rev Plant Boil 55:373–400
Arnon (1954) Criteria of essentiality of inorganic micronutrients for plants. In: Wallace T (ed) Trace elements and plant physiology. Chronica Botanica Waltham, MA, pp 31–39
Asada K (1994) Production and action of active oxygen species in photosynthetic tissues. In: Foyer CH, Mullineaux PM (eds) Causes of photooxidative stress and amelioration of defense systems in plants. CRC Press, Boca Raton, Florida, pp 77–104
Asada K (1997) The role of ascorbate peroxidase and monodehydroas corbate reductase in hydrogen peroxide scavenging in plants. In: Candalios JGS (ed) Oxidation states and the molecular biology of antioxidant defenses. Cold Sring Harbor Laboratory Press, pp 715–735
Baylock AD (1995) Navybean yield and maturity response to nitrogen and zinc. J Plant Nutr 18:163–178
Beak D, Cha JY, Kang S, Park B, Lee HJ, Hong H (2015) The *Arabidopsis* a zinc finger domain protein ARS1 is essential for seed germination and ROS homeostasis in response to ABA and oxidative stress. Front Plant Sci 6:963–970
Bell RW, McLay L, Plaskett D, Dell B, Loneragan JF (1989) Germination and vigour of black gram (*Vigna mungo L*., Hepper) seed from plants grown with and without boron. Aust J Agric Res 40:273–279
Berg JM, Shi Y (1996) The galvanization of biology: a growing appreciation for the role of zinc. Science 271:1081–1085
Bisht SS, Agarwala SC, Sharma CP (1991) Variable boron supply and sugarbeet metabolism. Proc Natl Acad Sci (India) 61:109–114
Bolanos I, Brewin NJ, Bonilla I (1996) Effects of boron on *Rhizobium-legume* cell-surface interactions and nodule development. Plant Physiol 110:1249–1256
Brive A, Sengupta AK, Beuchie D, Larsson J, Kennison JA, Rasmuson-Lestander A, Muller J (2001) Su (Z)12, a novel *Drosophila Polycombgroup* gene that is conserved in vertebrates and plants. Development 128:3371–3379
Brown PH, Bellaloui N, Wimmer MA, Bassil ES, Ruiz J, Hu H, Pfeffer H, Dannel F, Romheld V (2002) Boron in plant biology. Plant Boil 4:05–223
Brown PH, Welch RM, Cary EE (1987) Nickel: a micronutrient essential for higher plants. Plant Physiol 85(3):801–803
Caballero R, Arouzo M, Hernaiz M (1996) Accumulation and redistribution of mineral elements in common vetch during pod filling. Agron J 88:801–805
Cairns ALP, Kurtzinger JH (1992) The effect of molybdenum on seed dormancy in wheat. Plant Soil 145:295–297
Cakmak I (2000) Possible roles of zinc in protecting plant cells from damage by reactive oxygen spices. New Phytol 146:185–205
Cakmak I, Kurtz H, Marschner H (1995) Short-term effects of boron, germanium and high light intensity on membrane permeability in boron-deficient leaves of sunflower. Physiol Plant 95:11–18
Cakmak I, Marschner H, Bangerth F (1989) Effect of zinc nutritional status on growth, protein metabolism and levels of indole-3-acetic acid and other phytohormones in bean (*Phaseolus vulgaris* L.). J Exp Bot 40:405–412

Carpena RR, Carpena RO, Zoronza P, Collado G (1984) A possible role of boron in higher plants. J Plant Nutr 9:1341–1354
Casimiro A, Barroso J, Pais MS (1990) The effect of copper deficiency on photosynthetic electron transport in wheat plants. Physiol Plant 79:459–464
Chapple C (1998) Molecular genetic analysis of plant cytochrome P450-dependent monooxygenase. Annu Rev Plant Physiol Plant Mol Biol 49:311–343
Chatterjee C, Nautiyal N (2001) Molybdenum stress affects viability and vigour of wheat seeds. J Plant Nutr 24:1377–1386
Cooper TG, Filmer D, Wishnick M, Lane MD (1969) The active species of CO_2 utilized b ribulose diphosphate carboxylase. J Biol Chem 244:1081–1083
Cowles JR, Evans HJ, Russel S (1969) B_{12}Co-enzyme dependent ribonucleotide reductase in *Rhizobium* species and the effect of cobalt deficiency on the activity of enzyme. J Bacteriol 97:1460
Dell B (1981) Male sterility and anther wall structure in copper-deficient plants. Ann Bot (London) 48:599–608
Dembitsky VM, Smoum R, Al-Qantar AA, Ali HA, Pergament I, Srebnick M (2002) Natural occurrence of boron-containing compounds in plants, algae and microorganisms. Plant Sci 163:931–942
Dilworth MJ, Robson AD, Chatel DL (1979) Cobalt and nitrogen fixation in *Lupinusangustifolius* L. II. Nodule formation and fuction. New Phytol 83:63–79
Dordas C, Brown PH (2000) Permeability of boric acid across lipid bilayers and factors affecting it. J MembrBiol 175:95–105
Eide D, Broderius M, Feit J, Guerniot ML (1996) A novel iron-regulated metal transporter from plants identified by functional expression in yeast. Proc Natl Acad Sci USA 93:5624–5628
Elstner EF (1991) Mechanism of oxygen activation in different components of plants In: Active oxygen species, oxidative stress and plant metabolism. Am Soc Plant Physiol, 13–25 (Rockville MD)
Flowers TJ (1988) Chloride as a nutrient and as an osmoticum. In: Tinker B, Läuchli A (eds) Advances in plant nutrition. Praeger, New York, pp 55–78
Gerendas J, Polacco JC, Freyermuth SK, Sattelmacher B (1999) Significance of nickel for plant growth and metabolism. J Plant Nutr Soil Sci 162:241–256
Gorham J (1995) Betaines in higher plants-biosynthesis and role in stress metabolism. In: Walls-grove RM (ed) Aminoacids and their derivatives in higher plants. Cambridge University Press, Cambridge, pp 171–203
Gubler WD, Grogan RG, Osterli PP (1982) Yellows of melons caused by molybdenum deficiency in acid soil. Plant Dis 66:449–451
Gupta S, Pandey S (2019) Unravelling the biochemistry and genetics of ACC deaminase—an enzyme alleviating the biotic and abiotic stress in plants. Plant Gene 18. https://doi.org/10.1016/j.plgene.2019.10075
Gurley WB, Giddens J (1969) Factors affecting uptake, yield response and carry over of molybdenum on soyabean seed. Agron J 61:7–9
Halliwell B, Gutteridge JMC (1989) Free radicals in biology and medicine, 2nd edn. Clarendon Press, Oxford
Hatch MD, Burnell JN (1990) Carbonic anhydrase activity in leaves and its role in the first step of C_4 photosynthesis. Plant Physiol 93:825–828
Herrmann KM, Weaver LM (1999) The shikimate pathway. Annu Rev Plant Physiol Plant Mol Biol 50:473–503
Hewitt EJ, Bond G (1966) The cobalt requirement of non-legume root nodule plants. J Exp Bot 17:480–491
Hille R (1996) The mononuclear molybdenum enzymes. Chem Rev 96:2757–2816
Jaleel CA, Riadh K, Gopi R, Manivannan P (2009) Antioxidant defense responses: physiological plasticity in higher plants under abiotic constraints. Acta Physiol Plant 31:427–436

Kannaujia PK, Pandey SN (2013) Effect of zinc-stresses in alluvial soil on growth and yield of wheat. J Biol Chem Res 30:892–900

Kelley PM, Izawa S (1978) The role of chloride ion in photosynthesis II 1. Effects of chloride on photosystem II electron transport and hydroxylamine inhibition. Bio Chem Bio phys Acta 502:198–210

Khan Z, Upadhyaya H (2019) Impact of nanoparticles on abiotic stress responses in plants: an overview. Nanomater Plants Algae Microorgan 2:305–322

Kobayashi M, Mutosh T, Matoh T (2004) Boron nutrition of cultured tobacco by 2-cells. IV. Genes induced under low boron supply. J Exp Bot 55:1441–1443

Koshiba T, Saito H, Ono N, Yamarnoto N, Sato M (1996) Purification and properties of flavin-and molybdenum containing aldehyde oxidase from coleoptile of maize. Plant Biol 110:781–789

Li C, Dannel F, Pfeffer H, Romheld V, Bangerth F (2001) Effect of boron starvation on boron compartmentation and possibly hormone–mediated elongation growth and apical dominance in pea (*Pisumsativum*) plants. Physiol Plant 111:212–219

Loomis WD, Durst RW (1992) Chemistry and biology of boron. Bio Factors 3:229–239

Ma D, Sun D, Wang C, Ding H, Qin H, HouJ Huang X, Xie Y, Guo T (2017) Physiological responses and yield of wheat plants in Zinc-mediated alleviation of drought stress. Front Plant Sci. https://doi.org/10.3389/fpls.2017.00860

Maksymiec W (1997) Effect of copper on cellular processes in higher plants. Photosynthesis 34:321–342

Marschner H (1995) Mineral nutrition of higher plants. Academic Press, London

Marschner H, Cakmak I (1989) High light intensity enhances chlorosis and necrosis in leaves of zinc, potassium and magnesium deficient bean (*Phaseolus vulgaris*). J Plant Physiol 134:308–315

Mendel RR, Hansch (2002) Molybdoenzymes and molybdenum cofactor in plants. J Exp Bot 53:1689–1698

Muszynska E, Labudda M (2019) Dual role of metallic trace elements in stress biology—from negative to beneficial impact on plants. Int J Mol Sci 20:3117. https://doi.org/10.3390/ijms20133117

Nautiyal N, Chatterjee C (1999) Role of copper in improving the seed quality of sunflower (*Helianthus annuus*). Indian J Agric Sci 69:210–213

Nautiyal N, Chatterjee C, Sharma CP (1999) Copper stress effects grain filling in rice. Commun Soil Sci Plant Anal 30:1625–1632

Nautiyal N, Singh S, Chatterjee C (2005) Seed reserves of chickpea in relation to molybdenum supply. J Sci Food Agric 85:860–864

O'Neill MA, Ishii T, Albushein P, Darvill AG (2004) Rhamnogalacturonan II: structure and function of a borate cross-linked cell wall pectic polysaccharide. Annu Rev Plant Biol 55:109–139

Otegui MS, Capp R, Staehelin LA (2002) Developing seeds of *Arabidopsis* store different minerals in two types of vacuoles and in the endoplasmic reticulum. Plant cell 14:1311–1327

Pandey SN (2018) Biomolecular functions of micronutrients toward abiotic stress tolerance in plants. In: Hasanuzzaman M, Fujita M, Oku H, Nahar K, Hawrylak-Nowak B (eds) Plant nutrients and abiotic stress tolerance. Springer, Singapore, pp 153–170

Pandey SN, Gautam S (2009) Effects of zinc supply on its uptake, growth and biochemical constituents in lentil. Indian J Plant Physiol 14:67–70

Pandey N, Sharma CP (1989) Zinc deficiency effect on photosynthesis and transpiration in safflower and its reversal on making up the deficiency. Indian J Exp Biol 27:376–377

Pandey N, Sharma CP (1996) Copper effect on photosynthesis and transpiration in safflower. Indian J Exp Biol 34:821–822

Pandey SN, Verma P (2010) Effects on growth, nodulation and some physiological attributes of legumes with cobalt supply. J Eco Friend Agric 5:21–24

Pandey N, Gupta M, Sharma CP (1995) Ultrastural changes in pollen grains of green gram subjected to copper deficiency. Geophytology 25:147–150

Pandey N, Gupta M, Sharma CP (2000) Zinc deficience affects pollen structure and viability in green gram. Geophytology 28:31–34

Pandey N, Nautiyal BD, Sharma CP (2002) Cotton response to manganese stress and amelioration of stress. J Indian Bot Soc 81:93–95
Pandey N, Pathak GC, Sharma CP (2006) Zinc is critically required for pollen formation and fertilization in lentil. J Trace Elem Med Biol 20:89–96
Pandey SN, Abid M, Abid Ali Khan MM (2018) Diversity, functions, and stress responses of soil microorganisms. In: Egamberdieva D, Ahmad P (eds) Plant microbiome: stress response. Microorganisms for sustainability, vol 5. Springer, Singapore
Pathak GC, Gupta B, Pandey N (2012) Improving reproductive efficiency of chickpea by foliar application of zinc. Braz J Plant Physiol 24:173–180
Pilon-Smits EAH, Quinn CF, Tapken W, Malagoli M, Schiavon M (2009) Physiological functions of beneficial elements. Curr Opinion Plant Biol 12:267–274
Polar E (1970) The distribution of zinc in the cotyledons of *Vicia faba* and its translocation during the growth and maturation of the plant. Plant Soil 32:1–17
Prescott AG, John P (1996) Dioxygenases: molecular structure and role in plant metabolism. Annu Rev Plant Mol Biol 47:245–271
Purcell LC, King CA, Ball RA (2000) Soybean cultivar differences in ureides and relationship to drought-tolerant nitrogen fixation and manganese nutrition. Crop Sci 40:1062–1070
Ravichandran V, Pathmanabhan G (2014) Studies of nodulation and ethanol producing enzymes in copper under flood stress regime. J Plant Biol 33:75–80
Saikia J, Sarma RK, Dhandia R, Yadav A, Bharti R, Gupta VK, Saikia R (2018) Alleviation of drought stress in pulse crops with ACC deaminase producing rhizobacteria isolated from acidic soil of Northeast India. Sci Rep 8:3560. https://doi.org/10.1038/s41598-018-21921-w
Seo M, Koshiba T (2002) The complex regulation of ABA biosynthesis in plants. Trends Plant Sci 7:41–48
Sharma PN (1992) Pollen fertility in manganese deficient wheat. Tropical Agric 69:21–24
Sharma CP (2006) Plant micronutrients. Science Publishers, Enfield
Sharma CP, Sharma PN (1987) Mineral nutrient deficiencies affect plant water relations. J Plant Nutr 10:1637–1643
Sharma PN, Chatterjee C, Agarwala SC, Sharma CP (1990) Zinc deficiency and pollen fertility in maize (*Zea mays*). Plant Soil 124:221–225
Sharma CP, Khurana N, Chatterjee C (1995) Manganese stress changes physiology and oil content of linseed *Linumusitatissimum* L. Indian J Exp Biol 33:701–704
Sharma PN, Kumar N, Bisht SS (1996) Guard cell carbonic anhydrase activity and stomatal opening in zinc deficient faba bean, *Vicia faba*. Ind J Exp Biol 34:560–564
Shrotri CK, Tewari MN, Rathore VS (1978) Morphological and ultrastructural abnormalities in zinc deficient maize chloroplast. Plant Biochem J 5:89–96
Shrotri CK, Rathore VS, Mohanty P (1981) Studies on photosynthetic electron transport, photophosphorylation and CO_2 fixation in zinc deficient leaf cells of *Zea mays* L. J Plant Nutr 3:945–954
Sigel A, Sigel H (2002) Molybdenum and tungsten. Their roles in biological processes. Metal Ions in biological systems. Marcel Dekker, New York
Srivastava NK, Farooqi AHA, Bansal RP (1985) Response of opium poppy (*Papaver sommniferum*) to varying concentrations of boron in sand culture. Indian J Plant Nutr 4:91–94
Sun T-P, Gubler F (2004) Molecular mechanism of gibberellin signalling in plants. Annu Rev Plant Biol 55:197–224
Takahashi M, Terada Y, Nakai I, Nakanishi H, Yoshimura E, Mori S (2003) Role of nicotinamine in the intracellular delivery of metals and plant reproductive development. Plant Cell 15:1263–1280
Takatsuji H, Mori M, Benfey PN, Ren L, Cua NH (1992) Characterization of zinc finger DNA-binding protein expressed specifically in petunia petals and seedlings. EMBO J 11:241–249
Tang C, Robson AD, Dilworth MJ, Kuo J (1992) Microscopic evidence on how iron deficiency limits nodule initiation in *Lupinusaungstifolius*. New Phytol 121:457–467
Taylor IB (1991) Genetics of ABA synthesis. In: Davis WJ, Jones HG (eds) Abscisic acid, physiology and biochemistry. Bios Publishers, Oxford, pp 23–37

Terry N (1977) Photosynthesis, growth and role of chloride. Plant Physiol 60:69–75

Tewari RK, Kumar P, Sharma PN (2005) Signs of oxidative stress in the chlorotic leaves of iron starved plants. Plant Sci 169:1037–1045

Tewari RK, Hadacek F, Sassmann S, Lang I (2013) Iron deprivation-induced reactive oxygen species generation leads to non-autolytic PCD *in Brassica napus* leaves. Environ Exp Bot 91:74–83

Udavardi MK, Day DA (1997) Metabolic transport across symbiotic membranes of legume nodules. Annu Rev Plant Physiol Plant Mol Biol 48:493–523

UpadhyayA Tripathi S, Pandey SN (2013) Effects of soil sodicity on growth, nutrients uptake and biochemical responses of *Ammimajus* L. Res J Soil Biol 4:69–80

Wetzel CLR, Jenson WA (1992) Studies on pollen maturation in cotton: the storage reserve accumulation phase. Sex Plant Rep 5:117–127

Xu H, Huang Q, Shen K, Sten Z (1993) Anatomical studies on the effect of boron on the development of stamen and pistil of rape (*Brassica napus* L.). Zhiwa Xuebao 35:453–457

Xu G, Magen H, Tarchitzky J, Kafkafi U (2000) Advance in chloride nutrition. Adv Agron 66:96–150

Yu Q, Baluska F, Jasper F, Menzel D, Goldbach E (2002) Short term boron deprivation enhances levels of cytoskeletal proteins in maize, but not in zucchini, root apices. Physiol Plant 117:270–278

Yu Q, Hlavacka A, Matoh T, Volkmann D, Menzel D, Goldbach HE, Baluska F (2003) Short term boron dprivation inhibits endocytosis of wall pectins in meristematic cells of maize and wheat root apics. Plant Physiol 130:415–421

Response and Tolerance of Fabaceae Plants to Metal/Metalloid Toxicity

Jubayer Al Mahmud, M. H. M. Borhannuddin Bhuyan, Kamrun Nahar, Khursheda Parvin, and Mirza Hasanuzzaman

Abstract Plant of Fabaceae family or legumes are familiar for their nitrogen-fixing ability as well as their indispensable role in livestock and human health, in improving soil health and sustainable agricultural production. Excess toxic metals/metalloids in soil are one of the most important environmental reasons that hinder plant growth and productivity of a wide range of plant species including Fabaceae plant. On account of tolerance ability, member of Fabaceae plant varies and elevated level of metals/metalloids affected growth and yield of plant to the highest degree. Generally, Fabaceae plant can tolerate a small amount of metals/metalloids without compromising the growth and yield due to their well-established antioxidant defense system, osmolyte synthesis, hormone regulation, and chelation. Plant of Fabaceae family is also known for their phytoremediation capability. Plant scientists are trying to elucidate the underlying mechanisms of Fabaceae plant to metals/metalloids toxicity. In this chapter, we synthesized the recent research findings on diverse features of metals/metalloids stress tolerance approaches.

J. A. Mahmud
Department of Agroforestry and Environmental Science, Faculty of Agriculture, Sher-e-Bangla Agricultural University, Sher-e-Bangla Nagar, Dhaka 1207, Bangladesh

M. H. M. Borhannuddin Bhuyan
Citrus Research Station, Bangladesh Agricultural Research Institute, Jaintapur, Sylhet 3156, Bangladesh

K. Nahar
Department of Agricultural Botany, Faculty of Agriculture, Sher-e-Bangla Agricultural University, Sher-e-Bangla Nagar, Dhaka 1207, Bangladesh

K. Parvin
Laboratory of Plant Stress Responses, Department of Applied Biological Science, Faculty of Agriculture, Kagawa University, Miki-Cho, Kita-gun, Kagawa 761-0795, Japan

Department of Horticulture, Faculty of Agriculture, Sher-e-Bangla Agricultural University, Sher-e-Bangla Nagar, Dhaka 1207, Bangladesh

M. Hasanuzzaman (✉)
Department of Agronomy, Faculty of Agriculture, Sher-e-Bangla Agricultural University, Sher-e-Bangla Nagar, Dhaka 1207, Bangladesh
e-mail: mhzsauag@yahoo.com

M. Hasanuzzaman et al. (eds.), *The Plant Family Fabaceae*,
https://doi.org/10.1007/978-981-15-4752-2_17

Keywords Pulse crops · Soil pollution · Cadmium · Arsenic · Phytoremediation · Oxidative stress

1 Introduction

Fabaceae or Leguminosae, usually recognized as legume, pea or bean family, is the third largest angiosperm family but considered as the second most important plant family in the world agriculture following grass family called Poaceae (Gómez-Sagasti and Marino 2015; Ara et al. 2019). Members of Fabaceae family or legumes are familiar for their nitrogen-fixing ability. They improve the dietary value, which plays an imperative role in the health of livestock and human (Araújo et al. 2015). Legume root contains special organ called nodule which is host for some important symbiotic bacteria mostly rhizobia. These bacteria can fix atmospheric nitrogen and maintain a mutualistic relation with host plant. The introduction of legumes in crop rotation can retain the quality of degraded and polluted soils. So, legumes are also recognized for their significant function to retain soil health and confirm sustainable agricultural systems. Practice of crop rotation and intercrop (e.g., combined with cereals) with legumes diminishes the activity of pests, diseases, and weed populations while increasing the overall farm productivity (Matusso et al. 2014; Foyer et al. 2016). Similarly, other plant species of Fabaceae family also suffered from a different unavoidable environmental condition or abiotic stress including toxic metal/metalloid toxicity (Hasanuzzaman and Fujita 2012; Mahmud et al. 2018; Ara et al. 2019).

Rapid population expansion together with industrial insurgency and unplanned urbanization are contributing in enormous toxic metal/metalloids pollution to the environment. Metal/metalloids not only are big environmental concern but also create human health hazard due to their extensive occurrence in the air, water, and soil; perseverance in ecosystems; and noxious properties (Vijendra et al. 2016; Mahmud et al. 2019). Mining of ore, foundries and smelters, utilization of agrochemicals, metal-based industrial activities, emission from automobiles, blazing of fossil fuels, etc., are the main sources of releasing toxic metal/metalloids to the environment (Hasanuzzaman and Fujita 2012; Tchounwou et al. 2012; Rehman et al. 2017). Among the limiting factor of world agricultural production, toxic metals/metalloids are considered most important environmental contaminants as it enters into the plant body from the growing media and shrinks the quantity and quality of edible parts. Moreover, entry of toxic metal/metalloids to human food cycle from the agricultural commodity causes devastating effect on human health by crating different carcinogenic diseases including cancer (Mahmud et al. 2018, 2019). Different records confirm that lead (Pb), cadmium (Cd), chromium (Cr), mercury (Hg), aluminum (Al), arsenic (As), etc., are unnecessary and extremely toxic metal/metalloids for any sort of life including vegetation. Conversely, a number of metals/metalloids are indispensable for normal functioning and proper growth of plant in a trace amount but become lethal at higher concentration, e.g., copper (Cu), zinc (Zn), iron (Fe), boron (B), manganese (Mn), nickel (Ni), etc. (Tangahu et al. 2011; Hasanuzzaman and Fujita 2012; Zhou et al. 2014; Emamverdian et al. 2015; Mahmud et al. 2019). Toxic metal/metalloids severely demolish the usual cellular physiological and biochemical mechanisms including germination, photosynthesis, nutrient uptake, etc., and decrease growth and development of plant by creating nutritional disorder and oxidative stress (Sharma and Dietz 2009; Hasanuzzaman and Fujita 2012; Hossain et al. 2012; Sarwar et al. 2017; Mahmud et al. 2018).

Scientists are trying to explore the physiological and molecular basis of Fabaceae plant under toxic metal/metalloid stress. Under stress condition, production of reactive oxygen species (ROS; singlet oxygen, $^1O_2$; superoxide anion, $O_2^{\bullet -}$; hydrogen peroxide, H_2O_2; and hydroxyl radical, $OH^{\bullet}$) and methylglyoxal (MG) content increases tremendously which makes imbalance with components of antioxidant defense and glyoxalase system, respectively, and destroys cellular redox balance (Hasanuzzaman et al. 2012, 2017). Among the metal/metalloids, some (Fe, Cu, Cr, Co, etc.) can produce ROS directly through Haber–Weiss and Fenton reactions which are called redox-active. Redox-inactive metal/metalloids (Cd, Pb, Zn, Ni, Al, etc.) enhance ROS generation by reducing antioxidant defense system, provoking NADPH oxidase, relocating indispensable cations from definite binding enzyme site, triggering calcium (Ca)-reliant mechanisms, and influencing Fe-mediated systems (Hasanuzzaman and Fujita 2012; Mahmud et al. 2018, 2019). All types of plant try to scavenge production of excess ROS through antioxidant defense system consisting of non-enzymatic components (ascorbic acid, AsA; glutathione, GSH; phenolic compounds; alkaloids; α-tocopherol; and nonprotein amino acids) and enzymatic components (superoxide dismutase, SOD; catalase, CAT; ascorbate peroxidase, APX; glutathione reductase, GR; monodehydroascorbate reductase, MDHAR; dehydroascorbate reductase, DHAR; glutathione peroxidase, GPX; and glutathione S-transferase, GST) and excess MG through glyoxalase system containing glyoxalase I, Gly I, and Gly II enzyme (Hasanuzzaman et al. 2012, 2017; Stambulska et al. 2018). Moreover, few plants can increase its endogenous phytochelatins (PCs) and metallothioneins (MTs) content to combat stress. Phytochelatins and MTs make complex with metals/metalloids and transfer them in cell vacuole as inert form (Mahmud et al. 2019). Under metals/metalloids toxicity, soils interact with variety of microorganisms enhancing tolerance as well as growth of plant (Rajkumar et al. 2012). Fabaceae crops show significant variation broadly in their level of tolerance to toxic metal/metalloids. Plant tolerance-to-metal/metalloids stress is mostly determined by their genetics. Positive responses of different osmolytes, antioxidants, hormone regulation, metal chelation, and plant–microbe interaction of Fabaceae plant under metal/metalloid toxicity contribute in stress tolerance (Ara et al. 2019).

In this chapter, we will try to explore a general overview of physiological and molecular basis of Fabaceae plant under metal/metalloids toxicity. We hope these progresses will be able to support the development of adaptive Fabaceae plant for ensuring crop quality and quantity.

Table 1 Responses of Fabaceae plants to different metals/metalloids toxicity

Crops	Stress intensity and duration	Plant responses	References
Trigonella foenum-graceum L.	200 ppm Cu as $CuSO_4.5H_2O$; 7 days	• Reduced seed germination percentage by 45% with reduced seedling emergence rate • Decreased fresh weight of radical and plmule • Declined seed vigor index (SVI)	Menon et al. (2016)
T. foenum-graceum L.	200 ppm Cr as $K_2Cr_2O_7$; 7 d	• Lowered seed germination percentage by 60% • Reduced radical and plumule lengths and their fresh weight (FW) • Declined SVI	Menon et al. (2016)
T. foenum-graceum L.	200 ppm Zn as $ZnSO_4$; 7 d	• Reduced seed germination percentage by 20% • Decreased radical and plumule lengths and their FW • Decreased SVI	Menon et al. (2016)
T. foenum-graceum L.	3 and 4.5 g L^{-1} Cu as $CuSO_4$	• Decreased seed germination by 16% • Reduced radical length and its dry weight (DW) by 28 and 51%, respectively	Jalali et al. (2018)
Vicia faba L.	3 and 4.5 g L^{-1} Cu as $CuSO_4$	• Decreased seed germination by 47% • Reduced radical length and its DW by 70 and 26%, respectively	Jalali et al. (2018)
V. sativa L.	3 and 4.5 g L^{-1} Cu as $CuSO_4$	• Decreased seed germination by 17% • Reduced radical length and its DW by 30 and 35%, respectively	Jalali et al. (2018)
Medicago truncatula L.	3 and 4.5 g L^{-1} Cu as $CuSO_4$	• Reduced seed germination by 50% • Reduced radical length and its DW by 57 and 41%, respectively	Jalali et al. (2018)

(continued)

Table 1 (continued)

Crops	Stress intensity and duration	Plant responses	References
Vigna radiata L. cv. T-44	10 mM Al as $AlCl_3$; 14 d	• Caused a sharp decline in shoot and root lengths by 27 and 67%, respectively • Reduced plant DW and leaf area by 25 and 58%, respectively • Declined the content of chlorophyll (chl) and car by 48 and 66%, respectively • Decreased Ci content and carbonic anhydrase activity by 47 and 38%, respectively • Reduced stomatal conductance (gs), water-use efficiency (WUE), and photosynthetic rate (Pn) by 58, 37, and 43%, respectively • Lowered leaf relative water content (LRWC) by 40% with the elevation of proline (pro) content	Ali (2017)
V. radiata L.	400 mg Kg^{-1} Cr as $K_2Cr_2O_7$; 14 d	• Reduced 53% of seed germination • Inhibited stem length by 68% • Restricted biomass accumulation in above ground parts of plant	Hu et al. (2016)
V. radiata L. cv. NM-2006	250 μM Cr as $K_2Cr_2O_7$; 28 d	• Reduced plant height, root length, and number of leaves significantly • Caused significant reduction in shoot and root FW and DW • Decreased chl *a*, chl *b*, chl (*a* + *b*), and car contents	Jabeen et al. (2016)

(continued)

Table 1 (continued)

Crops	Stress intensity and duration	Plant responses	References
V. radiata L. cv. NM-2006	500 μM Cr as $K_2Cr_2O_7$; 28 d	• Drastically reduced plant height, root length, and number of leaves • Reduced shoot and root biomass accumulation • Decreased chl *a*, chl *b*, chl (*a* + *b*), and car contents	Jabeen et al. (2016)
V. radiata L. cv. BARI Mung-2	1.5 mM Cd as $CdCl_2$; 48 h	• Decreased plant height, root length, plant DW, and leaf area by 18, 28, 20, and 32%, respectively. • Reduced chl (*a* + *b*) content and LRWC • Increased pro contents	Nahar et al. (2016a, b)
V. radiata L.	5,7 and 9 μM Cd as $CdCl_2$; 24 h	• Reduced adventitious root number and FW • Increased pro content	Li et al. (2019)
V. radiata L.	5 mg L^{-1} Hg as $HgCl_2$; 4 d	• Decreased root length • Reduced chl *a* and chl *b* contents	Mao et al. (2018)
V. radiata L.	10 mg L^{-1} Cd $3CdSO_4.8H_2O$; 4 d	• Decreased root length • Reduced chl *a* and chl *b* contents • Decreased plant FW and DW	Mao et al. (2018)
V. radiata L.	15 mg L^{-1} Cr $K_2Cr_2O_7$; 4 d	• Decreased root length • Reduced biomass contents • Reduced chl *a* and chl *b* contents	Mao et al. (2018)
V. radiata L.	45 mg L^{-1} Cu as $CuSO_4.5H_2O$; 4 d	• Decreased root length • Reduced chl *a* and chl *b* contents	Mao et al. (2018)
V. radiaa L.	80 mg L^{-1} Pb as $(CH_3COO)_2.3H_2O$; 4 d	• Decreased root length • Reduced chl *a* and chl *b* contents	Mao et al. (2018)
Pisum sativum L. cv. Douce Provence	200 μM $CdCl_2$; 9 d	• Delayed embryo growth in seed germination • Reduced shoot and root lengths	Jaouani et al. (2018)

(continued)

Table 1 (continued)

Crops	Stress intensity and duration	Plant responses	References
Glycine max L.	0, 50, 100, and 150 mg L^{-1} of Cd as $CdSO_4 \bullet 8H_2O$; 35 d	• Decreased shoot length, root length, and their biomass content with the increase of Cd concentration • Showed highest reduction of LRWC under 150 mg L^{-1} Cd stress • Increased electrolyte leakage (EL) dramatically • Increased pro accumulation	Alyemeni et al. (2017)
G. max L. cv. Taekwan	100 μM $CuSO_4$; 5 d	• Reduced chl and car contents • Decreased the shoot and root lengths • Inhibited shoot and root FW and DW	Kang et al (2015)
G. max L. cv. Taekwan	100 μM $ZnSO_4$; 5 d	• Reduced chl and car contents • Decreased the shoot and root lengths • Inhibited shoot and root FW	Kang et al (2015)
G. max L. cv. BMX	4 mmol L^{-1} Al as $AlCl_3$; 48 h	• Decreased shoot and root DW • Decreased chl index (SPAD) • Decreased Ci and increased transpiration (E) • Decreased leaf xylem diameter • Caused leaf chlorosis and root tip lesion	Reis et al. (2018)
G. max L. cv. BMX	20 μmol L^{-1} Ni as $NiCl_2$; 7 d	• Decreased Pn, gs, and E • Increased Ci	Reis et al. (2017)
G. max L.	5 mg L^{-1} Hg as $HgCl_2$; 4 d	• Decreased root length • Reduced chl *a* and chl *b* contents	Mao et al. (2018)

(continued)

Table 1 (continued)

Crops	Stress intensity and duration	Plant responses	References
G. max L.	10 mg L^{-1} Cd as $3CdSO_4.8H_2O$; 4 d	• Decreased root length • Decreased plant FW and DW • Reduced chl *a* and chl *b* contents	Mao et al. (2018)
G. max L.	15 mg L^{-1} Cr as $K_2Cr_2O_7$; 4 d	• Decreased root length • Declined biomass accumulation • Reduced chl *a* and chl *b* contents	Mao et al. (2018)
G. max L.	45 mg L^{-1} Cu as $CuSO_4.5H_2O$; 4 d	• Decreased root length • Reduced chl *a* and chl *b* contents	Mao et al. (2018)
G. max L.	80 mg L^{-1} Pb as $(CH_3COO)_2.3H_2O$; 4 d	• Decreased root length • Reduced chl *a* and chl *b* contents	Mao et al. (2018)

2 Responses to Metal/Metalloid Toxicity

Metal/metalloid toxicity in plant-growing media hampers the usual growth and development of Fabaceae plant by creating nutrient imbalance and oxidative stress. Germination, emergence, cellular development, and physiological processes of plants are being disturbed by the toxicity of metal/metalloid (Stambulska et al. 2018). Moreover, these stresses also hinder nodulation process and reduce the quality and productivity of Fabaceae plant. In this subsection, we have discussed major responses of Fabaceae plant to metal/metalloid toxicity (Table 1).

2.1 Germination and Emergence

Metal stress restricted primary physiological process naming seed germination in plant by reducing the germination percentage of seeds later resulted in lower seedlings growth (Stambulska et al. 2018). With the increase of metal concentration, reduction of seed germination and seedling emergence rate are decreased gradually (Nath et al. 2008; Stambulska et al. 2018). Among 2, 105, 25, and 50 ppm Cr contamination, lowest seed germination of *Vigna sinensis* was recorded from 50 ppm concentration (Nath et al. 2008). Later, germinated seedlings were also suffered from inhibitions in plumule and radicle lengths under Cr toxicity. Stambulska et al. (2018) reported that Cr stress not only inhibited the seed germination but also delayed the required time for germination. As Cr toxicity causes the inhibition of amylase activities and subsequently decreases, the carbohydrates transport to the germ which explained the metal-induced restriction in seed germination process (Nath et al. 2008; Stambulska et al. 2018). Seed germination and emergence of *Trigonella foenum-graceum* were evaluated under Cr-, Cu-, and Zn-contaminated condition to study the effect of heavy metal (Menon et al. 2016). Seeds were kept to germinate on varying concentrated solution of each metal such as 1, 3, 5, 10, 50, 100, 200, 300, 500 ppm for 7 days. After that, germination percentage, seed vigor index (SVI), and length of radical and plumule with their fresh weight (FW) were measured. It was found that 500 ppm concentration of Cu, Cr, and Zn made maximum damage in each parameter. Hence, total germination was around 10, 25, and 60% from 500 ppm of Cr, Cu, and Zn, respectively, while 300 and 500 ppm caused the death of all seeds and 200 ppm of each metal showed lowest growth length and FW of radicle and plumule. Due to the metal-induced reduction of germination percentage, length of radical and plumule, FW of radical and plumule cumulatively reduced the SVI drastically. Seed germination and seedling emergence capacity of *Vicia faba*, *V. sativa*, *T. foenum-graceum*, and *Medicago truncatula* were tested under 3 and 4.5 g L^{-1} Cu toxicity for 5 days (Jalali et al. 2018). The Cu toxicity declined seed germination percentage in all legume species but varied depending on species and toxicity severity. Therefore, *V. sativa* and T. *foenum-graceum* showed germination percentage about 17 and 16%, respectively, which again reached by 47% and 50%, respectively, for *V. faba* and *M. truncatula*. Seedling emergence was also reduced by Cu stress, while *V. sativa* maintained comparatively less reduction of radical length and its dry weight (DW) which was 30 and 35%, respectively, but other three species showed higher rate of this reduction and then comparatively severe decreases revealed in *V. faba* (70% and 26%, respectively), *T. foenum-graceum* (28% and 51%, respectively), and *M. truncatula* (57% and 41%, respectively). Jalali et al. (2018) concluded that Cu stress-induced inhibitory effects on seed germination and emergence are associated with alteration in needed water absorption for required metabolic processes and later swerving of reserved mobilization process from cotyledon. A study was conducted on the effects of Cu, Mn, Pb, and Zn on the *C. arietinum* (Bhagyawant et al. 2019). It was revealed that seed germination and establishment hampered under stress condition indicated by retardation in radical length irrespective of metal exposure where most inhibition was recorded from Pb and least from Zn.

2.2 Growth and Development

Heavy metal stress illuminates the suppression in plant growth and development focused by the reduction in plant height, root length, adventitious root growth, leaf numbers, branch numbers, plant biomass accumulation, etc. Here, stress-induced growth inhibition is varied with the variation of plants, plant species, different types of metal, doses of metals, and durations. Nickel stress caused the hampers in normal growth of *Cicer arietinum* under varying doses of Ni including 25 mg L^{-1}, 50 mg L^{-1}, 100 mg L^{-1}, and 150 mg L^{-1} to soil (Batool 2018). Here, 150 mg L^{-1} Ni persuaded the highest reduction of shoot length, leaf number, branch number, and biomass content. In another study, growth inhibition rate (GIR) of *C. arietinum* was evaluated under Cu, Mn, Pb, and Zn toxicity (Bhagyawant et al. 2019). Where, the increasing of metal concentration, GIR was drastically increased upon all metals exposure. Six different genotypes of *C. arietinum* were exposed to vanadium (V; 25 mg L^{-1}) toxicity for 10 days (Imtiaz et al. 2015). Therefore, plant biomass, and shoot and root lengths were considered to evaluate the plant tolerance to V stress. The application of V lowered the plant growth and development measured by shoot length, root length, and their FW and DW. However, V toxicity resulted in gradual stunt of plant shoot and root growth along with restriction in lateral root development in all genotypes. Seven days old *V. radiata* seedlings were allowed to grow under 1 and 10 mM Al toxic media for 14 days (Ali 2017). After that, Al-treated seedlings showed the reduction of growth as dose-dependently while 10 mM Al caused highest damage. The Al toxicity caused severe inhibition in root length than shoot where decreased root length and shoot length are by 67 and 27%, respectively. Further plant DW and leaf area declined by 25 and 58%, respectively, upon 10 mM Al toxicity which might be correlated by Al-induced limited water and mineral uptakes due to extensive root damage. The inhibitory effects of Cr were evaluated on stem length and above ground FW of *V. radiata* plants by exposing to 50, 100, 200, 400, and 800 mg Kg^{-1} Cr (Hu et al. 2016). With the increasing Cr level, the stem length and above ground FW were reduced but 50 and 100 mg Kg^{-1} were not so significant in decreasing stem growth and biomass accumulation. *V. radiata* showed the symptoms of brown leaves and thin stem size at 400 mg kg^{-1} Cr, while 800 mg kg^{-1} Cr caused extensive growth retardation, stem blackening, and slight root decay. *Arachis hypogaea* seedlings were exposed to 50, 100 or 200 μM Cd; after two weeks of treatment, seedling growth was inhibited (Dong et al. 2016). Higher Cd concentration showed maximum reduction of seedling FW and DW around 45 and 47%, respectively. In one study, hydroponically grown 28 days old *M. sativa* seedlings were exposed to varying degree of Pb toxicity (0, 10, and 100 μM) for 2 and 7 days (Hattab et al. 2016). Two days treatment did not significantly restrict the plant growth indicated by shoot and root lengths and their FW and DW. Long-term exposure of *M. sativa* to Pd stress causes the reduction of shoot and root growth with 41 and 48% reduction of FW of shoot and root, respectively. The 100 μM Pb treatment reduced the shoot size by 43%. Here, remarkable decrease in root length was observed in 100 μM Pb toxicity, which was less affected by 10 μM Pb. Five genotypes (Pusa-37, Pusa-16, Pusa-40, PK-416, and Pusa-24) of *Glycine max* were evaluated upon varying degree of Cd toxicity (Alyemeni et al. 2017). Among 50, 100, and 150 mg L^{-1}, 150 mg L^{-1} showed the maximum toxicity effects in this study while plant growth was reduced gradually with the increasing Cd levels. All the genotypes suffered from reduction of shoot height, root length, and their FW and DW, although lower damage intensity showed the genotype tolerance to Cd.

2.3 Photosynthesis

The principal components of photosynthetic system are photosynthetic pigments including chlorophyll (chl) and carotenoid (car) by which dry matter accumulation and yield are governing, which decreased significantly in metal-treated plants. The 10 mM concentration of Al toxicity caused significant decline in chl and car pigmentation in *V. radiata* by 48 and 66%, respectively (Ali 2017). Again, stomatal conductance (gs), water-use efficiency (WUE), and finally photosynthetic rate (Pn) reduced by 58, 37, and 43%, respectively, in Al-treated *V. radiata*. Thus Al-induced 47 and 38% reduction of intercellular CO_2 concentration (Ci) and carbonic anhydrase (CA) activity supported the reduction of Pn. A pot experiment was conducted to study how Pb toxicity affects transpiration, photosynthetic pigments, *gs*, and net assimilation rate (NAR) of *V. radiata* by applying 250 and 500 mg Kg^{-1} Pb (Arif et al. 2019). Elevation of Pb toxicity enhanced the chl *a* degradation rate from 19 to 35% along with higher loss of chl *b* and car contents. Besides, higher Pb contamination excessively suppressed the NAR and gs. Leaf chl *a*, chl *b*, chl (*a* + *b*), and car contents notably decreased in Cr-stressed *V. radiata* in respective controls. Moreover, the decline was higher in higher doses of Cr (500 μM)-treated plants than lower dose (250 μM) (Jabeen et al. 2016). *A. hypogaea* suffered from chl degradation upon Cd toxicity indicated by decreased leaf chl *a*, chl *b*, and car contents, while the more decline was observed in higher Cd level (Dong et al. 2016). The Cd toxicity significantly reduced the Pn and transpiration (E), and maximum reduction was observed from 200 μM Cd-treated seedlings. In a study, individual application of 100 μM Cu and Zn caused the reduction in photosynthetic pigment synthesis in *G. max* (Kang et al. 2015). Here, total chl and car contents were reduced in both Cu- and Zn-stressed plants. A pot experiment was conducted to study the photosynthetic responses of five tree species of Fabaceae such as *Millettia peguensis, Pongamia pinnata, Albizia lebbeck, Bauhinia purpurea, Dalbergia sissoo* to Pb, Cu, Cr, and Mn contamination (Kanwal et al. 2019). Therefore, on the basis of photosynthetic rate these species can be arranged as follows *M. peguensis* > *P. pinnata* > *B. purpurea* > *A. lebbeck* > *D. sissoo*. In addition, as one of the contributing factors for photosynthesis, gs increased in *P. pinnata* and *M. peguensis* with the increase of metal contamination, while *A. lebbeck, D. sissoo*, and *B. purpurea* showed the decreasing trend. Thus upon metal toxicity, Fabaceae plants show the reduction in photosynthesis process which later contributes in lower yield production as well as further plant growth restriction.

2.4 Water Relation

Water is also considered one of the major factors for plant growth regulation due to its direct or indirect involvement in all growth processes. Plants are also suffered from water crisis upon raise in metal contamination owing to shallow root structure and poor physicochemical properties of soil, for which researchers become more interested in plant–water relation under heavy metal stress. Shah et al. (2010) demonstrated about the metal stress-induced series of events leading to water loss in plants including decrease in number and size of leaves, stomatal size, number and diameter of xylem vessels, with higher leaf rolling, leaf abscission, and root suberization. As like other abiotic stresses, heavy metal stress also restricts water uptakes by plants and hampers cell–water relation indicated by alteration in endogenous proline (pro) and water status. The Al toxicity increased the pro accumulation levels in *V. radiata* which indicated its involvement in maintaining osmoregulation under metal-induced water shortage condition (Ali 2017). In another study on Cd, modulated pro accumulation and water relation in *V. radiata* were investigated (Nahar et al. 2016a). In this study, 1.5 mM Cd toxicity enhanced the pro content as an indication of stress which was later confirmed by the reduction of tissue water content (LRWC) and leaf succulence. Here, it was summarized that Cd stress causes the disruption in water relationship as well as physiological drought in plants. The endogenous pro content was measured in leaves of Cd and Pb-exposed *Albizia julibrissin* after 7, 14, and 21 days of treatment (Baycu et al. 2017). After 21 days of treatment, Cd and Pb remarkably elevated the pro content about 25- to 40-fold than control which was insignificant in case of short-term exposure. Cu, Mn, Pb, and Zn toxicity accelerated the increase of pro content in *C. arietinum* which increased along with increase of metal concentration (Bhagyawant et al. 2019). This metal-induced higher pro concentration confirmed the stress condition in *C. arietinum* plants which was associated with pro-related gene expression. *A. hypogaea* seedlings also showed the gradual decrease in pro content with the increase of Cd toxicity from 50 to 200 μM levels and indicated the reduction of water status (Dong et al. 2016). *G. max* was suffered from the imbalance water relation upon Cd toxicity in both tolerant and sensitive genotypes (Alyemeni et al. 2017). Increased pro accumulation was higher in tolerant genotypes than sensitive genotypes, as Cd caused higher pro content compared to control. Consequently, LRWC was decreased in all genotypes of *G. max* with the gradual increase of Cd toxicity. Thus, the severity of pro and LRWC can be considered as the indicator of plant tolerance to stress. Pb-, Cu-, Cr-, and Mn-contaminated industrial waste water treatment to *M. peguensis, P. pinnata, A. lebbeck, B. purpurea, D. sissoo* increased endogenous pro contents significantly as stress indicator (Kanwal et al. 2019). According to maximum to minimum pro contents under stress conditions, these plants can be arranged as *A. lebbeck* > *M. peguensis* > *P. pinnata* > *D. sissoo* > *B. purpurea*. However, upon metal toxicity pro accumulation increased in all plant species.

2.5 Nodulation

Nodule is an important organ initiated on the roots of plant, principally in the member of Fabaceae family. Nodulation is a mutualistic relationship between host plant and soil bacteria, mainly between legumes and nitrogen-fixing bacteria, e.g., rhizobia. This interaction is very important for host plant as it gets extra atmospheric nitrogen through bacteria (Stambulska et al. 2018). Numerous scientific studies confirmed that toxic metal/metalloids destroy the normal function of nodulation of leguminous crop because nodule uptake more metal than the other parts of plant (Ghnaya et al. 2015). Haddad et al. (2015) carried out an experiment with three leguminous crops (clover, broad bean, and soybean) under different concentrations (0, 5, 25, 50, 100, and 250 mg kg^{-1} soil) of six heavy metals (Cd, Co, Cr, Cu, Ni, and Pb) and found that low concentrations of heavy metals did not notably affect process of nodulation and nodule number per plant, but under higher concentration those parameters become decreased notably (Table 2). In 2019, Navarro-Torre et al. recorded similar result from alfalfa plant under a different metal toxicity (Table 2). The familiar metalloid As also proved its damaging nature by hampering nodulation process of alfalfa and soybean (Melina et al. 2013; Pajuelo et al. 2008). Kong et al. (2017) checked the performance of *Medicago lupulina* L. under different concentrations of $CuSO_4$. They observed number of effective nodules, nodule FW, and nitrogenase activity decreased significantly with an increased amount of Cu in the medium. Recently, de Abreu et al. (2019) observed that Mn^{2+} did not inhibit the symbiotic process between isolated bacteria and plant species. In brief, it can be stated that toxic metal/metalloids hamper the overall nodulation process at higher concentration as nodule uptakes huge metal/metalloids from the growing media but is also a tolerance mechanism for legumes.

Table 2 Nodulation process affected by different metals/metalloids toxicity

Crops	Stress intensity and duration	Plant responses	References
Trifolium spp, *V. faba*, *G. max*	50, 100, and 250 mg $CdSO_4 \cdot 8H_2O$), $CoSO_4 \cdot 7H_2O$, $CuSO_4$, $Cr_2(SO_4)_3 \cdot 12H_2O$, $Pb(CH_3COO)_2$, $NiSO_4 \cdot 6H_2O$) kg^{-1} soil; 50 d	• Hampered nodulation process • Reduced nodule number per plant	Haddad et al. (2015)
M. sativa L.	650 μM As + 600 μM Cu + 30 μM Cd + 250 μM Pb + 750 μM Zn; 35 d	• Reduced nodule number per plant	Navarro-Torre et al. (2019)
M. lupulina L.	$CuSO_4$, 100 mg kg^{-1}, 200 mg kg^{-1}, 300 mg kg^{-1} growing medium; 50 d	• Decreased number of effective nodules • Decreased nodule FW • Decreased nitrogenase activity	Kong et al. (2017)

2.6 Reproductive Development, Yield, and Quality of Grain

Excessive heavy metal becomes responsible for producing lower yield in legumes. One hundred and fifty mg L^{-1} Ni contamination to soil medium led to inhibition in yield of *C. arietinum* by decreasing pod number and seed yield about 53 and 20%, respectively (Batool 2018). This Ni-mediated yield reduction might be correlated with shortage nutrient supply to the reproductive parts of *C. arietinum*. The yield of *C. arietinum* was investigated upon 200 μM of Cd toxicity (Ahmad et al. 2016). Cd stress significantly decreased the number of pods and seed yield by 50 and 53%, respectively. Moreover, Cd-mediated restriction in mineral and water uptake along with inhibition of photosynthetic pigment synthesis negatively affected this yield reduction togetherly. Both metal-tolerant (Sel 85 N) and metal-sensitive (P792) genotypes of *Cajanus cajan* were exposed to Cd and Zn toxicity at 15 days olds, where reproductive development and yield responses in metal-stressed plants were evaluated (Garg and Kaur 2013). Stress treatment delayed the flowering in both genotypes, while flower bud development drastically reduced in sensitive cultivars than tolerant one. Conversely, increasing the Cd toxicity had greater adverse effects in delaying the flowering process than Zn which indicated the more toxicity of Cd than Zn. Hence, metal stresses decreased the yield components in *C. cajan* indicated by lowered flower number, pod number, pod dry weight, seed number, seed dry weight, and 100-seed weight. In addition, yield responses to stress were lower in Cd-exposed plants than Zn exposure and thus revealed about the more toxicity of Cd to yield reduction than Zn (Garg and Kaur 2013).

Both Cd and Pb caused the reduction in yield of *A. hypogaea* and *G. max* (Fatoba et al. 2012). Hence, *A. hypogaea* and *G. max* were treated with 10, 20, 30, and 40 ppm of Cd and Pb, respectively, to observe the responses in yield attributes. Therefore, a significant reduction in pod numbers, seed numbers, and pod filling potentiality was found in both A. *hypogaea* and *G. max* upon higher toxic levels of Cd and Pb. *Phaseolus mungo* also showed significant reduction in yield potentiality upon Cd toxicity by decreasing of number of pods and seed yield (Siddhu and Khan 2012). The accumulated metal/metalloids in soil can alter the quality of produced grain in Fabaceae plants. Previous researchers studied about the negative effects of some of metals, viz. Cd, Cr, and Cu on grain quality, especially the protein content of mung bean grain, and found significant variation among the treatments in a dose-dependent way. Wani et al. (2007a) reported that Cr application can improve the gain protein in green gram. They also suggested that symbiosis did not get affected by the application of Cr, as a result then supply remains unchanged, and hence the protein content of grain was increased. Further, they added that grain protein content in green gram was significantly reduced by the application of Cd and Cu, when applied solely or in combination. On the other hand, when chickpea and mung bean were grown under metal toxic (Cd, Cr, Ni, Zn, Pb, and Cu) condition, the grain protein content decreased in a dose-dependent fashion, which increased after Cr and Pb application (Wani et al. 2007a, b; Table 3). Similarly, Cu-, Zn-, and Mn-contained sewage slims increased the protein content of grain as well as the nutrient (N and P) contents of

Table 3 Impact of toxic metal/metalloids toxicity on quality attributes of Fabaceae crops

Toxic metal/metalloids	Stress intensity	Crop	Grain quality indicator	Reference
Cd	0.011 (g kg^{-1} soil)	*Cicer arietinum* L.	11% decrease in grain protein	Wani et al. (2007a)
	0.023 (g kg^{-1} soil)	*C. arietinum* L.	22% decrease in grain protein	Wani et al. (2007a)
	0.006 (g kg^{-1} soil)	*Vigna radiata* (L.) R. Wilczek	4% decrease in grain protein	Wani et al. (2007b)
	0.012 (g kg^{-1} soil)	*V. radiata* (L.) R. Wilczek	6% decrease in grain protein	Wani et al. (2007b)
	0.024 (g kg^{-1} soil)	*V. radiata* (L.) R. Wilczek	8% decrease in grain protein	Wani et al. (2007b)
	0.1 mM	*Glycine max* (L.) Merr.	23% decrease in grain oil	Khan et al. (2013)
	0.5 mM	*G. max* (L.) Merr.	28% decrease in grain oil	Khan et al. (2013)
	1.0 mM	*G. max* (L.) Merr.	33% decrease in grain oil	Khan et al. (2013)
Cr	0.067 (g kg^{-1} soil)	*C. arietinum* L.	3% increase in grain protein	Wani et al. (2007a)
	0.135 (g kg^{-1} soil)	*C. arietinum* L.	2% decrease in grain protein	Wani et al. (2007a)
	0.034 (g kg^{-1} soil)	*V. radiata* (L.) R. Wilczek	5% increase in grain protein	Wani et al. (2007b)
	0.068 (g kg^{-1} soil)	*V. radiata* (L.) R. Wilczek	7% increase in grain protein	Wani et al. (2007b)
	0.14 (g kg^{-1} soil)	*V. radiata* (L.) R. Wilczek	11% increase in grain protein	Wani et al. (2007b)
Cu	0.066 (g kg^{-1} soil)	*C. arietinum* L.	9% decrease in grain protein	Wani et al. (2007a)
	0. 14 (g kg^{-1} soil)	*C. arietinum* L.	18% decrease in grain protein	Wani et al. (2007a)
	0. 33 (g kg^{-1} soil)	*V. radiata* (L.) R. Wilczek	4% decrease in grain protein	Wani et al. (2007b)
	0.66 (g kg^{-1} soil)	*V. radiata* (L.) R. Wilczek	5% decrease in grain protein	Wani et al. (2007b)
	1.38 (g kg^{-1} soil)	*V. radiata* (L.) R. Wilczek	6% decrease in grain protein	Wani et al. (2007b)
Ni	0. 29 (g kg^{-1} soil)	*C. arietinum* L.	2% decrease in grain protein	Wani et al. (2007a)
	0. 58 (g kg^{-1} soil)	*C. arietinum* L.	16% decrease in grain protein	Wani et al. (2007a)

(continued)

Table 3 (continued)

Toxic metal/metalloids	Stress intensity	Crop	Grain quality indicator	Reference
Pb	0.19 (g kg^{-1} soil)	*C. arietinum* L.	3% increase in grain protein	Wani et al. (2007a)
	0.39 (g kg^{-1} soil)	*C. arietinum* L.	6% increase in grain protein	Wani et al. (2007a)
Zn	4.8 (g kg^{-1} soil)	*C. arietinum* L.	10% increase in grain protein	Wani et al. (2007a)
	9.7 (g kg^{-1} soil)	*C. arietinum* L.	19% increase in grain protein	Wani et al. (2007a)
Hg	0.1 mM	*G. max* (L.) Merr.	38% decrease in grain oil	Khan et al. (2013)
	0.5 mM	*G. max* (L.) Merr.	58% decrease in grain oil	Khan et al. (2013)
	1.0 mM	*G. max* (L.) Merr.	68% decrease in grain oil	Khan et al. (2013)

lentil, whereas Zn application in chickpea decreased the content of grain protein content (Yamur et al. 2005). Further study suggested that toxic concentrations of metals, for example Cd and Hg, reduced the soybean grain oil content (Khan et al. 2013). They also suggested that metal toxicity decreased the essential fatty acid—oleic acid, and content in soybean decreased but palmitic, stearic, and linolenic acid increased markedly. They also added that the deleterious effect of Hg is greater than that of Cd on soybean oil content (Table 3).

2.7 *Oxidative Stress*

In recent era, the soil and water are contaminated by toxic metal/metalloids through industrial dumps, mining, sludges, and agrochemicals; as a result, the toxic buildup of metal/metalloids has been emerging as a potential threat to the plants as well as the human beings. Further, this heavy load of metal/metalloids is leached and dissolved with the groundwater aquifer, subsequently absorbed through the plant roots, and transferred into the aerial parts causing toxicity. This toxicity is extended from morphological to physicochemical level causing production of excessive ROS ($O_2^{\bullet-}$, $OH^\bullet$, $^1O_2$, and H_2O_2), which causes membrane lipid peroxidation, protein oxidation, enzyme inhibition, and damage to DNA and RNA (Anjum et al. 2012; Siddiqui 2013; Šiukšta et al. 2019). From biological point of view, metal/metalloids are of two groups, redox-active (Fe, Cu, Cr, Co) and redox-inactive (Cd, Pb, Zn, Ni, Al). The uptake and transport and distribution of metal/metalloids to organelles are guided to ROS generation, through redox activity by membrane-bound NADPH oxidase, while redox-inactive metal/metalloids are involved in the formation of $OH^\bullet$

from H_2O_2 via Haber–Weiss and Fenton reactions and initiate lipid peroxidation. Substantial indication proofed that Hg, Cu, Cr, Zn, Al, Pb, and Cd toxicity in Fabaceae plants can cause oxidative stress (Rui et al. 2016; Vijendra et al. 2016; Ashraf et al. 2016; Kapur and Singh 2019) (Table 4).

Nahar et al. (2017) reported that Al toxicity overproduced 83% of H_2O_2 and 110% of $O_2^{\bullet-}$, increased LOX activity up to 97%, and resulted in lipid peroxidation. They also observed a substantial increase (123%) in methylglyoxal content in the cell of mung bean plants. Anjum et al. (2011) reported that in *V. radiata* plants Cd level in soil (100 mg kg^{-1} soil) caused higher lipid peroxidation and H_2O_2 (69.6 and 113.9% increases in thiobarbituric acid reactive substances, TBARS, and H_2O_2 contents, respectively, over the control). The accumulation of H_2O_2 after Cd exposure has been detected in the leaves of different plant species, such as *Pisum sativum* (Romero-Puertas et al. 2004) and *V. mungo* (Singh et al. 2008). In another study, Bouazizi et al. (2010) assayed Cu toxicity in expanding *Phaseolus vulgaris* seedling leaves and found that the MDA content did not change significantly, but H_2O_2 level increased, when seedlings were exposed to 75 μM $CuSO_4$. They also observed that under 75 mM $CuSO_4$ stress GPX and CAT activities decreased, whereas GPX and CAT activities remained unchanged under 75 mM $CuSO_4$.

Mercury is another toxic metal for the leguminous crops, alfalfa root is very much sensitive to Hg^{2+}, and the root growth decreased in a dose-dependent way. Subsequently, the TBARS content increased, which was fourfold higher under 20 μM Hg^{2+}. In this context, maximum TBARS formation was accomplished by 12 h after Hg^{2+} exposure which was declined thereafter (Zhou et al. 2007). They also found that the activity of LOX increased, which is responsible for the peroxidation of biomolecules. Other antioxidant enzymes SOD and POD activities increased upon Hg exposure. But APX activity increased at a higher Hg concentration, whereas GR activity reduced at a lower Hg concentration (Zhou et al. 2007).

Thereafter, Anjum et al. (2014) also observed a significant increase of oxidative stress in roots and shoot of *V. radiata* in a dose-dependent way. The %EL, and H_2O_2 and TBARS illustrated a significant increase, where the maximum value occurred at 100 mg Cd kg^{-1} > 50 mg Cd kg^{-1}. Interestingly, the roots exhibited 1.6-, 1.3-, and 1.3-fold lower EL, H_2O_2, and TBARS, respectively, compared to the shoot. The enzyme GR, GPX, and GST exhibited varying responses, where GPX and GST activity decreased by Cd treatment, but GR activity increased. Among the non-enzymatic antioxidants, GSH content increase in *V. radiata* roots and shoot, it increased significantly, with reference to the control (Anjum et al. 2014). Similarly, Rui et al. (2016) examined the effects of Cd on growth, lipid peroxidation, reactive oxygen species (ROS) accumulation, antioxidant enzymatic activity, in the roots of two varieties of *V. sativa*. They observed that Cd treatment increased ROS and lipid peroxidation in the Cd-sensitive variety, where they found extracellular accumulation of H_2O_2 and $O_2^{\bullet-}$. Meanwhile, Vijendra et al. (2016) showed that Cd stress induced an increase $O_2^{\bullet-}$ in both leaf (up to fourfold) and root (up to 6.2-fold) tissues of *V. aconitifolia* L., compared to controls, which causes lipid peroxidation in leaves and roots. Compared to controls, significantly increased carbonyl content in leaves and roots was

Table 4 Impact of toxic metal/metalloids toxicity on oxidative stress of Fabaceae crops

Crops	Stress intensity and duration	Plant responses	References
Medicago sativa	40 μM Hg^{2+} as $HgCl_2$; 3 days	• Increased TBARS content in root and shoot • LOX activity increased • SOD isoform-1 increased remarkably • All the POD isoforms except VII and APX activity increased	Zhou et al. (2007)
Glycine max (L.) Merr.	Cd (0.3 mM as $CdSO_4.7H_2O$); reproductive stage	• MDA content increased by 26.81% • Proline content increased by 20.33% • H_2O_2 increased by 25.02% • AsA content dropped by 43.57% • SOD and APX activity increased but GR and CAT activity dropped	Kapur and Singh (2019)
G. max (L.) Merr.	Cd (0.6 mM as $CdSO_4.7H_2O$); reproductive stage	• MDA content increased by 26.81% • Proline content increased by 35.16% • H_2O_2 increased by 38.56% • AsA content dropped by 52.79% • SOD and APX activity increased but GR and CAT activity dropped	Kapur and Singh (2019)
Vigna aconitifolia L.	Cd 500 μM as $CdCl_2$; 15 d	• Increased $O_2^{\bullet -}$ and H_2O_2 in both shoot and root where 4.0-fold increase was observed in shoot compared with sevenfold increase in root • Both LOX and TBARS contents were increased in both shoot and root • AsA content was decreased, whereas GSH content increased in both leaves and root	Vijendra et al. (2016)

(continued)

Table 4 (continued)

Crops	Stress intensity and duration	Plant responses	References
V. radiata L. Wilczek cv. Pusa Ratna	Cd (50 and 100 mg Cd kg^{-1} soil); in the form of $CdCl_2$	• Roots exhibited 1.6-, 1.3-, and 1.3-fold lower levels of EL, H_2O_2, and TBARS, respectively • Compared to roots, the shoot displayed 2.6- and ≈3.0-fold higher activity of GPX and GST, respectively, and a twofold lower activity of GR	Anjum et al. (2014)
Cicer arietinum L.	3 and 4.5 g L^{-1} Cu as $CuSO_4.5H_2O$	• Lowest concentration showed nearly 16-fold higher tissue accumulation than the control at minimal dose and reaching more than 50-fold at highest dose of 250 mg L^{-1} dose	Bhagyawant et al. (2016)
C. arietinum L.	Pb 250 mL^{-1}as $Pb(NO_3)_2)$	• Lowest concentration showed nearly 16-fold higher tissue accumulation than the control at minimal dose and reaching more than 50-fold at highest dose of 250 mg L^{-1} dose	Bhagyawant et al. (2016)
C. arietinum L.	Zn 50 mL^{-1} as $ZnSO_4.7H_2O$, 7d	• At lower treatment doses, Zn accumulation in the tissues was nearly threefold	Bhagyawant et al. (2016)
C. arietinum L.	Mn 50 mL^{-1} as $MnSO_4.H_2O$, 7d	• At lower treatment doses, Mn accumulation in the tissues was nearly threefold	Bhagyawant et al. (2016)
V. radiata L. cv. BARI Mung-2	Al (0.5 mM) as $AlCl_3$; 48 and 72 h	• The content of H_2O_2, $O_2^{\bullet-}$ generation rate, and LOX activity increased by 83, 110, and 72%, which increased the lipid peroxidation by 97% under the Al stress of 48 h, compared to control • AsA content decreased, DHA content increased and AsA/DHA ratio decreased • Increased SOD, APX, MDHAR, DHAR, GPX, GST activity, whereas CAT activity decreased • Methylglyoxal level amplified by 86 and 123%	Nahar et al. (2017)

(continued)

Table 4 (continued)

Crops	Stress intensity and duration	Plant responses	References
Pisum sativum L.	Cd 50 μM as $CdCl_2$; 14 d	• Reduction of GR and GPX and, to a lower extent, of CAT, while total SOD activity showed a slight increase by the metal treatment • H_2O_2, $O_2^{\bullet-}$ content increased	Rodríguez-Serrano et al. (2006)
Vicia faba L.	Cd 10, 15, 20, 25 mg L^{-1} of substrate.	• Proline content increased in dose-dependent way	Alle et al. (2019)
V. faba L.	200, 400, 600, 800, 1,000 mg L^{-1} of substrate.	• Proline content increased in dose-dependent way	Alle et al. (2019)
V. mungo L. cv. Shekhar-2	0.25 mM Cr	• MDA content increased • SOD activity increased, while CAT decreased and POD remained unchanged	Srivastava and Shukla (2016)
V. mungo L. cv. Shekhar-2	0.25 mM Cd	• MDA content increased • SOD and POD activity increased, but CAT activity decreased	Srivastava and Shukla (2016)
V. mungo L. cv. Shekhar-2	0.25 mM Co	• MDA content increased • SOD and POD activity increased CAT activity decreased	Srivastava and Shukla (2016)
V. mungo L. cv. Shekhar-2	0.25 mM Pb	• MDA content increased • SOD, POD, and CAT activity decreased	Srivastava and Shukla (2016)

also recorded. Lipoxygenase activity was found to be increased in both leaves and roots of stressed plants.

In recent study, Kapur and Singh (2019) tried to find out the interaction between Cd and Zn in soybean [*G. max* (L.) Merr.]. They observed the Cd-induced higher activity of APX and SOD, accompanied by the other non-enzymatic antioxidants, but along with the stress severity H_2O_2, MDA, and proline content also raised creating oxidative stress of plants. Also, the CAT and GR activity reduced along with a decrease in AsA content with the increasing dose of Cd. They also found an antagonistic effect with Zn and concluded that when Zn is applied at a higher dose the activity of antioxidants increased. Moreover, the efficacy of Zn was more when applied in higher concentrations with low Cd. Thus, Zn plays a key role in plants to counter heavy metal stress by elevating antioxidative defense with higher activity of enzymes and reduced levels of non-enzymatic metabolites, and efficacy of Zn in combination is dose-dependent.

3 Tolerance Mechanism to Metal/Metalloid Toxicity

Tolerance of Fabaceae plant mostly depends on the type of metals/metalloids. Plant tries to adjust with stress condition by nutrient balance, metal chelation, and synthesis of different biomolecules including organic acids, phytohormones, osmolytes, polyamines, signaling molecules, etc. Here, we have discussed the tolerance mechanism of Fabaceae plant to combat metal/metaloid toxicity.

3.1 Nutrients

Different nutrients contributes in enhancing metals/metalloids stress tolerance in Fabaceae plant. Function of manures on heavy metals toxicity reduction and phytoremediation was studied in *Sinapis alba* L. and *Robinia pseudoacacia* L. Seed germination, biomass, root length, guaiacol peroxidase activity, and protein concentration were improved by manure application in soil contaminated with heavy metal. Those plants also showed better phytoremediation after manure application (Marta et al. 2019). Potassium (80 and 160 kg ha^{-1}) decreased heavy metal concentrations and *Lablab purpureus* productivity. This plant also showed increased plant biomass and enhanced root growth and mycorrhizal fungal colonization rates which were supposed to reduce metal content and toxicity in leaf (Ruthrof et al. 2018). Biochar and *Bradyrhizobium japonicum* were applied in metal-rich soil containing Ni, Mn, Cr, and Co. *Vigna mungo* plant growth, and N and P content of soil were upregulated by biochar and activity of those N-fixing bacteria. Soil bioavailable fractions of metals were also increased. 2.5% of biochar showed better performance than the 5% level of biochar (Seneviratne et al. 2017). Effect of individual application of Ca and K, and combined application of Ca and K to improve physiological performance of chickpea plants under Cd stress was studied. Addition of these macronutrients either alone or in combination maintained osmotic balance through Pro regulation in Cd-affected seedlings, total phenol, flavonoid, total protein, and enzymatic components of antioxidant defense system including SOD, CAT, APX, and GR activities upregulated by those macronutrients application under Cd stress. Reduction of H_2O_2 and MDA was the indication of Cd-induced oxidative stress reduction by Ca and K application. Regulation of root and shoot Cd accumulation together with mineral homeostasis maintenance (regulation of S, Mn, Mg, Ca, and K) was conferred by Ca and K, either single or dual application. Lastly augmenting the photosynthetic pigment (chl *a*, chl *b*, and total chl) and improving shoot and root growth, yield attributes and seed yield of chickpea Ca and K showed its additive effects in conferring Cd stress tolerance (Ahmad et al. 2016). Calcium and/or potassium were assessed for their role in Cd tolerance development in faba bean [*V. faba* L. (cv. TARA)]. Cadmium toxicity enforced faba bean plants toward physiological disorder. Calcium and potassium supplementation improved plant performance under Cd stress condition. Calcium and potassium improved the activity of antioxidant system enzyme activity including CAT, POD, and SOD. Exogenous addition of these nutrients also increased Pro content, decreased MDA level, and increased chl *a* and chl *b* contents. Root length, fresh and dry weight, shoot length, and fresh and dry weight plant^{-1} were also improved by Ca and K addition in *V. faba* under Cd stress. Magnesium-induced alleviation of Al toxicity was studied in rice bean [*V. umbellata* (Thunb.) Ohwi and Ohashi]. Upregulation of Al-induced citrate efflux also increased root growth. Beneficial role of Mg was through the citrate efflux in root. Vanadate (a H^+-ATPase inhibitor) helped to reduce Al-induced citrate efflux, but Mg reinstated the efflux. Supplementation of Mg enhanced activity of plasma membrane H^+-ATPase which was decreased by Al. Enhanced activity of plasma membrane H^+-ATPase by additional Mg application increased Mg and calcium contents in root apices (Yang et al. 2007). Function of calcium addition on chickpea (*C. arietinum* L.) germinating seeds exposed to cadmium stress was studied. Cadmium increased oxidative stress, but Ca supplementation decreased lipid peroxidation and protein carbonylation which indicates oxidative stress relaxation. Calcium addition decreased shoot and root Cd content. Shoot and root lengths decreased by Cd addition but increased by Ca supplementation with Cd. Under Cd toxicity, high increase of NAD^+/NADH and $NADP^+$/NADPH and activity of oxidative [NAD(P)H-oxidase] and dehydrogenase (glucose-6-phosphate dehydrogenase, 6-phosphogluconate dehydrogenase, and malate dehydrogenase) indicate oxidative stress which was alleviated by Ca. Redox balance was examined through the study of the redox state of pyridine nucleotide couples NAD^+/NADH and $NADP^+$/NADPH as well as their related oxidative [NAD(P)H-oxidase] and dehydrogenase (glucose-6-phosphate dehydrogenase, 6-phosphogluconate dehydrogenase, and malate dehydrogenase) enzyme activities. Cell death percentage highly increased by Cd toxicity which was reversed by Ca (Sakouhi et al. 2016). Faba bean (*V. faba* L.) seeds were treated with H_2O [nonprimed (NP)], or faba bean (*V. faba* L.) seeds were primed with $CaCl_2$, then these were exposed to $CdCl_2$, and also compared with nonprimed faba bean plants. Primed seedlings showed better performance in terms of gas exchange parameters, chl, and carotenoid contents than the nonprime seedlings under Cd stress. Cadmium acted as source of oxidative stress through increasing the H_2O_2 and $O_2^{\bullet-}$, increased electrolyte leakage which is an indication of membrane injury. In contrast, $CaCl_2$ priming alleviated ROS generation, oxidative damage, and decreased electrolyte leakage from the Cd-affected seedlings (Nouairi et al. 2019). Cadmium stress (0, 400, and 600 μM) caused oxidative stress to *Trifolium repens* L. plants and highly increased production of H_2O_2 content and thiobarbituric acid reactive substances (TBARS). The reason behind oxidative stress was disruption of antioxidant defense system. Cadmium also decreased growth of the plants. Application of $CaCl_2$ (5 mM) together with Cd treatment regulated several enzymes of antioxidant defense system such as activities of SOD, CAT, APX, and GPX which acted to enhance oxidative stress tolerance, decreased oxidative damage, and improved membrane integrity (compared to Cd-stressed plants without Ca supplementation). Calcium supplementation also increased the growth parameter of Cd-affected plants (Wang and Song 2009). Cadmium being a nonessential element is toxic for plants even in small amount. Zinc is essential nutrient element but becomes toxic at elevated levels. Silicon (Si) has been recognized as trace elements for plants having biological function and role in improving stress tolerance. Cadmium and zinc toxicity resulted in several deleterious effects on growth and development as well as physiology of pigeon pea plants. Silicon application and *Rhizophagus irregularis* inoculation recovered pigeon pea plants from those damages. Silicon was effective in upregulating cellular silicon content, and nitrogen and phosphorus content under Cd stress. Reduction of Cd or Zn content, silicon-mediated higher generation of non-protein thiol, and phytochelatin is correlated with the greater reduction of Cd and Zn content from the plant body. Si addition considerably increased the activity of SOD, CAT, and guaiacol peroxidase as well as GSH level of the pigeon pea plants under Cd or Zn stress. The enhancement of the antioxidant system thus declined the oxidative stress by decreasing H_2O_2 and MDA, percent of nodular membrane electrolyte leakage. Overall growth of pigeon pea was also improved significantly by Si addition under Cd or Zn stresses (Garg and Singh 2018)

3.2 Organic Acid

Organic acids proved their performance against diverse metals/metalloids toxicity in Fabaceae plant. Cadmium highly increased lipid peroxidation and protein carbonylation, NAD^+/NADH and $NADP^+$/NADPH, and activity of oxidative [NAD(P)H-oxidase] and dehydrogenase (glucose-6-phosphate dehydrogenase, 6-phosphogluconate dehydrogenase, and malate dehydrogenase) in chickpea (*C. arietinum* L.) plants which indicate oxidative stress. Shoot and root Cd content increased to a great extent under Cd stress. Shoot and root lengths decreased by Cd addition. Cadmium caused cell death. Ethylene glycol tetraacetic acid (EGTA) addition recovered damage of chickpea (*C. arietinum* L.) plants from cadmium stress. Addition of EGTA decreased Cd content of root and shoot, reduced oxidative stress, and percentage of cell death. The growth parameters measured by length of shoot and root were also increased by EGTA addition with Cd treatment, compared to Cd treatment only (Sakouhi et al. 2016). Organic acid-induced Al tolerance was inspected in different genotypes of soybean. Root tip of different genotypes of soybean released organic acid in exposure of Al. Citrate and malate secretion increased within 6 h of Al exposure. Higher amount of Al accumulated in root tips of the Al-sensitive genotype Young, compared to the Al-tolerant genotype PI 416937. Higher amount of organic acid accumulation specially the citrate accumulation was correlated the exclusion of Al and tolerance of Al. Lateral root showed higher susceptibility to Al, compared to taproots. Lateral root tips accumulated higher amount of Al which had lower content of citrate (Silva et al. 2001). Oxalate acid (50 and 100 μmol L^{-1}) containing nutrient media was made toxic with Cd (8 mg L^{-1}). Another nutrient solution was made toxic only with the same concentration of Cd. Control plants were grown only in the nutrient media. In bean plant, Cd in roots and shoots significantly increased due to the presence of Cd in the nutrient solution. The bean plants grown in oxalate acid solution with Cd performed better because the root and shoot Cd content was lower in this treatment, compared to the plants grown only in Cd containing nutrient media. Organic acid made a complex with Cd which was absorbed by roots and transported to shoot and became less toxic. Organic acid addition also improved root and shoot growth under Cd stress (Dongsen et al. 1995).

3.3 Osmolytes

Under metals/metalloids stress, a different osmolyte synthesis in cellular level is increased up to a certain extent which helps plant to survive. Function of glycinebetaine (GB) in augmenting Cr tolerance in mung bean was experimented. Different concentrations of Cr were applied alone or in combination with GB. Plant growth and biomass were reduced by Cr. Photosynthetic pigment contents, chl *a*, chl *b*, and carotenoids, were damaged by Cr. Chromium concentration within the mung bean plants and electrolyte leakage considerably and progressively increased in plants with rise in Cr levels. Activities of SOD, peroxidase (POD), and CAT were boosted up by exogenous GB which diminished electrolyte leakage. Reduction of Cr content inside plant and higher photosynthetic pigment contents were advantageous consequences of GB application in Cr-affected plants, and such effects promoted growth of Cr-affected plants (in contrast to Cr-affected plant without GB supplementation) (Jabeen et al. 2016). Thirty days aged chickpea plants were applied with 0, 25, 50, or 100 mg of Cd kg^{-1} of soil. The plants were sprayed with 20 mM proline. The control plants were sprayed with double distilled water. Proline spray improved plant performance against cadmium stress. Cadmium stress resulted in differential damage effects in chickpea plants that were severe with the increase of harshness of stress. Nodulation parameters, leghemoglobin and carbohydrate content, leaf nitrogen, and root nitrate content were increased by proline addition with Cd treatment. Activity of nitrogenase, nitrate reductase, glutamine synthetase, glutamate synthase, and glutamate dehydrogenase was increased in Pro-added Cd (all doses except 100 mg kg^{-1} of soil)-affected plants (Alyemeni et al. 2016). Increased amount of trehalose was confirmed in the nodules of Pusa 2002 than Pusa 991 cultivar of pigeon pea under Cd and Zn stress representing its function as osmoprotectant. Mycorrhization and Si supplementation further upregulated the activity of Tre biosynthetic enzymes trehalose-6-P synthase (T6PS) and trehalose-6-P phosphatase (T6PP) with concurrent decline in trehalase activity, which ensured improved endogenous Tre accumulation. This upregulated trehalose is supportive for mitigation of oxidative stress and lipid peroxidation under Cd and Zn stress (Garg and Singh 2018). Chickpea (*C. arietinum* L.) cv. Avarodhi plants were imposed with Cd (25, 50, or 100 mg per kg of soil) stress, and the role of exogenously sprayed 20 mM Pro was studied. Exogenous Pro application increased the endogenous level of Pro and increased leaf water potential. Carbonic anhydrase activity was amplified by Cd exposure, whereas Pro spray under Cd stress reduced its activity. Increase of CAT, peroxidase, and SOD activities was involved in defending oxidative damage of Cd-affected plants. Cadmium toxicity resulted in differential negative impact on photosynthetic parameters. When these plants were applied with Pro, the gs, Ci, E, WUE, and PN were significantly improved, compared to Cd-stressed plants. Proline also improved root and shoot lengths, and plant fresh and dry mass of Cd-affected plants. The overall improvement of physiological and growth parameters contributed to improve the yield attributes and yield of chickpea and seed quality of the plants in terms of pod number and number of seed pod^{-1}, seed yield $plant^{-1}$, 1000-seed weight, seed protein content (Hayat et al. 2013). Proline biosynthesis by SA addition in Cd-affected plants helped to confer osmoprotection. The same treatment also conferred oxidative stress protection as indicated by decreased lipid peroxidation and electrolyte leakage. Photosynthetic and growth performance were also improved (Popova et al. 2009). Selenium (Se), the trace elements, showed its toxic effects in hydroponically grown bean (*P. vulgaris* L.) seedlings at 4 and 6 ppm, but 1 and 2 ppm showed beneficial effect. Reduction of germination was recorded under Se stress. Root and shoot growth, chl content, leaf water content, and cellular respiration showed decreased pattern under Se toxicity. Elevation of lipid peroxidation and hydrogen peroxide which were due to disrupted enzymatic function including SOD, CAT, APX and GR together with reduced function of AsA and GSH. Proline content reduced considerably at 4 and 6 ppm levels. Exogenous application of Pro (50 μM) with the Se alleviated the injurious effect of Se and attributed better physiology in terms of all the studied parameters and growth of seedlings (Aggarwal et al. 2011).

3.4 Polyamine

Polyamine plays a significant role to overcome metal/metalloids toxicity in Fabaceae plant. Mung bean plants showed enhanced tolerance to metal stress when they were co-treated with polyamine. Spermine supplementation reduced Cd uptake and decreased further toxic effects in different physiological parameters in mung bean plant. Spermine decreased the accumulation and translocation of Cd from the growing media to root toward different parts of the mung bean plants. Spermine-treated seedlings showed higher phytochelatin content. Cadmium application increased oxidative damage in the mung bean plants which is indicated by the increase of H_2O_2 content, $O_2^{\bullet-}$ generation rate, LOX (lipoxygenase) activity, and lipid peroxidation level, whereas spermine supplementation decreased the oxidative damage by increasing the AsA and GSH contents and activities of antioxidant enzymes including SOD, CAT, GST, MDHAR, DHAR, and GR. Osmoregulation was maintained by increasing endogenous proline content induced by exogenous spermine for which water content of Cd-affected plants improved, compared to Cd-stressed plants without spermine application. Spermine treatment also increased the growth of seedlings of Cd-affected mung bean seedlings. Increasing the activities of glyoxalase system, the methylglyoxal was detoxified that also decreased Cd-induced toxicity in mung bean seedlings. The cytotoxicity of methylglyoxal (MG) was also reduced by exogenous Spm because it enhanced glyoxalase system enzymes and other component of this system (Nahar et al. 2016a). Putrescine (Put) increased Ni, Cd, and Pb accumulation in chickpea. Joined application of Put and *Bacillus subtilis, Bacillus thuringiensis*, and *Bacillus megaterium* increased the accumulation higher than the single application. Thus, Put showed its capacity to enhance phytoremediation capacity of chickpea (Khan and Bano 2017). Polyamine and nitric oxide both are recognized for their signaling function within the plants. They are interlinked in their biosynthesis pathways. Polyamines can be directly converted into nitric oxide, or polyamines induce NO production. Combined effect of polyamine and NO has been detected for putative effect on different abiotic stressed plants. Putrescine (0.2 mM) and nitric oxide (sodium nitroprusside, 1 mM) were applied concomitantly to defend Cd toxicity in mung bean seedlings. Restoration of root and shoot growth, chl repairment, reestablishment of osmotic adjustment (adjusted Pro), and leaf relative water content restoration were prominently noticed in mung bean plants as a result of combined application of putrescine and nitric oxide. Putrescine and nitric oxide either single or combined application alleviated oxidative damage by enhancing non-enzymatic and enzymatic components of antioxidant systems. Putrescine and nitric oxide modulated Gly I and Gly II activities and GSH content of glyoxalase system which reduced MG content and its adverse effect. One of the main reasons for the advantageous effect was reduction of Cd content within the root and shoot which was due to increase of GSH content that is involved in the formation of metal chelating agent phytochelatin. The research also suggested for probable membrane function that restricted entrance of Cd ion through the membrane (Nahar et al. 2016b). Aluminum toxicity aggravated reactive oxygen species (H_2O_2 and $O_2^{\bullet-}$), and generation, enforced lipoxygenase activity, and thus caused membrane lipid peroxidation in mung bean seedlings. The methylglyoxal content also increased significantly that is highly cytotoxic and oxidative in nature. Boosting the AsA and GSH content, AsA/dehydroascorbate (DHA) ratio, GSH/glutathione dissulfide (GSSG) ratio, activity of different enzymes of APX, DHAR, GR and CAT exogenous spermidine reduced reactive oxygen species production and oxidative stress induced by Al stress. Spermidine enhanced glyoxalase system and diminished methylglyoxal level. Reduction of chlorosis symptom by increasing chlorophyll content was observed in spermidine-supplemented Al-stressed seedlings, compared to the aluminum-stressed seedlings. Spermidine supplementation alleviated wilting symptoms of aluminum-affected mung bean seedlings by adjusting the osmolytic balance, i.e., the proline level which is also proved from higher leaf relative water content (Nahar et al. 2017).

3.5 *Phytohormone*

Different phytohormones confirmed their positive performance against metals/metalloids toxicity. Intonation of CuO nanoparticle toxicity by indole-3-acetic acid (IAA) was reported in green pea. Combination of different doses of IAA and CuO nanoparticles was tested. Some combinations did not show beneficial effect on green pea. Nano-CuO (10–100 nm, 74.3% Cu) and bulk CuO (bCuO, 100–10,000 nm, 79.7% Cu) were applied. Application of IAA at 10 and 100 μM on all Cu treatments reduced the number of plants. Supplementation of IAA at 10 μM, nano-CuO (100 mg/kg) and bCuO (50 mg/kg) increased stem Cu by ~84 and ~78%. On the other hand, when IAA was increased to 100 μM, nano-CuO and bCuO diminished stem Ca by 32 and 37%, and Mg by ~35% (Ochoa et al. 2017). Mung bean plants were exposed to Al stress, and exogenously applied salicylic acid was tested to alleviate adverse effect of Al. Decrease of length, and fresh and dry mass of root and shoot were recorded in aluminum-affected plants. Altered enzyme activities, osmolytic imbalance, and decreased water content were observed in aluminum-affected plants. Aluminum exposure destructed chlorophyll and declined rate of photosynthesis. Activity of carbonic anhydrase and proline content was regulated by exogenous salicylic acid. As a result, relative water content and water-use efficiency were higher in aluminum-affected and salicylic acid-supplemented mung bean plants. Salicylic acid also increased chlorophyll content and the rate of photosynthesis. Moreover, salicylic acid supplementation in Al-affected plants enhanced antioxidant defense system which is demonstrated from higher activities of antioxidant enzymes including catalase, peroxidase, and superoxide dismutase (Ali 2017). Non-enzymatic and enzymatic components of antioxidant systems. Putrescine and nitric oxide modulated Gly I and Gly II activities and GSH content of glyoxalase system which reduced MG content and its adverse effect. One of the main reasons for the advantageous effect was reduction of Cd content within the root and shoot which was due to increase of GSH content that is involved in the formation of metal chelating agent phytochelatin. The research also suggested for probable membrane function that restricted entrance of Cd ion through the membrane (Nahar et al. 2016b). Melatonin (*N*-acetyl-5-methoxytryptamine) is considered as a plant growth regulator and stress tolerance. In acidic soil, Al toxicity is considered as major constraint for plant's growth. Melatonin and gene encoding acetyltransferase NSI-like (nuclear shuttle protein-interacting) upregulation under Al stress were demonstrated to decrease Al-induced phytotoxicity in soybean (*G. max* L.). Further addition of melatonin (0.1 and 1 mM) conferred Al tolerance. Root growth inhibition and H_2O_2 generation were alleviated by melatonin addition with Al. Activities of the antioxidant enzymes (SOD, POD, CAT) and increased exudation of malate and citrate were recorded in melatonin-added Al treatment. Thus, melatonin enhanced Al tolerance in soybean. But high concentrations of melatonin (100 and 200 mM) had a negative effect and even reduced root growth in Al-imposed seedlings (Zhang et al. 2017). *Cassia tora* L. plant's root tip was applied with 20 μM Al. Aluminum-affected plants were damaged through production of H_2O_2 and $O_2^{\bullet -}$ in root tip. The oxidative stress damaged the plasma membrane. When the Al-exposed plants were supplemented with salicylic acid (SA), the plants showed reduced H_2O_2 and $O_2^{\bullet -}$, reduced oxidative stress, electrolyte leakage, and membrane damage. The reason behind the oxidative stress alleviation was modulation of antioxidant defense system. Activity of SOD, CAT, APX, GR, and guaiacol peroxidase was modulated which has been recognized as a key to enhance oxidative stress as well as Al tolerance (Wang et al. 2004). Role of auxin was inspected for enhancing Al tolerance in soybean. Auxin regulated organic acid production and membrane properties. Increase of indole acetic acid (IAA) content by 25 and 50 μM Al exposure to root of soybean plant increased citrate exudation and activation of the plasma membrane H^+-ATPase activity. In fact, in the presence of Al auxin stimulated citrate exudation through upregulation of *GmMATE*. Augmented phosphorylation of the plasma membrane H^+-ATPase in soybean roots was also correlated with higher auxin content in the presence of Al (Wang et al. 2016). Function of IAA (10 and 100 μM) in decreasing damaging effects of hexavalent chromium (Cr VI; 50, 100, and 250 μM) was studied in pea seedlings. Seed germination rate, nitrate reductase, nitrite reductase, glutamine synthetase, and glutamate synthase (GOGAT) activities were inhibited by Cr, but ammonium content and glutamate dehydrogenase activity were upregulated. Pea seedlings applied with IAA alleviated those damaging effects. Chromium accumulation was regulated by IAA. Superoxide radicals and reactive carbonyl groups were increased by Cr which is an indicator of oxidative stress, while IAA addition alleviated oxidative stress. Conversely, 100 μM IAA did not show positive effects on Cr-affected plants; moreover, this concentration of IAA showed toxic effects (Gangwar and Singh 2011).

Auxin induced and increased the production of adventitious roots under heavy metal stress in mung bean [*V. radiata* (L.) Wilczek]. Exposure of $CdCl_2$ inhibited generation and growth of adventitious roots, amplified soluble protein and MDA contents, and reduced the activity of peroxidase and APX, and content of GSH and phenols. Increase of CAT and SOD activities together with higher AsA content was demonstrated under Cd stress. Indole-3-butyric acid reverted inhibition in adventitious root development, controlled the antioxidant defense mechanism, and prevented oxidative stress of Cd-treated mung bean plants (Li et al. 2018). Abscisic acid (ABA) addition showed regulatory effect on antioxidative defense systems and the mitigation of Cd-stressed oxidative damage and physiology of mung bean plants. Regulation of SOD, APX, peroxidase, CAT activity, IAA oxidase activity, and AsA and GSH content was noticed in ABA-added Cd-stressed mung bean plants. Number and fresh weight of the adventitious roots of Cd-stressed seedlings were also increased by ABA. So, ABA induced Cd tolerance by the growth promotion, antioxidant system upregulation, and physiology of mung bean plants (Li et al. 2014). Pigeon pea [*Cajanus cajan* (L.) Millsp] was tested for its response to Al stress and role of 28-homobrassinolide (28-HBL) in alleviating Al stress. Seed germination and seedling growth parameters including root and shoot lengths, fresh weight, and dry weight decreased under Al stress. Inclusion of 28-HBL decreased the undesirable effects of Al on different growth parameters of seedling. Addition of 28-HBL augmented the Pro level and boosted the CAT, peroxidase, SOD, and APX activity of Al-stressed pigeon pea seedlings (Madhan et al. 2014). Mung bean seedlings were treated brassinolide to examine its role to eliminate the adverse effect of Al stress. Epicotyl length, dry weight of root and shoot, and chl content were decreased by Al, whereas brassinolide-added Cd-treated seedlings showed better performance in terms of the above-mentioned parameters (Abdullahi et al. 2003). Exposure of *P. vulgaris* L. to $CdCl_2$ increased the Cd content within the plant cell and increased membrane lipid peroxidation, and membrane stability index which caused electrolyte leakage. Cadmium also caused loss of water which is indicated by reduced relative water content. When the Cd and 24-epibrassinolide were applied concomitantly, the Cd content decreased in plant. The rising of proline level acted to increase the relative water content increased in Cd-affected plants after treatment with 24-epibrassinolide. The enhanced antioxidative enzyme activity boosted up by 24-epibrassinolide alleviated oxidative damage, improved membrane properties, and prevented electrolyte leakage. As a result, overall production of *P. vulgaris* L. increased under Cd stress. However, 24-epibrassinolide also performed better without stress condition (Rady 2011). Exogenous SA increased phytoremediation capacity of heavy metal of chickpea. The SA was applied singly or together with plant growth-promoting rhizobacteria (PGPR; *Bacillus subtilis, B. thuringiensis*, and *B. megaterium*). Either individual or together, SA and PGPR increased heavy metal (Ni, Cd, and Pb) accumulation in shoot. Addition of SA also increased macronutrient accumulation in shoot and root. Increase of Pro and decrease of lipid peroxidation level were also recorded in SA treatment in chickpea under heavy metal stress (Khan and Bano 2017). Biosynthesis of IAA and phosphate solubilization by the plant growth-promoting bacteria (PGPB) facilitated plant growth and development. Application of KM594397 increased the chickpea growth and dry matter under metal stress as compared to only metal-treated plants. Phosphorus availability upregulated the efficiency of biological nitrogen fixation. Inoculation with PGPB (*Rhizobium* spp.) improved the root and shoot growth and grain yield in chickpea plants (Gull et al. 2004). Salicylic acid improved pea plant performances in terms of shoot and root growth, CO_2 fixation, and the activity of ribulose-1,5-bisphosphate carboxylase (RuBPC). Reduction of lipid peroxidation and electrolyte leakage indicates oxidative stress reduction. Higher Pro level in SA-pretreated plants improved osmotic regulation under Cd stress. Adverse effect of Cd on photosynthesis, carboxylation reactions, and thermoluminescence characteristic parameters was also decreased in SA-pretreated plants. Overall improvement in the content of chl and growth was attributed by SA under Cd stress (Popova et al. 2009). Peanut plants were subjected to 200 μM $CdCl_2$ without or with 0.1 mM SA. It was clearly apparent that Cd accused oxidative injury whereas improving SOD, CAT, and peroxidase activity SA decreased oxidative damage. Application of SA reduced Cd distribution among different organalles. Uptake of nutrients including K, Ca, Mg, and Fe was improved by SA under Cd stress than the Cd treatment addition alone (Xu et al. 2014). Kinetin (KN; 10 and 100 μM) improved pea seedlings performance under Mn stress (50, 100 and 250 μM Mn; Gangwar et al. 2010). Growth, photosynthetic pigments, total protein, total nitrogen, ammonium (NH_4^+) content, NH_4^+ assimilating enzymes, and antioxidant system in pea seedlings decreased under Mn stress but improved by KN treatment. Kinetin increased the root and shoot, AsA and GSH content as well as activities of glutamine synthetase (GS), glutamate oxoglutarate aminotransferase (GOGAT), and CAT, GR, and DHAR activities in Mn-affected plants which were considered as a key to enhance Mn toxicity tolerance (Gangwar et al. 2010).

3.6 Signaling Molecule

Signaling molecule plays an important role to activate some enzyme and their related genes for enhancing metals/metalloids stress tolerance. Peanut (*Arachis hypogaea* L.) seedlings were experimented for their responses to sodium nitroprusside (SNP, a donor of NO) for defending cadmium toxicity effects or not. Cadmium imposition inhibited growth. Photosynthesis was hindered due to cadmium stress. Oxidative stress was also accelerated due to cadmium toxicity. Cadmium was accumulated mostly in root and leaf cell wall. When SNP was applied, the soluble fractions of Cd in cell organelles decreased whereas the root and leaf cell wall accumulation and deposition were higher, compared to without SNP addition under Cd stress. Improved photosynthesis, upregulated antioxidant system, and reduced Cd translocation from roots to shoots (as evidenced by decreased Cd accumulation) were the protective effects of SNP which conferred Cd tolerance (Dong et al. 2015). Application of nitric oxide (SNP, 1 mM) in Cd (1.5 mM $CdCl_2$)-affected mung bean seedlings improved growth, chl content, proline, and leaf relative water content, compared to Cd-affected plants without SNP addition. Cadmium increased H_2O_2 and $O_2^{\bullet-}$ content, LOX activity, and MDA content pointing toward oxidative stress, and SNP incorporation inverted the oxidative damage in Cd-stressed plants. Different components of antioxidant system like AsA, AsA/DHA, SOD, CAT, APX, MDHAR, etc., were modulated for ensuring oxidative safeguard of cellular components of SNP-supplemented Cd-affected plants. Glutathione metabolism was significantly modulated by SNP in Cd-distressed mung bean plants. The higher GSH content, GSH/GSSG, GR activity, and GST activity were acquired by SNP supplementation. The higher GSH is not only a component of antioxidant system but also structural components of phytochelatin and a component of glyoxalase system. The mung bean seedlings with SNP addition produced higher phytochelatin which chelated Cd and thus the reduction of root and shoot Cd content, biological concentration factor, translocation factor, and biological accumulation coefficient in mung bean seedlings. Increased GSH level together with modulation of Gly I and Gly II activity of glyoxalase system the toxic MG level was reduced by SNP in Cd-affected plants (Nahar et al. 2016b). Exogenous addition of SNP alleviated the injurious effects of Cd on adventitious rooting in mung bean hypocotyl cuttings in various ways. SNP-induced higher content of GSH, AsA, polyphenol, and Pro levels in Cd-affected plants, compared to Cd-treated plants without SNP. Augmentation of APX, POD, CAT, and SOD activities was also evident in SNP-supplemented Cd-affected plants. All these components were involved in oxidative stress alleviation. The Pro was not only involved in antioxidant defense but also involved in osmoregulation. Indole-3-acetic acid oxidase (IAAO) activity was reversed by SNP addition in Cd-stressed. All these components contributed to perform better in SNP-supplemented Cd-affected seedlings (Li et al. 2019). Cadmium reduced root growth and NO accumulation, and caused ROS burst in roots of in *M. truncatula*. Supplementation with NO resulted in inhibition of ROS generation and reversed root growth retardation in Cd-stressed plant. By slowing down IAA oxidase activity, NO prevented auxin degradation. Exogenous NO also enhanced the uptake of K^+ and Ca^{2+} . So, NO upregulated cadmium tolerance by reducing oxidative damage, maintaining the auxin equilibrium, and augmenting ion absorption (Xu et al. 2010). Cadmium in peanut (*A. hypogaea* L. cv. Huayu 22 and cv. Xiaobaisha) exerted negative effects. When 0.25 mM SNP was applied, improved better growth, chl content, photosynthesis, and mineral nutrition were noticed, compared to the Cd-affected plants. Addition of SNP decreased Cd-induced oxidative stress (H_2O_2, superoxide, and MDA content) by increasing activities of antioxidant enzymes SOD, CAT, and peroxidase activity, and AsA and GSH content. Root and shoot Cd contents were greatly decreased by SNP addition. Addition of SNP also maintained nutrient homeostasis increasing K, Ca, Mg, and Fe content in Cd-affected plants, compared to Cd-affected plants without SNP addition (Xu et al. 2014).

3.7 *Metal Chelation*

In general, plants are well equipped to reduce metals/metalloids toxicity at cellular level under certain extent of stress concentration. Among the tolerance strategies, synthesis of cysteine-rich/originated compound, phytochelatins and metallothioneins, is one of the important approaches which contribute in metal chelation. In this process, phytochelatins and metallothioneins make a complex with accumulated metals/metalloids and that are being transferred into cell vacuole as immobile/inert forms to reduce stress damage (Nahar et al. 2016a, b; Mahmud et al. 2019; Fig. 1). Metal/metalloid toxicity makes active the enzyme phytochelatin synthase which increases the synthesis of phytochelatin as well as phytochelatin metals/metalloids complex. On the contrary, cellular MTs are remarkably diverse and it is subdivided into three types based on the arrangement of the Cys residue (Zhou et al. 2006). Becana et al. (2010) reported that nodules are characterized by high levels of antioxidant defenses including nicotinamine, phytochelatins, and metallothioneins, which may contribute to detoxify Cd. Moreover, nodulation is a usual process where huge amount of Fe, Mg, Ca, and K are necessary. As a result, these metals are generally accumulated in the nodules for its normal functioning and contribute in reducing metal toxicity in Fabaceae plant which makes them more efficient metal accumulator than the plants of other family (Brear et al. 2013). Nahar et al. (2016a) reported that 1 mM and 1.5 mM Cd stress enhanced phytochelatin content of mung bean plant by 190 and 267%, respectively, while as a stress protectant exogenous spermine further increased phytochelatin content. In a separate experiment, Nahar et al. (2016b) confirmed similar result in mung bean plant under Cd stress condition but they used diverse polyamines as an exogenous protectant. On the other hand, Anjum et al. (2015) stated that Cu and Ni are moderate inducers while Pb and Zn are feeble inducers of phytochelatin. Alongside functioned as a metal detoxifier through chelating, MTs are concerned in metal homeostasis and an efficient protector of metal (Hossain et al. 2012). Synthesis of phytochelatin varies upon plant or metals/metalloids types where among the plant family Fabaceae and among the metals/metalloids Cd, Hg, As, and Fe proved higher phytochelatin synthesis in

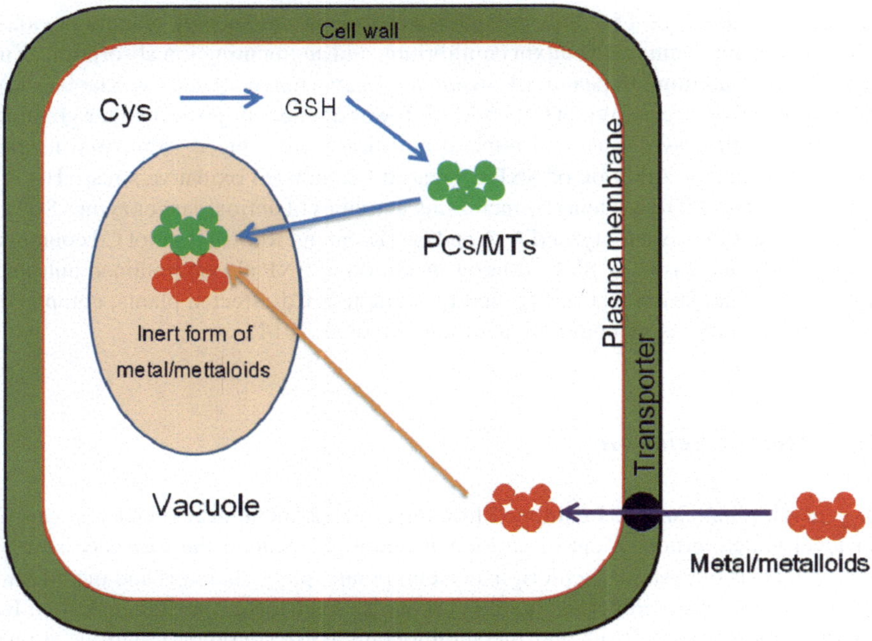

Fig. 1 Metal chelation by Fabaceae plant (Modified from Mahmud et al. 2019)

cellular level (Zhou et al. 2006; Becana et al. 2010). Furthermore, different exogenously applied phytoprotectors already confirmed their ability to further increase phytochelatins and metallothioneins.

4 Phytoremediation by Legume

Phytoremediation is a cost-effective plant-based bioremediation technique that employs a range of plants to eradicate, move, stabilize, and/or demolish contaminants in the soil and groundwater (Gómez-Sagasti and Marino 2015). A variety of vegetation has been recognized in the researches for their prospective in facilitating the phytoremediation process from toxic metal-/metalloid-polluted growing medium. Different grasses and plant of Brassicaceae and Fabaceae family showed their effectiveness in phytoremediation/rhizoremediation (Perveen et al. 2015; Navarro-Torre et al. 2019). Field crops including Fabaceae plant are also used in crop rotation process to improve soil mineral availability obviously for the legumes itself and following non-leguminous crop due to their symbiotic N-fixation, and the relationship between legumes and Rhizobium could be helpful for the rehabilitation of metal-contaminated soils by phytoextraction (Vamerali et al. 2010; Ghnaya et al. 2015). Symbiotic nitrogen fixations create Fabaceae family as perfect initiator to settle and renovate the worth and health of N-lacking surroundings, a widespread feature of metal-/metalloid-contaminated soils. This capacity together with legumes' deep-reaching root system and high biomass is ideal traits for efficient phytoremediation of Cd (Zaidi et al. 2012). Moreover, Merkl et al. (2005) confirmed that beside metal/metalloid remediation Fabaceae plant can work as phytoremediator of petroleum-contaminated soil. Several scientific researches established that toxic metal/metalloids demolish the usual function of nodulation of leguminous crop because nodule uptakes more metal and confirms phytoremediation (Ghnaya et al. 2015). However, Fabaceae are least investigated for their extensive utilization in phytoremediation technology (Zaidi et al. 2012).

5 Plant–Microbe Interaction in Fabaceae Plant to Mitigate Metal/Metalloid Toxicity

Metal/metalloid pollution not only causes environmental problems but also destructs ecological harmony to the agricultural soils, altered the structural and functional behavior of microbial community as well as their diversity. Moreover, metal/metalloid toxicity hampered the growth, physiology and yield of plants as well as contaminate the food chain. Plants under Fabaceae have an intimate relationship with microbes, which play a vital role for adaptation as well as survival of both plants and microbes under metal/metalloids stress, using inherited metabolic and genetic ability (Gopalakrishnan et al. 2015). Plants on metal-/metalloid-contaminated soils interact with variety of microorganisms, which can enhance growth and metal/metalloid toxicity tolerance (Rajkumar et al. 2012). Among the microorganisms, the plant growth promoting rhizobacterias (PGPRs) interact with the Fabaceae plants. A lot of previous reports suggested that PGPR have great effects on plants affected by metal/metalloid toxicity. Among the PGPR, bacteria from *Pseudomonas, Achromobacter, Azospirillum, Enterobacter, Variovorax, Bacillus, Azotobacter, Klebsiella, and Aeromonas* are reported to increase growth of plants under environment adversities, especially under metal/metalloid toxicity by various mechanisms including chelating agents (organic acids, siderophores, metallophores, bio-surfactants) and phytohormone production; facilitating nutrient availability; synthesizing 1-aminocyclopropane-1-carboxylic acid (ACC) deaminase; biomodification of root morphology and biomass; and acting as biocides (Glick 2010; Ma et al. 2016, 2011; Rajkumar et al. 2012). In a metal-/metalloid-contaminated environment, PGPR can adopt one or more strategies above to confer stress tolerance. However, the basic mechanisms lying in metal/metalloids toxicity tolerance are the mobilization and immobilization of metal/metalloids by bacterial proteins, which reduce their bioavailability as well as their absorption in pants. The PGPR have been reported to chelate the metal/metalloids by two ways; first, they actively produce several low molecular weight organic compounds (LMWOC) like gluconic acid, citric acid, oxalic acid, succinic acid, etc. On the other hand, some IAA-producing PGPR can increase exudation of LMWOC from roots by loosening the cell walls, like oxalic acid, malic acid, citric acid, acetic acid, etc. The organic acids in the root zone are reported to play various biochemical processes (Archana et al. 2012). Reports suggested that LMWOC-producing PGPR facilitate metal/metalloids stress tolerance by chelating the metal/metalloids ions form complex such as metaloxalate; those are less toxic to plants than from the free form, thus creating difficulties for the uptake to the stressed plants (Kavita et al. 2008; Gao et al. 2010; Rajkumar et al. 2012). Moreover, Gadd (2010) reported that LMWOC secreted by the microbes have more affinity toward toxic metal/metalloids than plant nutrients. Moreover, these organic acids are involved in mineral P solubilization, thus forming metal phosphate precipitates (e.g., Cd and Pb) (Patel et al. 2010; Sharma and Archana 2016). Park et al. (2011) reported that organic acid-producing and P-solubilizing *Pantoea* sp. and *Enterobacter* sp. are involved in Pb immobilizing in soil. Similarly in another study, Kavita et al. (2008) illustrated that gluconic acid-producing *E. asburiae* PSI3 can confer Cd toxicity tolerance in mung bean seedling. Bacterial siderophore (BSP) is one of the important mechanisms for iron (Fe)-chelating—secondary metabolites of 400–1000 Da (Burd et al. 2000; Glick 2010). Under metal/metalloids stress, Fe plays a vital role, which is mainly accomplished by BSP, increasing Fe absorption as well as decreasing free toxic metal/metalloids ions by binding them in siderophores, and leads to dilute toxicity of metal/metalloids, e.g., Cd, Ni, and Pb (Dimkpa et al. 2009a, b; Deicke et al. 2013). Tank and Saraf (2009) reported that decreased Ni uptake and increased plant growth in chickpea are attributed to *Pseudomonas* siderophore. Moreover, Becerra-Castro et al. (2011) reported that BSP could decrease the bioavailability of toxic Zn, Cd, and Ni. In addition, biosynthesis of chl is enhanced by the BSP by providing more Fe and nitrogen (N) (Rasouli-Sadaghiani et al. 2010). Soil pHs, Fe availability, soil nutrient condition, nutrient translocation from root to shoot, and root exudates are responsible for the generation of siderophore (Rajkumar et al. 2012). Another strategy followed by the microorganisms during metal/metalloids toxicity to save plants is producing the bio-surfactants—mainly lipid-containing complex, such as glycolipids, fatty acids, phospholipids, lipoprotein or lipopeptide, lipopeptides, and mycolic acid (Pacheco et al. 2010; Pacwa-Płociniczak et al. 2011). For example, di-rhamnolipid surfactant can remove 92% Cd and 88% Pb in Cd- and Pb-contaminated soils, respectively (Juwarkar et al. 2007). Moreover, *B. subtilis*, *P. aeruginosa*, *Torulopsis bombicola* produced bio-surfactants namely sophorolipids, di-rhamnolipids, and rhamnolipids, respectively, can remove metal/metalloid toxicity hence confers metal tolerance to Fabaceae plants (Juwarkar et al. 2007; Venkatesh and Vedaraman 2012). Among the phytohormones, abscisic acid (ABA), citokinins (CKs), brassinosteroids (BRs), auxins, and gibberellins (GA) and salicylic acid (SA) play a vital role reducing metal/metalloids toxicity (Peleg and Blumwald 2011). Gangwar et al. (2011) and Zhu et al. (2012) reported that GA reduces Cd toxicity as well as oxidative stress and lipid peroxidation as well as enhances antioxidant systems. Similarly, Ngumbi and Kloepper (2016) reported that microbes facilitate decreasing metal/metalloids toxicity, while Sharaf et al. (2009) reported that hormonal balance regulates proteases, catalase (CAT), and peroxidase (POD) activities.

In addition, PGPR also can produce phytohormones (IAA, GA, CK, ABA) and have been reported to stimulate growth under control as well as metal/metalloids toxicity (Ma et al. 2011; Etesami et al. 2015; Ullah et al. 2015). Microbial IAA has great importance increasing plant nutrient uptake, numbers of root hairs, length and surface of roots, as well as inducing physiological changes and tolerance to metals during metal/metalloids toxicity. Chmielowska-Bąk et al. (2014) reported that IAA reduced Cd adsorption and alleviated Cd toxicity in plants. Moreover, bacterial IAA can loosen the root cell walls, thus increasing root exudation, proton (H^+) release for acidification, producing organic/amino acid–metal/mineral complexes, increasing nutrient (e.g., N, P, K, Fe, Mn, Zn, Cu, etc.) mobility, prevention of toxic entering the cell symplast (Magdziak et al. 2013; Ma et al. 2016). Although most of the previous reports focused on microbial IAA, phytohormone GA and SA are also able to reduce metal/metalloids toxicity (Zhu et al. 2012). The conversion of insoluble phosphatic compounds is an important trait of phosphate-solubilizing bacteria (PSB), and 1–50% population of rhizospheric bacteria can dissolve phosphates (Zaidi et al. 2009). The PSB bind metal/metalloids to the cell wall forming complexes, which plays important roles in reducing toxic metal/metalloids uptake and translocation (Matusik et al. 2008; Park et al. 2010; Qiu et al. 2011), as reported in Pb immobilization via formation of pyromorphite ($Pb_5(PO_4)_3$) (Cao et al. 2008) and in Cd^{2+} immobilization by formation of insoluble Cd phosphate precipitates (Sharma and Archana 2016). During metal/metalloids toxicity, plant nutrients play a significant function to induce stress tolerance (Singh and Kumar 2015). Due to ionic similarity with nutrient elements, toxic metal/metalloids obstruct the uptake of essential plant nutrients like N, P, K, Mg, Ca, Fe, and Zn causing their deficiency and imbalance as well as alteration of their metabolic functions (Ouzounidou et al. 2006; Parida et al. 2003). DalCorso et al. (2013) reported that under As and Cd stress Ca^{2+}, Mg^{2+}, Zn^{2+}, and Fe^{2+} are competed by As and Cd^{2+} replacing the nutrient cations from their specific binding sites and disrupt their functions causing mineral deficiency in the plant. On the other hand, interaction with microbes can lead to reduce metal/metalloids toxicity by mobilizing the essential nutrients and making them essential for plants as well as plant's resistance to toxic metal/metalloids (Ma et al. 2016; Rajkumar et al. 2010; Saravanan et al. 2011; Sharma and Archana 2016). Plant growth-promoting bacteria like *Rhizobium*, *Bradyrhizobium*, *Mesorhizobium*, etc., provide N to the plant in two ways, first by mineralizing the organic N and second symbiotic nitrogen fixation, and thus gave tolerance to plants during metal/metalloids toxicity (Wani et al. 2008). Similarly, P-solubilizing PGPR are able to convert unavailable P to available forms by rhizosphere acidification, metal chelation, ion exchange, as well as releasing organic acids, protons, hydroxyl ions, CO_2, siderophore generation, phytohormone production, and ACC deaminase activity and extracellular phosphatases (Gyaneshwar et al. 2002; Glick 2012). Under metal/metalloids toxicity, *Pseudomonas* sp., *Pantoea* sp., and *Psychrobacter* sp. produce siderophores and dissolve P from $Ca_3(PO_4)_2$ and $Zn_3(PO_4)_2$ increase P uptake and ultimate plant growth (Rodriguez et al. 2004; Saravanan et al. 2007; Mia et al. 2010; Taurian et al. 2010). Similarly, some PGPR are reported to increase K and Si release from silicate minerals like feldspar, muscovite, and biotite mostly from the acidolysis by 2-ketogluconic acid, oxalic acid, propionic acid, tartaric acid, gluconic acid, citric acid, malic acid, succinic acid, fumaric acid, lactic acid, glycolic acid, malonic acid, etc. (In addition, these bacteria can also release H^+ ion to displace these ions and enhance Ca^{2+}, Mg^{2+}, and Mn^{2+} to plants.). Sulfate esters and sulfonates or amino acid sulfur are the main sources of S in soils, which are degraded by the microbes and increase S availability to plants. Sulfur is the main component to GSH in plants, which is involved in metal sequestration, antioxidant defense, and glyoxalase system for conferring metal/metalloids toxicity (Hasanuzzaman et al. 2018). Similarly, Etesami and Beattie (2017) reported that PGPR can increase micronutrient availability by altering soil pH and releasing chelating agents. Moreover, PGPR can solubilize Zn from Zn containing minerals (Saravanan et al. 2011). Under metal/metalloids toxicity, one of the most detrimental effects is production of phytohormone ET, which is involved in root growth retardation, ROS production as well as apostasies. Contrary PGPR under metal/metalloids toxicity produce ACC deaminase enzyme, which hydrolyzes the ET precursor ACC, thus reduce production of ET, and confer tolerance to plants from metal/metalloids toxicity (Glick 2010; Etesami et al. 2015). Reporters suggested that Fabaceae plants inoculated with ACC deaminase-releasing PGPR like *Achromobacter xylosoxidans*, *Actinobacteria* sp., *Arthrobacter* sp., *Bacillus sp. B. cereus*, *B. megaterium*, *Burkholderia* sp., *Enterobacter* sp., *E. aerogenes*, *E. intermedius*, *Firmicutes* sp., *Proteobacteria* sp., *Pseudomonas* sp., *P. tolaasii*, *P. fluorescens*, *P. aeruginosa*, *P. koreensis* are some of ACC deaminase-producing bacteria that could alleviate the toxicity of heavy metals in plants (Glick 2010; Ma et al. 2011; Rajkumar et al. 2012) (Fig. 2).

Bacterial extracellular polymeric substances (EPSs) decrease metal/metalloids toxicity by arresting the mobility of the metal/metalloids as well as their availability to plants (Kunito et al. 2001; Rajkumar et al. 2012). Basically, EPSs are homo- or heteropolysaccharides (includes proteins, humic substances, and mucopolysaccharides) and attach with bacterial cell surface in capsule or slime form. Moreover, neutral sugars, amino sugars, uronic acids, organic ester, and pyruvate ketals are also titled as EPS. Due to their acyl group, EPSs, which are anionic in nature, hence interact with other polysaccharides (Kaushal and Wani 2016), strongly bind with potentially toxic trace metal/metalloids, and subsequently increase trace

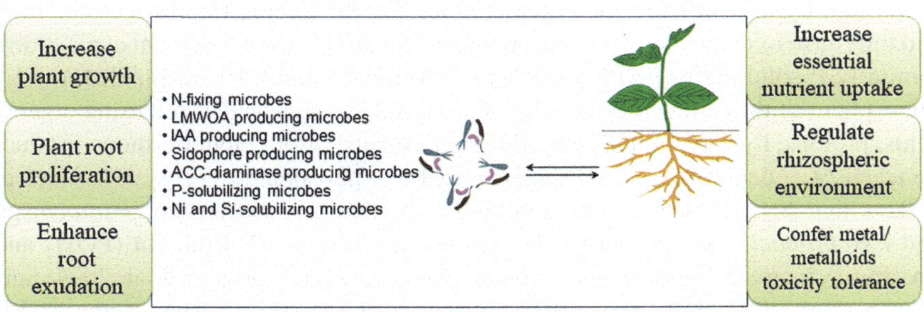

Fig. 2 Plant–microbe interaction in *Fabaceae* plant to mitigate metal/metalloid toxicity

element resistance (Xu et al. 2012). For example *P. putida* secreted EPS transformed toxic Cd^{2+} into organic complexes (Wei et al. 2011; Xu et al. 2012). Moreover, PSB are able to immobilize toxic metal/metalloids via EPS sorption of metals (Susilowati and Syekhfani 2014).

Therefore, further study is required on the metal/metalloids toxicity tolerance with the physiological status of Fabaceae plants as well investigating metal/metalloid-specific and crop-specific microbes for creating better tolerance in Fabaceae plant species.

6 Genetic Advances in Fabaceae Plant to Combat Metal/Metalloid Toxicity

Specialized genes possessed by the microbes can provide biological resistance in microbes and are a preferred technology due to their special metabolic pathway which can help them to convert toxic metal/metalloids into less toxic forms (Pieper and Reineke 2000; Furukawa 2003; Bondarenko et al. 2008; Jan et al. 2009; Ng et al. 2009; Hasin et al. 2010). Genetically engineered (GE) microorganisms producing metallothioneins (MTs) can enhance metal/metalloids accumulation in their cells (Pazirandeh et al. 1995). For example, *MerA* gene containing GE bacteria is able to remove Hg from Hg-contaminated soils by expressing MT and polyphosphate kinase (Barkay et al. 2003; Deckwer et al. 2004; De et al. 2006; Ruiz et al. 2011). In addition, GE microbes *Deinococcus radiodurans* and *Pseudomonas putida* are able to degrade chlorinated organic compounds and organophosphates in contaminated soils (Brim et al. 2000; Kumar et al. 1996). On the other hand, *Ralston metallidurans*, *Caulobacter* spp. strain JS4022/p723-6H, *E. coli* can be used for Cr, Cd, Ni removal, respectively (Liu et al. 2011; U 2005; Fulkerson et al. 1998). Meanwhile, GE plants can detoxify metal/metalloids and researchers have identified many genes that have been used to regulate metal/metalloids uptake, removal, translocation, and bioaccumulation. These plants may also be able for xenobiotics degradation and efficient for phytoremediation (Eapen et al. 2007; Bizily et al. 2000; Raskin 1996; Dietz and Schnoor 2001; Kurumata et al. 2005; Kumar 2012). Symbiosis is one of the promising efforts in plant community, which enables both the plant and the associated microorganisms to successfully sustain in the contaminated soils as well as removing the metals from the soils and can tolerate the metal/metalloids toxicity (Sriprang et al. 2002). To successfully accomplish this process, leguminous plants receive MT genes which will enable the plant to accumulate the toxic metal/metalloids in the nodules. On the other hand, *Rhizobia* inoculated with metallothionein (MTL4) and phytochelatin synthase gene can establish a successful relationship with Fabaceae plant *Astragalus sinicus*, hence, can accumulate a significant amount of Cd in the nodules, and can tolerate the Cd phytotoxicity (Ike et al. 2006). Although there are scopes for phyto- or bioremediation by microbes as well as plants using transgenic approaches in Fabaceae plants, there are also many challenges behind this such as transgenic pollen as well as transfer of plasmids horizontally, which will further act as hyper-accumulating plants, will accumulate higher amount of metals, and will be a powerful weapon for overcoming the present situation.

7 Conclusion

Toxic metal/metalloid contamination in the plant-growing medium hampers seed germination, cellular development, photosynthesis, growth and development of Fabaceae plant with a variety of morphological and physiological faults. The important event for legumes, nodulation, is negatively affected by the action of metal/metalloid stress because nodule uptakes huge amount of metal/metalloid and contributes in phytoextraction of them from growing medium. Moreover, it reduces the transportation of metal/metalloid to the upper parts of plant. Plant–microbe interaction with Fabaceae plant engages in tolerance capacity of plant up to certain extent. However, the defending efficiency of nodule bacteria depends on numerous factors, for instance category, nature, and concentrations of metals/metalloids, compatibleness of associates, virulence, adaptive capability, and nitrogen-fixing ability of bacteria. Toxicity of metal/metalloid in any plants is associated with the increased ROS generation which involve in creating oxidative stress resulting in protein oxidation, lipid peroxidation, and DNA damage. In most of the cases, Fabaceae plant can tolerate mild level of metal/metalloid toxicity due to their well-established antioxidant defense system but with the increase of stress concentration the system becomes incompetent. Scientists are trying to discover the approach to combat metal/metalloid toxicity of Fabaceae plant. Pretreatment of seed/seedling with nodule bacteria at low levels of oxidants could be prop up bacteria to defend against elevated concentration of toxic metals/metalloids in the surroundings. The production of nodule bacterial strains with higher resistance to environmental stresses may be a big prospect to enhance the advantage from their utilization in phytoremediation and cultivation at contaminated region. Our chapter tried to accumulate recent advances of Fabaceae plant under metal/metalloid toxicity which will help to develop research idea to overcome the problem.

References

Abdullahi BA, Gu XG, Gan QL, Yang YH (2003) Brassinolide amelioration of aluminium toxicity in mung bean seedling growth. J Plant Nutr 26:1725–1734

Aggarwal M, Sharma S, Kaur N, Pathania D, Bhandhari K, Kaushal N, Kaur R, Singh K, Srivastava A, Nayyar H (2011) Exogenous proline application reduces phytotoxic effects of selenium by minimising oxidative stress and improves growth in bean (*Phaseolus vulgaris* L.) seedlings. Biol Trace Elem Res 140:354–367

Ahmad P, Abdel Latef AA, Abd Allah EF, Hashem A, Sarwat M, Anjum NA, Gucel S (2016) Calcium and potassium supplementation enhanced growth, osmolyte secondary metabolite production, and

enzymatic antioxidant machinery in cadmium-exposed chickpea (*Cicer arietinum* L.). Front Plant Sci 7:513
Ali B (2017) Salicylic acid induced antioxidant system enhances the tolerence to aluminium in mung bean (*Vigna radiata* L. Wilczek) plants. Indian J Plant Physiol 22:178–189
Alle V, Osvalde A, Vikmane M, Kondratovics U (2019) The effect of cadmium and lead pollution on growth and physiological parameters of field beans (*Vicia faba*). Agron Res 17:1261–1272
Alyemeni MN, Hayat Q, Hayat S, Faizan M, Faraz A (2016) Exogenous proline application enhances the efficiency of nitrogen fixation and assimilation in chickpea plants exposed to cadmium. Legume Res 39:221–227
Alyemeni MN, Ahanger MA, Wijaya L, Alam P, Ahmad P (2017) Contrasting tolerance among soybean genotypes subjected to different levels of cadmium stress. Pak J Bot 49:903–911
Anjum SA, Xie XY, Wang LC, Saleem MF, Man C, Lei W (2011) Morphological, physiological and biochemical responses of plants to drought stress. Afr J Agric Res 6:2026–2032
Anjum NA, Ahmad I, Mohmood I, Pacheco M, Duarte AC, Pereira E, Umar S, Ahmad A, Khan NA, Iqbal M, Prasad MN (2012) Modulation of glutathione and its related enzymes in plants' responses to toxic metals and metalloids—a review. Environ Exp Bot 75:307–324
Anjum NA, Umar S, Iqbal M (2014) Assessment of cadmium accumulation, toxicity, and tolerance in Brassicaceae and Fabaceae plants—implications for phytoremediation. Environ Sci Pollut Res 21:10286–10293
Anjum NA, Hasanuzzaman M, Hossain MA, Thangavel P, Roychoudhury A, Gill SS, Rodrigo MAM, Adam V, Fujita M, Kizek R, Duarte AC, Pereira E, Ahmad I (2015) Jacks of metal/metalloid chelation trade in plants—an overview. Front Plant Sci 6:192
Ara A, Sofi PA, Rather MA, Rashid M, Gull M (2019) Abiotic stress tolerance in legumes—critical approaches. Int J Curr Microbiol App Sci 8:1991–2000
Araújo SS, Beebe S, Crespi M, Delbreil B, González EM, Gruber V, Lejeune-Henaut I, Link W, Monteros MJ, Prats E, Rao I, Vadez V, Vaz Patto MC (2015) Abiotic stress responses in legumes: strategies used to cope with environmental challenges. Crit Rev Plant Sci 34(1–3):237–280
Archana G, Buch A, Kumar GN (2012) Pivotal role of organic acid secretion by Rhizobacteria in plant growth promotion. In: Satyanarayana T, Johri BN, Prakash A (eds) Microorganisms in sustainable agriculture and biotechnology. Springer, Dordrecht, pp 35–53
Arif MS, Yasmeen T, Shahzad SM, Riaz M, Rizwan M, Iqbal S, Asif M, Soliman MH, Ali S (2019) Lead toxicity induced phytotoxic effects on mung bean can be relegated by lead tolerant *Bacillus subtilis* (PbRB3). Chemosphere 234:70–80
Ashraf MY, Roohi M, Iqbal Z, Ashraf M, Öztürk M, Gücel S (2016) Cadmium (Cd) and lead (Pb) induced changes in growth, some biochemical attributes, and mineral accumulation in two cultivars of mung bean [*Vigna radiata* (L.) Wilczek]. Commun Soil Sci Plant Anal 47:405–413
Barkay T, Miller SM, Summers AO (2003) Bacterial mercury resistance from atoms to ecosystems. FEMS Microbiol Rev 27:355–384
Batool S (2018) Impact of bioaccumulation of nickel on growth, seed yield and mineral uptake of chickpea (*Cicer arietinum* L.) varieties. Pak J Bot 50:2147–2150
Baycu G, Rognes SE, Özden H, Gören-Saglam N, Csatári I, Szabó S (2017) Abiotic stress effects on the antioxidative response profile of *Albizia julibrissin* Durazz. (Fabaceae). Braz J Bot 40:21–32
Becana M, Matamoros MA, Udvardi M, Dalton DA (2010) Recent insight into antioxidant defenses of legume root nodules. New Phytol 188:960–976
Becerra-Castro C, Prieto-Fernández A, Álvarez-Lopez V, Monterroso C, Cabello-Conejo MI, Acea MJ, Kidd PS (2011) Nickel solubilizing capacity and characterization of rhizobacteria isolated from hhyperaccumulating and non-hyperaccumulating subspecies of Alyssum serpyllifolium. Int J Phytorem 13(sup1):229–244
Bhagyawant SS, Narvekar DT, Gupta N, Bhadkaria A, Koul KK, Srivastava N (2019) Variations in the antioxidant and free radical scavenging under induced heavy metal stress expressed as proline content in chickpea. Physiol Mol Biol Plants 25:683–696
Bizily SP, Rugh CL, Meagher RB (2000) Phytodetoxification of hazardous organomercurials by genetically engineered plants. Nat Biotechnol 18:213–217

Bondarenko O, Rõlova T, Kahru A, Ivask A (2008) Bioavailability of Cd, Zn and Hg in soil to nine recombinant luminescent metal sensor bacteria. Sensors 8:6899–6923

Bouazizi H, Jouili H, Geitmann A, El Ferjani E (2010) Copper toxicity in expanding leaves of *Phaseolus vulgaris* L.: antioxidant enzyme response and nutrient element uptake. Ecotoxicol Environ Saf 73:1304–1308

Brear EM, Day DA, Smith PMC (2013) Iron: an essential micronutrient for the legume-rhizobium symbiosis. Front Plant Sci 4:359

Brim H, McFarlan SC, Fredrickson JK, Minton KW, Zhai M, Wackett LP, Daly MJ (2000) Engineering *Deinococcus radiodurans* for metal remediation in radioactive mixed waste environments. Nat Biotechnol 18:85–90

Burd GI, Dixon DG, Glick BR (2000) Plant growth-promoting bacteria that decrease heavy metal toxicity in plants. Can J Microbiol 46:237–245

Cao X, Ma LQ, Singh SP, Zhou Q (2008) Phosphate-induced lead immobilization from different lead minerals in soils under varying pH conditions. Environ Pollut 152:184–192

Chmielowska-Bąk J, Gzyl J, Rucińska-Sobkowiak R, Arasimowicz-Jelonek M, Deckert J (2014) The new insights into cadmium sensing. Front Plant Sci 5:245

DalCorso G, Manara A, Furini A (2013) An overview of heavy metal challenge in plants: from roots to shoots. Metallomics 5:1117–1132

de Abreu MGP, Filho PFM, Garcia KGV, da Silva Junior JMT, Albuquerque GHS, Araujo JM, Tavella LB (2019) Effect of manganese on growth and nodulation of *Mimosa caesalpiniaefolia* (Benth.). J Agric Sci https://doi.org/10.5539/jas.v11n2p339

De J, Sarker A, Rahman NS (2006) Bioremediation of toxic substances by mercury resistant marine bacteria. Ecotoxicology 15:385–389

Deckwer WD, Becker FU, Ledakowicz S, Wagner-Döbler I (2004) Microbial removal of ionic mercury in a three phase fluidzed bed reactor. Environ Sci Technol 38:1858–1865

Deicke M, Bellenger JP, Wichard T (2013) Direct quantification of bacterial molybdenum and iron metallophores with ultra-high-performance liquid chromatography coupled to time-of-flight mass spectrometry. J Chromatogr A 1298:50–60

Dietz AC, Schnoor JL (2001) Advances in phytoremediation. Environ Health Perspect 109:163–168

Dimkpa C, Weinand T, Asch F (2009a) Plant–rhizobacteria interactions alleviate abiotic stress conditions. Plant Cell Environ 32:1682–1694

Dimkpa CO, Merten D, Svatoš A, Büchel G, Kothe E (2009b) Siderophores mediate reduced and increased uptake of cadmium by *Streptomyces tendae* F4 and sunflower (*Helianthus annuus*), respectively. J Appl Microbiol 107:1687–1696

Dong YJ, Wang ZL, Zhang JW, Liu S, He ZL, He MR (2015) Interaction effects of nitric oxideand salicylic acid in alleviating salt stress of *Gossypium hirsutum* L. J Soil Sci Pant Nutr 15:561–573

Dong Y, Chen W, Xu L, Kong J, Liu S, He Z (2016) Nitric oxide can induce tolerance to oxidative stress of peanut seedlings under cadmium toxicity. Plant Growth Regul 79(1):19–28

Dongsen X, Robert Harrison RB, Henry CL (1995) Effect of organic acid on cadmium toxicity in tomato and bean growth. J Environ Sci 7:399–406

Eapen S, Singh S, D'Souza S (2007) Advances in development of transgenic plants for remediation of xenobiotic pollutants. Biotechnol Adv 25:442–451

Emamverdian A, Ding Y, Mokhberdoran F, Xie Y (2015) Heavy metal stress and some mechanisms of plant defense response. Sci World J 2015:756120

Etesami H, Beattie GA (2017) Plant-microbe interactions in adaptation of agricultural crops to abiotic stress conditions. In: Kumar V, Kumar M, Sharma S, Prasad R (eds) Probiotics and plant health. Springer, Singapore, pp 163–200

Etesami H, Alikhani HA, Hosseini HM (2015) Indole-3-Acetic Acid and 1-aminocyclopropane-1-carboxylate deaminase: bacterial traits required in rhizosphere, rhizoplane and/or endophytic competence by beneficial bacteria. In: Maheshwari DK (ed) Bacterial metabolites in sustainable agroecosystem. Springer, Cham, pp 183–258

Fatoba PO, Ogunkunle CO, Salihu BZ (2012) Toxic effects of cadmium (Cd) and Lead (Pb) on growth and productivity of *Arachis hypogaea* (L) and *Glycine max* (L). J Asian Sci Res 2:254–259

Foyer CH, Lam HM, Nguyen HT, Siddique KH, Varshney RK, Colmer TD, Cowling W, Bramley H, Mori TA, Hodgson JM, Cooper JW (2016) Neglecting legumes has compromised human health and sustainable food production. Nat plant 2:16112

Fulkerson JF, Garner RM, Mobley HLT (1998) Conserved residues and motifs in the nixA protein of Helicobacter pylori are critical for the high affinity transport of nickel ions. J Biol Chem 273:235–241

Furukawa K (2003) 'Super bugs' for bioremediation. Trends Biotechnol 21:187–190

Gadd GM (2010) Metals, minerals and microbes: geomicrobiology and bioremediation. Microbiology 156:609–643

Gangwar S, Singh VP (2011) Indole acetic acid differently changes growth and nitrogen metabolism in *Pisum sativum* L. seedlings under chromium (VI) phytotoxicity: implication of oxidative stress. Sci Hortic 129:321–328

Gangwar S, Singh VP, Prasad SM, Maurya JN (2010) Modulation of manganese toxicity in *Pisum sativum* L. seedlings by kinetin. Sci Hortic 126:467–474

Gangwar S, Singh VP, Srivastava PK, Maurya JN (2011) Modification of chromium (VI) phytotoxicity by exogenous gibberellic acid application in *Pisum sativum* (L.) seedlings. Acta Physiol Plant 33:1385–1397

Gao Y, Miao C, Mao L, Zhou P, Jin Z, Shi W (2010) Improvement of phytoextraction and antioxidative defense in *Solanum nigrum* L. under cadmium stress by application of cadmium-resistant strain and citric acid. J Hazard Mater 181:771–777

Garg N, Kaur H (2013) Impact of cadmium-zinc interactions on metal uptake, translocation and yield in pigeonpea genotypes colonized by arbuscular mycorrhizal fungi. J Plant Nutr 36(1):67–90

Garg N, Singh S (2018) Mycorrhizal inoculations and silicon fortifications improve rhizobial symbiosis, antioxidant defense, trehalose turnover in pigeon pea genotypes under cadmium and zinc stress. Plant Growth Regul 86:105–119

Ghnaya T, Mnassri M, Ghabriche R, Wali M, Poschenrieder C, Lutts S, Abdelly C (2015) Nodulation by Sinorhizobium meliloti originated from a mining soil alleviates Cd toxicity and increases Cd-phytoextraction in *Medicago sativa* L. Front Plant Sci 6:863

Glick BR (2010) Using soil bacteria to facilitate phytoremediation. Biotechnol Adv 28:367–374

Glick BR (2012) Plant growth-promoting bacteria: mechanisms and applications. Scientifica 2012:1–15

Gómez-Sagasti MT, Marino D (2015) PGPRs and nitrogen-fixing legumes: a perfect team for efficient Cd phytoremediation? Front Plant Sci 6:81

Gopalakrishnan S, Sathya A, Vijayabharathi R, Varshney RK, Gowda CL, Krishnamurthy L (2015) Plant growth promoting rhizobia: challenges and opportunities. 3 Biotech 5:355–377

Gull M, Hafeez F, Saleem M, Malik K (2004) Phosphorus uptake and growth promotion of chickpea by co-inoculation of mineral phosphate solubilising bacteria and a mixed rhizobial culture. Anim Prod Sci 44:623–628

Gyaneshwar P, Kumar GN, Parekh LJ, Poole PS (2002) Role of soil microorganisms in improving P nutrition of plants. Plant Soil 245:83–93

Haddad SA, Tabatabai MA, Abdel-Moneim A-MA, Loynachan TE (2015) Inhibition of nodulation and nitrogen nutrition of leguminous crops by selected heavy metals. Air, Soil Water Res 8:1–7

Hasanuzzaman M, Fujita M (2012) Heavy metals in the environment: current status, toxic effects on plants and possible phytoremediation. In: Anjum NA, Pereira MA, Ahmad I, Duarte AC, Umar S, Khan NA (eds) Phytotechnologies: remediation of environmental contaminants. CRC Press, Boca Raton, pp 7–73

Hasanuzzaman M, Hossain MA, da Silva JA, Fujita M (2012) Plant responses and tolerance to abiotic oxidative stress: antioxidant defense is a key factor. In: Bandi V, Shanker AK, Shanker C, Mandapaka M (eds) Crop stress and its management: perspectives and strategies. Springer, Berlin, pp 261–315

Hasanuzzaman M, Nahar K, Hossain M, Mahmud JA, Rahman A, Inafuku M, Oku H, Fujita M (2017) Coordinated actions of glyoxalase and antioxidant defense systems in conferring abiotic stress tolerance in plants. Int J Mol Sci 18:200

Hasanuzzaman M, Bhuyan MH, Mahmud JA, Nahar K, Mohsin SM, Parvin K, Fujita M (2018) Interaction of sulfur with phytohormones and signaling molecules in conferring abiotic stress tolerance to plants. Plant Signal Behav 13:e1477905

Hasin AA, Gurman SJ, Murphy LM et al (2010) Remediation of chromium (VI) by a methane-oxidizing bacterium. Environ Sci Technol 44:400–405

Hattab S, Hattab S, Flores-Casseres ML, Boussetta H, Doumas P, Hernandez LE, Banni M (2016) Characterisation of lead-induced stress molecular biomarkers in *Medicago sativa* plants. Environ Exp Bot 123:1–12

Hayat S, Hayat Q, Alyemeni MN, Ahmad A (2013) Proline enhances antioxidative enzyme activity, photosynthesis and yield of *Cicer arietinum* L. exposed to cadmium stress. Acta Bot Croat 72:323–335

Hossain MA, Piyatida P, da Silva JAT, Fujita M (2012) Molecular mechanism of heavy metal toxicity and tolerance in plants: central role of glutathione in detoxification of reactive oxygen species and methylglyoxal and in heavy metal chelation. J Bot 2012:872875

Hu S, Gu H, Cui C, Ji R (2016) Toxicity of combined chromium (VI) and phenanthrene pollution on the seed germination, stem lengths, and fresh weights of higher plants. Environ Sci Pollut Res 23:15227–15235

Ike M, Nagamatsu K, Shioya A, Nogawa M, Ogasawara W, Okada H, Morikawa Y (2006) Purification, characterization and gene cloning of 46 kDa chitinase (Chi46) from *Trichoderma reesei* PC-3-7 and its expression in *Escherichia coli*. Appl Microbiol Biotechnol 71:294–303

Imtiaz M, Tu S, Xie Z, Han D, Ashraf M, Rizwan MS (2015) Growth, V uptake, and antioxidant enzymes responses of chickpea (*Cicer arietinum* L.) genotypes under vanadium stress. Plant Soil 390:17–27

Jabeen N, Abbas Z, Iqbal M, Rizwan M, Jabbar A, Farid M, Ali S, Ibrahim M, Abbas F (2016) Glycinebetaine mediates chromium tolerance in mung bean through lowering of Cr uptake and improved antioxidant system. Arch Agron Soil Sci 62:648–662

Jalali K, Nouairi I, Kallala N, M'Sehli W, Zribi K, Mhadhbi H (2018) Germination, seedling growth, and antioxidant activity in four legume (fabaceae) species under copper sulphate fungicide treatment. Pak J Bot 50:1599–1606

Jan AT, Murtaza I, Ali A, Haq QM (2009) Mercury pollution: an emerging problem and potential bacterial remediation strategies. World J Microbiol Biotechnol 25:1529–1537

Jaouani K, Karmous I, Ostrowski M, El Ferjani E, Jakubowska A, Chaoui A (2018) Cadmium effects on embryo growth of pea seeds during germination: investigation of the mechanisms of interference of the heavy metal with protein mobilization-related factors. J Plant Physiol 226:64–76

Juwarkar AA, Nair A, Dubey KV, Singh SK, Devotta S (2007) Biosurfactant technology for remediation of cadmium and lead contaminated soils. Chemosphere 68:1996–2002

Kang SM, Radhakrishnan R, You YH, Khan AL, Lee KE, Lee JD, Lee IJ (2015) *Enterobacter asburiae* KE 17 association regulates physiological changes and mitigates the toxic effects of heavy metals in soybean. Plant Biol 17(5):1013–1022

Kanwal A, Ali S, Farhan M (2019) Heavy metal phytoextraction potential of indigenous tree species of the family fabaceae. Int J Phytorem 21(3):251–258

Kapur D, Singh KJ (2019) Zinc alleviates cadmium induced heavy metal stress by stimulating antioxidative defense in soybean [*Glycine max* (L.) Merr.] Crop J Appl Natur Sci 11:338–345

Kaushal M, Wani SP (2016) Rhizobacterial-plant interactions: strategies ensuring plant growth promotion under drought and salinity stress. Agric Ecosys Environ 231:68–78

Kavita B, Shukla S, Kumar GN, Archana G (2008) Amelioration of phytotoxic effects of Cd on mung bean seedlings by gluconic acid secreting rhizobacterium Enterobacter asburiae PSI3 and implication of role of organic acid. World J Microbiol Biotechnol 24:2965–2972

Khan N, Bano A (2017) Effects of exogenously applied salicylic acid and putrescine alone and in combination with rhizobacteria on the phytoremediation of heavy metals and chickpea growth in sandy soil. Int J Phytoremed 20:405–414

Khan R, Srivastava R, Abdin MZ, Manzoor N (2013) Effect of soil contamination with heavy metals on soybean seed oil quality. Eur Food Res Technol 236:707–714

Kong Z, Deng Z, Glick BR, Wei G, Chou M (2017) A nodule endophytic plant growth-promoting Pseudomonas and its effects on growth, nodulation and metal uptake in Medicago lupulina under copper stress. Ann Microbiol 49:58–67

Kumar S (2012) Phytoremediation of explosives using transgenic plants. J Pet Environ Biotechnol S4:1–2

Kumar S, Mukerji KG, Lal R (1996) Molecular aspects of pesticide degradation by microorganisms. Crit Rev Microbiol 22:1–26

Kunito T, Saeki K, Nagaoka K, Oyaizu H, Matsumoto S (2001) Characterization of copper-resistant bacterial community in rhizosphere of highly copper-contaminated soil. Eur J Soil Biol 37:95–102

Kurumata M, Takahashi M, Sakamoto A, Ramos JL, Nepovim A, Vanek T, Hirata T, Morikawa H (2005) Tolerance to and uptake and degradation of 2,4,6 trinitrotoluene (TNT) are enhanced by the expression of a bacterial nitroreductase gene in *Arabidopsis thaliana*. Z Naturforsch 60:272–278

Li SW, Leng Y, Feng L, Zeng XY (2014) Involvement of abscisic acid in regulating antioxidative defense systems and IAA-oxidase activity and improving adventitious rooting in mung bean [*Vigna radiata* (L.) Wilczek] seedlings under cadmium stress. Environ Sci Pollut Res 21:525–537

Li SW, Zeng XY, Leng Y, Feng L, Kang XH (2018) Indole-3-butyric acid mediates antioxidative defense systems to promote adventitious rooting in mung bean seedlings under cadmium and drought stresses. Ecotoxicol Environ Saf 161:332–341

Li SW, Li Y, Leng Y, Zeng XY, Ma YH (2019) Nitric oxide donor improves adventitious rooting in mung bean hypocotyl cuttings exposed to cadmium and osmotic stresses. Environ Exp Bot 164:114–123

Liu S, Zhang F, Chen J, Sun G (2011) Arsenic removal from contaminated soil via biovolatilization by genetically engineered bacteria under laboratory conditions. J Environ Sci 23:1544–1550

Ma Y, Prasad MN, Rajkumar M, Freitas H (2011) Plant growth promoting rhizobacteria and endophytes accelerate phytoremediation of metalliferous soils. Biotechnol Adv 29:248–258

Ma Y, Oliveira RS, Freitas H, Zhang C (2016) Biochemical and molecular mechanisms of plant-microbe-metal interactions: relevance for phytoremediation. Front Plant Sci 7:918

Madhan M, Mahesh K, Rao SSR (2014) Amelioration of aluminium toxicity on seed germination and early seedling growth of Pigeon Pea [*Cajanus cajan* (L.) Millsp.] by 28-Homobrassinolide. Int J Curr Microbiol App Sci 3:77–83

Magdziak Z, Mleczek M, Kaczmarek Z, Golinski P (2013) Influence of Ca/Mg ratio and Cd^{2+} and Pb^{2+} elements on low molecular weight organic acid secretion by *Salix viminalis* L. roots into the rhizosphere. Trees 27:663–673

Mahmud JA, Hasanuzzaman M, Nahar K, Bhuyan MHMB, Fujita M (2018) Insights into citric acid-induced cadmium tolerance and phytoremediation in *Brassica juncea* L.: coordinated functions of metal chelation, antioxidant defense and glyoxalase systems. Ecotoxicol Environ Saf 147:990–1001

Mahmud JA, Bhuyan MHMB, Anee TI, Nahar K, Fujita M, Hasanuzzaman M (2019) Reactive oxygen species metabolism and antioxidant defense in plants under metal/metalloid stress. In: Hasanuzzaman M, Hakeem K, Nahar K, Alharby H (eds) Plant Abiotic Stress Tolerance. Springer, Cham, pp 221–257

Mao F, Nan G, Cao M, Gao Y, Guo L, Meng X, Yang G (2018) The metal distribution and the change of physiological and biochemical process in soybean and mung bean plants under heavy metal stress. Int J Phytorem 20:1113–1120

Marta J, Rorat A, Grobelak A (2019) Enzymatic assays confirm the toxicity reduction after manure treatment of heavy metals contaminated soil. South Afr J Bot 124:47–53

Matusik J, Bajda T, Manecki M (2008) Immobilization of aqueous cadmium by addition of phosphates. J Hazard Mater 152:1332–1339

Matusso JMM, Mugwe JN, Mucheru-Muna M (2014) Potential role of cereal-legume intercropping systems in integrated soil fertility management in smallholder farming systems of Sub-Saharan. Afr Res J Agric Environ Manage 3:162–174

Melina AT, Romina BC, Paola SG, Elizabeth A (2013) Arsenic effect on the model crop symbiosis Bradyrhizobium-soybean. Plant Physiol Biochem 63:8–14

Merkl N, Schultze-Kraft R, Infante C (2005) Assessment of tropical grasses and legumes for phytoremediation of petroleum-contaminated soils. Water Air Soil Pollut 165(1–4):195–209

Menon P, Joshi N, Joshi A (2016) Effect of heavy metals on seed germination of *Trigonella foenum-graceum* L. Int J Life-Sci Sci Res 2(4):488–493

Mia MAB, Shamsuddin ZH, Wahab Z, Marziah M (2010) Effect of plant growth promoting rhizobacterial (PGPR) inoculation on growth and nitrogen incorporation of tissue-cultured Musa plantlets under nitrogen-free hydroponics condition. Aust J Crop Sci 4:85–90

Nahar K, Hasanuzzaman M, Alam MM, Rahman A, Suzuki T, Fujita M (2016a) Polyamine and nitric oxide crosstalk: antagonistic effects on cadmium toxicity in mung bean plants through upregulating the metal detoxification, antioxidant defense and methylglyoxal detoxification systems. Ecotoxicol Environ Saf 126:245–255

Nahar K, Rahman M, Hasanuzzaman M, Alam MM, Rahman A, Suzuki T, Fujita M (2016b) Physiological and biochemical mechanisms of spermine-induced cadmium stress tolerance in mung bean (*Vigna radiata* L.) seedlings. Environ Sci Pollut Res 23:21206–21218

Nahar K, Hasanuzzaman M, Suzuki T, Fujita M (2017) Polyamines-induced aluminum tolerance in mung bean: A study on antioxidant defense and methylglyoxal detoxification systems. Ecotoxicology 26:58–73

Nath K, Singh D, Shyam S, Sharma YK (2008) Effect of chromium and tannery effluent toxicity on metabolism and growth in cowpea (*Vigna sinensis* L. Saviex Hassk) seedling. Res Environ Life Sci 1:91–94

Navarro-Torre S, Rodríguez-Llorente ID, Doukkali B, Caviedes MA, Pajuelo E (2019) Competition for alfalfa nodulation under metal stress by the metal-tolerant strain *Ochrobactrum cytisi* Azn6.2. Ann Appl Biol 175:184–92. https://doi.org/10.1111/aab.12528

Ng SP, Davis B, Palombo EA, Bhave M (2009) Tn5051 like mer containing transposon identified in a heavy metal tolerant strain Achromobacter sp. AO22. BMC Res Notes 7:2–38

Ngumbi E, Kloepper J (2016) Bacterial-mediated drought tolerance: current and future prospects. Appl Soil Ecol 105:109–125

Nouairi I, Jalali K, Zribi F, Barhoumi F, Zribi K, Mhadhbi H (2019) Seed priming with calcium chloride improves the photosynthesis performance of faba bean plants subjected to cadmium stress. Photosynthetica 57(2):438–445

Ochoa L, Medina-Velo IA, Barrios AC, Bonilla-Bird NJ, Hernandez-Viezcas JA, Peralta-Videa JR, Gardea-Torresdey JL (2017) Modulation of CuO nanoparticles toxicity to green pea (*Pisum sativum* Fabaceae) by the phytohormone indole-3-acetic acid. Sci Total Environ 598:513–524

Ouzounidou G, Moustakas M, Symeonidis L, Karataglis S (2006) Response of wheat seedlings to Ni stress: effects of supplemental calcium. Arch Environ Contam Toxicol 50:346

Pacheco GJ, Ciapina EM, Gomes ED, Pereira Junior N (2010) Biosurfactant production by *Rhodococcus erythropolis* and its application to oil removal. Braz Microbiol 41:685–693

Pacwa-Płociniczak M, Płaza GA, Piotrowska-Seget Z, Cameotra SS (2011) Environmental applications of biosurfactants: recent advances. Int J Mol Sci 12:633–654

Pajuelo E, Rodríguez-Llorente ID, Dary M, Palomares AJ (2008) Toxic effects of arsenic on sinorhizobium-medicago sativa symbiotic interaction. Environ Pollut 154:203–211. https://doi.org/10.1016/j.envpol.2007.10.015

Parida BK, Chhibba IM, Nayyar VK (2003) Influence of nickel-contaminated soils on fenugreek (*Trigonella corniculata* L.) growth and mineral composition. Sci Hortic 98:113–119

Park J, Bolan N, Mallavarapu M, Naidu R (2010) Enhancing the solubility of insoluble phosphorus compounds by phosphate solubilizing bacteria. In: World Congress of Soil Science. Soil Solutions for a Changing World Brisbane. Published on DVD 66:1–6

Park JH, Bolan N, Megharaj M, Naidu R (2011) Isolation of phosphate solubilizing bacteria and their potential for lead immobilization in soil. J Hazard Mater 185:829–836

Patel KJ, Singh AK, Nareshkumar G, Archana G (2010) Organic-acid-producing, phytate-mineralizing rhizobacteria and their effect on growth of pigeon pea (*Cajanus cajan*). Appl Soil Ecol 44:252–261

Pazirandeh M, Chrisey LA, Mauro JM et al (1995) Expression of the *Neurospora crassa* metallothionein gene in *Escherichia coli* and its effect on heavy-metal uptake. Appl Microbiol Biotechnol 43:1112–1117

Peleg Z, Blumwald E (2011) Hormone balance and abiotic stress tolerance in crop plants. Curr Opin Plant Biol 14:290–295

Perveen R, Faizan S, Ansari AA (2015) Phytoremediation using leguminous plants: managing cadmium stress with app lications of Arbuscular Mycorrhiza (AM) fungi. In: Ansari AA, Gill SS, Gill R, Lanza GR, Newman L (eds) Phytoremediation. Springer, Cham, pp 131–142

Pieper DH, Reineke W (2000) Engineering bacteria for bioremediation. Curr Opin Biotechnol 11:262–270

Popova LP, Maslenkova LT, Yordanova RY, Ivanova AP, Krantev AP, Szalai G, Janda T (2009) Exogenous treatment with salicylic acid attenuates cadmium toxicity in pea seedlings. Plant Physiol Biochem 47:224–231

Rady MM (2011) Effect of 24-epibrassinolide on growth, yield, antioxidant system and cadmium content of bean (*Phaseolus vulgaris* L.) plants under salinity and cadmium stress. Sci Hortic 129:232–237

Rajkumar M, Ae N, Prasad MN, Freitas H (2010) Potential of siderophore-producing bacteria for improving heavy metal phytoextraction. Trends Biotechnol 28:142–149

Rajkumar M, Sandhya S, Prasad MN, Freitas H (2012) Perspectives of plant-associated microbes in heavy metal phytoremediation. Biotechnol Adv 30:1562–1574

Raskin I (1996) Plant genetic engineering may help with environmental cleanup. Proc Natl Acad Sci USA 93:3164–3166

Rasouli-Sadaghiani M, Hassani A, Barin M, Danesh YR, Sefidkon F (2010) Effects of AM fungi on growth, essential oil production and nutrients uptake in basil. J Med Plant Res 4:2222–2228

Rehman ZU, Khan S, Brusseau ML, Shah MT (2017) Lead and cadmium contamination and exposure risk assessment via consumption of vegetables grown in agricultural soils of five-selected regions of Pakistan. Chemosphere 168:1589–1596

Reis AR, de Queiroz Barcelos JP, de Souza Osório CRW, Santos EF, Lisboa LAM, Santini JMK, dos Santos MJD, Junior EF, Campos M, de Figueiredo PAM, Lavres J (2017) A glimpse into the physiological, biochemical and nutritional status of soybean plants under Ni-stress conditions. Environ Exp Bot 144:76–87

Reis AR, Lisboa LAM, Reis HPG, de Queiroz Barcelos JP, Santos EF, Santini JMK, Meyer-Sand BRV, Putti FF, Galindo FS, Kaneko FH, Barbosa JZ (2018) Depicting the physiological and ultrastructural responses of soybean plants to Al stress conditions. Plant Physiol Biochem 130:377–390

Rodriguez H, Gonzalez T, Goire I, Bashan Y (2004) Gluconic acid production and phosphate solubilization by the plant growth-promoting bacterium *Azospirillum* spp. Naturwissenschaften 91:552–555

Rodríguez-Serrano MA, Romero-Puertas MC, Zabalza AN, Corpas FJ, Gomez M, Del Rio LA, Sandalio LM (2006) Cadmium effect on oxidative metabolism of pea (*Pisum sativum* L.) roots. Imaging of reactive oxygen species and nitric oxide accumulation in vivo. Plant, Cell Environ 29:1532–1544

Romero-Puertas MC, Rodríguez-Serrano M, Corpas FJ, Gomez MD, Del Rio LA, Sandalio LM (2004) Cadmium-induced subcellular accumulation of O^{2-} and H_2O_2 in pea leaves. Plant, Cell Environ 27:1122–1134

Rui H, Chen C, Zhang X, Shen Z, Zhang F (2016) Cd-induced oxidative stress and lignification in the roots of two *Vicia sativa* L. varieties with different Cd tolerances. J Hazard Mater 301:304–313

Ruiz ON, Alvarez D, Gonzalez-Ruiz G, Torres C (2011) Characterization of mercury bioremediation by transgenic bacteria expressing metallothionein and polyphosphate kinase. BMC Biotechnol 11:1–8

Ruthrof KX, Fontaine JB, Hopkins AJ, McHenry MP, O'Hara G, McComb J, Hardy GE, Howieson J (2018) Potassium amendment increases biomass and reduces heavy metal concentrations in *Lablab purpureus* after phosphate mining. Land Degrad Dev 29:398–407

Sakouhi L, Rahoui S, Ben Massoud M, Munemasa S, Ferjani EL, Murata Y, Chaoui A (2016) Calcium and EGTA alleviate cadmium toxicity in germinating chickpea seeds. J Plant Growth Regul 35:1064–1073

Saravanan VS, Madhaiyan M, Thangaraju M (2007) Solubilization of zinc compounds by the diazotrophic, plant growth promoting bacterium *Gluconacetobacter diazotrophicus*.Chemosphere 66:1794–1798

Saravanan VS, Kumar MR, Sa TM (2011) Microbial zinc solubilization and their role on plants. In: Maheshwari DK (ed) Bacteria in agrobiology. Springer, Berlin, pp 47–63

Sarwar N, Imran M, Shaheen MR, Ishaque W, Kamran MA, Matloob A, Rehim A, Hussain S (2017) Phytoremediation strategies for soils contaminated with heavy metal: modifications and future perspectives. Chemosphere 171:710–721

Seneviratne M, Weerasundara L, Ok YS, Rinklebe J, Vithanage M (2017) Phytotoxicity attenuation in *Vigna radiata* under heavy metal stress at the presence of biochar and N fixing bacteria. J Environ Manage 186:293–300

Shah FUR, Ahmad N, Masood KR, Peralta-Videa JR (2010) Heavy metal toxicity in plants. In: Ashraf M, Ozturk M, Ahmad M (eds) Plant adaptation and phytoremediation. Springer, Dordrecht, pp 71–97

Sharaf AE, Farghal II, Sofy MR (2009) Role of gibberellic acid in abolishing the detrimental effects of Cd and Pb on broad bean and lupin plants. Res J Agric Biol Sci 5:668–673

Sharma RK, Archana G (2016) Cadmium minimization in food crops by cadmium resistant plant growth promoting rhizobacteria. Appl Soil Ecol 107:66–78

Sharma SS, Dietz KJ (2009) The relationship between metal toxicity and cellular redox imbalance. Trends Plant Sci 14:43–50

Siddhu G, Khan MAA (2012) Effects of cadmium on growth and metabolism of *Phaseolus mungo*. J Environ Biol 3:173–179

Siddiqui ZS (2013) Effects of double stress on antioxidant enzyme activity in Vigna radiata (L.) Wilczek. Acta Botanica Croatica 72(1):145–156

Siddiqui MH, Al-Whaibi MH, Sakran AM, Basalah MO, Ali HM (2012) Effect of calcium and potassium on antioxidant system of *Vicia faba* L. under cadmium stress. Int J Mol Sci 13:6604–6619

Silva IR, Smyth TJ, Raper CD, Carter TE, Rufty TW (2001) Differential aluminum tolerance in soybean: an evaluation of the role of organic acids. Physiol Plant 112:200–210

Singh S, Parihar P, Singh R, Singh VP, Prasad SM (2016) Heavy metal tolerance in plants: role of transcriptomics, proteomics, metabolomics, and ionomics. Front Plant Sci 6:1143

Singh S, Khan AN, Nazar R, Anjum NA (2008) Photosynthetic traits and activities of antioxidant enzymes in blackgram (*Vigna mungo* L. Hepper) under cadmium stress. Am J Plant Physiol 3(1):25–32

Singh M, Kumar J (2015) Adaptation strategies of plants against heavy metal toxicity: a short review. Biochem Pharmacol: Open Access 04(02)

Šiukšta R, Bondzinskaitė S, Kleizaitė V, Žvingila D, Taraškevičius R, Mockeliūnas L, Stapulionytė A, Mak K, Česnienė T (2019) Response of Tradescantia plants to oxidative stress induced by heavy metal pollution of soils from industrial areas. Environ Sci Pollut Res 26(1):44–61

Sriprang R, Hayashi M, Yamashita M, Ono H, Saeki K, Murooka Y (2002) A novel bioremediation system for heavy metals using the symbiosis between leguminous plant and genetically engineered Rhizobia. Biotechnol 99:279–293

Srivastava S, Shukla AK (2016) Differential response of black gram towards heavy metal stress. Environ Pollut Protect 2:9–96

Stambulska UY, Bayliak MM, Lushchak VI (2018) Chromium (VI) toxicity in legume plants: modulation effects of rhizobial symbiosis. Biomed Res Int 2018:8031213

Susilowati LE, Syekhfani S (2014) Characterization of phosphate solubilizing bacteria isolated from Pb contaminated soils and their potential for dissolving tricalcium phosphate. J Degrade Min Land Manage 1:57–62
Tangahu BV, Abdullah SRS, Basri H, Idris M, Anuar N, Mukhlisin M (2011) A review on heavy metals (As, Pb, and Hg) uptake by plants through phytoremediation. Int J Chem Eng 2011:939161
Tank N, Saraf M (2009) Enhancement of plant growth and decontamination of nickel-spiked soil using PGPR. J Basic Microbiol 49:195–204
Taurian T, Anzuay MS, Angelini JG, Tonelli ML, Ludueña L, Pena D, Ibáñez F, Fabra A (2010) Phosphate-solubilizing peanut associated bacteria: screening for plant growth-promoting activities. Plant Soil 329:421–431
Tchounwou PB, Yedjou CG, Patlolla AK, Sutton DJ (2012) Heavy metals toxicity and the environment. Mol Clinic Environ Toxicol 101:133–164
Ullah MA, Nadeem M, Hassan M, Ganter J, Muhammad B, Nawaz K, Shah AS, Hafeez FY (2015) Plant growth promoting rhizobacteria: an alternate way to improve yield and quality of wheat (Triticum aestivum). Int J Agric Biol 17:51–60
Vamerali T, Bandiera M, Mosca G (2010) Field crops for phytoremediation of metal ontaminated land—a review. Environ Chem Lett 8:1–17
Venkatesh NM, Vedaraman N (2012) Remediation of soil contaminated with copper using rhamnolipids produced from *Pseudomonas aeruginosa* MTCC 2297 using waste frying rice bran oil. Ann Microbiol 62:85–91
Vijendra PD, Huchappa KM, Lingappa R, Basappa G, Jayanna SG, Kumar V (2016) physiological and biochemical changes in moth bean (*Vigna aconitifolia* L.) under cadmium stress. J Bot 6403938
Wang CQ, Song H (2009) Calcium protects *Trifolium repens* L. seedlings against cadmium stress. Plant Cell Rep 28:1341–1349
Wang YS, Wang J, Yang ZM, Wang QY, Lu B, Li SQ, Lu YP, Wang SH, Sun X (2004) Salicylic acid modulates aluminum-induced oxidative stress in roots of *Cassia tora*. J Integ Plant Biol 46:819–828
Wang P, Yu W, Zhang J, Rengel Z, Xu J, Han Q, Chen L, Li K, Yu Y, Chen Q (2016) Auxin enhances aluminum-induced citrate exudation through upregulation of *GmMATE* and activation of the plasma membrane H^+-ATPase in soybean roots. Ann Bot 118:933–940
Wani PA, Khan MS, Zaidi A (2007a) Cadmium, chromium and copper in green gram plants. Agron Sustain Dev 27:145–153
Wani PA, Khan MS, Zaidi A (2007b) Impact of heavy metal toxicity on plant growth, symbiosis, seed yield and nitrogen and metal uptake in chickpea. Aust J Exp Agric 47:712–720
Wani PA, Khan MS, Zaidi A (2008) Chromium-reducing and plant growth-promoting mesorhizobium improves chickpea growth in chromium-amended soil. Biotechnol Lett 30:159–163
Wei X, Fang L, Cai P, Huang Q, Chen H, Liang W, Rong X (2011) Influence of extracellular polymeric substances (EPS) on Cd adsorption by bacteria. Environ Pollut 159:1369–1374
Xu J, Wang W, Yin H, Liu X, Sun H, Mi Q (2010) Exogenous nitric oxide improves antioxidative capacity and reduces auxin degradation in roots of *Medicago truncatula* seedlings under cadmium stress. Plant Soil 326:321–330
Xu X, Huang Q, Huang Q, Chen W (2012) Soil microbial augmentation by an EGFP tagged Pseudomonas putida X4 to reduce phytoavailable cadmium. Int Biodeter Biodegrad 71:55–60
Xu LL, Fan ZY, Dong YJ, Kong J, Bai XY (2014) Effects of exogenous salicylic acid and nitric oxide on physiological characteristics of two peanut cultivars under cadmium stress. Biol Plant 59:171–182
Yamur M, Kaydan D, Arvas O (2005) Effects of sewage biosolid application on seed protein ratios, seed NP contents, some morphological and yield characters in lentil (*Lens culinaris* Medic.). Res J Agric Biol Sci 1:308–314
Yang JL, You JF, Li YY, Wu P, Zheng SJ (2007) Magnesium enhances aluminum-induced citrate secretion in rice bean roots (*Vigna umbellata*) by restoring plasma membrane H^+-ATPase activity. Plant Cell Physiol 48:66–73

Zaidi A, Khan M, Ahemad M, Oves M (2009) Plant growth promotion by phosphate solubilizing bacteria. Acta Microbiol Imm H 56:263–284

Zaidi A, Wani PA, Khan MS (2012) Toxicity of heavy metals to legumes and bioremediation. In: Zaidi A, Wani PA, Khan MS (eds) Toxicity of heavy metals to legumes and bioremediation. Springer, Dordrecht

Zhang J, Zeng B, Mao Y, Kong X, Xinxun Wang, Yang Y, Zhang J, Xu J, Rengel Z, Chen Q (2017) Melatonin alleviates aluminium toxicity through modulating antioxidative enzymes and enhancing organic acid anion exudation in soybean. Function Plant Biol 44:961–968

Zhou G, Xu Y, Li J, Yang L, Liu J-Y (2006) Molecular analyses of the metallothionein gene family in rice (Oryza sativa L.). J Biochem Mol Biol 39:595–606

Zhou ZS, Huang SQ, Guo K, Mehta SK, Zhang PC, Yang ZM (2007) Metabolic adaptations to mercury-induced oxidative stress in roots of *Medicago sativa* L. J Inorg Biochem 101:1–9

Zhou B, Yao W, Wang S, Wang X, Jiang T (2014) The metallothionein gene, TaMT3, from Tamarix and rossowii confers Cd^{2+} tolerance in Tobacco. Int J Mol Sci 15:10398–10409

Zhu XF, Jiang T, Wang ZW, Lei GJ, Shi YZ, Li GX, Zheng SJ (2012) Gibberellic acid alleviates cadmium toxicity by reducing nitric oxide accumulation and expression of IRT1 in *Arabidopsis thaliana*. J Hazard Mater 239–240:302–307

Oxidative Stress and Antioxidant Defence in Fabaceae Plants Under Abiotic Stresses

Carla Gualtieri, Andrea Pagano, Anca Macovei, and Alma Balestrazzi

Abstract Legumes, grown worldwide under various climate conditions, are strongly appreciated for their nutritional value and impact in terms of environmental benefits and enhanced sustainability. Several environmental and anthropogenic factors can affect legume crop productivity, among which are the deleterious abiotic stresses, namely drought, salinity, temperature, heavy metals. Different abiotic stresses impair legume growth and performance by triggering a common scenario within the cell that is extensive oxidative damage. Thus, a better understanding of the molecular mechanisms underlying the oxidative stress response in legumes will lead to innovative agronomic and scientific developments, promoting the future competitiveness of the system. The chapter will present and discuss the state of the art concerning the hallmarks of oxidative damage and plant antioxidant response as well as the impact of oxidative injury on genome integrity. The focus will be on the DNA damage response and the way plants use this complex molecular network to cope with stress. Besides dissecting the cellular mechanisms, an in-depth evaluation of the several environmental and anthropogenic factors that are stress determinants is provided. In this context, the role of emerging players as miRNAs will be discussed. This chapter provides new insights on legume profiles of antioxidant stress response resulting from 'omics', covering issues of model legumes *versus* legume crops.

1 Hallmarks of Oxidative Damage

The evolution of photosynthesis has led to substantial changes in the composition of the Earth's atmosphere since carbon dioxide was progressively replaced by oxygen. The new living organisms were able to adapt to oxygen and its reduction intermediates, the cytotoxic reactive oxygen species (ROS), through the parallel evolution of antioxidant defence mechanisms that carefully regulate ROS levels within the cell (Apel and Hirt 2004; Gutteridge and Halliwell 2018; Xie et al. 2019). Oxidative

C. Gualtieri · A. Pagano · A. Macovei · A. Balestrazzi (✉)
Department of Biology and Biotechnology, "Lazzaro Spallanzani"-University of Pavia, via Ferrata 9, 27100, Pavia, Italy
e-mail: alma.balestrazzi@unipv.it

M. Hasanuzzaman et al. (eds.), *The Plant Family Fabaceae*,
https://doi.org/10.1007/978-981-15-4752-2_18

stress is a complex array of chemical and physiological events associated with all the different types of biotic (pathogen infection) and abiotic stresses (e.g. drought, salinity, extreme temperature, metal toxicity) that higher plants face during their life cycle (Xie et al. 2019). This condition results from the ROS overproduction and accumulation and compromises crop yields (Moller et al. 2007; Farooq et al. 2009; Challinor et al. 2014; Zorb et al. 2019).

Oxidative stress impacts the agronomic performance of grain legume crops that represent a fundamental source of dietary protein and essential mineral nutrients, particularly in the developing countries (Zhu et al. 2005; Bohra and Singh 2015; Bohra et al. 2015; Considine et al. 2017). In the context of global climate changes, adverse environmental conditions combined with improper agricultural land management are expected to challenge the global productivity of grain legume crops, thus limiting their benefits to soil fertility as well as their essential role in the cereal-legume-based cropping systems (Foyer et al. 2016; Considine et al. 2017).

Free radical species as O_2^- (superoxide radical), $OH^\cdot$ (hydroxyl radical), $HO_2^\cdot$(perhydroxyl radical) and non-radical molecules (H_2O_2, hydrogen peroxide; $^1O_2$, singlet oxygen) drive oxidative stress (Apel and Hirt 2004). The main sites of ROS production within the plant cell, in the presence of light, are chloroplasts and peroxisomes, whereas mitochondria represent the predominant source in the dark (Xia et al. 2015; Corpas et al. 2015). Other sites of ROS production have been identified, e.g. the plasma membrane where this event is mediated by the NADPH-dependent oxidase (Sharma et al. 2012), the endoplasmic reticulum (with the involvement of cytochrome P450) (Mittler 2002) and the apoplast (Hu et al. 2006). Excessive ROS levels cause oxidative injury to membrane lipids, proteins and nucleic acids, impairing the cell structure and metabolism, and ultimately lead to programmed cell death (Mittler and Blumwald 2015). Thus, the balance between ROS accumulation and scavenging is a critical parameter for plant survival and growth under environmental stress conditions. The hallmarks of oxidative stress damage can be precisely identified through the biochemical and molecular profiles of specific cellular components.

1.1 Lipid Peroxidation

At the level of cellular membranes, lipid peroxidation occurs when highly reactive ROS, such as hydroxyl radicals and singlet oxygen, abstract a hydrogen atom from polyunsaturated fatty acids to form lipid hydroperoxides. The latter are unstable and decompose to generate reactive aldehydes and ketones. Malondialdehyde (MDA) is a breakdown product of polyunsaturated fatty acid, and the MDA levels are regarded as a quantitative hallmark of lipid peroxidation induced by oxidative damage (Moller et al. 2007). Membrane lipids are sensitive targets, and the extent of ROS attack can be measured under adverse environmental conditions.

Lipid peroxidation is a reliable indicator of cell membrane damage, reported in several studies focused on the response to drought and heat stresses in alfalfa

(*Medicago sativa* L.) (Naya et al. 2007; Wang et al. 2009), white clover (*Trifolium repens* L.) (Lee et al. 2007), cowpea (*Vigna unguiculata* L.) (Nair et al. 2008) and *Lotus japonicus* L. (Sainz et al. 2010). Lipid peroxidation has been reported as one of the parameters used to assess the degree of oxidative damage triggered by heavy metals, as in the case of exposure to lead in pea (*Pisum sativum* L.) plants (Dias et al. 2019). MDA was also used as oxidative stress marker to monitor the response of chickpea (*Cicer arietinum* L.) roots to cadmium (Kar 2018). On the other hand, stress-tolerant genotypes can be selected, based on their ability to maintain low levels of lipid peroxidation, as reported for the drought-tolerant alfalfa varieties (Zhang et al. 2019). Similarly, lipid peroxidation has been used as hallmark to test the impact of salt stress on the *Medicago truncatula* L.-*Sinorhizobium meliloti* symbiosis and to evaluate the protective role of pre-treatments with the polyamines spermine and spermidine (Lopez-Gomez et al. 2017).

Another interesting use of this oxidative stress marker has been described in the context of seed germination. Doria et al. (2019) recently provided MDA profiles during the early step of water up-take by *M. truncatula* seeds in order to assess the impact of imbibitional damage on the lipid membranes. Lipid peroxidation helps the evaluation of an effective seed priming, the pre-sown treatment used by seed technologists to improve seed vigour (Bailly et al. 2000; Paparella et al. 2015). Polyamines are valuable priming agents, which are able to accelerate and increase germination of white clover seeds (Li et al. 2014). The treatment with polyamines triggered the pre-germinative metabolism, enhancing the antioxidant defence. Exogenous spermidine was able to improve the seed tolerance to water deficit during germination, and this was evidenced by the lower lipid peroxidation levels detected, compared to the untreated samples (Li et al. 2014).

1.2 Protein Oxidation

The irreversible protein oxidation is a major event that regulates the protein biological function and fate (Oracz et al. 2007). When free Fe^{2+} or Cu^{+} react with hydrogen peroxide, the resulting metal-catalysed oxidation generates highly reactive hydroxyl radicals (Fenton reaction; Halliwell 2006) that, in turn, oxidise amino acids. Protein carbonylation is induced through the interaction with reactive aldehydes generated by cell membrane lipid peroxides (Moller et al. 2011; Fedorova et al. 2014). Glycation represents a different type of oxidative post-translational modification, caused by the reaction of lysine and arginine residues with reducing sugars or their oxidation products. Legume root nodules contain high levels of Fe-proteins (nitrogenase, cytochromes and leghemoglobin) that can easily undergo oxidation (Becana et al. 2000). The occurrence of these oxidative post-translational modifications has been investigated in bean (*Phaseolus vulgaris* L.) nodule proteins (Matamoros et al. 2018). Metal-catalysed oxidation of amino acids led to malate dehydrogenase inactivation and leghemoglobin aggregation. The glycated proteins identified included the key nodule enzymes sucrose synthase, glutamine synthetase and glutamate synthase (Matamoros et al. 2018). Patterns of carbonylated proteins in maturing seeds of *Medicago truncatula* reported by Satour et al. (2018) revealed a correlation between carbonylation levels and seed deterioration.

2 The Plant Antioxidant Response

Plants have evolved highly complex defence mechanisms to limit the distructive effects of ROS. Such mechanisms contribute to the cellular redox homoeostasis through enzymatic and non-enzymatic antioxidants. The coordinated action of these players improves the cell ability to detoxify ROS, thus providing protection against oxidative damage (Mittler 2002; Gill and Tuteja 2010; Sharma et al. 2012).

The antioxidant enzymes mostly investigated include SOD (superoxide dismutase) which is able to catalyse the dismutation of O^{2-} radicals into O_2 and H_2O_2. Depending on the metal cofactor required to activate the catalytic site, they are classified as MnSODs, FeSODs and CuZnSODs. These classes can be distinguished based on different molecular features (e.g. their sensitivity to inhibitors) and subcellular locations. SOD isoforms are found in legume nodules where they play an essential role in removing the O^{2-} radicals released from the oxidation of legemoglobin, as shown for alfalfa MnSOD (Becana and Salin 1989) and soybean CuZnSOD (Puppo et al. 1982). The FeSOD isozyme has been identified first in cowpea nodules (Becana et al. 1989) and then in alfalfa, common bean and mung bean nodules (Becana et al. 2000). The tetrameric hemoproteins catalases are found as multiple isozymes, mainly located in peroxisomes and glyoxysomes (Scandalios et al. 1997). Catalase is located in the peroxisomes of determinate nodules, where it removes H_2O_2 generated by uricase and other oxidases (Kaneko and Newcomb 1987).

All the components of the ascorbate-gluthatione (AsA-GSH) cycle (Noctor and Foyer 1998; Asada 1999) are found in the cytosol of nodule cells (Dalton et al. 1986, 1992, 1993a, b; Dalton 1995). The AsA-GSH cycle includes four enzymes, namely ascorbate peroxidase (APX), monodehydroascorbate reductase (MDAR), dehydroascorbate reductase (DHAR) and glutathione reductase (GR) that act as a highly effective H_2O_2 detoxification route (Sofo et al. 2010). Non-enzymatic antioxidants consist of major cellular redox buffers, such as AsA, GSH, flavonoids, tocopherols, tocotrienols, carotenoids and alkaloids. AsA participates in several processes of plant growth and development (Noctor and Foyer 1998) and is found in legume nodules where it removes ROS (Matamoros et al. 1999). The thiol tripeptide GSH is a ROS scavenger, participating in the AsA-GSH cycle for H_2O_2 removal in chloroplasts and nodule cytosol (Dalton 1995), as well as in sulphur transport and storage, stress tolerance and heavy metal detoxification (Noctor and Foyer 1998). The leaves, roots and seeds of some legumes contain homoglutathione (hGSH) instead of or in addition to GSH (Klapheck et al. 1988). The relative abundance of GSH and hGSH in different legume tissues and species is determined by the corresponding thiol tripeptide synthetases. GSH and hGSH play protective roles during N_2 fixation

(Matamoros et al. 1999). Nodules possess other compounds with antioxidant properties, e.g. soybean nodules contain α-tocopherol (Evans et al. 1999). Polyamines are organic polycations involved in plant growth, but they are also powerful ROS scavengers (Fujihara et al. 1994). Uric acid, another potent antioxidant compound, is an intermediate of purine degradation and ureide synthesis found in peroxisomes (del Rio et al. 1998). Flavonoids and other phenolics are abundant in nodules, where they can inhibit lipid peroxidation by intercepting the peroxyl radicals formed in nodule membranes (Moran et al. 1997).

The antioxidant defence is essential for legume seed viability and successful germination under stress conditions, as reported in *M. truncatula* (Macovei et al. 2011a). The free radical scavenging activity of *M. truncatula* seeds was significantly higher at four and eight hours following rehydration, in presence and absence of PEG (polyethylene glycol)-mediated osmotic stress (Macovei et al. 2011a). *APX* and *SOD* genes, encoding cytosolic ascorbate peroxidise and superoxide dismutase, were up-regulated in *M. truncatula* seeds, as part of the antioxidant response (Balestrazzi et al. 2011a, b; Macovei et al. 2010, 2011b). The role played by antioxidant mechanisms in the response to specific abiotic stresses, namely drought stress, soil salinity, heat stress and heavy metal toxicity, will be described with more details in Paragraph 9.4.

3 Oxidative Injury and Genome Integrity: The DNA Damage Response

The integrity of genetic information is compromised under severe stress conditions that trigger oxidative DNA damage and impair genome stability. This results into reduced plant growth and crop productivity. To overcome genotoxic injury, plants have evolved an efficient repair machinery that responds to damage perception and signaling networks, the so-called DNA damage response (DDR) (Yoshiyama et al. 2013; Nikitaki et al. 2018). DDR has been mainly investigated in the model plant *Arabidopsis*; however, in recent years, *M. truncatula* has become a versatile system for the study of DNA damage sensing and repair, particularly, in relation to seed germination and seed quality issues (Pagano et al. 2017, 2019 and 2020), and hopefully, this knowledge will be translated to the most relevant legume crops (Parreira et al. 2018).

Macovei et al. (2011a) reported on the involvement of the *M. truncatulaOGG1* (8-oxoguanine DNA glycosylase/lyase) and *FPG* (formamidopyrimidine-DNAglycosylase) genes in the seed repair response, as part of the base-excision repair (BER) pathway (David et al. 2007). In *M. truncatula*, the *MtOGG1* and *MtFPG* genes were up-regulated during seed imbibition, with a peak at the timepoint of rehydration in which the highest levels of oxidative DNA damage, measured in terms of 8-oxoguanine accumulation, were recorded (Balestrazzi et al. 2011b). The contribution of nucleotide excision repair (NER) pathway to genome maintenance during

M. truncatula seed imbibition has been highlighted as well (Macovei et al. 2010, 2011b).

To date, *M. truncatula* is the plant system mostly used for the study of tyrosyl-DNA phosphodiesterases (TDPs) (Macovei et al. 2018a, b; Confalonieri et al. 2013; Faè et al. 2014; Sabatini et al. 2015, 2017; Araujo et al. 2016). The Tdp1 and Tdp2 DNA repair enzymes are involved in the processing of a wide range of 5′-and 3′-end blocking DNA lesions, among which the cytotoxicstabilised topoisomerase/DNA covalent complexes (Pommier et al. 2014). Differently from animals where Tdp1 is encoded by a single copy gene, plants possess a small *Tdp1* gene family, first identified in *M. truncatula* by Macovei et al. (2010). Both the *Tdp1α* and *Tdp1β* genes were up-regulated during seed imbibition, and *in planta*, in response to osmotic and heavy metal stress (Macovei et al. 2010). The involvement of plant *Tdp1* genes in the complex DDR response of *M. truncatula* cells has been evidenced using ionising radiation (Donà et al. 2014), antisense strategy and RNA-Seq-based analysis (Donà et al. 2013, 2017; Sabatini et al. 2015). Transgenic lines over expressing the *MtTdp2α* gene revealed enhanced tolerance to genotoxic stress caused by PEG6000 and copper (Confalonieri et al. 2013; Faè et al. 2014).

The knowledge concerning DNA damage response during seed development is still scarce. A transcriptomic profiling of the expression of genes related to DNA damage response/chromatin remodeling mechanisms was performed in *P. vulgaris* seeds at four distinct developmental stages (Parreira et al. 2018). In late embryogenesis, the observed up-regulation of genes related to DNA damage sensing and repair suggested for a tight control of DNA integrity. At the end of filling and onset of seed dehydration, the up-regulation of genes involved in sensing of DNA double-strand breaks indicated that genome integrity is challenged (Parreira et al. 2018).

4 Environmental and Anthropogenic Factors as Stress Determinants

4.1 Drought Stress

Water deficit, the main limiting factor for the successful establishment of crop cultivation, is playing a dramatic role in the current scenario of climate change (Boyer et al. 2013). Drought stress affects plant growth, reducing crop production, especially during grain filling and the reproductive phase. The extent of decrease in yield depends on the intensity and duration of drought stress, crop developmental stage and genotypic variability. Agricultural forage production requires adequate water supply, and this is problematic in semi-arid climates where insufficient water supply severely limits the production of forage legumes (Hopkins and Del Prado 2007; Nadeem et al. 2019a).

Drought resistance is achieved through reduced water loss combined with a more efficient water up-take, whereas drought tolerance relies on the action of osmoprotection and antioxidant mechanisms (Farooq et al. 2009; Fang and Xiong 2015). One of the biochemical changes occurring when plants experience these harmful conditions is ROS accumulation that is balanced through the antioxidant mechanisms. Enhanced antioxidant protection and declined lipid peroxidation are tightly associated with superior drought tolerance in legumes. Alfalfa avoids drought by reaching the deep moistured soil (Huang et al. 2018); however, several complex cultivar-specific morphological, physiological and molecular traits, including a robust antioxidant response, generally contribute to alfalfa drought resistance (Wang et al. 2009, 2011; Maghsoodi and Razmjoo 2015; Quan et al. 2016). Under oxidative stress, antioxidants may increase more during the recovery phase than in the stress phase, as observed in bean (Yasar et al. 2013), pea (Mittler and Zilinskas 1994; Osman 2015), soybean (Guler and Pehlivan 2016), chickpea (Patel et al. 2011) and cowpea (Carvalho et al. 2019). In the presence of drought stress, SOD, APX, GR, GST, GPX and POD activities are increased in resistant cultivars of common bean and horse gram (*Macrotyloma uniflorum* L.) (Saglam et al. 2011; Bhardwaj and Yadav 2012). The response of legumes at the onset of drought can vary; however, the final yield will significantly be reduced (Nadeem et al. 2019a).

4.2 Soil Salinity

Soil salinity exerts a deleterious impact on crops, including legumes, in terms of oxidative stress, genotoxicity, ionic imbalance, nutrition deficiency and osmotic stress (Murillo-Amador et al. 2007; He et al. 2015; Nadeem et al. 2019b). Composition of legume grains and grain yield is also altered (Manchanda and Garg 2008). This major constraint to crop production affects about 20% of the total irrigated land area in the world (FAO). Soil salinity is mainly caused by excess sodium chloride used for irrigation (Flowers and Flowers 2005).Salinity stress disrupts metabolic pathways by impairing key enzyme activities, while triggering ROS accumulation. Salt-tolerant legumes rely on effective antioxidant defence systems to scavenge the toxic-free radical species (Hernandez et al. 1999; Kukreja et al. 2005; Farooq et al. 2015). Under salinity stress, SOD, MDHAR, DHAR, GR and APX activities are significantly increased in salt-tolerant pea cultivars (Hernandez et al. 2000) and, similarly, a salt-tolerant common bean cultivar showed enhanced APX and CAT activities (Yasar et al. 2008).

4.3 Heat Stress

Temperature extremes (heat stress and cold stress) can severely damage crop plants at all stages of development, resulting in loss of productivity. Legumes, such as chickpea, lentil (*Lens culinaris* L.), mung bean (*Vigna radiata* L.), soybean and peas, are characterised by different degrees of sensitivity to high- and low-temperature stresses, and this compromises their performance at different developmental stages (germination, seedling emergence, vegetative phase, flowering and pod/seed filling phase) (HanumanthaRao et al. 2016; Sharma et al. 2016). In the case of mung bean, it has been reported that each degree rise in temperatures above optimum reduces the seed yield by 35–40% (Sharma et al. 2016). Increasing atmospheric CO_2 concentration along with temperature also limits plant growth, particularly in C_3 plant species, like mung bean (Nair et al. 2008). Heat stress decreases the activity of SOD, CAT and APX and increases ROS levels and membrane damage in the aerial parts of soybean, chickpea and mung bean plants (Djanaguiraman and Prasad 2010; Kumar et al. 2013; Nahar et al. 2015). Application of exogenous GSH enhanced heat stress tolerance in mung bean seedling by modulating antioxidant systems (Nahar et al. 2015), whereas higher levels of antioxidant metabolites have been detected in heat-tolerant chickpea and soybean genotypes under heat stress (Kumar et al. 2013; Chebrolu et al. 2016).

4.4 Heavy Metal Stress

Heavy metal environmental pollution has become a serious concern for living organisms and ecosystems. Heavy metals accumulated in soil are difficult to degrade and remove, and their further accumulation in plant tissues can cause higher plant-tissue toxicity (Nagajyoti et al. 2010). Heavy metal toxicity significantly affects plant growth as it suppresses antioxidant enzyme activity (Duan et al. 2018). The use of forages grown on metal-contaminated soil can increase the risk of heavy metals entering the food chain and affecting human health. The response to heavy metals has been investigated in legumes, highlighting the role of antioxidant players. *P. sativum* plants exposed to lead (Pb) in soil showed increased GR, APX and CAT activities in leaves and roots that were triggered by the phytohormone jasmonic acid (Dias et al. 2019). Similarly, in chickpea plants exposed to increasing vanadium (V) concentrations, the antioxidant enzyme activities (SOD, CAT, POD) were increased in a dose-dependent manner (Imtiaz et al. 2018).

5 Novel Players in the Response to Oxidative Stress: MicroRNAs

MicroRNAs (miRNAs) are small non-coding RNA molecules (21–22 nt) which are able to regulate gene expression at post-transcriptional level. They can silence the expression of a specific gene by matching to its messenger RNA (mRNA), triggering degradation or, in some cases, repressing the translation step (Yu et al. 2017; Tyagi et al. 2019). miRNAs have been retrieved in several legume species including *M. truncatula*, soybean, chickpea, common bean and *L. japonicas* (Subramanian et al. 2016; Szittya et al. 2008; Arenas-Huertero et al. 2009; Hu et al. 2013; Zheng et al. 2016; Pan et al. 2016; Wu et al. 2017). In a study performed on drought-sensitive and drought-tolerant soybean seedlings, Kulcheski et al. (2011) identified 256 miRNAs among the novel MIR07 and MIR11. The relative expression of MIR07 was increased under water deficit in both the drought-sensitive and drought-tolerant genotypes. In contrast, MIR11 expression showed a genotype-dependent response, being stable in the tolerant plants. Other soybean miRNAs involved in drought tolerance were also identified by Zheng et al. (2016). High-throughput sequencing was used by Jatan et al. (2019) to detect and characterise small RNAs in drought-tolerant chickpea roots in presence or absence of the rhizobacterium *Pseudomonas putida*, under water deficit. This study suggested that bacterial inoculation might play a crucial role in the modulation of miRNAs and their target genes in response to drought stress (Jatan et al. 2019). The genotype-dependent response of miRNAs to abiotic stresses was further assessed by Barrera-Figueroa et al. (2011) who investigated and compared the impact of water deficit on drought-tolerant and drought-sensitive cowpea cultivars. MicroRNAs also mediate the response to heat stress, as reported in common bean (Naya et al. 2014) and alfalfa (Matthews et al. 2019).

Members of the miR156 family control the *Squamosa promoter-binding protein-like* (*SPL*) genes in the context of plant growth under abiotic stresses. Alfalfa plants overexpressing miR156 and RNAi-mediated knockdown of the *SPL13* gene showed increased tolerance to heat stress associated with increased non-enzymatic antioxidant content (Matthews et al. 2019). A significant improvement in drought tolerance was also observed in alfalfa plants overexpressing miR156, concomitant with enhanced accumulation of proline, abscisic acid and antioxidants (Arshad et al. 2017).

The state of the art related to the miRNA word in legumes is rapidly expanding and databases resulting from the use of high-throughput technologies which are continuously enriched with huge amounts of information. The legume miRNAomes need to be integrated with tissue-specific transcriptomes and proteomes captured in response to abiotic stresses and/or ameliorating treatments (e.g. priming).

6 Insights on Legume Profiles of Antioxidant Stress Response Provided by 'Omics'

In the last decades, the approaches used to study biological systems and explain their behaviours have changed. The reductionist vision based on functional biology and aiming at model simplification is being progressively complemented and substituted by more holistic views based on system biology and aiming at huge dataset integration. Key functional and regulatory genes involved in abiotic stress resistance have been identified using next-generation technologies and related bioinformatic tools that enable the rapid and cost-effective analysis of whole genomes and transcriptomes in major crops.

In a recent review, Abdelrahman et al. (2018) provided an exhaustive overview of the knowledge gained on the molecular mechanisms underlying abiotic stress resistance in legume crops. Microarray-based gene expression analyses were initially used for transcriptome studies in legumes, as in the case of soybean (Le et al. 2012; Ha et al. 2015; Tripathi et al. 2015; Ramesh et al. 2019), *L. japonicus* (Asamizu et al. 2005), alfalfa (Gao et al. 2016), *M. truncatula* (Cheung et al. 2006), chickpea (Deokar et al. 2011) and candidate genes for drought, salinity, cold and heavy metal stress resistance were identified. Garg et al. (2016) performed a comparative transcriptome analysis of drought- and salinity-tolerant/sensitive chickpea genotypes in response to drought or salinity at different developmental stages. The study highlighted the extensive transcriptional reprogramming occurring in the different chickpea genotypes, showing enhanced drought sensitivity during the early reproductive stage, followed by pronounced salt stress sensitivity at the late reproductive stage (Garg et al. 2016).

High-throughput Illumina Hiseq 2500 sequencing allowed to identify drought-responsive miRNAs from alfalfa roots and leaves, providing 281 novel, predicted miRNAs (Li et al. 2017). Based on this study, drought-induced miRNA-related pathways were established, and the most representative families (miR166, miR159, miR482 and miR2118) with regulatory roles in legumes under drought were assessed (Li et al. 2017). Genome-wide analysis revealed auxin-responsive miRNAs differentially expressed in soybean roots in response to salt stress (Sun et al. 2016) as well as miRNAs involved in post-transcriptional regulation of gene expression in chickpea roots under salt and drought stresses (Khandal et al. 2017).

7 Model Legumes Versus Legume Crops

The choice of a model organism is essential to establish a representative experimental system, and it is functional for the definition of specific timepoints, physiological stages and treatments in which the phenomena of interest are to be studied. Two legume species in the galegoid clade (cool season legumes), *M. truncatula* and *L. japonicus*, which belong to the tribes Trifolieae and Loteae, respectively, were selected as model systems to investigate legume genomics and biology (Cook 1999; Stougaard 2001). Differently from the major crop legumes, *M. truncatula* and *L. japonicus* own a small genome, ideal for carrying forward and reverse genetic analyses and well suited for studying biological issues relevant to crop legumes, among the abiotic and biotic stress tolerances.

M. truncatula was originally chosen for the study of rhizobia-legume symbiosis and the molecular genetics of nitrogen fixation in legume root systems (Barker et al. 1990). Subsequently, the interest towards *M. truncatula* as a model organism has progressively increased because of other appealing features. Its phylogenetic relationship and genetic similarity with other legumes, along with his small sequenced genome (~500 Mbp) and its diploidy ($2n = 16$) makes it a suitable model system for molecular genetics. Furthermore, its autogamy, its short generation time (~3 months) and its in vitro regenerative capacities add practical advantages to its cultivation, both in field/greenhouse and in vitro (Frugoli and Harris 2001).

L. japonicus is a perennial temperate pasture species, closely related to birdsfoot trefoil (*Lotus corniculatus* L.) with features useful for genomics (e.g. a short life cycle of 2–3 months, self-fertility, diploidy ($2n = 12$) and a small genome of 472.1 Mb). Although both *L. japonicus* and *M. truncatula* belong to the 'temperate' or 'galegoid' legume group, *L. japonicus* forms determinate nodules as observed in 'tropical' or 'phaseoloid' legumes such as soybean and common bean. *L. japonicus* is widely used to study plant–microbe interactions, due to its ability to establish a range of different types of relationship with symbiotic and pathogenic microorganisms, and it has been demonstrated to be amenable to genetic analyses (Handberg et al. 1992; Sato and Tabata 2006; Mun et al. 2016).

Efforts and investments made to decipher *M. truncatula* and *L. japonicus* genomics have brought to significant advances in basic and applied research, facilitating knowledge transfer from the best-characterised models to related food and feed legumes (Zhu et al. 2005). Comparative genome mapping has revealed macrosynteny or the conserved gene order between species as well as microsynteny (conserved gene content and order at sequence level over a short, physically defined DNA contig) (Zhu et al. 2005). The conserved genome structure between *M. truncatula* and crop legumes has allowed for map-based cloning of genes required for nodulation in crop legumes (Zhu et al. 2005). Similarly, these approaches are currently used to speed up the identification of stress-tolerant phenotypes and design of new breeding strategies to ameliorate legume crop adaptation to adverse environments. In this context, a promising, integrated strategy aimed at the identification of stress-tolerant legume germplasm has been recently described by Menendez et al. (2019). The link between plant physiology and big data resulting from 'omics' was examined in order to highlight the gene-to-metabolite networks involved in the abiotic stress interactions dealing with the ROS scavengers polyamines. These approaches are expected to accelerate the identification of stress-tolerant phenotypes and the design of new biotechnological strategies to increase their yield and adaptation to marginal environments, making better use of available plant genetic resources.

8 New Avenues for Improving Abiotic Stress Tolerance in Legumes

Genome resources (physical maps, functional genomics tools) facilitate the isolation of key genes contributing to abiotic stress tolerance in several legume species, providing the opportunity for crop improvement. Thus, biotechnologists and breeders will be able to target more rapidly and precisely the desired agronomic traits. The molecular profiling of germplasm collections reflecting the global biodiversity will promote the discovery of novel players in abiotic stress tolerance that will be included breeding programs. In this context, the focus is directed towards the 'orphan' or underutilised legumes, suited to withstand harsh environments (e.g. arid regions) (Cullis et al. 2018). Orphan legumes include groundnut (*Arachis hypogaea*), grass pea (*Lathyrus sativus*), bambara groundnut (*Vigna subterranea*), cowpea (*Vigna unguiculata*) and marama bean (*Tylosema esculentum*). They are staple food crops in many developing countries, with little economic importance, not extensively improved by breeders (Foyer et al. 2016); however, these plants have evolved successful survival strategies using a combination of different traits and responses. To date, orphan legumes represent a unique source of information for breeders who are looking for traits able to allow for survival in extreme environments.

Genome editing stands as a powerful tool for legume crop improvement, whose potential is still underexploited. To date, a few studies are available that describe the use of CRISPR/Cas9 for editing drought-tolerance-related genes in legumes (Cai et al. 2015, 2018). The CRISPR-based approach requires a deep characterisation of the target gene(s), their function and regulatory mechanisms, and this might delay the use in legume crops that still own limited information on stress-related genes.

Due to their peculiar physiology, legumes are expected to play pivotal roles in the mitigation of the effects of climate changes as well as in the contribution to sustainable farming. Their broad genetic diversity will be the source of alleles with relevance in novel adaptive traits. Researchers from different disciplines in plant science must integrate their expertise to address the open questions and develop new tools. Traditional and modern breeding must join their efforts and integrate their tools in order to accelerate genetic improvement.

References

Abdelrahman M, Jogaiah S, Burritt DJ, Tran L-S (2018) Legume genetic resources and transcriptome dynamics under abiotic stress conditions. Plant Cell Environ 41:1972–1983

Apel K, Hirt H (2004) Reactive oxygen species: metabolism, oxidative stress, and signal transduction. Ann Rev Plant Biol 55:373–399

Araújo SS, Balestrazzi A, Faè M, Morano M, Carbonera D, Macovei A (2016) *MtTdp2α*-overexpression boosts the growth phase of *Medicago truncatula* cell suspension and increases the expression of key genes involved in antioxidant response and genome stability. Plant Cell Tissue Organ Cult 127:675–680

Araújo S, Pagano A, Dondi D, Lazzaroni S, Pinela E, Macovei A, Balestrazzi A (2019). Metabolic signatures of germination triggered by kinetin in *Medicago truncatula*. Sci. Rep. 9:10466

Arenas-Huertero C, Pérez B, Rabanal F, Blanco-Melo D, De la Rosa C, Estrada-Navarrete G, Sanchez F, Covarrubias AA, Reyes JL (2009) Conserved and novel miRNAs in the legume *Phaseolus vulgaris* in response to stress. Plant Mol Biol 70:385–401

Arshad M, Feyissa BA, Amyot L, Aung B, Hannoufa A (2017) MicroRNA156 improves drought stress tolerance in alfalfa (*Medicago sativa*) by silencing SPL13. Plant Sci 258:122–136

Asada K (1999) The water-water cycle in chloroplasts: scavenging of active oxygen and dissipation of excess photons. Annu Rev Plant Physiol Plant Mol Biol 50:601–639

Asamizu E, Nakamura Y, Sato S, Tabata S, Asamizu E, Nakamura Y, Sato S, Tabata S (2005) Comparison of the transcript profiles from the root and the nodulating root of the model legume *Lotus japonicus* by serial analysis of gene expression. Mol. Plant-Microbe Int. 18:487–498

Bailly C, Benamar A, Corbineau F, Come D (2000) Antioxidantsystems in sunflower (*Helianthus annuus* L.) seeds as affected bypriming. Seed Sci Res 10:35–42

Balestrazzi A, Confalonieri M, Macovei A, Donà M, Carbonera D (2011a) Genotoxic stress and DNA repair in plants: emerging functions and tools for improving crop productivity. Plant Cell Rep 30:287–295

Balestrazzi A, Confalonieri M, Macovei A, Carbonera D (2011b) Seed imbibition in *Medicago truncatula* Gaertn.: expression profiles ofDNArepair genes in relation to PEG-mediated stress. J Plant Physiol 168:706–713

Barker DG, Bianchi S, Blondon F, Datteé Y, Duc G, Essad S, Flament P, Gallusci P, Génier G, Guy P, Muel X, Tourneur J, Dénarié J, Huguet T (1990) *Medicago truncatula*, a model plant for studying the molecular genetics of the *Rhizobium*-legume symbiosis. Plant Mol. Biol Rep 8:40–49

Barrera-Figueroa BE, Gao L, Diop NN, Wu Z, Ehlers JD, Roberts PA, Close TJ, Zhu JK, Liu R (2011) Identification and comparative analysis of drought-associated microRNAs in two cowpea genotypes. BMC Plant Biol 11:127

Bhardwaj J, Yadav SK (2012) Comparative study on biochemical parameters and antioxidant enzymes in a drought tolerant and a sensitive variety of horsegram under drought stress. Am J Plant Physiol 7:17–29

Becana M, Salin ML (1989) Superoxide dismutases in nodules of leguminous plants. Can J Bot 67:415–421

Becana M, Paris FJ, Sandalio LM, del Rio LA (1989) Isoenzymes of superoxide dismutase in nodules of *Phaseolus vulgaris* L., *Pisum sativum* L., and *Vigna unguiculata* (L.) Walp. Plant Physiol 90:1286–1292

Becana M, Dalton DA, Moran JF, Iturbe-Ormaetxe I, Matamoros MA, Rubio MC (2000) Reactive oxygen species and antioxidants in legume nodules. Physiol Plant 109:372–381

Bohra A, Singh NP (2015) Whole genome sequences in pulse crops: a global community resource to expedite translational genomics and knowledge-based crop improvement. Biotechnol Lett 37:1529–1539

Bohra A, Sahrawat KL, Kumar S, Joshi R, Parihar AK, Singh U, Singh D, Singh NP (2015) Genetics- and genomics-based interventions for nutritional enhancement of grain legume crops: status and outlook. J Appl Genet 56:151–161

Boyer JS, Byrn P, Cassman KG, Cooper M, Delmer D, Greene T (2013) The US drought of 2012 in perspective: a call to action. Glob Food Secur 2:139–143

Cai Y, Chen L, Liu X, Sun S, Wu C, Jiang B, Han T, Hou W (2015) CRISPR/Cas9-mediated genome editing in soybean hairy roots. PLoS ONE 10:e0136064

Cai Y, Chen L, Liu X, Guo C, Sun S, Wu C, Jiang B, Han T, Hou W (2018) CRISPR/Cas9-mediated targeted mutagenesis of GmFT2a delays flowering time in soybean. Plant Biotechnol J 16:176–185

Carvalho M, Castro I, Moutinho-Pereira J, Correia C, Egea-Cortines M, Matos M, Rosa E, Carnide V, Lino-Neto T(2019) Evaluating stress responses in cowpea under drought stress. J Plant Physiol 241:153001

Challinor AJ, Watson J, Lobell DB, Howden SM, Smith DR, Chhetri N (2014) A meta-analysis of crop yield under climate change and adaptation. Nat Clim Change 4:287–291

Chebrolu KK, Fritschi FB, Ye SQ, Krishnan HB, Smith JR, Gillman JD (2016) Impact of heat stress during seed development on soybean seed metabolome. Metabolomics 12:1–14

Cheung F, Haas BJ, Goldberg SMD, May GD, Xiao Y, Town CD (2006) Sequencing *Medicago truncatula* expressed sequenced tags using 454 Life Sciences technology. BMC Genom 7:272

Confalonieri M, Faè M, Balestrazzi A, Donà M, Macovei A, Valassi A, Giraffa G, Carbonera D (2013) Enhanced osmotic stress tolerance in *Medicago truncatula* plants overexpressing the DNA repair gene *MtTdp2α* (tyrosyl-DNA phosphodiesterase 2). Plant Cell Tissue Organ Cult 116:187–203

Considine MJ, Siddique KHM, Foyer CH (2017) Nature's pulse power: legumes, food security and climate change. J Exp Bot 68:1815–1818

Cook DR (1999) *Medicago truncatula*-a model in the making! Curr. Opin. Plant Biol. 2:301–304

Corpas FJ, Gupta DK, Palma JM (2015) Production sites of reactive oxygen species (ROS) in organelles from plant cells. In: Gupta D, Palma J, Corpas F (eds) Reactive oxygen species and oxidative damage in plants under stress. Springer, Cham, Switzerland, pp 1–22

Cullis C, Chimwamurombe P, Barker N, Kunert K, Vorster J (2018) Orphan legumes growing in dry environments: marama bean as a case study. Front. Plant Sci. 9:1199

Dalton DA (1995) Antioxidant defenses of plants and fungi. In: Ahmad S (ed) Oxidative stress and antioxidant defenses in biology. Chapman and Hall, New York, NY, pp 298–355. ISBN 0-412-03971-0

Dalton DA, Russell SA, Hanus FJ, Pascoe GA, Evans HJ (1986) Enzymatic reactions of ascorbate and glutathione that prevent peroxide damage in soybean root nodules. Proc Natl Acad Sci USA 83:3811–3815

Dalton DA, Langeberg L, Robbins M (1992) Purification and characterization of monodehydroascorbate reductase from soybean root nodules. Arch Biochem Biophys 292:281–286

Dalton DA, Baird LM, Langeberg L, Taugher CY, Anyan WR, Vance CP, Sarath G (1993a) Subcellular localization of oxygen defense enzymes in soybean (*Glycine max* [L.] Merr.) root nodules. Plant Physiol 102:481–489

Dalton DA, Langeberg L, Treneman N (1993b) Correlations between the ascorbate–glutathione pathway and effectiveness in legume root nodules. Physiol Plant 87:365–370

David SS, O'Shea VL, Sucharita Kundu S (2007) Base-excisionrepair of oxidative DNA damage. Nature 447:941–950

del Rio LA, Sandalio LM, Corpas FJ, Lopez-Huertas E, Palma JM, Pastori GM (1998) Activated oxygen-mediated metabolic functions of leaf peroxisomes. Physiol Plant 104:673–680

Deokar AA, Kondawar V, Jain PK, Karuppayil SM, Raju NL, Vadez V, Varshney RK, Srinivasan R (2011) Comparative analysis of expressed sequence tags (ESTs) between drought-tolerant and -susceptible genotypes of chickpea under terminal drought stress. BMC Plant Biol 11:1–20

Dias MC, Mariz-Ponte N, Santos C (2019) Lead induces oxidative stress in *Pisum sativum* plants and changes the levels of phytohormones with antioxidant role. Plant Physiol Biochem 137:121–129

Djanaguiraman M, Prasad PVV (2010) Ethylene production under high temperature stress causes premature leaf senescence in soybean. Funct Plant Biol 37:1071–1084

Donà M, Confalonieri M, Minio A, Biggiogera M, Buttafava A, Raimondi E, Delledonne M, Ventura L, Sabatini ME, Macovei A, Giraffa G, Carbonera D, Balestrazzi A (2013) RNA-Seq analysis discloses early senescence and nucleolar dysfunction triggered by *Tdp1α* depletion in *Medicago truncatula*. J Exp Bot 64:1941–1951

Donà M, Ventura L, Balestrazzi A, Buttafava A, Carbonera D, Confalonieri M, Giraffa G, Macovei A (2014) Dose-dependent reactive species accumulation and preferential double strand breaks repair are featured in the gamma-ray response in *Medicago truncatula* cells. Plant Mol Biol Rep 32:129–141

Donà M, Sabatini ME, Biggiogera M, Confalonieri M, Minio A, Delledonne M, Giraffa G, Carbonera D, Araujo S, Balestrazzi A (2017) *MtTdp1alpha*-depleted *Medicago truncatula*plants show reduced cuticle permeability and altered expression of defense genes. Biol Plant 61:192–196

Doria E, Pagano A, Ferreri C, Larocca AV, Maçovei A, Araújo S, Balestrazzi A (2019) How does the seed pre-germinative metabolism fight against imbibition damage? Emerging roles of fatty acid cohort and antioxidant defence. Front. Plant Sci. In press

Duan C, Fang C, Yang W, Chen Y, Cui S (2018) Reveal the response of enzyme activities to heavy metals through in situ zymography. Ecotoxicol Environ Saf 156:106–115

Evans PJ, Gallesi D, Mathieu C, Hernandez MJ, de Felipe M, Halliwell B, Puppo A (1999) Oxidative stress occurs during soybean nodule senescence. Planta 208(1):73–79

Faè M, Balestrazzi A, Confalonieri M, Donà M, Macovei A, Valassi A, Giraffa G, Carbonera D (2014) Copper-mediated genotoxic stress is attenuated by the overexpression of the DNA repair gene *MtTdp2a* (tyrosyl-DNA phosphodiesterase 2 alpha) in *Medicago truncatula* plants. Plant Cell Rep 33:1071–1080

Fang Y, Xiong L (2015) General mechanisms of drought response and their application in drought resistance improvement in plants. Cell Mol Life Sci 72:673–689

Flowers TJ, Flowers SA (2005) Why does salinity pose such a difficult problem for plant breeders? Agric Water Manag 78:15–24

Farooq M, Wahid A, Kobayashi N, Fujita D, Basra SMA (2009) Plant drought stress: effects, mechanisms and management. Agron Sustain Dev 29:185–212

Farooq M, Hussain M, Wakeel A, Siddique KHM (2015) Salt stress in maize: effects, resistance mechanisms, and management. A Rev Agron Sustain Dev 35:461–481

Fedorova M, Bollineni RC, Hoffmann R (2014) Protein carbonylation as a major hallmark of oxidative damage: update of analytical strategies. Mass Spectrom Rev 33:79–97

Foyer CH, Lam HM, Nguyen HT, Siddique KH, Varshney RK, Colmer TD, Cowling W, Bramley H, Mori TA, Hodgson JM, Cooper JW, Miller AJ, Kunert K, Vorster J, Cullis C, Ozga JA, Wahlqvist ML, Liang Y, Shou H, Shi K, Yu J, Fodor N, Kaiser BN, Wong FL, Valliyodan B, Considine MJ (2016) Neglecting legumes has compromised human health and sustainable food production. Nat Plants 2:16112

Frugoli J, Harris J (2001) *Medicago truncatula* on the move! Plant Cell 13:458–463

Fujihara S, Abe H, Minakawa Y, Akao S, Yoneyama T (1994) Polyamines in nodules from various plant-microbe symbiotic associations. Plant Cell Physiol 35:1127–1134

Gao R, Austin RS, Amyot L, Hannoufa A (2016) Comparative transcriptome investigation of global gene expression changes caused by miR156 overexpression in *Medicago sativa*. BMC Genom 17:658

Garg R, Shankar R, Thakkar B, Kudapa H, Krishnamurthy L, Mantri N, Varshney RK, Bhatia S, Jain M (2016) Transcriptome analyses reveal genotype- and developmental stage-specific molecular responses to drought and salinity stresses in chickpea. Sci. Rep. 6:19228

Gill SS, Tuteja N (2010) Reactive oxygen species and antioxidant machinery in abiotic stress tolerance in crop plants. Plant Physiol Biochem 48:909–930

Guler NS, Pehlivan N (2016) Exogenous low-dose hydrogen peroxide enhances drought tolerance of soybean (*Glycine max* L.) through inducing antioxidant system. Acta Biol Hung 67:169–183

Gutteridge JMC, Halliwell B (2018) Mini-Review: oxidative stress, redox stress or redox success? Biochem Biophys Res Commun 502:183–186

Ha CV, Watanabe Y, Tran UT, Le DT, Tanaka M, Nguyen KH, Tran L-SP (2015) Comparative analysis of root transcriptomes from two contrasting drought-responsive Williams 82 and DT2008 soybean cultivars under normal and dehydration conditions. Front Plant Sci 6:551

Halliwell B (2006) Reactive species and antioxidants: redox biology is a fundamental theme of aerobic life. Plant Physiol 141:312–322

Handberg K, Stougaard J (1992) *Lotus japonicus*, an autogamous, diploid legume species for classical and molecular-genetics. Plant J. 2:487–496

HanumanthaRao B, Nair RM, Nayyar H (2016) Salinity and high temperature tolerance in mungbean [*Vigna radiata* (L.) Wilczwk] from a physiological perspective. Front Plant Sci 7:1–20

He Y, Fu J, Yu C, Wang X, Jiang Q, Hong J, Lu K, Xue G, Yan C, James A, Xu L, ChenJ Jiang D (2015) Increasing cyclic electron flow is related to Na^+ sequestration into vacuoles for salt tolerance in soybean. J Exp Bot 66:6877–6889

Hernandez JA, Campillo A, Jimenez A, Alarcon JJ, Sevilla F (1999) Response of antioxidant systems and leaf water relations to NaCl stress in pea plants. New Phytol 141:241–251

Hernandez JA, Jiménez A, Mullineaux P, Sevilla F (2000) Tolerance of pea (*Pisum sativum* L.) to long-term salt stress is associated with induction of antioxidant defences. Plant, Cell Environ 23:853–862

Hopkins A, Del Prado A (2007) Implications of climate change for grassland in Europe: impacts, adaptations and mitigation options: a review. Grass Forage Sci 62:118–126

Hu X, Zhang A, Zhang J, Jiang M (2006) Abscisic acid is a key inducer of hydrogen peroxide production in leaves of maize plants exposed to water stress. Plant Cell Physiol 47:1484–1495

Hu J, Zhang H, Ding Y (2013) Identification of conserved microRNAs and their targets in the model legume *Lotus japonicus*. J Biotechnol 164:520–524

Huang Z, Liu Y, Cui Z, Fang Y, He HH, Liu BR, Wu GL (2018) Soil water storage deficit of alfalfa (*Medicago sativa*) grasslands along ages in arid area (China). Field Crops Res. 221:1–6

Jatan R, Tiwari S, Asif MH, Lata C (2019) Genome-wide profiling reveals extensive alterations in *Pseudomonas putida*-mediated miRNAs expression during drought stress in chickpea (*Cicer arietinum* L.). Environ Exp Bot 157:217–227

Kaneko Y, Newcomb EH (1987) Cytochemical localization of uricase and catalase in developing root nodules of soybean. Protoplasma 140:1–12

Kar M (2018) Determination of the expression level of stress-related genes in *Cicer arietinum* root cell under Cd stress and the relationship to H_2O_2 concentrations. Ecotoxicology 27:1087–1094

Khandal H, Parween S, Roy R, Meena MK, Chattopadhyay D (2017) MicroRNA profiling provides insights into post-transcriptional regulation of gene expression in chickpea root apex under salinity and water deficiency. Sci Rep 7:4632

Klapheck S, Zopes H, Levels HG, Bergmann L (1988) Properties and localization of the homoglutathione synthetase from *Phaseolus coccineus* leaves. Physiol Plant 74:733–739

Kukreja S, Nandwal AS, Kumar N, Sharma SK, Unvi V, Sharma PK (2005) Plant water status, H_2O_2 scavenging enzymes, ethylene evolution and membrane integrity of *Cicer arietinum* roots as affected by salinity. Biol Plant 49:305–308

Kulcheski FR, de Oliveira LF, Molina LG, Almerão MP, Rodrigues FA, Marcolino J, Barbosa JF, Stolf-Moreira R, Nepomuceno AL, Marcelino-Guimarães FC, Abdelnoor RV, Nascimento LC, Carazzolle MC, Pereira GAG, Margis R (2011) Identification of novel soybean microRNAs involved in abiotic and biotic stresses. BMC Genom 12:307

Kumar S, Thakur P, Kaushal N, Malik JA, Gaur P, Nayyar H (2013) Effect of varying high temperatures during reproductive growth on reproductive function, oxidative stress and seed yield in chickpea genotypes differing in heat sensitivity. Arch Agron Soil Sci 59:823–843

Imtiaz M, Ashraf M, Rizwan MS, Nawaz MA, Rizwan M, Mehmood S, Yousaf B, Yuan Y, Ditta A, Mumtaz MA, Ali M, Mahmood S, Tu S (2018) Vanadium toxicity in chickpea (*Cicer arietinum* L.) grown in red soil: effects on cell death, ROS and antioxidative systems. Ecotoxicol Environ Saf 158:139–144

Le DT, Nishiyama R, Watanabe Y, Vankova R, Tanaka M, Seki M, Ham LH, Yamaguchi-Shinozaki K, Shinozaki K (2012) Tran L-SP Identification and expression analysis of cytokinin metabolic genes in soybean under normal and drought conditions in relation to cytokinin levels. PLoS One 7:e42411

Lee BR, Kim KY, Jung WJ, Avice JC, Ourry A, Kim TH (2007) Peroxidases and lignification in relation to the intensity of water-deficit stress in white clover (*Trifolium repens* L.). J Exp Bot 58:1271–1279

Li Z, Peng Y, Zhang XQ, Ma X, Huang LK, Yan YH (2014) Exogenous spermidine improves seed germination of white clover under water stress via involvement in starch metabolism, antioxidant defenses and relevant gene expression. Molecules 19:18003–18024

Li Y, Wan L, Bi S, Wan X, Li Z, Cao J, Tong Z, Xu H, He F, Li X (2017) Identification of drought-responsive microRNAs from roots and leaves of alfalfa by high-throughput sequencing. Genes (Basel) 8:E119

Lopez-Gomez M, Hidalgo-Castellanos J, Munoz-Sanchez JR, Mari-Pena AJ, Lluch C, Herrera-Cervera JA (2017) Polyamines contribute to salinity tolerance in the symbiosis *Medicago truncatula-Sinorhizobium meliloti* by preventing oxidative damage. Plant Physiol Biochem 116:9–17

Macovei A, Balestrazzi A, Confalonieri M, Carbonera D (2010) The tyrosyl-DNA phosphodiesterase gene family in *Medicago truncatula*Gaertn.: bioinformatic investigation and expression profiles inresponse to copper- and PEG-mediated stress. Planta 232:393–407

Macovei A, Balestrazzi A, Confalonieri M, Faè M, Carbonera D (2011a) New insights on the barrel medic *MtOGG1* and *MtFPG* functions in relation to oxidative stress response *in planta* and during seed imbibition. Plant Physiol Biochem 49:1040–1050

Macovei A, Balestrazzi A, Confalonieri M, Buttafava A, Carbonera D (2011b) The *TFIIS* and *TFIIS-like* gene from *Medicago truncatula* are involved in oxidative stress response. Gene 470:20–30

Macovei A, Faè M, Biggiogera M, de Sousa Araújo S, Carbonera D, Balestrazzi A (2018a) Ultrastructural and molecular analyses reveal enhanced nucleolar activity in medicago truncatula cells overexpressing the MtTdp2α Gene. Front. Plant Sci. 9:596

Macovei A, Pagano A, Sabatini ME, Grandi S, Balestrazzi A (2018b) The human tyrosyl-DNA phosphodiesterase 1 (hTdp1) inhibitor NSC120686 as an exploratory tool to investigate plant *Tdp1* genes. Genes 9:e186

Maghsoodi M, Razmjoo J (2015) Identify physiological markers for drought tolerance in alfalfa. Agron J 107:149–157

Manchanda G, Garg N (2008) Salinity and its effects on the functional biology of legumes. Acta Physiol Plant 30:595–618

Matamoros MA, Moran JF, Iturbe-Ormaetxe I, Rubio MC, Becana M (1999) Glutathione and homoglutathione synthesis in legume root nodules. Plant Physiol 121:879–888

Matamoros MA, Kim A, Penuelas M, Ihling C, Griesser E, Hoffmann R, Fedorova M, Frolov A, Becana M (2018) Protein carbonylation and glycation in legume nodules. Plant Physiol 177:1510–1528

Matthews C, Arshad M, Hannoufa A (2019) Alfalfa response to heat stress is modulated by microRNA156. Physiol Plant 165:830–842

Menéndez AB, Calzadilla PI, Sansberro PA, Espasandin FD, Gazquez A, Bordenave CD, Maiale SJ, Rodríguez AA, Maguire VG, Campestre MP, Garriz A, Rossi FR, Romero FM, Solmi L, Salloum MS, Monteoliva MI, Debat JH, Ruiz OA (2019) Polyamines and legumes: joint stories of stress, nitrogen fixation and environment. Front Plant Sci 10:1415

Mittler R, Zilinskas BA (1994) Regulation of pea cytosolic ascorbate peroxidase and other antioxidant enzymes during the progression of drought stress and following recovery from drought. Plant J 5:397–405

Mittler R (2002) Oxidative stress, antioxidants and stress tolerance. Trends Plant Sci 7:405–410

Mittler R, Blumwald E (2015) The roles of ROS and ABA in systemic acquired acclimation. Plant Cell 27:64–71

Moller IM, Jensen PE, Hansson A (2007) Oxidative modifications to cellular components in plants. Ann Rev Plant Biol 58:459–481

Moller IM, Rogowska-Wrzesinska A, Rao RSP (2011) Protein carbonylation and metal-catalyzed protein oxidation in a cellular perspective. J. Proteomics 74:2228–2242

Moran JF, Klucas RV, Grayer RJ, Abian J, Becana M (1997) Complexes of iron with phenolic compounds from soybean nodules and other legume tissues: prooxidant and antioxidant properties. Free Radic Biol Med 22:861–887

Mun T, Bachmann A, Gupta V, Stougaard J, Andersen SU (2016) *Lotus* Base: An integrated information portal for the model legume *Lotus japonicus*. Sci Rep 6:39447

Murillo-Amador B, Yamada S, Yamaguchi T, Rueda-Puente E, Ávila-Serrano N, García-Hernández JL, López-Aguilar R, Troyo-Diéguez E, Nieto-Garibay A (2007) Influence of calcium silicate on growth, physiological parameters and mineral nutrition in two legume species under salt stress. J Agron Crop Sci 193:413–421

Nadeem M, Li J, Yahya M, Sher A, Ma C, Wang X, Qiu L (2019a) Research progress and perspective on drought stress in legumes: a review. Int J Mol Sci 20:2541

Nadeem M, Li J, Yahya M, Wang M, Ali A, Cheng A, Wang X, Ma C (2019b) Grain legumes and fear of salt stress: focus onmechanisms and management strategies. Int J Mol Sci 20:799

Nagajyoti PC, Lee KD, Sreekanth TVM (2010) Heavy metals, occurrence and toxicity for plants: a review. Environ Chem Lett 8:199–216

Nahar K, Hasanuzzaman M, Alam MM, Fujita M (2015) Exogenous glutathione confers high temperature stress tolerance in mung bean (*Vigna radiata* L.) by modulating antioxidant defense and methylglyoxal detoxification system. Environ Exp Bot 112:44–54

Nair AS, Abraham TK, Jaya DS (2008) Studies on the changes in lipid peroxidation and antioxidants in drought stress induced cowpea (*Vigna unguicolata* L.) varieties. J Environ Biol 29:689–691

Naya L, Ladrera R, Ramos J, Gonzalez EM, Arrese-Igor C, Minchin FR, Becana M (2007) The response of carbon metabolism and antioxidant defenses of alfalfa nodules to drought stress and to the subsequent recovery of plants. Plant Physiol 144:1104–1114

Naya L, Paul S, Valdés-López O, Mendoza-Soto AB, Nova-Franco B, Sosa-Valencia G, Reyes JL, Hernández G (2014) Regulation of copper homeostasis and biotic interactions by microRNA 398b in common bean. PLoS One 9:e84416

Nikitaki Z, Holá M, Donà M, Pavlopoulou A, Michalopoulos I, Angelis KJ, Georgakilas AG, Macovei A, Balestrazzi A (2018) Integrating plant and animal biology for the search of novel DNA damage biomarkers. Mut Res Rev Mut Res 775:21–38

Noctor G, Foyer CH (1998) Ascorbate and glutathione: keeping active oxygen under control. Annu Rev Plant Physiol Plant Mol Biol 49:249–279

Oracz K, El-Maarouf Bouteau H, Farrant JM, Cooper K, Belghazi M, Job C, Job D, Corbineau F, Bailly C (2007) ROS production and protein oxidation as a novel mechanism for seed dormancy alleviation. Plant J. 50:452–465

Osman HS (2015) Enhancing antioxidant yield relationship of pea plant under drought at different growth stages by exogenously applied glycine betaine and proline. Ann Agric Sci 60:389–402

Pagano A, Araújo SDS, Macovei A, Leonetti P, Balestrazzi A (2017) The seed repair response during germination: disclosing correlations between DNA repair, antioxidant response, and chromatin remodeling in *Medicago truncatula*. Front Plant Sci 8:1972

Pagano A, de Sousa Araújo S, Macovei A, Dondi D, Lazzaroni S, Balestrazzi A (2019) Metabolic and gene expression hallmarks of seed germination uncovered by sodium butyrate in *Medicago truncatula*. Plant Cell Environ 42:259–269

Pagano A, Macovei A, Araujo S, Forti C, Balestrazzi A (2020) *Medicago truncatula*, an informative model to investigate the DNA damage response during seed germination. In: Set de Bruijn FJ (ed) The Model Legume Medicago truncatula, vol 2. Wiley. (ISBN 9781119409168)

Pan WJ, Tao JJ, Cheng T, Bian XH, Wei W, Zhang WK, Ma B, Chen SY, Zhang JS (2016) Soybean miR172a improves salt tolerance and can function as a long-distance signal. Mol Plant 9:1337–1340

Paparella S, Araujo S, Rossi G, Wijayasinghe M, Carbonera D, Balestrazzi A (2015) Seed priming: state of the art and new perspectives. Plant Cell Rep 34:1281–1293

Parreira JR, Balestrazzi A, Fevereiro P, De Sousa Araujo S (2018) Maintaining genome integrity during seed development in *Phaseolus vulgaris* L.: evidence from a transcriptomic profiling study. Genes 9:E463

Patel PK, Hemantaranjan A, Sarma BK, Singh R (2011) Growth and antioxidant system under drought stress in chickpea (*Cicer arietinum* L.) as sustained by salicylic acid. J Stress Physiol Biochem 7:130–144

Pommier Y, Huang SY, Gao R, Das BB, Murai J, Marchand C (2014) Tyrosyl-DNA-phosphodiesterases (TDP1 and TDP2). DNA Repair (Amst) 19:114–129

Puppo A, Dimitrijevic L, Rigaud J (1982) Possible involvement of nodule superoxide dismutase and catalase in leghemoglobin protection. Planta 156:374–379

Quan WL, Liu X, Wang HQ, Chan ZL (2016) Comparative physiological and transcriptional analyses of two contrasting drought tolerant alfalfa varieties. Front Plant Sci 6:1256

Ragland M, Theil EC (1993) Ferritin (mRNA, protein) and iron concentrations during soybean nodule development. Plant Mol Biol 21:555–560

Ramesh SV, Govindasamy V, Rajesh MK, Sabana AA, Praveen S (2019) Stress-responsive miRNAome of *Glycine max* (L.) Merrill: molecular insights and way forward. Planta 249:1267–1284

Sabatini ME, Pagano A, Araujo A, Balestrazzi A, Macovei A (2017) The tyrosyl-DNA phosphodiesterase 1β (Tdp1β) gene discloses an earlyresponse to abiotic stresses. Genes 8:305

Sabatini ME, Donà M, LeonettiP Minio A, Delledonne M, CarboneraD Confalonieri M, Giraffa G, Balestrazzi A (2015) Depletion oftyrosyl-DNA Phosphodiesterase 1 alpha (*MtTdp1alpha*) affects transposon expression in *Medicago truncatula*. J Int Plant Biol 58:618–622

Saglam A, Saruhan N, Terzi R, Kadioglu A (2011) The relations between antioxidant enzymes and chlorophyll fluorescence parameters in common bean cultivars differing in sensitivity to drought stress. Russ J Plant Physiol 58:60–68

Sainz M, Diaz P, Monza J, Borsani O (2010) Heat stress results in loss of chloroplast Cu/Zn superoxide dismutase and increased damage to Photosystem II in combined drought-heat stressed *Lotus Japonicus*. Physiol Plant 140:45–56

Sato S, Tabata S (2006) *Lotus japonicus* as a platform for legume research. Curr Opin Plant Biol 9:128–132

Satour P, Youssef C, Chatelain E, Vu BL, Teulat B, Job C, Job D, Montrichard F (2018) Patterns of protein carbonylation during *Medicago truncatula* seed maturation. Plant Cell Environ 41:2183–2194

Scandalios JG, Guan L, Polidoros AN (1997) Catalases in plants: Gene structure, properties, regulation, and expression. In: Scandalios JG (ed) Oxidative stress and the molecular biology of antioxidant defenses. Cold Spring Harbor, New York, NY, pp 343–406. ISBN 0-87969-502-1

Sharma PI, Jha AB, Dubey RS, Pessarakli M (2012) Reactive oxygen species, oxidative damage, and antioxidative defense mechanism in plants under stressful conditions. J Bot 2012:1–26

Sharma L, Priya M, Bindumadhava H, Nair RM, Nayyar H (2016) Influence of high temperature stress on growth, phenology and yield performance of mungbean (*Vigna radiata* (L.) Wilczek) under managed growth conditions. Sci Hort 213:379–391

Stougaard J (2001) Genetics and genomics of root symbiosis. Curr Opin Plant Biol 4:328–335

Sofo A, Cicco N, Paraggio M, Scopa A (2010) Regulation of the ascorbate–glutathione cycle in plants under drought stress. In: Anjum N, Chan MT, Umar S (eds) Ascorbate-glutathione pathway and stress tolerance in plants. Springer, Dordrecht

Sun Z, Wang Y, Mou F, Tian Y, Chen L, Zhang S, Jiang Q, Li X (2016) Genome-wide small RNA analysis of soybean reveals auxin-responsive microRNAs that are differentially expressed in response to salt stress in root apex. Front. Plant Sci 6:160

Subramanian S, Fu Y, Sunkar R, Barbazuk WB, Zhu JK, Yu O (2016) Novel and nodulation-regulated microRNAs in soybean roots. Front Plant Sci 6:1273

Szittya G, Moxo S, Santos DM, Jing R, Fevereiro MPS, Moulton V, Dalmay T (2008) High-throughput sequencing of *Medicago truncatula* short RNAs identifies eight new miRNA families. BMC Genom 9:593

Tripathi P, Rabara RC, Shulaev V, Shen QJ, Rushton PJ (2015) Understanding water-stress responses in soybean using hydroponics system-a systems biology perspective. Front Plant Sci 6:1145

Tyagi S, Sharma S, Ganie SA, Tahir M, Mir RR, Pandey R (2019) Plant microRNAs: biogenesis, gene silencing, web-based analysis tools and their use as molecular markers. Biotech 9:413

Valliyodan B, Ye H, Song L, Murphy M, Shannon JG, Nguyen HT (2017) Genetic diversity and genomic strategies for improving drought and waterlogging tolerance in soybeans. J Exp Bot 68:1835–1849

Wang WB, Kim YH, Lee HS, Kim KY, Deng XP, Kwak SS (2009) Analysis of antioxidant enzyme activity during germination of alfalfa under salt and drought stresses. Plant Physiol Biochem 47:570–577

Wang CQ, Xu HJ, Liu T (2011) Effect of selenium on ascorbate-glutathione metabolism during PEG-induced water deficit in *Trifolium repens* L. J Plant Growth Regul 30:436–444

Wu J, Wang L, Wang S (2017) MicroRNAs associated with drought response in the pulse crop common bean (*Phaseolus vulgaris* L.). Gene 628:78–86

Xia X-J, Zhou Y-H, Shi K, Zhou J, Foyer CH, Yu J-Q (2015) Interplay between reactive oxygen species and hormones in the control of plant development and stress tolerance. J Exp Bot 66:2839–2856

Xie X, He Z, Chen N, Tang Z, Wang Q, Cai Y (2019) The roles of environmental factors in regulation of oxidative stress in plant. Biomed Res Int 2019:9732325

Yasar F, Ellialtioglu S, Yildiz K (2008) Effect of salt stress on antioxidant defense systems, lipid peroxidation, and chlorophyll content in green bean. Russ J Plant Physiol 55:782–786

Yasar F, Uzal O, Yasar TO (2013) Investigation of the relationship between the tolerance to drought stress levels and antioxidant enzyme activities in green bean (*Phaseolus Vulgaris* L.) genotypes. Afr J Agric Res 8:5759–5763

Yoshiyama KO, Sakaguchi K, Kimura S (2013) DNA damage response in plants: conserved and variable response compared to animals. Biology 2:1338–1356

Yu Y, Jia T, Chen X (2017) The 'how' and 'where' of plant microRNAs. New Phytol 216:1002–1017

Zhang C, Shi S, Liu Z, Yang F, Yin G (2019) Drought tolerance in alfalfa (*Medicago sativa* L.) varieties ia associated with enhanced antioxidative protection and declined lipid peroxidation. J Plant Physiol 232:226–240

Zheng Y, Hivrale V, Zhang X, Valliyodan B, Lelandais-Brière C, Farmer AD, May GD, Crespi M, Nguyen HT, Sunkar R (2016) Small RNA profiles in soybean primary root tips under water deficit. BMC Syst Biol 10:126

Zhu H, Choi H, Cook DR, Shoemaker RC (2005) Bridging model and crop legumes through comparative genomics. Plant Physiol 137:1189–1196

Zorb C, Geifus C-M, Dietz K-J (2019) Salinity and crop yield. Plant Biol 21:31–38

Threat Imposed by O_3-Induced ROS on Defense, Nitrogen Fixation, Physiology, Biomass Allocation, and Yield of Legumes

Richa Rai

Abstract Fabaceae is the third largest family of flowering plants. Tropospheric O_3 is one of the most important secondary pollutants causing threat to agriculture productivity. The projected levels to which O_3 will increase are critically alarming and have caused a significant effect of productivity of legumes. The present chapter highlights about the effect of tropospheric O_3 on growth, development, and functioning of legumes. Tropospheric O_3 causes significant leaf injury, reductions in root biomass, and root nodule number. It causes reduction in photosynthesis and nitrogen-fixing enzymes. Changes in allocation of photosynthates between above and belowground due to O_3 also affect nitrogen fixation in legumes. Potential shortfalls in N-fixation will have severe economic and environmental consequences and it has not yet been considered on anagroecosystem basis. There is an utmost need for further investigation of tropospheric O_3 impacts of legume crops.

1 Introduction

The Leguminosae also known as Fabaceae is the third largest family of flowering plants. Legumes belong to the family Fabaceae/Leguminosae (with about 700 genera and 18000 sps) (Doyle 2001). Legumes rank third after cereals and oilseeds in world population and have major effects on the environment and health (Dita et al. 2006; Mantri et al. 2013). Legume crops can be divided into two groups according to their ability to grow in different seasons, namely cool season food legume and warm or tropical season food legumes (Miller et al. 2002; Toker and Yadav 2010). Legumes are a primary source of amino acids and provide around one-third (20–40%) of all dietary protein (Zhu et al. 2005; Kudapa et al. 2013). Legumes are an important source of protein oil fiber and micronutrients and play a vital role in cropping cycles

R. Rai (✉)
Department of Botany, Rameshwar College, Babasaheb Bhimrao Ambedkar Bihar University, Muzaffarpur, Bihar, India
e-mail: richarai81@gmail.com

M. Hasanuzzaman et al. (eds.), *The Plant Family Fabaceae*,
https://doi.org/10.1007/978-981-15-4752-2_19

due to their ability to fix atmospheric nitrogen (El-Enamy et al. 2013; Mantri et al. 2013). Legumes produce secondary metabolic compounds that can protect the plant against pathogens and pests (Kudapa et al. 2013; Ashraf et al. 2010). The present world population of 7.2 billion is expected to reach 9.6 billion by the middle of twenty-first century due to the high growth rate particularly in developing countries. There is no need to produce about 70% more food to feed this excessive population (Varshney et al. 2013). Climate change and a secondary pollutant O_3 are a threat to crop production.

Concentrations of tropospheric ozone (O_3) are a secondary pollutant and short-lived greenhouse gas formed through complex photochemical reaction of volatile organic compounds (VOCs), carbon monoxide (CO), and nitrogen oxides (NOx). The emissions of nitric oxide (NO) results from the reaction between nitrogen and o present in atmosphere under high temperature generally attained in the combustion chambers of engine. Nitric oxide once emitted scavenges singlet oxygen produced by dissociation of atmospheric O_2. NO_2 formed is easily dissociated by UV range of solar radiation into nitrogen monoxide and single oxygen ($^1O_2$). Singlet oxygen reacts with molecular O_2 to give rise to O_3. Under these conditions, the lifetime of O_3 is very short, because the O_3 produced reacts with NO to produce NO_2 and O_2 again; therefore, an equilibrium is formed between O_3 formation and degradation (Lorenzini and Saintanis 2003). But in rural area, O_3 scavengers NO is not available, hence O_3 concentrations are higher. Other precursors are non-methane hydrocarbons, volatile organic carbons (VOCs), and carbon monoxide (CO). The concentration of O_3 precursors (NOx, VOCs, CO etc.) is an important factor in determining the O_3 concentration in that region. The spatial and temporal distribution of O_3 and its precursors are mostly driven by the distribution of their emissions. These concentrations have increased tremendously across the globe since the start of the industrial era (IPCC 2013). Meehl et al. (2007) projected an increase of 20–25% in mean global O_3 concentration between 2015 and 2050 and of 40–60% by 2100. Ground-level O_3 concentrations (O_3) have more than doubled to 35–40 ppb since the Industrial Revolution and expected to increase in developing countries like India and China with peak values exceeding above 100 ppb when emissions of high O_3 precursor coincide with high temperatures and irradiance (The Royal Society 2008; IPCC 2013; Young et al. 2013).

Tropospheric O_3 causes damage to crops and affects crop yield which is a major challenge to global food supply in the present scenario. Studies have shown yield losses of wheat, rice, and legumes crop. Among Poaceae and Leguminosae (Fabaceae) members, Fabaceae members are more sensitive to tropospheric O_3 than Poaceae members. The present chapter highlights the impact of tropospheric O_3 on physiological machinery, N-metabolism, photosynthate allocation, growth and yield of Fabaceae members.

2 Uptake of O_3 and Defense Response in Legumes

Relationship between O_3 and plant responses is commonly based on the O_3 concentrations in the surrounding air. Tropospheric O_3 and CO_2 have a common route of entry within plant system and stomata play a fundamental role in determining the flux of O_3 in the apoplastic region of the plant. The flux of O_3 from the atmosphere into plants depends on a different level of resistance (inverse of conductance, 1/[conductance]) which exists at different morphological or physiological levels. Conductance is defined as a flux (concentration gradient) per unit driving force. The resistance to O_3 into plant cells consists of the following types: (i) aerodynamic resistance that depends upon atmospheric resistance; (ii) the boundary layer resistance caused by a layer of laminar air adjacent to leaves; (iii) the stomatal resistance exerted by the stomatal pores; and (iv) the internal resistance caused in the leaves after O_3 uptake through stomata (Guderian 1985). Internal structures of the plant such as cell wall and inter- and intracellular apparatus also contribute toward resistance against O_3. Various modeling results have indicated that a major factor affecting plant responses to O_3 appears to be stomatal conductance (g_s). The sensitivity of plants to O_3 is often correlated with their stomatal response because O_3 exposure generally results in a decline in stomatal aperture, and plants that show more rapid stomatal closure are reported to be resistant to O_3 in population-level studies (Yadav et al. 2019; Winner et al. 1991). However, O_3-induced declines in stomatal apertures may be of protective value as it may check uptake of O_3 and also it depends on the concentration of intercellular CO_2 within plant cell (Biswas et al. 2013). Studies have reported reductions in stomatal conductance (g_s) in *Vigna unguiculata* L. cultivars Blackeye and Asontem (Tetteh et al. 2015), soybean cultivars (Sun et al. 2014; Zhang et al. 2014; Rai et al. 2015), mung bean (Chaudhary and Agrawal 2015; Mishra and Agrawal 2015).

3 Oxidative Stress and Defense Response

Once O_3 enters the leaf through open stomata, it reacts instantaneously with the surrounding membranes leading to production on ROS, such as superoxide.O_2-, H_2O_2, and.OH- (Lamb and Dixon 1997). The production of ROS leads to alterations in the permeability of plasma membranes and lipid compositions (Heath 1987). Diara et al. (2005) reported that extracellular H_2O_2 accumulation is one of the earliest detectable responses to O_3 exposure. H_2O_2 disrupts photosynthesis and activates NAD(P)H-dependent oxidase (Park et al. 1998; Rao and Davis 1999) which leads to ROS accumulation. Alscher and Hess (1993) suggested that the superoxide radical is formed and produces leaf injury under O_3 exposure. Among other ROS, the hydroxyl radical is the most reactive oxygen species (ROS). It reacts rapidly with protein, lipids, and DNA and causes cell damage (Ibal et al. 1996). ROS leads to a chain of reactions which cause a significant effect on the cellular metabolism of the plants. ROS cause membrane damage and deleterious effects on the normal functioning of cells.

It has to be taken into account that ROS can directly affect the conformation of proteins and lipids. O_3-induced ROS formation activates several pathways, including Ca^{2+} and MAPK signaling pathways, in addition to ethylene-, salicylic acid-, and jasmonic acid-dependent pathways eventually leading to induction of defense reactions leading to programmed cell death (Samuel et al. 2000; Sandermann Jr. et al. 1998). ROS generated in the apoplast activates calcium (Ca^{2+}) channels and further induce the intracellular O_3-responsive signaling pathways. Clayton et al. (1999) showed a specific increase in the intracellular Ca^{2+} concentration, if atmospheric O_3 reached concentrations above 70 ppb. In terms of signaling, Ca^{2+} channels of the plasma membrane could also function as ROS sensors as they may affect the ion channel, redox-dependant alterations in ion conductance. Chaudhary and Agrawal (2015) reported an increase in lipid peroxidation (LPO) and solute leakage in all cultivars of mung bean (*Vigna radiata* L.) (HUM-1, HUM-2, HUM-6, HUM-23, HUM-24 and HUM-26) with the highest increase in HUM-1 and minimum in HUM-23 in mung bean exposed to elevated O_3 (70.9 ppb) than ambient O_3 (64 ppb).

The stress condition generated by O_3 results in an oxidative burst due to enhanced production of ROS, which causes negative effects on cellular components leading to damage to the lipids (peroxidation of the unsaturated fatty acids in membrane), protein (denaturation), carbohydrate, and nucleic acids (Blokhina et al. 2003).

To counteract the stress imposed by ROS, an array of antioxidant molecules is induced (Ashmore 2005; Caregnato et al. 2013) via non-enzymatic antioxidants such as ascorbic acid (AA), flavonoids, phenolics, tocopherol, peptides (glutathiones, carotenoids, and polyamines), and organic buffering systems or through enzymatic antioxidants (Blokhina et al. 2003) via superoxide dismutase (SOD), ascorbate peroxidase (APX), glutathione reductase (GR), catalase (CAT), and various types of peroxidases (POX) (Caregnato et al. 2013).

Non-enzymatic antioxidants, such as ascorbic acid (AA), protect critical macromolecules from oxidative damage by directly reacting with O_2 and H_2O_2-, regenerating α- tocopherol from tocopheroxyl and removing H_2O_2 via AA-GSH cycle (Pinto et al. 2003). Increase in AA pool has been observed in mung bean cultivars (Mishra and Agrawal 2015). Ascorbic acid (AA) is the most abundant low mol. wt. antioxidant which is synthesized in leaf cells (Castillo and Greppin 1988; Smirnoff 2000) and forms the first time of defense against O_3 exposure (Polle et al. 1995). The ability of AA to donate electrons in a wide range of enzymatic and non-enzymatic reaction signifies it as being the main ROS detoxifying compound. In addition, α-tocopherol is an important antioxidant for its ability to directly scavenge oxidizing radicals and protect chain propoagation during lipid peroxidation due to O_3 (Serbinov and Packer 1994).

Among enzymatic antioxidants, the primary one is superoxide dismutase (SOD), which represents a family of mettaloenzymes that catalyzes the dismutation of O_2- to H_2O_2 (Bowler et al. 1992; 1994). Another important enzyme is catalase (CAT) which is primarily in peroxisomes and catalyzes the disintegration of H_2O_2 (Scandalios et al. 1997). Ascorbate peroxidase (APX), glutathaione reductase (GR), monodehydroascorbate (MDHAR), and dehydroascorbate (DHAR) are enzymes of the ascorbate-glutathione cycle and help to regulate ROS formed in the presence of O_3 (Jimenez et al. 1997; Noctor and Foyer 1998). Similarly, O_2- (superoxide production rate) increased in elevated O_3-treated mung bean plants. Elevated O_3 altered the activities of antioxidative (APX, CAT, GR, and SOD) enzymes with varied magnitude in all the cultivars of *Vigna radiata* L. (Chaudhary and Agrawal 2015; Mishra and Agrawal 2015). Hassan et al. (2017) also recorded a significant increase in ascorbate peroxidase (APX), and glutathione peroxidase (GPX) in pea cultivars is exposed to ambient O_3 above 40 ppb. Level of antioxidative enzyme activities varied differentially between pea cultivars. An increase in APX and GPX activities was recorded in pea cultivar Little Marvel which indicates the adaptation to O_3 and increased ability of plants to scavenge toxic reactive oxygen species as a protective mechanism against oxidative stress (Hassan et al. 2017). Moreover, increasing activities of these enzymes is an indicator of oxidative stress and production of H_2O_2 against tropospheric O_3.

Both GR and APX are integral parts of the ascorbate glutathione cycle and efficient H_2O_2 scavengers; their induction is a strategic management of O_3-induced oxidative stress (Calatayud et al. 2004; Cheng et al. 2011). Their activities were increased in Little Marvel plants, while pea cultivar Victory showed insignificant response in the present study (Hassan et al. 2017).

4 Visible Injury

Exposure to elevated O_3 caused interveinal chlorosis on adaxial (upper) surface of the leaves in mung bean cultivars, with higher foliar injury percentage which was observed in sensitive cultivar *Vigna radiata* L. HUM-2 compared to resistant cultivar HUM-6 (Chaudhary et al. 2015). Taylor et al. (1982) recorded in soybean leaves the initial symptoms with dose of O_3 (250 ppb) were rapid desiccation bleaching, necrosis of leaves within 1–3 h of fumigation. After 12–24 h of fumigation, stippling (red-brown pigmentation) and silvering of leaf surfaces were observed with intermediate dose (Taylor et al. 1982). Burkey and Carter (2014) found foliar injury as stippling and necrosis depend on differential sensitivity among different cultivars, being much more pronounced on the sensitive genotype of soybean Forrest than tolerant cultivar exposed to 80 ppb for 7 h. Singh and Agrawal (2011) recorded the occurrence of injury symptoms like chlorotic spots on the adaxial surface of soybean cultivars Pusa 7712 and Pusa 9814 leaves grown under non-filtered air plus 20 ppb elevated O_3. With age, interveinal chlorosis was also observed in both cultivars. However, the percentage of injured leaves was slightly higher in O_3-sensitive cultivars Pusa 9712.

5 Impacts on Photosynthesis and Allocation of Photosynthates in Legumes

O_3 exposure results in decreased photosynthetic carbon assimilation (Morgan et al. 2003). In many cases, loss of assimilation capacity was shown to be done primarily to reduce carboxylation efficiency directly related to loss of RuBisCO activity. These losses of activity are thought to be due to decrease in activation state. M. Reid and Fiscus (1998); Reid et al. (1998) showed decline in RuBisCO content and activity, chlorophyll content and net photosynthesis rate with increased O_3 contents in ambient air.

Photosynthesis is a multistep process and alteration in Ps may be influenced by chl content, changes in internal CO_2, or by variations in the light reactions such as light energy utilization and conversion or dark reaction (carboxylation efficiency of RuBisCO). O_3 can alter light reaction decreasing the electron transfer between the two photosystems (Pellegrini et al. 2011; Calatayud et al. 2004). Sun et al. (2014) showed reductions in all chlorophyll fluorescence parameters including quenching of photochemical efficiency of PS II (qP), Fv/Fm ratio, PhiPSII, %P, (fraction of absorbed radiation dissipated in the antenna) %D, and (PSII antennae) %X and reduction in Chl decreased by 6.4% and carotenoid by 5.7%, light-saturated photosynthesis by 28% and also reduction in stomatal conductance in soybean plants exposed to 71 ppb for 9 h O_3 in a free-air concentration enrichment (FACE) experiment. Main causes of reduction in photosynthesis are due to reduction in RuBisCO carboxylation (Vc_{max}) and rates of electron transport (J_{max}) which are correlated with the reduction of RuBisCO gene expression.

The majority of the sugars decreased under elevated O_3 levels. Sucrose and starch are the major primary metabolites formed after photosynthesis so reductions in sugar levels or synthesis rates are the results of reduction in photosynthesis under elevated O_3 levels. On the other hand, reduction in rates of sucrose and starch synthesis would hence affect triose-p utilization and RuBP-regeneration-limited photosynthesis. Triose-p utilization, mainly the rate of starch and sucrose synthesis, affects Pi recycle and hence affecting RuBP regeneration. It is reported that O_3 stimulates PEP carboxylase (PEPC) activity in common bean *Phaseolus vulgaris* (Guidi et al. 2009). In anaplerotic reaction, PEP carboxylase (PEPC) catalyzes CO_2-fixation with PEP to produce oxaloacetate that can be subsequently reduced into malate and/or converted into citrate in TCA cycle. Such an increase could give rise to ATP and reducing power production by respiration to allow an enhancement of the constitutive detoxification systems activity in response to the release of reactive oxygen species (ROS) (Dizengremel et al. 2008).

6 Impact of O_3 on Nitrogen Fixation

Nitrogen is a vital element for plant growth and required for the synthesis of macromolecules such as amino acids and chlorophyll. About 78% of the atmospheric nitrogen cannot be used by living organism till it is not available ammonia or its reduced form. Biological nitrogen fixation (BNF) accounts for about 60% of nitrogen used in agriculture. Member of Fabaceae is capable of biological nitrogen fixation and meets their own need due to symbiotic association with diazotrophic Gram-bacteria known as *Rhizobium*. In legume-Rhizobium symbiosis, rhizobia induce nodules formation on the root of legumes in which N_2-fixation occurs. Legume nodules are found in a variety of shape, size, and structure. Some of the nodules have determinate growth with only a transient meristem, and these are relatively short-lived and most are indeterminate with a distinct apical meristem often showing branching. The rhizobia forming symbiosis with legume roots belong to five different genera: (i) *Rhizobium;* (ii) *Azorhizobium;* (iii) *Mesorhizobium;* (iv) *Sinorhizobium; and* (v) *Bradyrhizobium*.

In symbiotic relationship between legumes and Rhizobium, bacterial attachment to the root surface initiates root infection and formation of root nodules specialized structure which makes anaerobic environment and also leghaemoglobin produced by host cells to prevent the entry of O_2 into the infected cells for the activity of Nitrogenase, a key enzyme for nitrogen fixation. Efficient bacteroid respiration also restricts oxygen penetration into the cytoplasm and provides nitrogenise with the ATP and reductants required. In order to sustain N-fixation, the host plant must supply the bacteroids with a carbon source in the form of sucrose through phloem. (White et al. 2007). Bacteroids do not have enzymes for the assimilation of ammonia, so NH_3 obtained from N_2 reduction is released into the root cell from where the assimilation occurs via GS-GOGAT pathway in leaf of host cells through nitrite reductase, nitrate reductase, and glutamine synthase and leads to the production of glutamine, glutamate, and successively of the other transport compounds. Some of them are returned to bacteroid (White et al. 2007).

Symbiotic nitrogen fixation is crucial to the plant and it controls the number of nodules to balance the nitrogen gains and to avoid severe energy drain that would be imposed by having too many nitrogen-fixing nodules (Zahran 1999). There are many factors that affect biological nitrogen fixation such as temperature, water content, N concentration, root zone pH, plant nutrient status including C and N substrates in roots, and genetic variation in potential N-fixation capacity. It is also affected by plant nutritional status such as phosphorus (P) and potassium (K) levels that control nodule growth and nitrogenase activity directly or indirectly (Havelka et al. 1982). Photosynthate partitioned to roots supports nodule growth, provides energy for N-fixation, maintains a functional population of rhizobia, and allows the synthesis of amino compounds produced from N-fixation (Minchin and Pate 1973; Layzell et al. 1979; King et al. 1986).

Studies have shown tropospheric ozone reduced number, size, or mass of root nodules by 20% after exposure of legumes to 60 ppb (Hewitt et al. 2014; Zhao et al.

2012; Cheng et al. 2011). Several studies also report decreased N-fixation parameters with the majority of them focussing on negative effects in clover and soybean. Indeed O_3 induces suppression of N-fixation rates in peanut, chickpea, faba bean, mung bean, and cowpea within a range of 8–99% under different levels of O_3 exposure. Umponstira et al. (2009) found that nodule dry weight was significantly affected in cowpea by 61 and 90%, respectively, at vegetative stage (V3) and 42.3 and 55.6% at harvesting (R5) when exposed to ozone concentration at 40 and 70 ppb for 8 h per day. Nodules of size over 2 mm were recorded after the vegetative stage (V3). The number of nodules size over 2 mm found significantly decreased by 63.5 and 78.3% at vegetative stage (V3), 44.9 and 74.9% at flowering stage (R1), and 50.7 and 51.8% at harvesting (R5), respectively, after exposure to ozone concentration at 40 and 70 ppb. Ozone does not directly affect root nodules as it does not penetrate through the soil surface (Blum and Tingey 1977) but the leachates of ozone-treated plants may influence the growth and subsequent colonization of soil rhizobia and root fungi (Umponistira et al. 2009). A decreased supply of detritus may lead to a suppression of N-fixation in soil microbial communities in legume agroecosystems (Hewit et al. 2014). Stable isotope studies showed ozone-induced reductions in N-fixation rates which was associated with reduced availability and translocation of carbon assimilates to the root system during growth (Pausch et al. 1996; Cong et al. 2009) with a reduced sucrose flow to nodules the most likely cause (Udvardi and Poole 2013).

In peanut, seasonal exposure to treatments of 49–70 ppb is sufficient to reduce tissue N content and N-fixation rates (Cong et al. 2009), while mean 107 ppb exposure to soybean reduces N-fixation by 58% (Cheng et al. 2011), and in white clover, reductions recorded were reduced more than 50% in legumes exposed to 40–60 ppb. The nitrogen in plant shoot reduced at all stages like vegetative stage (V3), flowering stage (R1), and harvest (R5) exposed to ozone concentration at 40 and 70 ppb (Table 1).

Flavonoids are released by legume roots which serve as chemo-attractants for the attraction of bacteria Rhizobium and activate expression of rhizobial nod genes which are responsible for the production of nod factors (certain lipochitin oligosaccharides) that are perceived as a receptor in the legume host and trigger a sequence of events in host plant like curling of root hairs around the invading rhizobia, the entry of the rhizobia into the plant through infection threads and the induction of cell division in the root cortex that marks formation of the nodule primordium (Hirsch 1992). Galant et al. (2012) showed reduced flavonoid synthesis under O_3 treatment. The activity of nitrogen-fixing enzymes was also reported to reduce under O_3 exposure. Agrawal and Agrawal (1990) reported reduction in nitrogenase, nitrate reductase, and nitrite reductase activity in *Viciafaba* and *Cicerarientum*exposed to 92 ppb O_3 for 2 h. Rai et al. (2015) observed reductions in nitrate reductase activity in soybean exposed to ambient concentration of O_3 (42 ppb) for 12 h from germination to maturity. Reduction in nitrite reductase (NiR) could be attributed to its localization. Chloroplast is the site of localization of O_3 and also a primary place of action of O_3 after its absorption through stomata (Swanson et al. 1973). It is also known that O_3 alters the ultrastructure and functions of chloroplasts including electron flow

Table 1 Impact of O_3 on nitrogen fixation and nodulation in members of Fabaceae (Modified from Hewitt et al. 2015)

Plant name	O_3 concentration	Nitrogen fixation parameters	Percentage reduction (%)	References
Common bean	300 ppb	Nodulation	12	Blum and Heck (1980)
Chick pea Faba bean	100 ppb	Nodulation, N-fixing enzymes	12–34	Agrawal and Agrawal, (1990)
Soybean	58–107 ppb	N-fixation	27	Flagler et al. (1987)
Chickpea	85 ppb	Nodulation	18–21	Welfare et al. (2002)
Cowpea	40–70 ppb	Nodulation and N-fixation	31–90	Umponstira et al. (2009)
Soybean	65	N-fixation	37–40	Cheng et al. (2011)
Bean	100 ppb	Nitrate and nitrite reductase	25	Hassan and Anttonen (1999)
Soybean	42 ppb	Nitrate and nitrite reductase	28–31	Rai et al. (2015)
Cowpea	40–80 ppb	Nitrogenase enzyme activity	10–43	Malaiyandi and Natarajan (2014)

in photosystems I and II and various enzymes of the reductive pentose phosphate pathway (Nakamura and Saka 1978). It may be due to the O_3 treatment which resulted in the chloroplasts damage and thus reduced the NiR activity of leaves. On the other side, a decline in nitrate reductase (NR) may be due to reduction in enzyme synthesis or due to non-availability of an electron donor for proper enzyme activity (Agrawal and Agrawal 1990) (Table 1).

In leaves of soybean exposed to high O_3, Galant et al. (2012) describe an accumulation and/or oxidation of proteins implicated in amino acid biosynthesis and nitrogen homeostasis. The cytosolic form of glutamine synthase (GS), a central player in nitrogen sensing, increased in abundance, oxidation, and activity in the 116 ppb O_3 leaf sample of soybean (Galant et al. 2012). Indeed, increased expression of cytosolic GS is associated with leaf senescence and the recycling of ammonia during stress conditions (Brugie're et al. 2000).

7 Impact of O_3 on Plant Growth and Yield

Nitrogen fixation is also dependent on the availability of sugars from the host plant. As described before in this chapter, Ozone causes physiological changes in leaves that affect source strength, i.e., the amount of carbon available for allocation to sink tissues. Changes include decreased carbon assimilation, increased metabolic costs, and possibly decreased phloem loading. Several processes occur in O_3 exposed leaves that lead to increased carbon demand and sink strength in leaves and thus a reduction in carbon allocation below ground (Andersen 2003). Factors that increase the sink strength of leaves for carbon include repair processes, as well as the synthesis of antioxidants and other secondary compounds. Ozone stimulates biosynthetic pathways which involve carbon intermediates in support of defense and repair processes (Guderian 1985; Dizengremel 2001) and can lead to increased costs to construct tissues (Tingey et al. 1994). Synthesis of antioxidant compounds results in the diversion of carbohydrate from other metabolic processes (Dizengremel 2001). Amthor (1988) reported in beans (*Phaseolus vulgaris* L.) increased rate of metabolism to support repair and lesion formation O_3 exposure to 43 and 80 ppb. Such repair is an energy-consuming process which may lead to an increased rate of respiration. This increase in respiration for repair (maintenance) results in a diversion of a substrate from growth, and hence growth rate will decrease. Singh et al. (2015) reported in pea plants (*Pisum sativum* L.) under stress means low carbohydrate partitioning to the roots and nodules which led to a consequent reduction in the release of root exudates to the rhizosphere. This can, in turn, affect nodule formation and ultimately the biological N_2-fixation. Sucrose is oxidized to dicarboxylic acids and used as energy source by the bacteroids to fix atmospheric N, thus elevated rates of bacteroid respiration may lead to more formation of ROS (Dupont et al. 2012).

Increased costs associated with membrane repair, protein/antioxidant synthesis possible photosynthetic down-regulation, and decreased phloem loading could contribute to the reduction in allocation below ground. Roots are often dependent on the availability of photosynthate for their structural development (Andersen 2003), carbon-limiting stresses such as O_3 may have a rapid and significant effect on root growth. Decreased allocation below ground is often associated with decreased root: shoot ratio (RSR). Decreased RSR was recorded in cultivars of *Vigna mungo* exposed to 50 ppb ambient O_3 for 12 h (Singh and Agrawal 2011). Decreased allocation associated with ozone exposure alters N fixation in legumes which may result in a decrease in root nodule size or number or in the specific activity of the nitrogenases (Rees et al. 2005; Chaudhary and Agrawal (2015) found reductions in total biomass, number of leaves per plant and plant height exposed to elevated O_3 in mung bean and cultivar HUM-1 showed maximum reductions than HUM 23 which could be attributed to damage to fundamental metabolic processes of plants. Changes in the assimilation as reflected in the form of reduced plant biomass may also be due to disturbed partitioning which further altered the whole development.

The cumulative modifications in growth and biochemical parameters, biomass accumulation, and allocation under elevated O_3 concentration subsequently reflected in terms of negative impact on yield. Yield in Fabaceae members is influenced by the number of pods plant^{-1}, wt. of pods plant-1, number of seeds pod^{-1}, and number of seeds plant-1. Rai et al. (2015) reported the number and weight of pods plant^{-1} decreased significantly by 27.5 and 26.5%, respectively, after O3 exposure in soybean. Singh and Agrawal (2011) observed a reduction in yield by 40-44% in black gram (*Vigna mungo* L.).

8 Conclusions

The present chapter concludes that tropospheric O_3 affects physiology, nodulation, N-fixing enzymes, and photosynthate allocation which impact on growth, biomass, and yield of agricultural important legumes. Ozone within plant cell causes oxidative stress which diverts more of photosynthate toward repair and defense mechanism at the cost of below ground allocation which influences N-fixation and nodulation due to reduced availability of C assimilates. Different cultivars show differential responses to O_3. Multiple interacting environment stressors, such as ozone, drought, elevated CO_2, and N deposition will be increasingly important in determining the sustainability of legume-based agriculture and food production, particularly in parts of the world where severe issues with food security already exist.

Acknowledgements The author is thankful to funding agencies SERB, New Delhi, and CSIR, New Delhi, for providing research grant.

References

Agrawal M, Agrawal SB (1990) Effects of ozone exposure on enzymes and metabolites of nitrogen-metabolism. Sci Hortic 43:169–177

Alscher RG, Hess JL (1993) Antioxidants in higher plants. CRC, Boca Raton

Amthor JS (1988) Growth and maintenance respiration in leaves of bean (*Phaseolus vulgaris* L.) exposed to ozone in open-top chambers in the field. New Phytol 110:319–325

Andersen CP (2003) Source–sink balance and carbon allocation below ground in plants exposed to ozone. New Phytol 157:213–228

Ashmore MR (2005) Assessing the future global impacts of ozone on vegetation. Plant, Cell Environ 29:949–964

Ashraf MY, Ashraf M, Arshad M (2010) Major nutrients supply in legume crops under stress environments. In: Yadav SS, McNeil DL, Redden R, Patil SA (eds) Climate Change and Management of Cool Season Grain Legume Crops. Springer, Dordrecht, pp 155–170

Biswas DK, Xu H, Li YG, Ma BL, Jiang GM (2013) Modification of photosynthesis and growth responses to elevated CO_2 by ozone in two cultivars of winter wheat with different years of release. J Exp Bot 64(6):1485–1496

Blokhina O, Virolainen E, Fagerstedt KV (2003) Antioxidants, oxidative damage and oxygendeprivation stress: a review. Ann Bot 91:179–1943

Blum U, Tingey DT (1977) A study of the potential ways in which ozone could reduce root growth and nodulation of soybean. Atmos Environ 11:737–739

Bowler C, Montagu MV, Inze D (1992) Superoxide dismutase and stress tolerance. Annu Rev Plant Physiol 43:83–116

Bowler C, Van Camp W, Van Montagu M, Inze D (1994) Superoxide dismutase in plants. Crit Rev Plant Sci 13:199–218

Brugie're N, Dubois F, Masclaux C, Sangwan RS, Hirel B (2000) Immunolocalization of glutamine synthetase in senescing tobacco (*Nicotiana tabacum* L.) leaves suggests that ammonia assimilation is progressively shifted to the mesophyll cytosol. Planta 211: 519–527

Burkey KO, Carter TE (2014) Foliar resistance to ozone injury in the genetic base of U.S. and Canadian soybean and prediction of resistance in descendent cultivars using coefficient of parentage. Field Crop Res 111:207–217

Calatayud A, Barreno E (2004) Response to ozone in two lettuce varieties on chlorophyll a fluorescence, photosynthetic pigments and lipid peroxidation. Plant PhysiolBiochem 42:549–555
Castillo FJ, Greppin H (1988) Extracellular ascorbic acid and enzyme activities related to ascorbic acid metabolism in *Sedum album* L. leaves after ozone exposure. Environ Exp Bot 28:231–238
Chaudhary N, Agrawal SB (2015) The role of elevated ozone on growth, yield and seed quality amongst six cultivars of mung bean. Ecotoxicol Environ Safety 111:286–294
Cheng L, Booker FL, Burkey KO, Tu C, Shew HD, Rufty TW, Fiscus EL, Deforest JL, Hu SJ (2011) Soil microbial responses to elevated CO_2 and O_3 in a nitrogen-aggrading agroecosystem. PloS One 6
Cong T, Booker FL, Burkey KO, Hu S (2009) Elevated atmospheric carbon dioxide and O_3 differentially alter nitrogen acquisition in peanut. Crop Sci 49:1827–1836
Diara C, Castagna A, Baldan B, MensualiSodi A, Sahr T, Langebartels C, Sebastiani L, Ranieri A (2005) Differences in the kinetics and scale of signaling molecule production modulate the ozone sensitivity of hybrid poplar clones: the roles of H O, ethylene and salicylic acid. New Phytol 168:351–364
Dita MA, Rispail N, Prats E, Rubiales D, Singh KB (2006) Biotechnology approaches to overcome biotic and abiotic stress constraints in legumes. Euphytica 147:1–24
Dizengremel P (2001) Effects of ozone on the carbon metabolism of forest trees. Plant Physiol Biochem 39:729–742
Dizengremel P, Le Thiec D, Bagard M, Jolivet Y (2008) Ozone risk assessment for plants: central role of metabolism-dependant changes in reducing power. Environ Pollut 156:11–15
Doyle JJ (2001). Leguminosae. Encylopedia Genetics 1642:1081–1085
Dupont L, Alloing G, Pierre O, El Msehli S, Hopkins J, Herouart D, Frendo P (2012) The legume root nodule: from symbiotic nitrogen fixation to senescence. In: Nagata T (Ed) Senescence. Intechopen, Croatia, pp 137–68
El-Enany AE, Al-Anazi AD, Dief N, Al-Taisan WA (2013) Role of antioxidant enzymes in amelioration of water deficit and waterlogging stresses on vigna sinensis plants. J Biol Earth Sci 3:B144–B153
Flagler RB, Patterson RP, Hcagle AS, Heck WW (1987) Ozone and soil moisture deficit effects on nitrogen metabolism of soybean. Crop Sci 27:1177–1184
Galant A, Koester RP, Ainsworth EA, Hicks LM, Jex JM (2012) From climate change to molecular response: redox proteomics of ozone-induced responses insoybean. New Phytol 194:220–229
Guderian R (1985) Effects of pollutans combination. In: Guderian R (ed) Air pollution photochemical oxidants. Springer, Berlin, p 246
GuidiL., Degl'InnocentiE. F. Martinelli, M. Piras (2009) Ozone effects on carbonmetabolism in sensitive and insensitive phaseolus cultivars. Environ Exp Bot 66 117–125
Hassan IA, Anttonen S (1999) Changes in nitrogen metabolism in leaves of Bean (Phaseolus vulgaris L. cv. Giza 6) in response to Ozone. Pak J Biol Sci 2:327–331
Hassan IA, Haiba NS, Badr RH, Basahi JM, Almeelbi T, Ismail IM, Taia WK (2017) Effects of ambient ozone on reactive oxygen species and antioxidant metabolites in leaves of pea (*Pisum sativum* L.) plants. Pak J Bot 49:47–55
Havelka UD, Boyle MG, Hardy RWF (1982). Biological nitrogen fixation. In: Stevenson FJ (Ed), Nitrogen in Agricultural Soils ASA. Madison, Wisconsin USA, pp 365–422
Heath RL (1987) The biochemistry of ozone attack on the plasma membrane of plant cells. Rec Adv Phytochem 21:29–54
Hewitt DKL, Mills G, Hayes F, Wilkinson S, Davies W (2014) Highlighting the threat from current and near-future ozone pollution to clover in pasture. Environ Pollut 189:111–117
Hirsch AM (1992) Developmental biology of legume nodulation. New Phytol 122:211–237
IPCC (Intergovernmental Panel on Climate Change) (2013) Working Group I contribution to the IPCC fifth assessment report "Climate change 2013: the physical science basis". Final Draft Underlying Scientific-Technical Assessment

Iqbal M, Abdin M, Mahmooduzzafar Z, Yunus M, Agrawal M (1996) Resistance mechanisms in plants against air pollution. In: Iqbal M, Yunus M (eds) Plant response to air pollution. Wiley, New York, pp 195–240
Jimenez A, Hernandez JA, del Rio LA, Sevilla F (1997) Evidence for the presence of the ascorbate-glutathione cycle in mitochondria and peroxisomes of pea leaves. Plant Physiol 114(1):275–284
Kudapa H, Ramalingam A, Nayakoti S (2013) Functional genomics to study stress responses in crop legumes: progress and prospects. Funct Plant Biol 40:1221–1233
King BJ, Layzell DB, Canvin DT (1986) The role of dark carbon dioxide fixation in root nodules of soybean. Plant Physiol 81:200–205
Lamb C, Dixon RA (1997) The oxidative burst in plant disease resistance. Annu Rev Plant Physiol Plant Mol Biol 48:251–275
Layzell DB, Rainbird RM, Atkins CA, Pate JS (1979) Economy of photosynthate use in nitrogen-fixing legume nodules: observations on two contrasting symbioses. Plant Physiol 64:888–891
Lorenzini G, Saitanis C (2003) Ozone: a novel plant "pathogen". In: Toppi LSD, Pawlik-Skowronska B (eds) Abiotic stress in plants, Springer Science + Business Media, Dordrecht pp 205–229
Mantri N, Basker N, Ford R, Pang E, Pardeshi V (2013) The role of miRNAs in legumes with a focus on abiotic stress response. The Plant Genome. https://doi.org/10.3835/plantgenome2013.05.0013
Meehl GA, Stocker TF, Collins WD, Friedlingstein P, Gaye AT, Gregory JM, Kitoh A, Knutti R, Murphy JM, Noda A, Raper SCB, Watterson IG, Weaver AJ, Zhao Z-C (2007) Climate change 2007: The physical science basis, contribution of working group I to the fourth assessment report of the intergovernmental panel on climate change Global Climate Projections. Cambridge University Press, Cambridge, UK/New York, pp 747–846
Miller PR, McConkey BG, Clyton GW (2002) Pulse crop adaptation in Northern Great Plains. Agron J 94:261–272
Mishra AK, Agrawal SB (2015) Biochemical and physiological characteristics of tropical mung bean (Vigna radiata L.) cultivars against chronic ozone stress: an insight to cultivar-specific response. Protoplasma 252:797–811
Minchin FR, Pate JS (1973) The carbon balance of a legume and the functional economy of its root nodules. J. Exp. Bot. 24: 259–271.Noctor G, Foyer CH (1998) Ascorbate and glutathione: keeping active oxygen under control. Annu Rev Plant Physiol Plant Mol Biol 49:249–279
Morgan PB, Ainsworth EA, Long SP (2003) How does elevated ozone impact soybean? A meta-analysis of photosynthesis, growth and yield. Plant Cell Environ 26:1317–1328
Nakamura H, Saka H (1978) Photochemical oxidants injury in rice plants. III. Effect of ozone on physiological activities in rice plants. Jpn J Crop Sci 47:704–714
Noctor G, Foyer C (1998) Ascorbate and glutathione: keeping active oxygen under control. Annu Rev Plant Physiol Plant Mol Biol 49(1):249–279
Park JI, Grant CM, Davies MJ, Dawes IW (1998) The cytoplasmic Cu, Zn superoxide dismutase of Saccharomyces cerevisiae is required for resistance to freeze-thaw stress generation of free radicals during freezing and thawing. J Biol Chem 273:22921–22928
Pausch RC, Mulchi CL, Lee EH, Meisinger JJ (1996) Use of 13C and 15N isotopes to investigate O3 eff ects on C and N metabolism in soybeans: Part II. Nitrogen uptake, fi xation, and partitioning. Agric Ecosyst Environ 60(1):61–69
Pellegrini E, Francini A, Lorenzini G, Nali C (2011) PSII photochemistry and carboxylation efficiency in Liriodendron tulipifera under ozone exposure. Environ Exp Bot 70(2–3):217–226
Polle A, Wieser G, Havranek WM (1995) Quantification of ozone influx and apoplastic ascorbate content in needles of Norway spruce trees (Piceaabies L., Karst.) at high altitude. Plant. Cell Environ 18:681–688
Pinto E, Sigaud-Kutner TCS, Leitão MAS, Okamoto OK, Morse D, Colepicolo P (2003) Heavy metal-induced oxidative stress in algae. J Phycol 39:1008–1018
Rai R, Agrawal M, Choudhary KK, Agrawal SB, Lisa Emberson, Büker P (2015) Application of ethylene diurea (EDU) in assessing the response of a tropical soybean cultivar to ambient

O_3: nitrogen metabolism, antioxidants, reproductive development and yield. Ecotoxicol Environ Safety 112:29–38
Rao MV, Davis KR (1999) Ozone-induced cell death occurs via two distinct mechanisms in arabidopsis: the role of salicylic acid. Plant J 17:603–614
Rees DC, Tezcan FA, Haynes CA, Walton MY, Andrade S, Einsle O, Howard JB (2005) Structural basis of biological nitrogen fixation. Philos Trans R Soc A 363(1829):971–984
Reid CD, Fiscus EL, Burkey KO (1998) Combined effects of chronic ozone and elevated CO_2 on Rubisco activity and leaf components in soybean (*Glycine max*). J Exp Botany 49:1999–2011
Reid CD, Fiscus EL (1998) Effects of elevated CO_2 and/or ozone on limitations to CO_2 assimilation in soybean (*Glycine max*). J Exp Botany 49:885
Sandermann H Jr, Ernst D, Heller W, Langebartels C (1998) Ozone: an abiotic elicitor of plant defense reactions. Trends Plant Sci 3(2):47–50
Samuel MA, Miles GP, Ellis BE (2000) Ozone treatment rapidly activates MAP kinase signalling in plants. Plant J 22(4):367–376
Singh S, Agrawal SB (2011) Ambient ozone and two black gram cultivars: an assessment of amelioration by the use of ethylenediurea. Acta Physiologia Plantarum 33:2399–2411
Singh S, Agrawal SB, Agrawal M (2015) Responses of pea plants to elevated UV-B radiation at varying nutrient levels: N-metabolism, carbohydrate pool, total phenolics and yield. Funct Plant Biol 42(11):1045–1056
Swanson ES, Thomson WW, Mudd JB (1973) The effect of ozone on leaf cell membranes. Can J Bot 51(6):1213–1219
The Royal Society (2008) Ground-level Ozone in the 21st century: Future trends, impacts and policy implications. Royal society policy document 15/08, RS1276
Scandalios JG, Guan L, Polidoros AN (1997) Catalase in plants: gene structure, properties, regulation and expression. In: Scandalios JG (ed) Oxidative stress and molecular biology of antioxidant defenses. Cold Spring Harbor Laboratory Press, Cold Spring Harbor, pp 343–406
Serbinova EA, Packer L (1994) Antioxidant properties of α-tocopherol and α tocotrienol. Methods Enzymol 234:354–366
Smirnoff N (2000) Ascorbate biosynthesis and function in photoprotection. Philos Trans R Soc Lond B Biol Sci 355:1455–1464
Sun JD, Feng ZZ, Ort DR (2014) Impacts of rising tropospheric ozone on photosynthesis and metabolite levels on field grown soybean. Plant Sci 226:147–161
Taylor GE Jr, Tingey DT, Ratsch HC (1982) Ozone Flux In Glycine Max (L.) Merr. Sites of regulation and relationship to leave injury. Oecologia 53:179–186
Tetteh R, Yamaguchi M, Wada Y, Funada R, Izuta T (2015) Effects of ozone on growth, net photosynthesis and yield of two African varieties of *Vigna unguiculata*. Environ Pollut 196:230–238
Tingey DT, Hogsett WE, Rodecap KD, Lee EH, Moser TJ (1994) The impact of O3 on leaf construction cost and carbon isotope discrimination. Essener Ökologische Schriften 4:195–206
Toker C, Yadav SS (2010) Legume cultivars for stress environments. In: Yadav SS, McNeil DL, Redden R, Patil SA (eds) Climate change and management of cool season grain legume crops. Springer, Dordrecht, pp 351–376
Udvardi M, Poole PS (2013) Transport and metabolism in legume-rhizobia symbioses. Annu Rev Plant Biol 64(1):781–805. https://doi.og/10.1146/annurev-arplant-050312-120235http://doi.og/10.1146/annurev-arplant-050312-120235
Umponstira C, Kawayaskul S, Chuchaung S, Homhaul W (2009) Effect of ozone on nitrogen fixation, nitrogenase activity and rhizobium of cowpea (*Vigna unguiculata* (L.) Walp). Naresuan Univ J 17:213–220
Varshney RK, Roorkiwal M, Nguyen HN (2013) Legume genomics: from genomic resources to molecular breeding. Plant Genome 6:1–7
Welfare K, Yeo AR, Flowers TJ (2002) Effects of salinity and ozone, individually and in combination on the growth and ion contents of two chick pea (Cicer arietinum L.) varieties. Environ Pollut 120:397–403

White J, Prell J, Euan KJ, Philip P (2007) Nutrient sharing between symbionts. Plant Physiol 144:604–614

Winner EW, Coleman JS, Gillespie C, Mooney HA, Pell EJ (1991) Consequences of evolving resistance to air pollution. In: Taylor GE, Pitelka LF, Clegg MT (eds) Ecological genetics and air pollution. Springer, New York, pp 177–202

Yadav DS, Rai R, Mishra AK, Nivedita Chaudhary, Mukherjee Arideep Agrawal SB, Madhoolika Agrawal (2019) ROS production and its detoxification in early and late sown cultivars of wheat under future O_3 concentration. Sci Total Environ 659:200–210

Young PJ et al (2013) Pre-industrial to end 21st century projections of tropospheric ozone from the atmospheric chemistry and climate model intercomparison project (ACCMIP). Atmos Chem Phys 13:2063–2090

Zahran HH (1999) Rhizobium-Legume symbiosis and nitrogen fixation under severe conditions and in an arid climate. Microbiol Molecular Biol Rev 63:968–989

Zhang W, Wang G, Xiaobing L, Feng Z (2014) Effects of elevated O_3 exposure on seed yield, N concentration and photosynthesis of nine soybean cultivars (*Glycine max* (L.) Merr.) in Northeast China. Plant Sci 226:147–161

Zhao TH, Cao YH, Wang Y, Dai Z, Liu YO, Liu B (2012) Effects of ozone stress on root morphology and reactive oxygen species metabolism in soybean roots. Soybean Sci 12:1

Zhu H, Choi H, Cook DR, Shoemaker RC (2005) Bridging model and crop legumes through comparative genomics. Plant Physiol 137:1189–1196

Salinity Stress Responses in Three Popular Field Crops Belonging to *Fabaceae* Family: Current Status and Future Prospect

Debojyoti Moulick, Suman Samanta, Bedabrata Saha, Muhammed Khairujjaman Mazumder, Shainandni Dogra, Kishore C. S. Panigrahi, Saon Banerjee, Dibakar Ghosh, and Subhas Chandra Santra

Abstract Salinity stress is a constraint to the world crop production. The crop plants of *Fabaceae* family are a major source of dietary proteins for the population of third world countries. But salinity stress poses a serious threat in their yield. Several studies have been performed to elucidate the legume salinity stress responses as well as strategies to ameliorate salt stress. Here, in this chapter, we have tried to summarize the salt stress responses in as well as techniques employed to improve yield under salinity stress in three important crop plants of this family focusing on groundnut (*Arachis hypogaea* L.), mungbean (*Vigna radiata* L.) and blackgram (*Vigna mungo* L.).

Keywords Pulse crops · Salt stress · Osmolytes · Crop production · Nitrogen fixation · Beans

---

All authors share equal credit.

D. Moulick (✉) · M. K. Mazumder
Plant Stress and Metabolomics Laboratory, CIL, Assam University, Silchar, Assam 788011, India
e-mail: drubha31@gmail.com

S. Samanta · S. Banerjee
AICRP on Agrometeorology, Bidhan Chandra Krishi Viswavidyalaya, Kalyani, West Bengal 741235, India

B. Saha · K. C. S. Panigrahi
School of Biological Sciences, National Institute of Science Education and Research, Jatni, Bhubaneswar, Odisha 752050, India

S. Dogra
Punjab Biotechnology Incubator, Mohali, India

D. Ghosh
ICAR-Directorate of Weed Research, Jabalpur, Madhya Pradesh, India

S. C. Santra
Department of Environmental Science, University of Kalyani, Kalyani, West Bengal 741235, India

M. Hasanuzzaman et al. (eds.), *The Plant Family Fabaceae*,
https://doi.org/10.1007/978-981-15-4752-2_20

1 Introduction: Salinity Stress

Abiotic factors such as temperature, irregular rainfall, salinity stress, among others, strongly influence the soil quality which in turn results in the depletion of crop productivity. However, salinity is one of the most important limiting factors that inhibit the productivity of agricultural crops and plants through its adverse effects on germination, plant vigour and finally yield and yield attributing characters. USDA salinity laboratory has defined the saline soil as the soil having an electrical conductivity (EC) of solution extracted from the water-saturated soil paste ECe (EC of the extract) of 4 dS/m (deci Siemens per meter), where 4 dS/m $\approx$ 40 mM NaCl or more (Kotuby-Amacher et al. 2000; Chinnusamy et al. 2005). It is also observed that salt tolerance may vary depending on soil types and environmental factors (abiotic) such as vapour pressure deficit, radiation and temperature. Salinity is a concerning issue in arid and semi-arid areas as there are insufficient rain to leach salts and excess sodium ions out of the rhizosphere. Imadi et al. (2016) pointed out that almost 75 countries around the world are having sodic and saline-sodic soils and they have also denoted that their extent is increasing steadily among agricultural countries. Singh et al. (2015) also assumed that, by 2050, almost 50% of total arable land might lose its fertility due to this ever-increasing salinization procedure. Fertile lands across 100+ countries are facing the problem of salinization in varying intensities (Yadav 1993). As per the report of Munns and James (2003), globally, nearly 100 million ha of arable land that accounts for 6–7% of total arable land is affected due to salinity. However, as per the recent findings of Singh (2018), it has been observed that worldwide currently almost 953 million ha of land area comes under the severe threat of salinization. According to the above-mentioned report, about 6.73 million ha of productive lands of India are exposed to the severe threat of being saline-prone. Out of 29 states, fifteen states of the country contain saline and sodic components in their productive lands. Gujrat topped the list with 2.2 million ha of saline and sodic soil while Uttar Pradesh ranked second with 1.3 million ha of saline prone areas. Arid and semi-arid regions are supposed to be more vulnerable due to the utilization of brackish water for irrigation purpose that induces secondary salinization resulting a great threat to the sustainable productivity. Soils of any origin contain salts in different amount stretching from humid tropics to the polar regions. Minerals, in the form of salts, are very common and the most significant part of soil and are considered as the essential component for plant nutrition. Irrigation waters, from either canals or underground pumping, contain some amount of salt, which starts secondary soil salinization procedure through crop irrigation. Lauchli et al. (2008) presented alarming statistics about soil salinization. They described that more than 45 million ha of irrigated land (~20% of total irrigated land acreage) and almost 2% of the lands farmed by dryland agriculture have been severely damaged or lost its productivity due to high salinity level. It is evident that irrigation water containing calcium (Ca^{2+}), magnesium (Mg^{2+}) and sodium (Na^{+}) produce carbonates after evaporation, leaving Na^{+} dominant in the soil (Serrano et al. 1999). This residual Na^{+} often exceeds the quantity of aggregated macronutrient as well as micronutrient by one or two orders of magnitude. Grattana and Grieveb (1999) observed that these higher concentrations of Na^+ in soil may reduce the nutrient-ion activities by producing extreme ratios of Na^+/Ca^{2+} or Na^+/K^+.

2 Consequences of Salinity Stress on Field Crops

Physiological processes of plants are directly related to the soil salinity level which reduce the crop growth and its vigour development through processes such as water stress, ion toxicity, nutritional disorders, oxidative stress, alteration of metabolic processes, membrane disorganization, reduction of cell division and expansion, genotoxicity (Munns 2002; Zhu 2007). Soil salinity is one of the most important abiotic factors that drastically limits the crop production, depending on the magnitude of salinity and the tolerance limit at the critical phenological stages of a particular crop (Hasanuzzaman et al. 2013a, b; Parihar et al. 2015; Shrivastava and Kumar 2015; Negrao et al. 2017; Parvin et al. 2019; Hasanuzzaman et al. 2018a, b, 2019; Mahmud et al. 2020). Presence of excessive amount of salt in the soil strongly obstructs the germination process and retards the growth by making water less available resulting a physiological drought (Moreno-Casasola 2008). Plant morphology and physiology are adversely affected due to the presence of salt. Under high salinity stress, the enhanced uptake of Na ions negatively influence the plant growth due to the result of five different reasons and they are as follows:

i. Decreased uptake of essential nutrients such as potassium by the plant.
ii. Reduced water use efficiency
iii. Toxic effects of Na^+ and Cl^- on plant morphological traits such as root system size.
iv. Toxic effects of both the ions on plant physiology including the activity of enzymes, the function of cell membranes and the production of plant hormones (Alam 1994; Munns and James 2003);
v. Increased oxidative stress caused by high Na^+ and Cl^- levels (Munns 2002; Dolatabadian et al. 2011; Golpayegani and Tilebeni 2011; Bothe 2012).
vi. In short, shoot growth is affected more by salinity than root (Omisun et al. 2018).

However, at low concentrations, crops and their productivity levels might hardly be affected (Maggio et al. 2001). Plants such as glycophytes are very sensitive to salinization as the presence of 100–200 mM NaCl in soil might be lethal for the plant. On the contrary, halophytes are more susceptible to high salt concentration (up to 300–400 mM). For some species, a relation between leaf area and number and salinity was described, where leaf number may increase with the increasing salt concentration (Chen et al. 2007; Rui et al. 2009; Gama et al. 2007; Ha et al. 2008). Recent studies also proved that fresh and dry weights of the shoot systems are affected (either positive or negative) depending on salts concentration, type of salt and importantly the type of plant species (Taffouo et al. 2010; Memon et al. 2010).

3 Salinity Stress on Selected Field Crops from *Fabaceae* Family

3.1 Groundnut

Groundnut (*Arachis hypogaea* L.), also known as peanut, is an important oil, food and feed legume crop cultivated in more than 100 countries of tropics and subtropics stretching from 40° N to 40° S latitudes. Since it is rich in protein and fill with high-energy value, it is considered as the fourth most important source of edible oil and third most source of vegetable protein in the world. China holds the topmost position by producing almost 1.4×10^{10} kg of groundnut (Fig. 1) (FAOSTAT 2017). India is the second largest producer (80–85 lakh MT) having the highest annual all-season acreage of 70 lakh hectares (GOI 2018). Groundnut can be cultivated throughout the year, i.e. in monsoon season (commonly known as *Kharif)*, winter season (commonly known as Rabi) and summer. However, in India, nearly 80% of annual acreage and production comes from *Kharif* season. According to the Directorate of Economics and Statistics, GOI, all India *Kharif* 2018 groundnut acreage was 3,890,000 hectares. Gujarat is leading the table with 40.1% of national production followed by Rajasthan which produced almost 21.6%. The national average yield was estimated as 1336 kg ha^{-1}.

Groundnut, a self-pollinated, tropical annual legume, belongs to genus Arachis in subtribe Stylosanthinae of tribe Aeschynomenea of family Leguminosae. Mainly two subtypes, hypogaea and fastigiata, are being cultivated across the globe with two

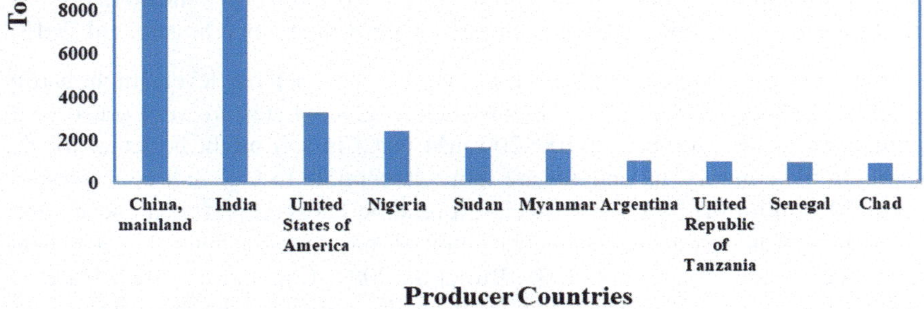

Fig. 1 Top ten producers of groundnuts and their production (in tonnes) in 2017. *Source* FAOSTAT (accessed on 26.11.2019)

botanical varieties (var. hypogaea and var. aequatoriana). Each of these botanical varieties can be differentiated by their distinct features in terms of plant, pod and seed morphology. However, most of the commercially cultivated varieties belong to the hypogaea (common name/market type: Virginia or runner), fastigiata (Valencia) and vulgaris (Spanish) botanical variety groups.

Now if we consider the economic scenario, we can state that it is a very important agricultural export commodity. Groundnut kernels contain 48–50% of oil and 26–28% of protein along with antioxidants, biologically active polyphenols, flavonoids, isoflavones, minerals and vitamins (Janila et al. 2013). On equal weight basis (Kg for Kg), groundnuts contain more protein than meat and about two and a half times more than eggs. In addition to protein, groundnuts are a good source of calcium, phosphorus, iron, zinc and boron. The groundnuts also contain vitamin 'E' and small amounts of vitamin 'B' complex. It has been observed that its growth and development is at the optimum level when the mean temperature remains between 21 and 26.5 °C along with ≥50 cm rainfall. A warm and dry condition is assumed as the best weather during its ripening period. It has been noticed that 100 gm of groundnut kernel can produce 564 kcal of energy (Jambunathan 1991; Pasupuleti and Nigan 2013). Birthal et al. (2010) described that over 60% of global groundnut production is crushed for extraction of edible oil as well as industrial uses. Rest of the amount, i.e. 40% is being used for direct consumption as nut along with seed storage (for next year). Most of the Asian countries utilized groundnut oil as the cooking medium because of its high smoking point (Singh and Diwakar 1993). The best part of it is the cake (crushed portion after oil extraction) which is utilized for feeding animal industry and nowadays, with advance technologies, they have transformed them into enriched easily digestible food for children and aged persons. Generally, the cake contains protein (43–65%) and fat (6–20%) along with vitamin B depending upon the method of extraction. Its nutritious quality is so high that NASA has selected groundnut as a possible food for the advanced life support system for extended space missions. Fermented dough and kisra, prepared from groundnut, are very popular among Sudanese (Singh 1992). Groundnut cake meal has been used to prepare bakery products such as bread, biscuits and cookies which can replenish protein demand of malnourished people in the developing countries. Nowadays, people of USA, Canada, Australia, etc., are using peanut butter as the most popular product. Each and every portion of groundnut plant is very useful and rich in nutrient values. Not only the kernels but also the haulms contain protein @8–15%, lipids (1–5%), minerals (9–17%) and carbohydrate (38–45%) which constitute a nutritious fodder for livestock.

The most interesting part of this crop is that, being a legume crop, it can improve the soil health and fertility by leaving behind N_2 (through nitrogen fixation at root zone) and organic matter in the soil. Due to these superqualities, the projected demand of peanut in Asia alone by 2020 is assumed to be 1.6 times more than the level of production in 2000 (Birthal et al. 2010). As a whole, its cultivation contributes to the sustainability of mixed crop–livestock production systems, the most predominant system of the semi-arid areas.

3.1.1 Impact of Salinity Stress on Groundnut's Physiology, Agronomic Performance and Nutritional Quality (2010–Till Now)

Rapid urbanization for better lifestyle signals higher incomes, higher opportunity cost of time and therefore greater demand for convenience foods. To contribute to the feeding of the 2/3 portion of the world population living in the developing countries of Asia and Africa, a rapid growth in groundnut production is needed. However, the groundnut genotypes are very sensitive to salt and drought stress as they affect both productivity and quality of peanut in semi-arid tropics (Reddy and Anbumozhi 2003; Krishna et al. 2015). Salt stress sharply reduces the plants' ability to uptake minerals and its corresponding growth and development through 'osmotic effect' (Singh et al. 1989; Janila et al. 1999). The impact of salinity stress on the physiological and biochemical changes of groundnut was examined by two different ways and they are as follows:

3.1.2 Laboratory-Based Study (In Situ Experiments)

Nigeria is the fourth largest producer of groundnut in the world, and due to its economic as well as nutritious value, it continues to be the predominant crop of this country. Mensah et al. (2006) tested five genotypes (RMOP 91, Ex Dakar, RMP 12, RRB 12 and Esan Local) over two different approaches. For germination tests, seeds were placed in Petri dishes, whereas plastic buckets were used for the rest of the growing period. They also recorded a delay in germination as well as the reduction in agronomic characters such as plant height and number of leaves per plant under EC greater than 2.60 mS/cm. It is also noticed that the selected varieties were more salt-tolerant during the germination stage during the vegetative stage.

Numerous studies proved that under saline conditions, leaves of the plant contain more Na^+ and Cl^- ions which are toxic to them (Dogar et al. 2012). In addition, the absorption of K^+, NO_3^- and $H_2PO_4^-$ ions are regulated rather reduced due to the presence of these two ions (White and Broadley 2001). Presence of substantial amount of Na^+ also induces nutrient imbalance (Rogers et al. 2003; Hu and Schmidhalter 2005) and inhibits the activity of Ca^{2+}, K^+ and Mg^{2+} due to high Na^+/Ca^{2+} or Na^+/K^+ (Hu and Schmidhalter 1997; Grattan and Grieve 1999) and N, P and K translocation (Hirpara et al. 2005). Mungala et al. (2008) reported that the germination of groundnut seeds is reduced by 10% when grown on media containing 2.5% NaCl. They have also noticed a consecutive reduction in groundnut yield by 25%, 50% and 100% when cultivated under the salinity level of 4.1, 4.9 and 6.5 mmhos/cm. On the other side, Yadav et al. (2011) demonstrated that EC 3.2 dS/m is the critical value for yield reduction. Salinity level of 9.2 dS/m might capable of reducing the chlorophyll content which, in turn, able to inhibit/retard the productivity in groundnut (Hammad et al. 2010). Seedling, seed development and flowering stages are considered as the most critical phenological stages that can be severely influenced by the soil salinity (Mudgal 2004; Cuartero et al. 2006; Amin 2011). Taufiq et al. (2016) studied the growth of two groundnut varieties, namely Domba (Spanish type) and

Hypoma 1 (Valencia type) under saline condition where EC varies from 2.65 to 4.08 dS/m. Results revealed that above the critical level (3.2 dS/m), the number of pod per plant and pod yield both reduced. In another study, Taufiq et al. (2015) cultured ten groundnut varieties in a glasshouse and the results revealed that the Indonesian cultivars failed to produce pods on soil with EC value of 2.0 dS/m.

Groundnut is considered as moderately sensitive to salinity stress (Azad et al. 2013). Researchers from Kerala raised groundnuts under in vitro and in vivo conditions where different concentrations of NaCl (No NaCl, 100 mM, 200 mM) were used as the stress-inducing salt (Mohan and Shashidharan 2018). A delayed response was noticed under both types of cultures due to the adaptive feature that plants adopt to survive under stress affecting multiple resources (e.g. building blocks and energy). Mild increase in protein content was observed in the 200 mM salt. Plants are also capable of synthesizing new proteins under a stressed situation that may provide a storage form of nitrogen which can be reutilized when stress is over (Garratt et al. 2002). The response to salinity is directly regulated with the upregulation of key enzymes that regulates reactive oxygen species levels such as peroxidases catalase, superoxide dismutase and maintenance of H_2O_2 at concentrations required for cell signalling (Munns and Tester 2008). Mohan and Shashidharan (2018) also noticed a sudden increase in the concentration of phenolic compounds under the 100 mM solution of both in vitro and in vivo conditions. However, this drastic change is mostly limited up to 100 mM as no further increment was recorded with the increment of the saline dose.

Neighbouring countries like Bangladesh are also facing the concerning issue of salinization of arable land. Around the coastal belts of Bangladesh, approximately 1.02 million ha of cultivated lands are facing the disastrous effect of varying degree of salinity (2.0 → 16 dS/m) which changes according to the seasons (Karim and Iqbal 2001; SRDI 2003). Thus, these places are not suitable for growing high water-demanding crops during the dry period of November to May as suitable irrigation water could not be supplied that time. Farmers like to cultivate groundnut in the said period as the crop demands only 350 mm water to complete its life cycle, which can be accomplished through rainfall (Reddy and Kaul 1996). Azad et al. (2014) tested the growth of five genotypes of groundnut in the glasshouse of Bangladesh Institute of Nuclear Agriculture to study the impact of salinity on those cultivars. It was observed that under 7–9 dS/m salinity stress, the 'Dacca-1' variety showed more resistance during flowering to harvest in the time frame of August to January and it produced more pods and kernels compared to unstressed treatment. However, a contrasting scenario was also noticed when the variety was exposed to 8 dS/m during April to August where they could not produce any pod due to high maximum temperature (>50 °C).

In a recent study conducted under the laboratory condition of Plant Physiology Department of Bidhan Chandra Krishi Viswavidyalaya (BCKV), West Bengal, 26 genotypes of groundnut were investigated to identify the salt-tolerant varieties and their physiological changes in response to stress (Pal and Pal 2017). The same study confirmed that high salinity causes fluctuation of biochemical parameters such as reduction of leaf sugar content in the range of 2.70–70.32% (varying with genotypes) over control, increase in leaf proline content (531.5–780.16%) in tolerant genotypes. Five genotypes, namely KDG-197, R2001-2, VG-315, TCGS1157 and TG-51 were the best performers under the salinity stress (up to 200 mM NaCl), whereas genotype Girnar 3 was found as the mostly affected one in this study.

3.1.3 Field-Based Study (Ex Situ Experiment)

Very few ex situ trials were done to examine the impact of salinity on the productivity and growth of groundnut. One of these works was conducted by Chakraborty et al. (2016) where two Spanish bunch type varieties (TG 37A and GG 2) have been tested in a split-split design to define the role of external K^+ application in alleviating the salinity stress and their responses towards conjoint saline-K environment. Imposition of salinity stress results in varying degree of response and adaptability in two different genotypes. Being a stress-tolerant variety, GG 2 represents the lesser degree of reduction (11–20%) of relative water content (RWC) in the leaves compared to TG 37A (16–28%). However, external application of K^+ improved the resistance capability of both genotypes, specifically in TG 37A. Leaf water potential also dropped significantly with concurrent fall in RWC. Under control condition, i.e. without stress, the stress-tolerant cultivar (GG 2) produced less biomass compared to stress susceptible cultivar. This may be due to the higher metabolic cost of imparting inherent tolerance character in GG 2 resulted in sacrificing total biomass under control condition. Exogenously applied K^+ helped both the cultivars to maintain better physiological status by retaining tissue K^+ or through higher uptake of K^+. Previous studies also pointed out the same points (Ikeda 2005; Akram et al. 2009).

3.2 Mungbean

Vigna radiata (L.) or mungbean is a vital legume crop, widely cultivated in Asian countries. Mungbeans are a member of the genus *Vigna* that comprise 150+ species originated mostly in Asia and Africa. The tropical regions of Asia have the highest degree of genetic diversity (USDA-ARS 2012). Besides complementing cereal-based diets, it is also used as an indispensable ingredient in various salads consumed in developed and developing countries. Cultivated species of mungbeans are also known as green gram, moong, golden gram, oregon pea, chickasono pea, chickasaw pea and chiroko in India. Agronomically, mungbean belongs to short-duration field crops with crop duration of 70–110 days and can be cultivated in a wide range of agroecosystem (except in frost-free land mass) (Imrie 1995). Morphologically, mungbean has small structure (<1.25 m), nodule bearing roots (for N_2 fixation) with erect to suberect branches, trifoliate leaves, bears yellow flowers with butterfly-like structure, which may vary from variety to variety. Seeds of mungbean have a wide range of appearances depending upon the colour of testa as well as absence or presence of texture layer (Watt et al. 1977). From morphological point of view, seeds are either having a dull or buff appearance with a golden or green shiny appearance. India has the credit to be the world's largest producer of mungbean (>50%) (Vijayalakshmi et al. 2003), whereas Thailand is the largest exporter of mungbean in the world (Srinives 1991). During 2012–2017 (under Twelfth Plan) in India, 15.91 lakh tonnes of mungbean were produced from 34.50 lakh hectares' agricultural land. In India, in terms of production per unit land, the order lies Andhra Pradesh (696 kg/ha) < Jharkhand (704 kg/ha) < Punjab (845 kg/ha), respectively, with national yield average of 461 kg/ha, whereas if we consider agricultural land used for mungbean cultivation, the top three positions were occupied by Rajasthan, Maharashtra and Karnataka, respectively (Kumar and Pandey 2018).

3.2.1 Salinity Stress on Mungbean

Salinity stress has been documented as the main abiotic stressor for the field crops. Hasanuzzaman et al. (2013a) reported that due to salinity stress, as much as 70% yield reduction of mungbean is observed in the presence of 50 mM NaCl. Authors such as Kandil et al. (2012) and Karthikeyan et al. (2012) were of the opinion that 1–3% of agriculture land has gradually become saline every year due to both anthropogenic influence as well as naturally. The authors further added that this soil salinization, by the middle of twenty-first century, will significantly modulate the production of agronomically important crops throughout the world. Regarding the adverse effects of soil salinity on field crops, mungbean is no exception. Under field conditions, mungbean experiences a wide range of adverse situation along with salinity stress, such as high temperature, pod shattering, attack from pests, among others, which stimulates significant yield reduction. From agronomically or plant biology point of view, the adverse effects of salinity on mungbean depend on various factors. The magnitude of salinity induced injuries is governed by (a) duration of exposure, (b) concentration of salt, (c) growth stage where it exposed to salinity, (d) nature of plant and finally and (e) how salinity was applied. Main obstacles behind proper categorization or documentation of salinity-induced responses in mungbean are (1) complex nature of salinity stress and (2) polygenic nature of stress-induced responses (Mahajan and Tuteja 2005). Other authors, as Mahdavi and Sanavy (2007), were of the view that unavailability of adequate technology to standardize the stress-induced responses in grain legumes can hinder the development of transgenic or tolerant lines.

In the plant–soil system context, salinity stress inhibits germination and a reduction in root and shoot elongation was observed. It is reported that under saline stress environment, Na^+ and Cl^- create an osmotic potential on the exterior of root and thus preventing water and dissolved mineral uptake can be taken as a possible mode of action of excess salinity stress (Moose and Mumm 2008; Muhammad and Majid 2013; Munns and James 2003). Nafees et al. (2010) and Natr and Lawlor (2005) reported that salinity-induced reduction in pigment (chlorophyll content) in mungbean and attributed this to inhibition of chlorophyll biogenesis or modulating chloroplast structure by accumulated Na^+ and Cl^- in mungbean's leaf. Sehrawat et al. (2013,

2014a, b, 2015) have investigated the consequences of salinity stress on mungbean and found that salinity stress can impose inhibitory or adverse effects irrespective of growth stage, but especially damage at post-vegetative stage, i.e. flowering and pod-filling stage. To elaborate the possible mechanism associated with drastic yield loss, authors like Tavakkoli et al. (2010) and Hasanuzzaman et al. (2012) found that greater accumulation of Na^+ in saline soil and its subsequent accumulation in different plant parts may lead to yield reduction.

3.3 *Blackgram*

Blackgram (*Vigna mungo* (L.) Hepper) is one of the important pulse mainly consumed because it is a cheap source of dietary protein (21%), in the underdeveloped and developing countries. Blackgram plants can also maintain nitrogen level in the rhizosphere by fixing atmospheric nitrogen (Chowra et al. 2017). It is mainly cultivated in Asian countries including India, Pakistan, Myanmar and parts of southern Asia. About 70% of world's blackgram production comes from India (Khariff crop). The seeds of blackgram are small and have black seed coat.

Blackgram is considered to be highly sensitive to salt and as such is prone to salt from airborne spray and tidal immersion in coastal areas. Despite its high susceptibility, very few works on blackgram and salinity stress responses have been reported till date. The first report studied the physiological impact of sodium chloride treatment in water relations, proline and chlorophyll contents, whereas the second report a year later by the same group discussed the differential tolerance of nine varieties (Ashraf 1989; Ashraf and Karim 1990). In the second report, they showed that there is considerable variability in tolerance among the varieties through assessing chlorophyll content and ionic status upon NaCl treatment (Ashraf and Karim 1990). Dash and Panda 2001, showed dose-dependent response in growth and antioxidant enzyme activity in blackgram during germination. Again after a decade gap, Win et al (2011) studied the variation in response on salt treatment in several Vigna genotypes, being some of them *Vigna mungo*. Among other aspects, this study evidenced the relevance of genetically diverse accessions to study salt response mechanisms and usefulness as a good resource for breeding programs.

4 Salinity Stress Management

From the previous discussions and literature surveys, it is clear that salinity stress changes the physiological as well as biochemical properties of a plant. Scientists projected that the rate at which fertile agricultural lands are affected by high soil salinity, the amount of available arable land will reduce by 30% in the next 25 years and up to 50% by the year 2050 (Rozema and Flowers 2008). Such kind of alarming situation strictly demands the development of salt-tolerant varieties as well as the more advanced techniques to mitigate or at least slow down the soil salinization procedures. There are two ways to eradicate the impact of salinity in legume crop production and that are discussed further.

4.1 Methods so Far Applied in Groundnut to Mitigate Salinity Stress

4.1.1 Techniques or Procedures to Mitigate Salt Stress in the Soil

Several substances have been applied to soils in a tentative to ameliorate crop yield in soils with high levels of salt. One of these examples is the application of gypsum @3–5 tons per acre (recommended rate) in the field during its preparation. Taufiq et al. (2016) have proved that combination of gypsum and organic manure is a better option to ameliorate saline soil compared to the individual application of K fertilizer, dolomite, gypsum and organic manure. These authors found that the combination of gypsum and organic manure was consistent in reducing exchangeable Na, Na saturation and soil EC and the groundnut varieties, cultivated under this treatment, produced higher yield compared to other treatments. Other approaches include:

i. Incorporation of Dhaincha (*Sesbania bispinosa*) (6.25 t ha^{-1}) in soil before planting (Cuevas et al. 2019).
ii. Maintenance of high K/Na ratio by applying potash and Ca fertilization. Application of exogenous K^+ might enhance the resistance power against salinity in terms of plant water status, biomass produced under stress, osmotic adjustment and better ionic balance (Chakraborty et al. 2016). However, the response may differ depending upon the salt sensitivity of the genotype studied.
iii. Split application of N and K fertilizers.

4.1.2 Techniques to Minimize or Eradicate the Influence of Salinity in the Plant Itself

With the advancement of genetic engineering, worldwide researchers are focusing mainly on the development of stress-resilient transgenic crops by using recombinant DNA techniques (Asif et al. 2011; Iqbal et al. 2012). Azad et al. (2013) found that one of the mutant varieties had increased total sugar contents when exposed to salinity stresses, namely during flowering and pod-filling stage, as well as, increased free amino acid during pod-filling stage. These physiological responses contributed to maintain turgor of guard cell and intake of CO_2 (through open stomata). This modulates photosynthesis rate and contributing to the mobilization of assimilates to reproductive organs, particularly kernel. The ectopic expression of a peroxisomal ascorbate peroxidase gene (*Sb*APX) from the extreme halophyte Salicornia brachiate was found to improve salinity responses in *A. hypogaea* (Singh et al. 2014).

Traits as growth, chlorophyll content, relative water content and electrolyte leakage under 150 mM NaCl stress condition were found improved by the described approach (Singh et al. 2014). In a recent experiment, Banavath et al. (2018) have also developed one *A. hypogaea* transgenic genotype expressing an *Arabidopsis* homeodomain-leucine zipper transcription factor (AtHDG11) under stress-inducible rd29A promoter. Such approach influenced few biochemical and physiological attributes like: efficient uptake of water from deeper layers of soil due to enhanced root length.

Reduction in water loss through stomata by reducing its density; boosted water use efficiency traits and elevated photosynthetic rates to lower down the yield loss under stress conditions and; contributed for the control of cellular salt tolerance with enhanced accumulation of different enzymes and efficient ROS detoxification mechanism.

Nevertheless, other suitable approaches could be used also to improve the plant response to salinity. Nowadays, researchers also prefer the use of plant growth-promoting rhizobacteria to mitigate the salt effects in plants. One of these examples is the use of diazotropic salt-tolerant bacterial isolates that under stress condition can contribute to maintain ion homeostasis in groundnut cultivars, which also showed less ROS accumulation and even enhanced growth in under salt stress condition (Sharma et al. 2016).

i. Seed hardening with NaCl (10 mM concentration) (TNAU Agritech Portal 2020).
ii. Foliar spray of 0.5 ppm brassinolode for increasing photosynthetic activity.
iii. Foliar spray of 2% DAP + 1% KCl during critical stages.
iv. Foliar application of ascorbic acid could improve the number of leaves and leaf area, while in combination with zinc sulphate, it can increase the plant height and total plant biomass.

4.2 Salinity Management in Mungbean

A lot of work has been done on mungbean stress mitigation through priming, pre-treatment and co-treatment during this decade. Since legume crops are relatively recalcitrant to genetic engineering, very few studies have been performed. Exogenous application of proline, ethylene and glycinebetaine reduced ROS accumulation and ROS-induced damages, whereas boosted the antioxidant machinery, photosynthesis and glyoxylase system in mungbean under salt stress (Hossain and Fujita 2010; Hossain et al. 2011; Khan et al. 2014). Adverse effect of salinity was also found to be reduced by exogenous application of thiourea and glutathione through modulation of cellular metabolites and methylglyoxal detoxification system (Perveen et al. 2016; Nahar et al. 2015). Another interesting study found that pre-treatment with sublethal dose of salt (NaCl) stimulated the antioxidant defence against salinity stress as well as DNA damage at latter stages of seedling development (Saha et al. 2010, 2015). Rhizobia, rhizobacteria and *Pseudomonas* strain co-inoculation also ameliorates salt stress response in mungbean and co-inoculated plants have increased chlorophyll content, relative water content, transpiration rate and nutrient balance (Ahmad et al. 2011, 2012, 2013; Tittabutr et al. 2013; Panwar et al. 2016). Salicylic acid treatment has been reported to enhance photosynthetic efficiency under saline condition by enhancing nitrogen and sulphur assimilation (Nazar et al. 2011; Shakeel and Mansoor 2012; Akhtar et al. 2013; Kaur and Nayyar 2015; Ghassemi-Golezani and Lotfi 2015). Co-application of phyto-hormones (gibberellic acid and brassinosteroid) and polyamines (spermine, spermidine and putrescine) resulted in alleviation of salt stress and stressed seedlings showing better physiology, nutrient homeostasis, antioxidant activity and reduced sodium uptake (Ghosh et al. 2015; Nahar et al. 2016; Lalotra et al. 2017). Biochar is an eco-friendly alternative for amelioration of environmental stressors, which has been experimented in different crops under different conditions. In mungbean, biochar also effectively attenuates salinity stress by regulating H^+-ATPase and H^+-PPase activities in root tonoplast among other physiological and morphological aspects (Torabian et al. 2018; Nikpour-Rashidabad et al. 2019). Even elements like sulphur, calcium, zinc and silica have been used to ameliorate salinity stress in mungbean, and like other agents, they both boost up the antioxidant activity leading to better phenotype than the untreated controls (Hussain et al. 2019; Ahmad et al. 2019; Ashraf et al. 2019). Very recently, a low dosage acetic acid has been found to be a cost-effective way to minimize toxicity in mungbean due to sea water (Rahman et al. 2019). The treated plants showed better morpho-physiological aspects and photosynthetic efficiency under the influence of sea water with respect to untreated controls.

Trait improvement for salinity tolerance in mungbean through genetix engineering approach has been reported by a single laboratory till date. An *Arabidopsis* NHX1, a Na^+/H^+ exchanger protein localized in the vacuolar membrane was constitutively overexpressed in mungbean (Sahoo et al. 2016). NHX1 is known to compartmentalize Na^+ in vacuoles to protect the cytoplasmic macromolecules and organelles from the toxic effects. The transgenic mungbean plants showed better adaptability to saline condition as depicted from the soil salinity stress assays depicted in the manuscript (Sahoo et al. 2016).

4.3 Salinity Stress Management in Blackgram

Genetic engineering has long been used as an alternative tool for crop improvement, and blackgram has not been an exception to this. Blackgram is recalcitrant to genetic transformation but despite this bottleneck attempts have been made to engineer candidate genes into blackgram for trait improvement (Saini et al. 2003; Chopra and Saini 2014; Das 2018). The blackgram recalcitrance to generate transgenic calli was reported by Karthikeyan et al. (1996). Seven years after, the recovery of transgenic blackgram was reported by Saini et al. (2003) and Balaji et al. (2003). Constitutive overexpression of Glyoxalase I gene in blackgram resulted in plants with enhanced tolerance to salt stress (Bhomkar et al. 2008). Glyoxalase I belongs to metalloglutathione transferase superfamily, thus implicated in a decisive part in the detoxification of cytotoxic methylglyoxal. Perhaps this was the first report for engineering of abiotic stress tolerance in blackgram. Latter from the same group, another report for the improvement of multiple abiotic stress tolerance in blackgram through genetic engineering was published (Singh et al. 2016). In this study, the overexpression of ALDRXV4 was the approach defined. This enzyme belongs to the aldo–keto reductase superfamily which catalyse the reduction of carbonyl metabolites in the cells thus consititute a key player for osmoprotection and remediation of the reactive carbonyl species. Meantime, a considerable number of genomic and transcriptomic resources have been generated, which in the future might contribute to the development of improved varieties to abiotic stresses including salinity. Using high throughput Illumina sequencing technology, RNA libraries were constructed from leaves of blackgram paving the way for identification 13 novel miRNA (Paul et al. 2014). Blackgram genetic diversity analysis and linkage maps were also generated using AFLP, RAPD and ISSR markers (Souframanien and Gopalakrishna 2004; Chaitieng et al. 2006; Gupta et al. 2008; Gupta and Gopalakrishna 2009) and may constitute a relevant tool for marker-assisted breeding approaches in this species.

Besides genomic and genetic approaches, seed and seedling treatments including the well-described seed priming have also been used as an excellent tool to enhance the tolerance/avoidance capacity of crop plants against abiotic stress like drought, salinity, cold, heavy metal, etc. (Saha et al. 2019; Sahoo et al. 2019). For example, the effectivity of seed priming as an alternative mean to ameliorate against other abiotic stresses like arsenic stress in rice has been widely studied using selenium and zinc seed treatments (Moulick et al. 2016, 2017, 2018a, b, c, 2019). Diverse group of agents has been used to ameliorate stress in *Vigna* species including polyamines and hormones, but being a neglected crop, very few studies have been done on blackgram. A triazol compound, triadimefon, was one of the first agents to be experimented with as co-treatment along with salinity (Jaleel et al. 2009). Triadimefon drastically alleviated salinity in blackgram by triggering enhanced antioxidant enzyme activity. Devi et al. (2012) demonstrated the use of IAA as an ameliorating agent to salt stress. In this work, stressed blackgram plants treated with IAA showed better growth reflected on shoot–root length, biomass, leaf area and chlorophyll content. Very recently, the impact of halotolerant plant growth-promoting rhizobacteria (H-PGPR) and kinetin was studied on salt tolerance of blackgram (Yasin et al. 2018). In this work, 15 rhizobacterial isolates were screened for their potential to enhance growth in blackgram. Two of the selected strains were then applied in combination with kinetin doses, and plant performance was evaluated in field condition (brackish water) for improving tolerance to blackram. The plants treated with kinetin and H-PGPR exhibited enhanced chlorophyll content, growth, yield and reduced electrolyte leakage.

5 Conclusion

Keeping in mind the economic importance of *Fabaceae* family, its susceptibility to salinity stress and impact on productivity and nutritional quality, we are still far from having a complete picture of the physiological and molecular mechanisms governing legume response to salinity. More research efforts in terms of funding and research projects should be intensified. Moreover, reluctances of some crops to genomic alternation, lack of standardization in research experiments and the existence of a large number of orphan crops perhaps hampered the generation of knowledge. Indeed, it would be relevant to extend research from the questions of particular region to global level, in which large collaborative research consortium should be established to maximize the efforts to improve legume performance and yield in the current scenario of climate change. Large-scale omic studies will go a long way in this regard. Besides adopting modern methodologies to mitigate salinity stress in the abovementioned field crops, a parallel attention should also have directed towards developing farmer-friendly mitigation options too. Speaking in terms of research, basic research must be encouraging at any cost but at the same time strategic research also encouraged considerably or else the gap between the basic and applied (i.e. application of findings) remains as it was.

Acknowledgements DM wish to thank Department of Biotechnology (DBT), Government of India for providing financial support (DBT—RA Award; Dated 29/5/2018; File No. AUS/CIL/DBT–RA/DM 2019). Furthermore, DM wants to also acknowledge Dr. Shuvasish Choudhury and Assam University for mentorship and providing laboratory facility. BS is grateful to Science and Engineering Research Board (SERB), India, for National Post-Doctoral Fellowship (NPDF, PDF/2018/003216).

References

Ahmad M, Zahir ZA, Asghar HN, Asghar M (2011) Inducing salt tolerance in mung bean through coinoculation with rhizobia and plant-growth-promoting rhizobacteria containing 1-aminocyclopropane-1-carboxylate deaminase. Canad J Microbiol 57(7):578–589

Ahmad M, Zahir ZA, Asghar HN, Arshad M (2012) The combined application of rhizobial strains and plant growth promoting rhizobacteria improves growth and productivity of mung bean (*Vigna radiata* L.) under salt-stressed conditions. Ann Microbiol 62(3):1321–1330

Ahmad M, Zahir ZA, Khalid M, Nazli F, Arshad M (2013) Efficacy of rhizobium and pseudomonas strains to improve physiology, ionic balance and quality of mung bean under salt-affected conditions on farmer's fields. Plant Physiol Biochem 63:170–176

Ahmad P, Ahanger MA, Alam P, Alyemeni MN, Wijaya L, Ali S, Ashraf M (2019) Silicon (Si) supplementation alleviates NaCl toxicity in mung bean (*Vigna radiata* (L.) Wilczek) through the modifications of physio-biochemical attributes and key antioxidant enzymes. J Plant Growth Regul 38(1):70–82

Akhtar J, Ahmad R, Ashraf MY, Tanveer A, Waraich EA, Oraby H (2013) Influence of exogenous application of salicylic acid on salt-stressed mungbean (*Vigna radiata*): growth and nitrogen metabolism. Pak J Bot 45(1):119–125

Akram MS, Ashraf M, Akram NA (2009) Effectiveness of potassium sulfate in mitigating salt-induced adverse effects on different physio-biochemical attributes in sunflower (*Helianthus annuus* L.). Flora-Morpho Distr Func Ecol Plants 204(6):471–483

Alam S (1994) Nutrient uptake by plants under stress condition. In: Pessarakli M (ed) Handbook of plant and crop stress. Marcel Dekker, New York, pp 227–243

Amin M (2011) Adaptation of Suitable Crops in Saline Soils of Noakhali District. Technical Bull 02: 5 pp

Ashraf M (1989) The effect of NaCl on water relations, chlorophyll, and protein and proline contents of two cultivars of blackgram (*Vigna mungo* L.). Plant Soil 119(2):205–210

Ashraf M, Karim F (1990) Differences between blackgram (*Vigna mungo* L.) cultivars in NaCl resistance. Zeitschrift für Pflanzenernährung und Bodenkunde 153(1):15–20

Ashraf MY, Tariq S, Saleem M, Khan MA, Hassan SWU, Sadef Y (2019) Calcium and zinc mediated growth and physio-biochemical changes in mungbean grown under saline conditions. J Plant Nutr: 1–14

Asif MA, Zafar Y, Iqbal J, Iqbal MM, Rashid U, Ali GM, Arif A, Nazir F (2011) Enhanced expression of AtNHX1, in transgenic groundnut (*Arachis hypogaea* L.) improves salt and drought tolerence. Mol Biotechnol 49(3):250–256

Azad MAK, Alam MS, Hamid MA (2013) Modification of salt tolerance level in groundnut (*Arachis hypogaea* L.) through induced mutation. Leg Res 36:224–233

Azad MAK, Shah-E-Alam M, Hamid MA, Rafii MY, Malek MA (2014) Combining ability of pod yield and related traits of groundnut (*Arachis hypogaea* L.) under salinity stress. Sci World J

Balaji V, Rajamuni P, Sridevi G, Veluthambi K (2003) *Agrobacterium*-mediated transformation efficiency in blackgram and rice enhanced by multiple copies of pTiBo542 virB and virG

Banavath JN, Chakradhar T, Pandit V, Konduru S, Guduru KK, Akila CS, Podha S, Puli COR (2018) Stress inducible overexpression of AtHDG11 leads to improved drought and salt stress tolerance in peanut (*Arachis hypogaea* L.). Front Chem 6(34):1–22. https://doi.org/10.3389/fchem.2018.00034

Bhomkar P, Upadhyay CP, Saxena M, Muthusamy A, Prakash NS, Pooggin M, Sarin NB (2008) Salt stress alleviation in transgenic *Vigna mungo* L. Hepper (blackgram) by overexpression of the glyoxalase I gene using a novel Cestrum yellow leaf curling virus (CmYLCV) promoter. Mol Breed 22(2):169–181

Birthal PS, Rao PP, Nigam SN, Bantilan CS, Bhagavatulu S (2010) Groundnut and soybean economies in Asia: facts, trends and outlook. International Crops Research Institute for the Semi-Arid Tropics, Patancheru

Bothe H (2012) Arbuscular mycorrhiza and salt tolerance of plants. Symbiosis 58:7–16

Chaitieng B, Kaga A, Tomooka N, Isemura T, Kuroda Y, Vaughan DA (2006) Development of a black gram (*Vigna mungo* (L.) Hepper) linkage map and its comparison with an azuki bean (*Vigna angularis* (Willd.) Ohwi and Ohashi) linkage map. Theor Appl Genet 113(7):1261–1269

Chakraborty K, Bhaduria D, Meena HN, Kalariya K (2016) External potassium (K^+) application improves salinity tolerance by promoting Na^+-exclusion, K^+-accumulation and osmotic adjustment in contrasting peanut cultivars. Plant Physiol Biochem. https://doi.org/10.1016/j.plaphy.2016.02.039

Chen C, Tao C, Peng H, Ding Y (2007) Genetic analysis of salt stress responses in asparagus bean (*Vigna unguiculata* L. ssp. Sesquipedalis verdc.). J Hered 98(7):655–665

Chinnusamy V, Jagendorf A, Zhu JK (2005) Understanding and improving salt tolerance in plants. Crop Sci 45:437–448

Chopra R, Saini R (2014) Transformation of blackgram (*Vigna mungo* (L.) Hepper) by barley chitinase and ribosome-inactivating protein genes towards improving resistance to Corynespora leaf spot fungal disease. Appl Biochem Biotechnol 174(8):2791–2800

Chowra U, Yanase E, Koyama H, Panda SK (2017) Aluminium-induced excessive ROS causes cellular damage and metabolic shifts in black gram *Vigna mungo* (L.) Hepper. Protoplasma 254(1):293–302

Cuartero J, Bolarin MC, Asins MJ, Moreno V (2006) Increasing salt tolerance in the tomato. J Exp Bot 57(5):1045–1058

Cuevas J, Daliakopoulos IN, Moral FD, Hueso JJ, Tsanis IK (2019) A review of soil-improving cropping systems for soil salinization. Agronomy 9:295–316. https://doi.org/10.3390/agronomy9060295

Das DK (2018) Expression of a Bacterial Chitinase (ChiB) gene enhances resistance against *Erysiphae polygoni* induced powdery mildew disease in the transgenic black gram (*Vigna mungo* L.) (cv. T9). Amer J Plant Sci 9(08):1759

Dash M, Panda SK (2001) Salt stress induced changes in growth and enzyme activities in germinating Phaseolus mungo seeds. Biol Plant 44(4):587–589

Devi RG, Pandiyarajan V, Gurusaravanan P (2012) Alleviating effect of IAA on salt stressed *Phaseolus mungo* (L.) with reference to growth and biochemical characteristics. Rec Res Sci Technol 4(3)

Dogar UF, Naila N, Maira A, Iqra A, Maryam I, Khalid H, Khalid N, Ejaz HS, Khizar HB (2012) Noxious effects of NaCl salinity on plants. Bot Res Int 5(1):20–23

Dolatabadian A, Sanavy SAMM, Ghanati F (2011) Effect of salinity on growth, xylem structure and anatomical characteristics of soybean. Not Sci Biol 3(1):41–45

FAO (2017) FAOSTAT agricultural data. Agricultural production, crops, primary. http://www.fao.org/faostat/en/#data/QC/visualize. Accessed on 25 Nov 2019

Gama PBS, Inanaga S, Tanaka K, Nakazawa R (2007) Physiological response of common bean (*Phaseolus vulgaris* L.) seedlings to salinity stress. Afr J Biotechnol 6(2):79–88

Garratt LC, Janagoudar BS, Lowe KC, Anthony P, Power JB, Davey MR (2002) Salinity tolerance and antioxidant status in cotton cultures. Free Radical Biol Med 33:502–511

Ghassemi-Golezani K, Lotfi R (2015) The impact of salicylic acid and silicon on chlorophyll a fluorescence in mung bean under salt stress. Russian J Plant Physiol 62(5):611–616

Ghosh S, Mitra S, Paul A (2015). Physiochemical studies of sodium chloride on mungbean (*Vigna radiata* L. Wilczek) and its possible recovery with spermine and gibberellic acid. The Sci World J

GOI (2018) Directorate of economics and statistics, Government of India. All India *Kharif* 2018. https://apeda.gov.in/apedawebsite/HACCP/2018_Groundnut_Survey_Report.pdf. Accessed on 17 Nov 2019

Golpayegani A, Tilebeni H (2011) Effect of biological fertilizers on biochemical and physiological parameters of basil (*Ociumum basilicm* L.) medicine plant. Amer Eur J Agric Environ Sci 11:411–416

Grattan SR, Grieve CM (1999) Salinity-mineral nutrient relations in horticultural crops. Sci Hortic 78:127–157

Gupta SK, Gopalakrishna T (2009) Genetic diversity analysis in blackgram (*Vigna mungo* (L.) Hepper) using AFLP and transferable microsatellite markers from azuki bean (*Vigna angularis* (Willd.) Ohwi and Ohashi). Genome 52(2):120–129

Gupta SK, Souframanien J, Gopalakrishna T (2008) Construction of a genetic linkage map of black gram, *Vigna mungo* (L.) Hepper, based on molecular markers and comparative studies. Genome 51(8):628–637

Ha E, Ikhajiagba B, Bamidele JF, Ogic-odia E (2008) Salinity effects on young healthy seedling of kyllingia peruviana collected from escravos, delta state. Global J Environ Res 2(2):74–88

Hammad SAR, Shaban KA, Tantawy MF (2010) Studies on salinity tolerance of two peanut cultivars in relation to growth, leaf water content. Some chemical aspects and yield. J Appl Sci Res 6(10):1517–1526

Hasanuzzaman M, Hossain MA, Silva JAT, Fujita M (2012) Plant responses and tolerance to abiotic oxidative stress: antioxidant defenses is a key factors. In: Bandi V, Shanker AK, Shanker C, Mandapaka M (eds) Crop stress and its management: perspectives and strategies. Springer, Berlin, pp 261–316

Hasanuzzaman M, Nahar K, Fujita M (2013a) Plant response to salt stress and role of exogenous protectants to mitigate salt-induced damages. In: Ahmad P, Azooz MM, Prasad MNV (eds) Ecophysiology and responses of plants under salt stress. Springer, New York, pp 25–87

Hasanuzzama M, Nahar K, Fujita M, Ahmad P, Chandna R, Prasad MNV, Ozturk M (2013b) Enhancing plant productivity under salt stress—Relevance of poly-omics. In: Ahmad P, Azooz MM, Prasad MNV (eds) Salt stress in plants: omics, signaling and responses. Springer, Berlin. pp 113–156

Hasanuzzaman M, Nahar K, Rohman MM, Anee TI, Huang Y, Fujita M (2018a) Exogenous silicon protects *Brassica napus* plants from salinity-induced oxidative stress through the modulation of AsA-GSH pathway, thiol-dependent antioxidant enzymes and glyoxalase systems. Gesunde Pflanzen 70:185–194

Hasanuzzaman M, Oku H, Nahar K, Bhuyan MHMB, Mahmud JA, Baluska F, Fujita M (2018b) Nitric oxide-induced salt stress tolerance in plants: ROS metabolism, signaling, and molecular interactions. Plant Biotechnol Rep. https://doi.org/10.1007/s11816-018-0480-0

Hasanuzzaman M, Bhuyan MHMB, Anee TI, Parvin K, Nahar K, Mahmud JA, Fujita M (2019) Regulation of ascorbate-glutathione pathway in mitigating oxidative damage in plants under abiotic stress. Antioxidants 8(9):384. https://doi.org/10.3390/antiox8090384

Hirpara KD, Prakash JR, Ashish DP, Amar NP (2005) Effect of salinisation of soil on growth and macro- and micro-nutrient accumulation in seedlings of Butea monosperma (Fabaceae). Anales de Biologia 27:3–14

Hossain MA, Fujita M (2010) Evidence for a role of exogenous glycinebetaine and proline in antioxidant defense and methylglyoxal detoxification systems in mung bean seedlings under salt stress. Physiol Mol Biol Plants 16(1):19–29

Hossain MA, Hasanuzzaman M, Fujita M (2011) Coordinate induction of antioxidant defense and glyoxalase system by exogenous proline and glycinebetaine is correlated with salt tolerance in mung bean. Front Agric China 5(1):1–14

Hu Y, Schmidhalter U (1997) Interactive effects of salinity and macronutrient level on wheat. 2. Composition. J Plant Nutr 20:1169–1182

Hu Y, Schmidhalter U (2005) Drought and salinity: a comparison of their effects on mineral nutrition of plants. J Plant Nutr Soil Sci 168:541–549

Hussain SJ, Masood A, Anjum NA, Khan NA (2019) Sulfur-mediated control of salinity impact on photosynthesis and growth in mungbean cultivars screened for salt tolerance involves glutathione and proline metabolism, and glucose sensitivity. Acta Physiolog Planta 41(8):129

Ikeda M (2005) Distribution of K, Na and Cl in root and leaf cells of soybean and cucumber plants grown under salinity conditions. Soil Sci Plant Nutri 51(7):1053–1057

Imadi SR, Shah SW, Kazi AG, Azooz MM, Ahmad P (2016) Phytoremediation of saline soils for sustainable agricultural productivity plant metal interaction emerging remediation techniques, pp 455–468. https://doi.org/10.1016/B978-0-12-803158-2.00018-7

Imrie BC (1995) Register of Australian grain legume cultivars. *Vigna radiata* (L.) Wilczek. (mungbean) cv. Emerald. Aust J Exp Agric 35(5): 677–677

Iqbal MM, Nazir F, Ali S, Asif MA, Zafar Y, Iqbal J, Ali GM (2012) Over expression of rice chitinase gene in transgenic peanut (*Arachis hypogaea* L.) improves resistance against leaf spot. Mol Biotechnol 50(2):129–136

Jaleel CA, Iqbal M, Panneerselvam R (2009) Triadimefon Protects Blackgram (*Vigna mungo* L. Hepper) Plants from Sodium Chloride Stress. Plant Stress 3(1):13–16

Jambunathan R (1991) Groundnut quality characteristics. In: Uses of tropical grain legumes: proceedings of a consultants meeting. ICRISAT, Patancheru, pp 267–275, 27–30 Mar 1989

Janila P, Rao TN, Kumar AA (1999) Germination and early seedling growth of groundnut (*Arachis hypogaea* L.) varieties under salt stress. Ann Agril Res 20:180–182

Janila P, Nigam SN, Pandey MK, Nagesh P, Varshney RK (2013) Groundnut improvement: use of genetic and genomic tools. Front Plant Sci 4:1–16. https://doi.org/10.3389/fpls.2013.00023

Kandil AA, Arafa AA, Sharief AE, Ramadan AN (2012) Genotypic differences between two mungbean varieties in response to salt stress at seedling stage. Int J Agric Sci 4(7): 278–283. ISSN: 0975-3710 and E-ISSN: 0975-9107

Karim Z, Iqbal A (2001) Impact of land degradation in Bangladesh: changing scenario in agricultural land use. Bangladesh Agricultural Research Council, Dhaka, Bangladesh

Karthikeyan A, Valarmathi R, Nandini S, Nandhakumar MR (2012) Genetically modified crops: insect resistance. Biotechnology 11:119–126

Kaur S, Nayyar H (2015) Selenium fertilization to salt-stressed mungbean (*Vigna radiata* L. Wilczek) plants reduces sodium uptake, improves reproductive function, pod set and seed yield. Sci Hortic 197:304–317

Khan MIR, Asgher M, Khan NA (2014) Alleviation of salt-induced photosynthesis and growth inhibition by salicylic acid involves glycinebetaine and ethylene in mungbean (*Vigna radiata* L.). Plant Physiol Biochem 80:67–74

Kotuby-Amacher J, Koenig K, Kitchen B (2000) Salinity and plant tolerance. Available at https://extension.usu.edu/files/publications/publication/AG-SO-03.pdf

Krishna G, Singh BK, Kim EK, Morya VK, Ramteke PR (2015) Progress in genetic engineering of peanut (*Arachis hypogaea* L.)—a review. Plant Biotechnol J 13:147–162. https://doi.org/10.1111/pbi.12339

Kumar V, Pandey SK (2018) Current status of mungbean in Madhya Pradesh—a review. Int J Curr Microbiol App Sci 7(11):1062–1072

Lalotra S, Hemantaranjan A, Kumar S, Kant R (2017) Effect of brassinosteroid (brassinolide) on seedling traits, morphology and metabolism in mung bean under salinity stress. Annu Res Rev Biol 12:1–8

Lauchli A, James RA, Huang CX, McCully M, Munns R (2008) Cell-specific localization of Na^+ in roots of durum wheat and possible control points for salt exclusion. Plant Cell Environ 31(11):1565–1574

Maggio A, Hasegawa PM, Bressan RA, Consiglio MF, Joly RJ (2001) Unravelling the functional relationship between root anatomy and stress tolerance. Func Plant Biol 28(10):999–1004

Mahajan S, Tuteja N (2005) Cold, salinity and drought stresses: an overview. Arch Biochem Biophys 444:139–158

Mahdavi B, Sanavy SA (2007) Germination and seedling growth in grass pea (*Lathyrus sativus*) cultivars under salinity conditions. Pak J Biol Sci 10:273–279

Mahmud JA, Hasanuzzaman M, Khan MIR, Nahar K, Fujita M (2020) β-aminobutyric acid pretreatment confers salt stress tolerance in *Brassica napus* L. by modulating reactive oxygen species metabolism and methylglyoxal detoxification. Plants 9(2): 241. https://doi.org/10.3390/plants9020241

Memon SA, Hou X, Wang LJ (2010) Morphological analysis of salt stress response of pak Choi. EJEAFChe 9(1):248–254

Mensah JK, Akomeah PA, Ikhajiagbe B, Ekpekurede EO (2006) Effects of salinity on germination, growth and yield of five groundnut genotypes. Afr J Biotechnol 5(20):1973–1979

Mohan D, Shashidharan A (2018) An in vitro system for studying the effect of salt stress in groundnut (*Arachis hypogaea* L.). Current Biotechnol 7:464–471

Moose SP, Mumm RH (2008) Molecular plant breeding as the foundation for 21st century crop improvement. Plant Physiol 147:969–977

Moreno-Casasola P (2008) Dunes. Encyclopedia of ecology, pp 971–976. https://doi.org/10.1016/B978-008045405-4.00328-1

Moulick D, Ghosh D, Santra SC (2016) Evaluation of effectiveness of seed priming with selenium in rice during germination under arsenic stress. Plant Physiol Biochem 109:571–578

Moulick D, Santra SC, Ghosh D (2017) Seed priming with se alleviate as induced phytotoxicity during germination and seedling growth by restricting as translocation in rice (*Oryza sativa* L. cv IET-4094). Ecotoxicol Environ Saf 145:449–456

Moulick D, Santra SC, Ghosh D (2018a) Effect of selenium induced seed priming on arsenic accumulation in rice plant and subsequent transmission in human food chain. Ecotoxicol Environ Saf 152:67–77

Moulick D, Santra SC, Ghosh D (2018b) Rice seed priming with se: a novel approach to mitigate as induced adverse consequences on growth, yield and as load in brown rice. J Hazard Mater 355:187–196

Moulick D, Santra SC, Ghosh D (2018c) Seed priming with se mitigates as-induced phytotoxicity in rice seedlings by enhancing essential micronutrient uptake and translocation and reducing as translocation. Environ Sci Pollut Res. https://doi.org/10.1007/s11356-018-2711-x

Moulick D, Santra SC, Ghosh D, Panda SK. (2019) An assessment of efficiency of zinc priming in rice (cv. MTU-7029) during germination and early seedling growth. In: Priming and pretreatment of seeds and seedlings. Springer, Singapore, pp 495–507

Mudgal V (2004) Physiological studies on growth and nitrogen metabolism in *Cicer arietinum* L. under saline conditions. In: Ph.D. thesis. Rohilkhand University, India

Muhammad A, Majid RF (2013) Crop breeding for salt tolerance in the era of molecular markers and marker-assisted selection. Plant Breeding 132:10–20

Mungala AJ, Radhakrishnan T, Junagadh JRD (2008) In vitro screening of 123 Indian peanut cultivars for sodium chloride induced salinity tolerance. World J Agri Sci 4(5):574–582

Munns R (2002) Comparative physiology of salt and water stress. Plant Cell Environ 25(2):239–250

Munns R, James RA (2003) Screening methods for salinity tolerance: a case study with tetraploid wheat. Plant Soil 253:201–218

Munns R, Tester M (2008) Mechanisms of salinity tolerance. Annual Rev Plant Biol 59:651–681

Nafees A, Shabina S, Asim M, Rahat N, Noushina I (2010) Application of salicylic acid increases contents of nutrients and antioxidative metabolism in mungbean and alleviates adverse effects of salinity stress. Int J Plant Biol 1:e.1

Nahar K, Hasanuzzaman M, Alam MM, Fujita M (2015) Roles of exogenous glutathione in antioxidant defense system and methylglyoxal detoxification during salt stress in mung bean. Biologia Planta 59(4):745–756

Nahar K, Hasanuzzaman M, Rahman A, Alam M, Mahmud JA, Suzuki T, Fujita M (2016) Polyamines confer salt tolerance in mung bean (*Vigna radiata* L.) by reducing sodium uptake, improving nutrient homeostasis, antioxidant defense, and methylglyoxal detoxification systems. Front Plant Sci 7:1104

Natr L, Lawlor DW (2005) Photosynthetic plant productivity, 2nd ed. In: Pessarakli M (ed) Hand book photosynthesis. CRC Press, New York, USA, pp 501–524

Nazar R, Iqbal N, Syeed S, Khan NA (2011) Salicylic acid alleviates decreases in photosynthesis under salt stress by enhancing nitrogen and sulfur assimilation and antioxidant metabolism differentially in two mungbean cultivars. J Plant Physiol 168(8):807–815

Negrao S, Schmockel SM, Tester M (2017) Evaluating physiological responses of plants to salinity stress. Ann Bot 119(1):1–11

Nikpour-Rashidabad N, Tavasolee A, Torabian S, Farhangi-Abriz S (2019) The effect of biochar on the physiological, morphological and anatomical characteristics of mung bean roots after exposure to salt stress. Arch Biol Sci 71(2):321–327

Omisun T, Sahoo S, Saha B, Panda SK (2018) Relative salinity tolerance of rice cultivars native to North East India: a physiological, biochemical and molecular perspective. Protoplasma 255(1):193–202. https://doi.org/10.1007/s00709-017-1142-8

Pal A, Pal AK (2017) Physiological Basis of Salt Tolerance in Groundnut (*Arachis hypogaea* L.). Int J Curr Microbiol App Sci 6(9):2157–2171

Panwar M, Tewari R, Nayyar H (2016) Native halo-tolerant plant growth promoting rhizobacteria Enterococcus and Pantoea sp. improve seed yield of Mungbean (*Vigna radiata* L.) under soil salinity by reducing sodium uptake and stress injury. Physiol Mol Biol Plants 22(4):445–459

Parihar P, Singh S, Singh R, Singh VP, Prasad SM (2015) Effect of salinity stress on plants and its tolerance strategies: a review. Environ Sci Pollut Control Ser 22(6):4056–4075

Parvin K, Hasanuzzaman M, Bhuyan MHMB, Mohsin SM, Fujita M (2019) Quercetin mediated salt tolerance in tomato through the enhancement of plant antioxidant defense and glyoxalase systems. Plants 8(8):247. https://doi.org/10.3390/plants8080247

Pasupuleti J, Nigan SN (2013) Phenotyping for groundnut (*Arachis hyogea* L.) Improvement. In: Pangulur SK, Kumar AA (eds) Phenotyping for Plant Breeding. Springer, Berlin

Paul S, Kundu A, Pal A (2014) Identification and expression profiling of *Vigna mungo* microRNAs from leaf small RNA transcriptome by deep sequencing. J Int Plant Biol 56(1):15–23

Perveen S, Farooq R, Shahbaz M (2016) Thiourea-induced metabolic changes in two mung bean [*Vigna radiata* (L.) Wilczek] (Fabaceae) varieties under salt stress. Braz J Bot 39(1):41–54

Rahman MM, Mostofa MG, Rahman MA, Islam MR, Keya SS, Das AK, Tabassum B (2019) Acetic acid: a cost-effective agent for mitigation of seawater-induced salt toxicity in mung bean. Sci Rep 9(1):1–15

Reddy TY, Anbumozhi VRV (2003) Physiological responses of groundnut (*Arachis hypogea* L.) to drought stress and its amelioration. A critical review. Plant Growth Regul 41:75–88. https://doi.org/10.1023/A:1027353430164

Reddy LJ, Kaul AK (1996) Status and prospects of groundnut in Bangladesh. Bangladesh Agricultural Research Council, Dhaka, Bangladesh

Rogers ME, Grieve CM, Shannon MC (2003) Plant growth and ion relations in Lucerne (*Medicago sativa* L.) in response to the combined effects of NaCl and P. Plant Soil 253:187–194

Rozema J, Flowers T (2008) Ecology. Crops for a salinized world. Science 322:1478–1480

Rui L, Wei S, Mu-xiang C, Cheng-jun J, Min W, Bo-ping Y (2009) Leaf anatomical changes of *Burguiera gymnorrhiza* seedlings under salt stress. J Trop Subtrop Bot 17(2):169–175

Saha P, Chatterjee P, Biswas AK (2010) NaCl pretreatment alleviates salt stress by enhancement of antioxidant defense system and osmolyte accumulation in mungbean (*Vigna radiata* L. Wilczek). Indian J Exp Biol 48 (6):593–600

Saha P, Mukherjee A, Biswas AK (2015) Modulation of NaCl induced DNA damage and oxidative stress in mungbean by pretreatment with sublethal dose. Biologia Planta 59(1):139–146

Saha B, Chowardhara B, Kar S, Devi SS, Awasthi JP, Moulick D, Tanti B, Panda SK (2019) Advances in heavy metal-induced stress alleviation with respect to exogenous amendments in crop plants. In: Priming and pretreatment of seeds and seedlings. Springer, Singapore, pp 313–332

Sahoo S, Borgohain P, Saha B, Moulick D, Tanti, B, Panda SK (2019) Seed priming and seedling pre-treatment induced tolerance to drought and salt stress: recent advances. In: Priming and pretreatment of seeds and seedlings. Springer, Singapore, pp 253–263

Sahoo DP, Kumar S, Mishra S, Kobayashi Y, Panda SK, Sahoo L (2016) Enhanced salinity tolerance in transgenic mungbean overexpressing *Arabidopsis antiporter* (NHX1) gene. Mol Breed 36(10):144

Saini R, Jaiwal S, Jaiwal PK (2003) Stable genetic transformation of *Vigna mungo* L. Hepper via *Agrobacterium tumefaciens*. Plant Cell Reports 21(9):851–859

Sehrawat N, Bhat KV, Sairam RK, Toomoka N, Kaga A, Shu Y, Jaiwal PK (2013) Diversity analysis and confirmation of intra-specific hybrids for salt tolerance in mungbean (*Vigna radiata* L. Wilczek). Int J Integ Biol 14:65–73

Sehrawat N, Bhat KV, Sairam RK, Jaiwal PK (2014a) Screening of mungbean (*Vigna radiata* (L.) Wilczek) genotypes for salt tolerance. Int J Plan Environ Sci 4:36–43

Sehrawat N, Yadav M, Bhat KV, Sairam RK, Jaiwal PK (2014b) Evaluation of mungbean genotypes for salt tolerance at early seedling growth stage. Biocatal Agric Biotechnol 3:108–113

Sehrawat N, Yadav M, Bhat KV, Sairam RK, Jaiwal PK (2015) Effect of salinity stress on mungbean (*Vigna radiata* (L.) Wilczek) during consecutive summer and spring seasons. J Agri Sci Belgrade 60(1):23–32

Serrano R, Culianz-Macia F, Moreno V (1999) Genetic engineering of salt and drought tolerance with yeast regulatory genes. Sci Hortic 78:261–269

Shakeel S, Mansoor S (2012) Salicylic acid prevents the damaging action of salt in mung bean [(*Vigna radiata* L.) Wilczek] seedlings. Pak J Bot 44(2):559–562

Sharma S, Kulkarni J, Jha B (2016) Halotolerant rhizobacteria promote growth and enhance salinity tolerance in peanut. Front Microbiol 7(1600):1–11

Shrivastava P, Kumar R (2015) Soil salinity: a serious environmental issue and plant growth promoting bacteria as one of the tolls for its alleviation. Saudi J Biol Sci 22:123–131

Singh B (1992) Aspects of groundnut utilization and possible improvements of Indiginous food in some countries of semi-arid tropical Africa. In: Nigam SN (ed) Groundnut a global prespective. International Crop Research Institute for the Semi-arid Tropics, Patancheru, Andhra Pradesh, India, pp 257–263

Singh G (2018) Climate change and sustainable management of salinity in agriculture. Res Med Eng Sci 6(2):608–614. https://doi.org/10.31031/RMES.2018.06.000635

Singh F, Diwakar B (1993) Nutritive value and uses of Pigeonpea and groundnut. Manual. International Crops Research Institute for the Semi-Arid Tropics, Patancheru

Singh BG, Rao PV, Hiremath SM (1989) Effect of salinity on germination and seedling growth in three varieties of groundnut (*Arachis hypogaea* L.) J Res APAU 17: 432–433

Singh N, Mishra A, Jha B (2014) Ectopic over-expression of peroxisomal ascorbate peroxidase (SbpPAX) gene confers salt stress tolerance in transgenic peanut (*Arachis hypogaea*). Gene 547(1):119–125

Singh RP, Jha P, Jha PN (2015) The plant-growth-promoting bacterium Klebsiella sp. SBP-8 confers induced systemic tolerance in wheat (*Triticum aestivum*) under salt stress. J Plant Physiol 184:57–67

Singh P, Kumar D, Sarin NB (2016) Multiple abiotic stress tolerance in *Vigna mungo* is altered by overexpression of ALDRXV4 gene via reactive carbonyl detoxification. Plant Mol Biol 91(3):257–273

Souframanien J, Gopalakrishna T (2004) A comparative analysis of genetic diversity in blackgram genotypes using RAPD and ISSR markers. Theor Appl Genet 109(8):1687–1693

SRDI (2003) Soil salinity in Bangladesh 2000. Soil Resource Development Institute, Ministry of Agriculture, Dhaka, Bangladesh

Srinives P (1991) Mungbean breeding and genetic resources in Thailand. In: Proceedings of the mungbean meeting 90. Department of Agriculture

Taffouo VD, Wamba OF, Yombi E, Nono GV, Akoe A (2010) Growth, yield, water status and ionic distribution response of three Bambara groundnut (*Vigna subterranean* (L.) verdc.) landraces grown under saline conditions. Int J Bot 6(1):53–58

Taufiq A, Kristiono A, dan Harnowo D (2015) Respon varietas unggul kacang tanah terhadap cekaman salinitas. Jurnal Penelitian Pertanian Tanaman Pangan 34(2):153–164

Taufiq A, Wijanarko A, Kristiono A (2016) Effect of amelioration on growth and yield of two groundnut varieties on saline soil. J Degrad Min Lands Manage 3(4):639–647

Tavakkoli E, Rengasamy P, McDonald GK (2010) High concentrations of Na+ and Cl—ions in soil solution have simultaneous detrimental effects on growth of faba bean under salinity stress. J Exp Bot 61:4449–4459

Tittabutr P, Piromyou P, Longtonglang A, Noisa-Ngiam R, Boonkerd N, Teaumroong N (2013) Alleviation of the effect of environmental stresses using co-inoculation of mungbean by Bradyrhizobium and rhizobacteria containing stress-induced ACC deaminase enzyme. Soil Sci Plant Nutr 59(4):559–571

TNAU Agritech Portal (2020) Tamil Nadu Agricultural University. http://agritech.tnau.ac.in/agriculture/agri_salinity_management.html. Accessed on 23 Jan 2020

Torabian S, Farhangi-Abriz S, Rathjen J (2018) Biochar and lignite affect H^+-ATPase and H^+-PPase activities in root tonoplast and nutrient contents of mung bean under salt stress. Plant Physiol Biochem 129:141–149

Vijayalakshmi P, Amirthaveni S, Devadas RP, Weinberger K, Tsou SCS, Shanmugasundaram S (2003) Enhanced bioavailability of iron from mungbeans and its effects on health of schoolchildren. AVRDC-World Vegetable Center

Watt EE, Poehlman JM, Cumbie BG (1977) Origin and composition of a texture layer on seeds of mungbean 1. Crop Sci 17(1):121–125

White PJ, Broadley MR (2001) Chloride in soils and its uptake and movement within the plant: a review. Ann Botany 88:967–988

Win KT, Aung ZO, Hirasawa T, Ookawa T, Yutaka H (2011) Genetic analysis of Myanmar *Vigna* species in responses to salt stress at the seedling stage. Afr J Biotechnol 10(9):1615–1624

Yadav JPS (1993) Salt affected soils and their management with special reference to Uttar Pradesh. J Indian Soc Soil Sci 41:623–629

Yadav S, Mohammad I, Aqi A, Shamsul H (2011) Causes of salinity and plant manifestations to salt stress: a review. J Environ Biol 32:667–685

Yasin NA, Khan WU, Ahmad SR, Ali A, Ahmad A, Akram W (2018) Imperative roles of halotolerant plant growth-promoting rhizobacteria and kinetin in improving salt tolerance and growth of black gram (*Phaseolus mungo*). Environ Sci Pollut Res 25(5):4491–4505

Zhu JK (2007) Plant salt stress. Wiley

The manufacturer's authorised representative in the EU is Springer Nature Customer Service Centre GmbH, Europaplatz 3, 69115 Heidelberg, Germany. If you have any concerns regarding our products, please contact ProductSafety@springernature.com

Printed and bound by CPI Group (UK) Ltd, Croydon, CR0 4YY
30/09/2026
02225557-0014